Student Solutions Manual

for

SINGLE VARIABLE CALCULUS
EARLY TRANSCENDENTALS

FIFTH EDITION

DANIEL ANDERSON
University of Iowa

JEFFERY A. COLE
Anoka-Ramsey Community College

DANIEL DRUCKER
Wayne State University

THOMSON
™
BROOKS/COLE

Australia ▫ Canada ▫ Mexico ▫ Singapore ▫ Spain ▫ United Kingdom ▫ United States

Printed in Canada

5 6 7 8 9 10   08 07 06 05

Printer: Webcom

ISBN 0-534-39333-0

For more information about our products, contact us at:
Thomson Learning Academic Resource Center
1-800-423-0563

For permission to use material from this text, contact us by:
Phone: 1-800-730-2214
Fax: 1-800-730-2215
Web: http://www.thomsonrights.com

Brooks/Cole–Thomson Learning
10 Davis Drive
Belmont, CA 94002
USA

Asia
Thomson Learning
5 Shenton Way #01-01
UIC Building
Singapore 068808

Australia/New Zealand
Nelson Thomson Learning
102 Dodds Street
Southbank, Victoria 3006
Australia

Canada
Nelson
1120 Birchmount Road
Toronto, Ontario M1K 5G4
Canada

Europe/Middle East/Africa
Thomson Learning
High Holborn House
50/51 Bedford Row
London WC1R 4LR
United Kingdom

Latin America
Thomson Learning
Seneca, 53
Colonia Polanco
11560 Mexico D.F.
Mexico

Spain
Paraninfo
Calle/Magallanes, 25
28015 Madrid, Spain

◻ PREFACE

This *Student Solutions Manual* contains strategies for solving and solutions to selected exercises in the text *Single Variable Calculus, Early Transcendentals, Fifth Edition*, by James Stewart. It contains solutions to the odd-numbered exercises in each section, the review sections, the True-False Quizzes, and the Problem Solving sections, as well as solutions to all the exercises in the Concept Checks.

We use some non-standard notation in order to save space. If you see a symbol which you don't recognize, refer to the Table of Abbreviations and symbols on page iv.

This manual is a text supplement and should be read along with the text. You should read all exercise solutions in this manual because many concept explanations are given and then used in subsequent solutions. All concepts necessary to solve a particular problem are not reviewed for every exercise. If you are having difficulty with a previously covered concept, refer back to the section where it was covered for more complete help.

A significant number of today's students are involved in various outside activities, and find it difficult, if not impossible, to attend all class sessions; this manual should help meet the needs of these students. In addition, it is our hope that this manual's solutions will enhance the understanding of all readers of the material and provide insights to solving other exercises.

We appreciate feedback concerning errors, solution correctness or style, and manual style. Any comments may be sent directly to jeff.cole@anokaramsey.edu, or in care of the publisher: Brooks/Cole Publishing Company, 10 Davis Drive, Belmont CA 94002-3098.

We would like to thank Brian Betsill, Stephanie Kuhns, and Kathi Townes, of TECHarts, for their production services; and Stacy Green, of Wadsworth/Brookscole Publishing Company, for her patience and support. All of these people have provided invaluable help in creating this manual.

Jeffery A. Cole
Anoka-Ramsey Community College

James Stewart
McMaster University

Daniel Drucker
Wayne State University

Daniel Anderson
University of Iowa

☐ ABBREVIATIONS AND SYMBOLS

CD	concave downward
CU	concave upward
D	the domain of f
FDT	First Derivative Test
HA	horizontal asymptote(s)
I	interval of convergence
IP	inflection points(s)
R	radius of convergence
VA	vertical asymptote(s)
$\overset{CAS}{=}$	indicates the use of a computer algebra system
$\overset{H}{=}$	indicates the use of l'Hospital's Rule
$\overset{j}{=}$	indicates the use of Formula j in the Table of Integrals given on Reference Pages 6–10
$\overset{s}{=}$	indicates the use of the substitution $\{u = \sin x,\ du = \cos x\,dx\}$
$\overset{c}{=}$	indicates the use of the substitution $\{u = \cos x,\ du = -\sin x\,dx\}$

CONTENTS

□ 11 INFINITE SEQUENCES AND SERIES   443

□ APPENDIXES

1 ☐ FUNCTIONS AND MODELS

1.1 Four Ways to Represent a Function

In exercises requiring estimations or approximations, your answers may vary slightly from the answers given here.

1. (a) The point $(-1, -2)$ is on the graph of f, so $f(-1) = -2$.

(b) When $x = 2$, y is about 2.8, so $f(2) \approx 2.8$.

(c) $f(x) = 2$ is equivalent to $y = 2$. When $y = 2$, we have $x = -3$ and $x = 1$.

(d) Reasonable estimates for x when $y = 0$ are $x = -2.5$ and $x = 0.3$.

(e) The domain of f consists of all x-values on the graph of f. For this function, the domain is $-3 \le x \le 3$, or $[-3, 3]$. The range of f consists of all y-values on the graph of f. For this function, the range is $-2 \le y \le 3$, or $[-2, 3]$.

(f) As x increases from -1 to 3, y increases from -2 to 3. Thus, f is increasing on the interval $[-1, 3]$.

3. From Figure 1 in the text, the lowest point occurs at about $(t, a) = (12, -85)$. The highest point occurs at about $(17, 115)$. Thus, the range of the vertical ground acceleration is $-85 \le a \le 115$. In Figure 11, the range of the north-south acceleration is approximately $-325 \le a \le 485$. In Figure 12, the range of the east-west acceleration is approximately $-210 \le a \le 200$.

5. No, the curve is not the graph of a function because a vertical line intersects the curve more than once. Hence, the curve fails the Vertical Line Test.

7. Yes, the curve is the graph of a function because it passes the Vertical Line Test. The domain is $[-3, 2]$ and the range is $[-3, -2) \cup [-1, 3]$.

9. The person's weight increased to about 160 pounds at age 20 and stayed fairly steady for 10 years. The person's weight dropped to about 120 pounds for the next 5 years, then increased rapidly to about 170 pounds. The next 30 years saw a gradual increase to 190 pounds. Possible reasons for the drop in weight at 30 years of age: diet, exercise, health problems.

11. The water will cool down almost to freezing as the ice melts. Then, when the ice has melted, the water will slowly warm up to room temperature.

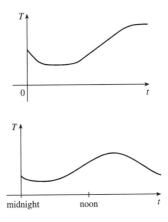

13. Of course, this graph depends strongly on the geographical location!

15.

17. (a)

(b) From the graph, we estimate the number of cell-phone subscribers in Malaysia to be about 540 in 1994 and 1450 in 1996.

19. $f(x) = 3x^2 - x + 2$.

$f(2) = 3(2)^2 - 2 + 2 = 12 - 2 + 2 = 12$.

$f(-2) = 3(-2)^2 - (-2) + 2 = 12 + 2 + 2 = 16$.

$f(a) = 3a^2 - a + 2$.

$f(-a) = 3(-a)^2 - (-a) + 2 = 3a^2 + a + 2$.

$f(a+1) = 3(a+1)^2 - (a+1) + 2 = 3(a^2 + 2a + 1) - a - 1 + 2 = 3a^2 + 6a + 3 - a + 1 = 3a^2 + 5a + 4$.

$2f(a) = 2 \cdot f(a) = 2(3a^2 - a + 2) = 6a^2 - 2a + 4$.

$f(2a) = 3(2a)^2 - (2a) + 2 = 3(4a^2) - 2a + 2 = 12a^2 - 2a + 2$.

$f(a^2) = 3(a^2)^2 - (a^2) + 2 = 3(a^4) - a^2 + 2 = 3a^4 - a^2 + 2$.

$[f(a)]^2 = [3a^2 - a + 2]^2 = (3a^2 - a + 2)(3a^2 - a + 2)$

$\qquad = 9a^4 - 3a^3 + 6a^2 - 3a^3 + a^2 - 2a + 6a^2 - 2a + 4 = 9a^4 - 6a^3 + 13a^2 - 4a + 4$.

$f(a+h) = 3(a+h)^2 - (a+h) + 2 = 3(a^2 + 2ah + h^2) - a - h + 2 = 3a^2 + 6ah + 3h^2 - a - h + 2$.

21. $f(x) = x - x^2$, so

$f(2+h) = 2 + h - (2+h)^2 = 2 + h - (4 + 4h + h^2) = 2 + h - 4 - 4h - h^2 = -(h^2 + 3h + 2)$,

$f(x+h) = x + h - (x+h)^2 = x + h - (x^2 + 2xh + h^2) = x + h - x^2 - 2xh - h^2$, and

$\dfrac{f(x+h) - f(x)}{h} = \dfrac{x + h - x^2 - 2xh - h^2 - x + x^2}{h} = \dfrac{h - 2xh - h^2}{h} = \dfrac{h(1 - 2x - h)}{h} = 1 - 2x - h$.

23. $f(x) = x/(3x - 1)$ is defined for all x except when $0 = 3x - 1 \quad \Leftrightarrow \quad x = \frac{1}{3}$, so the domain is $\left\{ x \in \mathbb{R} \mid x \neq \frac{1}{3} \right\} = \left(-\infty, \frac{1}{3} \right) \cup \left(\frac{1}{3}, \infty \right)$.

25. $f(t) = \sqrt{t} + \sqrt[3]{t}$ is defined when $t \geq 0$. These values of t give real number results for $\sqrt{t}$, whereas any value of t gives a real number result for $\sqrt[3]{t}$. The domain is $[0, \infty)$.

27. $h(x) = 1 \big/ \sqrt[4]{x^2 - 5x}$ is defined when $x^2 - 5x > 0$ $\Leftrightarrow$ $x(x - 5) > 0$. Note that $x^2 - 5x \neq 0$ since that would result in division by zero. The expression $x(x - 5)$ is positive if $x < 0$ or $x > 5$. (See Appendix A for methods for solving inequalities.) Thus, the domain is $(-\infty, 0) \cup (5, \infty)$.

29. $f(x) = 5$ is defined for all real numbers, so the domain is $\mathbb{R}$, or $(-\infty, \infty)$. The graph of f is a horizontal line with y-intercept 5.

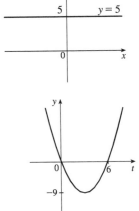

31. $f(t) = t^2 - 6t$ is defined for all real numbers, so the domain is $\mathbb{R}$, or $(-\infty, \infty)$. The graph of f is a parabola opening upward since the coefficient of t^2 is positive. To find the t-intercepts, let $y = 0$ and solve for t. $0 = t^2 - 6t = t(t - 6)$ $\Rightarrow$ $t = 0$ and $t = 6$. The t-coordinate of the vertex is halfway between the t-intercepts, that is, at $t = 3$. Since $f(3) = 3^2 - 6 \cdot 3 = -9$, the vertex is $(3, -9)$.

33. $g(x) = \sqrt{x - 5}$ is defined when $x - 5 \geq 0$ or $x \geq 5$, so the domain is $[5, \infty)$. Since $y = \sqrt{x - 5}$ $\Rightarrow$ $y^2 = x - 5$ $\Rightarrow$ $x = y^2 + 5$, we see that g is the top half of a parabola.

35. $G(x) = \dfrac{3x + |x|}{x}$. Since $|x| = \begin{cases} x & \text{if } x \geq 0 \\ -x & \text{if } x < 0 \end{cases}$, we have

$$G(x) = \begin{cases} \dfrac{3x + x}{x} & \text{if } x > 0 \\ \dfrac{3x - x}{x} & \text{if } x < 0 \end{cases} = \begin{cases} \dfrac{4x}{x} & \text{if } x > 0 \\ \dfrac{2x}{x} & \text{if } x < 0 \end{cases} = \begin{cases} 4 & \text{if } x > 0 \\ 2 & \text{if } x < 0 \end{cases}$$

Note that G is not defined for $x = 0$. The domain is $(-\infty, 0) \cup (0, \infty)$.

37. $f(x) = \begin{cases} x & \text{if } x \leq 0 \\ x + 1 & \text{if } x > 0 \end{cases}$

Domain is $\mathbb{R}$, or $(-\infty, \infty)$.

39. $f(x) = \begin{cases} x + 2 & \text{if } x \leq -1 \\ x^2 & \text{if } x > -1 \end{cases}$

Note that for $x = -1$, both $x + 2$ and x^2 are equal to 1. Domain is $\mathbb{R}$.

41. Recall that the slope m of a line between the two points (x_1, y_1) and (x_2, y_2) is $m = \dfrac{y_2 - y_1}{x_2 - x_1}$ and an equation of

the line connecting those two points is $y - y_1 = m(x - x_1)$. The slope of this line segment is $\dfrac{-6 - 1}{4 - (-2)} = -\dfrac{7}{6}$, so

an equation is $y - 1 = -\frac{7}{6}(x + 2)$. The function is $f(x) = -\frac{7}{6}x - \frac{4}{3}$, $-2 \le x \le 4$.

43. We need to solve the given equation for y. $x + (y - 1)^2 = 0 \iff (y - 1)^2 = -x \iff y - 1 = \pm\sqrt{-x} \iff$

$y = 1 \pm \sqrt{-x}$. The expression with the positive radical represents the top half of the parabola, and the one with the

negative radical represents the bottom half. Hence, we want $f(x) = 1 - \sqrt{-x}$. Note that the domain is $x \le 0$.

45. For $-1 \le x \le 2$, the graph is the line with slope 1 and y-intercept 1, that is, the line $y = x + 1$. For $2 < x \le 4$, the

graph is the line with slope $-\frac{3}{2}$ and x-intercept 4 [which corresponds to the point $(4, 0)$], so

$y - 0 = -\frac{3}{2}(x - 4) = -\frac{3}{2}x + 6$. So the function is $f(x) = \begin{cases} x + 1 & \text{if } -1 \le x \le 2 \\ -\frac{3}{2}x + 6 & \text{if } 2 < x \le 4 \end{cases}$

47. Let the length and width of the rectangle be L and W. Then the perimeter is $2L + 2W = 20$ and the area is

$A = LW$. Solving the first equation for W in terms of L gives $W = \dfrac{20 - 2L}{2} = 10 - L$. Thus,

$A(L) = L(10 - L) = 10L - L^2$. Since lengths are positive, the domain of A is $0 < L < 10$. If we further restrict

L to be larger than W, then $5 < L < 10$ would be the domain.

49. Let the length of a side of the equilateral triangle be x. Then by the Pythagorean Theorem, the height y of the

triangle satisfies $y^2 + \left(\frac{1}{2}x\right)^2 = x^2$, so that $y^2 = x^2 - \frac{1}{4}x^2 = \frac{3}{4}x^2$ and $y = \frac{\sqrt{3}}{2}x$. Using the formula for the area A

of a triangle, $A = \frac{1}{2}(\text{base})(\text{height})$, we obtain $A(x) = \frac{1}{2}(x)\left(\frac{\sqrt{3}}{2}x\right) = \frac{\sqrt{3}}{4}x^2$, with domain $x > 0$.

51. Let each side of the base of the box have length x, and let the height of the box be h. Since the volume is 2, we

know that $2 = hx^2$, so that $h = 2/x^2$, and the surface area is $S = x^2 + 4xh$. Thus,

$S(x) = x^2 + 4x(2/x^2) = x^2 + (8/x)$, with domain $x > 0$.

53. The height of the box is x and the length and width are $L = 20 - 2x$, $W = 12 - 2x$. Then $V = LWx$ and so

$V(x) = (20 - 2x)(12 - 2x)(x) = 4(10 - x)(6 - x)(x) = 4x(60 - 16x + x^2) = 4x^3 - 64x^2 + 240x$.

The sides L, W, and x must be positive. Thus, $L > 0 \iff 20 - 2x > 0 \iff x < 10$; $W > 0 \iff$

$12 - 2x > 0 \iff x < 6$; and $x > 0$. Combining these restrictions gives us the domain $0 < x < 6$.

55. (a)

(b) On \$14,000, tax is assessed on \$4000, and 10%(\$4000) = \$400.

On \$26,000, tax is assessed on \$16,000, and

10%(\$10,000) + 15%(\$6000) = \$1000 + \$900 = \$1900.

(c) As in part (b), there is \$1000 tax assessed on \$20,000 of

income, so the graph of T is a line segment from $(10,000, 0)$

to $(20,000, 1000)$. The tax on \$30,000 is \$2500, so the graph

of T for $x > 20,000$ is the ray with initial point

$(20,000, 1000)$ that passes through $(30,000, 2500)$.

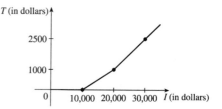

57. f is an odd function because its graph is symmetric about the origin. g is an even function because its graph is symmetric with respect to the y-axis.

59. (a) Because an even function is symmetric with respect to the y-axis, and the point $(5, 3)$ is on the graph of this even function, the point $(-5, 3)$ must also be on its graph.

(b) Because an odd function is symmetric with respect to the origin, and the point $(5, 3)$ is on the graph of this odd function, the point $(-5, -3)$ must also be on its graph.

61. $f(x) = x^{-2}$.

$$f(-x) = (-x)^{-2} = \frac{1}{(-x)^2} = \frac{1}{x^2}$$

$$= x^{-2} = f(x)$$

So f is an even function.

63. $f(x) = x^2 + x$, so $f(-x) = (-x)^2 + (-x) = x^2 - x$. Since this is neither $f(x)$ nor $-f(x)$, the function f is neither even nor odd.

65. $f(x) = x^3 - x$.

$$f(-x) = (-x)^3 - (-x) = -x^3 + x$$

$$= -(x^3 - x) = -f(x)$$

So f is odd.

1.2 Mathematical Models: A Catalog of Essential Functions

1. (a) $f(x) = \sqrt[5]{x}$ is a root function with $n = 5$.

(b) $g(x) = \sqrt{1 - x^2}$ is an algebraic function because it is a root of a polynomial.

(c) $h(x) = x^9 + x^4$ is a polynomial of degree 9.

(d) $r(x) = \dfrac{x^2 + 1}{x^3 + x}$ is a rational function because it is a ratio of polynomials.

(e) $s(x) = \tan 2x$ is a trigonometric function.

(f) $t(x) = \log_{10} x$ is a logarithmic function.

3. We notice from the figure that g and h are even functions (symmetric with respect to the y-axis) and that f is an odd function (symmetric with respect to the origin). So (b) $\left[y = x^5\right]$ must be f. Since g is flatter than h near the origin, we must have (c) $\left[y = x^8\right]$ matched with g and (a) $\left[y = x^2\right]$ matched with h.

5. (a) An equation for the family of linear functions with slope 2 is

$y = f(x) = 2x + b$, where b is the

y-intercept.

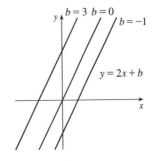

(b) $f(2) = 1$ means that the point $(2, 1)$ is on the graph of f. We can use the point-slope form of a line to obtain an equation for the family of linear functions through the point $(2, 1)$.

$y - 1 = m(x - 2)$, which is equivalent to

$y = mx + (1 - 2m)$ in slope-intercept form.

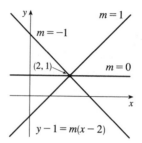

(c) To belong to both families, an equation must have slope $m = 2$, so the equation in part (b), $y = mx + (1 - 2m)$, becomes $y = 2x - 3$. It is the *only* function that belongs to both families.

7. All members of the family of linear functions $f(x) = c - x$ have graphs that are lines with slope -1. The y-intercept is c.

9. (a)

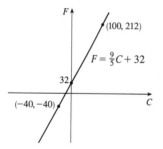

(b) The slope of $\frac{9}{5}$ means that F increases $\frac{9}{5}$ degrees for each increase of $1\,°C$. (Equivalently, F increases by 9 when C increases by 5 and F decreases by 9 when C decreases by 5.) The F-intercept of 32 is the Fahrenheit temperature corresponding to a Celsius temperature of 0.

11. (a) Using N in place of x and T in place of y, we find the slope to be $\dfrac{T_2 - T_1}{N_2 - N_1} = \dfrac{80 - 70}{173 - 113} = \dfrac{10}{60} = \dfrac{1}{6}$. So a linear equation is $T - 80 = \frac{1}{6}(N - 173) \Leftrightarrow T - 80 = \frac{1}{6}N - \frac{173}{6} \Leftrightarrow T = \frac{1}{6}N + \frac{307}{6}$ $\left[\frac{307}{6} = 51.1\overline{6}\right]$.

(b) The slope of $\frac{1}{6}$ means that the temperature in Fahrenheit degrees increases one-sixth as rapidly as the number of cricket chirps per minute. Said differently, each increase of 6 cricket chirps per minute corresponds to an increase of $1\,°F$.

(c) When $N = 150$, the temperature is given approximately by $T = \frac{1}{6}(150) + \frac{307}{6} = 76.1\overline{6}\,°F \approx 76\,°F$.

13. (a) We are given $\dfrac{\text{change in pressure}}{10\ \text{feet change in depth}} = \dfrac{4.34}{10} = 0.434$. Using P for pressure and d for depth with the point $(d, P) = (0, 15)$, we have the slope-intercept form of the line, $P = 0.434d + 15$.

(b) When $P = 100$, then $100 = 0.434d + 15 \Leftrightarrow 0.434d = 85 \Leftrightarrow d = \frac{85}{0.434} \approx 195.85$ feet. Thus, the pressure is 100 lb/in^2 at a depth of approximately 196 feet.

15. (a) The data appear to be periodic and a sine or cosine function would make the best model. A model of the form $f(x) = a\cos(bx) + c$ seems appropriate.

(b) The data appear to be decreasing in a linear fashion. A model of the form $f(x) = mx + b$ seems appropriate.

Some values are given to many decimal places. These are the results given by several computer algebra systems — rounding is left to the reader.

17. (a)

A linear model does seem appropriate.

(b) Using the points $(4000, 14.1)$ and $(60{,}000, 8.2)$, we obtain

$$y - 14.1 = \frac{8.2 - 14.1}{60{,}000 - 4000}(x - 4000) \text{ or, equivalently,}$$

$$y \approx -0.000105357x + 14.521429.$$

(c) Using a computing device, we obtain the least squares regression line $y = -0.0000997855x + 13.950764$. The following commands and screens illustrate how to find the least squares regression line on a TI-83 Plus. Enter the data into list one (L1) and list two (L2). Press $\boxed{\text{STAT}}\ \boxed{1}$ to enter the editor.

Find the regession line and store it in Y_1. Press $\boxed{\text{2nd}}\ \boxed{\text{QUIT}}\ \boxed{\text{STAT}}\ \boxed{\blacktriangleright}\ \boxed{4}\ \boxed{\text{VARS}}\ \boxed{\blacktriangleright}\ \boxed{1}\ \boxed{1}\ \boxed{\text{ENTER}}$.

Note from the last figure that the regression line has been stored in Y_1 and that Plot1 has been turned on (Plot1 is highlighted). You can turn on Plot1 from the Y= menu by placing the cursor on Plot1 and pressing $\boxed{\text{ENTER}}$ or by pressing $\boxed{\text{2nd}}\ \boxed{\text{STAT PLOT}}\ \boxed{1}\ \boxed{\text{ENTER}}$.

Now press $\boxed{\text{ZOOM}}$ $\boxed{9}$ to produce a graph of the data and the regression line. Note that choice 9 of the ZOOM menu automatically selects a window that displays all of the data.

(d) When $x = 25{,}000$, $y \approx 11.456$; or about 11.5 per 100 population.

(e) When $x = 80{,}000$, $y \approx 5.968$; or about a 6% chance.

(f) When $x = 200{,}000$, y is negative, so the model does not apply.

19. (a)   (b)

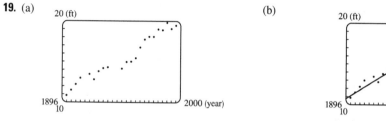

A linear model does seem appropriate.

Using a computing device, we obtain the least squares regression line $y = 0.089119747x - 158.2403249$, where x is the year and y is the height in feet.

(c) When $x = 2000$, the model gives $y \approx 20.00$ ft. Note that the actual winning height for the 2000 Olympics is *less than* the winning height for 1996—so much for that prediction.

(d) When $x = 2100$, $y \approx 28.91$ ft. This would be an increase of 9.49 ft from 1996 to 2100. Even though there was an increase of 8.59 ft from 1900 to 1996, it is unlikely that a similar increase will occur over the next 100 years.

21.

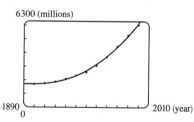

Using a computing device, we obtain the cubic function $y = ax^3 + bx^2 + cx + d$ with $a = 0.0012937$, $b = -7.06142$, $c = 12{,}823$, and $d = -7{,}743{,}770$. When $x = 1925$, $y \approx 1914$ (million).

1.3 New Functions from Old Functions

1. (a) If the graph of f is shifted 3 units upward, its equation becomes $y = f(x) + 3$.

(b) If the graph of f is shifted 3 units downward, its equation becomes $y = f(x) - 3$.

(c) If the graph of f is shifted 3 units to the right, its equation becomes $y = f(x - 3)$.

(d) If the graph of f is shifted 3 units to the left, its equation becomes $y = f(x + 3)$.

(e) If the graph of f is reflected about the x-axis, its equation becomes $y = -f(x)$.

(f) If the graph of f is reflected about the y-axis, its equation becomes $y = f(-x)$.

(g) If the graph of f is stretched vertically by a factor of 3, its equation becomes $y = 3f(x)$.

(h) If the graph of f is shrunk vertically by a factor of 3, its equation becomes $y = \frac{1}{3}f(x)$.

3. (a) (graph 3) The graph of f is shifted 4 units to the right and has equation $y = f(x - 4)$.

(b) (graph 1) The graph of f is shifted 3 units upward and has equation $y = f(x) + 3$.

(c) (graph 4) The graph of f is shrunk vertically by a factor of 3 and has equation $y = \frac{1}{3}f(x)$.

(d) (graph 5) The graph of f is shifted 4 units to the left and reflected about the x-axis. Its equation is
$y = -f(x + 4)$.

(e) (graph 2) The graph of f is shifted 6 units to the left and stretched vertically by a factor of 2. Its equation is
$y = 2f(x + 6)$.

5. (a) To graph $y = f(2x)$ we shrink the graph of f horizontally by a factor of 2.

(b) To graph $y = f\left(\frac{1}{2}x\right)$ we stretch the graph of f horizontally by a factor of 2.

The point $(4, -1)$ on the graph of f corresponds to the point $\left(\frac{1}{2} \cdot 4, -1\right) = (2, -1)$.

The point $(4, -1)$ on the graph of f corresponds to the point $(2 \cdot 4, -1) = (8, -1)$.

(c) To graph $y = f(-x)$ we reflect the graph of f about the y-axis.

(d) To graph $y = -f(-x)$ we reflect the graph of f about the y-axis, then about the x-axis.

The point $(4, -1)$ on the graph of f corresponds to the point $(-1 \cdot 4, -1) = (-4, -1)$.

The point $(4, -1)$ on the graph of f corresponds to the point $(-1 \cdot 4, -1 \cdot -1) = (-4, 1)$.

7. The graph of $y = f(x) = \sqrt{3x - x^2}$ has been shifted 4 units to the left, reflected about the x-axis, and shifted downward 1 unit. Thus, a function describing the graph is

$$y = \underbrace{-1 \cdot}_{\substack{\text{reflect} \\ \text{about} \\ x\text{-axis}}} \underbrace{f(x + 4)}_{\substack{\text{shift} \\ 4 \text{ units} \\ \text{left}}} \underbrace{-1}_{\substack{\text{shift} \\ 1 \text{ unit} \\ \text{down}}}$$

This function can be written as

$$y = -f(x + 4) - 1 = -\sqrt{3(x + 4) - (x + 4)^2} - 1 = -\sqrt{3x + 12 - (x^2 + 8x + 16)} - 1$$
$$= -\sqrt{-x^2 - 5x - 4} - 1$$

9. $y = -x^3$: Start with the graph of $y = x^3$ and reflect about the x-axis. Note: Reflecting about the y-axis gives the same result since substituting $-x$ for x gives us $y = (-x)^3 = -x^3$.

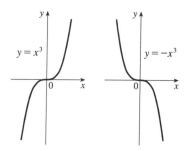

11. $y = (x + 1)^2$: Start with the graph of $y = x^2$ and shift 1 unit to the left.

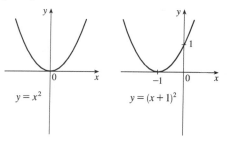

13. $y = 1 + 2\cos x$: Start with the graph of $y = \cos x$, stretch vertically by a factor of 2, and then shift 1 unit upward.

15. $y = \sin(x/2)$: Start with the graph of $y = \sin x$ and stretch horizontally by a factor of 2.

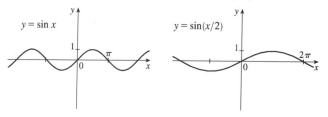

17. $y = \sqrt{x + 3}$: Start with the graph of $y = \sqrt{x}$ and shift 3 units to the left.

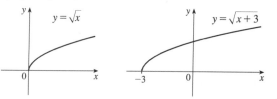

19. $y = \frac{1}{2}(x^2 + 8x) = \frac{1}{2}(x^2 + 8x + 16) - 8 = \frac{1}{2}(x + 4)^2 - 8$: Start with the graph of $y = x^2$, compress vertically by a factor of 2, shift 4 units to the left, and then shift 8 units downward.

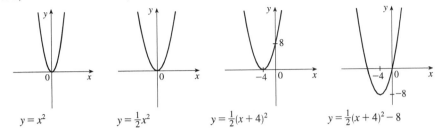

$$y = x^2 \qquad y = \tfrac{1}{2}x^2 \qquad y = \tfrac{1}{2}(x+4)^2 \qquad y = \tfrac{1}{2}(x+4)^2 - 8$$

21. $y = 2/(x + 1)$: Start with the graph of $y = 1/x$, shift 1 unit to the left, and then stretch vertically by a factor of 2.

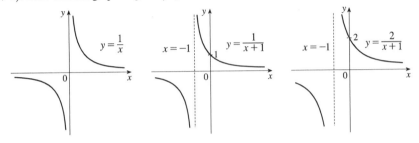

23. $y = |\sin x|$: Start with the graph of $y = \sin x$ and reflect all the parts of the graph below the x-axis about the x-axis.

25. This is just like the solution to Example 4 except the amplitude of the curve (the $30°$N curve in Figure 9 on June 21) is $14 - 12 = 2$. So the function is $L(t) = 12 + 2\sin\left[\frac{2\pi}{365}(t - 80)\right]$. March 31 is the 90th day of the year, so the model gives $L(90) \approx 12.34$ h. The daylight time (5:51 A.M. to 6:18 P.M.) is 12 hours and 27 minutes, or 12.45 h. The model value differs from the actual value by $\frac{12.45 - 12.34}{12.45} \approx 0.009$, less than 1%.

27. (a) To obtain $y = f(|x|)$, the portion of the graph of $y = f(x)$ to the right of the y-axis is reflected about the y-axis.

(b) $y = \sin|x|$ (c) $y = \sqrt{|x|}$

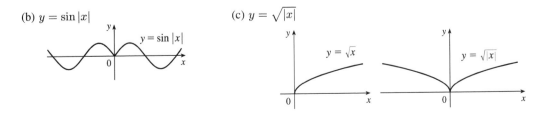

29. Assuming that successive horizontal and vertical gridlines are a unit apart, we can make a table of approximate values as follows.

x	0	1	2	3	4	5	6
$f(x)$	2	1.7	1.3	1.0	0.7	0.3	0
$g(x)$	2	2.7	3	2.8	2.4	1.7	0
$f(x) + g(x)$	4	4.4	4.3	3.8	3.1	2.0	0

Connecting the points $(x, f(x) + g(x))$ with a smooth curve gives an approximation to the graph of $f + g$. Extra points can be plotted between those listed above if necessary.

31. $f(x) = x^3 + 2x^2; g(x) = 3x^2 - 1$. $D = \mathbb{R}$ for both f and g.

$(f + g)(x) = (x^3 + 2x^2) + (3x^2 - 1) = x^3 + 5x^2 - 1$, $D = \mathbb{R}$.

$(f - g)(x) = (x^3 + 2x^2) - (3x^2 - 1) = x^3 - x^2 + 1$, $D = \mathbb{R}$.

$(fg)(x) = (x^3 + 2x^2)(3x^2 - 1) = 3x^5 + 6x^4 - x^3 - 2x^2$, $D = \mathbb{R}$.

$\left(\dfrac{f}{g}\right)(x) = \dfrac{x^3 + 2x^2}{3x^2 - 1}$, $D = \left\{x \mid x \neq \pm\dfrac{1}{\sqrt{3}}\right\}$ since $3x^2 - 1 \neq 0$.

33. $f(x) = x$, $g(x) = 1/x$

35. $f(x) = 2x^2 - x; g(x) = 3x + 2$. $D = \mathbb{R}$ for both f and g, and hence for their composites.

$(f \circ g)(x) = f(g(x)) = f(3x + 2) = 2(3x + 2)^2 - (3x + 2) = 2(9x^2 + 12x + 4) - 3x - 2 = 18x^2 + 21x + 6$.

$(g \circ f)(x) = g(f(x)) = g(2x^2 - x) = 3(2x^2 - x) + 2 = 6x^2 - 3x + 2$.

$(f \circ f)(x) = f(f(x)) = f(2x^2 - x) = 2(2x^2 - x)^2 - (2x^2 - x) = 2(4x^4 - 4x^3 + x^2) - 2x^2 + x = 8x^4 - 8x^3 + x$.

$(g \circ g)(x) = g(g(x)) = g(3x + 2) = 3(3x + 2) + 2 = 9x + 6 + 2 = 9x + 8$.

37. $f(x) = \sin x$, $D = \mathbb{R}$; $g(x) = 1 - \sqrt{x}$, $D = [0, \infty)$.

$(f \circ g)(x) = f(g(x)) = f(1 - \sqrt{x}) = \sin(1 - \sqrt{x}), D = [0, \infty)$.

$(g \circ f)(x) = g(f(x)) = g(\sin x) = 1 - \sqrt{\sin x}$. For $\sqrt{\sin x}$ to be defined, we must have

$\sin x \geq 0 \iff x \in [0, \pi] \cup [2\pi, 3\pi] \cup [-2\pi, -\pi] \cup [4\pi, 5\pi] \cup [-4\pi, -3\pi] \cup \dots$, so

$D = \{x \mid x \in [2n\pi, \pi + 2n\pi]$, where n is an integer$\}$.

$(f \circ f)(x) = f(f(x)) = f(\sin x) = \sin(\sin x), D = \mathbb{R}$.

$(g \circ g)(x) = g(g(x)) = g(1 - \sqrt{x}) = 1 - \sqrt{1 - \sqrt{x}}$,

$D = \{x \geq 0 \mid 1 - \sqrt{x} \geq 0\} = \{x \geq 0 \mid 1 \geq \sqrt{x}\} = \{x \geq 0 \mid \sqrt{x} \leq 1\} = [0, 1]$.

39. $f(x) = x + \dfrac{1}{x}$, $D = \{x \mid x \neq 0\}$; $g(x) = \dfrac{x+1}{x+2}$, $D = \{x \mid x \neq -2\}$.

$$(f \circ g)(x) = f(g(x)) = f\left(\frac{x+1}{x+2}\right) = \frac{x+1}{x+2} + \frac{1}{\frac{x+1}{x+2}} = \frac{x+1}{x+2} + \frac{x+2}{x+1}$$

$$= \frac{(x+1)(x+1) + (x+2)(x+2)}{(x+2)(x+1)} = \frac{(x^2 + 2x + 1) + (x^2 + 4x + 4)}{(x+2)(x+1)} = \frac{2x^2 + 6x + 5}{(x+2)(x+1)}$$

Since $g(x)$ is not defined for $x = -2$ and $f(g(x))$ is not defined for $x = -2$ and $x = -1$, the domain of $(f \circ g)(x)$ is $D = \{x \mid x \neq -2, -1\}$.

$$(g \circ f)(x) = g(f(x)) = g\left(x + \frac{1}{x}\right) = \frac{\left(x + \frac{1}{x}\right) + 1}{\left(x + \frac{1}{x}\right) + 2} = \frac{\frac{x^2 + 1 + x}{x}}{\frac{x^2 + 1 + 2x}{x}} = \frac{x^2 + x + 1}{x^2 + 2x + 1} = \frac{x^2 + x + 1}{(x+1)^2}.$$

Since $f(x)$ is not defined for $x = 0$ and $g(f(x))$ is not defined for $x = -1$, the domain of $(g \circ f)(x)$ is $D = \{x \mid x \neq -1, 0\}$.

$$(f \circ f)(x) = f(f(x)) = f\left(x + \frac{1}{x}\right) = \left(x + \frac{1}{x}\right) + \frac{1}{x + \frac{1}{x}} = x + \frac{1}{x} + \frac{1}{\frac{x^2+1}{x}} = x + \frac{1}{x} + \frac{x}{x^2 + 1}$$

$$= \frac{x(x)(x^2 + 1) + 1\,(x^2 + 1) + x(x)}{x(x^2 + 1)} = \frac{x^4 + x^2 + x^2 + 1 + x^2}{x(x^2 + 1)}$$

$$= \frac{x^4 + 3x^2 + 1}{x(x^2 + 1)},   D = \{x \mid x \neq 0\}.$$

$$(g \circ g)(x) = g(g(x)) = g\left(\frac{x+1}{x+2}\right) = \frac{\frac{x+1}{x+2} + 1}{\frac{x+1}{x+2} + 2} = \frac{\frac{x+1+1(x+2)}{x+2}}{\frac{x+1+2(x+2)}{x+2}} = \frac{x+1+x+2}{x+1+2x+4} = \frac{2x+3}{3x+5}.$$

Since $g(x)$ is not defined for $x = -2$ and $g(g(x))$ is not defined for $x = -\frac{5}{3}$, the domain of $(g \circ g)(x)$ is $D = \left\{x \mid x \neq -2, -\frac{5}{3}\right\}$.

41. $(f \circ g \circ h)(x) = f(g(h(x))) = f(g(x-1)) = f(2(x-1))$
$\qquad = 2(x-1) + 1 = 2x - 1$

43. $(f \circ g \circ h)(x) = f(g(h(x))) = f(g(x+3)) = f((x+3)^2 + 2)$
$\qquad = f(x^2 + 6x + 11) = \sqrt{(x^2 + 6x + 11) - 1} = \sqrt{x^2 + 6x + 10}$

45. Let $g(x) = x^2 + 1$ and $f(x) = x^{10}$. Then $(f \circ g)(x) = f(g(x)) = (x^2 + 1)^{10} = F(x)$.

47. Let $g(x) = x^2$ and $f(x) = \dfrac{x}{x+4}$. Then $(f \circ g)(x) = f(g(x)) = \dfrac{x^2}{x^2 + 4} = G(x)$.

49. Let $g(t) = \cos t$ and $f(t) = \sqrt{t}$. Then $(f \circ g)(t) = f(g(t)) = \sqrt{\cos t} = u(t)$.

51. Let $h(x) = x^2$, $g(x) = 3^x$, and $f(x) = 1 - x$. Then $(f \circ g \circ h)(x) = 1 - 3^{x^2} = H(x)$.

53. Let $h(x) = \sqrt{x}$, $g(x) = \sec x$, and $f(x) = x^4$. Then $(f \circ g \circ h)(x) = (\sec \sqrt{x})^4 = \sec^4(\sqrt{x}) = H(x)$.

55. (a) $g(2) = 5$, because the point $(2, 5)$ is on the graph of g. Thus, $f(g(2)) = f(5) = 4$, because the point $(5, 4)$ is on the graph of f.

(b) $g(f(0)) = g(0) = 3$

(c) $(f \circ g)(0) = f(g(0)) = f(3) = 0$

(d) $(g \circ f)(6) = g(f(6)) = g(6)$. This value is not defined, because there is no point on the graph of g that has x-coordinate 6.

(e) $(g \circ g)(-2) = g(g(-2)) = g(1) = 4$

(f) $(f \circ f)(4) = f(f(4)) = f(2) = -2$

57. (a) Using the relationship *distance* = *rate* · *time* with the radius r as the distance, we have $r(t) = 60t$.

(b) $A = \pi r^2 \Rightarrow (A \circ r)(t) = A(r(t)) = \pi(60t)^2 = 3600\pi t^2$. This formula gives us the extent of the rippled area (in cm^2) at any time t.

59. (a)   (b)

$$H(t) = \begin{cases} 0 & \text{if } t < 0 \\ 1 & \text{if } t \geq 0 \end{cases}$$

$$V(t) = \begin{cases} 0 & \text{if } t < 0 \\ 120 & \text{if } t \geq 0 \end{cases} \quad \text{so } V(t) = 120H(t).$$

(c)

V, 240 graph

Starting with the formula in part (b), we replace 120 with 240 to reflect the different voltage. Also, because we are starting 5 units to the right of $t = 0$, we replace t with $t - 5$. Thus, the formula is $V(t) = 240H(t - 5)$.

61. (a) By examining the variable terms in g and h, we deduce that we must square g to get the terms $4x^2$ and $4x$ in h. If we let $f(x) = x^2 + c$, then $(f \circ g)(x) = f(g(x)) = f(2x + 1) = (2x + 1)^2 + c = 4x^2 + 4x + (1 + c)$. Since $h(x) = 4x^2 + 4x + 7$, we must have $1 + c = 7$. So $c = 6$ and $f(x) = x^2 + 6$.

(b) We need a function g so that $f(g(x)) = 3(g(x)) + 5 = h(x)$. But $h(x) = 3x^2 + 3x + 2 = 3(x^2 + x) + 2 = 3(x^2 + x - 1) + 5$, so we see that $g(x) = x^2 + x - 1$.

63. We need to examine $h(-x)$.

$$h(-x) = (f \circ g)(-x) = f(g(-x)) = f(g(x)) \quad \text{[because } g \text{ is even]} \quad = h(x)$$

Because $h(-x) = h(x)$, h is an even function.

1.4 Graphing Calculators and Computers

1. $f(x) = x^4 + 2$

(a) $[-2, 2]$ by $[-2, 2]$ (b) $[0, 4]$ by $[0, 4]$ (c) $[-4, 4]$ by $[-4, 4]$

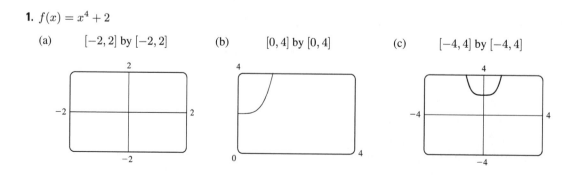

(d) $[-8, 8]$ by $[-4, 40]$ (e) $[-40, 40]$ by $[-80, 800]$

The most appropriate graph is produced in viewing rectangle (d).

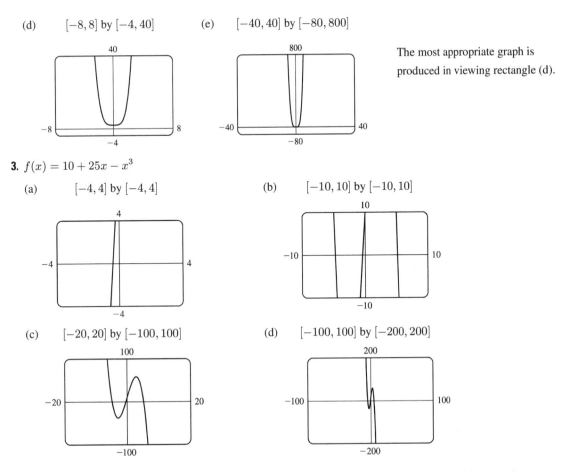

3. $f(x) = 10 + 25x - x^3$

(a) $[-4, 4]$ by $[-4, 4]$ (b) $[-10, 10]$ by $[-10, 10]$

(c) $[-20, 20]$ by $[-100, 100]$ (d) $[-100, 100]$ by $[-200, 200]$

The most appropriate graph is produced in viewing rectangle (c) because the maximum and minimum points are fairly easy to see and estimate.

5. Since the graph of $f(x) = 5 + 20x - x^2$ is a parabola opening downward, an appropriate viewing rectangle should include the maximum point.

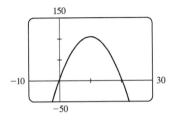

7. $f(x) = 0.01x^3 - x^2 + 5$. Graphing f in a standard viewing rectangle, $[-10, 10]$ by $[-10, 10]$, shows us what appears to be a parabola. But since this is a cubic polynomial, we know that a larger viewing rectangle will reveal a minimum point as well as the maximum point. After some trial and error, we choose the viewing rectangle $[-50, 150]$ by $[-2000, 2000]$.

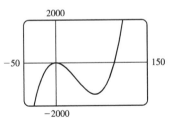

9. $f(x) = \sqrt[4]{81 - x^4}$ is defined when

$81 - x^4 \geq 0 \iff x^4 \leq 81 \iff |x| \leq 3$, so

the domain of f is $[-3, 3]$. Also

$0 \leq \sqrt[4]{81 - x^4} \leq \sqrt[4]{81} = 3$, so the range is $[0, 3]$.

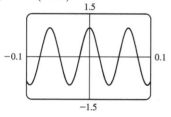

11. The graph of $f(x) = x^2 + (100/x)$ has a
vertical asymptote of $x = 0$. As you zoom
out, the graph of f looks more and more like
that of $y = x^2$.

13. $f(x) = \cos(100x)$

15. $f(x) = \sin(x/40)$

17. $y = 3^{\cos(x^2)}$

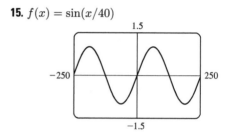

19. We must solve the given equation for y to obtain equations
for the upper and lower halves of the ellipse.

$4x^2 + 2y^2 = 1 \iff 2y^2 = 1 - 4x^2 \iff y^2 = \dfrac{1 - 4x^2}{2}$

$\iff y = \pm\sqrt{\dfrac{1 - 4x^2}{2}}$

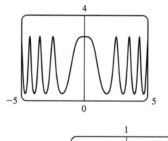

21. From the graph of $f(x) = x^3 - 9x^2 - 4$, we see that there is
one solution of the equation $f(x) = 0$ and it is slightly larger
than 9. By zooming in or using a root or zero feature, we
obtain $x \approx 9.05$.

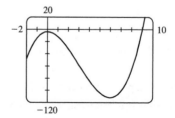

23. We see that the graphs of $f(x) = x^2$ and $g(x) = \sin x$

intersect twice. One solution is $x = 0$. The other solution of

$f = g$ is the x-coordinate of the point of intersection in the

first quadrant. Using an intersect feature or zooming in, we

find this value to be approximately 0.88. Alternatively, we

could find that value by finding the positive zero of

$h(x) = x^2 - \sin x.$

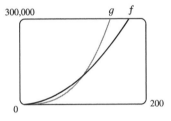

Note: After producing the graph on a TI-83 Plus, we can find the approximate value 0.88 by using the following

keystrokes: 2nd CALC 5 ENTER ENTER 1 ENTER. The "1" is just a guess for 0.88.

25. $g(x) = x^3/10$ is larger than $f(x) = 10x^2$ whenever $x > 100$.

27.

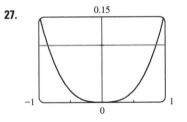

We see from the graphs of $y = |\sin x - x|$ and $y = 0.1$ that there are two

solutions to the equation $|\sin x - x| = 0.1$: $x \approx -0.85$ and $x \approx 0.85$.

The condition $|\sin x - x| < 0.1$ holds for any x lying between these two

values.

29. (a) The root functions $y = \sqrt{x}$,
  $y = \sqrt[4]{x}$ and $y = \sqrt[6]{x}$

  (b) The root functions $y = x$,
  $y = \sqrt[3]{x}$ and $y = \sqrt[5]{x}$

  (c) The root functions $y = \sqrt{x}$,
  $y = \sqrt[3]{x}, y = \sqrt[4]{x}$ and $y = \sqrt[5]{x}$

(d) • For any n, the nth root of 0 is 0 and the nth root of 1 is 1; that is, all nth root functions pass through the
  points $(0, 0)$ and $(1, 1)$.

  • For odd n, the domain of the nth root function is $\mathbb{R}$, while for even n, it is $\{x \in \mathbb{R} \mid x \geq 0\}$.

  • Graphs of even root functions look similar to that of $\sqrt{x}$, while those of odd root functions resemble that
  of $\sqrt[3]{x}$.

  • As n increases, the graph of $\sqrt[n]{x}$ becomes steeper near 0 and flatter for $x > 1$.

31. $f(x) = x^4 + cx^2 + x$. If $c < 0$, there are three humps: two minimum points and a maximum point. These humps get flatter as c increases, until at $c = 0$ two of the humps disappear and there is only one minimum point. This single hump then moves to the right and approaches the origin as c increases.

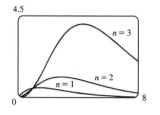

33. $y = x^n 2^{-x}$. As n increases, the maximum of the function moves further from the origin, and gets larger. Note, however, that regardless of n, the function approaches 0 as $x \to \infty$.

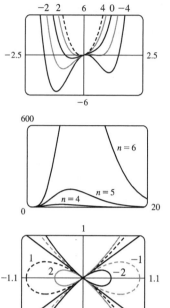

35. $y^2 = cx^3 + x^2$

If $c < 0$, the loop is to the right of the origin, and if c is positive, it is to the left. In both cases, the closer c is to 0, the larger the loop is. (In the limiting case, $c = 0$, the loop is "infinite", that is, it doesn't close.) Also, the larger $|c|$ is, the steeper the slope is on the loopless side of the origin.

37. The graphing window is 95 pixels wide and we want to start with $x = 0$ and end with $x = 2\pi$. Since there are 94 "gaps" between pixels, the distance between pixels is $\frac{2\pi - 0}{94}$. Thus, the x-values that the calculator actually plots are $x = 0 + \frac{2\pi}{94} \cdot n$, where $n = 0, 1, 2, \ldots, 93, 94$. For $y = \sin 2x$, the actual points plotted by the calculator are $\left(\frac{2\pi}{94} \cdot n, \sin\left(2 \cdot \frac{2\pi}{94} \cdot n\right)\right)$ for $n = 0, 1, \ldots, 94$. For $y = \sin 96x$, the points plotted are $\left(\frac{2\pi}{94} \cdot n, \sin\left(96 \cdot \frac{2\pi}{94} \cdot n\right)\right)$ for $n = 0, 1, \ldots, 94$. But

$$\sin\left(96 \cdot \tfrac{2\pi}{94} \cdot n\right) = \sin\left(94 \cdot \tfrac{2\pi}{94} \cdot n + 2 \cdot \tfrac{2\pi}{94} \cdot n\right) = \sin\left(2\pi n + 2 \cdot \tfrac{2\pi}{94} \cdot n\right)$$
$$= \sin\left(2 \cdot \tfrac{2\pi}{94} \cdot n\right) \quad \text{[by periodicity of sine]}, \quad n = 0, 1, \ldots, 94$$

So the y-values, and hence the points, plotted for $y = \sin 96x$ are identical to those plotted for $y = \sin 2x$. *Note:* Try graphing $y = \sin 94x$. Can you see why all the y-values are zero?

1.5 Exponential Functions

1. (a) $f(x) = a^x$, $a > 0$ (b) $\mathbb{R}$

(c) $(0, \infty)$ (d) See Figures 6(c), 6(b), and 6(a), respectively.

3. All of these graphs approach 0 as $x \to -\infty$, all of them pass through the point $(0, 1)$, and all of them are increasing and approach ∞ as $x \to \infty$. The larger the base, the faster the function increases for $x > 0$, and the faster it approaches 0 as $x \to -\infty$.

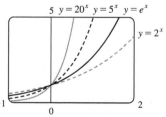

5. The functions with bases greater than 1 (3^x and 10^x) are
increasing, while those with bases less than 1
$\left[\left(\frac{1}{3}\right)^x \text{ and } \left(\frac{1}{10}\right)^x\right]$ are decreasing. The graph of $\left(\frac{1}{3}\right)^x$ is the
reflection of that of 3^x about the y-axis, and the graph of
$\left(\frac{1}{10}\right)^x$ is the reflection of that of 10^x about the y-axis. The
graph of 10^x increases more quickly than that of 3^x for
$x > 0$, and approaches 0 faster as $x \to -\infty$.

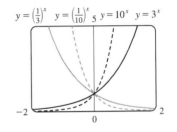

7. We start with the graph of $y = 4^x$ (Figure 3) and then shift 3 units downward. This shift doesn't affect the domain, but the range of $y = 4^x - 3$ is $(-3, \infty)$. There is a horizontal asymptote of $y = -3$.

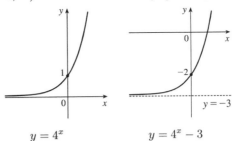

9. We start with the graph of $y = 2^x$
(Figure 2), reflect it about the
y-axis, and then about the x-axis
(or just rotate $180°$ to handle both
reflections) to obtain the graph of
$y = -2^{-x}$. In each graph, $y = 0$
is the horizontal asymptote.

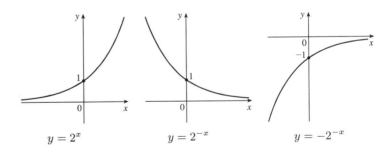

11. We start with the graph of $y = e^x$ (Figure 13), reflect it about the x-axis, and then shift 3 units upward. Note the horizontal asymptote of $y = 3$.

13. (a) To find the equation of the graph that results from shifting the graph of $y = e^x$ 2 units downward, we subtract 2 from the original function to get $y = e^x - 2$.

(b) To find the equation of the graph that results from shifting the graph of $y = e^x$ 2 units to the right, we replace x with $x - 2$ in the original function to get $y = e^{(x-2)}$.

(c) To find the equation of the graph that results from reflecting the graph of $y = e^x$ about the x-axis, we multiply the original function by -1 to get $y = -e^x$.

(d) To find the equation of the graph that results from reflecting the graph of $y = e^x$ about the y-axis, we replace x with $-x$ in the original function to get $y = e^{-x}$.

(e) To find the equation of the graph that results from reflecting the graph of $y = e^x$ about the x-axis and then about the y-axis, we first multiply the original function by -1 (to get $y = -e^x$) and then replace x with $-x$ in this equation to get $y = -e^{-x}$.

15. (a) The denominator $1 + e^x$ is never equal to zero because $e^x > 0$, so the domain of $f(x) = 1/(1 + e^x)$ is $\mathbb{R}$.

(b) $1 - e^x = 0 \Leftrightarrow e^x = 1 \Leftrightarrow x = 0$, so the domain of $f(x) = 1/(1 - e^x)$ is $(-\infty, 0) \cup (0, \infty)$.

17. Use $y = Ca^x$ with the points $(1, 6)$ and $(3, 24)$. $6 = Ca^1$ $\left[C = \frac{6}{a} \right]$ and $24 = Ca^3 \Rightarrow 24 = \left(\frac{6}{a} \right) a^3 \Rightarrow$

$4 = a^2 \Rightarrow a = 2$ [since $a > 0$] and $C = \frac{6}{2} = 3$. The function is $f(x) = 3 \cdot 2^x$.

19. If $f(x) = 5^x$, then $\dfrac{f(x+h) - f(x)}{h} = \dfrac{5^{x+h} - 5^x}{h} = \dfrac{5^x 5^h - 5^x}{h} = \dfrac{5^x (5^h - 1)}{h} = 5^x \left(\dfrac{5^h - 1}{h} \right)$.

21. 2 ft = 24 in, $f(24) = 24^2$ in = 576 in = 48 ft. $g(24) = 2^{24}$ in = $2^{24}/(12 \cdot 5280)$ mi ≈ 265 mi

23. The graph of g finally surpasses that of f at $x \approx 35.8$.

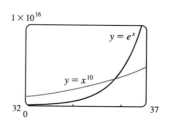

25. (a) Fifteen hours represents 5 doubling periods (one doubling period is three hours).

$100 \cdot 2^5 = 3200$

(b) In t hours, there will be $t/3$ doubling periods. The initial population is 100, so the population y at time t is $y = 100 \cdot 2^{t/3}$.

(c) $t = 20 \Rightarrow y = 100 \cdot 2^{20/3} \approx 10{,}159$

(d) We graph $y_1 = 100 \cdot 2^{x/3}$ and $y_2 = 50{,}000$. The two curves intersect at $x \approx 26.9$, so the population reaches 50,000 in about 26.9 hours.

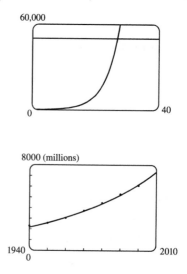

27. An exponential model is $y = ab^t$, where $a = 3.154832569 \times 10^{-12}$ and $b = 1.017764706$. This model gives $y(1993) \approx 5498$ million and $y(2010) \approx 7417$ million.

1.6 Inverse Functions and Logarithms

1. (a) See Definition 1.

(b) It must pass the Horizontal Line Test.

3. f is not one-to-one because $2 \neq 6$, but $f(2) = 2.0 = f(6)$.

5. No horizontal line intersects the graph of f more than once. Thus, by the Horizontal Line Test, f is one-to-one.

7. The horizontal line $y = 0$ (the x-axis) intersects the graph of f in more than one point. Thus, by the Horizontal Line Test, f is not one-to-one.

9. The graph of $f(x) = \frac{1}{2}(x+5)$ is a line with slope $\frac{1}{2}$. It passes the Horizontal Line Test, so f is one-to-one.

Algebraic solution: If $x_1 \neq x_2$, then $x_1 + 5 \neq x_2 + 5$ $\Rightarrow$ $\frac{1}{2}(x_1+5) \neq \frac{1}{2}(x_2+5)$ $\Rightarrow$ $f(x_1) \neq f(x_2)$, so f is one-to-one.

11. $g(x) = |x|$ $\Rightarrow$ $g(-1) = 1 = g(1)$, so g is not one-to-one.

13. A football will attain every height h up to its maximum height twice: once on the way up, and again on the way down. Thus, even if t_1 does not equal t_2, $f(t_1)$ may equal $f(t_2)$, so f is not 1-1.

15. f does not pass the Horizontal Line Test, so f is not 1-1.

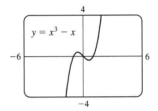

17. Since $f(2) = 9$ and f is 1-1, we know that $f^{-1}(9) = 2$. Remember, if the point $(2, 9)$ is on the graph of f, then the point $(9, 2)$ is on the graph of f^{-1}.

19. First, we must determine x such that $g(x) = 4$. By inspection, we see that if $x = 0$, then $g(x) = 4$. Since g is 1-1 (g is an increasing function), it has an inverse, and $g^{-1}(4) = 0$.

21. We solve $C = \frac{5}{9}(F - 32)$ for F: $\frac{9}{5}C = F - 32$ $\Rightarrow$ $F = \frac{9}{5}C + 32$. This gives us a formula for the inverse function, that is, the Fahrenheit temperature F as a function of the Celsius temperature C. $F \geq -459.67$ $\Rightarrow$ $\frac{9}{5}C + 32 \geq -459.67$ $\Rightarrow$ $\frac{9}{5}C \geq -491.67$ $\Rightarrow$ $C \geq -273.15$, the domain of the inverse function.

23. $f(x) = \sqrt{10 - 3x}$ $\Rightarrow$ $y = \sqrt{10 - 3x}$ $(y \geq 0)$ $\Rightarrow$ $y^2 = 10 - 3x$ $\Rightarrow$ $3x = 10 - y^2$ $\Rightarrow$ $x = -\frac{1}{3}y^2 + \frac{10}{3}$. Interchange x and y: $y = -\frac{1}{3}x^2 + \frac{10}{3}$. So $f^{-1}(x) = -\frac{1}{3}x^2 + \frac{10}{3}$. Note that the domain of f^{-1} is $x \geq 0$.

25. $f(x) = e^{x^3}$ $\Rightarrow$ $y = e^{x^3}$ $\Rightarrow$ $\ln y = x^3$ $\Rightarrow$ $x = \sqrt[3]{\ln y}$. Interchange x and y: $y = \sqrt[3]{\ln x}$. So $f^{-1}(x) = \sqrt[3]{\ln x}$.

27. $y = \ln(x+3)$ $\Rightarrow$ $x + 3 = e^y$ $\Rightarrow$ $x = e^y - 3$. Interchange x and y: $y = e^x - 3$. So $f^{-1}(x) = e^x - 3$.

29. $y = f(x) = 1 - \dfrac{2}{x^2} \;\Rightarrow\; 1 - y = \dfrac{2}{x^2} \;\Rightarrow\; x^2 = \dfrac{2}{1-y} \;\Rightarrow\;$

$x = \sqrt{\dfrac{2}{1-y}}$, since $x > 0$. Interchange x and y: $y = \sqrt{\dfrac{2}{1-x}}$.

So $f^{-1}(x) = \sqrt{\dfrac{2}{1-x}}$.

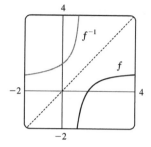

31. The function f is one-to-one, so its inverse exists and the graph of its inverse can be obtained by reflecting the graph of f about the line $y = x$.

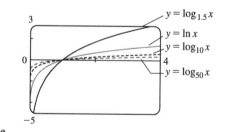

33. (a) It is defined as the inverse of the exponential function with base a, that is, $\log_a x = y \;\Leftrightarrow\; a^y = x$.

   (b) $(0, \infty)$ (c) $\mathbb{R}$ (d) See Figure 11.

35. (a) $\log_2 64 = 6$ since $2^6 = 64$.   (b) $\log_6 \frac{1}{36} = -2$ since $6^{-2} = \frac{1}{36}$.

37. (a) $\log_{10} 1.25 + \log_{10} 80 = \log_{10} (1.25 \cdot 80) = \log_{10} 100 = \log_{10} 10^2 = 2$

   (b) $\log_5 10 + \log_5 20 - 3 \log_5 2 = \log_5 (10 \cdot 20) - \log_5 2^3 = \log_5 \frac{200}{8} = \log_5 25 = \log_5 5^2 = 2$

39. $2 \ln 4 - \ln 2 = \ln 4^2 - \ln 2 = \ln 16 - \ln 2 = \ln \frac{16}{2} = \ln 8$

41. $\ln(1 + x^2) + \frac{1}{2} \ln x - \ln \sin x = \ln(1 + x^2) + \ln x^{1/2} - \ln \sin x = \ln[(1 + x^2)\sqrt{x}] - \ln \sin x = \ln \dfrac{(1 + x^2)\sqrt{x}}{\sin x}$

43. To graph these functions, we use $\log_{1.5} x = \dfrac{\ln x}{\ln 1.5}$ and

$\log_{50} x = \dfrac{\ln x}{\ln 50}$. These graphs all approach $-\infty$ as $x \to 0^+$, and

they all pass through the point $(1, 0)$. Also, they are all increasing,

and all approach ∞ as $x \to \infty$. The functions with larger bases

increase extremely slowly, and the ones with smaller bases do so

somewhat more quickly. The functions with large bases approach the

y-axis more closely as $x \to 0^+$.

45. $3 \text{ ft} = 36 \text{ in}$, so we need x such that $\log_2 x = 36 \;\Leftrightarrow\; x = 2^{36} = 68{,}719{,}476{,}736$. In miles, this is

$68{,}719{,}476{,}736 \text{ in} \cdot \dfrac{1 \text{ ft}}{12 \text{ in}} \cdot \dfrac{1 \text{ mi}}{5280 \text{ ft}} \approx 1{,}084{,}587.7 \text{ mi}.$

47. (a) Shift the graph of $y = \log_{10} x$ five units to the left to obtain the graph of $y = \log_{10}(x + 5)$. Note the vertical asymptote of $x = -5$.

$$y = \log_{10} x \qquad y = \log_{10}(x + 5)$$

(b) Reflect the graph of $y = \ln x$ about the x-axis to obtain the graph of $y = -\ln x$.

$$y = \ln x \qquad y = -\ln x$$

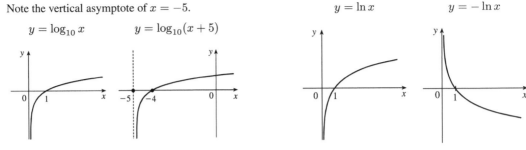

49. (a) $2 \ln x = 1 \quad \Rightarrow \quad \ln x = \frac{1}{2} \quad \Rightarrow \quad x = e^{1/2} = \sqrt{e}$

(b) $e^{-x} = 5 \quad \Rightarrow \quad -x = \ln 5 \quad \Rightarrow \quad x = -\ln 5$

51. (a) $2^{x-5} = 3 \quad \Leftrightarrow \quad \log_2 3 = x - 5 \quad \Leftrightarrow \quad x = 5 + \log_2 3$.

Or: $2^{x-5} = 3 \quad \Leftrightarrow \quad \ln\left(2^{x-5}\right) = \ln 3 \quad \Leftrightarrow \quad (x - 5)\ln 2 = \ln 3 \quad \Leftrightarrow \quad x - 5 = \dfrac{\ln 3}{\ln 2} \quad \Leftrightarrow \quad x = 5 + \dfrac{\ln 3}{\ln 2}$

(b) $\ln x + \ln(x - 1) = \ln(x(x - 1)) = 1 \quad \Leftrightarrow \quad x(x - 1) = e^1 \quad \Leftrightarrow \quad x^2 - x - e = 0$. The quadratic formula (with $a = 1$, $b = -1$, and $c = -e$) gives $x = \frac{1}{2}\left(1 \pm \sqrt{1 + 4e}\right)$, but we reject the negative root since the natural logarithm is not defined for $x < 0$. So $x = \frac{1}{2}\left(1 + \sqrt{1 + 4e}\right)$.

53. (a) $e^x < 10 \quad \Rightarrow \quad \ln e^x < \ln 10 \quad \Rightarrow \quad x < \ln 10 \quad \Rightarrow \quad x \in (-\infty, \ln 10)$

(b) $\ln x > -1 \quad \Rightarrow \quad e^{\ln x} > e^{-1} \quad \Rightarrow \quad x > e^{-1} \quad \Rightarrow \quad x \in (1/e, \infty)$

55. (a) For $f(x) = \sqrt{3 - e^{2x}}$, we must have $3 - e^{2x} \geq 0 \quad \Rightarrow \quad e^{2x} \leq 3 \quad \Rightarrow \quad 2x \leq \ln 3 \quad \Rightarrow \quad x \leq \frac{1}{2}\ln 3$. Thus, the domain of f is $(-\infty, \frac{1}{2}\ln 3]$.

(b) $y = f(x) = \sqrt{3 - e^{2x}}$ [note that $y \geq 0$] $\quad \Rightarrow \quad y^2 = 3 - e^{2x} \quad \Rightarrow \quad e^{2x} = 3 - y^2 \quad \Rightarrow \quad 2x = \ln(3 - y^2)$ $\quad \Rightarrow \quad x = \frac{1}{2}\ln(3 - y^2)$. Interchange x and y: $y = \frac{1}{2}\ln(3 - x^2)$. So $f^{-1}(x) = \frac{1}{2}\ln(3 - x^2)$. For the domain of f^{-1}, we must have $3 - x^2 > 0 \quad \Rightarrow \quad x^2 < 3 \quad \Rightarrow \quad |x| < \sqrt{3} \quad \Rightarrow \quad -\sqrt{3} < x < \sqrt{3} \quad \Rightarrow \quad 0 \leq x < \sqrt{3}$ since $x \geq 0$. Note that the domain of f^{-1}, $[0, \sqrt{3})$, equals the range of f.

57. We see that the graph of $y = f(x) = \sqrt{x^3 + x^2 + x + 1}$ is increasing, so f is 1-1. Enter $x = \sqrt{y^3 + y^2 + y + 1}$ and use your CAS to solve the equation for y. Using Derive, we get two (irrelevant) solutions involving imaginary expressions, as well as one which can be simplified to the following:

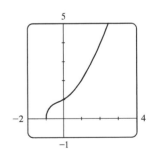

$$y = f^{-1}(x) = -\frac{\sqrt[3]{4}}{6}\left(\sqrt[3]{D - 27x^2 + 20} - \sqrt[3]{D + 27x^2 - 20} + \sqrt[3]{2}\right)$$

where $D = 3\sqrt{3}\sqrt{27x^4 - 40x^2 + 16}$. Maple and Mathematica each give two complex expressions and one real expression, and the real expression is equivalent to that given by Derive. For example, Maple's expression simplifies

to $\dfrac{1}{6} \dfrac{M^{2/3} - 8 - 2M^{1/3}}{2M^{1/3}}$, where $M = 108x^2 + 12\sqrt{48 - 120x^2 + 81x^4} - 80$.

59. (a) $n = 100 \cdot 2^{t/3}$ $\Rightarrow$ $\dfrac{n}{100} = 2^{t/3}$ $\Rightarrow$ $\log_2\left(\dfrac{n}{100}\right) = \dfrac{t}{3}$ $\Rightarrow$ $t = 3\log_2\left(\dfrac{n}{100}\right)$. Using formula (10), we

can write this as $t = 3 \cdot \dfrac{\ln(n/100)}{\ln 2}$. This function tells us how long it will take to obtain n bacteria (given the

number n).

(b) $n = 50{,}000$ $\Rightarrow$ $t = 3\log_2\dfrac{50{,}000}{100} = 3\log_2 500 = 3\left(\dfrac{\ln 500}{\ln 2}\right) \approx 26.9$ hours

61. (a) To find the equation of the graph that results from shifting the graph of $y = \ln x$ 3 units upward, we add 3 to the

original function to get $y = \ln x + 3$.

(b) To find the equation of the graph that results from shifting the graph of $y = \ln x$ 3 units to the left, we replace x

with $x + 3$ in the original function to get $y = \ln(x + 3)$.

(c) To find the equation of the graph that results from reflecting the graph of $y = \ln x$ about the x-axis, we multiply

the original equation by -1 to get $y = -\ln x$.

(d) To find the equation of the graph that results from reflecting the graph of $y = \ln x$ about the y-axis, we replace x

with $-x$ in the original equation to get $y = \ln(-x)$.

(e) To find the equation of the graph that results from reflecting the graph of $y = \ln x$ about the line $y = x$, we

interchange x and y in the original equation to get $x = \ln y$ $\Leftrightarrow$ $y = e^x$.

(f) To find the equation of the graph that results from reflecting the graph of $y = \ln x$ about the x-axis and then

about the line $y = x$, we first multiply the original equation by -1 [to get $y = -\ln x$] and then interchange x

and y in this equation to get $x = -\ln y$ $\Leftrightarrow$ $\ln y = -x$ $\Leftrightarrow$ $y = e^{-x}$.

(g) To find the equation of the graph that results from reflecting the graph of $y = \ln x$ about the y-axis and then

about the line $y = x$, we first replace x with $-x$ in the original equation [to get $y = \ln(-x)$] and then

interchange x and y to get $x = \ln(-y)$ $\Leftrightarrow$ $-y = e^x$ $\Leftrightarrow$ $y = -e^x$.

(h) To find the equation of the graph that results from shifting the graph of $y = \ln x$ 3 units to the left and then

reflecting it about the line $y = x$, we first replace x with $x + 3$ in the original equation [to get $y = \ln(x + 3)$]

and then interchange x and y in this equation to get $x = \ln(y + 3)$ $\Leftrightarrow$ $y + 3 = e^x$ $\Leftrightarrow$ $y = e^x - 3$.

63. (a) $\sin^{-1}\left(\dfrac{\sqrt{3}}{2}\right) = \dfrac{\pi}{3}$ since $\sin\dfrac{\pi}{3} = \dfrac{\sqrt{3}}{2}$ and $\dfrac{\pi}{3}$ is in $\left[-\dfrac{\pi}{2}, \dfrac{\pi}{2}\right]$.

(b) $\cos^{-1}(-1) = \pi$ since $\cos \pi = -1$ and π is in $[0, \pi]$.

65. (a) $\tan^{-1}\sqrt{3} = \dfrac{\pi}{3}$ since $\tan\dfrac{\pi}{3} = \sqrt{3}$ and $\dfrac{\pi}{3}$ is in $\left(-\dfrac{\pi}{2}, \dfrac{\pi}{2}\right)$.

(b) $\arcsin\left(-\dfrac{1}{\sqrt{2}}\right) = -\dfrac{\pi}{4}$ since $\sin\left(-\dfrac{\pi}{4}\right) = -\dfrac{1}{\sqrt{2}}$ and $-\dfrac{\pi}{4}$ is in $\left[-\dfrac{\pi}{2}, \dfrac{\pi}{2}\right]$.

67. (a) $\sin(\sin^{-1} 0.7) = 0.7$ since 0.7 is in $[-1, 1]$.

(b) $\tan^{-1}\left(\tan\dfrac{4\pi}{3}\right) = \tan^{-1}\sqrt{3} = \dfrac{\pi}{3}$ since $\dfrac{\pi}{3}$ is in $\left[-\dfrac{\pi}{2}, \dfrac{\pi}{2}\right]$.

69. Let $y = \sin^{-1} x$. Then $-\dfrac{\pi}{2} \le y \le \dfrac{\pi}{2}$ $\Rightarrow$ $\cos y \ge 0$, so $\cos(\sin^{-1} x) = \cos y = \sqrt{1 - \sin^2 y} = \sqrt{1 - x^2}$

71. Let $y = \tan^{-1} x$. Then $\tan y = x$, so from the triangle we see that

$$\sin(\tan^{-1} x) = \sin y = \frac{x}{\sqrt{1 + x^2}}.$$

73.

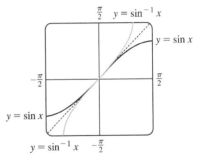

The graph of $\sin^{-1} x$ is the reflection of the graph of $\sin x$ about the line $y = x$.

75. $g(x) = \sin^{-1}(3x + 1)$.

Domain $(g) = \{x \mid -1 \le 3x + 1 \le 1\} = \{x \mid -2 \le 3x \le 0\} = \{x \mid -\frac{2}{3} \le x \le 0\} = \left[-\frac{2}{3}, 0\right]$.

Range $(g) = \{y \mid -\frac{\pi}{2} \le y \le \frac{\pi}{2}\} = \left[-\frac{\pi}{2}, \frac{\pi}{2}\right]$.

1 Review

CONCEPT CHECK

1. (a) A **function** f is a rule that assigns to each element x in a set A exactly one element, called $f(x)$, in a set B. The set A is called the **domain** of the function. The **range** of f is the set of all possible values of $f(x)$ as x varies throughout the domain.

(b) If f is a function with domain A, then its **graph** is the set of ordered pairs $\{(x, f(x)) \mid x \in A\}$.

(c) Use the Vertical Line Test on page 17.

2. The four ways to represent a function are: verbally, numerically, visually, and algebraically. An example of each is given below.

Verbally: An assignment of students to chairs in a classroom (a description in words)

Numerically: A tax table that assigns an amount of tax to an income (a table of values)

Visually: A graphical history of the Dow Jones average (a graph)

Algebraically: A relationship between distance, rate, and time: $d = rt$ (an explicit formula)

3. (a) An **even function** f satisfies $f(-x) = f(x)$ for every number x in its domain. It is symmetric with respect to the y-axis.

(b) An **odd function** g satisfies $g(-x) = -g(x)$ for every number x in its domain. It is symmetric with respect to the origin.

4. A function f is called **increasing** on an interval I if $f(x_1) < f(x_2)$ whenever $x_1 < x_2$ in I.

5. A **mathematical model** is a mathematical description (often by means of a function or an equation) of a real-world phenomenon.

6. (a) Linear function: $f(x) = 2x + 1$, $f(x) = ax + b$

(b) Power function: $f(x) = x^2$, $f(x) = x^a$

(c) Exponential function: $f(x) = 2^x$, $f(x) = a^x$

(d) Quadratic function: $f(x) = x^2 + x + 1$,

$f(x) = ax^2 + bx + c$

(e) Polynomial of degree 5: $f(x) = x^5 + 2$

(f) Rational function: $f(x) = \dfrac{x}{x+2}$, $f(x) = \dfrac{P(x)}{Q(x)}$ where

$P(x)$ and $Q(x)$ are polynomials

7.

8. (a)

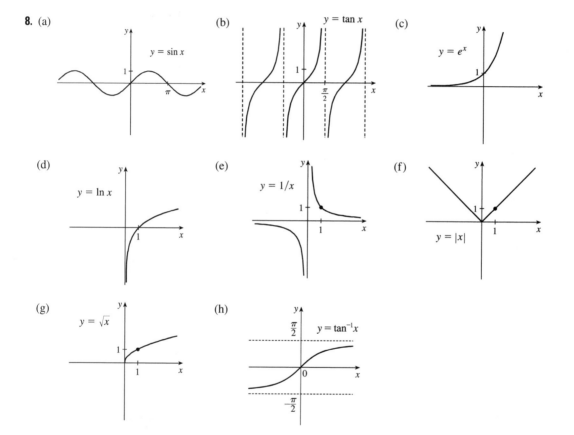

9. (a) The domain of $f + g$ is the intersection of the domain of f and the domain of g; that is, $A \cap B$.

(b) The domain of fg is also $A \cap B$.

(c) The domain of f/g must exclude values of x that make g equal to 0; that is, $\{x \in A \cap B \mid g(x) \neq 0\}$.

10. Given two functions f and g, the **composite** function $f \circ g$ is defined by $(f \circ g)(x) = f(g(x))$. The domain of $f \circ g$ is the set of all x in the domain of g such that $g(x)$ is in the domain of f.

11. (a) If the graph of f is shifted 2 units upward, its equation becomes $y = f(x) + 2$.

(b) If the graph of f is shifted 2 units downward, its equation becomes $y = f(x) - 2$.

(c) If the graph of f is shifted 2 units to the right, its equation becomes $y = f(x - 2)$.

(d) If the graph of f is shifted 2 units to the left, its equation becomes $y = f(x + 2)$.

(e) If the graph of f is reflected about the x-axis, its equation becomes $y = -f(x)$.

(f) If the graph of f is reflected about the y-axis, its equation becomes $y = f(-x)$.

(g) If the graph of f is stretched vertically by a factor of 2, its equation becomes $y = 2f(x)$.

(h) If the graph of f is shrunk vertically by a factor of 2, its equation becomes $y = \frac{1}{2}f(x)$.

(i) If the graph of f is stretched horizontally by a factor of 2, its equation becomes $y = f(\frac{1}{2})x$.

(j) If the graph of f is shrunk horizontally by a factor of 2, its equation becomes $y = f(2x)$.

12. (a) A function f is called a *one-to-one function* if it never takes on the same value twice; that is, if $f(x_1) \neq f(x_2)$ whenever $x_1 \neq x_2$. (Or, f is 1-1 if each output corresponds to only one input.)

Use the Horizontal Line Test: A function is one-to-one if and only if no horizontal line intersects its graph more than once.

(b) If f is a one-to-one function with domain A and range B, then its *inverse function* f^{-1} has domain B and range A and is defined by

$$f^{-1}(y) = x \quad \Leftrightarrow \quad f(x) = y$$

for any y in B. The graph of f^{-1} is obtained by reflecting the graph of f about the line $y = x$.

13. (a) The inverse sine function $f(x) = \sin^{-1} x$ is defined as follows:

$$\sin^{-1} x = y \quad \Leftrightarrow \quad \sin y = x \quad and \quad -\frac{\pi}{2} \le y \le \frac{\pi}{2}$$

Its domain is $-1 \le x \le 1$ and its range is $-\frac{\pi}{2} \le y \le \frac{\pi}{2}$.

(b) The inverse cosine function $f(x) = \cos^{-1} x$ is defined as follows:

$$\cos^{-1} x = y \quad \Leftrightarrow \quad \cos y = x \quad and \quad 0 \le y \le \pi$$

Its domain is $-1 \le x \le 1$ and its range is $0 \le y \le \pi$.

(c) The inverse tangent function $f(x) = \tan^{-1} x$ is defined as follows:

$$\tan^{-1} x = y \quad \Leftrightarrow \quad \tan y = x \quad and \quad -\frac{\pi}{2} < y < \frac{\pi}{2}$$

Its domain is $\mathbb{R}$ and its range is $-\frac{\pi}{2} < y < \frac{\pi}{2}$.

——————————————————— TRUE-FALSE QUIZ ———————————————————

1. False. Let $f(x) = x^2$, $s = -1$, and $t = 1$. Then $f(s + t) = (-1 + 1)^2 = 0^2 = 0$, but
$f(s) + f(t) = (-1)^2 + 1^2 = 2 \neq 0 = f(s + t)$.

3. False. Let $f(x) = x^2$. Then $f(3x) = (3x)^2 = 9x^2$ and $3f(x) = 3x^2$. So $f(3x) \neq 3f(x)$.

5. True. See the Vertical Line Test.

7. False. Let $f(x) = x^3$. Then f is one-to-one and $f^{-1}(x) = \sqrt[3]{x}$. But $1/f(x) = 1/x^3$, which is not equal to $f^{-1}(x)$.

9. True. The function $\ln x$ is an increasing function on $(0, \infty)$.

11. False. Let $x = e^2$ and $a = e$. Then $\dfrac{\ln x}{\ln a} = \dfrac{\ln e^2}{\ln e} = \dfrac{2 \ln e}{\ln e} = 2$ and $\ln \dfrac{x}{a} = \ln \dfrac{e^2}{e} = \ln e = 1$, so in general the statement is false. What *is* true, however, is that $\ln \dfrac{x}{a} = \ln x - \ln a$.

--- EXERCISES ---

1. (a) When $x = 2$, $y \approx 2.7$. Thus, $f(2) \approx 2.7$.

(b) $f(x) = 3 \Rightarrow x \approx 2.3, 5.6$

(c) The domain of f is $-6 \le x \le 6$, or $[-6, 6]$.

(d) The range of f is $-4 \le y \le 4$, or $[-4, 4]$.

(e) f is increasing on $[-4, 4]$, that is, on $-4 \le x \le 4$.

(f) f is not one-to-one since it fails the Horizontal Line Test.

(g) f is odd since its graph is symmetric about the origin.

3. (a)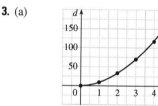

(b) From the graph, we see that the distance traveled after 4.5 seconds is slightly less than 150 feet.

5. $f(x) = \sqrt{4 - 3x^2}$. Domain: $4 - 3x^2 \ge 0 \Rightarrow 3x^2 \le 4 \Rightarrow x^2 \le \frac{4}{3} \Rightarrow |x| \le \frac{2}{\sqrt{3}}$. Range: $y \ge 0$ and $y \le \sqrt{4} \Rightarrow 0 \le y \le 2$.

7. $y = 1 + \sin x$. Domain: $\mathbb{R}$. Range: $-1 \le \sin x \le 1 \Rightarrow 0 \le 1 + \sin x \le 2 \Rightarrow 0 \le y \le 2$.

9. (a) To obtain the graph of $y = f(x) + 8$, we shift the graph of $y = f(x)$ up 8 units.

(b) To obtain the graph of $y = f(x + 8)$, we shift the graph of $y = f(x)$ left 8 units.

(c) To obtain the graph of $y = 1 + 2f(x)$, we stretch the graph of $y = f(x)$ vertically by a factor of 2, and then shift the resulting graph 1 unit upward.

(d) To obtain the graph of $y = f(x - 2) - 2$, we shift the graph of $y = f(x)$ right 2 units (for the "-2" inside the parentheses), and then shift the resulting graph 2 units downward.

(e) To obtain the graph of $y = -f(x)$, we reflect the graph of $y = f(x)$ about the x-axis.

(f) To obtain the graph of $y = f^{-1}(x)$, we reflect the graph of $y = f(x)$ about the line $y = x$ (assuming f is one-to-one).

11. $y = -\sin 2x$: Start with the graph of $y = \sin x$, compress horizontally by a factor of 2, and reflect about the x-axis.

13. $y = (1 + e^x)/2$: Start with the graph of $y = e^x$, shift 1 unit upward, and compress vertically by a factor of 2.

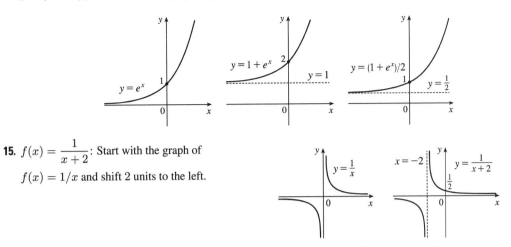

15. $f(x) = \dfrac{1}{x+2}$: Start with the graph of

$f(x) = 1/x$ and shift 2 units to the left.

17. (a) The terms of f are a mixture of odd and even powers of x, so f is neither even nor odd.

(b) The terms of f are all odd powers of x, so f is odd.

(c) $f(-x) = e^{-(-x)^2} = e^{-x^2} = f(x)$, so f is even.

(d) $f(-x) = 1 + \sin(-x) = 1 - \sin x$. Now $f(-x) \neq f(x)$ and $f(-x) \neq -f(x)$, so f is neither even nor odd.

19. $f(x) = \ln x$, $\quad D = (0, \infty)$; $\quad g(x) = x^2 - 9$, $\quad D = \mathbb{R}$.

$(f \circ g)(x) = f(g(x)) = f(x^2 - 9) = \ln(x^2 - 9)$.

$\quad$ Domain: $x^2 - 9 > 0 \;\; \Rightarrow \;\; x^2 > 9 \;\; \Rightarrow \;\; |x| > 3 \;\; \Rightarrow \;\; x \in (-\infty, -3) \cup (3, \infty)$

$(g \circ f)(x) = g(f(x)) = g(\ln x) = (\ln x)^2 - 9.$ $\quad$ Domain: $x > 0$, or $(0, \infty)$

$(f \circ f)(x) = f(f(x)) = f(\ln x) = \ln(\ln x).$ $\quad$ Domain: $\ln x > 0 \;\; \Rightarrow \;\; x > e^0 = 1$, or $(1, \infty)$

$(g \circ g)(x) = g(g(x)) = g(x^2 - 9) = (x^2 - 9)^2 - 9.$ $\quad$ Domain: $x \in \mathbb{R}$, or $(-\infty, \infty)$

21.

Many models appear to be plausible. Your choice depends on whether you think medical advances will keep increasing life expectancy, or if there is bound to be a natural leveling-off of life expectancy. A linear model, $y = 0.2493x - 423.4818$ gives us an estimate of 77.6 years for the year 2010.

23. We need to know the value of x such that $f(x) = 2x + \ln x = 2$. Since $x = 1$ gives us $y = 2$, $f^{-1}(2) = 1$.

25. (a) $e^{2 \ln 3} = \left(e^{\ln 3} \right)^2 = 3^2 = 9$

(b) $\log_{10} 25 + \log_{10} 4 = \log_{10}(25 \cdot 4) = \log_{10} 100 = \log_{10} 10^2 = 2$

(c) $\tan\left(\arcsin \frac{1}{2} \right) = \tan \frac{\pi}{6} = \frac{1}{\sqrt{3}}$

(d) Let $\theta = \cos^{-1} \frac{4}{5}$, so $\cos \theta = \frac{4}{5}$. Then $\sin\left(\cos^{-1} \frac{4}{5} \right) = \sin \theta = \sqrt{1 - \cos^2 \theta} = \sqrt{1 - \left(\frac{4}{5} \right)^2} = \sqrt{\frac{9}{25}} = \frac{3}{5}$.

27. (a) After 4 days, $\frac{1}{2}$ gram remains; after 8 days, $\frac{1}{4}$ g; after 12 days, $\frac{1}{8}$ g; after 16 days, $\frac{1}{16}$ g.

(b) $m(4) = \dfrac{1}{2}$, $m(8) = \dfrac{1}{2^2}$, $m(12) = \dfrac{1}{2^3}$, $m(16) = \dfrac{1}{2^4}$. From the pattern, we see that $m(t) = \dfrac{1}{2^{t/4}}$, or $2^{-t/4}$.

(c) $m = 2^{-t/4} \implies \log_2 m = -t/4 \implies t = -4\log_2 m$; this is the time elapsed when there are m grams of ^{100}Pd.

(d) $m = 0.01 \implies t = -4\log_2 0.01 = -4\left(\dfrac{\ln 0.01}{\ln 2}\right) \approx 26.6$ days

29. $f(x) = \ln(x^2 - c)$. If $c < 0$, the domain of f is $\mathbb{R}$. If $c = 0$, the domain of f is $(-\infty, 0) \cup (0, \infty)$. If $c > 0$, the domain of f is $(-\infty, -\sqrt{c}) \cup (\sqrt{c}, \infty)$. As c increases, the dip at $x = 0$ becomes deeper. For $c \geq 0$, the graph has asymptotes at $x = \pm\sqrt{c}$.

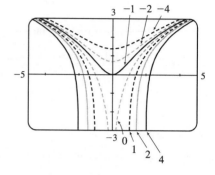

☐ PRINCIPLES OF PROBLEM SOLVING

1.

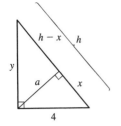

4

By using the area formula for a triangle, $\frac{1}{2}$ (base) (height), in two ways,

we see that $\frac{1}{2}\,(4)\,(y) = \frac{1}{2}\,(h)\,(a)$, so $a = \dfrac{4y}{h}$. Since $4^2 + y^2 = h^2$,

$y = \sqrt{h^2 - 16}$, and $a = \dfrac{4\sqrt{h^2 - 16}}{h}$.

3. $|2x - 1| = \begin{cases} 2x - 1 & \text{if } x \geq \frac{1}{2} \\ 1 - 2x & \text{if } x < \frac{1}{2} \end{cases}$ and $|x + 5| = \begin{cases} x + 5 & \text{if } x \geq -5 \\ -x - 5 & \text{if } x < -5 \end{cases}$

Therefore, we consider the three cases $x < -5$, $-5 \leq x < \frac{1}{2}$, and $x \geq \frac{1}{2}$.

If $x < -5$, we must have $1 - 2x - (-x - 5) = 3 \Leftrightarrow x = 3$, which is false, since we are considering $x < -5$.

If $-5 \leq x < \frac{1}{2}$, we must have $1 - 2x - (x + 5) = 3 \Leftrightarrow x = -\frac{7}{3}$.

If $x \geq \frac{1}{2}$, we must have $2x - 1 - (x + 5) = 3 \Leftrightarrow x = 9$.

So the two solutions of the equation are $x = -\frac{7}{3}$ and $x = 9$.

5. $f(x) = \left|x^2 - 4\,|x| + 3\right|$. If $x \geq 0$, then $f(x) = \left|x^2 - 4x + 3\right| = |(x - 1)(x - 3)|$.

Case (i):   If $0 < x \leq 1$, then $f(x) = x^2 - 4x + 3$.

Case (ii):   If $1 < x \leq 3$, then $f(x) = -(x^2 - 4x + 3) = -x^2 + 4x - 3$.

Case (iii):   If $x > 3$, then $f(x) = x^2 - 4x + 3$.

This enables us to sketch the graph for $x \geq 0$. Then we use the fact that f is an

even function to reflect this part of the graph about the y-axis to obtain the

entire graph. Or, we could consider also the cases $x < -3$, $-3 \leq x < -1$,

and $-1 \leq x < 0$.

7. Remember that $|a| = a$ if $a \geq 0$ and that $|a| = -a$ if $a < 0$. Thus,

$$x + |x| = \begin{cases} 2x & \text{if } x \geq 0 \\ 0 & \text{if } x < 0 \end{cases} \quad \text{and} \quad y + |y| = \begin{cases} 2y & \text{if } y \geq 0 \\ 0 & \text{if } y < 0 \end{cases}$$

We will consider the equation $x + |x| = y + |y|$ in four cases.

(1) $x \geq 0, y \geq 0$	(2) $x \geq 0, y < 0$	(3) $x < 0, y \geq 0$	(4) $x < 0, y < 0$
$2x = 2y$	$2x = 0$	$0 = 2y$	$0 = 0$
$x = y$	$x = 0$	$0 = y$	

Case 1 gives us the line $y = x$ with nonnegative x and y.

Case 2 gives us the portion of the y-axis with y negative.

Case 3 gives us the portion of the x-axis with x negative.

Case 4 gives us the entire third quadrant.

9. $|x| + |y| \leq 1$. The boundary of the region has equation $|x| + |y| = 1$.

In quadrants I, II, III, and IV, this becomes the lines $x + y = 1$,

$-x + y = 1$, $-x - y = 1$, and $x - y = 1$ respectively.

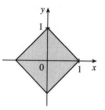

11. $(\log_2 3)(\log_3 4)(\log_4 5) \cdots (\log_{31} 32) = \left(\dfrac{\ln 3}{\ln 2}\right)\left(\dfrac{\ln 4}{\ln 3}\right)\left(\dfrac{\ln 5}{\ln 4}\right) \cdots \left(\dfrac{\ln 32}{\ln 31}\right) = \dfrac{\ln 32}{\ln 2} = \dfrac{\ln 2^5}{\ln 2} = \dfrac{5\ln 2}{\ln 2} = 5$

13. $\ln(x^2 - 2x - 2) \leq 0 \;\Rightarrow\; x^2 - 2x - 2 \leq e^0 = 1 \;\Rightarrow\; x^2 - 2x - 3 \leq 0 \;\Rightarrow\; (x - 3)(x + 1) \leq 0 \;\Rightarrow$

$x \in [-1, 3]$. Since the argument must be positive, $x^2 - 2x - 2 > 0 \;\Rightarrow\; \left[x - (1 - \sqrt{3})\right]\left[x - (1 + \sqrt{3})\right] > 0$

$\Rightarrow\; x \in (-\infty, 1 - \sqrt{3}) \cup (1 + \sqrt{3}, \infty)$. The intersection of these intervals is $[-1, 1 - \sqrt{3}) \cup (1 + \sqrt{3}, 3]$.

15. Let d be the distance traveled on each half of the trip. Let t_1 and t_2 be the times taken for the first and second halves
of the trip.

For the first half of the trip we have $t_1 = d/30$ and for the second half we have $t_2 = d/60$. Thus, the average speed

for the entire trip is $\dfrac{\text{total distance}}{\text{total time}} = \dfrac{2d}{t_1 + t_2} = \dfrac{2d}{\dfrac{d}{30} + \dfrac{d}{60}} \cdot \dfrac{60}{60} = \dfrac{120d}{2d + d} = \dfrac{120d}{3d} = 40$. The average speed for

the entire trip is 40 mi/h.

17. Let S_n be the statement that $7^n - 1$ is divisible by 6.

- S_1 is true because $7^1 - 1 = 6$ is divisible by 6.

- Assume S_k is true, that is, $7^k - 1$ is divisible by 6. In other words, $7^k - 1 = 6m$ for some positive integer m.
  Then $7^{k+1} - 1 = 7^k \cdot 7 - 1 = (6m + 1) \cdot 7 - 1 = 42m + 6 = 6(7m + 1)$, which is divisible by 6, so S_{k+1}
  is true.

- Therefore, by mathematical induction, $7^n - 1$ is divisible by 6 for every positive integer n.

19. $f_0(x) = x^2$ and $f_{n+1}(x) = f_0(f_n(x))$ for $n = 0, 1, 2, \ldots$.

$f_1(x) = f_0(f_0(x)) = f_0(x^2) = (x^2)^2 = x^4$, $f_2(x) = f_0(f_1(x)) = f_0(x^4) = (x^4)^2 = x^8$,

$f_3(x) = f_0(f_2(x)) = f_0(x^8) = (x^8)^2 = x^{16}, \ldots$. Thus, a general formula is $f_n(x) = x^{2^{n+1}}$.

2 ☐ LIMITS AND DERIVATIVES

2.1 The Tangent and Velocity Problems

1. (a) Using $P(15, 250)$, we construct the following table:

t	Q	slope $= m_{PQ}$
5	$(5, 694)$	$\frac{694-250}{5-15} = -\frac{444}{10} = -44.4$
10	$(10, 444)$	$\frac{444-250}{10-15} = -\frac{194}{5} = -38.8$
20	$(20, 111)$	$\frac{111-250}{20-15} = -\frac{139}{5} = -27.8$
25	$(25, 28)$	$\frac{28-250}{25-15} = -\frac{222}{10} = -22.2$
30	$(30, 0)$	$\frac{0-250}{30-15} = -\frac{250}{15} = -16.\overline{6}$

(b) Using the values of t that correspond to the points closest to P ($t = 10$ and $t = 20$), we have

$$\frac{-38.8 + (-27.8)}{2} = -33.3$$

(c) From the graph, we can estimate the slope of the tangent line at P to be $\frac{-300}{9} = -33.\overline{3}$.

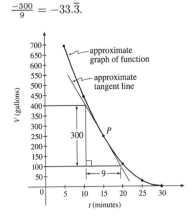

3. For the curve $y = x/(1+x)$ and the point $P\left(1, \frac{1}{2}\right)$:

(a)

	x	Q	m_{PQ}
(i)	0.5	$(0.5, 0.333333)$	0.333333
(ii)	0.9	$(0.9, 0.473684)$	0.263158
(iii)	0.99	$(0.99, 0.497487)$	0.251256
(iv)	0.999	$(0.999, 0.499750)$	0.250125
(v)	1.5	$(1.5, 0.6)$	0.2
(vi)	1.1	$(1.1, 0.523810)$	0.238095
(vii)	1.01	$(1.01, 0.502488)$	0.248756
(viii)	1.001	$(1.001, 0.500250)$	0.249875

(b) The slope appears to be $\frac{1}{4}$.

(c) $y - \frac{1}{2} = \frac{1}{4}(x - 1)$ or
$$y = \frac{1}{4}x + \frac{1}{4}.$$

5. (a) $y = y(t) = 40t - 16t^2$. At $t = 2$, $y = 40(2) - 16(2)^2 = 16$. The average velocity between times 2 and $2 + h$

is $v_{\text{ave}} = \dfrac{y(2 + h) - y(2)}{(2 + h) - 2} = \dfrac{[40(2 + h) - 16(2 + h)^2] - 16}{h} = \dfrac{-24h - 16h^2}{h} = -24 - 16h$, if $h \neq 0$.

(i) $[2, 2.5]$: $h = 0.5$, $v_{\text{ave}} = -32$ ft/s   (ii) $[2, 2.1]$: $h = 0.1$, $v_{\text{ave}} = -25.6$ ft/s

(iii) $[2, 2.05]$: $h = 0.05$, $v_{\text{ave}} = -24.8$ ft/s   (iv) $[2, 2.01]$: $h = 0.01$, $v_{\text{ave}} = -24.16$ ft/s

(b) The instantaneous velocity when $t = 2$ (h approaches 0) is -24 ft/s.

7. $s = s(t) = t^3/6$. Average velocity between times 1 and $1 + h$ is

$$v_{\text{ave}} = \frac{s(1+h) - s(1)}{(1+h) - 1} = \frac{(1+h)^3/6 - 1/6}{h} = \frac{h^3 + 3h^2 + 3h}{6h} = \frac{h^2 + 3h + 3}{6} \text{ if } h \neq 0.$$

(a)   (i) $[1, 3]$: $h = 2$, $v_{\text{ave}} = \frac{13}{6}$ ft/s   (ii) $[1, 2]$: $h = 1$, $v_{\text{ave}} = \frac{7}{6}$ ft/s

(iii) $[1, 1.5]$: $h = 0.5$, $v_{\text{ave}} = \frac{19}{24}$ ft/s   (iv) $[1, 1.1]$: $h = 0.1$, $v_{\text{ave}} = \frac{331}{600}$ ft/s

(b) As h approaches 0, the velocity approaches $\frac{3}{6} = \frac{1}{2}$ ft/s.

(c)

(d)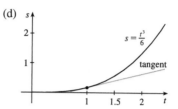

9. For the curve $y = \sin(10\pi/x)$ and the point $P(1, 0)$:

(a)

x	Q	m_{PQ}
2	$(2, 0)$	0
1.5	$(1.5, 0.8660)$	1.7321
1.4	$(1.4, -0.4339)$	-1.0847
1.3	$(1.3, -0.8230)$	-2.7433
1.2	$(1.2, 0.8660)$	4.3301
1.1	$(1.1, -0.2817)$	-2.8173

x	Q	m_{PQ}
0.5	$(0.5, 0)$	0
0.6	$(0.6, 0.8660)$	-2.1651
0.7	$(0.7, 0.7818)$	-2.6061
0.8	$(0.8, 1)$	-5
0.9	$(0.9, -0.3420)$	3.4202

As x approaches 1, the slopes do not appear to be approaching any particular value.

(b)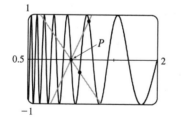

We see that problems with estimation are caused by the frequent oscillations of the graph. The tangent is so steep at P that we need to take x-values much closer to 1 in order to get accurate estimates of its slope.

(c) If we choose $x = 1.001$, then the point Q is $(1.001, -0.0314)$ and $m_{PQ} \approx -31.3794$. If $x = 0.999$, then Q is $(0.999, 0.0314)$ and $m_{PQ} = -31.4422$. The average of these slopes is -31.4108. So we estimate that the slope of the tangent line at P is about -31.4.

2.2 The Limit of a Function

1. As x approaches 2, $f(x)$ approaches 5. [Or, the values of $f(x)$ can be made as close to 5 as we like by taking x sufficiently close to 2 (but $x \neq 2$).] Yes, the graph could have a hole at $(2, 5)$ and be defined such that $f(2) = 3$.

3. (a) $\lim\limits_{x \to -3} f(x) = \infty$ means that the values of $f(x)$ can be made arbitrarily large (as large as we please) by taking x sufficiently close to -3 (but not equal to -3).

(b) $\lim\limits_{x \to 4+} f(x) = -\infty$ means that the values of $f(x)$ can be made arbitrarily large negative by taking x sufficiently close to 4 through values larger than 4.

5. (a) $f(x)$ approaches 2 as x approaches 1 from the left, so $\lim\limits_{x \to 1^-} f(x) = 2$.

(b) $f(x)$ approaches 3 as x approaches 1 from the right, so $\lim\limits_{x \to 1+} f(x) = 3$.

(c) $\lim\limits_{x \to 1} f(x)$ does not exist because the limits in part (a) and part (b) are not equal.

(d) $f(x)$ approaches 4 as x approaches 5 from the left and from the right, so $\lim\limits_{x \to 5} f(x) = 4$.

(e) $f(5)$ is not defined, so it doesn't exist.

7. (a) $\lim\limits_{t \to 0^-} g(t) = -1$ (b) $\lim\limits_{t \to 0+} g(t) = -2$

(c) $\lim\limits_{t \to 0} g(t)$ does not exist because the limits in part (a) and part (b) are not equal.

(d) $\lim\limits_{t \to 2^-} g(t) = 2$ (e) $\lim\limits_{t \to 2+} g(t) = 0$

(f) $\lim\limits_{t \to 2} g(t)$ does not exist because the limits in part (d) and part (e) are not equal.

(g) $g(2) = 1$ (h) $\lim\limits_{t \to 4} g(t) = 3$

9. (a) $\lim\limits_{x \to -7} f(x) = -\infty$ (b) $\lim\limits_{x \to -3} f(x) = \infty$ (c) $\lim\limits_{x \to 0} f(x) = \infty$

(d) $\lim\limits_{x \to 6^-} f(x) = -\infty$ (e) $\lim\limits_{x \to 6+} f(x) = \infty$

(f) The equations of the vertical asymptotes are $x = -7$, $x = -3$, $x = 0$, and $x = 6$.

11.

(a) $\lim\limits_{x \to 0^-} f(x) = 1$

(b) $\lim\limits_{x \to 0+} f(x) = 0$

(c) $\lim\limits_{x \to 0} f(x)$ does not exist because the limits in part (a) and part (b) are not equal.

13. $\lim\limits_{x \to 3^+} f(x) = 4$, $\quad \lim\limits_{x \to 3^-} f(x) = 2$,

$\lim\limits_{x \to -2} f(x) = 2$, $\quad f(3) = 3$, $\quad f(-2) = 1$

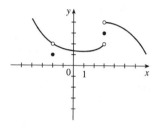

15. For $f(x) = \dfrac{x^2 - 2x}{x^2 - x - 2}$:

x	$f(x)$	x	$f(x)$
2.5	0.714286	1.9	0.655172
2.1	0.677419	1.95	0.661017
2.05	0.672131	1.99	0.665552
2.01	0.667774	1.995	0.666110
2.005	0.667221	1.999	0.666556
2.001	0.666778		

It appears that $\lim\limits_{x \to 2} \dfrac{x^2 - 2x}{x^2 - x - 2} = 0.\bar{6} = \frac{2}{3}$.

17. For $f(x) = \dfrac{e^x - 1 - x}{x^2}$:

x	$f(x)$	x	$f(x)$
1	0.718282	−1	0.367879
0.5	0.594885	−0.5	0.426123
0.1	0.517092	−0.1	0.483742
0.05	0.508439	−0.05	0.491770
0.01	0.501671	−0.01	0.498337

It appears that $\lim\limits_{x \to 0} \dfrac{e^x - 1 - x}{x^2} = 0.5 = \frac{1}{2}$.

19. For $f(x) = \dfrac{\sqrt{x + 4} - 2}{x}$:

x	$f(x)$	x	$f(x)$
1	0.236068	−1	0.267949
0.5	0.242641	−0.5	0.258343
0.1	0.248457	−0.1	0.251582
0.05	0.249224	−0.05	0.250786
0.01	0.249844	−0.01	0.250156

It appears that $\lim\limits_{x \to 0} \dfrac{\sqrt{x + 4} - 2}{x} = 0.25 = \frac{1}{4}$.

21. For $f(x) = \dfrac{x^6 - 1}{x^{10} - 1}$:

x	$f(x)$	x	$f(x)$
0.5	0.985337	1.5	0.183369
0.9	0.719397	1.1	0.484119
0.95	0.660186	1.05	0.540783
0.99	0.612018	1.01	0.588022
0.999	0.601200	1.001	0.598800

It appears that $\lim\limits_{x \to 1} \dfrac{x^6 - 1}{x^{10} - 1} = 0.6 = \frac{3}{5}$.

23. $\lim\limits_{x \to 5^+} \dfrac{6}{x - 5} = \infty$ since $(x - 5) \to 0$ as $x \to 5^+$ and $\dfrac{6}{x - 5} > 0$ for $x > 5$.

25. $\lim\limits_{x \to 1} \dfrac{2 - x}{(x - 1)^2} = \infty$ since the numerator is positive and the denominator approaches 0 through positive values

as $x \to 1$.

27. $\lim\limits_{x\to-2+} \dfrac{x-1}{x^2(x+2)} = -\infty$ since $(x+2) \to 0$ as $x \to -2^+$ and $\dfrac{x-1}{x^2(x+2)} < 0$ for $-2 < x < 0$.

29. $\lim\limits_{x\to(-\pi/2)^-} \sec x = \lim\limits_{x\to(-\pi/2)^-} (1/\cos x) = -\infty$ since $\cos x \to 0$ as $x \to (-\pi/2)^-$ and $\cos x < 0$ for

$-\pi < x < -\pi/2$.

31. (a) $f(x) = 1/(x^3 - 1)$

x	$f(x)$
0.5	-1.14
0.9	-3.69
0.99	-33.7
0.999	-333.7
0.9999	-3333.7
0.99999	$-33{,}333.7$

x	$f(x)$
1.5	0.42
1.1	3.02
1.01	33.0
1.001	333.0
1.0001	3333.0
1.00001	33,333.3

From these calculations, it seems that $\lim\limits_{x\to1^-} f(x) = -\infty$ and $\lim\limits_{x\to1^+} f(x) = \infty$.

(b) If x is slightly smaller than 1, then $x^3 - 1$ will be a negative number close to 0, and the reciprocal of $x^3 - 1$, that is, $f(x)$, will be a negative number with large absolute value. So $\lim\limits_{x\to1^-} f(x) = -\infty$.

If x is slightly larger than 1, then $x^3 - 1$ will be a small positive number, and its reciprocal, $f(x)$, will be a large positive number. So $\lim\limits_{x\to1^+} f(x) = \infty$.

(c) It appears from the graph of f that $\lim\limits_{x\to1^-} f(x) = -\infty$ and

$\lim\limits_{x\to1^+} f(x) = \infty$.

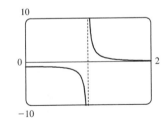

33. (a) Let $h(x) = (1+x)^{1/x}$.

(b)

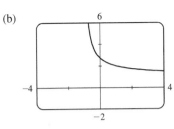

x	$h(x)$
-0.001	2.71964
-0.0001	2.71842
-0.00001	2.71830
-0.000001	2.71828
0.000001	2.71828
0.00001	2.71827
0.0001	2.71815
0.001	2.71692

It appears that $\lim\limits_{x\to0} (1+x)^{1/x} \approx 2.71828$, which is approximately e.

In Section 3.8 we will see that the value of the limit is exactly e.

35. For $f(x) = x^2 - (2^x/1000)$:

(a)

x	$f(x)$
1	0.998000
0.8	0.638259
0.6	0.358484
0.4	0.158680
0.2	0.038851
0.1	0.008928
0.05	0.001465

It appears that $\lim\limits_{x \to 0} f(x) = 0$.

(b)

x	$f(x)$
0.04	0.000572
0.02	−0.000614
0.01	−0.000907
0.005	−0.000978
0.003	−0.000993
0.001	−0.001000

It appears that $\lim\limits_{x \to 0} f(x) = -0.001$.

37. No matter how many times we zoom in toward the origin, the graphs of $f(x) = \sin(\pi/x)$ appear to consist of almost-vertical lines. This indicates more and more frequent oscillations as $x \to 0$.

39.

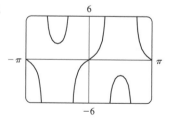

There appear to be vertical asymptotes of the curve $y = \tan(2 \sin x)$ at $x \approx \pm 0.90$ and $x \approx \pm 2.24$. To find the exact equations of these asymptotes, we note that the graph of the tangent function has vertical asymptotes at $x = \frac{\pi}{2} + \pi n$. Thus, we must have $2 \sin x = \frac{\pi}{2} + \pi n$, or equivalently, $\sin x = \frac{\pi}{4} + \frac{\pi}{2} n$. Since $-1 \leq \sin x \leq 1$, we must have $\sin x = \pm \frac{\pi}{4}$ and so $x = \pm \sin^{-1} \frac{\pi}{4}$ (corresponding to $x \approx \pm 0.90$).

Just as $150°$ is the reference angle for $30°$, $\pi - \sin^{-1} \frac{\pi}{4}$ is the reference angle for $\sin^{-1} \frac{\pi}{4}$. So $x = \pm \left(\pi - \sin^{-1} \frac{\pi}{4}\right)$ are also equations of the vertical asymptotes (corresponding to $x \approx \pm 2.24$).

2.3 Calculating Limits Using the Limit Laws

1. (a) $\lim\limits_{x \to a} [f(x) + h(x)] = \lim\limits_{x \to a} f(x) + \lim\limits_{x \to a} h(x)$

$$= -3 + 8 = 5$$

(b) $\lim\limits_{x \to a} [f(x)]^2 = \left[\lim\limits_{x \to a} f(x)\right]^2 = (-3)^2 = 9$

(c) $\lim\limits_{x \to a} \sqrt[3]{h(x)} = \sqrt[3]{\lim\limits_{x \to a} h(x)} = \sqrt[3]{8} = 2$

(d) $\lim\limits_{x \to a} \dfrac{1}{f(x)} = \dfrac{1}{\lim\limits_{x \to a} f(x)} = \dfrac{1}{-3} = -\dfrac{1}{3}$

(e) $\lim\limits_{x \to a} \dfrac{f(x)}{h(x)} = \dfrac{\lim\limits_{x \to a} f(x)}{\lim\limits_{x \to a} h(x)} = \dfrac{-3}{8} = -\dfrac{3}{8}$

(f) $\lim\limits_{x \to a} \dfrac{g(x)}{f(x)} = \dfrac{\lim\limits_{x \to a} g(x)}{\lim\limits_{x \to a} f(x)} = \dfrac{0}{-3} = 0$

(g) The limit does not exist, since $\lim\limits_{x \to a} g(x) = 0$ but $\lim\limits_{x \to a} f(x) \neq 0$.

(h) $\lim\limits_{x \to a} \dfrac{2f(x)}{h(x) - f(x)} = \dfrac{2\lim\limits_{x \to a} f(x)}{\lim\limits_{x \to a} h(x) - \lim\limits_{x \to a} f(x)} = \dfrac{2(-3)}{8 - (-3)} = -\dfrac{6}{11}$

3. $\lim\limits_{x \to -2} (3x^4 + 2x^2 - x + 1) = \lim\limits_{x \to -2} 3x^4 + \lim\limits_{x \to -2} 2x^2 - \lim\limits_{x \to -2} x + \lim\limits_{x \to -2} 1$ [Limit Laws 1 and 2]

$$= 3 \lim\limits_{x \to -2} x^4 + 2 \lim\limits_{x \to -2} x^2 - \lim\limits_{x \to -2} x + \lim\limits_{x \to -2} 1 \qquad [3]$$

$$= 3(-2)^4 + 2(-2)^2 - (-2) + (1) \qquad [9, 8, \text{ and } 7]$$

$$= 48 + 8 + 2 + 1 = 59$$

5. $\lim\limits_{x \to 3} (x^2 - 4)(x^3 + 5x - 1) = \lim\limits_{x \to 3} (x^2 - 4) \cdot \lim\limits_{x \to 3} (x^3 + 5x - 1)$ [Limit Law 4]

$$= \left(\lim\limits_{x \to 3} x^2 - \lim\limits_{x \to 3} 4\right) \cdot \left(\lim\limits_{x \to 3} x^3 + 5 \lim\limits_{x \to 3} x - \lim\limits_{x \to 3} 1\right) \qquad [2, 1, \text{ and } 3]$$

$$= (3^2 - 4) \cdot (3^3 + 5 \cdot 3 - 1) \qquad [7, 8, \text{ and } 9]$$

$$= 5 \cdot 41 = 205$$

7. $\lim\limits_{x \to 1} \left(\dfrac{1 + 3x}{1 + 4x^2 + 3x^4}\right)^3 = \left(\lim\limits_{x \to 1} \dfrac{1 + 3x}{1 + 4x^2 + 3x^4}\right)^3$ [6]

$$= \left[\dfrac{\lim\limits_{x \to 1} (1 + 3x)}{\lim\limits_{x \to 1} (1 + 4x^2 + 3x^4)}\right]^3 \qquad [5]$$

$$= \left[\dfrac{\lim\limits_{x \to 1} 1 + 3 \lim\limits_{x \to 1} x}{\lim\limits_{x \to 1} 1 + 4 \lim\limits_{x \to 1} x^2 + 3 \lim\limits_{x \to 1} x^4}\right]^3 \qquad [2, 1, \text{ and } 3]$$

$$= \left[\dfrac{1 + 3(1)}{1 + 4(1)^2 + 3(1)^4}\right]^3 = \left[\dfrac{4}{8}\right]^3 = \left(\dfrac{1}{2}\right)^3 = \dfrac{1}{8} \qquad [7, 8, \text{ and } 9]$$

9. $\lim\limits_{x \to 4^-} \sqrt{16 - x^2} = \sqrt{\lim\limits_{x \to 4^-} (16 - x^2)}$ [11]

$$= \sqrt{\lim\limits_{x \to 4^-} 16 - \lim\limits_{x \to 4^-} x^2} \qquad [2]$$

$$= \sqrt{16 - (4)^2} = 0 \qquad [7 \text{ and } 9]$$

11. $\lim\limits_{x \to 2} \dfrac{x^2 + x - 6}{x - 2} = \lim\limits_{x \to 2} \dfrac{(x + 3)(x - 2)}{x - 2} = \lim\limits_{x \to 2} (x + 3) = 2 + 3 = 5$

13. $\lim\limits_{x \to 2} \dfrac{x^2 - x + 6}{x - 2}$ does not exist since $x - 2 \to 0$ but $x^2 - x + 6 \to 8$ as $x \to 2$.

15. $\lim\limits_{t \to -3} \dfrac{t^2 - 9}{2t^2 + 7t + 3} = \lim\limits_{t \to -3} \dfrac{(t+3)(t-3)}{(2t+1)(t+3)} = \lim\limits_{t \to -3} \dfrac{t-3}{2t+1} = \dfrac{-3-3}{2(-3)+1} = \dfrac{-6}{-5} = \dfrac{6}{5}$

17. $\lim\limits_{h \to 0} \dfrac{(4+h)^2 - 16}{h} = \lim\limits_{h \to 0} \dfrac{(16 + 8h + h^2) - 16}{h} = \lim\limits_{h \to 0} \dfrac{8h + h^2}{h} = \lim\limits_{h \to 0} \dfrac{h(8+h)}{h} = \lim\limits_{h \to 0} (8+h) = 8 + 0 = 8$

19. $\lim\limits_{h \to 0} \dfrac{(1+h)^4 - 1}{h} = \lim\limits_{h \to 0} \dfrac{\left(1 + 4h + 6h^2 + 4h^3 + h^4\right) - 1}{h} = \lim\limits_{h \to 0} \dfrac{4h + 6h^2 + 4h^3 + h^4}{h}$

$\qquad = \lim\limits_{h \to 0} \dfrac{h(4 + 6h + 4h^2 + h^3)}{h} = \lim\limits_{h \to 0} \left(4 + 6h + 4h^2 + h^3\right) = 4 + 0 + 0 + 0 = 4$

21. $\lim\limits_{t \to 9} \dfrac{9 - t}{3 - \sqrt{t}} = \lim\limits_{t \to 9} \dfrac{\left(3 + \sqrt{t}\right)\left(3 - \sqrt{t}\right)}{3 - \sqrt{t}} = \lim\limits_{t \to 9} \left(3 + \sqrt{t}\right) = 3 + \sqrt{9} = 6$.

23. $\lim\limits_{x \to 7} \dfrac{\sqrt{x+2} - 3}{x - 7} = \lim\limits_{x \to 7} \dfrac{\sqrt{x+2} - 3}{x - 7} \cdot \dfrac{\sqrt{x+2} + 3}{\sqrt{x+2} + 3} = \lim\limits_{x \to 7} \dfrac{(x+2) - 9}{(x-7)\left(\sqrt{x+2} + 3\right)}$

$\qquad = \lim\limits_{x \to 7} \dfrac{x - 7}{(x-7)\left(\sqrt{x+2} + 3\right)} = \lim\limits_{x \to 7} \dfrac{1}{\sqrt{x+2} + 3} = \dfrac{1}{\sqrt{9} + 3} = \dfrac{1}{6}$

25. $\lim\limits_{x \to -4} \dfrac{\dfrac{1}{4} + \dfrac{1}{x}}{4 + x} = \lim\limits_{x \to -4} \dfrac{\dfrac{x+4}{4x}}{4 + x} = \lim\limits_{x \to -4} \dfrac{x + 4}{4x(4 + x)} = \lim\limits_{x \to -4} \dfrac{1}{4x} = \dfrac{1}{4(-4)} = -\dfrac{1}{16}$

27. $\lim\limits_{x \to 9} \dfrac{x^2 - 81}{\sqrt{x} - 3} = \lim\limits_{x \to 9} \dfrac{(x-9)(x+9)}{\sqrt{x} - 3} = \lim\limits_{x \to 9} \dfrac{(\sqrt{x} - 3)(\sqrt{x} + 3)(x+9)}{\sqrt{x} - 3}$ $\quad \begin{bmatrix} \text{factor } x - 9 \text{ as a} \\ \text{difference of squares} \end{bmatrix}$

$\qquad = \lim\limits_{x \to 9} \left[(\sqrt{x} + 3)(x+9)\right] = (\sqrt{9} + 3)(9 + 9) = 6 \cdot 18 = 108$

29. $\lim\limits_{t \to 0} \left(\dfrac{1}{t\sqrt{1+t}} - \dfrac{1}{t} \right) = \lim\limits_{t \to 0} \dfrac{1 - \sqrt{1+t}}{t\sqrt{1+t}} = \lim\limits_{t \to 0} \dfrac{\left(1 - \sqrt{1+t}\right)\left(1 + \sqrt{1+t}\right)}{t\sqrt{t+1}\left(1 + \sqrt{1+t}\right)} = \lim\limits_{t \to 0} \dfrac{-t}{t\sqrt{1+t}\left(1 + \sqrt{1+t}\right)}$

$\qquad = \lim\limits_{t \to 0} \dfrac{-1}{\sqrt{1+t}\left(1 + \sqrt{1+t}\right)} = \dfrac{-1}{\sqrt{1+0}\left(1 + \sqrt{1+0}\right)} = -\dfrac{1}{2}$

31. (a)

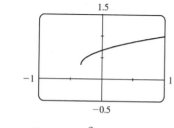

$\lim\limits_{x \to 0} \dfrac{x}{\sqrt{1+3x} - 1} \approx \dfrac{2}{3}$

(b)

x	$f(x)$
-0.001	0.6661663
-0.0001	0.6666167
-0.00001	0.6666617
-0.000001	0.6666662
0.000001	0.6666672
0.00001	0.6666717
0.0001	0.6667167
0.001	0.6671663

The limit appears to be $\dfrac{2}{3}$.

(c) $\lim\limits_{x \to 0} \left(\dfrac{x}{\sqrt{1+3x}-1} \cdot \dfrac{\sqrt{1+3x}+1}{\sqrt{1+3x}+1} \right) = \lim\limits_{x \to 0} \dfrac{x\left(\sqrt{1+3x}+1\right)}{(1+3x)-1} = \lim\limits_{x \to 0} \dfrac{x\left(\sqrt{1+3x}+1\right)}{3x}$

$\qquad\qquad\qquad = \dfrac{1}{3} \lim\limits_{x \to 0} \left(\sqrt{1+3x}+1\right)$ [Limit Law 3]

$\qquad\qquad\qquad = \dfrac{1}{3} \left[\sqrt{\lim\limits_{x \to 0} (1+3x)} + \lim\limits_{x \to 0} 1 \right]$ [1 and 11]

$\qquad\qquad\qquad = \dfrac{1}{3} \left(\sqrt{\lim\limits_{x \to 0} 1 + 3 \lim\limits_{x \to 0} x} + 1 \right)$ [1, 3, and 7]

$\qquad\qquad\qquad = \dfrac{1}{3} \left(\sqrt{1 + 3 \cdot 0} + 1 \right)$ [7 and 8]

$\qquad\qquad\qquad = \dfrac{1}{3}(1+1) = \dfrac{2}{3}$

33. Let $f(x) = -x^2$, $g(x) = x^2 \cos 20\pi x$ and $h(x) = x^2$. Then

$\qquad -1 \le \cos 20\pi x \le 1 \quad \Rightarrow \quad -x^2 \le x^2 \cos 20\pi x \le x^2 \quad \Rightarrow$

$\qquad f(x) \le g(x) \le h(x)$. So since $\lim\limits_{x \to 0} f(x) = \lim\limits_{x \to 0} h(x) = 0$, by the

Squeeze Theorem we have $\lim\limits_{x \to 0} g(x) = 0$.

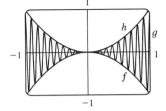

35. $1 \le f(x) \le x^2 + 2x + 2$ for all x. Now $\lim\limits_{x \to -1} 1 = 1$ and

$\qquad \lim\limits_{x \to -1} (x^2 + 2x + 2) = \lim\limits_{x \to -1} x^2 + 2 \lim\limits_{x \to -1} x + \lim\limits_{x \to -1} 2 = (-1)^2 + 2(-1) + 2 = 1$. Therefore, by the Squeeze

Theorem, $\lim\limits_{x \to -1} f(x) = 1$.

37. $-1 \le \cos(2/x) \le 1 \quad \Rightarrow \quad -x^4 \le x^4 \cos(2/x) \le x^4$. Since $\lim\limits_{x \to 0} \left(-x^4 \right) = 0$ and $\lim\limits_{x \to 0} x^4 = 0$, we have

$\qquad \lim\limits_{x \to 0} \left[x^4 \cos(2/x) \right] = 0$ by the Squeeze Theorem.

39. If $x > -4$, then $|x + 4| = x + 4$, so $\lim\limits_{x \to -4^+} |x + 4| = \lim\limits_{x \to -4^+} (x + 4) = -4 + 4 = 0$.

If $x < -4$, then $|x + 4| = -(x + 4)$, so $\lim\limits_{x \to -4^-} |x + 4| = \lim\limits_{x \to -4^-} -(x + 4) = -(-4 + 4) = 0$.

Since the right and left limits are equal, $\lim\limits_{x \to -4} |x + 4| = 0$.

41. If $x > 2$, then $|x - 2| = x - 2$, so $\lim\limits_{x \to 2^+} \dfrac{|x - 2|}{x - 2} = \lim\limits_{x \to 2^+} \dfrac{x - 2}{x - 2} = \lim\limits_{x \to 2^+} 1 = 1$. If $x < 2$, then

$|x - 2| = -(x - 2)$, so $\lim\limits_{x \to 2^-} \dfrac{|x - 2|}{x - 2} = \lim\limits_{x \to 2^-} \dfrac{-(x - 2)}{x - 2} = \lim\limits_{x \to 2^-} -1 = -1$. The right and left limits are

different, so $\lim\limits_{x \to 2} \dfrac{|x - 2|}{x - 2}$ does not exist.

43. Since $|x| = -x$ for $x < 0$, we have $\lim\limits_{x \to 0^-} \left(\dfrac{1}{x} - \dfrac{1}{|x|} \right) = \lim\limits_{x \to 0^-} \left(\dfrac{1}{x} - \dfrac{1}{-x} \right) = \lim\limits_{x \to 0^-} \dfrac{2}{x}$, which does not exist

since the denominator approaches 0 and the numerator does not.

45. (a)

(b) (i) Since sgn $x = 1$ for $x > 0$, $\lim\limits_{x \to 0^+} \text{sgn}\, x = \lim\limits_{x \to 0^+} 1 = 1$.

    (ii) Since sgn $x = -1$ for $x < 0$, $\lim\limits_{x \to 0^-} \text{sgn}\, x = \lim\limits_{x \to 0^-} -1 = -1$.

    (iii) Since $\lim\limits_{x \to 0^-} \text{sgn}\, x \neq \lim\limits_{x \to 0^+} \text{sgn}\, x$, $\lim\limits_{x \to 0} \text{sgn}\, x$ does not exist.

    (iv) Since $|\text{sgn}\, x| = 1$ for $x \neq 0$, $\lim\limits_{x \to 0} |\text{sgn}\, x| = \lim\limits_{x \to 0} 1 = 1$.

47. (a) (i) $\lim\limits_{x \to 1^+} \dfrac{x^2 - 1}{|x - 1|} = \lim\limits_{x \to 1^+} \dfrac{x^2 - 1}{x - 1} = \lim\limits_{x \to 1^+} (x + 1) = 2$

    (ii) $\lim\limits_{x \to 1^-} \dfrac{x^2 - 1}{|x - 1|} = \lim\limits_{x \to 1^-} \dfrac{x^2 - 1}{-(x - 1)} = \lim\limits_{x \to 1^-} -(x + 1) = -2$

(c)

(b) No, $\lim\limits_{x \to 1} F(x)$ does not exist since $\lim\limits_{x \to 1^+} F(x) \neq \lim\limits_{x \to 1^-} F(x)$.

49. (a) (i) $[\![x]\!] = -2$ for $-2 \leq x < -1$, so $\lim\limits_{x \to -2^+} [\![x]\!] = \lim\limits_{x \to -2^+} (-2) = -2$

    (ii) $[\![x]\!] = -3$ for $-3 \leq x < -2$, so $\lim\limits_{x \to -2^-} [\![x]\!] = \lim\limits_{x \to -2^-} (-3) = -3$. The right and left limits are different,

    so $\lim\limits_{x \to -2} [\![x]\!]$ does not exist.

    (iii) $[\![x]\!] = -3$ for $-3 \leq x < -2$, so $\lim\limits_{x \to -2.4} [\![x]\!] = \lim\limits_{x \to -2.4} (-3) = -3$.

  (b) (i) $[\![x]\!] = n - 1$ for $n - 1 \leq x < n$, so $\lim\limits_{x \to n^-} [\![x]\!] = \lim\limits_{x \to n^-} (n - 1) = n - 1$.

    (ii) $[\![x]\!] = n$ for $n \leq x < n + 1$, so $\lim\limits_{x \to n^+} [\![x]\!] = \lim\limits_{x \to n^+} n = n$.

  (c) $\lim\limits_{x \to a} [\![x]\!]$ exists $\Leftrightarrow$ a is not an integer.

51. The graph of $f(x) = [\![x]\!] + [\![-x]\!]$ is the same as the graph of $g(x) = -1$ with holes at each integer, since $f(a) = 0$
for any integer a. Thus, $\lim\limits_{x \to 2^-} f(x) = -1$ and $\lim\limits_{x \to 2^+} f(x) = -1$, so $\lim\limits_{x \to 2} f(x) = -1$. However,
$f(2) = [\![2]\!] + [\![-2]\!] = 2 + (-2) = 0$, so $\lim\limits_{x \to 2} f(x) \neq f(2)$.

53. Since $p(x)$ is a polynomial, $p(x) = a_0 + a_1 x + a_2 x^2 + \cdots + a_n x^n$. Thus, by the Limit Laws,

$$\lim_{x \to a} p(x) = \lim_{x \to a} \left(a_0 + a_1 x + a_2 x^2 + \cdots + a_n x^n\right)$$

$$= a_0 + a_1 \lim_{x \to a} x + a_2 \lim_{x \to a} x^2 + \cdots + a_n \lim_{x \to a} x^n$$

$$= a_0 + a_1 a + a_2 a^2 + \cdots + a_n a^n = p(a)$$

Thus, for any polynomial p, $\lim\limits_{x \to a} p(x) = p(a)$.

55. Observe that $0 \leq f(x) \leq x^2$ for all x, and $\lim\limits_{x \to 0} 0 = 0 = \lim\limits_{x \to 0} x^2$. So, by the Squeeze Theorem, $\lim\limits_{x \to 0} f(x) = 0$.

57. Let $f(x) = H(x)$ and $g(x) = 1 - H(x)$, where H is the Heaviside function defined in Exercise 1.3.59.
Thus, either f or g is 0 for any value of x. Then $\lim\limits_{x \to 0} f(x)$ and $\lim\limits_{x \to 0} g(x)$ do not exist, but

$\lim\limits_{x \to 0} [f(x)g(x)] = \lim\limits_{x \to 0} 0 = 0$.

59. Since the denominator approaches 0 as $x \to -2$, the limit will exist only if the numerator also approaches 0 as $x \to -2$. In order for this to happen, we need $\displaystyle\lim_{x \to -2} \left(3x^2 + ax + a + 3\right) = 0 \iff$

$3(-2)^2 + a(-2) + a + 3 = 0 \iff 12 - 2a + a + 3 = 0 \iff a = 15$. With $a = 15$, the limit becomes

$$\lim_{x \to -2} \frac{3x^2 + 15x + 18}{x^2 + x - 2} = \lim_{x \to -2} \frac{3(x+2)(x+3)}{(x-1)(x+2)} = \lim_{x \to -2} \frac{3(x+3)}{x-1} = \frac{3(-2+3)}{-2-1} = \frac{3}{-3} = -1.$$

2.4 The Precise Definition of a Limit

1. (a) To have $5x + 3$ within a distance of 0.1 of 13, we must have $12.9 \le 5x + 3 \le 13.1 \implies 9.9 \le 5x \le 10.1$ $\implies 1.98 \le x \le 2.02$. Thus, x must be within 0.02 units of 2 so that $5x + 3$ is within 0.1 of 13.

(b) Use 0.01 in place of 0.1 in part (a) to obtain 0.002.

3. On the left side of $x = 2$, we need $|x - 2| < \left|\frac{10}{7} - 2\right| = \frac{4}{7}$. On the right side, we need $|x - 2| < \left|\frac{10}{3} - 2\right| = \frac{4}{3}$. For both of these conditions to be satisfied at once, we need the more restrictive of the two to hold, that is, $|x - 2| < \frac{4}{7}$. So we can choose $\delta = \frac{4}{7}$, or any smaller positive number.

5. The leftmost question mark is the solution of $\sqrt{x} = 1.6$ and the rightmost, $\sqrt{x} = 2.4$. So the values are $1.6^2 = 2.56$ and $2.4^2 = 5.76$. On the left side, we need $|x - 4| < |2.56 - 4| = 1.44$. On the right side, we need $|x - 4| < |5.76 - 4| = 1.76$. To satisfy both conditions, we need the more restrictive condition to hold—namely, $|x - 4| < 1.44$. Thus, we can choose $\delta = 1.44$, or any smaller positive number.

7. $\left|\sqrt{4x+1} - 3\right| < 0.5 \iff 2.5 < \sqrt{4x+1} < 3.5$. We plot the three parts of this inequality on the same screen and identify the x-coordinates of the points of intersection using the cursor. It appears that the inequality holds for $1.3125 \le x \le 2.8125$. Since $|2 - 1.3125| = 0.6875$ and $|2 - 2.8125| = 0.8125$, we choose $0 < \delta < \min\{0.6875, 0.8125\} = 0.6875$.

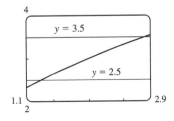

9. For $\varepsilon = 1$, the definition of a limit requires that we find δ such that $\left|\left(4 + x - 3x^3\right) - 2\right| < 1 \iff 1 < 4 + x - 3x^3 < 3$ whenever $0 < |x - 1| < \delta$. If we plot the graphs of $y = 1$, $y = 4 + x - 3x^3$ and $y = 3$ on the same screen, we see that we need $0.86 \le x \le 1.11$. So since $|1 - 0.86| = 0.14$ and $|1 - 1.11| = 0.11$, we choose $\delta = 0.11$ (or any smaller positive number). For $\varepsilon = 0.1$, we must find δ such that $\left|\left(4 + x - 3x^3\right) - 2\right| < 0.1 \iff 1.9 < 4 + x - 3x^3 < 2.1$ whenever $0 < |x - 1| < \delta$. From the graph, we see that we need $0.988 \le x \le 1.012$. So since $|1 - 0.988| = 0.012$ and $|1 - 1.012| = 0.012$, we choose $\delta = 0.012$ (or any smaller positive number) for the inequality to hold.

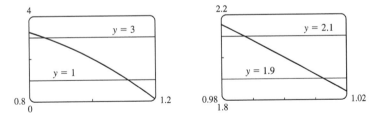

11. From the graph, we see that $\dfrac{x}{(x^2+1)(x-1)^2} > 100$ whenever

$0.93 \le x \le 1.07$. So since $|1 - 0.93| = 0.07$ and

$|1 - 1.07| = 0.07$, we can take $\delta = 0.07$ (or any smaller positive

number).

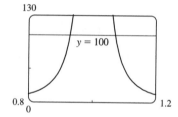

13. (a) $A = \pi r^2$ and $A = 1000 \text{ cm}^2 \quad \Rightarrow \quad \pi r^2 = 1000 \quad \Rightarrow \quad r^2 = \frac{1000}{\pi} \quad \Rightarrow$

$r = \sqrt{\dfrac{1000}{\pi}} \quad [r > 0] \quad \approx 17.8412 \text{ cm}.$

(b) $|A - 1000| \le 5 \quad \Rightarrow \quad -5 \le \pi r^2 - 1000 \le 5 \quad \Rightarrow \quad 1000 - 5 \le \pi r^2 \le 1000 + 5 \quad \Rightarrow$

$\sqrt{\dfrac{995}{\pi}} \le r \le \sqrt{\dfrac{1005}{\pi}} \quad \Rightarrow \quad 17.7966 \le r \le 17.8858.$ $\sqrt{\dfrac{1000}{\pi}} - \sqrt{\dfrac{995}{\pi}} \approx 0.04466$ and

$\sqrt{\dfrac{1005}{\pi}} - \sqrt{\dfrac{1000}{\pi}} \approx 0.04455.$ So if the machinist gets the radius within 0.0445 cm of 17.8412, the area will be

within 5 cm^2 of 1000.

(c) x is the radius, $f(x)$ is the area, a is the target radius given in part (a), L is the target area (1000), ε is the

tolerance in the area (5), and δ is the tolerance in the radius given in part (b).

15. Given $\varepsilon > 0$, we need $\delta > 0$ such that if $0 < |x - 1| < \delta$, then

$|(2x+3) - 5| < \varepsilon.$ But $|(2x+3) - 5| < \varepsilon \quad \Leftrightarrow \quad |2x - 2| < \varepsilon$

$\Leftrightarrow \quad 2|x - 1| < \varepsilon \quad \Leftrightarrow \quad |x - 1| < \varepsilon/2.$ So if we choose $\delta = \varepsilon/2$,

then $0 < |x - 1| < \delta \quad \Rightarrow \quad |(2x+3) - 5| < \varepsilon.$ Thus,

$\lim\limits_{x \to 1} (2x + 3) = 5$ by the definition of a limit.

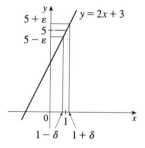

17. Given $\varepsilon > 0$, we need $\delta > 0$ such that if $0 < |x - (-3)| < \delta$, then

$|(1 - 4x) - 13| < \varepsilon.$ But $|(1 - 4x) - 13| < \varepsilon \quad \Leftrightarrow$

$|-4x - 12| < \varepsilon \quad \Leftrightarrow \quad |-4| \, |x + 3| < \varepsilon \quad \Leftrightarrow \quad |x - (-3)| < \varepsilon/4.$

So if we choose $\delta = \varepsilon/4$, then $0 < |x - (-3)| < \delta \quad \Rightarrow$

$|(1 - 4x) - 13| < \varepsilon.$ Thus, $\lim\limits_{x \to -3} (1 - 4x) = 13$ by the definition of

a limit.

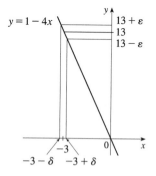

19. Given $\varepsilon > 0$, we need $\delta > 0$ such that if $0 < |x - 3| < \delta$, then $\left| \dfrac{x}{5} - \dfrac{3}{5} \right| < \varepsilon \quad \Leftrightarrow \quad \frac{1}{5} |x - 3| < \varepsilon \quad \Leftrightarrow$

$|x - 3| < 5\varepsilon.$ So choose $\delta = 5\varepsilon.$ Then $0 < |x - 3| < \delta \quad \Rightarrow \quad |x - 3| < 5\varepsilon \quad \Rightarrow \quad \dfrac{|x - 3|}{5} < \varepsilon \quad \Rightarrow$

$\left| \dfrac{x}{5} - \dfrac{3}{5} \right| < \varepsilon.$ By the definition of a limit, $\lim\limits_{x \to 3} \dfrac{x}{5} = \dfrac{3}{5}.$

21. Given $\varepsilon > 0$, we need $\delta > 0$ such that if $0 < |x - (-5)| < \delta$, then $\left|\left(4 - \frac{3}{5}x\right) - 7\right| < \varepsilon$ $\Leftrightarrow$

$\left|-\frac{3}{5}x - 3\right| < \varepsilon$ $\Leftrightarrow$ $\frac{3}{5}|x + 5| < \varepsilon$ $\Leftrightarrow$ $|x - (-5)| < \frac{5}{3}\varepsilon$. So choose $\delta = \frac{5}{3}\varepsilon$. Then $|x - (-5)| < \delta$ $\Rightarrow$

$\left|\left(4 - \frac{3}{5}x\right) - 7\right| < \varepsilon$. Thus, $\lim\limits_{x \to -5} \left(4 - \frac{3}{5}x\right) = 7$ by the definition of a limit.

23. Given $\varepsilon > 0$, we need $\delta > 0$ such that if $0 < |x - a| < \delta$, then $|x - a| < \varepsilon$. So $\delta = \varepsilon$ will work.

25. Given $\varepsilon > 0$, we need $\delta > 0$ such that if $0 < |x - 0| < \delta$, then $|x^2 - 0| < \varepsilon$ $\Leftrightarrow$ $x^2 < \varepsilon$ $\Leftrightarrow$ $|x| < \sqrt{\varepsilon}$. Take

$\delta = \sqrt{\varepsilon}$. Then $0 < |x - 0| < \delta$ $\Rightarrow$ $|x^2 - 0| < \varepsilon$. Thus, $\lim\limits_{x \to 0} x^2 = 0$ by the definition of a limit.

27. Given $\varepsilon > 0$, we need $\delta > 0$ such that if $0 < |x - 0| < \delta$, then $||x| - 0| < \varepsilon$. But $||x|| = |x|$. So this is true if we

pick $\delta = \varepsilon$. Thus, $\lim\limits_{x \to 0} |x| = 0$ by the definition of a limit.

29. Given $\varepsilon > 0$, we need $\delta > 0$ such that if $0 < |x - 2| < \delta$, then $\left|(x^2 - 4x + 5) - 1\right| < \varepsilon$ $\Leftrightarrow$

$\left|x^2 - 4x + 4\right| < \varepsilon$ $\Leftrightarrow$ $\left|(x - 2)^2\right| < \varepsilon$. So take $\delta = \sqrt{\varepsilon}$. Then $0 < |x - 2| < \delta$ $\Leftrightarrow$ $|x - 2| < \sqrt{\varepsilon}$ $\Leftrightarrow$

$\left|(x - 2)^2\right| < \varepsilon$. Thus, $\lim\limits_{x \to 2} (x^2 - 4x + 5) = 1$ by the definition of a limit.

31. Given $\varepsilon > 0$, we need $\delta > 0$ such that if $0 < |x - (-2)| < \delta$, then $\left|(x^2 - 1) - 3\right| < \varepsilon$ or upon simplifying we

need $\left|x^2 - 4\right| < \varepsilon$ whenever $0 < |x + 2| < \delta$. Notice that if $|x + 2| < 1$, then $-1 < x + 2 < 1$ $\Rightarrow$

$-5 < x - 2 < -3$ $\Rightarrow$ $|x - 2| < 5$. So take $\delta = \min\{\varepsilon/5, 1\}$. Then $0 < |x + 2| < \delta$ $\Rightarrow$ $|x - 2| < 5$ and

$|x + 2| < \varepsilon/5$, so $\left|(x^2 - 1) - 3\right| = |(x + 2)(x - 2)| = |x + 2| \, |x - 2| < (\varepsilon/5)(5) = \varepsilon$. Thus, by the definition

of a limit, $\lim\limits_{x \to -2} (x^2 - 1) = 3$.

33. Given $\varepsilon > 0$, we let $\delta = \min\left\{2, \frac{\varepsilon}{8}\right\}$. If $0 < |x - 3| < \delta$, then $|x - 3| < 2$ $\Rightarrow$ $-2 < x - 3 < 2$ $\Rightarrow$

$4 < x + 3 < 8$ $\Rightarrow$ $|x + 3| < 8$. Also $|x - 3| < \frac{\varepsilon}{8}$, so $|x^2 - 9| = |x + 3| \, |x - 3| < 8 \cdot \frac{\varepsilon}{8} = \varepsilon$. Thus,

$\lim\limits_{x \to 3} x^2 = 9$.

35. (a) The points of intersection in the graph are $(x_1, 2.6)$ and $(x_2, 3.4)$

with $x_1 \approx 0.891$ and $x_2 \approx 1.093$. Thus, we can take δ to be the

smaller of $1 - x_1$ and $x_2 - 1$. So $\delta = x_2 - 1 \approx 0.093$.

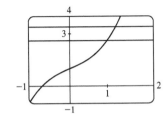

(b) Solving $x^3 + x + 1 = 3 + \varepsilon$ gives us two nonreal complex roots and one real root, which is

$$x(\varepsilon) = \frac{\left(216 + 108\varepsilon + 12\sqrt{336 + 324\varepsilon + 81\varepsilon^2}\,\right)^{2/3} - 12}{6\left(216 + 108\varepsilon + 12\sqrt{336 + 324\varepsilon + 81\varepsilon^2}\,\right)^{1/3}}. \text{ Thus, } \delta = x(\varepsilon) - 1.$$

(c) If $\varepsilon = 0.4$, then $x(\varepsilon) \approx 1.093\,272\,342$ and $\delta = x(\varepsilon) - 1 \approx 0.093$, which agrees with our answer in part (a).

37. *1. Guessing a value for δ* Given $\varepsilon > 0$, we must find $\delta > 0$ such that $|\sqrt{x} - \sqrt{a}| < \varepsilon$ whenever

$0 < |x - a| < \delta$. But $|\sqrt{x} - \sqrt{a}| = \dfrac{|x - a|}{\sqrt{x} + \sqrt{a}} < \varepsilon$ (from the hint). Now if we can find a positive constant C such

that $\sqrt{x} + \sqrt{a} > C$ then $\dfrac{|x - a|}{\sqrt{x} + \sqrt{a}} < \dfrac{|x - a|}{C} < \varepsilon$, and we take $|x - a| < C\varepsilon$. We can find this number by

restricting x to lie in some interval centered at a. If $|x - a| < \frac{1}{2}a$, then $-\frac{1}{2}a < x - a < \frac{1}{2}a$ $\Rightarrow$ $\frac{1}{2}a < x < \frac{3}{2}a$

$\Rightarrow$ $\sqrt{x} + \sqrt{a} > \sqrt{\frac{1}{2}a} + \sqrt{a}$, and so $C = \sqrt{\frac{1}{2}a} + \sqrt{a}$ is a suitable choice for the constant. So

$|x - a| < \left(\sqrt{\frac{1}{2}a} + \sqrt{a}\right)\varepsilon$. This suggests that we let $\delta = \min\left\{\frac{1}{2}a, \left(\sqrt{\frac{1}{2}a} + \sqrt{a}\right)\varepsilon\right\}$.

2. Showing that δ works Given $\varepsilon > 0$, we let $\delta = \min\left\{\frac{1}{2}a, \left(\sqrt{\frac{1}{2}a} + \sqrt{a}\right)\varepsilon\right\}$. If $0 < |x - a| < \delta$, then

$|x - a| < \frac{1}{2}a$ $\Rightarrow$ $\sqrt{x} + \sqrt{a} > \sqrt{\frac{1}{2}a} + \sqrt{a}$ (as in part 1). Also $|x - a| < \left(\sqrt{\frac{1}{2}a} + \sqrt{a}\right)\varepsilon$, so

$|\sqrt{x} - \sqrt{a}| = \dfrac{|x - a|}{\sqrt{x} + \sqrt{a}} < \dfrac{\left(\sqrt{a/2} + \sqrt{a}\right)\varepsilon}{\left(\sqrt{a/2} + \sqrt{a}\right)} = \varepsilon$. Therefore, $\lim\limits_{x \to a} \sqrt{x} = \sqrt{a}$ by the definition of a limit.

39. Suppose that $\lim\limits_{x \to 0} f(x) = L$. Given $\varepsilon = \frac{1}{2}$, there exists $\delta > 0$ such that $0 < |x| < \delta$ $\Rightarrow$ $|f(x) - L| < \frac{1}{2}$. Take

any rational number r with $0 < |r| < \delta$. Then $f(r) = 0$, so $|0 - L| < \frac{1}{2}$, so $L \le |L| < \frac{1}{2}$. Now take any irrational

number s with $0 < |s| < \delta$. Then $f(s) = 1$, so $|1 - L| < \frac{1}{2}$. Hence, $1 - L < \frac{1}{2}$, so $L > \frac{1}{2}$. This contradicts

$L < \frac{1}{2}$, so $\lim\limits_{x \to 0} f(x)$ does not exist.

41. $\dfrac{1}{(x + 3)^4} > 10{,}000$ $\Leftrightarrow$ $(x + 3)^4 < \dfrac{1}{10{,}000}$ $\Leftrightarrow$ $|x + 3| < \dfrac{1}{\sqrt[4]{10{,}000}}$ $\Leftrightarrow$ $|x - (-3)| < \dfrac{1}{10}$

43. Given $M < 0$ we need $\delta > 0$ so that $\ln x < M$ whenever $0 < x < \delta$; that is, $x = e^{\ln x} < e^M$ whenever $0 < x < \delta$.

This suggests that we take $\delta = e^M$. If $0 < x < e^M$, then $\ln x < \ln e^M = M$. By the definition of a limit,

$\lim\limits_{x \to 0^+} \ln x = -\infty$.

2.5 Continuity

1. From Definition 1, $\lim\limits_{x \to 4} f(x) = f(4)$.

3. (a) The following are the numbers at which f is discontinuous and the type of discontinuity at that number:

-4 (removable), -2 (jump), 2 (jump), 4 (infinite).

(b) f is continuous from the left at -2 since $\lim\limits_{x \to -2^-} f(x) = f(-2)$. f is continuous from the right at 2 and 4 since

$\lim\limits_{x \to 2^+} f(x) = f(2)$ and $\lim\limits_{x \to 4^+} f(x) = f(4)$. It is continuous from neither side at -4 since $f(-4)$ is undefined.

5. The graph of $y = f(x)$ must have a discontinuity at $x = 3$ and must show that $\lim\limits_{x \to 3^-} f(x) = f(3)$.

7. (a)

(b) There are discontinuities at times $t = 1, 2, 3$, and 4. A person parking in the lot would want to keep in mind that the charge will jump at the beginning of each hour.

9. Since f and g are continuous functions,

$$\lim_{x\to 3}[2f(x) - g(x)] = 2\lim_{x\to 3}f(x) - \lim_{x\to 3}g(x) \qquad \text{[by Limit Laws 2 and 3]}$$
$$= 2f(3) - g(3) \qquad \text{[by continuity of } f \text{ and } g \text{ at } x = 3]$$
$$= 2 \cdot 5 - g(3) = 10 - g(3)$$

Since it is given that $\lim_{x\to 3}[2f(x) - g(x)] = 4$, we have $10 - g(3) = 4$, so $g(3) = 6$.

11. $\lim_{x\to -1}f(x) = \lim_{x\to -1}(x + 2x^3)^4 = \left(\lim_{x\to -1}x + 2\lim_{x\to -1}x^3\right)^4 = \left[-1 + 2(-1)^3\right]^4 = (-3)^4 = 81 = f(-1)$.

By the definition of continuity, f is continuous at $a = -1$.

13. For $a > 2$, we have $\lim_{x\to a}f(x) = \lim_{x\to a}\dfrac{2x + 3}{x - 2} = \dfrac{\lim_{x\to a}(2x + 3)}{\lim_{x\to a}(x - 2)}$ [Limit Law 5] $= \dfrac{2\lim_{x\to a}x + \lim_{x\to a}3}{\lim_{x\to a}x - \lim_{x\to a}2}$

[1, 2, and 3] $= \dfrac{2a + 3}{a - 2}$ [7 and 8] $= f(a)$. Thus, f is continuous at $x = a$ for every a in $(2, \infty)$; that is, f is continuous on $(2, \infty)$.

15. $f(x) = \ln|x - 2|$ is discontinuous at 2 since $f(2) = \ln 0$ is not defined.

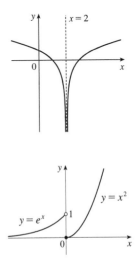

17. $f(x) = \begin{cases} e^x & \text{if } x < 0 \\ x^2 & \text{if } x \geq 0 \end{cases}$

The left-hand limit of f at $a = 0$ is $\lim_{x\to 0^-}f(x) = \lim_{x\to 0^-}e^x = 1$. The right-hand limit of f at $a = 0$ is $\lim_{x\to 0^+}f(x) = \lim_{x\to 0^+}x^2 = 0$. Since these limits are not equal, $\lim_{x\to 0}f(x)$ does not exist and f is discontinuous at 0.

19. $f(x) = \begin{cases} \dfrac{x^2 - x - 12}{x + 3} & \text{if } x \neq -3 \\ -5 & \text{if } x = -3 \end{cases} = \begin{cases} x - 4 & \text{if } x \neq -3 \\ -5 & \text{if } x = -3 \end{cases}$

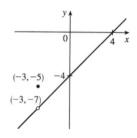

So $\lim\limits_{x \to -3} f(x) = \lim\limits_{x \to -3} (x - 4) = -7$ and $f(-3) = -5$.

Since $\lim\limits_{x \to -3} f(x) \neq f(-3)$, f is discontinuous at -3.

21. $F(x) = \dfrac{x}{x^2 + 5x + 6}$ is a rational function. So by Theorem 5 (or Theorem 7), F is continuous at every number in

its domain, $\{x \mid x^2 + 5x + 6 \neq 0\} = \{x \mid (x + 3)(x + 2) \neq 0\} = \{x \mid x \neq -3, \, -2\}$ or

$(-\infty, -3) \cup (-3, -2) \cup (-2, \infty)$.

23. By Theorem 5, the polynomials x^2 and $2x - 1$ are continuous on $(-\infty, \infty)$. By Theorem 7, the root function $\sqrt{x}$ is

continuous on $[0, \infty)$. By Theorem 9, the composite function $\sqrt{2x - 1}$ is continuous on its domain, $\left[\frac{1}{2}, \infty\right)$. By

part 1 of Theorem 4, the sum $R(x) = x^2 + \sqrt{2x - 1}$ is continuous on $\left[\frac{1}{2}, \infty\right)$.

25. By Theorem 5, the polynomial $5x$ is continuous on $(-\infty, \infty)$. By Theorems 9 and 7, $\sin 5x$ is continuous on

$(-\infty, \infty)$. By Theorem 7, e^x is continuous on $(-\infty, \infty)$. By part 4 of Theorem 4, the product of e^x and $\sin 5x$ is

continuous at all numbers which are in both of their domains, that is, on $(-\infty, \infty)$.

27. By Theorem 5, the polynomial $t^4 - 1$ is continuous on $(-\infty, \infty)$. By Theorem 7, $\ln x$ is continuous on its

domain, $(0, \infty)$. By Theorem 9, $\ln(t^4 - 1)$ is continuous on its domain, which is

$\{t \mid t^4 - 1 > 0\} = \{t \mid t^4 > 1\} = \{t \mid |t| > 1\} = (-\infty, -1) \cup (1, \infty)$.

29. The function $y = \dfrac{1}{1 + e^{1/x}}$ is discontinuous at $x = 0$ because the

left- and right-hand limits at $x = 0$ are different.

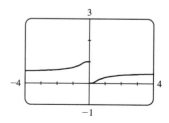

31. Because we are dealing with root functions, $5 + \sqrt{x}$ is continuous on $[0, \infty)$, $\sqrt{x + 5}$ is continuous on $[-5, \infty)$, so

the quotient $f(x) = \dfrac{5 + \sqrt{x}}{\sqrt{5 + x}}$ is continuous on $[0, \infty)$. Since f is continuous at $x = 4$, $\lim\limits_{x \to 4} f(x) = f(4) = \frac{7}{3}$.

33. Because $x^2 - x$ is continuous on $\mathbb{R}$, the composite function $f(x) = e^{x^2 - x}$ is continuous on $\mathbb{R}$, so

$\lim\limits_{x \to 1} f(x) = f(1) = e^{1 - 1} = e^0 = 1$.

35. $f(x) = \begin{cases} x^2 & \text{if } x < 1 \\ \sqrt{x} & \text{if } x \geq 1 \end{cases}$

By Theorem 5, since $f(x)$ equals the polynomial x^2 on $(-\infty, 1)$, f is continuous on $(-\infty, 1)$. By Theorem 7,

since $f(x)$ equals the root function $\sqrt{x}$ on $(1, \infty)$, f is continuous on $(1, \infty)$. At $x = 1$,

$\lim\limits_{x \to 1^-} f(x) = \lim\limits_{x \to 1^-} x^2 = 1$ and $\lim\limits_{x \to 1^+} f(x) = \lim\limits_{x \to 1^+} \sqrt{x} = 1$. Thus, $\lim\limits_{x \to 1} f(x)$ exists and equals 1. Also,

$f(1) = \sqrt{1} = 1$. Thus, f is continuous at $x = 1$. We conclude that f is continuous on $(-\infty, \infty)$.

37. $f(x) = \begin{cases} 1 + x^2 & \text{if } x \le 0 \\ 2 - x & \text{if } 0 < x \le 2 \\ (x-2)^2 & \text{if } x > 2 \end{cases}$

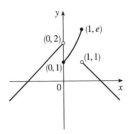

f is continuous on $(-\infty, 0)$, $(0, 2)$, and $(2, \infty)$ since it is a polynomial on

each of these intervals. Now $\lim\limits_{x \to 0^-} f(x) = \lim\limits_{x \to 0^-} (1 + x^2) = 1$ and

$\lim\limits_{x \to 0^+} f(x) = \lim\limits_{x \to 0^+} (2 - x) = 2$, so f is discontinuous at 0. Since $f(0) = 1$, f is continuous from the left at 0.

Also, $\lim\limits_{x \to 2^-} f(x) = \lim\limits_{x \to 2^-} (2 - x) = 0$, $\lim\limits_{x \to 2^+} f(x) = \lim\limits_{x \to 2^+} (x - 2)^2 = 0$, and $f(2) = 0$, so f is continuous at 2.

The only number at which f is discontinuous is 0.

39. $f(x) = \begin{cases} x + 2 & \text{if } x < 0 \\ e^x & \text{if } 0 \le x \le 1 \\ 2 - x & \text{if } x > 1 \end{cases}$

f is continuous on $(-\infty, 0)$ and $(1, \infty)$ since on each of these intervals it is

a polynomial; it is continuous on $(0, 1)$ since it is an exponential. Now

$\lim\limits_{x \to 0^-} f(x) = \lim\limits_{x \to 0^-} (x + 2) = 2$ and $\lim\limits_{x \to 0^+} f(x) = \lim\limits_{x \to 0^+} e^x = 1$, so f is

discontinuous at 0. Since $f(0) = 1$, f is continuous from the right at 0. Also $\lim\limits_{x \to 1^-} f(x) = \lim\limits_{x \to 1^-} e^x = e$ and

$\lim\limits_{x \to 1^+} f(x) = \lim\limits_{x \to 1^+} (2 - x) = 1$, so f is discontinuous at 1. Since $f(1) = e$, f is continuous from the left at 1.

41. f is continuous on $(-\infty, 3)$ and $(3, \infty)$. Now $\lim\limits_{x \to 3^-} f(x) = \lim\limits_{x \to 3^-} (cx + 1) = 3c + 1$ and

$\lim\limits_{x \to 3^+} f(x) = \lim\limits_{x \to 3^+} (cx^2 - 1) = 9c - 1$. So f is continuous $\Leftrightarrow$ $3c + 1 = 9c - 1$ $\Leftrightarrow$ $6c = 2$ $\Leftrightarrow$ $c = \frac{1}{3}$.

Thus, for f to be continuous on $(-\infty, \infty)$, $c = \frac{1}{3}$.

43. (a) $f(x) = \dfrac{x^2 - 2x - 8}{x + 2} = \dfrac{(x-4)(x+2)}{x+2}$ has a removable discontinuity at -2 because $g(x) = x - 4$ is

continuous on $\mathbb{R}$ and $f(x) = g(x)$ for $x \ne -2$. [The discontinuity is removed by defining $f(-2) = -6$.]

(b) $f(x) = \dfrac{x - 7}{|x - 7|}$ $\Rightarrow$ $\lim\limits_{x \to 7^-} f(x) = -1$ and $\lim\limits_{x \to 7^+} f(x) = 1$. Thus, $\lim\limits_{x \to 7} f(x)$ does not exist, so the

discontinuity is not removable. (It is a jump discontinuity.)

(c) $f(x) = \dfrac{x^3 + 64}{x + 4} = \dfrac{(x+4)(x^2 - 4x + 16)}{x+4}$ has a removable discontinuity at -4 because

$g(x) = x^2 - 4x + 16$ is continuous on $\mathbb{R}$ and $f(x) = g(x)$ for $x \ne -4$. [The discontinuity is removed by

defining $f(-4) = 48$.]

(d) $f(x) = \dfrac{3 - \sqrt{x}}{9 - x} = \dfrac{3 - \sqrt{x}}{(3 - \sqrt{x})(3 + \sqrt{x})}$ has a removable discontinuity at 9 because $g(x) = \dfrac{1}{3 + \sqrt{x}}$ is

continuous on $\mathbb{R}$ and $f(x) = g(x)$ for $x \ne 9$. [The discontinuity is removed by defining $f(9) = \frac{1}{6}$.]

45. $f(x) = x^3 - x^2 + x$ is continuous on the interval $[2, 3]$, $f(2) = 6$, and $f(3) = 21$. Since $6 < 10 < 21$, there is a number c in $(2, 3)$ such that $f(c) = 10$ by the Intermediate Value Theorem.

47. $f(x) = x^4 + x - 3$ is continuous on the interval $[1, 2]$, $f(1) = -1$, and $f(2) = 15$. Since $-1 < 0 < 15$, there is a number c in $(1, 2)$ such that $f(c) = 0$ by the Intermediate Value Theorem. Thus, there is a root of the equation $x^4 + x - 3 = 0$ in the interval $(1, 2)$.

49. $f(x) = \cos x - x$ is continuous on the interval $[0, 1]$, $f(0) = 1$, and $f(1) = \cos 1 - 1 \approx -0.46$. Since $-0.46 < 0 < 1$, there is a number c in $(0, 1)$ such that $f(c) = 0$ by the Intermediate Value Theorem. Thus, there is a root of the equation $\cos x - x = 0$, or $\cos x = x$, in the interval $(0, 1)$.

51. (a) $f(x) = e^x + x - 2$ is continuous on the interval $[0, 1]$, $f(0) = -1 < 0$, and $f(1) = e - 1 \approx 1.72 > 0$. Since $-1 < 0 < 1.72$, there is a number c in $(0, 1)$ such that $f(c) = 0$ by the Intermediate Value Theorem. Thus, there is a root of the equation $e^x + x - 2 = 0$, or $e^x = 2 - x$, in the interval $(0, 1)$.

(b) $f(0.44) \approx -0.007 < 0$ and $f(0.45) \approx 0.018 > 0$, so there is a root between 0.44 and 0.45.

53. (a) Let $f(x) = x^5 - x^2 - 4$. Then $f(1) = 1^5 - 1^2 - 4 = -4 < 0$ and $f(2) = 2^5 - 2^2 - 4 = 24 > 0$. So by the Intermediate Value Theorem, there is a number c in $(1, 2)$ such that $f(c) = c^5 - c^2 - 4 = 0$.

(b) We can see from the graphs that, correct to three decimal places, the root is $x \approx 1.434$.

55. ($\Rightarrow$) If f is continuous at a, then by Theorem 8 with $g(h) = a + h$, we have

$$\lim_{h \to 0} f(a + h) = f\left(\lim_{h \to 0} (a + h)\right) = f(a).$$

($\Leftarrow$) Let $\varepsilon > 0$. Since $\lim_{h \to 0} f(a + h) = f(a)$, there exists $\delta > 0$ such that $0 < |h| < \delta \Rightarrow$ $|f(a + h) - f(a)| < \varepsilon$. So if $0 < |x - a| < \delta$, then $|f(x) - f(a)| = |f(a + (x - a)) - f(a)| < \varepsilon$. Thus, $\lim_{x \to a} f(x) = f(a)$ and so f is continuous at a.

57. As in the previous exercise, we must show that $\lim_{h \to 0} \cos(a + h) = \cos a$ to prove that the cosine function is continuous.

$$\lim_{h \to 0} \cos(a + h) = \lim_{h \to 0} (\cos a \cos h - \sin a \sin h)$$

$$= \lim_{h \to 0} (\cos a \cos h) - \lim_{h \to 0} (\sin a \sin h)$$

$$= \left(\lim_{h \to 0} \cos a\right)\left(\lim_{h \to 0} \cos h\right) - \left(\lim_{h \to 0} \sin a\right)\left(\lim_{h \to 0} \sin h\right)$$

$$= (\cos a)(1) - (\sin a)(0) = \cos a$$

59. $f(x) = \begin{cases} 0 & \text{if } x \text{ is rational} \\ 1 & \text{if } x \text{ is irrational} \end{cases}$ is continuous nowhere. For, given any number a and any $\delta > 0$, the interval

$(a - \delta, a + \delta)$ contains both infinitely many rational and infinitely many irrational numbers. Since $f(a) = 0$ or 1,

there are infinitely many numbers x with $0 < |x - a| < \delta$ and $|f(x) - f(a)| = 1$. Thus, $\lim\limits_{x \to a} f(x) \neq f(a)$.

[In fact, $\lim\limits_{x \to a} f(x)$ does not even exist.]

61. If there is such a number, it satisfies the equation $x^3 + 1 = x \quad \Leftrightarrow \quad x^3 - x + 1 = 0$. Let the left-hand side of this

equation be called $f(x)$. Now $f(-2) = -5 < 0$, and $f(-1) = 1 > 0$. Note also that $f(x)$ is a polynomial, and

thus continuous. So by the Intermediate Value Theorem, there is a number c between -2 and -1 such that

$f(c) = 0$, so that $c = c^3 + 1$.

63. Define $u(t)$ to be the monk's distance from the monastery, as a function of time, on the first day, and define $d(t)$ to

be his distance from the monastery, as a function of time, on the second day. Let D be the distance from the

monastery to the top of the mountain. From the given information we know that $u(0) = 0$, $u(12) = D$, $d(0) = D$

and $d(12) = 0$. Now consider the function $u - d$, which is clearly continuous. We calculate that $(u - d)(0) = -D$

and $(u - d)(12) = D$. So by the Intermediate Value Theorem, there must be some time t_0 between 0 and 12 such

that $(u - d)(t_0) = 0 \quad \Leftrightarrow \quad u(t_0) = d(t_0)$. So at time t_0 after 7:00 A.M., the monk will be at the same place on

both days.

2.6 Limits at Infinity; Horizontal Asymptotes

1. (a) As x becomes large, the values of $f(x)$ approach 5.

(b) As x becomes large negative, the values of $f(x)$ approach 3.

3. (a) $\lim\limits_{x \to 2} f(x) = \infty$ (b) $\lim\limits_{x \to -1^-} f(x) = \infty$

(c) $\lim\limits_{x \to -1^+} f(x) = -\infty$ (d) $\lim\limits_{x \to \infty} f(x) = 1$

(e) $\lim\limits_{x \to -\infty} f(x) = 2$ (f) Vertical: $x = -1$, $x = 2$; Horizontal: $y = 1$, $y = 2$

5. $f(0) = 0$, $f(1) = 1$, $\lim\limits_{x \to \infty} f(x) = 0$,

f is odd

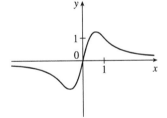

7. $\lim\limits_{x \to 2} f(x) = -\infty, \quad \lim\limits_{x \to \infty} f(x) = \infty,$

$\lim\limits_{x \to -\infty} f(x) = 0, \quad \lim\limits_{x \to 0^+} f(x) = \infty,$

$\lim\limits_{x \to 0^-} f(x) = -\infty$

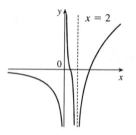

9. If $f(x) = x^2/2^x$, then a calculator gives $f(0) = 0$, $f(1) = 0.5$, $f(2) = 1$, $f(3) = 1.125$, $f(4) = 1$,

$f(5) = 0.78125$, $f(6) = 0.5625$, $f(7) = 0.3828125$, $f(8) = 0.25$, $f(9) = 0.158203125$, $f(10) = 0.09765625$,

$f(20) \approx 0.00038147$, $f(50) \approx 2.2204 \times 10^{-12}$, $f(100) \approx 7.8886 \times 10^{-27}$.

It appears that $\lim\limits_{x \to \infty} \left(x^2/2^x\right) = 0$.

11. $\lim\limits_{x \to \infty} \dfrac{3x^2 - x + 4}{2x^2 + 5x - 8} = \lim\limits_{x \to \infty} \dfrac{(3x^2 - x + 4)/x^2}{(2x^2 + 5x - 8)/x^2}$

[divide both the numerator and denominator by x^2 (the highest power of x that appears in the denominator)]

$= \dfrac{\lim\limits_{x \to \infty} (3 - 1/x + 4/x^2)}{\lim\limits_{x \to \infty} (2 + 5/x - 8/x^2)}$ [Limit Law 5]

$= \dfrac{\lim\limits_{x \to \infty} 3 - \lim\limits_{x \to \infty} (1/x) + \lim\limits_{x \to \infty} (4/x^2)}{\lim\limits_{x \to \infty} 2 + \lim\limits_{x \to \infty} (5/x) - \lim\limits_{x \to \infty} (8/x^2)}$ [Limit Laws 1 and 2]

$= \dfrac{3 - \lim\limits_{x \to \infty} (1/x) + 4 \lim\limits_{x \to \infty} (1/x^2)}{2 + 5 \lim\limits_{x \to \infty} (1/x) - 8 \lim\limits_{x \to \infty} (1/x^2)}$ [Limit Laws 7 and 3]

$= \dfrac{3 - 0 + 4(0)}{2 + 5(0) - 8(0)}$ [Theorem 5 of Section 2.5]

$= \dfrac{3}{2}$

13. $\lim\limits_{x \to \infty} \dfrac{1}{2x + 3} = \lim\limits_{x \to \infty} \dfrac{1/x}{(2x + 3)/x} = \dfrac{\lim\limits_{x \to \infty} (1/x)}{\lim\limits_{x \to \infty} (2 + 3/x)} = \dfrac{\lim\limits_{x \to \infty} (1/x)}{\lim\limits_{x \to \infty} 2 + 3 \lim\limits_{x \to \infty} (1/x)} = \dfrac{0}{2 + 3(0)} = \dfrac{0}{2} = 0$

15. $\lim\limits_{x \to -\infty} \dfrac{1 - x - x^2}{2x^2 - 7} = \lim\limits_{x \to -\infty} \dfrac{(1 - x - x^2)/x^2}{(2x^2 - 7)/x^2} = \dfrac{\lim\limits_{x \to -\infty} (1/x^2 - 1/x - 1)}{\lim\limits_{x \to -\infty} (2 - 7/x^2)}$

$= \dfrac{\lim\limits_{x \to -\infty} (1/x^2) - \lim\limits_{x \to -\infty} (1/x) - \lim\limits_{x \to -\infty} 1}{\lim\limits_{x \to -\infty} 2 - 7 \lim\limits_{x \to -\infty} (1/x^2)} = \dfrac{0 - 0 - 1}{2 - 7(0)} = -\dfrac{1}{2}$

17. Divide both the numerator and denominator by x^3 (the highest power of x that occurs in the denominator).

$$\lim_{x\to\infty}\frac{x^3+5x}{2x^3-x^2+4}=\lim_{x\to\infty}\frac{\dfrac{x^3+5x}{x^3}}{\dfrac{2x^3-x^2+4}{x^3}}=\lim_{x\to\infty}\frac{1+\dfrac{5}{x^2}}{2-\dfrac{1}{x}+\dfrac{4}{x^3}}=\frac{\lim\limits_{x\to\infty}\left(1+\dfrac{5}{x^2}\right)}{\lim\limits_{x\to\infty}\left(2-\dfrac{1}{x}+\dfrac{4}{x^3}\right)}$$

$$=\frac{\lim\limits_{x\to\infty}1+5\lim\limits_{x\to\infty}\dfrac{1}{x^2}}{\lim\limits_{x\to\infty}2-\lim\limits_{x\to\infty}\dfrac{1}{x}+4\lim\limits_{x\to\infty}\dfrac{1}{x^3}}=\frac{1+5(0)}{2-0+4(0)}=\frac{1}{2}$$

19. First, multiply the factors in the denominator. Then divide both the numerator and denominator by u^4.

$$\lim_{u\to\infty}\frac{4u^4+5}{(u^2-2)(2u^2-1)}=\lim_{u\to\infty}\frac{4u^4+5}{2u^4-5u^2+2}=\lim_{u\to\infty}\frac{\dfrac{4u^4+5}{u^4}}{\dfrac{2u^4-5u^2+2}{u^4}}=\lim_{u\to\infty}\frac{4+\dfrac{5}{u^4}}{2-\dfrac{5}{u^2}+\dfrac{2}{u^4}}$$

$$=\frac{\lim\limits_{u\to\infty}\left(4+\dfrac{5}{u^4}\right)}{\lim\limits_{u\to\infty}\left(2-\dfrac{5}{u^2}+\dfrac{2}{u^4}\right)}=\frac{\lim\limits_{u\to\infty}4+5\lim\limits_{u\to\infty}\dfrac{1}{u^4}}{\lim\limits_{u\to\infty}2-5\lim\limits_{u\to\infty}\dfrac{1}{u^2}+2\lim\limits_{u\to\infty}\dfrac{1}{u^4}}=\frac{4+5(0)}{2-5(0)+2(0)}$$

$$=\frac{4}{2}=2$$

21. $\lim\limits_{x\to\infty}\dfrac{\sqrt{9x^6-x}}{x^3+1}=\lim\limits_{x\to\infty}\dfrac{\sqrt{9x^6-x}\,/x^3}{(x^3+1)/x^3}=\dfrac{\lim\limits_{x\to\infty}\sqrt{(9x^6-x)/x^6}}{\lim\limits_{x\to\infty}(1+1/x^3)}$ [since $x^3=\sqrt{x^6}$ for $x>0$]

$$=\frac{\lim\limits_{x\to\infty}\sqrt{9-1/x^5}}{\lim\limits_{x\to\infty}1+\lim\limits_{x\to\infty}(1/x^3)}=\frac{\sqrt{\lim\limits_{x\to\infty}9-\lim\limits_{x\to\infty}(1/x^5)}}{1+0}$$

$$=\sqrt{9-0}=3$$

23. $\lim\limits_{x\to\infty}\left(\sqrt{9x^2+x}-3x\right)=\lim\limits_{x\to\infty}\dfrac{\left(\sqrt{9x^2+x}-3x\right)\left(\sqrt{9x^2+x}+3x\right)}{\sqrt{9x^2+x}+3x}=\lim\limits_{x\to\infty}\dfrac{\left(\sqrt{9x^2+x}\right)^2-(3x)^2}{\sqrt{9x^2+x}+3x}$

$$=\lim\limits_{x\to\infty}\frac{(9x^2+x)-9x^2}{\sqrt{9x^2+x}+3x}=\lim\limits_{x\to\infty}\frac{x}{\sqrt{9x^2+x}+3x}\cdot\frac{1/x}{1/x}$$

$$=\lim\limits_{x\to\infty}\frac{x/x}{\sqrt{9x^2/x^2+x/x^2}+3x/x}=\lim\limits_{x\to\infty}\frac{1}{\sqrt{9+1/x}+3}=\frac{1}{\sqrt{9}+3}=\frac{1}{3+3}=\frac{1}{6}$$

25. $\lim\limits_{x\to\infty}\left(\sqrt{x^2+ax}-\sqrt{x^2+bx}\,\right)=\lim\limits_{x\to\infty}\dfrac{\left(\sqrt{x^2+ax}-\sqrt{x^2+bx}\,\right)\left(\sqrt{x^2+ax}+\sqrt{x^2+bx}\,\right)}{\sqrt{x^2+ax}+\sqrt{x^2+bx}}$

$$=\lim\limits_{x\to\infty}\frac{(x^2+ax)-(x^2+bx)}{\sqrt{x^2+ax}+\sqrt{x^2+bx}}=\lim\limits_{x\to\infty}\frac{[(a-b)x]/x}{\left(\sqrt{x^2+ax}+\sqrt{x^2+bx}\,\right)/\sqrt{x^2}}$$

$$=\lim\limits_{x\to\infty}\frac{a-b}{\sqrt{1+a/x}+\sqrt{1+b/x}}=\frac{a-b}{\sqrt{1+0}+\sqrt{1+0}}=\frac{a-b}{2}$$

27. $\sqrt{x}$ is large when x is large, so $\lim\limits_{x\to\infty}\sqrt{x}=\infty$.

29. $\lim\limits_{x\to\infty}(x-\sqrt{x})=\lim\limits_{x\to\infty}\sqrt{x}\,(\sqrt{x}-1)=\infty$ since $\sqrt{x}\to\infty$ and $\sqrt{x}-1\to\infty$ as $x\to\infty$.

31. $\lim\limits_{x\to-\infty}(x^4+x^5)=\lim\limits_{x\to-\infty}x^5(\frac{1}{x}+1)$ [factor out the largest power of x] $=-\infty$ because $x^5\to-\infty$ and

$1/x+1\to1$ as $x\to-\infty$.

33. $\lim\limits_{x\to\infty}\dfrac{x+x^3+x^5}{1-x^2+x^4}=\lim\limits_{x\to\infty}\dfrac{(x+x^3+x^5)/x^4}{(1-x^2+x^4)/x^4}$ [divide by the highest power of x in the denominator]

$\qquad\qquad=\lim\limits_{x\to\infty}\dfrac{1/x^3+1/x+x}{1/x^4-1/x^2+1}=\infty$

because $(1/x^3+1/x+x)\to\infty$ and $(1/x^4-1/x^2+1)\to1$ as $x\to\infty$.

35. (a)

From the graph of $f(x)=\sqrt{x^2+x+1}+x$, we estimate the value of $\lim\limits_{x\to-\infty}f(x)$ to be -0.5.

(b)

x	$f(x)$
$-10{,}000$	-0.4999625
$-100{,}000$	-0.4999962
$-1{,}000{,}000$	-0.4999996

From the table, we estimate the limit to be -0.5.

(c) $\lim\limits_{x\to-\infty}\left(\sqrt{x^2+x+1}+x\right)=\lim\limits_{x\to-\infty}\left(\sqrt{x^2+x+1}+x\right)\left[\dfrac{\sqrt{x^2+x+1}-x}{\sqrt{x^2+x+1}-x}\right]=\lim\limits_{x\to-\infty}\dfrac{(x^2+x+1)-x^2}{\sqrt{x^2+x+1}-x}$

$\qquad\qquad=\lim\limits_{x\to-\infty}\dfrac{(x+1)(1/x)}{(\sqrt{x^2+x+1}-x)(1/x)}=\lim\limits_{x\to-\infty}\dfrac{1+(1/x)}{-\sqrt{1+(1/x)+(1/x^2)}-1}$

$\qquad\qquad=\dfrac{1+0}{-\sqrt{1+0+0}-1}=-\dfrac{1}{2}$

Note that for $x<0$, we have $\sqrt{x^2}=|x|=-x$, so when we divide the radical by x, with $x<0$, we get

$\dfrac{1}{x}\sqrt{x^2+x+1}=-\dfrac{1}{\sqrt{x^2}}\sqrt{x^2+x+1}=-\sqrt{1+(1/x)+(1/x^2)}$.

37. $\lim\limits_{x\to\pm\infty}\dfrac{x}{x+4}=\lim\limits_{x\to\pm\infty}\dfrac{1}{1+4/x}=\dfrac{1}{1+0}=1$, so $y=1$ is a horizontal

asymptote. $\lim\limits_{x\to-4^-}\dfrac{x}{x+4}=\infty$ and $\lim\limits_{x\to-4^+}\dfrac{x}{x+4}=-\infty$, so $x=-4$ is

a vertical asymptote. The graph confirms these calculations.

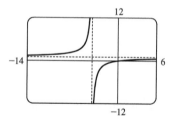

39. $\displaystyle\lim_{x\to\pm\infty}\frac{x^3}{x^2+3x-10}=\lim_{x\to\pm\infty}\frac{x}{1+(3/x)-(10/x^2)}=\pm\infty$, so there is

no horizontal asymptote.

$$\lim_{x\to2^+}\frac{x^3}{x^2+3x-10}=\lim_{x\to2^+}\frac{x^3}{(x+5)(x-2)}=\infty,\text{ since}$$

$\dfrac{x^3}{(x+5)(x-2)}>0$ for $x>2$. Similarly, $\displaystyle\lim_{x\to2^-}\frac{x^3}{x^2+3x-10}=-\infty$,

$\displaystyle\lim_{x\to-5^-}\frac{x^3}{x^2+3x-10}=-\infty,$ and $\displaystyle\lim_{x\to-5^+}\frac{x^3}{x^2+3x-10}=\infty,$ so

$x=2$ and $x=-5$ are vertical asymptotes. The graph confirms these

calculations.

41. $\displaystyle\lim_{x\to\infty}\frac{x}{\sqrt[4]{x^4+1}}\cdot\frac{1/x}{1/\sqrt[4]{x^4}}=\lim_{x\to\infty}\frac{1}{\sqrt[4]{1+\dfrac{1}{x^4}}}=\frac{1}{\sqrt[4]{1+0}}=1$ and

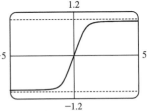

$$\lim_{x\to-\infty}\frac{x}{\sqrt[4]{x^4+1}}\cdot\frac{1/x}{-1/\sqrt[4]{x^4}}=\lim_{x\to-\infty}\frac{1}{-\sqrt[4]{1+\dfrac{1}{x^4}}}=\frac{1}{-\sqrt[4]{1+0}}=-1,$$

so $y=\pm1$ are horizontal asymptotes. There is no vertical asymptote.

43. Let's look for a rational function.

(1) $\displaystyle\lim_{x\to\pm\infty}f(x)=0\quad\Rightarrow\quad$ degree of numerator < degree of denominator

(2) $\displaystyle\lim_{x\to0}f(x)=-\infty\quad\Rightarrow\quad$ there is a factor of x^2 in the denominator (not just x, since that would produce a

sign change at $x=0$), and the function is negative near $x=0$.

(3) $\displaystyle\lim_{x\to3^-}f(x)=\infty$ and $\displaystyle\lim_{x\to3^+}f(x)=-\infty\quad\Rightarrow\quad$ vertical asymptote at $x=3$; there is a factor of $(x-3)$ in

the denominator.

(4) $f(2)=0\quad\Rightarrow\quad2$ is an x-intercept; there is at least one factor of $(x-2)$ in the numerator.

Combining all of this information and putting in a negative sign to give us the desired left- and right-hand limits

gives us $f(x)=\dfrac{2-x}{x^2(x-3)}$ as one possibility.

45. $y=f(x)=x^2(x-2)(1-x)$. The y-intercept is $f(0)=0$, and the x-intercepts

occur when $y=0\quad\Rightarrow\quad x=0,1,$ and 2. Notice that, since x^2 is always positive, the

graph does not cross the x-axis at 0, but does cross the x-axis at 1 and 2.

$\displaystyle\lim_{x\to\infty}x^2(x-2)(1-x)=-\infty$, since the first two factors are large positive and the

third large negative when x is large positive. $\displaystyle\lim_{x\to-\infty}x^2(x-2)(1-x)=-\infty$ because

the first and third factors are large positive and the second large negative as $x\to-\infty$.

47. $y = f(x) = (x + 4)^5(x - 3)^4$. The y-intercept is $f(0) = 4^5(-3)^4 = 82{,}944$.

The x-intercepts occur when $y = 0 \;\Rightarrow\; x = -4, 3$. Notice that the graph does

not cross the x-axis at 3 because $(x - 3)^4$ is always positive, but does cross the

x-axis at -4. $\lim\limits_{x \to \infty} (x + 4)^5(x - 3)^4 = \infty$ since both factors are large positive

when x is large positive. $\lim\limits_{x \to -\infty} (x + 4)^5(x - 3)^4 = -\infty$ since the first factor is

large negative and the second factor is large positive when x is large negative.

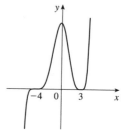

49. (a) Since $-1 \le \sin x \le 1$ for all x, $-\dfrac{1}{x} \le \dfrac{\sin x}{x} \le \dfrac{1}{x}$ for $x > 0$. As $x \to \infty$, $-1/x \to 0$ and $1/x \to 0$, so by the

Squeeze Theorem, $(\sin x)/x \to 0$. Thus, $\lim\limits_{x \to \infty} \dfrac{\sin x}{x} = 0$.

(b) From part (a), the horizontal asymptote is $y = 0$. The function

$y = (\sin x)/x$ crosses the horizontal asymptote whenever $\sin x = 0$;

that is, at $x = \pi n$ for every integer n. Thus, the graph crosses the

asymptote *an infinite number of times.*

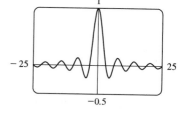

51. Divide the numerator and the denominator by the highest power of x in $Q(x)$.

(a) If $\deg P < \deg Q$, then the numerator $\to 0$ but the denominator doesn't. So $\lim\limits_{x \to \infty} [P(x)/Q(x)] = 0$.

(b) If $\deg P > \deg Q$, then the numerator $\to \pm\infty$ but the denominator doesn't, so $\lim\limits_{x \to \infty} [P(x)/Q(x)] = \pm\infty$

(depending on the ratio of the leading coefficients of P and Q).

53. $\lim\limits_{x \to \infty} \dfrac{4x - 1}{x} = \lim\limits_{x \to \infty} \left(4 - \dfrac{1}{x}\right) = 4$, and $\lim\limits_{x \to \infty} \dfrac{4x^2 + 3x}{x^2} = \lim\limits_{x \to \infty} \left(4 + \dfrac{3}{x}\right) = 4$. Therefore, by the Squeeze

Theorem, $\lim\limits_{x \to \infty} f(x) = 4$.

55. (a) $\lim\limits_{t \to \infty} v(t) = \lim\limits_{t \to \infty} v^*\left(1 - e^{-gt/v^*}\right) = v^*(1 - 0) = v^*$

(b) We graph $v(t) = 1 - e^{-9.8t}$ and $v(t) = 0.99v^*$, or in this case,

$v(t) = 0.99$. Using an intersect feature or zooming in on the point

of intersection, we find that $t \approx 0.47$ s.

57. $\left| \dfrac{6x^2 + 5x - 3}{2x^2 - 1} - 3 \right| < 0.2 \;\Leftrightarrow\; 2.8 < \dfrac{6x^2 + 5x - 3}{2x^2 - 1} < 3.2$. So

we graph the three parts of this inequality on the same screen, and

find that the curve $y = \dfrac{6x^2 + 5x - 3}{2x^2 - 1}$ seems to lie between the lines

$y = 2.8$ and $y = 3.2$ whenever $x > 12.8$. So we can choose $N = 13$

(or any larger number) so that the inequality holds whenever $x \ge N$.

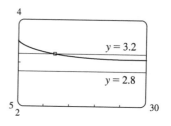

59. For $\varepsilon = 0.5$, we need to find N such that $\left| \dfrac{\sqrt{4x^2 + 1}}{x + 1} - (-2) \right| < 0.5$

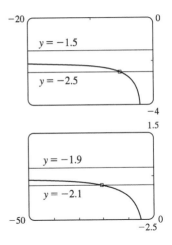

$\Leftrightarrow \quad -2.5 < \dfrac{\sqrt{4x^2 + 1}}{x + 1} < -1.5$ whenever $x \leq N$. We graph the

three parts of this inequality on the same screen, and see that the

inequality holds for $x \leq -6$. So we choose $N = -6$ (or any smaller

number).

For $\varepsilon = 0.1$, we need $-2.1 < \dfrac{\sqrt{4x^2 + 1}}{x + 1} < -1.9$ whenever

$x \leq N$. From the graph, it seems that this inequality holds for

$x \leq -22$. So we choose $N = -22$ (or any smaller number).

61. (a) $1/x^2 < 0.0001 \quad \Leftrightarrow \quad x^2 > 1/0.0001 = 10{,}000 \quad \Leftrightarrow \quad x > 100 \quad (x > 0)$

(b) If $\varepsilon > 0$ is given, then $1/x^2 < \varepsilon \quad \Leftrightarrow \quad x^2 > 1/\varepsilon \quad \Leftrightarrow \quad x > 1/\sqrt{\varepsilon}$. Let $N = 1/\sqrt{\varepsilon}$.

Then $x > N \quad \Rightarrow \quad x > \dfrac{1}{\sqrt{\varepsilon}} \quad \Rightarrow \quad \left| \dfrac{1}{x^2} - 0 \right| = \dfrac{1}{x^2} < \varepsilon$, so $\lim\limits_{x \to \infty} \dfrac{1}{x^2} = 0$.

63. For $x < 0$, $|1/x - 0| = -1/x$. If $\varepsilon > 0$ is given, then $-1/x < \varepsilon \quad \Leftrightarrow \quad x < -1/\varepsilon$.

Take $N = -1/\varepsilon$. Then $x < N \quad \Rightarrow \quad x < -1/\varepsilon \quad \Rightarrow \quad |(1/x) - 0| = -1/x < \varepsilon$, so $\lim\limits_{x \to -\infty} (1/x) = 0$.

65. Given $M > 0$, we need $N > 0$ such that $x > N \quad \Rightarrow \quad e^x > M$. Now $e^x > M \quad \Leftrightarrow \quad x > \ln M$, so take

$N = \max(1, \ln M)$. (This ensures that $N > 0$.) Then $x > N = \max(1, \ln M) \quad \Rightarrow \quad e^x > \max(e, M) \geq M$,

so $\lim\limits_{x \to \infty} e^x = \infty$.

67. Suppose that $\lim\limits_{x \to \infty} f(x) = L$. Then for every $\varepsilon > 0$ there is a corresponding positive number N such that

$|f(x) - L| < \varepsilon$ whenever $x > N$. If $t = 1/x$, then $x > N \quad \Leftrightarrow \quad 0 < 1/x < 1/N \quad \Leftrightarrow \quad 0 < t < 1/N$. Thus,

for every $\varepsilon > 0$ there is a corresponding $\delta > 0$ (namely $1/N$) such that $|f(1/t) - L| < \varepsilon$ whenever $0 < t < \delta$.

This proves that $\lim\limits_{t \to 0^+} f(1/t) = L = \lim\limits_{x \to \infty} f(x)$.

Now suppose that $\lim\limits_{x \to -\infty} f(x) = L$. Then for every $\varepsilon > 0$ there is a corresponding negative number N such that

$|f(x) - L| < \varepsilon$ whenever $x < N$. If $t = 1/x$, then $x < N \quad \Leftrightarrow \quad 1/N < 1/x < 0 \quad \Leftrightarrow \quad 1/N < t < 0$. Thus, for

every $\varepsilon > 0$ there is a corresponding $\delta > 0$ (namely $-1/N$) such that $|f(1/t) - L| < \varepsilon$ whenever $-\delta < t < 0$.

This proves that $\lim\limits_{t \to 0^-} f(1/t) = L = \lim\limits_{x \to -\infty} f(x)$.

2.7 Tangents, Velocities, and Other Rates of Change

1. (a) This is just the slope of the line through two points: $m_{PQ} = \dfrac{\Delta y}{\Delta x} = \dfrac{f(x) - f(3)}{x - 3}$.

(b) This is the limit of the slope of the secant line PQ as Q approaches P: $m = \lim\limits_{x \to 3} \dfrac{f(x) - f(3)}{x - 3}$.

3. The slope at D is the largest positive slope, followed by the positive slope at E. The slope at C is zero. The slope at B is steeper than at A (both are negative). In decreasing order, we have the slopes at: D, E, C, A, and B.

5. (a) (i) Using Definition 1,

$$m = \lim_{x \to a} \frac{f(x) - f(a)}{x - a} \quad \lim_{x \to -3} \frac{f(x) - f(-3)}{x - (-3)} = \lim_{x \to -3} \frac{(x^2 + 2x) - (3)}{x - (-3)} = \lim_{x \to -3} \frac{(x + 3)(x - 1)}{x + 3}$$

$$= \lim_{x \to -3}(x - 1) = -4$$

(ii) Using Equation 2,

$$m = \lim_{h \to 0} \frac{f(a + h) - f(a)}{h} = \lim_{h \to 0} \frac{f(-3 + h) - f(-3)}{h} = \lim_{h \to 0} \frac{[(-3 + h)^2 + 2(-3 + h)] - (3)}{h}$$

$$= \lim_{h \to 0} \frac{9 - 6h + h^2 - 6 + 2h - 3}{h} = \lim_{h \to 0} \frac{h(h - 4)}{h} = \lim_{h \to 0}(h - 4) = -4$$

(b) Using the point-slope form of the equation of a line, an equation of the tangent line is $y - 3 = -4(x + 3)$.

Solving for y gives us $y = -4x - 9$, which is the slope-intercept form of the equation of the tangent line.

(c)

7. Using (2) with $f(x) = 1 + 2x - x^3$ and $P(1, 2)$,

$$m = \lim_{h \to 0} \frac{f(a + h) - f(a)}{h} = \lim_{h \to 0} \frac{f(1 + h) - f(1)}{h} = \lim_{h \to 0} \frac{[1 + 2(1 + h) - (1 + h)^3] - 2}{h}$$

$$= \lim_{h \to 0} \frac{1 + 2 + 2h - (1 + 3h + 3h^2 + h^3) - 2}{h} = \lim_{h \to 0} \frac{-h^3 - 3h^2 - h}{h}$$

$$= \lim_{h \to 0} \frac{h(-h^2 - 3h - 1)}{h} = \lim_{h \to 0}(-h^2 - 3h - 1) = -1$$

Tangent line: $y - 2 = -1(x - 1) \Leftrightarrow y - 2 = -x + 1 \Leftrightarrow y = -x + 3$

9. Using (1) with $f(x) = \dfrac{x - 1}{x - 2}$ and $P(3, 2)$,

$$m = \lim_{x \to a} \frac{f(x) - f(a)}{x - a} = \lim_{x \to 3} \frac{\dfrac{x - 1}{x - 2} - 2}{x - 3} = \lim_{x \to 3} \frac{\dfrac{x - 1 - 2(x - 2)}{x - 2}}{x - 3} = \lim_{x \to 3} \frac{3 - x}{(x - 2)(x - 3)}$$

$$= \lim_{x \to 3} \frac{-1}{x - 2} = \frac{-1}{1} = -1.$$

Tangent line: $y - 2 = -1(x - 3) \Leftrightarrow y - 2 = -x + 3 \Leftrightarrow y = -x + 5$

11. (a) $m = \lim\limits_{x \to a} \dfrac{f(x) - f(a)}{x - a} = \lim\limits_{x \to a} \dfrac{2/(x + 3) - 2/(a + 3)}{x - a} = \lim\limits_{x \to a} \dfrac{2(a + 3) - 2(x + 3)}{(x - a)(x + 3)(a + 3)}$

$\quad = \lim\limits_{x \to a} \dfrac{2(a - x)}{(x - a)(x + 3)(a + 3)} = \lim\limits_{x \to a} \dfrac{-2}{(x + 3)(a + 3)} = \dfrac{-2}{(a + 3)^2}$

(b) (i) $a = -1 \;\Rightarrow\; m = \dfrac{-2}{(-1 + 3)^2} = -\dfrac{1}{2}$ (ii) $a = 0 \;\Rightarrow\; m = \dfrac{-2}{(0 + 3)^2} = -\dfrac{2}{9}$

(iii) $a = 1 \;\Rightarrow\; m = \dfrac{-2}{(1 + 3)^2} = -\dfrac{1}{8}$

13. (a) Using (1),

$$m = \lim\limits_{x \to a} \dfrac{(x^3 - 4x + 1) - (a^3 - 4a + 1)}{x - a} = \lim\limits_{x \to a} \dfrac{(x^3 - a^3) - 4(x - a)}{x - a}$$

$$= \lim\limits_{x \to a} \dfrac{(x - a)(x^2 + ax + a^2) - 4(x - a)}{x - a} = \lim\limits_{x \to a} (x^2 + ax + a^2 - 4) = 3a^2 - 4$$

(b) At $(1, -2)$: $m = 3(1)^2 - 4 = -1$, so an equation of the tangent line is $y - (-2) = -1(x - 1) \;\Leftrightarrow$
$y = -x - 1$. At $(2, 1)$: $m = 3(2)^2 - 4 = 8$, so an equation of the tangent line is $y - 1 = 8(x - 2) \;\Leftrightarrow$
$y = 8x - 15$.

(c)

15. (a) Since the slope of the tangent at $t = 0$ is 0, the car's initial velocity was 0.

(b) The slope of the tangent is greater at C than at B, so the car was going faster at C.

(c) Near A, the tangent lines are becoming steeper as x increases, so the velocity was increasing, so the car was
speeding up. Near B, the tangent lines are becoming less steep, so the car was slowing down. The steepest
tangent near C is the one at C, so at C the car had just finished speeding up, and was about to start slowing
down.

(d) Between D and E, the slope of the tangent is 0, so the car did not move during that time.

17. Let $s(t) = 40t - 16t^2$.

$$v(2) = \lim\limits_{t \to 2} \dfrac{s(t) - s(2)}{t - 2} = \lim\limits_{t \to 2} \dfrac{(40t - 16t^2) - 16}{t - 2} = \lim\limits_{t \to 2} \dfrac{-16t^2 + 40t - 16}{t - 2} = \lim\limits_{t \to 2} \dfrac{-8(2t^2 - 5t + 2)}{t - 2}$$

$$= \lim\limits_{t \to 2} \dfrac{-8(t - 2)(2t - 1)}{t - 2} = -8 \lim\limits_{t \to 2} (2t - 1) = -8(3) = -24$$

Thus, the instantaneous velocity when $t = 2$ is -24 ft/s.

19. $v(a) = \lim\limits_{h \to 0} \dfrac{s(a + h) - s(a)}{h} = \lim\limits_{h \to 0} \dfrac{4(a + h)^3 + 6(a + h) + 2 - (4a^3 + 6a + 2)}{h}$

$\quad = \lim\limits_{h \to 0} \dfrac{4a^3 + 12a^2 h + 12ah^2 + 4h^3 + 6a + 6h + 2 - 4a^3 - 6a - 2}{h}$

$\quad = \lim\limits_{h \to 0} \dfrac{12a^2 h + 12ah^2 + 4h^3 + 6h}{h} = \lim\limits_{h \to 0} (12a^2 + 12ah + 4h^2 + 6) = (12a^2 + 6)$ m/s

So $v(1) = 12(1)^2 + 6 = 18$ m/s, $v(2) = 12(2)^2 + 6 = 54$ m/s, and $v(3) = 12(3)^2 + 6 = 114$ m/s.

21. The sketch shows the graph for a room temperature of $72°$ and a refrigerator temperature of $38°$. The initial rate of change is greater in magnitude than the rate of change after an hour.

23. (a) (i) $[20, 23]$: $\dfrac{7.9 - 11.5}{23 - 20} = -1.2\,°\mathrm{C/h}$

  (ii) $[20, 22]$: $\dfrac{9.0 - 11.5}{22 - 20} = -1.25\,°\mathrm{C/h}$

  (iii) $[20, 21]$: $\dfrac{10.2 - 11.5}{21 - 20} = -1.3\,°\mathrm{C/h}$

(b) In the figure, we estimate A to be $(18, 15.5)$ and B as $(23, 6)$. So the slope is

$\dfrac{6 - 15.5}{23 - 18} = -1.9\,°\mathrm{C/h}$ at 8:00 P.M.

25. (a) (i) $[1995, 1997]$: $\dfrac{N(1997) - N(1995)}{1997 - 1995} = \dfrac{2461 - 873}{2} = \dfrac{1588}{2} = 794$ thousand/year

  (ii) $[1995, 1996]$: $\dfrac{N(1996) - N(1995)}{1996 - 1995} = \dfrac{1513 - 873}{1} = 640$ thousand/year

  (iii) $[1994, 1995]$: $\dfrac{N(1995) - N(1994)}{1995 - 1994} = \dfrac{873 - 572}{1} = 301$ thousand/year

(b) Using the values from (ii) and (iii), we have $\dfrac{640 + 301}{2} = \dfrac{941}{2} = 470.5$ thousand/year.

(c) Estimating A as $(1994, 420)$ and B as $(1996, 1275)$, the slope at 1995 is $\dfrac{1275 - 420}{1996 - 1994} = \dfrac{855}{2} = 427.5$ thousand/year

27. (a) (i) $\dfrac{\Delta C}{\Delta x} = \dfrac{C(105) - C(100)}{105 - 100} = \dfrac{6601.25 - 6500}{5} = \$20.25/\text{unit}.$

  (ii) $\dfrac{\Delta C}{\Delta x} = \dfrac{C(101) - C(100)}{101 - 100} = \dfrac{6520.05 - 6500}{1} = \$20.05/\text{unit}.$

(b) $\dfrac{C(100 + h) - C(100)}{h} = \dfrac{\left[5000 + 10(100 + h) + 0.05(100 + h)^2\right] - 6500}{h} = \dfrac{20h + 0.05h^2}{h}$

$= 20 + 0.05h,\ h \neq 0$

So the instantaneous rate of change is $\displaystyle\lim_{h \to 0} \dfrac{C(100 + h) - C(100)}{h} = \lim_{h \to 0}(20 + 0.05h) = \$20/\text{unit}.$

2.8 Derivatives

1.

The line from $P(2, f(2))$ to $Q(2 + h, f(2 + h))$

is the line that has slope $\dfrac{f(2 + h) - f(2)}{h}$.

3. $g'(0)$ is the only negative value. The slope at $x = 4$ is smaller than the slope at $x = 2$ and both are smaller than the slope at $x = -2$. Thus, $g'(0) < 0 < g'(4) < g'(2) < g'(-2)$.

5. We begin by drawing a curve through the origin at a slope of 3 to satisfy $f(0) = 0$ and $f'(0) = 3$. Since $f'(1) = 0$, we will round off our figure so that there is a horizontal tangent directly over $x = 1$. Lastly, we make sure that the curve has a slope of -1 as we pass over $x = 2$. Two of the many possibilities are shown.

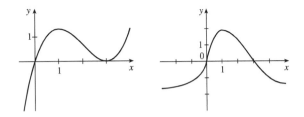

7. Using Definition 2 with $f(x) = 3x^2 - 5x$ and the point $(2, 2)$, we have

$$f'(2) = \lim_{h \to 0} \frac{f(2 + h) - f(2)}{h} = \lim_{h \to 0} \frac{[3(2 + h)^2 - 5(2 + h)] - 2}{h}$$

$$= \lim_{h \to 0} \frac{(12 + 12h + 3h^2 - 10 - 5h) - 2}{h} = \lim_{h \to 0} \frac{3h^2 + 7h}{h} = \lim_{h \to 0} (3h + 7) = 7.$$

So an equation of the tangent line at $(2, 2)$ is $y - 2 = 7(x - 2)$ or $y = 7x - 12$.

9. (a) Using Definition 2 with $F(x) = x^3 - 5x + 1$ and the point $(1, -3)$, we have

$$F'(1) = \lim_{h \to 0} \frac{F(1 + h) - F(1)}{h} = \lim_{h \to 0} \frac{[(1 + h)^3 - 5(1 + h) + 1] - (-3)}{h}$$

$$= \lim_{h \to 0} \frac{(1 + 3h + 3h^2 + h^3 - 5 - 5h + 1) + 3}{h} = \lim_{h \to 0} \frac{h^3 + 3h^2 - 2h}{h}$$

$$= \lim_{h \to 0} \frac{h(h^2 + 3h - 2)}{h} = \lim_{h \to 0} (h^2 + 3h - 2) = -2$$

So an equation of the tangent line at $(1, -3)$ is $y - (-3) = -2(x - 1) \quad \Leftrightarrow \quad y = -2x - 1$.

(b)

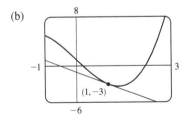

11. (a) $f'(1) = \lim\limits_{h \to 0} \dfrac{f(1+h) - f(1)}{h} = \lim\limits_{h \to 0} \dfrac{3^{1+h} - 3^1}{h}$.

(b)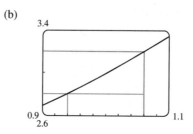

So let $F(h) = \dfrac{3^{1+h} - 3}{h}$. We calculate:

h	$F(h)$	h	$F(h)$
0.1	3.484	−0.1	3.121
0.01	3.314	−0.01	3.278
0.001	3.298	−0.001	3.294
0.0001	3.296	−0.0001	3.296

We estimate that $f'(1) \approx 3.296$.

From the graph, we estimate that the slope of the tangent is about

$$\frac{3.2 - 2.8}{1.06 - 0.94} = \frac{0.4}{0.12} \approx 3.3.$$

13. Use Definition 2 with $f(x) = 3 - 2x + 4x^2$.

$$f'(a) = \lim_{h \to 0} \frac{f(a+h) - f(a)}{h} = \lim_{h \to 0} \frac{\left[3 - 2(a+h) + 4(a+h)^2\right] - \left(3 - 2a + 4a^2\right)}{h}$$

$$= \lim_{h \to 0} \frac{\left(3 - 2a - 2h + 4a^2 + 8ah + 4h^2\right) - \left(3 - 2a + 4a^2\right)}{h}$$

$$= \lim_{h \to 0} \frac{-2h + 8ah + 4h^2}{h} = \lim_{h \to 0} \frac{h(-2 + 8a + 4h)}{h} = \lim_{h \to 0}(-2 + 8a + 4h) = -2 + 8a$$

15. Use Definition 2 with $f(t) = (2t + 1)/(t + 3)$.

$$f'(a) = \lim_{h \to 0} \frac{f(a+h) - f(a)}{h} = \lim_{h \to 0} \frac{\dfrac{2(a+h) + 1}{(a+h) + 3} - \dfrac{2a + 1}{a + 3}}{h}$$

$$= \lim_{h \to 0} \frac{(2a + 2h + 1)(a + 3) - (2a + 1)(a + h + 3)}{h(a + h + 3)(a + 3)}$$

$$= \lim_{h \to 0} \frac{\left(2a^2 + 6a + 2ah + 6h + a + 3\right) - \left(2a^2 + 2ah + 6a + a + h + 3\right)}{h(a + h + 3)(a + 3)}$$

$$= \lim_{h \to 0} \frac{5h}{h(a + h + 3)(a + 3)} = \lim_{h \to 0} \frac{5}{(a + h + 3)(a + 3)} = \frac{5}{(a + 3)^2}$$

17. Use Definition 2 with $f(x) = 1/\sqrt{x + 2}$.

$$f'(a) = \lim_{h \to 0} \frac{f(a+h) - f(a)}{h} = \lim_{h \to 0} \frac{\dfrac{1}{\sqrt{(a+h) + 2}} - \dfrac{1}{\sqrt{a + 2}}}{h}$$

$$= \lim_{h \to 0} \frac{\dfrac{\sqrt{a + 2} - \sqrt{a + h + 2}}{\sqrt{a + h + 2}\sqrt{a + 2}}}{h} = \lim_{h \to 0} \left[\frac{\sqrt{a + 2} - \sqrt{a + h + 2}}{h\sqrt{a + h + 2}\sqrt{a + 2}} \cdot \frac{\sqrt{a + 2} + \sqrt{a + h + 2}}{\sqrt{a + 2} + \sqrt{a + h + 2}}\right]$$

$$= \lim_{h \to 0} \frac{(a + 2) - (a + h + 2)}{h\sqrt{a + h + 2}\sqrt{a + 2}\left(\sqrt{a + 2} + \sqrt{a + h + 2}\right)}$$

$$= \lim_{h \to 0} \frac{-h}{h\sqrt{a + h + 2}\sqrt{a + 2}\left(\sqrt{a + 2} + \sqrt{a + h + 2}\right)}$$

$$= \lim_{h \to 0} \frac{-1}{\sqrt{a + h + 2}\sqrt{a + 2}\left(\sqrt{a + 2} + \sqrt{a + h + 2}\right)}$$

$$= \frac{-1}{\left(\sqrt{a + 2}\right)^2 \left(2\sqrt{a + 2}\right)} = -\frac{1}{2(a + 2)^{3/2}}$$

Note that the answers to Exercises 19–24 are not unique.

19. By Definition 2, $\lim\limits_{h \to 0} \dfrac{(1+h)^{10} - 1}{h} = f'(1)$, where $f(x) = x^{10}$ and $a = 1$.

Or: By Definition 2, $\lim\limits_{h \to 0} \dfrac{(1+h)^{10} - 1}{h} = f'(0)$, where $f(x) = (1+x)^{10}$ and $a = 0$.

21. By Equation 3, $\lim\limits_{x \to 5} \dfrac{2^x - 32}{x - 5} = f'(5)$, where $f(x) = 2^x$ and $a = 5$.

23. By Definition 2, $\lim\limits_{h \to 0} \dfrac{\cos(\pi + h) + 1}{h} = f'(\pi)$, where $f(x) = \cos x$ and $a = \pi$.

Or: By Definition 2, $\lim\limits_{h \to 0} \dfrac{\cos(\pi + h) + 1}{h} = f'(0)$, where $f(x) = \cos(\pi + x)$ and $a = 0$.

25. $v(2) = f'(2) = \lim\limits_{h \to 0} \dfrac{f(2+h) - f(2)}{h} = \lim\limits_{h \to 0} \dfrac{\left[(2+h)^2 - 6(2+h) - 5\right] - \left[2^2 - 6(2) - 5\right]}{h}$

$= \lim\limits_{h \to 0} \dfrac{(4 + 4h + h^2 - 12 - 6h - 5) - (-13)}{h} = \lim\limits_{h \to 0} \dfrac{h^2 - 2h}{h} = \lim\limits_{h \to 0} (h - 2) = -2 \text{ m/s}$

27. (a) $f'(x)$ is the rate of change of the production cost with respect to the number of ounces of gold produced. Its units are dollars per ounce.

(b) After 800 ounces of gold have been produced, the rate at which the production cost is increasing is $17/ounce. So the cost of producing the 800th (or 801st) ounce is about $17.

(c) In the short term, the values of $f'(x)$ will decrease because more efficient use is made of start-up costs as x increases. But eventually $f'(x)$ might increase due to large-scale operations.

29. (a) $f'(v)$ is the rate at which the fuel consumption is changing with respect to the speed. Its units are (gal/h)/(mi/h).

(b) The fuel consumption is decreasing by 0.05 (gal/h)/(mi/h) as the car's speed reaches 20 mi/h. So if you increase your speed to 21 mi/h, you could expect to decrease your fuel consumption by about 0.05 (gal/h)/(mi/h).

31. $T'(10)$ is the rate at which the temperature is changing at 10:00 A.M. To estimate the value of $T'(10)$, we will average the difference quotients obtained using the times $t = 8$ and $t = 12$. Let

$A = \dfrac{T(8) - T(10)}{8 - 10} = \dfrac{72 - 81}{-2} = 4.5$ and $B = \dfrac{T(12) - T(10)}{12 - 10} = \dfrac{88 - 81}{2} = 3.5$. Then

$T'(10) = \lim\limits_{t \to 10} \dfrac{T(t) - T(10)}{t - 10} \approx \dfrac{A + B}{2} = \dfrac{4.5 + 3.5}{2} = 4°\text{F/h}$.

33. (a) $S'(T)$ is the rate at which the oxygen solubility changes with respect to the water temperature. Its units are $(\text{mg/L})/°\text{C}$.

(b) For $T = 16°\text{C}$, it appears that the tangent line to the curve goes through the points $(0, 14)$ and $(32, 6)$. So

$S'(16) \approx \dfrac{6 - 14}{32 - 0} = -\dfrac{8}{32} = -0.25 \ (\text{mg/L})/°\text{C}$. This means that as the temperature increases past $16°\text{C}$, the oxygen solubility is decreasing at a rate of $0.25 \ (\text{mg/L})/°\text{C}$.

35. Since $f(x) = x \sin(1/x)$ when $x \neq 0$ and $f(0) = 0$, we have

$$f'(0) = \lim_{h \to 0} \frac{f(0+h) - f(0)}{h} = \lim_{h \to 0} \frac{h \sin(1/h) - 0}{h} = \lim_{h \to 0} \sin(1/h). \text{ This limit does not exist since } \sin(1/h)$$

takes the values -1 and 1 on any interval containing 0. (Compare with Example 4 in Section 2.2.)

2.9 The Derivative as a Function

1. *Note:* Your answers may vary depending on your estimates. By estimating the slopes of tangent lines on the graph of f, it appears that

(a) $f'(1) \approx -2$ (b) $f'(2) \approx 0.8$

(c) $f'(3) \approx -1$ (d) $f'(4) \approx -0.5$

3. It appears that f is an odd function, so f' will be an even function—that is, $f'(-a) = f'(a)$.

(a) $f'(-3) \approx 1.5$ (b) $f'(-2) \approx 1$

(c) $f'(-1) \approx 0$ (d) $f'(0) \approx -4$

(e) $f'(1) \approx 0$ (f) $f'(2) \approx 1$

(g) $f'(3) \approx 1.5$

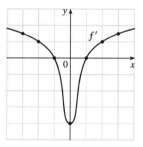

Hints for Exercises 5–13: First plot x-intercepts on the graph of f' for any horizontal tangents on the graph of f. Look for any corners on the graph of f — there will be a discontinuity on the graph of f'. On any interval where f has a tangent with positive (or negative) slope, the graph of f' will be positive (or negative). If the graph of the function is linear, the graph of f' will be a horizontal line.

5.

7.

9.

11.

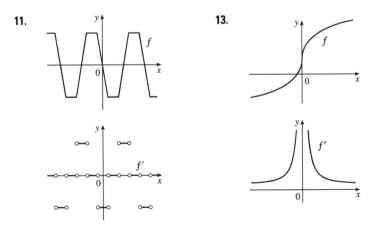

13.

15. It appears that there are horizontal tangents on the graph of M for $t = 1963$ and $t = 1971$. Thus, there are zeros for those values of t on the graph of M'. The derivative is negative for the years 1963 to 1971.

17.

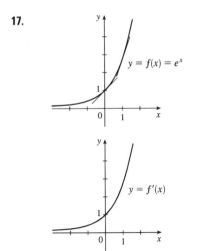

The slope at 0 appears to be 1 and the slope at 1 appears to be 2.7. As x decreases, the slope gets closer to 0. Since the graphs are so similar, we might guess that $f'(x) = e^x$.

19. (a) By zooming in, we estimate that $f'(0) = 0$, $f'\left(\frac{1}{2}\right) = 1$, $f'(1) = 2$, and $f'(2) = 4$.

(b) By symmetry, $f'(-x) = -f'(x)$. So $f'\left(-\frac{1}{2}\right) = -1$, $f'(-1) = -2$, and $f'(-2) = -4$.

(c) It appears that $f'(x)$ is twice the value of x, so we guess that $f'(x) = 2x$.

(d) $f'(x) = \lim_{h \to 0} \dfrac{f(x+h) - f(x)}{h} = \lim_{h \to 0} \dfrac{(x+h)^2 - x^2}{h}$

$= \lim_{h \to 0} \dfrac{(x^2 + 2hx + h^2) - x^2}{h} = \lim_{h \to 0} \dfrac{2hx + h^2}{h} = \lim_{h \to 0} \dfrac{h(2x + h)}{h} = \lim_{h \to 0} (2x + h) = 2x$

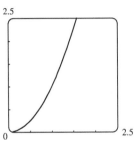

21. $f'(x) = \lim\limits_{h \to 0} \dfrac{f(x+h) - f(x)}{h} = \lim\limits_{h \to 0} \dfrac{37 - 37}{h} = \lim\limits_{h \to 0} \dfrac{0}{h} = \lim\limits_{h \to 0} 0 = 0$

Domain of f = domain of $f' = \mathbb{R}$.

23. $f'(x) = \lim\limits_{h \to 0} \dfrac{f(x+h) - f(x)}{h} = \lim\limits_{h \to 0} \dfrac{[1 - 3(x+h)^2] - (1 - 3x^2)}{h}$

$= \lim\limits_{h \to 0} \dfrac{[1 - 3(x^2 + 2xh + h^2)] - (1 - 3x^2)}{h} = \lim\limits_{h \to 0} \dfrac{1 - 3x^2 - 6xh - 3h^2 - 1 + 3x^2}{h}$

$= \lim\limits_{h \to 0} \dfrac{-6xh - 3h^2}{h} = \lim\limits_{h \to 0} \dfrac{h(-6x - 3h)}{h} = \lim\limits_{h \to 0} (-6x - 3h) = -6x$

Domain of f = domain of $f' = \mathbb{R}$.

25. $f'(x) = \lim\limits_{h \to 0} \dfrac{f(x+h) - f(x)}{h} = \lim\limits_{h \to 0} \dfrac{[(x+h)^3 - 3(x+h) + 5] - (x^3 - 3x + 5)}{h}$

$= \lim\limits_{h \to 0} \dfrac{(x^3 + 3x^2h + 3xh^2 + h^3 - 3x - 3h + 5) - (x^3 - 3x + 5)}{h}$

$= \lim\limits_{h \to 0} \dfrac{3x^2h + 3xh^2 + h^3 - 3h}{h} = \lim\limits_{h \to 0} \dfrac{h(3x^2 + 3xh + h^2 - 3)}{h}$

$= \lim\limits_{h \to 0} (3x^2 + 3xh + h^2 - 3) = 3x^2 - 3$

Domain of f = domain of $f' = \mathbb{R}$.

27. $g'(x) = \lim\limits_{h \to 0} \dfrac{g(x+h) - g(x)}{h} = \lim\limits_{h \to 0} \dfrac{\sqrt{1 + 2(x+h)} - \sqrt{1 + 2x}}{h} \left[\dfrac{\sqrt{1 + 2(x+h)} + \sqrt{1 + 2x}}{\sqrt{1 + 2(x+h)} + \sqrt{1 + 2x}} \right]$

$= \lim\limits_{h \to 0} \dfrac{(1 + 2x + 2h) - (1 + 2x)}{h \left[\sqrt{1 + 2(x+h)} + \sqrt{1 + 2x} \right]} = \lim\limits_{h \to 0} \dfrac{2}{\sqrt{1 + 2x + 2h} + \sqrt{1 + 2x}} = \dfrac{2}{2\sqrt{1 + 2x}} = \dfrac{1}{\sqrt{1 + 2x}}$

Domain of $g = [-\frac{1}{2}, \infty)$, domain of $g' = (-\frac{1}{2}, \infty)$.

29. $G'(t) = \lim\limits_{h \to 0} \dfrac{G(t+h) - G(t)}{h} = \lim\limits_{h \to 0} \dfrac{\dfrac{4(t+h)}{(t+h)+1} - \dfrac{4t}{t+1}}{h} = \lim\limits_{h \to 0} \dfrac{\dfrac{4(t+h)(t+1) - 4t(t+h+1)}{(t+h+1)(t+1)}}{h}$

$= \lim\limits_{h \to 0} \dfrac{(4t^2 + 4ht + 4t + 4h) - (4t^2 + 4ht + 4t)}{h(t+h+1)(t+1)}$

$= \lim\limits_{h \to 0} \dfrac{4h}{h(t+h+1)(t+1)} = \lim\limits_{h \to 0} \dfrac{4}{(t+h+1)(t+1)} = \dfrac{4}{(t+1)^2}$

Domain of G = domain of $G' = (-\infty, -1) \cup (-1, \infty)$.

31. $f'(x) = \lim\limits_{h \to 0} \dfrac{f(x+h) - f(x)}{h} = \lim\limits_{h \to 0} \dfrac{(x+h)^4 - x^4}{h} = \lim\limits_{h \to 0} \dfrac{(x^4 + 4x^3h + 6x^2h^2 + 4xh^3 + h^4) - x^4}{h}$

$= \lim\limits_{h \to 0} \dfrac{4x^3h + 6x^2h^2 + 4xh^3 + h^4}{h} = \lim\limits_{h \to 0} (4x^3 + 6x^2h + 4xh^2 + h^3) = 4x^3$

Domain of f = domain of $f' = \mathbb{R}$.

33. (a) $f'(x) = \lim\limits_{h \to 0} \dfrac{f(x+h) - f(x)}{h} = \lim\limits_{h \to 0} \dfrac{\left[x + h - \left(\dfrac{2}{x+h}\right)\right] - \left[x - \left(\dfrac{2}{x}\right)\right]}{h}$

$= \lim\limits_{h \to 0} \left[\dfrac{h - \dfrac{2}{(x+h)} + \dfrac{2}{x}}{h}\right] = \lim\limits_{h \to 0} \left[1 + \dfrac{-2x + 2(x+h)}{hx(x+h)}\right] = \lim\limits_{h \to 0} \left[1 + \dfrac{2h}{hx(x+h)}\right]$

$= \lim\limits_{h \to 0} \left[1 + \dfrac{2}{x(x+h)}\right] = 1 + \dfrac{2}{x^2}$

(b) Notice that when f has steep tangent lines, $f'(x)$ is very large.
When f is flatter, $f'(x)$ is smaller.

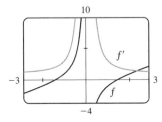

35. (a) $U'(t)$ is the rate at which the unemployment rate is changing with respect to time. Its units are percent per year.

(b) To find $U'(t)$, we use $\lim\limits_{h \to 0} \dfrac{U(t+h) - U(t)}{h} \approx \dfrac{U(t+h) - U(t)}{h}$ for small values of h.

For 1991: $U'(1991) = \dfrac{U(1992) - U(1991)}{1992 - 1991} = \dfrac{7.5 - 6.8}{1} = 0.70$

For 1992: We estimate $U'(1992)$ by using $h = -1$ and $h = 1$, and then average the two results to obtain a final estimate.

$h = -1 \quad \Rightarrow \quad U'(1992) \approx \dfrac{U(1991) - U(1992)}{1991 - 1992} = \dfrac{6.8 - 7.5}{-1} = 0.70;$

$h = 1 \quad \Rightarrow \quad U'(1992) \approx \dfrac{U(1993) - U(1992)}{1993 - 1992} = \dfrac{6.9 - 7.5}{1} = -0.60.$

So we estimate that $U'(1992) \approx \frac{1}{2}[0.70 + (-0.60)] = 0.05$.

t	1991	1992	1993	1994	1995	1996	1997	1998	1999	2000
$U'(t)$	0.70	0.05	-0.70	-0.65	-0.35	-0.35	-0.45	-0.35	-0.25	-0.20

37. f is not differentiable at $x = -1$ or at $x = 11$ because the graph has vertical tangents at those points; at $x = 4$, because there is a discontinuity there; and at $x = 8$, because the graph has a corner there.

39. As we zoom in toward $(-1, 0)$, the curve appears more and more like a straight line, so $f(x) = x + \sqrt{|x|}$ is differentiable at $x = -1$. But no matter how much we zoom in toward the origin, the curve doesn't straighten out—we can't eliminate the sharp point (a cusp). So f is not differentiable at $x = 0$.

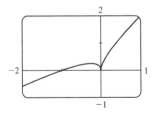

41. (a) Note that we have factored $x - a$ as the difference of two cubes in the third step.

$$f'(a) = \lim_{x \to a} \frac{f(x) - f(a)}{x - a} = \lim_{x \to a} \frac{x^{1/3} - a^{1/3}}{x - a} = \lim_{x \to a} \frac{x^{1/3} - a^{1/3}}{(x^{1/3} - a^{1/3})(x^{2/3} + x^{1/3}a^{1/3} + a^{2/3})}$$

$$= \lim_{x \to a} \frac{1}{x^{2/3} + x^{1/3}a^{1/3} + a^{2/3}} = \frac{1}{3a^{2/3}} \text{ or } \tfrac{1}{3}a^{-2/3}$$

(b) $f'(0) = \lim_{h \to 0} \dfrac{f(0 + h) - f(0)}{h} = \lim_{h \to 0} \dfrac{\sqrt[3]{h} - 0}{h} = \lim_{h \to 0} \dfrac{1}{h^{2/3}}$. This function increases without bound, so the

limit does not exist, and therefore $f'(0)$ does not exist.

(c) $\lim_{x \to 0} |f'(x)| = \lim_{x \to 0} \dfrac{1}{3x^{2/3}} = \infty$ and f is continuous at $x = 0$ (root function), so f has a vertical tangent

at $x = 0$.

43. $f(x) = |x - 6| = \begin{cases} -(x - 6) & \text{if } x < 6 \\ x - 6 & \text{if } x \geq 6 \end{cases} = \begin{cases} 6 - x & \text{if } x < 6 \\ x - 6 & \text{if } x \geq 6 \end{cases}$

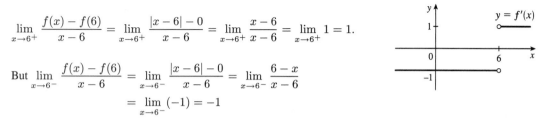

$$\lim_{x \to 6^+} \frac{f(x) - f(6)}{x - 6} = \lim_{x \to 6^+} \frac{|x - 6| - 0}{x - 6} = \lim_{x \to 6^+} \frac{x - 6}{x - 6} = \lim_{x \to 6^+} 1 = 1.$$

But $\lim_{x \to 6^-} \dfrac{f(x) - f(6)}{x - 6} = \lim_{x \to 6^-} \dfrac{|x - 6| - 0}{x - 6} = \lim_{x \to 6^-} \dfrac{6 - x}{x - 6}$

$$= \lim_{x \to 6^-} (-1) = -1$$

So $f'(6) = \lim_{x \to 6} \dfrac{f(x) - f(6)}{x - 6}$ does not exist. However, $f'(x) = \begin{cases} -1 & \text{if } x < 6 \\ 1 & \text{if } x > 6 \end{cases}$

Another way of writing the answer is $f'(x) = \dfrac{x - 6}{|x - 6|}$.

45. (a) $f(x) = x |x| = \begin{cases} x^2 & \text{if } x \geq 0 \\ -x^2 & \text{if } x < 0 \end{cases}$

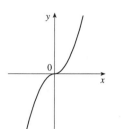

(b) Since $f(x) = x^2$ for $x \geq 0$, we have $f'(x) = 2x$ for $x > 0$. [See Exercise 2.9.19(d).] Similarly, since $f(x) = -x^2$ for $x < 0$, we have $f'(x) = -2x$ for $x < 0$. At $x = 0$, we have

$$f'(0) = \lim_{x \to 0} \frac{f(x) - f(0)}{x - 0} = \lim_{x \to 0} \frac{x |x|}{x} = \lim_{x \to 0} |x| = 0.$$

So f is differentiable at 0. Thus, f is differentiable for all x.

(c) From part (b), we have $f'(x) = \begin{cases} 2x & \text{if } x \geq 0 \\ -2x & \text{if } x < 0 \end{cases} = 2|x|$.

47. (a) If f is even, then

$$f'(-x) = \lim_{h \to 0} \frac{f(-x + h) - f(-x)}{h} = \lim_{h \to 0} \frac{f[-(x - h)] - f(-x)}{h} = \lim_{h \to 0} \frac{f(x - h) - f(x)}{h}$$

$$= -\lim_{h \to 0} \frac{f(x - h) - f(x)}{-h} \quad [\text{let } \Delta x = -h] \quad = -\lim_{\Delta x \to 0} \frac{f(x + \Delta x) - f(x)}{\Delta x} = -f'(x)$$

Therefore, f' is odd.

(b) If f is odd, then

$$f'(-x) = \lim_{h \to 0} \frac{f(-x+h) - f(-x)}{h} = \lim_{h \to 0} \frac{f[-(x-h)] - f(-x)}{h} = \lim_{h \to 0} \frac{-f(x-h) + f(x)}{h}$$

$$= \lim_{h \to 0} \frac{f(x-h) - f(x)}{-h} \quad [\text{let } \Delta x = -h] \quad = \lim_{\Delta x \to 0} \frac{f(x+\Delta x) - f(x)}{\Delta x} = f'(x)$$

Therefore, f' is even.

49.

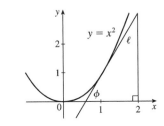

In the right triangle in the diagram, let Δy be the side opposite angle ϕ and Δx the side adjacent angle ϕ. Then the slope of the tangent line ℓ is $m = \Delta y / \Delta x = \tan \phi$. Note that $0 < \phi < \frac{\pi}{2}$. We know (see Exercise 19) that the derivative of $f(x) = x^2$ is $f'(x) = 2x$. So the slope of the tangent to the curve at the point $(1, 1)$ is 2. Thus, ϕ is the angle between 0 and $\frac{\pi}{2}$ whose tangent is 2; that is, $\phi = \tan^{-1} 2 \approx 63°$.

2 Review

———————————————— CONCEPT CHECK ————————————————

1. (a) $\lim_{x \to a} f(x) = L$: See Definition 2.2.1 and Figures 1 and 2 in Section 2.2.

   (b) $\lim_{x \to a^+} f(x) = L$: See the paragraph after Definition 2.2.2 and Figure 9(b) in Section 2.2.

   (c) $\lim_{x \to a^-} f(x) = L$: See Definition 2.2.2 and Figure 9(a) in Section 2.2.

   (d) $\lim_{x \to a} f(x) = \infty$: See Definition 2.2.4 and Figure 12 in Section 2.2.

   (e) $\lim_{x \to \infty} f(x) = L$: See Definition 2.6.1 and Figure 2 in Section 2.6.

2. In general, the limit of a function fails to exist when the function does not approach a fixed number. For each of the following functions, the limit fails to exist at $x = 2$.

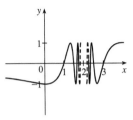

The left- and right-hand
limits are not equal.

There is an
infinite discontinuity.

There are an infinite
number of oscillations.

3. (a)–(g) See the statements of Limit Laws 1–6 and 11 in Section 2.3.

4. See Theorem 3 in Section 2.3.

5. (a) See Definition 2.2.6 and Figures 12–14 in Section 2.2.

(b) See Definition 2.6.3 and Figures 3 and 4 in Section 2.6.

6. (a) $y = x^4$: No asymptote

(b) $y = \sin x$: No asymptote

(c) $y = \tan x$: Vertical asymptotes $x = \frac{\pi}{2} + \pi n$,
n an integer

(d) $y = \tan^{-1} x$: Horizontal asymptotes $y = \pm\frac{\pi}{2}$

(e) $y = e^x$: Horizontal asymptote $y = 0$
($\lim\limits_{x \to -\infty} e^x = 0$)

(f) $y = \ln x$: Vertical asymptote $x = 0$
($\lim\limits_{x \to 0^+} \ln x = -\infty$)

(g) $y = 1/x$: Vertical asymptote $x = 0$,
horizontal asymptote $y = 0$

(h) $y = \sqrt{x}$: No asymptote

7. (a) A function f is continuous at a number a if $f(x)$ approaches $f(a)$ as x approaches a; that is,
$\lim\limits_{x \to a} f(x) = f(a)$.

(b) A function f is continuous on the interval $(-\infty, \infty)$ if f is continuous at every real number a. The graph of such a function has no breaks and every vertical line crosses it.

8. See Theorem 2.5.10.

9. See Definition 2.7.1.

10. See the paragraph containing Formula 3 in Section 2.7.

11. (a) The average rate of change of y with respect to x over the interval $[x_1, x_2]$ is $\dfrac{f(x_2) - f(x_1)}{x_2 - x_1}$.

(b) The instantaneous rate of change of y with respect to x at $x = x_1$ is $\lim\limits_{x_2 \to x_1} \dfrac{f(x_2) - f(x_1)}{x_2 - x_1}$.

12. See Definition 2.8.2. The pages following the definition discuss interpretations of $f'(a)$ as the slope of a tangent line to the graph of f at $x = a$ and as an instantaneous rate of change of $f(x)$ with respect to x when $x = a$.

13. (a) A function f is differentiable at a number a if its derivative f' exists at $x = a$; that is, if $f'(a)$ exists.

(b) See Theorem 2.9.4. This theorem also tells us that if f is *not* continuous at a, then f is *not* differentiable at a.

(c)

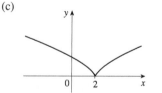

14. See the discussion and Figure 8 on page 172.

—————————————————————— TRUE-FALSE QUIZ ——————————————————————

1. False.   Limit Law 2 applies only if the individual limits exist (these don't).

3. True.   Limit Law 5 applies.

5. False.   Consider $\lim\limits_{x \to 5} \dfrac{x(x - 5)}{x - 5}$ or $\lim\limits_{x \to 5} \dfrac{\sin(x - 5)}{x - 5}$. The first limit exists and is equal to 5. By Example 3 in Section 2.2, we know that the latter limit exists (and is equal to 1).

7. True. A polynomial is continuous everywhere, so $\lim\limits_{x \to b} p(x)$ exists and is equal to $p(b)$.

9. True. See Figure 4 in Section 2.6.

11. False. Consider $f(x) = \begin{cases} 1/(x-1) & \text{if } x \neq 1 \\ 2 & \text{if } x = 1 \end{cases}$

13. True. Use Theorem 2.5.8 with $a = 2$, $b = 5$, and $g(x) = 4x^2 - 11$. Note that $f(4) = 3$ is not needed.

15. True, by the definition of a limit with $\varepsilon = 1$.

17. False. See the note after Theorem 4 in Section 2.9.

─────────────────────── EXERCISES ───────────────────────

1. (a) (i) $\lim\limits_{x \to 2^+} f(x) = 3$

(ii) $\lim\limits_{x \to -3^+} f(x) = 0$

(iii) $\lim\limits_{x \to -3} f(x)$ does not exist since the left and right limits are not equal. (The left limit is -2.)

(iv) $\lim\limits_{x \to 4} f(x) = 2$

(v) $\lim\limits_{x \to 0} f(x) = \infty$

(vi) $\lim\limits_{x \to 2^-} f(x) = -\infty$

(vii) $\lim\limits_{x \to \infty} f(x) = 4$

(viii) $\lim\limits_{x \to -\infty} f(x) = -1$

(b) The equations of the horizontal asymptotes are $y = -1$ and $y = 4$.

(c) The equations of the vertical asymptotes are $x = 0$ and $x = 2$.

(d) f is discontinuous at $x = -3, 0, 2$, and 4. The discontinuities are jump, infinite, infinite, and removable, respectively.

3. Since the exponential function is continuous, $\lim\limits_{x \to 1} e^{x^3 - x} = e^{1-1} = e^0 = 1$.

5. $\lim\limits_{x \to -3} \dfrac{x^2 - 9}{x^2 + 2x - 3} = \lim\limits_{x \to -3} \dfrac{(x+3)(x-3)}{(x+3)(x-1)} = \lim\limits_{x \to -3} \dfrac{x-3}{x-1} = \dfrac{-3-3}{-3-1} = \dfrac{-6}{-4} = \dfrac{3}{2}$

7. $\lim\limits_{h \to 0} \dfrac{(h-1)^3 + 1}{h} = \lim\limits_{h \to 0} \dfrac{(h^3 - 3h^2 + 3h - 1) + 1}{h} = \lim\limits_{h \to 0} \dfrac{h^3 - 3h^2 + 3h}{h} = \lim\limits_{h \to 0} (h^2 - 3h + 3) = 3$

Another solution: Factor the numerator as a sum of two cubes and then simplify.

$$\lim\limits_{h \to 0} \dfrac{(h-1)^3 + 1}{h} = \lim\limits_{h \to 0} \dfrac{(h-1)^3 + 1^3}{h} = \lim\limits_{h \to 0} \dfrac{[(h-1)+1]\left[(h-1)^2 - 1(h-1) + 1^2\right]}{h}$$

$$= \lim\limits_{h \to 0} \left[(h-1)^2 - h + 2\right] = 1 - 0 + 2 = 3$$

9. $\lim\limits_{r \to 9} \dfrac{\sqrt{r}}{(r-9)^4} = \infty$ since $(r-9)^4 \to 0$ as $r \to 9$ and $\dfrac{\sqrt{r}}{(r-9)^4} > 0$ for $r \neq 9$.

11. $\lim\limits_{s \to 16} \dfrac{4 - \sqrt{s}}{s - 16} = \lim\limits_{s \to 16} \dfrac{4 - \sqrt{s}}{(\sqrt{s}+4)(\sqrt{s}-4)} = \lim\limits_{s \to 16} \dfrac{-1}{\sqrt{s}+4} = \dfrac{-1}{\sqrt{16}+4} = -\dfrac{1}{8}$

13. $\dfrac{|x-8|}{x-8} = \begin{cases} \dfrac{x-8}{x-8} & \text{if } x - 8 > 0 \\ \dfrac{-(x-8)}{x-8} & \text{if } x - 8 < 0 \end{cases} = \begin{cases} 1 & \text{if } x > 8 \\ -1 & \text{if } x < 8 \end{cases}$

Thus, $\lim\limits_{x \to 8^-} \dfrac{|x-8|}{x-8} = \lim\limits_{x \to 8^-} (-1) = -1$.

15. $\lim\limits_{x \to 0} \dfrac{1 - \sqrt{1 - x^2}}{x} \cdot \dfrac{1 + \sqrt{1 - x^2}}{1 + \sqrt{1 - x^2}} = \lim\limits_{x \to 0} \dfrac{1 - (1 - x^2)}{x(1 + \sqrt{1 - x^2})} = \lim\limits_{x \to 0} \dfrac{x^2}{x(1 + \sqrt{1 - x^2})} = \lim\limits_{x \to 0} \dfrac{x}{1 + \sqrt{1 - x^2}} = 0$

17. $\lim\limits_{x \to \infty} \dfrac{1 + 2x - x^2}{1 - x + 2x^2} = \lim\limits_{x \to \infty} \dfrac{(1 + 2x - x^2)/x^2}{(1 - x + 2x^2)/x^2} = \lim\limits_{x \to \infty} \dfrac{1/x^2 + 2/x - 1}{1/x^2 - 1/x + 2} = \dfrac{0 + 0 - 1}{0 - 0 + 2} = -\dfrac{1}{2}$

19. Since x is positive, $\sqrt{x^2} = |x| = x$.

$$\lim\limits_{x \to \infty} \dfrac{\sqrt{x^2 - 9}}{2x - 6} = \lim\limits_{x \to \infty} \dfrac{\sqrt{x^2 - 9}/\sqrt{x^2}}{(2x - 6)/x} = \lim\limits_{x \to \infty} \dfrac{\sqrt{1 - 9/x^2}}{2 - 6/x} = \dfrac{\sqrt{1 - 0}}{2 - 0} = \dfrac{1}{2}$$

21. $\lim\limits_{x \to \infty} e^{-3x} = 0$ since $-3x \to -\infty$ as $x \to \infty$ and $\lim\limits_{t \to -\infty} e^t = 0$.

23. From the graph of $y = (\cos^2 x)/x^2$, it appears that $y = 0$ is the horizontal

asymptote and $x = 0$ is the vertical asymptote. Now $0 \le (\cos x)^2 \le 1$

$\Rightarrow \quad \dfrac{0}{x^2} \le \dfrac{\cos^2 x}{x^2} \le \dfrac{1}{x^2} \quad \Rightarrow \quad 0 \le \dfrac{\cos^2 x}{x^2} \le \dfrac{1}{x^2}$. But $\lim\limits_{x \to \pm\infty} 0 = 0$ and

$\lim\limits_{x \to \pm\infty} \dfrac{1}{x^2} = 0$, so by the Squeeze Theorem,

$\lim\limits_{x \to \pm\infty} \dfrac{\cos^2 x}{x^2} = 0$. Thus, $y = 0$ is the horizontal asymptote. $\lim\limits_{x \to 0} \dfrac{\cos^2 x}{x^2} = \infty$ because $\cos^2 x \to 1$ and $x^2 \to 0$ as

$x \to 0$, so $x = 0$ is the vertical asymptote.

25. Since $2x - 1 \le f(x) \le x^2$ for $0 < x < 3$ and $\lim\limits_{x \to 1} (2x - 1) = 1 = \lim\limits_{x \to 1} x^2$, we have $\lim\limits_{x \to 1} f(x) = 1$ by the

Squeeze Theorem.

27. Given $\varepsilon > 0$, we need $\delta > 0$ so that if $0 < |x - 5| < \delta$, then $|(7x - 27) - 8| < \varepsilon \quad \Leftrightarrow \quad |7x - 35| < \varepsilon \quad \Leftrightarrow$

$|x - 5| < \varepsilon/7$. So take $\delta = \varepsilon/7$. Then $0 < |x - 5| < \delta \quad \Rightarrow \quad |(7x - 27) - 8| < \varepsilon$. Thus, $\lim\limits_{x \to 5} (7x - 27) = 8$

by the definition of a limit.

29. Given $\varepsilon > 0$, we need $\delta > 0$ so that if $0 < |x - 2| < \delta$, then $|x^2 - 3x - (-2)| < \varepsilon$. First, note that if $|x - 2| < 1$,

then $-1 < x - 2 < 1$, so $0 < x - 1 < 2 \quad \Rightarrow \quad |x - 1| < 2$. Now let $\delta = \min\{\varepsilon/2, 1\}$. Then $0 < |x - 2| < \delta$

$\Rightarrow \quad |x^2 - 3x - (-2)| = |(x - 2)(x - 1)| = |x - 2|\,|x - 1| < (\varepsilon/2)(2) = \varepsilon$.

Thus, $\lim\limits_{x \to 2} (x^2 - 3x) = -2$ by the definition of a limit.

31. (a) $f(x) = \sqrt{-x}$ if $x < 0$, $f(x) = 3 - x$ if $0 \le x < 3$, $f(x) = (x - 3)^2$ if $x > 3$.

    (i) $\lim\limits_{x \to 0^+} f(x) = \lim\limits_{x \to 0^+} (3 - x) = 3$ (ii) $\lim\limits_{x \to 0^-} f(x) = \lim\limits_{x \to 0^-} \sqrt{-x} = 0$

    (iii) Because of (i) and (ii), $\lim\limits_{x \to 0} f(x)$ does not exist.   (iv) $\lim\limits_{x \to 3^-} f(x) = \lim\limits_{x \to 3^-} (3 - x) = 0$

    (v) $\lim\limits_{x \to 3^+} f(x) = \lim\limits_{x \to 3^+} (x - 3)^2 = 0$ (vi) Because of (iv) and (v), $\lim\limits_{x \to 3} f(x) = 0$.

  (b) f is discontinuous at 0 since $\lim\limits_{x \to 0} f(x)$ does not exist.   (c)

      f is discontinuous at 3 since $f(3)$ does not exist.

33. $\sin x$ is continuous on $\mathbb{R}$ by Theorem 7 in Section 2.5. Since e^x is continuous on $\mathbb{R}$, $e^{\sin x}$ is continuous on $\mathbb{R}$ by

Theorem 9 in Section 2.5. Lastly, x is continuous on $\mathbb{R}$ since it's a polynomial and the product $xe^{\sin x}$ is continuous

on its domain $\mathbb{R}$ by Theorem 4 in Section 2.5.

35. $f(x) = 2x^3 + x^2 + 2$ is a polynomial, so it is continuous on $[-2, -1]$ and $f(-2) = -10 < 0 < 1 = f(-1)$. So

by the Intermediate Value Theorem there is a number c in $(-2, -1)$ such that $f(c) = 0$, that is, the equation

$2x^3 + x^2 + 2 = 0$ has a root in $(-2, -1)$.

37. (a) The slope of the tangent line at $(2, 1)$ is

$$\lim_{x \to 2} \frac{f(x) - f(2)}{x - 2} = \lim_{x \to 2} \frac{9 - 2x^2 - 1}{x - 2} = \lim_{x \to 2} \frac{8 - 2x^2}{x - 2} = \lim_{x \to 2} \frac{-2(x^2 - 4)}{x - 2} = \lim_{x \to 2} \frac{-2(x - 2)(x + 2)}{x - 2}$$

$$= \lim_{x \to 2} [-2(x + 2)] = -2 \cdot 4 = -8$$

(b) An equation of this tangent line is $y - 1 = -8(x - 2)$ or $y = -8x + 17$.

39. (a) $s = s(t) = 1 + 2t + t^2/4$. The average velocity over the time interval $[1, 1 + h]$ is

$$v_{\text{ave}} = \frac{s(1 + h) - s(1)}{(1 + h) - 1} = \frac{1 + 2(1 + h) + (1 + h)^2/4 - 13/4}{h} = \frac{10h + h^2}{4h} = \frac{10 + h}{4}.$$

So for the following intervals the average velocities are:

(i) $[1, 3]$: $h = 2$, $v_{\text{ave}} = (10 + 2)/4 = 3$ m/s

(ii) $[1, 2]$: $h = 1$, $v_{\text{ave}} = (10 + 1)/4 = 2.75$ m/s

(iii) $[1, 1.5]$: $h = 0.5$, $v_{\text{ave}} = (10 + 0.5)/4 = 2.625$ m/s

(iv) $[1, 1.1]$: $h = 0.1$, $v_{\text{ave}} = (10 + 0.1)/4 = 2.525$ m/s

(b) When $t = 1$, the instantaneous velocity is $\lim_{h \to 0} \frac{s(1 + h) - s(1)}{h} = \lim_{h \to 0} \frac{10 + h}{4} = \frac{10}{4} = 2.5$ m/s.

41. (a) $f'(2) = \lim_{x \to 2} \frac{f(x) - f(2)}{x - 2} = \lim_{x \to 2} \frac{x^3 - 2x - 4}{x - 2}$ (c)

$$= \lim_{x \to 2} \frac{(x - 2)(x^2 + 2x + 2)}{x - 2} = \lim_{x \to 2} (x^2 + 2x + 2) = 10$$

(b) $y - 4 = 10(x - 2)$ or $y = 10x - 16$

43. (a) $f'(r)$ is the rate at which the total cost changes with respect to the interest rate. Its units are

dollars/(percent per year).

(b) The total cost of paying off the loan is increasing by $1200/(percent per year) as the interest rate reaches 10%.

So if the interest rate goes up from 10% to 11%, the cost goes up approximately $1200.

(c) As r increases, C increases. So $f'(r)$ will always be positive.

For Exercises 44–46, see the hints before Exercise 5 in Section 2.9.

45.

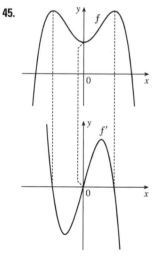

47. (a) $f'(x) = \lim\limits_{h \to 0} \dfrac{f(x+h) - f(x)}{h} = \lim\limits_{h \to 0} \dfrac{\sqrt{3 - 5(x+h)} - \sqrt{3 - 5x}}{h} \cdot \dfrac{\sqrt{3 - 5(x+h)} + \sqrt{3 - 5x}}{\sqrt{3 - 5(x+h)} + \sqrt{3 - 5x}}$

$\qquad = \lim\limits_{h \to 0} \dfrac{[3 - 5(x+h)] - (3 - 5x)}{h \left(\sqrt{3 - 5(x+h)} + \sqrt{3 - 5x} \right)} = \lim\limits_{h \to 0} \dfrac{-5}{\sqrt{3 - 5(x+h)} + \sqrt{3 - 5x}} = \dfrac{-5}{2\sqrt{3 - 5x}}$

(b) Domain of f: (the radicand must be nonnegative) $3 - 5x \ge 0 \;\Rightarrow\; 5x \le 3 \;\Rightarrow\; x \in \left(-\infty, \frac{3}{5}\right]$

Domain of f': exclude $\frac{3}{5}$ because it makes the denominator zero; $x \in \left(-\infty, \frac{3}{5}\right)$

(c) Our answer to part (a) is reasonable because $f'(x)$ is always

negative and f is always decreasing.

49. f is not differentiable: at $x = -4$ because f is not continuous, at $x = -1$ because f has a corner, at $x = 2$ because f is not continuous, and at $x = 5$ because f has a vertical tangent.

51. $B'(1990)$ is the rate at which the total value of U.S. banknotes in circulation is changing in billions of dollars per

year. To estimate the value of $B'(1990)$, we will average the difference quotients obtained using the times $t = 1985$

and $t = 1995$. Let $A = \dfrac{B(1985) - B(1990)}{1985 - 1990} = \dfrac{182.0 - 268.2}{-5} = 17.24$ and

$C = \dfrac{B(1995) - B(1990)}{1995 - 1990} = \dfrac{401.5 - 268.2}{5} = 26.66.$ Then

$B'(1990) = \lim\limits_{t \to 1990} \dfrac{B(t) - B(1990)}{t - 1990} \approx \dfrac{A + C}{2} = \dfrac{17.24 + 26.66}{2} = 21.95$ billions of dollars/year.

53. $|f(x)| \le g(x) \;\Leftrightarrow\; -g(x) \le f(x) \le g(x)$ and $\lim\limits_{x \to a} g(x) = 0 = \lim\limits_{x \to a} -g(x)$. Thus, by the Squeeze Theorem,

$\lim\limits_{x \to a} f(x) = 0.$

□ PROBLEMS PLUS

1. Let $t = \sqrt[6]{x}$, so $x = t^6$. Then $t \to 1$ as $x \to 1$, so

$$\lim_{x \to 1} \frac{\sqrt[3]{x} - 1}{\sqrt{x} - 1} = \lim_{t \to 1} \frac{t^2 - 1}{t^3 - 1} = \lim_{t \to 1} \frac{(t-1)(t+1)}{(t-1)(t^2+t+1)} = \lim_{t \to 1} \frac{t+1}{t^2+t+1} = \frac{1+1}{1^2+1+1} = \frac{2}{3}.$$

Another method: Multiply both the numerator and the denominator by $(\sqrt{x}+1)\left(\sqrt[3]{x^2} + \sqrt[3]{x} + 1\right)$.

3. For $-\frac{1}{2} < x < \frac{1}{2}$, we have $2x - 1 < 0$ and $2x + 1 > 0$, so $|2x - 1| = -(2x-1)$ and $|2x+1| = 2x+1$.

Therefore, $\displaystyle \lim_{x \to 0} \frac{|2x-1| - |2x+1|}{x} = \lim_{x \to 0} \frac{-(2x-1) - (2x+1)}{x} = \lim_{x \to 0} \frac{-4x}{x} = \lim_{x \to 0} (-4) = -4.$

5. Since $[\![x]\!] \le x < [\![x]\!] + 1$, we have $\dfrac{[\![x]\!]}{[\![x]\!]} \le \dfrac{x}{[\![x]\!]} < \dfrac{[\![x]\!]+1}{[\![x]\!]} \quad \Rightarrow \quad 1 \le \dfrac{x}{[\![x]\!]} < 1 + \dfrac{1}{[\![x]\!]}$ for $x \ge 1$. As $x \to \infty$,

$[\![x]\!] \to \infty$, so $\dfrac{1}{[\![x]\!]} \to 0$ and $1 + \dfrac{1}{[\![x]\!]} \to 1$. Thus, $\displaystyle \lim_{x \to \infty} \frac{x}{[\![x]\!]} = 1$ by the Squeeze Theorem.

7. f is continuous on $(-\infty, a)$ and (a, ∞). To make f continuous on $\mathbb{R}$, we must have continuity at a. Thus,

$$\lim_{x \to a^+} f(x) = \lim_{x \to a^-} f(x) \quad \Rightarrow \quad \lim_{x \to a^+} x^2 = \lim_{x \to a^-} (x+1) \quad \Rightarrow \quad a^2 = a + 1 \quad \Rightarrow \quad a^2 - a - 1 = 0 \quad \Rightarrow$$

[by the quadratic formula] $a = \left(1 \pm \sqrt{5}\right)/2 \approx 1.618$ or -0.618.

9. $\displaystyle \lim_{x \to a} f(x) = \lim_{x \to a} \left(\frac{1}{2}[f(x) + g(x)] + \frac{1}{2}[f(x) - g(x)]\right)$

$\qquad = \frac{1}{2} \lim_{x \to a} [f(x) + g(x)] + \frac{1}{2} \lim_{x \to a} [f(x) - g(x)]$

$\qquad = \frac{1}{2} \cdot 2 + \frac{1}{2} \cdot 1 = \frac{3}{2}$, and

$\displaystyle \lim_{x \to a} g(x) = \lim_{x \to a} \Big([f(x) + g(x)] - f(x) \Big) = \lim_{x \to a} [f(x) + g(x)] - \lim_{x \to a} f(x) = 2 - \frac{3}{2} = \frac{1}{2}.$

So $\displaystyle \lim_{x \to a} [f(x)g(x)] = \left[\lim_{x \to a} f(x)\right]\left[\lim_{x \to a} g(x)\right] = \frac{3}{2} \cdot \frac{1}{2} = \frac{3}{4}.$

Another solution: Since $\displaystyle \lim_{x \to a} [f(x) + g(x)]$ and $\displaystyle \lim_{x \to a} [f(x) - g(x)]$ exist, we must have

$$\lim_{x \to a} [f(x) + g(x)]^2 = \left(\lim_{x \to a} [f(x) + g(x)]\right)^2 \text{ and } \lim_{x \to a} [f(x) - g(x)]^2 = \left(\lim_{x \to a} [f(x) - g(x)]\right)^2, \text{ so}$$

$$\lim_{x \to a} [f(x) \, g(x)] = \lim_{x \to a} \frac{1}{4}\left([f(x) + g(x)]^2 - [f(x) - g(x)]^2\right) \quad \text{[because all of the } f^2 \text{ and } g^2 \text{ cancel]}$$

$$= \frac{1}{4}\left(\lim_{x \to a} [f(x) + g(x)]^2 - \lim_{x \to a} [f(x) - g(x)]^2\right) = \frac{1}{4}(2^2 - 1^2) = \frac{3}{4}.$$

11. (a) Consider $G(x) = T(x + 180°) - T(x)$. Fix any number a. If $G(a) = 0$, we are done:

Temperature at a = Temperature at $a + 180°$. If $G(a) > 0$, then $G(a + 180°) = T(a + 360°) - T(a + 180°)$

$= T(a) - T(a + 180°) = -G(a) < 0$. Also, G is continuous since temperature varies continuously. So, by the

Intermediate Value Theorem, G has a zero on the interval $[a, a + 180°]$. If $G(a) < 0$, then a similar argument

applies.

(b) Yes. The same argument applies.

(c) The same argument applies for quantities that vary continuously, such as barometric pressure. But one could

argue that altitude above sea level is sometimes discontinuous, so the result might not always hold for that

quantity.

13. (a) Put $x = 0$ and $y = 0$ in the equation: $f(0 + 0) = f(0) + f(0) + 0^2 \cdot 0 + 0 \cdot 0^2$ $\Rightarrow$ $f(0) = 2f(0)$.

Subtracting $f(0)$ from each side of this equation gives $f(0) = 0$.

(b) $f'(0) = \lim_{h \to 0} \dfrac{f(0 + h) - f(0)}{h} = \lim_{h \to 0} \dfrac{\left[f(0) + f(h) + 0^2 h + 0h^2\right] - f(0)}{h}$

$= \lim_{h \to 0} \dfrac{f(h)}{h} = \lim_{x \to 0} \dfrac{f(x)}{x} = 1$

(c) $f'(x) = \lim_{h \to 0} \dfrac{f(x + h) - f(x)}{h} = \lim_{h \to 0} \dfrac{\left[f(x) + f(h) + x^2 h + xh^2\right] - f(x)}{h}$

$= \lim_{h \to 0} \dfrac{f(h) + x^2 h + xh^2}{h} = \lim_{h \to 0} \left[\dfrac{f(h)}{h} + x^2 + xh\right] = 1 + x^2$

3 ☐ DIFFERENTIATION RULES

3.1 Derivatives of Polynomials and Exponential Functions

1. (a) e is the number such that $\lim\limits_{h \to 0} \dfrac{e^h - 1}{h} = 1$.

(b)

x	$(2.7^x - 1)/x$
-0.001	0.9928
-0.0001	0.9932
0.001	0.9937
0.0001	0.9933

x	$(2.8^x - 1)/x$
-0.001	1.0291
-0.0001	1.0296
0.001	1.0301
0.0001	1.0297

From the tables (to two decimal places), $\lim\limits_{h \to 0} \dfrac{2.7^h - 1}{h} = 0.99$ and $\lim\limits_{h \to 0} \dfrac{2.8^h - 1}{h} = 1.03$. Since $0.99 < 1 < 1.03$, $2.7 < e < 2.8$.

3. $f(x) = 186.5$ is a constant function, so its derivative is 0, that is, $f'(x) = 0$.

5. $f(x) = 5x - 1 \quad \Rightarrow \quad f'(x) = 5 - 0 = 5$

7. $f(x) = x^2 + 3x - 4 \quad \Rightarrow \quad f'(x) = 2x^{2-1} + 3 - 0 = 2x + 3$

9. $f(t) = \frac{1}{4}(t^4 + 8) \quad \Rightarrow \quad f'(t) = \frac{1}{4}(t^4 + 8)' = \frac{1}{4}(4t^{4-1} + 0) = t^3$

11. $y = x^{-2/5} \quad \Rightarrow \quad y' = -\frac{2}{5}x^{(-2/5)-1} = -\frac{2}{5}x^{-7/5} = -\dfrac{2}{5x^{7/5}}$

13. $V(r) = \frac{4}{3}\pi r^3 \quad \Rightarrow \quad V'(r) = \frac{4}{3}\pi(3r^2) = 4\pi r^2$

15. $Y(t) = 6t^{-9} \quad \Rightarrow \quad Y'(t) = 6(-9)t^{-10} = -54t^{-10}$

17. $G(x) = \sqrt{x} - 2e^x = x^{1/2} - 2e^x \quad \Rightarrow \quad G'(x) = \frac{1}{2}x^{-1/2} - 2e^x = \dfrac{1}{2\sqrt{x}} - 2e^x$

19. $F(x) = (\frac{1}{2}x)^5 = (\frac{1}{2})^5 x^5 = \frac{1}{32}x^5 \quad \Rightarrow \quad F'(x) = \frac{1}{32}(5x^4) = \frac{5}{32}x^4$

21. $g(x) = x^2 + \dfrac{1}{x^2} = x^2 + x^{-2} \quad \Rightarrow \quad g'(x) = 2x + (-2)x^{-3} = 2x - \dfrac{2}{x^3}$

23. $y = \dfrac{x^2 + 4x + 3}{\sqrt{x}} = x^{3/2} + 4x^{1/2} + 3x^{-1/2} \quad \Rightarrow$

$y' = \frac{3}{2}x^{1/2} + 4(\frac{1}{2})x^{-1/2} + 3(-\frac{1}{2})x^{-3/2} = \frac{3}{2}\sqrt{x} + \dfrac{2}{\sqrt{x}} - \dfrac{3}{2x\sqrt{x}}$

$\left[\text{note that } x^{3/2} = x^{2/2} \cdot x^{1/2} = x\sqrt{x} \right]$

25. $y = 4\pi^2 \quad \Rightarrow \quad y' = 0$ since $4\pi^2$ is a constant.

27. $y = ax^2 + bx + c \quad \Rightarrow \quad y' = 2ax + b$

29. $v = t^2 - \dfrac{1}{\sqrt[4]{t^3}} = t^2 - t^{-3/4} \quad \Rightarrow \quad v' = 2t - (-\frac{3}{4})t^{-7/4} = 2t + \dfrac{3}{4t^{7/4}} = 2t + \dfrac{3}{4t\sqrt[4]{t^3}}$

31. $z = \dfrac{A}{y^{10}} + Be^y = Ay^{-10} + Be^y \quad \Rightarrow \quad z' = -10Ay^{-11} + Be^y = -\dfrac{10A}{y^{11}} + Be^y$

33. $f(x) = e^x - 5x \Rightarrow f'(x) = e^x - 5$.

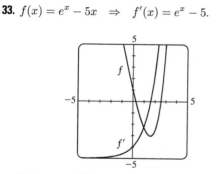

Notice that $f'(x) = 0$ when f has a horizontal tangent, f' is positive when f is increasing, and f' is negative when f is decreasing.

35. $f(x) = 3x^{15} - 5x^3 + 3 \Rightarrow$
$$f'(x) = 45x^{14} - 15x^2.$$

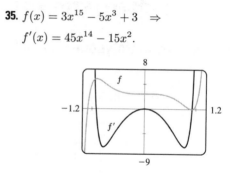

Notice that $f'(x) = 0$ when f has a horizontal tangent, f' is positive when f is increasing, and f' is negative when f is decreasing.

37. To graphically estimate the value of $f'(1)$ for $f(x) = 3x^2 - x^3$, we'll graph f in the viewing rectangle

$[1 - 0.1, 1 + 0.1]$ by $[f(0.9), f(1.1)]$, as shown in the figure. [When assigning values to the window variables, it is

convenient to use $Y_1(0.9)$ for $Y_{\min}$ and $Y_1(1.1)$ for $Y_{\max}$.] If we have sufficiently zoomed in on the graph of f, we

should obtain a graph that looks like a diagonal line; if not, graph again with $1 - 0.01$ and $1 + 0.01$, etc.

Estimated value:

$$f'(1) \approx \frac{2.299 - 1.701}{1.1 - 0.9} = \frac{0.589}{0.2} = 2.99.$$

Exact value: $f(x) = 3x^2 - x^3 \Rightarrow f'(x) = 6x - 3x^2$,

so $f'(1) = 6 - 3 = 3$.

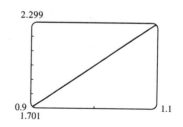

39. $y = x^4 + 2e^x \Rightarrow y' = 4x^3 + 2e^x$. At $(0, 2)$, $y' = 2$ and an equation of the tangent line is $y - 2 = 2(x - 0)$

or $y = 2x + 2$.

41. $y = 3x^2 - x^3 \Rightarrow y' = 6x - 3x^2$. At $(1, 2)$, $y' = 6 - 3 = 3$,

so an equation of the tangent line is $y - 2 = 3(x - 1)$,

or $y = 3x - 1$.

43. (a)

(b)

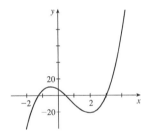

From the graph in part (a), it appears that f' is zero at $x_1 \approx -1.25$, $x_2 \approx 0.5$, and $x_3 \approx 3$. The slopes are negative (so f' is negative) on $(-\infty, x_1)$ and (x_2, x_3). The slopes are positive (so f' is positive) on (x_1, x_2) and (x_3, ∞).

(c) $f(x) = x^4 - 3x^3 - 6x^2 + 7x + 30 \quad \Rightarrow$
$f'(x) = 4x^3 - 9x^2 - 12x + 7$

45. The curve $y = 2x^3 + 3x^2 - 12x + 1$ has a horizontal tangent when $y' = 6x^2 + 6x - 12 = 0 \quad \Leftrightarrow$

$6(x^2 + x - 2) = 0 \quad \Leftrightarrow \quad 6(x + 2)(x - 1) = 0 \quad \Leftrightarrow \quad x = -2$ or $x = 1$. The points on the curve are $(-2, 21)$

and $(1, -6)$.

47. $y = 6x^3 + 5x - 3 \quad \Rightarrow \quad m = y' = 18x^2 + 5$, but $x^2 \geq 0$ for all x, so $m \geq 5$ for all x.

49.

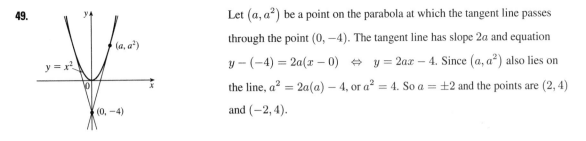

Let (a, a^2) be a point on the parabola at which the tangent line passes through the point $(0, -4)$. The tangent line has slope $2a$ and equation

$y - (-4) = 2a(x - 0) \quad \Leftrightarrow \quad y = 2ax - 4$. Since (a, a^2) also lies on the line, $a^2 = 2a(a) - 4$, or $a^2 = 4$. So $a = \pm 2$ and the points are $(2, 4)$ and $(-2, 4)$.

51. $y = f(x) = 1 - x^2 \quad \Rightarrow \quad f'(x) = -2x$, so the tangent line at $(2, -3)$ has

slope $f'(2) = -4$. The normal line has slope $-\frac{1}{-4} = \frac{1}{4}$ and equation

$y + 3 = \frac{1}{4}(x - 2)$ or $y = \frac{1}{4}x - \frac{7}{2}$.

53. $f'(x) = \lim\limits_{h \to 0} \dfrac{f(x+h) - f(x)}{h} = \lim\limits_{h \to 0} \dfrac{\dfrac{1}{x+h} - \dfrac{1}{x}}{h} = \lim\limits_{h \to 0} \dfrac{x - (x+h)}{hx(x+h)}$

$= \lim\limits_{h \to 0} \dfrac{-h}{hx(x+h)} = \lim\limits_{h \to 0} \dfrac{-1}{x(x+h)} = -\dfrac{1}{x^2}$

55. $f(x) = 2 - x$ if $x \le 1$ and $f(x) = x^2 - 2x + 2$ if $x > 1$. Now we compute the right- and left-hand derivatives

defined in Exercise 2.9.46:

$f'_-(1) = \lim\limits_{h \to 0^-} \dfrac{f(1+h) - f(1)}{h} = \lim\limits_{h \to 0^-} \dfrac{2 - (1+h) - 1}{h} = \lim\limits_{h \to 0^-} \dfrac{-h}{h} = \lim\limits_{h \to 0^-} -1 = -1$ and

$f'_+(1) = \lim\limits_{h \to 0^+} \dfrac{f(1+h) - f(1)}{h} = \lim\limits_{h \to 0^+} \dfrac{(1+h)^2 - 2(1+h) + 2 - 1}{h} = \lim\limits_{h \to 0^+} \dfrac{h^2}{h} = \lim\limits_{h \to 0^+} h = 0.$

Thus, $f'(1)$ does not exist since $f'_-(1) \ne f'_+(1)$,

so f is not differentiable at 1. But $f'(x) = -1$

for $x < 1$ and $f'(x) = 2x - 2$ if $x > 1$.

57. (a) Note that $x^2 - 9 < 0$ for $x^2 < 9 \iff |x| < 3 \iff -3 < x < 3$. So

$$f(x) = \begin{cases} x^2 - 9 & \text{if } x \le -3 \\ -x^2 + 9 & \text{if } -3 < x < 3 \\ x^2 - 9 & \text{if } x \ge 3 \end{cases} \Rightarrow$$

$$f'(x) = \begin{cases} 2x & \text{if } x < -3 \\ -2x & \text{if } -3 < x < 3 \\ 2x & \text{if } x > 3 \end{cases} = \begin{cases} 2x & \text{if } |x| > 3 \\ -2x & \text{if } |x| < 3 \end{cases}$$

To show that $f'(3)$ does not exist we investigate $\lim\limits_{h \to 0} \dfrac{f(3+h) - f(3)}{h}$ by computing the left- and right-hand

derivatives defined in Exercise 2.9.46.

$f'_-(3) = \lim\limits_{h \to 0^-} \dfrac{f(3+h) - f(3)}{h} = \lim\limits_{h \to 0^-} \dfrac{[-(3+h)^2 + 9] - 0}{h} = \lim\limits_{h \to 0^-} (-6 - h) = -6$ and

$f'_+(3) = \lim\limits_{h \to 0^+} \dfrac{f(3+h) - f(3)}{h} = \lim\limits_{h \to 0^+} \dfrac{[(3+h)^2 - 9] - 0}{h} = \lim\limits_{h \to 0^+} \dfrac{6h + h^2}{h} = \lim\limits_{h \to 0^+} (6 + h) = 6.$

Since the left and right limits are different,

$\lim\limits_{h \to 0} \dfrac{f(3+h) - f(3)}{h}$ does not exist, that is,

$f'(3)$ does not exist. Similarly, $f'(-3)$ does not

exist. Therefore, f is not differentiable at 3 or

at -3.

(b)

59. $y = f(x) = ax^2$ $\Rightarrow$ $f'(x) = 2ax$. So the slope of the tangent to the parabola at $x = 2$ is $m = 2a(2) = 4a$. The slope of the given line, $2x + y = b$ $\Leftrightarrow$ $y = -2x + b$, is seen to be -2, so we must have $4a = -2$ $\Leftrightarrow$ $a = -\frac{1}{2}$. So when $x = 2$, the point in question has y-coordinate $-\frac{1}{2} \cdot 2^2 = -2$. Now we simply require that the given line, whose equation is $2x + y = b$, pass through the point $(2, -2)$: $2(2) + (-2) = b$ $\Leftrightarrow$ $b = 2$. So we must have $a = -\frac{1}{2}$ and $b = 2$.

61. $y = f(x) = ax^3 + bx^2 + cx + d$ $\Rightarrow$ $f'(x) = 3ax^2 + 2bx + c$. The point $(-2, 6)$ is on f, so $f(-2) = 6$ $\Rightarrow$ $-8a + 4b - 2c + d = 6$ **(1)**. The point $(2, 0)$ is on f, so $f(2) = 0$ $\Rightarrow$ $8a + 4b + 2c + d = 0$ **(2)**. Since there are horizontal tangents at $(-2, 6)$ and $(2, 0)$, $f'(\pm 2) = 0$. $f'(-2) = 0$ $\Rightarrow$ $12a - 4b + c = 0$ **(3)** and $f'(2) = 0$ $\Rightarrow$ $12a + 4b + c = 0$ **(4)**. Subtracting equation **(3)** from **(4)** gives $8b = 0$ $\Rightarrow$ $b = 0$. Adding **(1)** and **(2)** gives $8b + 2d = 6$, so $d = 3$ since $b = 0$. From **(3)** we have $c = -12a$, so **(2)** becomes $8a + 4(0) + 2(-12a) + 3 = 0$ $\Rightarrow$ $3 = 16a$ $\Rightarrow$ $a = \frac{3}{16}$. Now $c = -12a = -12\left(\frac{3}{16}\right) = -\frac{9}{4}$ and the desired cubic function is $y = \frac{3}{16}x^3 - \frac{9}{4}x + 3$.

63. *Solution 1:*   Let $f(x) = x^{1000}$. Then, by the definition of a derivative,

$$f'(1) = \lim_{x \to 1} \frac{f(x) - f(1)}{x - 1} = \lim_{x \to 1} \frac{x^{1000} - 1}{x - 1}.$$ But this is just the limit we want to find, and we know (from the

Power Rule) that $f'(x) = 1000x^{999}$, so $f'(1) = 1000(1)^{999} = 1000$. So $\lim_{x \to 1} \dfrac{x^{1000} - 1}{x - 1} = 1000$.

Solution 2:   Note that $\left(x^{1000} - 1\right) = (x - 1)\left(x^{999} + x^{998} + x^{997} + \cdots + x^2 + x + 1\right)$. So

$$\lim_{x \to 1} \frac{x^{1000} - 1}{x - 1} = \lim_{x \to 1} \frac{(x - 1)\left(x^{999} + x^{998} + x^{997} + \cdots + x^2 + x + 1\right)}{x - 1}$$

$$= \lim_{x \to 1} \left(x^{999} + x^{998} + x^{997} + \cdots + x^2 + x + 1\right) = \underbrace{1 + 1 + 1 + \cdots + 1 + 1 + 1}_{1000 \text{ ones}}$$

$$= 1000, \text{ as above.}$$

3.2 The Product and Quotient Rules

1. Product Rule: $y = \left(x^2 + 1\right)\left(x^3 + 1\right)$ $\Rightarrow$

$\qquad y' = \left(x^2 + 1\right)\left(3x^2\right) + \left(x^3 + 1\right)(2x) = 3x^4 + 3x^2 + 2x^4 + 2x = 5x^4 + 3x^2 + 2x.$

Multiplying first: $y = \left(x^2 + 1\right)\left(x^3 + 1\right) = x^5 + x^3 + x^2 + 1$ $\Rightarrow$ $y' = 5x^4 + 3x^2 + 2x$ (equivalent).

3. By the Product Rule, $f(x) = x^2 e^x$ $\Rightarrow$ $f'(x) = x^2 \dfrac{d}{dx}\left(e^x\right) + e^x \dfrac{d}{dx}\left(x^2\right) = x^2 e^x + e^x(2x) = xe^x(x + 2)$.

5. By the Quotient Rule, $y = \dfrac{e^x}{x^2}$ $\Rightarrow$

$$y' = \frac{x^2 \dfrac{d}{dx}\left(e^x\right) - e^x \dfrac{d}{dx}\left(x^2\right)}{\left(x^2\right)^2} = \frac{x^2\left(e^x\right) - e^x(2x)}{x^4} = \frac{xe^x(x - 2)}{x^4} = \frac{e^x(x - 2)}{x^3}$$

The notations $\overset{PR}{\Rightarrow}$ and $\overset{QR}{\Rightarrow}$ indicate the use of the Product and Quotient Rules, respectively.

7. $g(x) = \dfrac{3x - 1}{2x + 1}$ $\overset{QR}{\Rightarrow}$ $g'(x) = \dfrac{(2x + 1)(3) - (3x - 1)(2)}{(2x + 1)^2} = \dfrac{6x + 3 - 6x + 2}{(2x + 1)^2} = \dfrac{5}{(2x + 1)^2}$

9. $V(x) = (2x^3 + 3)(x^4 - 2x)$ $\overset{PR}{\Rightarrow}$
$V'(x) = (2x^3 + 3)(4x^3 - 2) + (x^4 - 2x)(6x^2) = (8x^6 + 8x^3 - 6) + (6x^6 - 12x^3) = 14x^6 - 4x^3 - 6$

11. $F(y) = \left(\dfrac{1}{y^2} - \dfrac{3}{y^4}\right)(y + 5y^3) = (y^{-2} - 3y^{-4})(y + 5y^3)$ $\overset{PR}{\Rightarrow}$
$F'(y) = (y^{-2} - 3y^{-4})(1 + 15y^2) + (y + 5y^3)(-2y^{-3} + 12y^{-5})$
$\quad = (y^{-2} + 15 - 3y^{-4} - 45y^{-2}) + (-2y^{-2} + 12y^{-4} - 10 + 60y^{-2})$
$\quad = 5 + 14y^{-2} + 9y^{-4}$ or $5 + 14/y^2 + 9/y^4$

13. $y = \dfrac{t^2}{3t^2 - 2t + 1}$ $\overset{QR}{\Rightarrow}$
$y' = \dfrac{(3t^2 - 2t + 1)(2t) - t^2(6t - 2)}{(3t^2 - 2t + 1)^2} = \dfrac{2t[3t^2 - 2t + 1 - t(3t - 1)]}{(3t^2 - 2t + 1)^2}$
$\quad = \dfrac{2t(3t^2 - 2t + 1 - 3t^2 + t)}{(3t^2 - 2t + 1)^2} = \dfrac{2t(1 - t)}{(3t^2 - 2t + 1)^2}$

15. $y = (r^2 - 2r)e^r$ $\overset{PR}{\Rightarrow}$ $y' = (r^2 - 2r)(e^r) + e^r(2r - 2) = e^r(r^2 - 2r + 2r - 2) = e^r(r^2 - 2)$

17. $y = \dfrac{v^3 - 2v\sqrt{v}}{v} = v^2 - 2\sqrt{v} = v^2 - 2v^{1/2}$ $\Rightarrow$ $y' = 2v - 2\left(\tfrac{1}{2}\right)v^{-1/2} = 2v - v^{-1/2}$.

We can change the form of the answer as follows: $2v - v^{-1/2} = 2v - \dfrac{1}{\sqrt{v}} = \dfrac{2v\sqrt{v} - 1}{\sqrt{v}} = \dfrac{2v^{3/2} - 1}{\sqrt{v}}$

19. $y = \dfrac{1}{x^4 + x^2 + 1}$ $\Rightarrow$ $y' = \dfrac{(x^4 + x^2 + 1)(0) - 1(4x^3 + 2x)}{(x^4 + x^2 + 1)^2} = -\dfrac{2x(2x^2 + 1)}{(x^4 + x^2 + 1)^2}$

21. $f(x) = \dfrac{x}{x + c/x}$ $\Rightarrow$
$f'(x) = \dfrac{(x + c/x)(1) - x(1 - c/x^2)}{\left(x + \dfrac{c}{x}\right)^2} = \dfrac{x + c/x - x + c/x}{\left(\dfrac{x^2 + c}{x}\right)^2} = \dfrac{2c/x}{\dfrac{(x^2 + c)^2}{x^2}} \cdot \dfrac{x^2}{x^2} = \dfrac{2cx}{(x^2 + c)^2}$

23. $y = \dfrac{2x}{x + 1}$ $\Rightarrow$ $y' = \dfrac{(x + 1)(2) - (2x)(1)}{(x + 1)^2} = \dfrac{2}{(x + 1)^2}$. At $(1, 1)$, $y' = \tfrac{1}{2}$, and an equation of the tangent line
is $y - 1 = \tfrac{1}{2}(x - 1)$, or $y = \tfrac{1}{2}x + \tfrac{1}{2}$.

25. $y = 2xe^x$ $\Rightarrow$ $y' = 2(x \cdot e^x + e^x \cdot 1) = 2e^x(x + 1)$. At $(0, 0)$, $y' = 2e^0(0 + 1) = 2 \cdot 1 \cdot 1 = 2$, and an
equation of the tangent line is $y - 0 = 2(x - 0)$, or $y = 2x$.

27. (a) $y = f(x) = \dfrac{1}{1 + x^2}$ $\Rightarrow$

$f'(x) = \dfrac{(1 + x^2)(0) - 1(2x)}{(1 + x^2)^2} = \dfrac{-2x}{(1 + x^2)^2}$. So the slope of the

tangent line at the point $\left(-1, \tfrac{1}{2}\right)$ is $f'(-1) = \dfrac{2}{2^2} = \tfrac{1}{2}$ and its

equation is $y - \tfrac{1}{2} = \tfrac{1}{2}(x + 1)$ or $y = \tfrac{1}{2}x + 1$.

(b)

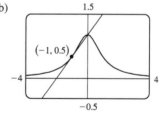

29. (a) $f(x) = \dfrac{e^x}{x^3}$ $\Rightarrow$ $f'(x) = \dfrac{x^3(e^x) - e^x(3x^2)}{(x^3)^2} = \dfrac{x^2 e^x(x-3)}{x^6} = \dfrac{e^x(x-3)}{x^4}$

(b)

$f' = 0$ when f has a horizontal tangent line, f' is negative when f is decreasing, and f' is positive when f is increasing.

31. We are given that $f(5) = 1$, $f'(5) = 6$, $g(5) = -3$, and $g'(5) = 2$.

(a) $(fg)'(5) = f(5)g'(5) + g(5)f'(5) = (1)(2) + (-3)(6) = 2 - 18 = -16$

(b) $\left(\dfrac{f}{g}\right)'(5) = \dfrac{g(5)f'(5) - f(5)g'(5)}{[g(5)]^2} = \dfrac{(-3)(6) - (1)(2)}{(-3)^2} = -\dfrac{20}{9}$

(c) $\left(\dfrac{g}{f}\right)'(5) = \dfrac{f(5)g'(5) - g(5)f'(5)}{[f(5)]^2} = \dfrac{(1)(2) - (-3)(6)}{(1)^2} = 20$

33. $f(x) = e^x g(x)$ $\Rightarrow$ $f'(x) = e^x g'(x) + g(x)e^x = e^x[g'(x) + g(x)]$.
$f'(0) = e^0[g'(0) + g(0)] = 1(5 + 2) = 7$

35. (a) From the graphs of f and g, we obtain the following values: $f(1) = 2$ since the point $(1, 2)$ is on the graph of f; $g(1) = 1$ since the point $(1, 1)$ is on the graph of g; $f'(1) = 2$ since the slope of the line segment between $(0, 0)$ and $(2, 4)$ is $\dfrac{4 - 0}{2 - 0} = 2$; $g'(1) = -1$ since the slope of the line segment between $(-2, 4)$ and $(2, 0)$ is $\dfrac{0 - 4}{2 - (-2)} = -1$. Now $u(x) = f(x)g(x)$, so $u'(1) = f(1)g'(1) + g(1)f'(1) = 2 \cdot (-1) + 1 \cdot 2 = 0$.

(b) $v(x) = f(x)/g(x)$, so $v'(5) = \dfrac{g(5)f'(5) - f(5)g'(5)}{[g(5)]^2} = \dfrac{2\left(-\frac{1}{3}\right) - 3 \cdot \frac{2}{3}}{2^2} = \dfrac{-\frac{8}{3}}{4} = -\dfrac{2}{3}$

37. (a) $y = xg(x)$ $\Rightarrow$ $y' = xg'(x) + g(x) \cdot 1 = xg'(x) + g(x)$

(b) $y = \dfrac{x}{g(x)}$ $\Rightarrow$ $y' = \dfrac{g(x) \cdot 1 - xg'(x)}{[g(x)]^2} = \dfrac{g(x) - xg'(x)}{[g(x)]^2}$

(c) $y = \dfrac{g(x)}{x}$ $\Rightarrow$ $y' = \dfrac{xg'(x) - g(x) \cdot 1}{(x)^2} = \dfrac{xg'(x) - g(x)}{x^2}$

39. If $P(t)$ denotes the population at time t and $A(t)$ the average annual income, then $T(t) = P(t)A(t)$ is the total personal income. The rate at which $T(t)$ is rising is given by $T'(t) = P(t)A'(t) + A(t)P'(t)$ $\Rightarrow$
$T'(1999) = P(1999)A'(1999) + A(1999)P'(1999) = (961,400)(\$1400/\text{yr}) + (\$30,593)(9200/\text{yr})$
$= \$1,345,960,000/\text{yr} + \$281,455,600/\text{yr} = \$1,627,415,600/\text{yr}$

So the total personal income was rising by about $1.627 billion per year in 1999.

The term $P(t)A'(t) \approx \$1.346$ billion represents the portion of the rate of change of total income due to the existing population's increasing income. The term $A(t)P'(t) \approx \$281$ million represents the portion of the rate of change of total income due to increasing population.

41. If $y = f(x) = \dfrac{x}{x+1}$, then $f'(x) = \dfrac{(x+1)(1) - x(1)}{(x+1)^2} = \dfrac{1}{(x+1)^2}$. When $x = a$, the equation of the tangent

line is $y - \dfrac{a}{a+1} = \dfrac{1}{(a+1)^2}(x-a)$. This line passes through $(1,2)$ when $2 - \dfrac{a}{a+1} = \dfrac{1}{(a+1)^2}(1-a)$ $\Leftrightarrow$

$2(a+1)^2 - a(a+1) = 1 - a$ $\Leftrightarrow$ $2a^2 + 4a + 2 - a^2 - a - 1 + a = 0$ $\Leftrightarrow$ $a^2 + 4a + 1 = 0$.

The quadratic formula gives the roots of this equation as $a = \dfrac{-4 \pm \sqrt{4^2 - 4(1)(1)}}{2(1)} = \dfrac{-4 \pm \sqrt{12}}{2} = -2 \pm \sqrt{3}$,

so there are two such tangent lines. Since

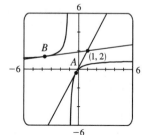

$$f\left(-2 \pm \sqrt{3}\right) = \frac{-2 \pm \sqrt{3}}{-2 \pm \sqrt{3} + 1} = \frac{-2 \pm \sqrt{3}}{-1 \pm \sqrt{3}} \cdot \frac{-1 \mp \sqrt{3}}{-1 \mp \sqrt{3}}$$

$$= \frac{2 \pm 2\sqrt{3} \mp \sqrt{3} - 3}{1 - 3} = \frac{-1 \pm \sqrt{3}}{-2} = \frac{1 \mp \sqrt{3}}{2},$$

the lines touch the curve at $A\left(-2 + \sqrt{3}, \frac{1-\sqrt{3}}{2}\right) \approx (-0.27, -0.37)$ and

$B\left(-2 - \sqrt{3}, \frac{1+\sqrt{3}}{2}\right) \approx (-3.73, 1.37)$.

We will sometimes use the form $f'g + fg'$ rather than the form $fg' + gf'$ for the Product Rule.

43. (a) $(fgh)' = [(fg)h]' = (fg)'h + (fg)h' = (f'g + fg')h + (fg)h' = f'gh + fg'h + fgh'$

(b) Putting $f = g = h$ in part (a), we have

$$\frac{d}{dx}[f(x)]^3 = (fff)' = f'ff + ff'f + fff' = 3fff' = 3[f(x)]^2 f'(x).$$

(c) $\dfrac{d}{dx}\left(e^{3x}\right) = \dfrac{d}{dx}\left(e^x\right)^3 = 3(e^x)^2 e^x = 3e^{2x}e^x = 3e^{3x}$

3.3 Rates of Change in the Natural and Social Sciences

1. (a) $s = f(t) = t^2 - 10t + 12$ $\Rightarrow$ $v(t) = f'(t) = 2t - 10$

(b) $v(3) = 2(3) - 10 = -4$ ft/s

(c) The particle is at rest when $v(t) = 0$ $\Leftrightarrow$ $2t - 10 = 0$ $\Leftrightarrow$ $t = 5$ s.

(d) The particle is moving in the positive direction when $v(t) > 0$ $\Leftrightarrow$ $2t - 10 > 0$ $\Leftrightarrow$ $2t > 10$ $\Leftrightarrow$ $t > 5$.

(e) Since the particle is moving in the positive direction   (f)
and in the negative direction, we need to calculate the
distance traveled in the intervals $[0, 5]$ and $[5, 8]$
separately. $|f(5) - f(0)| = |-13 - 12| = 25$ ft and
$|f(8) - f(5)| = |-4 - (-13)| = 9$ ft. The total
distance traveled during the first 8 s is $25 + 9 = 34$ ft.

3. (a) $s = f(t) = t^3 - 12t^2 + 36t$ $\Rightarrow$ $v(t) = f'(t) = 3t^2 - 24t + 36$

(b) $v(3) = 27 - 72 + 36 = -9$ ft/s

(c) The particle is at rest when $v(t) = 0$. $3t^2 - 24t + 36 = 0$ $\Rightarrow$ $3(t-2)(t-6) = 0$ $\Rightarrow$ $t = 2$ s or 6 s.

(d) The particle is moving in the positive direction when $v(t) > 0$. $3(t-2)(t-6) > 0$ $\Leftrightarrow$ $0 \le t < 2$ or $t > 6$.

(e) Since the particle is moving in the positive direction and in the negative direction, we need to calculate the distance traveled in the intervals $[0, 2]$, $[2, 6]$, and $[6, 8]$ separately.

$$|f(2) - f(0)| = |32 - 0| = 32.$$
$$|f(6) - f(2)| = |0 - 32| = 32.$$
$$|f(8) - f(6)| = |32 - 0| = 32.$$

The total distance is $32 + 32 + 32 = 96$ ft.

(f)

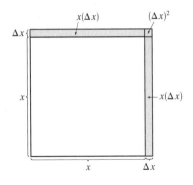

5. (a) $s = \dfrac{t}{t^2 + 1}$ $\Rightarrow$ $v(t) = s'(t) = \dfrac{(t^2 + 1)(1) - t(2t)}{(t^2 + 1)^2} = \dfrac{1 - t^2}{(t^2 + 1)^2}$

(b) $v(3) = \dfrac{1 - (3)^2}{(3^2 + 1)^2} = \dfrac{1 - 9}{10^2} = \dfrac{-8}{100} = -\dfrac{2}{25}$ ft/s

(c) It is at rest when $v = 0$ $\Leftrightarrow$ $1 - t^2 = 0$ $\Leftrightarrow$ $t = 1$ s $[t \neq -1$ since $t \geq 0]$.

(d) It moves in the positive direction when $v > 0$ $\Leftrightarrow$ $1 - t^2 > 0$ $\Leftrightarrow$ $t^2 < 1$ $\Leftrightarrow$ $0 \leq t < 1$.

(e) Distance in positive direction $= |s(1) - s(0)| = \left|\frac{1}{2} - 0\right| = \frac{1}{2}$ ft

Distance in negative direction $= |s(8) - s(1)| = \left|\frac{8}{65} - \frac{1}{2}\right| = \frac{49}{130}$ ft

Total distance traveled $= \frac{1}{2} + \frac{49}{130} = \frac{57}{65}$ ft

(f)

7. $s(t) = t^3 - 4.5t^2 - 7t$ $\Rightarrow$ $v(t) = s'(t) = 3t^2 - 9t - 7 = 5$ $\Leftrightarrow$ $3t^2 - 9t - 12 = 0$ $\Leftrightarrow$
$3(t - 4)(t + 1) = 0$ $\Leftrightarrow$ $t = 4$ or -1. Since $t \geq 0$, the particle reaches a velocity of 5 m/s at $t = 4$ s.

9. (a) $h = 10t - 0.83t^2$ $\Rightarrow$ $v(t) = \dfrac{dh}{dt} = 10 - 1.66t$, so $v(3) = 10 - 1.66(3) = 5.02$ m/s.

(b) $h = 25$ $\Rightarrow$ $10t - 0.83t^2 = 25$ $\Rightarrow$ $0.83t^2 - 10t + 25 = 0$ $\Rightarrow$ $t = \dfrac{10 \pm \sqrt{17}}{1.66} \approx 3.54$ or 8.51.

The value $t_1 = (10 - \sqrt{17})/1.66$ corresponds to the time it takes for the stone to rise 25 m and
$t_2 = (10 + \sqrt{17})/1.66$ corresponds to the time when the stone is 25 m high on the way down. Thus,
$v(t_1) = 10 - 1.66[(10 - \sqrt{17})/1.66] = \sqrt{17} \approx 4.12$ m/s.

11. (a) $A(x) = x^2$ $\Rightarrow$ $A'(x) = 2x$. $A'(15) = 30$ mm^2/mm is the rate at which the area is increasing with respect to the side length as x reaches 15 mm.

(b) The perimeter is $P(x) = 4x$, so $A'(x) = 2x = \frac{1}{2}(4x) = \frac{1}{2}P(x)$.
The figure suggests that if Δx is small, then the change in the area of the square is approximately half of its perimeter (2 of the 4 sides) times Δx. From the figure, $\Delta A = 2x(\Delta x) + (\Delta x)^2$. If Δx is small, then $\Delta A \approx 2x(\Delta x)$ and so $\Delta A/\Delta x \approx 2x$.

13. (a) Using $A(r) = \pi r^2$, we find that the average rate of change is:

(i) $\dfrac{A(3) - A(2)}{3 - 2} = \dfrac{9\pi - 4\pi}{1} = 5\pi$ (ii) $\dfrac{A(2.5) - A(2)}{2.5 - 2} = \dfrac{6.25\pi - 4\pi}{0.5} = 4.5\pi$

(iii) $\dfrac{A(2.1) - A(2)}{2.1 - 2} = \dfrac{4.41\pi - 4\pi}{0.1} = 4.1\pi$

(b) $A(r) = \pi r^2 \;\Rightarrow\; A'(r) = 2\pi r$, so $A'(2) = 4\pi$.

(c) The circumference is $C(r) = 2\pi r = A'(r)$. The figure suggests that if Δr is small, then the change in the area of the circle (a ring around the outside) is approximately equal to its circumference times Δr. Straightening out this ring gives us a shape that is approximately rectangular with length $2\pi r$ and width Δr, so $\Delta A \approx 2\pi r(\Delta r)$. Algebraically,

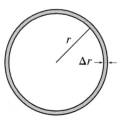

$\Delta A = A(r + \Delta r) - A(r) = \pi(r + \Delta r)^2 - \pi r^2 = 2\pi r(\Delta r) + \pi(\Delta r)^2$.

So we see that if Δr is small, then $\Delta A \approx 2\pi r(\Delta r)$ and therefore, $\Delta A / \Delta r \approx 2\pi r$.

15. $S(r) = 4\pi r^2 \;\Rightarrow\; S'(r) = 8\pi r \;\Rightarrow$

(a) $S'(1) = 8\pi$ ft^2/ft (b) $S'(2) = 16\pi$ ft^2/ft (c) $S'(3) = 24\pi$ ft^2/ft

As the radius increases, the surface area grows at an increasing rate. In fact, the rate of change is linear with respect to the radius.

17. The mass is $f(x) = 3x^2$, so the linear density at x is $\rho(x) = f'(x) = 6x$.

(a) $\rho(1) = 6$ kg/m (b) $\rho(2) = 12$ kg/m (c) $\rho(3) = 18$ kg/m

Since ρ is an increasing function, the density will be the highest at the right end of the rod and lowest at the left end.

19. The quantity of charge is $Q(t) = t^3 - 2t^2 + 6t + 2$, so the current is $Q'(t) = 3t^2 - 4t + 6$.

(a) $Q'(0.5) = 3(0.5)^2 - 4(0.5) + 6 = 4.75$ A (b) $Q'(1) = 3(1)^2 - 4(1) + 6 = 5$ A

The current is lowest when Q' has a minimum. $Q''(t) = 6t - 4 < 0$ when $t < \frac{2}{3}$. So the current decreases when $t < \frac{2}{3}$ and increases when $t > \frac{2}{3}$. Thus, the current is lowest at $t = \frac{2}{3}$ s.

21. (a) To find the rate of change of volume with respect to pressure, we first solve for V in terms of P.

$$PV = C \;\Rightarrow\; V = \frac{C}{P} \;\Rightarrow\; \frac{dV}{dP} = -\frac{C}{P^2}.$$

(b) From the formula for dV/dP in part (a), we see that as P increases, the absolute value of dV/dP decreases. Thus, the volume is decreasing more rapidly at the beginning.

(c) $\beta = -\dfrac{1}{V}\dfrac{dV}{dP} = -\dfrac{1}{V}\left(-\dfrac{C}{P^2}\right) = \dfrac{C}{(PV)P} = \dfrac{C}{CP} = \dfrac{1}{P}$

23. (a) **1920:** $m_1 = \dfrac{1860 - 1750}{1920 - 1910} = \frac{110}{10} = 11,\; m_2 = \dfrac{2070 - 1860}{1930 - 1920} = \frac{210}{10} = 21,$

$(m_1 + m_2)/2 = (11 + 21)/2 = 16$ million/year

1980: $m_1 = \dfrac{4450 - 3710}{1980 - 1970} = \frac{740}{10} = 74,\; m_2 = \dfrac{5280 - 4450}{1990 - 1980} = \frac{830}{10} = 83,$

$(m_1 + m_2)/2 = (74 + 83)/2 = 78.5$ million/year

(b) $P(t) = at^3 + bt^2 + ct + d$ (in millions of people), where $a \approx 0.0012937063$, $b \approx -7.061421911$, $c \approx 12{,}822.97902$, and $d \approx -7{,}743{,}770.396$.

(c) $P(t) = at^3 + bt^2 + ct + d \Rightarrow P'(t) = 3at^2 + 2bt + c$ (in millions of people per year)

(d) $P'(1920) = 3(0.0012937063)(1920)^2 + 2(-7.061421911)(1920) + 12,822.97902$

≈ 14.48 million/year [smaller than the answer in part (a), but close to it]

$P'(1980) \approx 75.29$ million/year (smaller, but close)

(e) $P'(1985) \approx 81.62$ million/year, so the rate of growth in 1985 was about 81.62 million/year.

25. (a) $[C] = \dfrac{a^2 kt}{akt + 1} \Rightarrow$

$\text{rate of reaction} = \dfrac{d\,[C]}{dt} = \dfrac{(akt + 1)(a^2 k) - (a^2 kt)(ak)}{(akt + 1)^2} = \dfrac{a^2 k(akt + 1 - akt)}{(akt + 1)^2} = \dfrac{a^2 k}{(akt + 1)^2}$

(b) If $x = [C]$, then $a - x = a - \dfrac{a^2 kt}{akt + 1} = \dfrac{a^2 kt + a - a^2 kt}{akt + 1} = \dfrac{a}{akt + 1}$.

So $k(a - x)^2 = k\left(\dfrac{a}{akt + 1}\right)^2 = \dfrac{a^2 k}{(akt + 1)^2} = \dfrac{d\,[C]}{dt}$ [from part (a)] $= \dfrac{dx}{dt}$.

(c) As $t \to \infty$, $[C] = \dfrac{a^2 kt}{akt + 1} = \dfrac{(a^2 kt)/t}{(akt + 1)/t} = \dfrac{a^2 k}{ak + (1/t)} \to \dfrac{a^2 k}{ak} = a$ moles/L.

(d) As $t \to \infty$, $\dfrac{d[C]}{dt} = \dfrac{a^2 k}{(akt + 1)^2} \to 0$.

(e) As t increases, nearly all of the reactants A and B are converted into product C. In practical terms, the reaction virtually stops.

27. (a) Using $v = \dfrac{P}{4\eta l}(R^2 - r^2)$ with $R = 0.01$, $l = 3$, $P = 3000$, and $\eta = 0.027$, we have v as a function of r:

$v(r) = \dfrac{3000}{4(0.027)3}(0.01^2 - r^2)$. $v(0) = 0.\overline{925}$ cm/s, $v(0.005) = 0.69\overline{4}$ cm/s, $v(0.01) = 0$.

(b) $v(r) = \dfrac{P}{4\eta l}(R^2 - r^2) \Rightarrow v'(r) = \dfrac{P}{4\eta l}(-2r) = -\dfrac{Pr}{2\eta l}$. When $l = 3$, $P = 3000$, and $\eta = 0.027$, we have

$v'(r) = -\dfrac{3000r}{2(0.027)3}$. $v'(0) = 0$, $v'(0.005) = -92.\overline{592}$ (cm/s)/cm, and $v'(0.01) = -185.\overline{185}$ (cm/s)/cm.

(c) The velocity is greatest where $r = 0$ (at the center) and the velocity is changing most where $r = R = 0.01$ cm (at the edge).

29. (a) $C(x) = 2000 + 3x + 0.01x^2 + 0.0002x^3 \Rightarrow C'(x) = 3 + 0.02x + 0.0006x^2$

(b) $C'(100) = 3 + 0.02(100) + 0.0006(10,000) = 3 + 2 + 6 = \$11$/pair. $C'(100)$ is the rate at which the cost is increasing as the 100th pair of jeans is produced. It predicts the cost of the 101st pair.

(c) The cost of manufacturing the 101st pair of jeans is

$C(101) - C(100) = (2000 + 303 + 102.01 + 206.0602) - (2000 + 300 + 100 + 200)$

$= 11.0702 \approx \$11.07$

31. (a) $A(x) = \dfrac{p(x)}{x} \Rightarrow A'(x) = \dfrac{xp'(x) - p(x) \cdot 1}{x^2} = \dfrac{xp'(x) - p(x)}{x^2}$. $A'(x) > 0 \Rightarrow A(x)$ is increasing; that is, the average productivity increases as the size of the workforce increases.

(b) $p'(x)$ is greater than the average productivity $\Rightarrow p'(x) > A(x) \Rightarrow p'(x) > \dfrac{p(x)}{x} \Rightarrow$

$xp'(x) > p(x) \Rightarrow xp'(x) - p(x) > 0 \Rightarrow \dfrac{xp'(x) - p(x)}{x^2} > 0 \Rightarrow A'(x) > 0$.

33. $PV = nRT \;\Rightarrow\; T = \dfrac{PV}{nR} = \dfrac{PV}{(10)(0.0821)} = \dfrac{1}{0.821}(PV)$. Using the Product Rule, we have

$$\frac{dT}{dt} = \frac{1}{0.821}\left[P(t)V'(t) + V(t)P'(t)\right] = \frac{1}{0.821}\left[(8)(-0.15) + (10)(0.10)\right] \approx -0.2436 \text{ K/min}.$$

35. (a) If the populations are stable, then the growth rates are neither positive nor negative; that is,

$$\frac{dC}{dt} = 0 \text{ and } \frac{dW}{dt} = 0.$$

(b) "The caribou go extinct" means that the population is zero, or mathematically, $C = 0$.

(c) We have the equations $\dfrac{dC}{dt} = aC - bCW$ and $\dfrac{dW}{dt} = -cW + dCW$. Let $dC/dt = dW/dt = 0$, $a = 0.05$,

$b = 0.001$, $c = 0.05$, and $d = 0.0001$ to obtain $0.05C - 0.001CW = 0$ **(1)** and

$-0.05W + 0.0001CW = 0$ **(2)**. Adding 10 times **(2)** to **(1)** eliminates the CW-terms and gives us

$0.05C - 0.5W = 0 \;\Rightarrow\; C = 10W$. Substituting $C = 10W$ into **(1)** results in

$0.05(10W) - 0.001(10W)W = 0 \;\Leftrightarrow\; 0.5W - 0.01W^2 = 0 \;\Leftrightarrow\; 50W - W^2 = 0 \;\Leftrightarrow\;$

$W(50 - W) = 0 \;\Leftrightarrow\; W = 0$ or 50. Since $C = 10W$, $C = 0$ or 500. Thus, the population pairs (C, W)

that lead to stable populations are $(0, 0)$ and $(500, 50)$. So it is possible for the two species to live in harmony.

3.4 Derivatives of Trigonometric Functions

1. $f(x) = x - 3\sin x \;\Rightarrow\; f'(x) = 1 - 3\cos x$

3. $y = \sin x + 10\tan x \;\Rightarrow\; y' = \cos x + 10\sec^2 x$

5. $g(t) = t^3 \cos t \;\Rightarrow\; g'(t) = t^3(-\sin t) + (\cos t) \cdot 3t^2 = 3t^2 \cos t - t^3 \sin t$ or $t^2(3\cos t - t\sin t)$

7. $h(\theta) = \csc\theta + e^\theta \cot\theta \;\Rightarrow\;$
$h'(\theta) = -\csc\theta\cot\theta + e^\theta\left(-\csc^2\theta\right) + (\cot\theta)e^\theta = -\csc\theta\cot\theta + e^\theta\left(\cot\theta - \csc^2\theta\right)$

9. $y = \dfrac{x}{\cos x} \;\Rightarrow\; y' = \dfrac{(\cos x)(1) - (x)(-\sin x)}{(\cos x)^2} = \dfrac{\cos x + x\sin x}{\cos^2 x}$

11. $f(\theta) = \dfrac{\sec\theta}{1 + \sec\theta} \;\Rightarrow\;$

$$f'(\theta) = \frac{(1 + \sec\theta)(\sec\theta\tan\theta) - (\sec\theta)(\sec\theta\tan\theta)}{(1 + \sec\theta)^2} = \frac{(\sec\theta\tan\theta)[(1 + \sec\theta) - \sec\theta]}{(1 + \sec\theta)^2} = \frac{\sec\theta\tan\theta}{(1 + \sec\theta)^2}$$

13. $y = \dfrac{\sin x}{x^2} \;\Rightarrow\; y' = \dfrac{x^2\cos x - (\sin x)(2x)}{(x^2)^2} = \dfrac{x(x\cos x - 2\sin x)}{x^4} = \dfrac{x\cos x - 2\sin x}{x^3}$

15. $y = \sec\theta\tan\theta \;\Rightarrow\; y' = \sec\theta\left(\sec^2\theta\right) + \tan\theta\left(\sec\theta\tan\theta\right) = \sec\theta\left(\sec^2\theta + \tan^2\theta\right)$
Using the identity $1 + \tan^2\theta = \sec^2\theta$, we can write alternative forms of the answer as

$$\sec\theta\left(1 + 2\tan^2\theta\right) \qquad \text{or} \qquad \sec\theta\left(2\sec^2\theta - 1\right)$$

17. $\dfrac{d}{dx}(\csc x) = \dfrac{d}{dx}\left(\dfrac{1}{\sin x}\right) = \dfrac{(\sin x)(0) - 1(\cos x)}{\sin^2 x} = \dfrac{-\cos x}{\sin^2 x} = -\dfrac{1}{\sin x} \cdot \dfrac{\cos x}{\sin x} = -\csc x\cot x$

19. $\dfrac{d}{dx}(\cot x) = \dfrac{d}{dx}\left(\dfrac{\cos x}{\sin x}\right) = \dfrac{(\sin x)(-\sin x) - (\cos x)(\cos x)}{\sin^2 x} = -\dfrac{\sin^2 x + \cos^2 x}{\sin^2 x} = -\dfrac{1}{\sin^2 x} = -\csc^2 x$

21. $y = \tan x \;\Rightarrow\; y' = \sec^2 x \;\Rightarrow\;$ the slope of the tangent line at $\left(\frac{\pi}{4}, 1\right)$ is $\sec^2\frac{\pi}{4} = \left(\sqrt{2}\right)^2 = 2$ and an equation
of the tangent line is $y - 1 = 2\left(x - \frac{\pi}{4}\right)$ or $y = 2x + 1 - \frac{\pi}{2}$.

23. $y = x + \cos x$ $\Rightarrow$ $y' = 1 - \sin x$. At $(0, 1)$, $y' = 1$, and an equation of the tangent line is $y - 1 = 1(x - 0)$, or $y = x + 1$.

25. (a) $y = x \cos x$ $\Rightarrow$ $y' = x(-\sin x) + \cos x(1) = \cos x - x \sin x$.
So the slope of the tangent at the point $(\pi, -\pi)$ is
$\cos \pi - \pi \sin \pi = -1 - \pi(0) = -1$, and an equation is
$y + \pi = -(x - \pi)$ or $y = -x$.

(b)

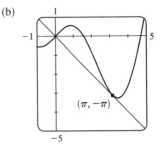

27. (a) $f(x) = 2x + \cot x$ $\Rightarrow$ $f'(x) = 2 - \csc^2 x$

(b)

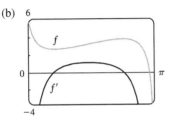

Notice that $f'(x) = 0$ when f has a horizontal tangent.
f' is positive when f is increasing and f' is negative when f
is decreasing. Also, $f'(x)$ is large negative when the graph
of f is steep.

29. $f(x) = x + 2 \sin x$ has a horizontal tangent when $f'(x) = 0$ $\Leftrightarrow$ $1 + 2 \cos x = 0$ $\Leftrightarrow$ $\cos x = -\frac{1}{2}$ $\Leftrightarrow$
$x = \frac{2\pi}{3} + 2\pi n$ or $\frac{4\pi}{3} + 2\pi n$, where n is an integer. Note that $\frac{4\pi}{3}$ and $\frac{2\pi}{3}$ are $\pm\frac{\pi}{3}$ units from π. This allows us to
write the solutions in the more compact equivalent form $(2n + 1)\pi \pm \frac{\pi}{3}$, n an integer.

31. (a) $x(t) = 8 \sin t$ $\Rightarrow$ $v(t) = x'(t) = 8 \cos t$

(b) The mass at time $t = \frac{2\pi}{3}$ has position $x\left(\frac{2\pi}{3}\right) = 8 \sin \frac{2\pi}{3} = 8\left(\frac{\sqrt{3}}{2}\right) = 4\sqrt{3}$ and velocity
$v\left(\frac{2\pi}{3}\right) = 8 \cos \frac{2\pi}{3} = 8\left(-\frac{1}{2}\right) = -4$. Since $v\left(\frac{2\pi}{3}\right) < 0$, the particle is moving to the left.

33.

From the diagram we can see that $\sin \theta = x/10$ $\Leftrightarrow$ $x = 10 \sin \theta$. We want to find the
rate of change of x with respect to θ; that is, $dx/d\theta$. Taking the derivative of the above
expression, $dx/d\theta = 10(\cos \theta)$. So when $\theta = \frac{\pi}{3}$,
$$dx/d\theta = 10 \cos \frac{\pi}{3} = 10\left(\frac{1}{2}\right) = 5 \text{ ft/rad}$$

35. $\displaystyle\lim_{x \to 0} \frac{\sin 3x}{x} = \lim_{x \to 0} \frac{3 \sin 3x}{3x}$ [multiply numerator and denominator by 3]

$\displaystyle = 3 \lim_{3x \to 0} \frac{\sin 3x}{3x}$ [as $x \to 0$, $3x \to 0$]

$\displaystyle = 3 \lim_{\theta \to 0} \frac{\sin \theta}{\theta}$ [let $\theta = 3x$]

$= 3(1)$ [Equation 2]

$= 3$

37. $\displaystyle\lim_{t \to 0} \frac{\tan 6t}{\sin 2t} = \lim_{t \to 0} \left(\frac{\sin 6t}{t} \cdot \frac{1}{\cos 6t} \cdot \frac{t}{\sin 2t} \right) = \lim_{t \to 0} \frac{6 \sin 6t}{6t} \cdot \lim_{t \to 0} \frac{1}{\cos 6t} \cdot \lim_{t \to 0} \frac{2t}{2 \sin 2t}$

$\displaystyle = 6 \lim_{t \to 0} \frac{\sin 6t}{6t} \cdot \lim_{t \to 0} \frac{1}{\cos 6t} \cdot \frac{1}{2} \lim_{t \to 0} \frac{2t}{\sin 2t} = 6(1) \cdot \frac{1}{1} \cdot \frac{1}{2}(1) = 3$

39. $\displaystyle\lim_{\theta\to 0}\frac{\sin(\cos\theta)}{\sec\theta} = \frac{\sin\left(\lim\limits_{\theta\to 0}\cos\theta\right)}{\lim\limits_{\theta\to 0}\sec\theta} = \frac{\sin 1}{1} = \sin 1$

41. $\displaystyle\lim_{x\to 0}\frac{\cot 2x}{\csc x} = \lim_{x\to 0}\frac{\cos 2x\sin x}{\sin 2x} = \lim_{x\to 0}\cos 2x\left[\frac{(\sin x)/x}{(\sin 2x)/x}\right] = \lim_{x\to 0}\cos 2x\left[\frac{\lim\limits_{x\to 0}[(\sin x)/x]}{2\lim\limits_{x\to 0}[(\sin 2x)/2x]}\right]$

$\displaystyle = 1\cdot\frac{1}{2\cdot 1} = \frac{1}{2}$

43. Divide numerator and denominator by θ. ($\sin\theta$ also works.)

$\displaystyle\lim_{\theta\to 0}\frac{\sin\theta}{\theta + \tan\theta} = \lim_{\theta\to 0}\frac{\dfrac{\sin\theta}{\theta}}{1 + \dfrac{\sin\theta}{\theta}\cdot\dfrac{1}{\cos\theta}} = \frac{\lim\limits_{\theta\to 0}\dfrac{\sin\theta}{\theta}}{1 + \lim\limits_{\theta\to 0}\dfrac{\sin\theta}{\theta}\lim\limits_{\theta\to 0}\dfrac{1}{\cos\theta}} = \frac{1}{1 + 1\cdot 1} = \frac{1}{2}$

45. (a) $\dfrac{d}{dx}\tan x = \dfrac{d}{dx}\dfrac{\sin x}{\cos x} \Rightarrow \sec^2 x = \dfrac{\cos x\cos x - \sin x\,(-\sin x)}{\cos^2 x} = \dfrac{\cos^2 x + \sin^2 x}{\cos^2 x}$. So $\sec^2 x = \dfrac{1}{\cos^2 x}$.

(b) $\dfrac{d}{dx}\sec x = \dfrac{d}{dx}\dfrac{1}{\cos x} \Rightarrow \sec x\tan x = \dfrac{(\cos x)(0) - 1(-\sin x)}{\cos^2 x}$. So $\sec x\tan x = \dfrac{\sin x}{\cos^2 x}$.

(c) $\dfrac{d}{dx}(\sin x + \cos x) = \dfrac{d}{dx}\dfrac{1 + \cot x}{\csc x} \Rightarrow$

$\cos x - \sin x = \dfrac{\csc x\,(-\csc^2 x) - (1 + \cot x)(-\csc x\cot x)}{\csc^2 x} = \dfrac{\csc x\left[-\csc^2 x + (1 + \cot x)\cot x\right]}{\csc^2 x}$

$= \dfrac{-\csc^2 x + \cot^2 x + \cot x}{\csc x} = \dfrac{-1 + \cot x}{\csc x}$

So $\cos x - \sin x = \dfrac{\cot x - 1}{\csc x}$.

47. By the definition of radian measure, $s = r\theta$, where r is the radius of the circle.

By drawing the bisector of the angle θ, we can see that $\sin\dfrac{\theta}{2} = \dfrac{d/2}{r} \Rightarrow d = 2r\sin\dfrac{\theta}{2}$.

So $\displaystyle\lim_{\theta\to 0^+}\frac{s}{d} = \lim_{\theta\to 0^+}\frac{r\theta}{2r\sin(\theta/2)} = \lim_{\theta\to 0^+}\frac{2\cdot(\theta/2)}{2\sin(\theta/2)} = \lim_{\theta\to 0}\frac{\theta/2}{\sin(\theta/2)} = 1$. [This is just the reciprocal of the limit

$\displaystyle\lim_{x\to 0}\frac{\sin x}{x} = 1$ combined with the fact that as $\theta\to 0$, $\frac{\theta}{2}\to 0$ also.]

3.5 The Chain Rule

1. Let $u = g(x) = 4x$ and $y = f(u) = \sin u$. Then $\dfrac{dy}{dx} = \dfrac{dy}{du}\dfrac{du}{dx} = (\cos u)(4) = 4\cos 4x$.

3. Let $u = g(x) = 1 - x^2$ and $y = f(u) = u^{10}$. Then $\dfrac{dy}{dx} = \dfrac{dy}{du}\dfrac{du}{dx} = (10u^9)(-2x) = -20x(1 - x^2)^9$.

5. Let $u = g(x) = \sqrt{x}$ and $y = f(u) = e^u$.

Then $\dfrac{dy}{dx} = \dfrac{dy}{du}\dfrac{du}{dx} = (e^u)\left(\frac{1}{2}x^{-1/2}\right) = e^{\sqrt{x}}\cdot\dfrac{1}{2\sqrt{x}} = \dfrac{e^{\sqrt{x}}}{2\sqrt{x}}$.

7. $F(x) = (x^3 + 4x)^7 \Rightarrow F'(x) = 7(x^3 + 4x)^6(3x^2 + 4)$ [or $7x^6(x^2 + 4)^6(3x^2 + 4)$]

9. $F(x) = \sqrt[4]{1 + 2x + x^3} = (1 + 2x + x^3)^{1/4} \Rightarrow$

$$F'(x) = \tfrac{1}{4}(1 + 2x + x^3)^{-3/4} \cdot \frac{d}{dx}(1 + 2x + x^3) = \frac{1}{4(1 + 2x + x^3)^{3/4}} \cdot (2 + 3x^2)$$

$$= \frac{2 + 3x^2}{4(1 + 2x + x^3)^{3/4}} = \frac{2 + 3x^2}{4\sqrt[4]{(1 + 2x + x^3)^3}}$$

11. $g(t) = \dfrac{1}{(t^4 + 1)^3} = (t^4 + 1)^{-3} \Rightarrow g'(t) = -3(t^4 + 1)^{-4}(4t^3) = -12t^3(t^4 + 1)^{-4} = \dfrac{-12t^3}{(t^4 + 1)^4}$

13. $y = \cos(a^3 + x^3) \Rightarrow y' = -\sin(a^3 + x^3) \cdot 3x^2$ [a^3 is just a constant] $= -3x^2 \sin(a^3 + x^3)$

15. $y = e^{-mx} \Rightarrow y' = e^{-mx}\dfrac{d}{dx}(-mx) = e^{-mx}(-m) = -me^{-mx}$

17. $g(x) = (1 + 4x)^5(3 + x - x^2)^8 \Rightarrow$

$$g'(x) = (1 + 4x)^5 \cdot 8(3 + x - x^2)^7(1 - 2x) + (3 + x - x^2)^8 \cdot 5(1 + 4x)^4 \cdot 4$$

$$= 4(1 + 4x)^4(3 + x - x^2)^7\left[2(1 + 4x)(1 - 2x) + 5(3 + x - x^2)\right]$$

$$= 4(1 + 4x)^4(3 + x - x^2)^7\left[(2 + 4x - 16x^2) + (15 + 5x - 5x^2)\right]$$

$$= 4(1 + 4x)^4(3 + x - x^2)^7\left(17 + 9x - 21x^2\right)$$

19. $y = (2x - 5)^4(8x^2 - 5)^{-3} \Rightarrow$

$$y' = 4(2x - 5)^3(2)(8x^2 - 5)^{-3} + (2x - 5)^4(-3)(8x^2 - 5)^{-4}(16x)$$

$$= 8(2x - 5)^3(8x^2 - 5)^{-3} - 48x(2x - 5)^4(8x^2 - 5)^{-4}$$

[This simplifies to $8(2x - 5)^3(8x^2 - 5)^{-4}(-4x^2 + 30x - 5)$.]

21. $y = xe^{-x^2} \Rightarrow y' = xe^{-x^2}(-2x) + e^{-x^2} \cdot 1 = e^{-x^2}(-2x^2 + 1) = e^{-x^2}(1 - 2x^2)$

23. $y = e^{x\cos x} \Rightarrow y' = e^{x\cos x} \cdot \dfrac{d}{dx}(x\cos x) = e^{x\cos x}[x(-\sin x) + (\cos x) \cdot 1] = e^{x\cos x}(\cos x - x\sin x)$

25. $F(z) = \sqrt{\dfrac{z - 1}{z + 1}} = \left(\dfrac{z - 1}{z + 1}\right)^{1/2} \Rightarrow$

$$F'(z) = \frac{1}{2}\left(\frac{z - 1}{z + 1}\right)^{-1/2} \cdot \frac{d}{dz}\left(\frac{z - 1}{z + 1}\right) = \frac{1}{2}\left(\frac{z + 1}{z - 1}\right)^{1/2} \cdot \frac{(z + 1)(1) - (z - 1)(1)}{(z + 1)^2}$$

$$= \frac{1}{2}\frac{(z + 1)^{1/2}}{(z - 1)^{1/2}} \cdot \frac{z + 1 - z + 1}{(z + 1)^2} = \frac{1}{2}\frac{(z + 1)^{1/2}}{(z - 1)^{1/2}} \cdot \frac{2}{(z + 1)^2} = \frac{1}{(z - 1)^{1/2}(z + 1)^{3/2}}$$

27. $y = \dfrac{r}{\sqrt{r^2 + 1}}$ ⇒

$$y' = \frac{\sqrt{r^2 + 1}\,(1) - r \cdot \frac{1}{2}(r^2 + 1)^{-1/2}(2r)}{\left(\sqrt{r^2 + 1}\right)^2} = \frac{\sqrt{r^2 + 1} - \dfrac{r^2}{\sqrt{r^2 + 1}}}{\left(\sqrt{r^2 + 1}\right)^2} = \frac{\dfrac{\sqrt{r^2 + 1}\,\sqrt{r^2 + 1} - r^2}{\sqrt{r^2 + 1}}}{\left(\sqrt{r^2 + 1}\right)^2}$$

$$= \frac{(r^2 + 1) - r^2}{\left(\sqrt{r^2 + 1}\right)^3} = \frac{1}{(r^2 + 1)^{3/2}} \text{ or } (r^2 + 1)^{-3/2}$$

Another solution: Write y as a product and make use of the Product Rule. $y = r(r^2 + 1)^{-1/2}$ ⇒

$$y' = r \cdot -\tfrac{1}{2}(r^2 + 1)^{-3/2}(2r) + (r^2 + 1)^{-1/2} \cdot 1$$

$$= (r^2 + 1)^{-3/2}\left[-r^2 + (r^2 + 1)^1\right] = (r^2 + 1)^{-3/2}(1) = (r^2 + 1)^{-3/2}$$

The step that students usually have trouble with is factoring out $(r^2 + 1)^{-3/2}$. But this is no different than factoring out x^2 from $x^2 + x^5$; that is, we are just factoring out a factor with the *smallest* exponent that appears on it. In this case, $-\tfrac{3}{2}$ is smaller than $-\tfrac{1}{2}$.

29. $y = \tan(\cos x)$ ⇒ $y' = \sec^2(\cos x) \cdot (-\sin x) = -\sin x \sec^2(\cos x)$

31. Using Formula 5 and the Chain Rule, $y = 2^{\sin \pi x}$ ⇒

$$y' = 2^{\sin \pi x}(\ln 2) \cdot \frac{d}{dx}(\sin \pi x) = 2^{\sin \pi x}(\ln 2) \cdot \cos \pi x \cdot \pi = 2^{\sin \pi x}(\pi \ln 2)\cos \pi x$$

33. $y = (1 + \cos^2 x)^6$ ⇒ $y' = 6(1 + \cos^2 x)^5 \cdot 2\cos x\,(-\sin x) = -12\cos x \sin x\,(1 + \cos^2 x)^5$

35. $y = \sec^2 x + \tan^2 x = (\sec x)^2 + (\tan x)^2$ ⇒
$$y' = 2(\sec x)(\sec x \tan x) + 2(\tan x)(\sec^2 x) = 2\sec^2 x \tan x + 2\sec^2 x \tan x = 4\sec^2 x \tan x$$

37. $y = \cot^2(\sin \theta) = [\cot(\sin \theta)]^2$ ⇒
$$y' = 2\,[\cot(\sin \theta)] \cdot \frac{d}{d\theta}\,[\cot(\sin \theta)] = 2\cot(\sin \theta) \cdot [-\csc^2(\sin \theta) \cdot \cos \theta] = -2\cos \theta \cot(\sin \theta)\csc^2(\sin \theta)$$

39. $y = \sqrt{x + \sqrt{x}}$ ⇒ $y' = \tfrac{1}{2}(x + \sqrt{x})^{-1/2}\left(1 + \tfrac{1}{2}x^{-1/2}\right) = \dfrac{1}{2\sqrt{x + \sqrt{x}}}\left(1 + \dfrac{1}{2\sqrt{x}}\right)$

41. $y = \sin\left(\tan \sqrt{\sin x}\right)$ ⇒

$$y' = \cos\left(\tan \sqrt{\sin x}\right) \cdot \frac{d}{dx}\left(\tan \sqrt{\sin x}\right) = \cos\left(\tan \sqrt{\sin x}\right)\sec^2 \sqrt{\sin x} \cdot \frac{d}{dx}(\sin x)^{1/2}$$

$$= \cos\left(\tan \sqrt{\sin x}\right)\sec^2 \sqrt{\sin x} \cdot \tfrac{1}{2}(\sin x)^{-1/2} \cdot \cos x$$

$$= \cos\left(\tan \sqrt{\sin x}\right)\left(\sec^2 \sqrt{\sin x}\right)\left(\frac{1}{2\sqrt{\sin x}}\right)(\cos x)$$

43. $y = (1 + 2x)^{10}$ ⇒ $y' = 10(1 + 2x)^9 \cdot 2 = 20(1 + 2x)^9$. At $(0, 1)$, $y' = 20(1 + 0)^9 = 20$, and an equation of the tangent line is $y - 1 = 20(x - 0)$, or $y = 20x + 1$.

45. $y = \sin(\sin x)$ ⇒ $y' = \cos(\sin x) \cdot \cos x$. At $(\pi, 0)$, $y' = \cos(\sin \pi) \cdot \cos \pi = \cos(0) \cdot (-1) = 1(-1) = -1$, and an equation of the tangent line is $y - 0 = -1(x - \pi)$, or $y = -x + \pi$.

47. (a) $y = \dfrac{2}{1+e^{-x}}$ $\Rightarrow$ $y' = \dfrac{(1+e^{-x})(0) - 2(-e^{-x})}{(1+e^{-x})^2} = \dfrac{2e^{-x}}{(1+e^{-x})^2}$.

At $(0,1)$, $y' = \dfrac{2e^0}{(1+e^0)^2} = \dfrac{2(1)}{(1+1)^2} = \dfrac{2}{2^2} = \dfrac{1}{2}$.

So an equation of the tangent line is $y - 1 = \frac{1}{2}(x - 0)$ or $y = \frac{1}{2}x + 1$.

(b)

49. (a) $f(x) = \dfrac{\sqrt{1-x^2}}{x}$ $\Rightarrow$

$f'(x) = \dfrac{x \cdot \frac{1}{2}(1-x^2)^{-1/2}(-2x) - \sqrt{1-x^2}\,(1)}{x^2} \cdot \dfrac{\sqrt{1-x^2}}{\sqrt{1-x^2}}$

$= \dfrac{-x^2 - (1-x^2)}{x^2\sqrt{1-x^2}} = \dfrac{-1}{x^2\sqrt{1-x^2}}$

(b)

Notice that all tangents to the graph of f have negative slopes and $f'(x) < 0$ always.

51. For the tangent line to be horizontal, $f'(x) = 0$. $f(x) = 2\sin x + \sin^2 x$ $\Rightarrow$

$f'(x) = 2\cos x + 2\sin x \cos x = 0$ $\Leftrightarrow$ $2\cos x\,(1+\sin x) = 0$ $\Leftrightarrow$ $\cos x = 0$ or $\sin x = -1$, so

$x = \frac{\pi}{2} + 2n\pi$ or $\frac{3\pi}{2} + 2n\pi$, where n is any integer. Now $f\left(\frac{\pi}{2}\right) = 3$ and $f\left(\frac{3\pi}{2}\right) = -1$, so the points on the curve

with a horizontal tangent are $\left(\frac{\pi}{2} + 2n\pi, 3\right)$ and $\left(\frac{3\pi}{2} + 2n\pi, -1\right)$, where n is any integer.

53. $F(x) = f(g(x))$ $\Rightarrow$ $F'(x) = f'(g(x)) \cdot g'(x)$,

so $F'(3) = f'(g(3)) \cdot g'(3) = f'(6) \cdot g'(3) = 7 \cdot 4 = 28$. Notice that we did not use $f'(3) = 2$.

55. (a) $h(x) = f(g(x))$ $\Rightarrow$ $h'(x) = f'(g(x)) \cdot g'(x)$, so $h'(1) = f'(g(1)) \cdot g'(1) = f'(2) \cdot 6 = 5 \cdot 6 = 30$.

(b) $H(x) = g(f(x))$ $\Rightarrow$ $H'(x) = g'(f(x)) \cdot f'(x)$, so $H'(1) = g'(f(1)) \cdot f'(1) = g'(3) \cdot 4 = 9 \cdot 4 = 36$.

57. (a) $u(x) = f(g(x))$ $\Rightarrow$ $u'(x) = f'(g(x))g'(x)$. So $u'(1) = f'(g(1))g'(1) = f'(3)g'(1)$. To find $f'(3)$, note

that f is linear from $(2,4)$ to $(6,3)$, so its slope is $\dfrac{3-4}{6-2} = -\dfrac{1}{4}$. To find $g'(1)$, note that g is linear from $(0,6)$

to $(2,0)$, so its slope is $\dfrac{0-6}{2-0} = -3$. Thus, $f'(3)g'(1) = \left(-\frac{1}{4}\right)(-3) = \frac{3}{4}$.

(b) $v(x) = g(f(x))$ $\Rightarrow$ $v'(x) = g'(f(x))f'(x)$. So $v'(1) = g'(f(1))f'(1) = g'(2)f'(1)$, which does not exist

since $g'(2)$ does not exist.

(c) $w(x) = g(g(x))$ $\Rightarrow$ $w'(x) = g'(g(x))g'(x)$. So $w'(1) = g'(g(1))g'(1) = g'(3)g'(1)$. To find $g'(3)$, note

that g is linear from $(2,0)$ to $(5,2)$, so its slope is $\dfrac{2-0}{5-2} = \dfrac{2}{3}$. Thus, $g'(3)g'(1) = \left(\frac{2}{3}\right)(-3) = -2$.

59. $h(x) = f(g(x))$ $\Rightarrow$ $h'(x) = f'(g(x))g'(x)$. So $h'(0.5) = f'(g(0.5))g'(0.5) = f'(0.1)g'(0.5)$. We can

estimate the derivatives by taking the average of two secant slopes.

For $f'(0.1)$: $m_1 = \dfrac{14.8 - 12.6}{0.1 - 0} = 22$, $m_2 = \dfrac{18.4 - 14.8}{0.2 - 0.1} = 36$. So $f'(0.1) \approx \dfrac{m_1 + m_2}{2} = \dfrac{22 + 36}{2} = 29$.

For $g'(0.5)$: $m_1 = \dfrac{0.10 - 0.17}{0.5 - 0.4} = -0.7$, $m_2 = \dfrac{0.05 - 0.10}{0.6 - 0.5} = -0.5$. So $g'(0.5) \approx \dfrac{m_1 + m_2}{2} = -0.6$.

Hence, $h'(0.5) = f'(0.1)g'(0.5) \approx (29)(-0.6) = -17.4$.

61. (a) $F(x) = f(e^x) \quad \Rightarrow \quad F'(x) = f'(e^x)\dfrac{d}{dx}(e^x) = f'(e^x)e^x$

(b) $G(x) = e^{f(x)} \quad \Rightarrow \quad G'(x) = e^{f(x)}\dfrac{d}{dx}f(x) = e^{f(x)}f'(x)$

63. (a) $f(x) = L(x^4) \quad \Rightarrow \quad f'(x) = L'(x^4) \cdot 4x^3 = (1/x^4) \cdot 4x^3 = 4/x$ for $x > 0$.

(b) $g(x) = L(4x) \quad \Rightarrow \quad g'(x) = L'(4x) \cdot 4 = (1/(4x)) \cdot 4 = 1/x$ for $x > 0$.

(c) $F(x) = [L(x)]^4 \quad \Rightarrow \quad F'(x) = 4[L(x)]^3 \cdot L'(x) = 4[L(x)]^3 \cdot (1/x) = 4[L(x)]^3/x$

(d) $G(x) = L(1/x) \quad \Rightarrow \quad G'(x) = L'(1/x) \cdot (-1/x^2) = (1/(1/x)) \cdot (-1/x^2) = x \cdot (-1/x^2) = -1/x$
for $x > 0$.

65. $s(t) = 10 + \frac{1}{4}\sin(10\pi t) \quad \Rightarrow \quad$ the velocity after t seconds is
$v(t) = s'(t) = \frac{1}{4}\cos(10\pi t)(10\pi) = \frac{5\pi}{2}\cos(10\pi t)$ cm/s.

67. (a) $B(t) = 4.0 + 0.35\sin\dfrac{2\pi t}{5.4} \quad \Rightarrow \quad \dfrac{dB}{dt} = \left(0.35\cos\dfrac{2\pi t}{5.4}\right)\left(\dfrac{2\pi}{5.4}\right) = \dfrac{0.7\pi}{5.4}\cos\dfrac{2\pi t}{5.4} = \dfrac{7\pi}{54}\cos\dfrac{2\pi t}{5.4}$

(b) At $t = 1$, $\dfrac{dB}{dt} = \dfrac{7\pi}{54}\cos\dfrac{2\pi}{5.4} \approx 0.16$.

69. $s(t) = 2e^{-1.5t}\sin 2\pi t \quad \Rightarrow$
$v(t) = s'(t) = 2\left[e^{-1.5t}(\cos 2\pi t)(2\pi) + (\sin 2\pi t)e^{-1.5t}(-1.5)\right] = 2e^{-1.5t}(2\pi\cos 2\pi t - 1.5\sin 2\pi t)$

Graph of position

Graph of velocity

71. (a) Using a calculator or CAS, we obtain the model $Q = ab^t$ with $a = 100.0124369$ and $b = 0.000045145933$. We can change this model to one with base e and exponent $\ln b$ [$b^t = e^{t\ln b}$ from precalculus mathematics or from Section 7.3]: $Q = ae^{t\ln b} = 100.012437e^{-10.005531t}$.

(b) Use $Q'(t) = ab^t \ln b$ or the calculator command `nDeriv(Y₁,X,.04)` with $Y_1 = ab^x$ to get $Q'(0.04) \approx -670.63$ μA. The result of Example 2 in Section 2.1 was -670 μA.

73. (a) Derive gives $g'(t) = \dfrac{45(t-2)^8}{(2t+1)^{10}}$ without simplifying. With either Maple or Mathematica, we first get

$g'(t) = 9\dfrac{(t-2)^8}{(2t+1)^9} - 18\dfrac{(t-2)^9}{(2t+1)^{10}}$, and the simplification command results in the above expression.

(b) Derive gives $y' = 2(x^3 - x + 1)^3(2x + 1)^4(17x^3 + 6x^2 - 9x + 3)$ without simplifying. With either Maple or Mathematica, we first get

$y' = 10(2x + 1)^4(x^3 - x + 1)^4 + 4(2x + 1)^5(x^3 - x + 1)^3(3x^2 - 1)$. If we use Mathematica's `Factor` or `Simplify`, or Maple's `factor`, we get the above expression, but Maple's `simplify` gives the polynomial expansion instead. For locating horizontal tangents, the factored form is the most helpful.

75. (a) If f is even, then $f(x) = f(-x)$. Using the Chain Rule to differentiate this equation, we get

$f'(x) = f'(-x)\dfrac{d}{dx}(-x) = -f'(-x)$. Thus, $f'(-x) = -f'(x)$, so f' is odd.

(b) If f is odd, then $f(x) = -f(-x)$. Differentiating this equation, we get $f'(x) = -f'(-x)(-1) = f'(-x)$, so f' is even.

77. (a) $\dfrac{d}{dx}(\sin^n x \cos nx) = n\sin^{n-1} x \cos x \cos nx + \sin^n x\,(-n\sin nx)$ [Product Rule]

$$= n\sin^{n-1} x\,(\cos nx \cos x - \sin nx \sin x) \qquad \text{[factor out } n\sin^{n-1} x\text{]}$$
$$= n\sin^{n-1} x \cos(nx + x) \qquad \text{[Addition Formula for cosine]}$$
$$= n\sin^{n-1} x \cos[(n+1)x] \qquad \text{[factor out } x\text{]}$$

(b) $\dfrac{d}{dx}(\cos^n x \cos nx) = n\cos^{n-1} x\,(-\sin x)\cos nx + \cos^n x\,(-n\sin nx)$ [Product Rule]

$$= -n\cos^{n-1} x\,(\cos nx \sin x + \sin nx \cos x) \qquad \text{[factor out } -n\cos^{n-1} x\text{]}$$
$$= -n\cos^{n-1} x \sin(nx + x) \qquad \text{[Addition Formula for sine]}$$
$$= -n\cos^{n-1} x \sin[(n+1)x] \qquad \text{[factor out } x\text{]}$$

79. Since $\theta° = \left(\frac{\pi}{180}\right)\theta$ rad, we have $\dfrac{d}{d\theta}(\sin \theta°) = \dfrac{d}{d\theta}\left(\sin \frac{\pi}{180}\theta\right) = \frac{\pi}{180}\cos \frac{\pi}{180}\theta = \frac{\pi}{180}\cos \theta°$.

81. First note that products and differences of polynomials are polynomials and that the derivative of a polynomial is also a polynomial. When $n = 1$, $f^{(1)}(x) = \left(\dfrac{P(x)}{Q(x)}\right)' = \dfrac{Q(x)P'(x) - P(x)Q'(x)}{[Q(x)]^2} = \dfrac{A_1(x)}{[Q(x)]^{1+1}}$, where $A_1(x) = Q(x)P'(x) - P(x)Q'(x)$. Suppose the result is true for $n = k$, where $k \geq 1$. Then $f^{(k)}(x) = \dfrac{A_k(x)}{[Q(x)]^{k+1}}$, so

$$f^{(k+1)}(x) = \left(\dfrac{A_k(x)}{[Q(x)]^{k+1}}\right)' = \dfrac{[Q(x)]^{k+1}A_k'(x) - A_k(x)\cdot(k+1)[Q(x)]^k \cdot Q'(x)}{\{[Q(x)]^{k+1}\}^2}$$

$$= \dfrac{[Q(x)]^{k+1}A_k'(x) - (k+1)A_k(x)[Q(x)]^k Q'(x)}{[Q(x)]^{2k+2}}$$

$$= \dfrac{[Q(x)]^k\{[Q(x)]^1 A_k'(x) - (k+1)A_k(x)Q'(x)\}}{[Q(x)]^k[Q(x)]^{k+2}} = \dfrac{Q(x)A_k'(x) - (k+1)A_k(x)Q'(x)}{[Q(x)]^{k+2}}$$

$$= A_{k+1}(x)/[Q(x)]^{k+2}, \text{ where } A_{k+1}(x) = Q(x)A_k'(x) - (k+1)A_k(x)Q'(x).$$

We have shown that the formula holds for $n = 1$, and that when it holds for $n = k$ it also holds for $n = k + 1$. Thus, by mathematical induction, the formula holds for all positive integers n.

3.6 Implicit Differentiation

1. (a) $\dfrac{d}{dx}(xy + 2x + 3x^2) = \dfrac{d}{dx}(4) \;\Rightarrow\; (x \cdot y' + y \cdot 1) + 2 + 6x = 0 \;\Rightarrow\; xy' = -y - 2 - 6x \;\Rightarrow$

$y' = \dfrac{-y - 2 - 6x}{x}$ or $y' = -6 - \dfrac{y+2}{x}$.

(b) $xy + 2x + 3x^2 = 4 \;\Rightarrow\; xy = 4 - 2x - 3x^2 \;\Rightarrow\; y = \dfrac{4 - 2x - 3x^2}{x} = \dfrac{4}{x} - 2 - 3x$, so $y' = -\dfrac{4}{x^2} - 3$.

(c) From part (a), $y' = \dfrac{-y - 2 - 6x}{x} = \dfrac{-(4/x - 2 - 3x) - 2 - 6x}{x} = \dfrac{-4/x - 3x}{x} = -\dfrac{4}{x^2} - 3$.

3. (a) $\dfrac{d}{dx}\left(\dfrac{1}{x}+\dfrac{1}{y}\right)=\dfrac{d}{dx}\,(1)$ $\Rightarrow$ $-\dfrac{1}{x^2}-\dfrac{1}{y^2}y'=0$ $\Rightarrow$ $-\dfrac{1}{y^2}\,y'=\dfrac{1}{x^2}$ $\Rightarrow$ $y'=-\dfrac{y^2}{x^2}$

(b) $\dfrac{1}{x}+\dfrac{1}{y}=1$ $\Rightarrow$ $\dfrac{1}{y}=1-\dfrac{1}{x}=\dfrac{x-1}{x}$ $\Rightarrow$ $y=\dfrac{x}{x-1}$, so $y'=\dfrac{(x-1)(1)-(x)(1)}{(x-1)^2}=\dfrac{-1}{(x-1)^2}$.

(c) $y'=-\dfrac{y^2}{x^2}=-\dfrac{[x/(x-1)]^2}{x^2}=-\dfrac{x^2}{x^2(x-1)^2}=-\dfrac{1}{(x-1)^2}$

5. $\dfrac{d}{dx}\left(x^2+y^2\right)=\dfrac{d}{dx}\,(1)$ $\Rightarrow$ $2x+2yy'=0$ $\Rightarrow$ $2yy'=-2x$ $\Rightarrow$ $y'=-\dfrac{x}{y}$

7. $\dfrac{d}{dx}\left(x^3+x^2y+4y^2\right)=\dfrac{d}{dx}\,(6)$ $\Rightarrow$ $3x^2+(x^2y'+y\cdot 2x)+8yy'=0$ $\Rightarrow$ $x^2y'+8yy'=-3x^2-2xy$

$\Rightarrow$ $(x^2+8y)y'=-3x^2-2xy$ $\Rightarrow$ $y'=-\dfrac{3x^2+2xy}{x^2+8y}=-\dfrac{x(3x+2y)}{x^2+8y}$

9. $\dfrac{d}{dx}\left(x^2y+xy^2\right)=\dfrac{d}{dx}\,(3x)$ $\Rightarrow$ $(x^2y'+y\cdot 2x)+(x\cdot 2yy'+y^2\cdot 1)=3$ $\Rightarrow$

$x^2y'+2xyy'=3-2xy-y^2$ $\Rightarrow$ $y'(x^2+2xy)=3-2xy-y^2$ $\Rightarrow$ $y'=\dfrac{3-2xy-y^2}{x^2+2xy}$

11. $\dfrac{d}{dx}\left(x^2y^2+x\sin y\right)=\dfrac{d}{dx}\,(4)$ $\Rightarrow$ $x^2\cdot 2yy'+y^2\cdot 2x+x\cos y\cdot y'+\sin y\cdot 1=0$ $\Rightarrow$

$2x^2yy'+x\cos y\cdot y'=-2xy^2-\sin y$ $\Rightarrow$ $(2x^2y+x\cos y)y'=-2xy^2-\sin y$ $\Rightarrow$ $y'=\dfrac{-2xy^2-\sin y}{2x^2y+x\cos y}$

13. $\dfrac{d}{dx}\left(4\cos x\sin y\right)=\dfrac{d}{dx}\,(1)$ $\Rightarrow$ $4\left[\cos x\cdot\cos y\cdot y'+\sin y\cdot(-\sin x)\right]=0$ $\Rightarrow$

$y'(4\cos x\cos y)=4\sin x\sin y$ $\Rightarrow$ $y'=\dfrac{4\sin x\sin y}{4\cos x\cos y}=\tan x\tan y$

15. $\dfrac{d}{dx}\left(e^{x^2y}\right)=\dfrac{d}{dx}\,(x+y)$ $\Rightarrow$ $e^{x^2y}(x^2y'+y\cdot 2x)=1+y'$ $\Rightarrow$ $x^2e^{x^2y}y'+2xye^{x^2y}=1+y'$ $\Rightarrow$

$x^2e^{x^2y}y'-y'=1-2xye^{x^2y}$ $\Rightarrow$ $y'(x^2e^{x^2y}-1)=1-2xye^{x^2y}$ $\Rightarrow$ $y'=\dfrac{1-2xye^{x^2y}}{x^2e^{x^2y}-1}$

17. $\sqrt{xy}=1+x^2y$ $\Rightarrow$ $\tfrac{1}{2}(xy)^{-1/2}(xy'+y\cdot 1)=0+x^2y'+y\cdot 2x$ $\Rightarrow$ $\dfrac{x}{2\sqrt{xy}}\,y'+\dfrac{y}{2\sqrt{xy}}=x^2y'+2xy$

$\Rightarrow$ $y'\left(\dfrac{x}{2\sqrt{xy}}-x^2\right)=2xy-\dfrac{y}{2\sqrt{xy}}$ $\Rightarrow$ $y'\left(\dfrac{x-2x^2\sqrt{xy}}{2\sqrt{xy}}\right)=\dfrac{4xy\sqrt{xy}-y}{2\sqrt{xy}}$ $\Rightarrow$ $y'=\dfrac{4xy\sqrt{xy}-y}{x-2x^2\sqrt{xy}}$

19. $xy=\cot(xy)$ $\Rightarrow$ $y+xy'=-\csc^2(xy)(y+xy')$ $\Rightarrow$ $(y+xy')\left[1+\csc^2(xy)\right]=0$ $\Rightarrow$

$y+xy'=0$ [since $1+\csc^2(xy)>0$] $\Rightarrow$ $y'=-y/x$

21. $\dfrac{d}{dx}\left\{1+f(x)+x^2[f(x)]^3\right\}=\dfrac{d}{dx}\,(0)$ $\Rightarrow$ $f'(x)+x^2\cdot 3[f(x)]^2\cdot f'(x)+[f(x)]^3\cdot 2x=0$. If $x=1$, we have

$f'(1)+1^2\cdot 3[f(1)]^2\cdot f'(1)+[f(1)]^3\cdot 2(1)=0$ $\Rightarrow$ $f'(1)+1\cdot 3\cdot 2^2\cdot f'(1)+2^3\cdot 2=0$ $\Rightarrow$

$f'(1)+12f'(1)=-16$ $\Rightarrow$ $13f'(1)=-16$ $\Rightarrow$ $f'(1)=-\tfrac{16}{13}$.

23. $y^4 + x^2y^2 + yx^4 = y + 1$ $\Rightarrow$ $4y^3 + \left(x^2 \cdot 2y + y^2 \cdot 2x \dfrac{dx}{dy}\right) + \left(y \cdot 4x^3 \dfrac{dx}{dy} + x^4 \cdot 1\right) = 1$ $\Rightarrow$

$2xy^2 \dfrac{dx}{dy} + 4x^3y \dfrac{dx}{dy} = 1 - 4y^3 - 2x^2y - x^4$ $\Rightarrow$ $\dfrac{dx}{dy} = \dfrac{1 - 4y^3 - 2x^2y - x^4}{2xy^2 + 4x^3y}$

25. $x^2 + xy + y^2 = 3$ $\Rightarrow$ $2x + xy' + y \cdot 1 + 2yy' = 0$ $\Rightarrow$ $xy' + 2yy' = -2x - y$ $\Rightarrow$

$y'(x + 2y) = -2x - y$ $\Rightarrow$ $y' = \dfrac{-2x - y}{x + 2y}$. When $x = 1$ and $y = 1$, we have $y' = \dfrac{-2 - 1}{1 + 2 \cdot 1} = \dfrac{-3}{3} = -1$, so

an equation of the tangent line is $y - 1 = -1(x - 1)$ or $y = -x + 2$.

27. $x^2 + y^2 = (2x^2 + 2y^2 - x)^2$ $\Rightarrow$ $2x + 2yy' = 2(2x^2 + 2y^2 - x)(4x + 4yy' - 1)$. When $x = 0$ and $y = \frac{1}{2}$, we

have $0 + y' = 2(\frac{1}{2})(2y' - 1)$ $\Rightarrow$ $y' = 2y' - 1$ $\Rightarrow$ $y' = 1$, so an equation of the tangent line is

$y - \frac{1}{2} = 1(x - 0)$ or $y = x + \frac{1}{2}$.

29. $2(x^2 + y^2)^2 = 25(x^2 - y^2)$ $\Rightarrow$ $4(x^2 + y^2)(2x + 2yy') = 25(2x - 2yy')$ $\Rightarrow$

$4(x + yy')(x^2 + y^2) = 25(x - yy')$ $\Rightarrow$ $4yy'(x^2 + y^2) + 25yy' = 25x - 4x(x^2 + y^2)$ $\Rightarrow$

$y' = \dfrac{25x - 4x(x^2 + y^2)}{25y + 4y(x^2 + y^2)}$. When $x = 3$ and $y = 1$, we have $y' = \frac{75 - 120}{25 + 40} = -\frac{45}{65} = -\frac{9}{13}$, so an equation of the

tangent line is $y - 1 = -\frac{9}{13}(x - 3)$ or $y = -\frac{9}{13}x + \frac{40}{13}$.

31. (a) $y^2 = 5x^4 - x^2$ $\Rightarrow$ $2yy' = 5(4x^3) - 2x$ $\Rightarrow$ $y' = \dfrac{10x^3 - x}{y}$. **(b)**

So at the point $(1, 2)$ we have $y' = \dfrac{10(1)^3 - 1}{2} = \dfrac{9}{2}$, and an

equation of the tangent line is $y - 2 = \frac{9}{2}(x - 1)$ or $y = \frac{9}{2}x - \frac{5}{2}$.

33. (a)

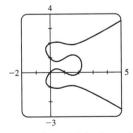

There are eight points with horizontal tangents:
four at $x \approx 1.57735$ and four at $x \approx 0.42265$.

(b) $y' = \dfrac{3x^2 - 6x + 2}{2(2y^3 - 3y^2 - y + 1)}$ $\Rightarrow$ $y' = -1$ at

$(0, 1)$ and $y' = \frac{1}{3}$ at $(0, 2)$.

Equations of the tangent lines are $y = -x + 1$

and $y = \frac{1}{3}x + 2$.

(c) $y' = 0$ $\Rightarrow$ $3x^2 - 6x + 2 = 0$ $\Rightarrow$

$x = 1 \pm \frac{1}{3}\sqrt{3}$

(d) By multiplying the right side of the equation by $x - 3$, we obtain the first graph.

By modifying the equation in other ways, we can generate the other graphs.

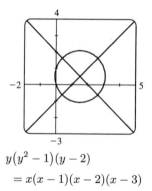

$$y(y^2 - 1)(y - 2)$$
$$= x(x - 1)(x - 2)(x - 3)$$

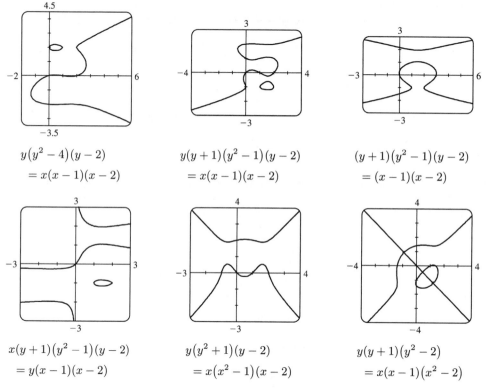

$$y(y^2 - 4)(y - 2)$$
$$= x(x - 1)(x - 2)$$

$$y(y + 1)(y^2 - 1)(y - 2)$$
$$= x(x - 1)(x - 2)$$

$$(y + 1)(y^2 - 1)(y - 2)$$
$$= (x - 1)(x - 2)$$

$$x(y + 1)(y^2 - 1)(y - 2)$$
$$= y(x - 1)(x - 2)$$

$$y(y^2 + 1)(y - 2)$$
$$= x(x^2 - 1)(x - 2)$$

$$y(y + 1)(y^2 - 2)$$
$$= x(x - 1)(x^2 - 2)$$

35. From Exercise 29, a tangent to the lemniscate will be horizontal if $y' = 0 \Rightarrow 25x - 4x(x^2 + y^2) = 0 \Rightarrow x[25 - 4(x^2 + y^2)] = 0 \Rightarrow x^2 + y^2 = \frac{25}{4}$ **(1)**. (Note that when x is 0, y is also 0, and there is no horizontal tangent at the origin.) Substituting $\frac{25}{4}$ for $x^2 + y^2$ in the equation of the lemniscate, $2(x^2 + y^2)^2 = 25(x^2 - y^2)$, we get $x^2 - y^2 = \frac{25}{8}$ **(2)**. Solving **(1)** and **(2)**, we have $x^2 = \frac{75}{16}$ and $y^2 = \frac{25}{16}$, so the four points are $\left(\pm\frac{5\sqrt{3}}{4}, \pm\frac{5}{4}\right)$.

37. $\dfrac{x^2}{a^2} - \dfrac{y^2}{b^2} = 1 \Rightarrow \dfrac{2x}{a^2} - \dfrac{2yy'}{b^2} = 0 \Rightarrow y' = \dfrac{b^2 x}{a^2 y} \Rightarrow$ an equation of the tangent line at (x_0, y_0) is

$y - y_0 = \dfrac{b^2 x_0}{a^2 y_0}(x - x_0)$. Multiplying both sides by $\dfrac{y_0}{b^2}$ gives $\dfrac{y_0 y}{b^2} - \dfrac{y_0^2}{b^2} = \dfrac{x_0 x}{a^2} - \dfrac{x_0^2}{a^2}$. Since (x_0, y_0) lies on the

hyperbola, we have $\dfrac{x_0 x}{a^2} - \dfrac{y_0 y}{b^2} = \dfrac{x_0^2}{a^2} - \dfrac{y_0^2}{b^2} = 1$.

39. If the circle has radius r, its equation is $x^2 + y^2 = r^2 \;\Rightarrow\; 2x + 2yy' = 0 \;\Rightarrow\; y' = -\dfrac{x}{y}$, so the slope of the

tangent line at $P(x_0, y_0)$ is $-\dfrac{x_0}{y_0}$. The negative reciprocal of that slope is $\dfrac{-1}{-x_0/y_0} = \dfrac{y_0}{x_0}$, which is the slope of

OP, so the tangent line at P is perpendicular to the radius OP.

41. $y = \tan^{-1}\sqrt{x} \;\Rightarrow\; y' = \dfrac{1}{1 + (\sqrt{x})^2} \cdot \dfrac{d}{dx}(\sqrt{x}) = \dfrac{1}{1+x}\left(\tfrac{1}{2}x^{-1/2}\right) = \dfrac{1}{2\sqrt{x}\,(1+x)}$

43. $y = \sin^{-1}(2x + 1) \;\Rightarrow$

$$y' = \frac{1}{\sqrt{1 - (2x+1)^2}} \cdot \frac{d}{dx}(2x + 1) = \frac{1}{\sqrt{1 - (4x^2 + 4x + 1)}} \cdot 2 = \frac{2}{\sqrt{-4x^2 - 4x}} = \frac{1}{\sqrt{-x^2 - x}}$$

45. $H(x) = (1 + x^2)\arctan x \;\Rightarrow\; H'(x) = (1 + x^2)\dfrac{1}{1 + x^2} + (\arctan x)(2x) = 1 + 2x\arctan x$

47. $h(t) = \cot^{-1}(t) + \cot^{-1}(1/t) \;\Rightarrow$

$$h'(t) = -\frac{1}{1 + t^2} - \frac{1}{1 + (1/t)^2} \cdot \frac{d}{dt}\frac{1}{t} = -\frac{1}{1 + t^2} - \frac{t^2}{t^2 + 1}\cdot\left(-\frac{1}{t^2}\right) = -\frac{1}{1 + t^2} + \frac{1}{t^2 + 1} = 0.$$

Note that this makes sense because $h(t) = \dfrac{\pi}{2}$ for $t > 0$ and $h(t) = -\dfrac{\pi}{2}$ for $t < 0$.

49. $y = \cos^{-1}(e^{2x}) \;\Rightarrow\; y' = -\dfrac{1}{\sqrt{1 - (e^{2x})^2}} \cdot \dfrac{d}{dx}(e^{2x}) = -\dfrac{2e^{2x}}{\sqrt{1 - e^{4x}}}$

51. $f(x) = e^x - x^2 \arctan x \;\Rightarrow$

$$f'(x) = e^x - \left[x^2\left(\frac{1}{1 + x^2}\right) + (\arctan x)(2x)\right]$$

$$= e^x - \frac{x^2}{1 + x^2} - 2x\arctan x$$

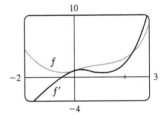

This is reasonable because the graphs show that f is increasing when f' is

positive, and f' is zero when f has a minimum.

53. Let $y = \cos^{-1} x$. Then $\cos y = x$ and $0 \le y \le \pi \;\Rightarrow\; -\sin y\,\dfrac{dy}{dx} = 1 \;\Rightarrow$

$$\frac{dy}{dx} = -\frac{1}{\sin y} = -\frac{1}{\sqrt{1 - \cos^2 y}} = -\frac{1}{\sqrt{1 - x^2}}. \quad \text{(Note that } \sin y \ge 0 \text{ for } 0 \le y \le \pi.\text{)}$$

55. $2x^2 + y^2 = 3$ and $x = y^2$ intersect when $2x^2 + x - 3 = 0 \;\Leftrightarrow\; (2x + 3)(x - 1) = 0 \;\Leftrightarrow\; x = -\tfrac{3}{2}$ or 1, but

$-\tfrac{3}{2}$ is extraneous since $x = y^2$ is nonnegative. When $x = 1$, $1 = y^2 \;\Rightarrow\; y = \pm 1$, so there are two points of

intersection: $(1, \pm 1)$. $2x^2 + y^2 = 3 \;\Rightarrow\; 4x + 2yy' = 0 \;\Rightarrow\; y' = -2x/y$, and $x = y^2 \;\Rightarrow\; 1 = 2yy' \;\Rightarrow$

$y' = 1/(2y)$. At $(1, 1)$, the slopes are $m_1 = -2(1)/1 = -2$ and $m_2 = 1/(2 \cdot 1) = \tfrac{1}{2}$, so the curves are orthogonal

(since m_1 and m_2 are negative reciprocals of each other). By symmetry, the curves are also orthogonal at $(1, -1)$.

57.

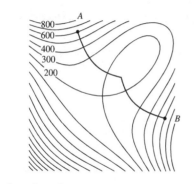

59. $x^2 + y^2 = r^2$ is a circle with center O and $ax + by = 0$ is a line through O.

$x^2 + y^2 = r^2 \implies 2x + 2yy' = 0 \implies y' = -x/y$, so the slope of the tangent

line at $P_0\,(x_0, y_0)$ is $-x_0/y_0$. The slope of the line OP_0 is y_0/x_0, which is the

negative reciprocal of $-x_0/y_0$. Hence, the curves are orthogonal, and the families

of curves are orthogonal trajectories of each other.

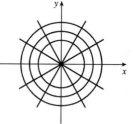

61. $y = cx^2 \implies y' = 2cx$ and $x^2 + 2y^2 = k \implies 2x + 4yy' = 0 \implies$

$2yy' = -x \implies y' = -\dfrac{x}{2(y)} = -\dfrac{x}{2(cx^2)} = -\dfrac{1}{2cx}$, so the curves are

orthogonal.

63. To find the points at which the ellipse $x^2 - xy + y^2 = 3$ crosses the x-axis, let $y = 0$ and solve for x.

$y = 0 \implies x^2 - x(0) + 0^2 = 3 \iff x = \pm\sqrt{3}$. So the graph of the ellipse crosses the x-axis at the points

$(\pm\sqrt{3}, 0)$. Using implicit differentiation to find y', we get $2x - xy' - y + 2yy' = 0 \implies y'(2y - x) = y - 2x$

$\iff y' = \dfrac{y - 2x}{2y - x}$. So y' at $(\sqrt{3}, 0)$ is $\dfrac{0 - 2\sqrt{3}}{2(0) - \sqrt{3}} = 2$ and y' at $(-\sqrt{3}, 0)$ is $\dfrac{0 + 2\sqrt{3}}{2(0) + \sqrt{3}} = 2$. Thus, the tangent

lines at these points are parallel.

65. $x^2 y^2 + xy = 2 \implies x^2 \cdot 2yy' + y^2 \cdot 2x + x \cdot y' + y \cdot 1 = 0 \iff y'(2x^2 y + x) = -2xy^2 - y \iff$

$y' = -\dfrac{2xy^2 + y}{2x^2 y + x}$. So $\dfrac{2xy^2 + y}{2x^2 y + x} = -1 \iff 2xy^2 + y = 2x^2 y + x \iff y(2xy + 1) = x(2xy + 1) \iff$

$y(2xy + 1) - x(2xy + 1) = 0 \iff (2xy + 1)(y - x) = 0 \iff xy = -\tfrac{1}{2}$ or $y = x$. But $xy = -\tfrac{1}{2} \implies$

$x^2 y^2 + xy = \tfrac{1}{4} - \tfrac{1}{2} \neq 2$, so we must have $x = y$. Then $x^2 y^2 + xy = 2 \implies x^4 + x^2 = 2 \iff$

$x^4 + x^2 - 2 = 0 \iff (x^2 + 2)(x^2 - 1) = 0$. So $x^2 = -2$, which is impossible, or $x^2 = 1 \iff x = \pm 1$.

Since $x = y$, the points on the curve where the tangent line has a slope of -1 are $(-1, -1)$ and $(1, 1)$.

67. (a) If $y = f^{-1}(x)$, then $f(y) = x$. Differentiating implicitly with respect to x and remembering that y is a function

of x, we get $f'(y)\dfrac{dy}{dx} = 1$, so $\dfrac{dy}{dx} = \dfrac{1}{f'(y)} \implies \left(f^{-1}\right)'(x) = \dfrac{1}{f'(f^{-1}(x))}$.

(b) $f(4) = 5 \;\Rightarrow\; f^{-1}(5) = 4$. By part (a), $\left(f^{-1}\right)'(5) = 1/f'\left(f^{-1}(5)\right) = 1/f'(4) = 1/\left(\frac{2}{3}\right) = \frac{3}{2}$.

69. $x^2 + 4y^2 = 5 \;\Rightarrow\; 2x + 4(2yy') = 0 \;\Rightarrow\; y' = -\dfrac{x}{4y}$. Now let h be the height of the lamp, and let (a, b) be the

point of tangency of the line passing through the points $(3, h)$ and $(-5, 0)$. This line has slope

$(h - 0)/[3 - (-5)] = \frac{1}{8}h$. But the slope of the tangent line through the point (a, b) can be expressed as $y' = -\dfrac{a}{4b}$,

or as $\dfrac{b - 0}{a - (-5)} = \dfrac{b}{a + 5}$ [since the line passes through $(-5, 0)$ and (a, b)], so $-\dfrac{a}{4b} = \dfrac{b}{a + 5} \;\Leftrightarrow$

$4b^2 = -a^2 - 5a \;\Leftrightarrow\; a^2 + 4b^2 = -5a$. But $a^2 + 4b^2 = 5$ [since (a, b) is on the ellipse], so $5 = -5a \;\Leftrightarrow$

$a = -1$. Then $4b^2 = -a^2 - 5a = -1 - 5(-1) = 4 \;\Rightarrow\; b = 1$, since the point is on the top half of the ellipse.

So $\dfrac{h}{8} = \dfrac{b}{a + 5} = \dfrac{1}{-1 + 5} = \dfrac{1}{4} \;\Rightarrow\; h = 2$. So the lamp is located 2 units above the x-axis.

3.7 Higher Derivatives

1. $a = f$, $b = f'$, $c = f''$. We can see this because where a has a horizontal tangent, $b = 0$, and where b has a
horizontal tangent, $c = 0$. We can immediately see that c can be neither f nor f', since at the points where c has a
horizontal tangent, neither a nor b is equal to 0.

3. We can immediately see that a is the graph of the acceleration function, since at the points where a has a horizontal
tangent, neither c nor b is equal to 0. Next, we note that $a = 0$ at the point where b has a horizontal tangent, so b
must be the graph of the velocity function, and hence, $b' = a$. We conclude that c is the graph of the position
function.

5. $f(x) = x^5 + 6x^2 - 7x \;\Rightarrow\; f'(x) = 5x^4 + 12x - 7 \;\Rightarrow\; f''(x) = 20x^3 + 12$

7. $y = \cos 2\theta \;\Rightarrow\; y' = -2\sin 2\theta \;\Rightarrow\; y'' = -4\cos 2\theta$

9. $F(t) = (1 - 7t)^6 \;\Rightarrow\; F'(t) = 6(1 - 7t)^5(-7) = -42(1 - 7t)^5 \;\Rightarrow$
$F''(t) = -42 \cdot 5(1 - 7t)^4(-7) = 1470(1 - 7t)^4$

11. $h(u) = \dfrac{1 - 4u}{1 + 3u} \;\Rightarrow\; h'(u) = \dfrac{(1 + 3u)(-4) - (1 - 4u)(3)}{(1 + 3u)^2} = \dfrac{-4 - 12u - 3 + 12u}{(1 + 3u)^2} = \dfrac{-7}{(1 + 3u)^2}$ or

$-7(1 + 3u)^{-2} \;\Rightarrow\; h''(u) = -7(-2)(1 + 3u)^{-3}(3) = 42(1 + 3u)^{-3}$ or $\dfrac{42}{(1 + 3u)^3}$

13. $h(x) = \sqrt{x^2 + 1} \;\Rightarrow\; h'(x) = \dfrac{1}{2}(x^2 + 1)^{-1/2}(2x) = \dfrac{x}{\sqrt{x^2 + 1}} \;\Rightarrow$

$h''(x) = \dfrac{\sqrt{x^2 + 1} \cdot 1 - x\left[\frac{1}{2}(x^2 + 1)^{-1/2}(2x)\right]}{\left(\sqrt{x^2 + 1}\right)^2} = \dfrac{(x^2 + 1)^{-1/2}\left[(x^2 + 1) - x^2\right]}{(x^2 + 1)^1} = \dfrac{1}{(x^2 + 1)^{3/2}}$

15. $y = (x^3 + 1)^{2/3} \;\Rightarrow\; y' = \frac{2}{3}(x^3 + 1)^{-1/3}(3x^2) = 2x^2(x^3 + 1)^{-1/3} \;\Rightarrow$
$y'' = 2x^2\left(-\frac{1}{3}\right)(x^3 + 1)^{-4/3}(3x^2) + (x^3 + 1)^{-1/3}(4x) = 4x(x^3 + 1)^{-1/3} - 2x^4(x^3 + 1)^{-4/3}$

17. $H(t) = \tan 3t \;\Rightarrow\; H'(t) = 3\sec^2 3t \;\Rightarrow$
$H''(t) = 2 \cdot 3\sec 3t \dfrac{d}{dt}(\sec 3t) = 6\sec 3t\,(3\sec 3t \tan 3t) = 18\sec^2 3t \tan 3t$

19. $g(t) = t^3 e^{5t} \Rightarrow g'(t) = t^3 e^{5t} \cdot 5 + e^{5t} \cdot 3t^2 = t^2 e^{5t}(5t + 3) \Rightarrow$
$g''(t) = (2t)e^{5t}(5t + 3) + t^2\left(5e^{5t}\right)(5t + 3) + t^2 e^{5t}(5)$
$\quad = te^{5t}\left[2(5t + 3) + 5t(5t + 3) + 5t\right] = te^{5t}\left(25t^2 + 30t + 6\right)$

21. (a) $f(x) = 2\cos x + \sin^2 x \Rightarrow f'(x) = 2(-\sin x) + 2\sin x(\cos x) = \sin 2x - 2\sin x \Rightarrow$
$f''(x) = 2\cos 2x - 2\cos x = 2(\cos 2x - \cos x)$

(b)

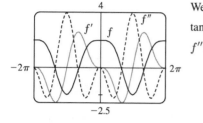

We can see that our answers are plausible, since f has horizontal tangents where $f'(x) = 0$, and f' has horizontal tangents where $f''(x) = 0$.

23. $y = \sqrt{2x + 3} = (2x + 3)^{1/2} \Rightarrow y' = \frac{1}{2}(2x + 3)^{-1/2} \cdot 2 = (2x + 3)^{-1/2} \Rightarrow$
$y'' = -\frac{1}{2}(2x + 3)^{-3/2} \cdot 2 = -(2x + 3)^{-3/2} \Rightarrow y''' = \frac{3}{2}(2x + 3)^{-5/2} \cdot 2 = 3(2x + 3)^{-5/2}$

25. $f(t) = t\cos t \Rightarrow f'(t) = t(-\sin t) + \cos t \cdot 1 \Rightarrow f''(t) = t(-\cos t) - \sin t \cdot 1 - \sin t \Rightarrow$
$f'''(t) = t\sin t - \cos t \cdot 1 - \cos t - \cos t = t\sin t - 3\cos t$, so $f'''(0) = 0 - 3 = -3$.

27. $f(\theta) = \cot\theta \Rightarrow f'(\theta) = -\csc^2\theta \Rightarrow f''(\theta) = -2\csc\theta\left(-\csc\theta\cot\theta\right) = 2\csc^2\theta\cot\theta \Rightarrow$
$f'''(\theta) = 2\left(-2\csc^2\theta\cot\theta\right)\cot\theta + 2\csc^2\theta\left(-\csc^2\theta\right) = -2\csc^2\theta\left(2\cot^2\theta + \csc^2\theta\right) \Rightarrow$
$f'''\left(\frac{\pi}{6}\right) = -2(2)^2\left[2\left(\sqrt{3}\right)^2 + (2)^2\right] = -80$

29. $9x^2 + y^2 = 9 \Rightarrow 18x + 2yy' = 0 \Rightarrow 2yy' = -18x \Rightarrow y' = -9x/y \Rightarrow$
$y'' = -9\left(\dfrac{y \cdot 1 - x \cdot y'}{y^2}\right) = -9\left(\dfrac{y - x(-9x/y)}{y^2}\right) = -9 \cdot \dfrac{y^2 + 9x^2}{y^3} = -9 \cdot \dfrac{9}{y^3}$ [since x and y must satisfy the original equation, $9x^2 + y^2 = 9$]. Thus, $y'' = -81/y^3$.

31. $x^3 + y^3 = 1 \Rightarrow 3x^2 + 3y^2 y' = 0 \Rightarrow y' = -\dfrac{x^2}{y^2} \Rightarrow$
$y'' = -\dfrac{y^2(2x) - x^2 \cdot 2yy'}{(y^2)^2} = -\dfrac{2xy^2 - 2x^2 y\left(-x^2/y^2\right)}{y^4} = -\dfrac{2xy^4 + 2x^4 y}{y^6} = -\dfrac{2xy(y^3 + x^3)}{y^6} = -\dfrac{2x}{y^5}$,
since x and y must satisfy the original equation, $x^3 + y^3 = 1$.

33. $f(x) = x^n \Rightarrow f'(x) = nx^{n-1} \Rightarrow f''(x) = n(n - 1)x^{n-2} \Rightarrow \cdots \Rightarrow$
$f^{(n)}(x) = n(n - 1)(n - 2)\cdots 2 \cdot 1x^{n-n} = n!$

35. $f(x) = e^{2x} \Rightarrow f'(x) = 2e^{2x} \Rightarrow f''(x) = 2 \cdot 2e^{2x} = 2^2 e^{2x} \Rightarrow$
$f'''(x) = 2^2 \cdot 2e^{2x} = 2^3 e^{2x} \Rightarrow \cdots \Rightarrow f^{(n)}(x) = 2^n e^{2x}$

37. $f(x) = 1/(3x^3) = \frac{1}{3}x^{-3} \Rightarrow f'(x) = \frac{1}{3}(-3)x^{-4} \Rightarrow f''(x) = \frac{1}{3}(-3)(-4)x^{-5} \Rightarrow$
$f'''(x) = \frac{1}{3}(-3)(-4)(-5)x^{-6} \Rightarrow \cdots \Rightarrow$
$f^{(n)}(x) = \frac{1}{3}(-3)(-4)\cdots[-(n + 2)]x^{-(n+3)} = \dfrac{(-1)^n \cdot 3 \cdot 4 \cdot 5 \cdots (n + 2)}{3x^{n+3}} \cdot \dfrac{2}{2} = \dfrac{(-1)^n(n + 2)!}{6x^{n+3}}$

39. Let $f(x) = \cos x$. Then $Df(2x) = 2f'(2x)$, $D^2 f(2x) = 2^2 f''(2x)$, $D^3 f(2x) = 2^3 f'''(2x), \ldots,$
$D^{(n)} f(2x) = 2^n f^{(n)}(2x)$. Since the derivatives of $\cos x$ occur in a cycle of four, and since $103 = 4(25) + 3$, we
have $f^{(103)}(x) = f^{(3)}(x) = \sin x$ and $D^{103}\cos 2x = 2^{103} f^{(103)}(2x) = 2^{103}\sin 2x$.

41. By measuring the slope of the graph of $s = f(t)$ at $t = 0, 1, 2, 3, 4$, and 5, and using the method of Example 1 in Section 2.9, we plot the graph of the velocity function $v = f'(t)$ in the first figure. The acceleration when $t = 2$ s is $a = f''(2)$, the slope of the tangent line to the graph of f' when $t = 2$. We estimate the slope of this tangent line to be $a(2) = f''(2) = v'(2) \approx \frac{27}{3} = 9$ ft/s². Similar measurements enable us to graph the acceleration function in the second figure.

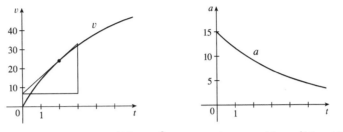

43. (a) $s = 2t^3 - 15t^2 + 36t + 2 \Rightarrow v(t) = s'(t) = 6t^2 - 30t + 36 \Rightarrow a(t) = v'(t) = 12t - 30$

(b) $a(1) = 12 \cdot 1 - 30 = -18$ m/s²

(c) $v(t) = 6(t^2 - 5t + 6) = 6(t-2)(t-3) = 0$ when $t = 2$ or 3 and $a(2) = 24 - 30 = -6$ m/s², $a(3) = 36 - 30 = 6$ m/s².

45. (a) $s = \sin\left(\frac{\pi}{6}t\right) + \cos\left(\frac{\pi}{6}t\right), 0 \le t \le 2$. $v(t) = s'(t) = \cos\left(\frac{\pi}{6}t\right) \cdot \frac{\pi}{6} - \sin\left(\frac{\pi}{6}t\right) \cdot \frac{\pi}{6} = \frac{\pi}{6}\left[\cos\left(\frac{\pi}{6}t\right) - \sin\left(\frac{\pi}{6}t\right)\right]$

$\Rightarrow a(t) = v'(t) = \frac{\pi}{6}\left[-\sin\left(\frac{\pi}{6}t\right) \cdot \frac{\pi}{6} - \cos\left(\frac{\pi}{6}t\right) \cdot \frac{\pi}{6}\right] = -\frac{\pi^2}{36}\left[\sin\left(\frac{\pi}{6}t\right) + \cos\left(\frac{\pi}{6}t\right)\right]$

(b) $a(1) = -\frac{\pi^2}{36}\left[\sin\left(\frac{\pi}{6} \cdot 1\right) + \cos\left(\frac{\pi}{6} \cdot 1\right)\right] = -\frac{\pi^2}{36}\left[\frac{1}{2} + \frac{\sqrt{3}}{2}\right] = -\frac{\pi^2}{72}\left(1 + \sqrt{3}\right) \approx -0.3745$ m/s²

(c) $v(t) = 0$ for $0 \le t \le 2 \Rightarrow \cos\left(\frac{\pi}{6}t\right) = \sin\left(\frac{\pi}{6}t\right) \Rightarrow 1 = \dfrac{\sin\left(\frac{\pi}{6}t\right)}{\cos\left(\frac{\pi}{6}t\right)} \Rightarrow$

$\tan\left(\frac{\pi}{6}t\right) = 1 \Rightarrow \frac{\pi}{6}t = \tan^{-1} 1 \Rightarrow t = \frac{6}{\pi} \cdot \frac{\pi}{4} = \frac{3}{2} = 1.5$ s. Thus,

$a\left(\frac{3}{2}\right) = -\frac{\pi^2}{36}\left[\sin\left(\frac{\pi}{6} \cdot \frac{3}{2}\right) + \cos\left(\frac{\pi}{6} \cdot \frac{3}{2}\right)\right] = -\frac{\pi^2}{36}\left[\frac{\sqrt{2}}{2} + \frac{\sqrt{2}}{2}\right] = -\frac{\pi^2}{36}\sqrt{2} \approx -0.3877$ m/s².

47. (a) $s(t) = t^4 - 4t^3 + 2 \Rightarrow v(t) = s'(t) = 4t^3 - 12t^2 \Rightarrow a(t) = v'(t) = 12t^2 - 24t = 12t(t-2) = 0$ when $t = 0$ or 2.

(b) $s(0) = 2$ m, $v(0) = 0$ m/s, $s(2) = -14$ m, $v(2) = -16$ m/s

49. (a) $s = f(t) = t^3 - 12t^2 + 36t, t \ge 0 \Rightarrow v(t) = f'(t) = 3t^2 - 24t + 36$.

$a(t) = v'(t) = 6t - 24$. $a(3) = 6(3) - 24 = -6$ (m/s)/s or m/s².

(b)

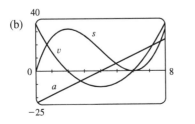

(c) The particle is speeding up when v and a have the same sign. This occurs when $2 < t < 4$ and when $t > 6$. It is slowing down when v and a have opposite signs; that is, when $0 \le t < 2$ and when $4 < t < 6$.

51. (a) $y(t) = A \sin \omega t$ $\Rightarrow$ $v(t) = y'(t) = A\omega \cos \omega t$ $\Rightarrow$ $a(t) = v'(t) = -A\omega^2 \sin \omega t$

(b) $a(t) = -A\omega^2 \sin \omega t = -\omega^2(A \sin \omega t) = -\omega^2 y(t)$, so $a(t)$ is proportional to $y(t)$.

(c) speed $= |v(t)| = A\omega |\cos \omega t|$ is a maximum when $\cos \omega t = \pm 1$. But when $\cos \omega t = \pm 1$, we have $\sin \omega t = 0$, and $a(t) = -A\omega^2 \sin \omega t = -A\omega^2(0) = 0$.

53. Let $P(x) = ax^2 + bx + c$. Then $P'(x) = 2ax + b$ and $P''(x) = 2a$. $P''(2) = 2$ $\Rightarrow$ $2a = 2$ $\Rightarrow$ $a = 1$.
$P'(2) = 3$ $\Rightarrow$ $2(1)(2) + b = 3$ $\Rightarrow$ $4 + b = 3$ $\Rightarrow$ $b = -1$.
$P(2) = 5$ $\Rightarrow$ $1(2)^2 + (-1)(2) + c = 5$ $\Rightarrow$ $2 + c = 5$ $\Rightarrow$ $c = 3$. So $P(x) = x^2 - x + 3$.

55. $y = A \sin x + B \cos x$ $\Rightarrow$ $y' = A \cos x - B \sin x$ $\Rightarrow$ $y'' = -A \sin x - B \cos x$. Substituting into
$y'' + y' - 2y = \sin x$ gives us $(-3A - B) \sin x + (A - 3B) \cos x = 1 \sin x$, so we must have $-3A - B = 1$
and $A - 3B = 0$. Solving for A and B, we add the first equation to three times the second to get $B = -\frac{1}{10}$
and $A = -\frac{3}{10}$.

57. $y = e^{rx}$ $\Rightarrow$ $y' = re^{rx}$ $\Rightarrow$ $y'' = r^2 e^{rx}$, so
$y'' + 5y' - 6y = r^2 e^{rx} + 5re^{rx} - 6e^{rx} = e^{rx}(r^2 + 5r - 6) = e^{rx}(r + 6)(r - 1) = 0$ $\Rightarrow$
$(r + 6)(r - 1) = 0$ $\Rightarrow$ $r = 1$ or -6.

59. $f(x) = xg(x^2)$ $\Rightarrow$ $f'(x) = x \cdot g'(x^2) \cdot 2x + g(x^2) \cdot 1 = g(x^2) + 2x^2 g'(x^2)$ $\Rightarrow$
$f''(x) = g'(x^2) \cdot 2x + 2x^2 \cdot g''(x^2) \cdot 2x + g'(x^2) \cdot 4x = 6xg'(x^2) + 4x^3 g''(x^2)$

61. $f(x) = g(\sqrt{x})$ $\Rightarrow$ $f'(x) = g'(\sqrt{x}) \cdot \frac{1}{2}x^{-1/2} = \dfrac{g'(\sqrt{x})}{2\sqrt{x}}$ $\Rightarrow$

$f''(x) = \dfrac{2\sqrt{x} \cdot g''(\sqrt{x}) \cdot \frac{1}{2}x^{-1/2} - g'(\sqrt{x}) \cdot 2 \cdot \frac{1}{2}x^{-1/2}}{(2\sqrt{x})^2} = \dfrac{x^{-1/2}\left[\sqrt{x}\, g''(\sqrt{x}) - g'(\sqrt{x})\right]}{4x}$

$= \dfrac{\sqrt{x}\, g''(\sqrt{x}) - g'(\sqrt{x})}{4x\sqrt{x}}$

63. (a) $f(x) = \dfrac{1}{x^2 + x}$ $\Rightarrow$ $f'(x) = \dfrac{-(2x + 1)}{(x^2 + x)^2}$ $\Rightarrow$

$f''(x) = \dfrac{(x^2 + x)^2(-2) + (2x + 1)(2)(x^2 + x)(2x + 1)}{(x^2 + x)^4} = \dfrac{2(3x^2 + 3x + 1)}{(x^2 + x)^3}$ $\Rightarrow$

$f'''(x) = \dfrac{(x^2 + x)^3(2)(6x + 3) - 2(3x^2 + 3x + 1)(3)(x^2 + x)^2(2x + 1)}{(x^2 + x)^6}$

$= \dfrac{-6(4x^3 + 6x^2 + 4x + 1)}{(x^2 + x)^4}$ $\Rightarrow$

$f^{(4)}(x) = \dfrac{(x^2 + x)^4(-6)(12x^2 + 12x + 4) + 6(4x^3 + 6x^2 + 4x + 1)(4)(x^2 + x)^3(2x + 1)}{(x^2 + x)^8}$

$= \dfrac{24(5x^4 + 10x^3 + 10x^2 + 5x + 1)}{(x^2 + x)^5}$

$f^{(5)}(x) = ?$

(b) $f(x) = \dfrac{1}{x(x + 1)} = \dfrac{1}{x} - \dfrac{1}{x + 1}$ $\Rightarrow$ $f'(x) = -x^{-2} + (x + 1)^{-2}$ $\Rightarrow$ $f''(x) = 2x^{-3} - 2(x + 1)^{-3}$ $\Rightarrow$

$f'''(x) = (-3)(2)x^{-4} + (3)(2)(x + 1)^{-4}$ $\Rightarrow$ $\cdots$ $\Rightarrow$ $f^{(n)}(x) = (-1)^n n! \left[x^{-(n+1)} - (x + 1)^{-(n+1)}\right]$

65. For $f(x) = x^2 e^x$, $f'(x) = x^2 e^x + e^x(2x) = e^x(x^2 + 2x)$. Similarly, we have

$$f''(x) = e^x(x^2 + 4x + 2)$$

$$f'''(x) = e^x(x^2 + 6x + 6)$$

$$f^{(4)}(x) = e^x(x^2 + 8x + 12)$$

$$f^{(5)}(x) = e^x(x^2 + 10x + 20)$$

It appears that the coefficient of x in the quadratic term increases by 2 with each differentiation. The pattern for the constant terms seems to be $0 = 1 \cdot 0, 2 = 2 \cdot 1, 6 = 3 \cdot 2, 12 = 4 \cdot 3, 20 = 5 \cdot 4$. So a reasonable guess is that $f^{(n)}(x) = e^x[x^2 + 2nx + n(n-1)]$.

Proof: Let S_n be the statement that $f^{(n)}(x) = e^x[x^2 + 2nx + n(n-1)]$.

1. S_1 is true because $f'(x) = e^x(x^2 + 2x)$.

2. Assume that S_k is true; that is, $f^{(k)}(x) = e^x[x^2 + 2kx + k(k-1)]$. Then

$$f^{(k+1)}(x) = \frac{d}{dx}\left[f^{(k)}(x)\right] = e^x(2x + 2k) + [x^2 + 2kx + k(k-1)]e^x$$

$$= e^x[x^2 + (2k+2)x + (k^2 + k)] = e^x[x^2 + 2(k+1)x + (k+1)k]$$

This shows that S_{k+1} is true.

3. Therefore, by mathematical induction, S_n is true for all n; that is, $f^{(n)}(x) = e^x[x^2 + 2nx + n(n-1)]$ for every positive integer n.

67. The Chain Rule says that $\dfrac{dy}{dx} = \dfrac{dy}{du}\dfrac{du}{dx}$, so

$$\frac{d^2y}{dx^2} = \frac{d}{dx}\left(\frac{dy}{dx}\right) = \frac{d}{dx}\left(\frac{dy}{du}\frac{du}{dx}\right) = \left[\frac{d}{dx}\left(\frac{dy}{du}\right)\right]\frac{du}{dx} + \frac{dy}{du}\frac{d}{dx}\left(\frac{du}{dx}\right) \quad \text{[Product Rule]}$$

$$= \left[\frac{d}{du}\left(\frac{dy}{du}\right)\frac{du}{dx}\right]\frac{du}{dx} + \frac{dy}{du}\frac{d^2u}{dx^2} = \frac{d^2y}{du^2}\left(\frac{du}{dx}\right)^2 + \frac{dy}{du}\frac{d^2u}{dx^2}$$

69. We will show that for each positive integer n, the nth derivative $f^{(n)}$ exists and equals one of $f, f', f'', f''', \ldots$, $f^{(p-1)}$. Since $f^{(p)} = f$, the first p derivatives of f are $f', f'', f''', \ldots, f^{(p-1)}$, and f. In particular, our statement is true for $n = 1$. Suppose that k is an integer, $k \geq 1$, for which f is k-times differentiable with $f^{(k)}$ in the set $S = \{f, f', f'', \ldots, f^{(p-1)}\}$. Since f is p-times differentiable, every member of S [including $f^{(k)}$] is differentiable, so $f^{(k+1)}$ exists and equals the derivative of some member of S. Thus, $f^{(k+1)}$ is in the set $\{f', f'', f''', \ldots, f^{(p)}\}$, which equals S since $f^{(p)} = f$. We have shown that the statement is true for $n = 1$ and that its truth for $n = k$ implies its truth for $n = k + 1$. By mathematical induction, the statement is true for all positive integers n.

3.8 Derivatives of Logarithmic Functions

1. The differentiation formula for logarithmic functions, $\dfrac{d}{dx}(\log_a x) = \dfrac{1}{x \ln a}$, is simplest when $a = e$ because $\ln e = 1$.

3. $f(\theta) = \ln(\cos \theta) \quad \Rightarrow \quad f'(\theta) = \dfrac{1}{\cos \theta}\dfrac{d}{d\theta}(\cos \theta) = \dfrac{-\sin \theta}{\cos \theta} = -\tan \theta$

5. $f(x) = \log_2(1 - 3x) \Rightarrow f'(x) = \dfrac{1}{(1 - 3x)\ln 2} \dfrac{d}{dx}(1 - 3x) = \dfrac{-3}{(1 - 3x)\ln 2}$ or $\dfrac{3}{(3x - 1)\ln 2}$

7. $f(x) = \sqrt[5]{\ln x} = (\ln x)^{1/5} \Rightarrow f'(x) = \frac{1}{5}(\ln x)^{-4/5}\dfrac{d}{dx}(\ln x) = \dfrac{1}{5(\ln x)^{4/5}} \cdot \dfrac{1}{x} = \dfrac{1}{5x\sqrt[5]{(\ln x)^4}}$

9. $f(x) = \sqrt{x}\,\ln x \Rightarrow f'(x) = \sqrt{x}\left(\dfrac{1}{x}\right) + (\ln x)\cdot\dfrac{1}{2\sqrt{x}} = \dfrac{1}{\sqrt{x}} + \dfrac{\ln x}{2\sqrt{x}} = \dfrac{2 + \ln x}{2\sqrt{x}}$

11. $F(t) = \ln\dfrac{(2t + 1)^3}{(3t - 1)^4} = \ln(2t + 1)^3 - \ln(3t - 1)^4 = 3\ln(2t + 1) - 4\ln(3t - 1) \Rightarrow$

$F'(t) = 3\cdot\dfrac{1}{2t + 1}\cdot 2 - 4\cdot\dfrac{1}{3t - 1}\cdot 3 = \dfrac{6}{2t + 1} - \dfrac{12}{3t - 1}$, or combined, $\dfrac{-6(t + 3)}{(2t + 1)(3t - 1)}$.

13. $g(x) = \ln\dfrac{a - x}{a + x} = \ln(a - x) - \ln(a + x) \Rightarrow$

$g'(x) = \dfrac{1}{a - x}(-1) - \dfrac{1}{a + x} = \dfrac{-(a + x) - (a - x)}{(a - x)(a + x)} = \dfrac{-2a}{a^2 - x^2}$

15. $f(u) = \dfrac{\ln u}{1 + \ln(2u)} \Rightarrow$

$f'(u) = \dfrac{[1 + \ln(2u)]\cdot\frac{1}{u} - \ln u\cdot\frac{1}{2u}\cdot 2}{[1 + \ln(2u)]^2} = \dfrac{\frac{1}{u}[1 + \ln(2u) - \ln u]}{[1 + \ln(2u)]^2}$

$= \dfrac{1 + (\ln 2 + \ln u) - \ln u}{u\,[1 + \ln(2u)]^2} = \dfrac{1 + \ln 2}{u\,[1 + \ln(2u)]^2}$

17. $y = \ln|2 - x - 5x^2| \Rightarrow y' = \dfrac{1}{2 - x - 5x^2}\cdot(-1 - 10x) = \dfrac{-10x - 1}{2 - x - 5x^2}$ or $\dfrac{10x + 1}{5x^2 + x - 2}$

19. $y = \ln(e^{-x} + xe^{-x}) = \ln(e^{-x}(1 + x)) = \ln(e^{-x}) + \ln(1 + x) = -x + \ln(1 + x) \Rightarrow$

$y' = -1 + \dfrac{1}{1 + x} = \dfrac{-1 - x + 1}{1 + x} = -\dfrac{x}{1 + x}$

21. $y = x\ln x \Rightarrow y' = x(1/x) + (\ln x)\cdot 1 = 1 + \ln x \Rightarrow y'' = 1/x$

23. $y = \log_{10} x \Rightarrow y' = \dfrac{1}{x\ln 10} = \dfrac{1}{\ln 10}\left(\dfrac{1}{x}\right) \Rightarrow y'' = \dfrac{1}{\ln 10}\left(-\dfrac{1}{x^2}\right) = -\dfrac{1}{x^2\ln 10}$

25. $f(x) = \dfrac{x}{1 - \ln(x - 1)} \Rightarrow$

$f'(x) = \dfrac{[1 - \ln(x - 1)]\cdot 1 - x\cdot\dfrac{-1}{x - 1}}{[1 - \ln(x - 1)]^2} = \dfrac{\dfrac{(x - 1)[1 - \ln(x - 1)] + x}{x - 1}}{[1 - \ln(x - 1)]^2} = \dfrac{x - 1 - (x - 1)\ln(x - 1) + x}{(x - 1)[1 - \ln(x - 1)]^2}$

$= \dfrac{2x - 1 - (x - 1)\ln(x - 1)}{(x - 1)[1 - \ln(x - 1)]^2}$

$\text{Dom}(f) = \{x \mid x - 1 > 0 \text{ and } 1 - \ln(x - 1) \neq 0\} = \{x \mid x > 1 \text{ and } \ln(x - 1) \neq 1\}$

$= \{x \mid x > 1 \text{ and } x - 1 \neq e^1\} = \{x \mid x > 1 \text{ and } x \neq 1 + e\} = (1, 1 + e)\cup(1 + e, \infty)$

27. $f(x) = x^2\ln(1 - x^2) \Rightarrow f'(x) = 2x\ln(1 - x^2) + \dfrac{x^2(-2x)}{1 - x^2} = 2x\ln(1 - x^2) - \dfrac{2x^3}{1 - x^2}$.

$\text{Dom}(f) = \{x \mid 1 - x^2 > 0\} = \{x \mid |x| < 1\} = (-1, 1)$.

29. $f(x) = \dfrac{x}{\ln x} \Rightarrow f'(x) = \dfrac{\ln x - x(1/x)}{(\ln x)^2} = \dfrac{\ln x - 1}{(\ln x)^2} \Rightarrow f'(e) = \dfrac{1 - 1}{1^2} = 0$

31. $y = f(x) = \ln \ln x$ $\Rightarrow$ $f'(x) = \dfrac{1}{\ln x}\left(\dfrac{1}{x}\right)$ $\Rightarrow$ $f'(e) = \dfrac{1}{e}$, so an equation of the tangent line at $(e, 0)$ is

$y - 0 = \dfrac{1}{e}(x - e)$, or $y = \dfrac{1}{e}x - 1$, or $x - ey = e$.

33. $f(x) = \sin x + \ln x$ $\Rightarrow$ $f'(x) = \cos x + 1/x$. This is reasonable,

because the graph shows that f increases when f' is positive, and

$f'(x) = 0$ when f has a horizontal tangent.

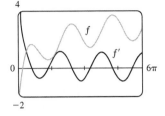

35. $y = (2x + 1)^5 (x^4 - 3)^6$ $\Rightarrow$ $\ln y = \ln\left((2x + 1)^5 (x^4 - 3)^6\right)$ $\Rightarrow$

$\ln y = 5\ln(2x + 1) + 6\ln(x^4 - 3)$ $\Rightarrow$ $\dfrac{1}{y}y' = 5\cdot\dfrac{1}{2x + 1}\cdot 2 + 6\cdot\dfrac{1}{x^4 - 3}\cdot 4x^3$ $\Rightarrow$

$y' = y\left(\dfrac{10}{2x + 1} + \dfrac{24x^3}{x^4 - 3}\right) = (2x + 1)^5 (x^4 - 3)^6 \left(\dfrac{10}{2x + 1} + \dfrac{24x^3}{x^4 - 3}\right)$.

[The answer could be simplified to $y' = 2(2x + 1)^4 (x^4 - 3)^5 (29x^4 + 12x^3 - 15)$, but this is unnecessary.]

37. $y = \dfrac{\sin^2 x \tan^4 x}{(x^2 + 1)^2}$ $\Rightarrow$ $\ln y = \ln\left(\sin^2 x \tan^4 x\right) - \ln\left(x^2 + 1\right)^2$ $\Rightarrow$

$\ln y = \ln(\sin x)^2 + \ln(\tan x)^4 - \ln\left(x^2 + 1\right)^2$ $\Rightarrow$ $\ln y = 2\ln|\sin x| + 4\ln|\tan x| - 2\ln\left(x^2 + 1\right)$ $\Rightarrow$

$\dfrac{1}{y}y' = 2\cdot\dfrac{1}{\sin x}\cdot\cos x + 4\cdot\dfrac{1}{\tan x}\cdot\sec^2 x - 2\cdot\dfrac{1}{x^2 + 1}\cdot 2x$ $\Rightarrow$

$y' = \dfrac{\sin^2 x \tan^4 x}{(x^2 + 1)^2}\left(2\cot x + \dfrac{4\sec^2 x}{\tan x} - \dfrac{4x}{x^2 + 1}\right)$

39. $y = x^x$ $\Rightarrow$ $\ln y = \ln x^x$ $\Rightarrow$ $\ln y = x\ln x$ $\Rightarrow$ $y'/y = x(1/x) + (\ln x)\cdot 1$ $\Rightarrow$
$y' = y(1 + \ln x)$ $\Rightarrow$ $y' = x^x(1 + \ln x)$

41. $y = x^{\sin x}$ $\Rightarrow$ $\ln y = \ln x^{\sin x}$ $\Rightarrow$ $\ln y = \sin x \ln x$ $\Rightarrow$ $\dfrac{y'}{y} = (\sin x)\cdot\dfrac{1}{x} + (\ln x)(\cos x)$ $\Rightarrow$

$y' = y\left(\dfrac{\sin x}{x} + \ln x \cos x\right)$ $\Rightarrow$ $y' = x^{\sin x}\left(\dfrac{\sin x}{x} + \ln x \cos x\right)$

43. $y = (\ln x)^x$ $\Rightarrow$ $\ln y = \ln(\ln x)^x$ $\Rightarrow$ $\ln y = x\ln\ln x$ $\Rightarrow$ $\dfrac{y'}{y} = x\cdot\dfrac{1}{\ln x}\cdot\dfrac{1}{x} + (\ln\ln x)\cdot 1$ $\Rightarrow$

$y' = y\left(\dfrac{x}{x\ln x} + \ln\ln x\right)$ $\Rightarrow$ $y' = (\ln x)^x\left(\dfrac{1}{\ln x} + \ln\ln x\right)$

45. $y = x^{e^x}$ $\Rightarrow$ $\ln y = e^x \ln x$ $\Rightarrow$ $\dfrac{y'}{y} = e^x\cdot\dfrac{1}{x} + (\ln x)\cdot e^x$ $\Rightarrow$ $y' = x^{e^x}e^x\left(\ln x + \dfrac{1}{x}\right)$

47. $y = \ln\left(x^2 + y^2\right)$ $\Rightarrow$ $y' = \dfrac{1}{x^2 + y^2}\dfrac{d}{dx}\left(x^2 + y^2\right)$ $\Rightarrow$ $y' = \dfrac{2x + 2yy'}{x^2 + y^2}$ $\Rightarrow$ $x^2 y' + y^2 y' = 2x + 2yy'$

$\Rightarrow$ $x^2 y' + y^2 y' - 2yy' = 2x$ $\Rightarrow$ $\left(x^2 + y^2 - 2y\right)y' = 2x$ $\Rightarrow$ $y' = \dfrac{2x}{x^2 + y^2 - 2y}$

49. $f(x) = \ln(x-1) \;\Rightarrow\; f'(x) = 1/(x-1) = (x-1)^{-1} \;\Rightarrow\; f''(x) = -(x-1)^{-2} \;\Rightarrow$
$f'''(x) = 2(x-1)^{-3} \;\Rightarrow\; f^{(4)}(x) = -2 \cdot 3(x-1)^{-4} \;\Rightarrow\; \cdots \;\Rightarrow$
$f^{(n)}(x) = (-1)^{n-1} \cdot 2 \cdot 3 \cdot 4 \cdot \; \cdots \; \cdot (n-1)(x-1)^{-n} = (-1)^{n-1} \dfrac{(n-1)!}{(x-1)^n}$

51. If $f(x) = \ln(1+x)$, then $f'(x) = \dfrac{1}{1+x}$, so $f'(0) = 1$.

Thus, $\lim\limits_{x \to 0} \dfrac{\ln(1+x)}{x} = \lim\limits_{x \to 0} \dfrac{f(x)}{x} = \lim\limits_{x \to 0} \dfrac{f(x) - f(0)}{x - 0} = f'(0) = 1$.

3.9 Hyperbolic Functions

1. (a) $\sinh 0 = \frac{1}{2}\left(e^0 - e^0\right) = 0$ \qquad\qquad (b) $\cosh 0 = \frac{1}{2}\left(e^0 + e^0\right) = \frac{1}{2}(1+1) = 1$

3. (a) $\sinh(\ln 2) = \dfrac{e^{\ln 2} - e^{-\ln 2}}{2} = \dfrac{e^{\ln 2} - \left(e^{\ln 2}\right)^{-1}}{2} = \dfrac{2 - 2^{-1}}{2} = \dfrac{2 - \frac{1}{2}}{2} = \dfrac{3}{4}$

  (b) $\sinh 2 = \frac{1}{2}\left(e^2 - e^{-2}\right) \approx 3.62686$

5. (a) $\operatorname{sech} 0 = \dfrac{1}{\cosh 0} = \dfrac{1}{1} = 1$ \qquad\qquad (b) $\cosh^{-1} 1 = 0$ because $\cosh 0 = 1$.

7. $\sinh(-x) = \frac{1}{2}\left[e^{-x} - e^{-(-x)}\right] = \frac{1}{2}\left(e^{-x} - e^x\right) = -\frac{1}{2}\left(e^x - e^{-x}\right) = -\sinh x$

9. $\cosh x + \sinh x = \frac{1}{2}\left(e^x + e^{-x}\right) + \frac{1}{2}\left(e^x - e^{-x}\right) = \frac{1}{2}(2e^x) = e^x$

11. $\sinh x \cosh y + \cosh x \sinh y = \left[\frac{1}{2}\left(e^x - e^{-x}\right)\right]\left[\frac{1}{2}\left(e^y + e^{-y}\right)\right] + \left[\frac{1}{2}\left(e^x + e^{-x}\right)\right]\left[\frac{1}{2}\left(e^y - e^{-y}\right)\right]$
$= \frac{1}{4}\left[\left(e^{x+y} + e^{x-y} - e^{-x+y} - e^{-x-y}\right) + \left(e^{x+y} - e^{x-y} + e^{-x+y} - e^{-x-y}\right)\right]$
$= \frac{1}{4}\left(2e^{x+y} - 2e^{-x-y}\right) = \frac{1}{2}\left[e^{x+y} - e^{-(x+y)}\right] = \sinh(x+y)$

13. Divide both sides of the identity $\cosh^2 x - \sinh^2 x = 1$ by $\sinh^2 x$:
$\dfrac{\cosh^2 x}{\sinh^2 x} - \dfrac{\sinh^2 x}{\sinh^2 x} = \dfrac{1}{\sinh^2 x} \;\Leftrightarrow\; \coth^2 x - 1 = \operatorname{csch}^2 x$.

15. Putting $y = x$ in the result from Exercise 11, we have
$\sinh 2x = \sinh(x+x) = \sinh x \cosh x + \cosh x \sinh x = 2\sinh x \cosh x$.

17. $\tanh(\ln x) = \dfrac{\sinh(\ln x)}{\cosh(\ln x)} = \dfrac{\left(e^{\ln x} - e^{-\ln x}\right)/2}{\left(e^{\ln x} + e^{-\ln x}\right)/2} = \dfrac{x - \left(e^{\ln x}\right)^{-1}}{x + \left(e^{\ln x}\right)^{-1}} = \dfrac{x - x^{-1}}{x + x^{-1}}$
$= \dfrac{x - 1/x}{x + 1/x} = \dfrac{(x^2 - 1)/x}{(x^2 + 1)/x} = \dfrac{x^2 - 1}{x^2 + 1}$

19. By Exercise 9, $(\cosh x + \sinh x)^n = (e^x)^n = e^{nx} = \cosh nx + \sinh nx$.

21. $\tanh x = \frac{4}{5} > 0$, so $x > 0$. $\coth x = 1/\tanh x = \frac{5}{4}$, $\operatorname{sech}^2 x = 1 - \tanh^2 x = 1 - \left(\frac{4}{5}\right)^2 = \frac{9}{25} \;\Rightarrow$
$\operatorname{sech} x = \frac{3}{5}$ (since $\operatorname{sech} x > 0$), $\cosh x = 1/\operatorname{sech} x = \frac{5}{3}$, $\sinh x = \tanh x \cosh x = \frac{4}{5} \cdot \frac{5}{3} = \frac{4}{3}$, and
$\operatorname{csch} x = 1/\sinh x = \frac{3}{4}$.

23. (a) $\lim\limits_{x \to \infty} \tanh x = \lim\limits_{x \to \infty} \dfrac{e^x - e^{-x}}{e^x + e^{-x}} \cdot \dfrac{e^{-x}}{e^{-x}} = \lim\limits_{x \to \infty} \dfrac{1 - e^{-2x}}{1 + e^{-2x}} = \dfrac{1 - 0}{1 + 0} = 1$

  (b) $\lim\limits_{x \to -\infty} \tanh x = \lim\limits_{x \to -\infty} \dfrac{e^x - e^{-x}}{e^x + e^{-x}} \cdot \dfrac{e^x}{e^x} = \lim\limits_{x \to -\infty} \dfrac{e^{2x} - 1}{e^{2x} + 1} = \dfrac{0 - 1}{0 + 1} = -1$

  (c) $\lim\limits_{x \to \infty} \sinh x = \lim\limits_{x \to \infty} \dfrac{e^x - e^{-x}}{2} = \infty$

(d) $\displaystyle\lim_{x\to-\infty} \sinh x = \lim_{x\to-\infty} \frac{e^x - e^{-x}}{2} = -\infty$

(e) $\displaystyle\lim_{x\to\infty} \operatorname{sech} x = \lim_{x\to\infty} \frac{2}{e^x + e^{-x}} = 0$

(f) $\displaystyle\lim_{x\to\infty} \coth x = \lim_{x\to\infty} \frac{e^x + e^{-x}}{e^x - e^{-x}} \cdot \frac{e^{-x}}{e^{-x}} = \lim_{x\to\infty} \frac{1 + e^{-2x}}{1 - e^{-2x}} = \frac{1+0}{1-0} = 1$ [*Or:* Use part (a)]

(g) $\displaystyle\lim_{x\to 0^+} \coth x = \lim_{x\to 0^+} \frac{\cosh x}{\sinh x} = \infty$, since $\sinh x \to 0$ through positive values and $\cosh x \to 1$.

(h) $\displaystyle\lim_{x\to 0^-} \coth x = \lim_{x\to 0^-} \frac{\cosh x}{\sinh x} = -\infty$, since $\sinh x \to 0$ through negative values and $\cosh x \to 1$.

(i) $\displaystyle\lim_{x\to-\infty} \operatorname{csch} x = \lim_{x\to-\infty} \frac{2}{e^x - e^{-x}} = 0$

25. Let $y = \sinh^{-1} x$. Then $\sinh y = x$ and, by Example 1(a), $\cosh^2 y - \sinh^2 y = 1 \;\Rightarrow\;$ [with $\cosh y > 0$]
$\cosh y = \sqrt{1 + \sinh^2 y} = \sqrt{1 + x^2}$. So by Exercise 9, $e^y = \sinh y + \cosh y = x + \sqrt{1 + x^2} \;\Rightarrow\;$
$y = \ln\!\left(x + \sqrt{1 + x^2}\,\right)$.

27. (a) Let $y = \tanh^{-1} x$. Then $x = \tanh y = \dfrac{\sinh y}{\cosh y} = \dfrac{(e^y - e^{-y})/2}{(e^y + e^{-y})/2} \cdot \dfrac{e^y}{e^y} = \dfrac{e^{2y} - 1}{e^{2y} + 1} \;\Rightarrow\;$

$xe^{2y} + x = e^{2y} - 1 \;\Rightarrow\; 1 + x = e^{2y} - xe^{2y} \;\Rightarrow\; 1 + x = e^{2y}(1 - x) \;\Rightarrow\;$

$e^{2y} = \dfrac{1 + x}{1 - x} \;\Rightarrow\; 2y = \ln\!\left(\dfrac{1 + x}{1 - x}\right) \;\Rightarrow\; y = \tfrac{1}{2}\ln\!\left(\dfrac{1 + x}{1 - x}\right)$.

(b) Let $y = \tanh^{-1} x$. Then $x = \tanh y$, so from Exercise 18 we have

$e^{2y} = \dfrac{1 + \tanh y}{1 - \tanh y} = \dfrac{1 + x}{1 - x} \;\Rightarrow\; 2y = \ln\!\left(\dfrac{1 + x}{1 - x}\right) \;\Rightarrow\; y = \tfrac{1}{2}\ln\!\left(\dfrac{1 + x}{1 - x}\right)$.

29. (a) Let $y = \cosh^{-1} x$. Then $\cosh y = x$ and $y \geq 0 \;\Rightarrow\; \sinh y \dfrac{dy}{dx} = 1 \;\Rightarrow\;$

$\dfrac{dy}{dx} = \dfrac{1}{\sinh y} = \dfrac{1}{\sqrt{\cosh^2 y - 1}} = \dfrac{1}{\sqrt{x^2 - 1}}$ (since $\sinh y \geq 0$ for $y \geq 0$). *Or:* Use Formula 4.

(b) Let $y = \tanh^{-1} x$. Then $\tanh y = x \;\Rightarrow\; \operatorname{sech}^2 y \dfrac{dy}{dx} = 1 \;\Rightarrow\; \dfrac{dy}{dx} = \dfrac{1}{\operatorname{sech}^2 y} = \dfrac{1}{1 - \tanh^2 y} = \dfrac{1}{1 - x^2}$.

Or: Use Formula 5.

(c) Let $y = \operatorname{csch}^{-1} x$. Then $\operatorname{csch} y = x \;\Rightarrow\; -\operatorname{csch} y \coth y \dfrac{dy}{dx} = 1 \;\Rightarrow\; \dfrac{dy}{dx} = -\dfrac{1}{\operatorname{csch} y \coth y}$.

By Exercise 13, $\coth y = \pm\sqrt{\operatorname{csch}^2 y + 1} = \pm\sqrt{x^2 + 1}$. If $x > 0$, then $\coth y > 0$, so $\coth y = \sqrt{x^2 + 1}$.
If $x < 0$, then $\coth y < 0$, so $\coth y = -\sqrt{x^2 + 1}$. In either case we have

$\dfrac{dy}{dx} = -\dfrac{1}{\operatorname{csch} y \coth y} = -\dfrac{1}{|x|\sqrt{x^2 + 1}}$.

(d) Let $y = \operatorname{sech}^{-1} x$. Then $\operatorname{sech} y = x \;\Rightarrow\; -\operatorname{sech} y \tanh y \dfrac{dy}{dx} = 1 \;\Rightarrow\;$

$\dfrac{dy}{dx} = -\dfrac{1}{\operatorname{sech} y \tanh y} = -\dfrac{1}{\operatorname{sech} y \sqrt{1 - \operatorname{sech}^2 y}} = -\dfrac{1}{x\sqrt{1 - x^2}}$. (Note that $y > 0$ and so $\tanh y > 0$.)

(e) Let $y = \coth^{-1} x$. Then $\coth y = x \;\Rightarrow\; -\operatorname{csch}^2 y \dfrac{dy}{dx} = 1 \;\Rightarrow\;$

$\dfrac{dy}{dx} = -\dfrac{1}{\operatorname{csch}^2 y} = \dfrac{1}{1 - \coth^2 y} = \dfrac{1}{1 - x^2}$ by Exercise 13.

31. $f(x) = x \cosh x \Rightarrow f'(x) = x(\cosh x)' + (\cosh x)(x)' = x \sinh x + \cosh x$

33. $h(x) = \sinh(x^2) \Rightarrow h'(x) = \cosh(x^2) \cdot 2x = 2x \cosh(x^2)$

35. $G(x) = \dfrac{1 - \cosh x}{1 + \cosh x} \Rightarrow$

$$G'(x) = \frac{(1 + \cosh x)(-\sinh x) - (1 - \cosh x)(\sinh x)}{(1 + \cosh x)^2}$$

$$= \frac{-\sinh x - \sinh x \cosh x - \sinh x + \sinh x \cosh x}{(1 + \cosh x)^2} = \frac{-2 \sinh x}{(1 + \cosh x)^2}$$

37. $h(t) = \coth \sqrt{1 + t^2} \Rightarrow h'(t) = -\operatorname{csch}^2 \sqrt{1 + t^2} \cdot \frac{1}{2}(1 + t^2)^{-1/2}(2t) = -\dfrac{t \operatorname{csch}^2 \sqrt{1 + t^2}}{\sqrt{1 + t^2}}$

39. $H(t) = \tanh(e^t) \Rightarrow H'(t) = \operatorname{sech}^2(e^t) \cdot e^t = e^t \operatorname{sech}^2(e^t)$

41. $y = e^{\cosh 3x} \Rightarrow y' = e^{\cosh 3x} \cdot \sinh 3x \cdot 3 = 3e^{\cosh 3x} \sinh 3x$

43. $y = \tanh^{-1} \sqrt{x} \Rightarrow y' = \dfrac{1}{1 - (\sqrt{x})^2} \cdot \frac{1}{2} x^{-1/2} = \dfrac{1}{2 \sqrt{x}(1 - x)}$

45. $y = x \sinh^{-1}(x/3) - \sqrt{9 + x^2} \Rightarrow$

$$y' = \sinh^{-1}\left(\frac{x}{3}\right) + x \frac{1/3}{\sqrt{1 + (x/3)^2}} - \frac{2x}{2\sqrt{9 + x^2}} = \sinh^{-1}\left(\frac{x}{3}\right) + \frac{x}{\sqrt{9 + x^2}} - \frac{x}{\sqrt{9 + x^2}} = \sinh^{-1}\left(\frac{x}{3}\right)$$

47. $y = \coth^{-1} \sqrt{x^2 + 1} \Rightarrow y' = \dfrac{1}{1 - (x^2 + 1)} \dfrac{2x}{2\sqrt{x^2 + 1}} = -\dfrac{1}{x \sqrt{x^2 + 1}}$

49. (a) $y = 20 \cosh(x/20) - 15 \Rightarrow y' = 20 \sinh(x/20) \cdot \frac{1}{20} = \sinh(x/20)$. Since the right pole is positioned at

$x = 7$, we have $y'(7) = \sinh \frac{7}{20} \approx 0.3572$.

(b) If α is the angle between the tangent line and the x-axis, then $\tan \alpha = $ slope of the line $= \sinh \frac{7}{20}$, so

$\alpha = \tan^{-1}\left(\sinh \frac{7}{20}\right) \approx 0.343$ rad $\approx 19.66°$. Thus, the angle between the line and the pole is

$\theta = 90° - \alpha \approx 70.34°$.

51. (a) $y = A \sinh mx + B \cosh mx \Rightarrow y' = mA \cosh mx + mB \sinh mx \Rightarrow$

$y'' = m^2 A \sinh mx + m^2 B \cosh mx = m^2(A \sinh mx + B \cosh mx) = m^2 y$

(b) From part (a), a solution of $y'' = 9y$ is $y(x) = A \sinh 3x + B \cosh 3x$. So

$-4 = y(0) = A \sinh 0 + B \cosh 0 = B$, so $B = -4$. Now $y'(x) = 3A \cosh 3x - 12 \sinh 3x \Rightarrow$

$6 = y'(0) = 3A \Rightarrow A = 2$, so $y = 2 \sinh 3x - 4 \cosh 3x$.

53. The tangent to $y = \cosh x$ has slope 1 when $y' = \sinh x = 1 \Rightarrow x = \sinh^{-1} 1 = \ln(1 + \sqrt{2})$, by Equation 3.

Since $\sinh x = 1$ and $y = \cosh x = \sqrt{1 + \sinh^2 x}$, we have $\cosh x = \sqrt{2}$. The point is $\left(\ln(1 + \sqrt{2}), \sqrt{2}\right)$.

55. If $ae^x + be^{-x} = \alpha \cosh(x + \beta)$ [or $\alpha \sinh(x + \beta)$], then

$ae^x + be^{-x} = \frac{\alpha}{2}\left(e^{x+\beta} \pm e^{-x-\beta}\right) = \frac{\alpha}{2}\left(e^x e^\beta \pm e^{-x} e^{-\beta}\right) = \left(\frac{\alpha}{2} e^\beta\right) e^x \pm \left(\frac{\alpha}{2} e^{-\beta}\right) e^{-x}$. Comparing coefficients

of e^x and e^{-x}, we have $a = \frac{\alpha}{2} e^\beta$ **(1)** and $b = \pm \frac{\alpha}{2} e^{-\beta}$ **(2)**. We need to find α and β. Dividing equation **(1)** by equation **(2)** gives us $\dfrac{a}{b} = \pm e^{2\beta}$ ⇒ **(∗)** $2\beta = \ln(\pm \frac{a}{b})$ ⇒ $\beta = \frac{1}{2}\ln(\pm \frac{a}{b})$. Solving equations **(1)** and **(2)**

for e^β gives us $e^\beta = \dfrac{2a}{\alpha}$ and $e^\beta = \pm \dfrac{\alpha}{2b}$, so $\dfrac{2a}{\alpha} = \pm \dfrac{\alpha}{2b}$ ⇒ $\alpha^2 = \pm 4ab$ ⇒ $\alpha = 2\sqrt{\pm ab}$.

(∗) If $\dfrac{a}{b} > 0$, we use the $+$ sign and obtain a cosh function, whereas if $\dfrac{a}{b} < 0$, we use the $-$ sign and obtain a

sinh function.

In summary, if a and b have the same sign, we have $ae^x + be^{-x} = 2\sqrt{ab}\cosh(x + \frac{1}{2}\ln\frac{a}{b})$, whereas, if a and b

have the opposite sign, then $ae^x + be^{-x} = 2\sqrt{-ab}\sinh(x + \frac{1}{2}\ln(-\frac{a}{b}))$.

3.10 Related Rates

1. $V = x^3$ ⇒ $\dfrac{dV}{dt} = \dfrac{dV}{dx}\dfrac{dx}{dt} = 3x^2 \dfrac{dx}{dt}$

3. $y = x^3 + 2x$ ⇒ $\dfrac{dy}{dt} = \dfrac{dy}{dx}\dfrac{dx}{dt} = (3x^2 + 2)(5) = 5(3x^2 + 2)$. When $x = 2$, $\dfrac{dy}{dt} = 5(14) = 70$.

5. $z^2 = x^2 + y^2$ ⇒ $2z\dfrac{dz}{dt} = 2x\dfrac{dx}{dt} + 2y\dfrac{dy}{dt}$ ⇒ $\dfrac{dz}{dt} = \dfrac{1}{z}\left(x\dfrac{dx}{dt} + y\dfrac{dy}{dt}\right)$. When $x = 5$ and $y = 12$,

$z^2 = 5^2 + 12^2$ ⇒ $z^2 = 169$ ⇒ $z = \pm 13$. For $\dfrac{dx}{dt} = 2$ and $\dfrac{dy}{dt} = 3$, $\dfrac{dz}{dt} = \dfrac{1}{\pm 13}(5 \cdot 2 + 12 \cdot 3) = \pm\dfrac{46}{13}$.

7. (a) Given: a plane flying horizontally at an altitude of 1 mi and a speed of 500 mi/h passes directly over a radar
station. If we let t be time (in hours) and x be the horizontal distance traveled by the plane (in mi), then we are
given that $dx/dt = 500$ mi/h.

(b) Unknown: the rate at which the distance from the plane to the station is
increasing when it is 2 mi from the station. If we let y be the distance from
the plane to the station, then we want to find dy/dt when $y = 2$ mi.

(c)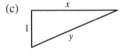

(d) By the Pythagorean Theorem, $y^2 = x^2 + 1$ ⇒ $2y(dy/dt) = 2x(dx/dt)$.

(e) $\dfrac{dy}{dt} = \dfrac{x}{y}\dfrac{dx}{dt} = \dfrac{x}{y}(500)$. Since $y^2 = x^2 + 1$, when $y = 2$, $x = \sqrt{3}$, so $\dfrac{dy}{dt} = \dfrac{\sqrt{3}}{2}(500) = 250\sqrt{3} \approx 433$ mi/h.

9. (a) Given: a man 6 ft tall walks away from a street light mounted on a 15-ft-tall pole at a rate of 5 ft/s. If we let t be
time (in s) and x be the distance from the pole to the man (in ft), then we are given that $dx/dt = 5$ ft/s.

(b) Unknown: the rate at which the tip of his shadow is moving when he is
40 ft from the pole. If we let y be the distance from the man to the tip of
his shadow (in ft), then we want to find $\dfrac{d}{dt}(x + y)$ when $x = 40$ ft.

(c)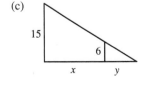

(d) By similar triangles, $\dfrac{15}{6} = \dfrac{x + y}{y}$ ⇒ $15y = 6x + 6y$ ⇒ $9y = 6x$ ⇒ $y = \frac{2}{3}x$.

(e) The tip of the shadow moves at a rate of $\dfrac{d}{dt}(x + y) = \dfrac{d}{dt}\left(x + \frac{2}{3}x\right) = \dfrac{5}{3}\dfrac{dx}{dt} = \frac{5}{3}(5) = \frac{25}{3}$ ft/s.

11.

We are given that $\dfrac{dx}{dt} = 60$ mi/h and $\dfrac{dy}{dt} = 25$ mi/h. $z^2 = x^2 + y^2 \Rightarrow$

$$2z \frac{dz}{dt} = 2x \frac{dx}{dt} + 2y \frac{dy}{dt} \Rightarrow z \frac{dz}{dt} = x \frac{dx}{dt} + y \frac{dy}{dt} \Rightarrow \frac{dz}{dt} = \frac{1}{z}\left(x \frac{dx}{dt} + y \frac{dy}{dt}\right).$$

After 2 hours, $x = 2(60) = 120$ and $y = 2(25) = 50 \Rightarrow z = \sqrt{120^2 + 50^2} = 130$,

so $\dfrac{dz}{dt} = \dfrac{1}{z}\left(x \dfrac{dx}{dt} + y \dfrac{dy}{dt}\right) = \dfrac{120(60) + 50(25)}{130} = 65$ mi/h.

13.

We are given that $\dfrac{dx}{dt} = 4$ ft/s and $\dfrac{dy}{dt} = 5$ ft/s. $z^2 = (x+y)^2 + 500^2 \Rightarrow$

$$2z \frac{dz}{dt} = 2(x+y)\left(\frac{dx}{dt} + \frac{dy}{dt}\right). \text{ 15 minutes after the woman starts, we have}$$

$x = (4 \text{ ft/s})(20 \text{ min})(60 \text{ s/min}) = 4800 \text{ ft}$ and $y = 5 \cdot 15 \cdot 60 = 4500 \Rightarrow$

$z = \sqrt{(4800 + 4500)^2 + 500^2} = \sqrt{86{,}740{,}000}$, so

$\dfrac{dz}{dt} = \dfrac{x+y}{z}\left(\dfrac{dx}{dt} + \dfrac{dy}{dt}\right) = \dfrac{4800 + 4500}{\sqrt{86{,}740{,}000}}(4+5) = \dfrac{837}{\sqrt{8674}} \approx 8.99$ ft/s.

15. $A = \frac{1}{2}bh$, where b is the base and h is the altitude. We are given that $\dfrac{dh}{dt} = 1$ cm/min and $\dfrac{dA}{dt} = 2$ cm^2/min.

Using the Product Rule, we have $\dfrac{dA}{dt} = \dfrac{1}{2}\left(b \dfrac{dh}{dt} + h \dfrac{db}{dt}\right)$. When $h = 10$ and $A = 100$, we have

$100 = \frac{1}{2}b(10) \Rightarrow \frac{1}{2}b = 10 \Rightarrow b = 20$, so $2 = \dfrac{1}{2}\left(20 \cdot 1 + 10 \dfrac{db}{dt}\right) \Rightarrow 4 = 20 + 10 \dfrac{db}{dt} \Rightarrow$

$\dfrac{db}{dt} = \dfrac{4-20}{10} = -1.6$ cm/min.

17.

We are given that $\dfrac{dx}{dt} = 35$ km/h and $\dfrac{dy}{dt} = 25$ km/h. $z^2 = (x+y)^2 + 100^2$

$\Rightarrow 2z \dfrac{dz}{dt} = 2(x+y)\left(\dfrac{dx}{dt} + \dfrac{dy}{dt}\right)$. At 4:00 P.M., $x = 4(35) = 140$ and

$y = 4(25) = 100 \Rightarrow z = \sqrt{(140+100)^2 + 100^2} = \sqrt{67{,}600} = 260$, so

$\dfrac{dz}{dt} = \dfrac{x+y}{z}\left(\dfrac{dx}{dt} + \dfrac{dy}{dt}\right) = \dfrac{140+100}{260}(35+25) = \dfrac{720}{13} \approx 55.4$ km/h.

19.

If $C =$ the rate at which water is pumped in, then $\dfrac{dV}{dt} = C - 10{,}000$, where

$V = \frac{1}{3}\pi r^2 h$ is the volume at time t. By similar triangles, $\dfrac{r}{2} = \dfrac{h}{6} \Rightarrow$

$r = \frac{1}{3}h \Rightarrow V = \frac{1}{3}\pi\left(\frac{1}{3}h\right)^2 h = \frac{\pi}{27}h^3 \Rightarrow \dfrac{dV}{dt} = \frac{\pi}{9}h^2 \dfrac{dh}{dt}$.

When $h = 200$ cm, $\dfrac{dh}{dt} = 20$ cm/min, so $C - 10{,}000 = \frac{\pi}{9}(200)^2(20) \Rightarrow$

$C = 10{,}000 + \frac{800{,}000}{9}\pi \approx 289{,}253$ cm^3/min.

21.

The figure is labeled in meters. The area A of a trapezoid is $\frac{1}{2}(\text{base}_1 + \text{base}_2)(\text{height})$, and the volume V of the 10-meter-long trough is $10A$. Thus, the volume of the trapezoid with height h is

$$V = (10)\tfrac{1}{2}\,[0.3 + (0.3 + 2a)]\,h. \text{ By similar triangles, } \frac{a}{h} = \frac{0.25}{0.5} = \frac{1}{2}, \text{ so}$$

$$2a = h \;\Rightarrow\; V = 5(0.6 + h)h = 3h + 5h^2. \text{ Now } \frac{dV}{dt} = \frac{dV}{dh}\frac{dh}{dt} \;\Rightarrow\; 0.2 = (3 + 10h)\frac{dh}{dt} \;\Rightarrow$$

$$\frac{dh}{dt} = \frac{0.2}{3 + 10h}. \text{ When } h = 0.3, \frac{dh}{dt} = \frac{0.2}{3 + 10(0.3)} = \frac{0.2}{6} \text{ m/min} = \frac{1}{30} \text{ m/min or } \frac{10}{3} \text{ cm/min}.$$

23.

We are given that $\frac{dV}{dt} = 30 \text{ ft}^3/\text{min}. \; V = \frac{1}{3}\pi r^2 h = \frac{1}{3}\pi\left(\frac{h}{2}\right)^2 h = \frac{\pi h^3}{12}$

$$\Rightarrow \frac{dV}{dt} = \frac{dV}{dh}\frac{dh}{dt} \;\Rightarrow\; 30 = \frac{\pi h^2}{4}\frac{dh}{dt} \;\Rightarrow\; \frac{dh}{dt} = \frac{120}{\pi h^2}. \text{ When}$$

$$h = 10 \text{ ft}, \frac{dh}{dt} = \frac{120}{10^2\pi} = \frac{6}{5\pi} \approx 0.38 \text{ ft/min}.$$

25.

$A = \frac{1}{2}bh$, but $b = 5$ m and $\sin\theta = \frac{h}{4} \;\Rightarrow\; h = 4\sin\theta$, so

$A = \frac{1}{2}(5)(4\sin\theta) = 10\sin\theta.$ We are given $\frac{d\theta}{dt} = 0.06 \text{ rad/s}$, so

$$\frac{dA}{dt} = \frac{dA}{d\theta}\frac{d\theta}{dt} = (10\cos\theta)(0.06) = 0.6\cos\theta. \text{ When } \theta = \tfrac{\pi}{3},$$

$$\frac{dA}{dt} = 0.6\left(\cos\tfrac{\pi}{3}\right) = (0.6)\left(\tfrac{1}{2}\right) = 0.3 \text{ m}^2/\text{s}.$$

27. Differentiating both sides of $PV = C$ with respect to t and using the Product Rule gives us $P\frac{dV}{dt} + V\frac{dP}{dt} = 0$

$\Rightarrow \frac{dV}{dt} = -\frac{V}{P}\frac{dP}{dt}.$ When $V = 600$, $P = 150$ and $\frac{dP}{dt} = 20$, so we have $\frac{dV}{dt} = -\frac{600}{150}(20) = -80.$ Thus, the volume is decreasing at a rate of 80 cm³/min.

29. With $R_1 = 80$ and $R_2 = 100$, $\frac{1}{R} = \frac{1}{R_1} + \frac{1}{R_2} = \frac{1}{80} + \frac{1}{100} = \frac{180}{8000} = \frac{9}{400}$, so $R = \frac{400}{9}$. Differentiating

$\frac{1}{R} = \frac{1}{R_1} + \frac{1}{R_2}$ with respect to t, we have $-\frac{1}{R^2}\frac{dR}{dt} = -\frac{1}{R_1^2}\frac{dR_1}{dt} - \frac{1}{R_2^2}\frac{dR_2}{dt} \;\Rightarrow$

$\frac{dR}{dt} = R^2\left(\frac{1}{R_1^2}\frac{dR_1}{dt} + \frac{1}{R_2^2}\frac{dR_2}{dt}\right).$ When $R_1 = 80$ and $R_2 = 100$,

$\frac{dR}{dt} = \frac{400^2}{9^2}\left[\frac{1}{80^2}(0.3) + \frac{1}{100^2}(0.2)\right] = \frac{107}{810} \approx 0.132 \ \Omega/\text{s}.$

31.

We are given that $\frac{dx}{dt} = 2 \text{ ft/s}. \; \sin\theta = \frac{x}{10} \;\Rightarrow\; x = 10\sin\theta \;\Rightarrow$

$\frac{dx}{dt} = 10\cos\theta\frac{d\theta}{dt}.$ When $\theta = \frac{\pi}{4}$, $2 = 10\cos\frac{\pi}{4}\frac{d\theta}{dt} \;\Rightarrow$

$\frac{d\theta}{dt} = \frac{2}{10(1/\sqrt{2})} = \frac{\sqrt{2}}{5} \text{ rad/s}.$

33. (a)

By the Pythagorean Theorem, $4000^2 + y^2 = \ell^2$. Differentiating with respect to t, we obtain $2y\dfrac{dy}{dt} = 2\ell\dfrac{d\ell}{dt}$. We know that $\dfrac{dy}{dt} = 600$ ft/s, so when

$y = 3000$ ft, $\ell = \sqrt{4000^2 + 3000^2} = \sqrt{25{,}000{,}000} = 5000$ ft and

$\dfrac{d\ell}{dt} = \dfrac{y}{\ell}\dfrac{dy}{dt} = \dfrac{3000}{5000}(600) = \dfrac{1800}{5} = 360$ ft/s.

(b) Here $\tan\theta = \dfrac{y}{4000}$ $\Rightarrow$ $\dfrac{d}{dt}(\tan\theta) = \dfrac{d}{dt}\left(\dfrac{y}{4000}\right)$ $\Rightarrow$ $\sec^2\theta\,\dfrac{d\theta}{dt} = \dfrac{1}{4000}\dfrac{dy}{dt}$ $\Rightarrow$ $\dfrac{d\theta}{dt} = \dfrac{\cos^2\theta}{4000}\dfrac{dy}{dt}$.

When $y = 3000$ ft, $\dfrac{dy}{dt} = 600$ ft/s, $\ell = 5000$ and $\cos\theta = \dfrac{4000}{\ell} = \dfrac{4000}{5000} = \dfrac{4}{5}$, so

$\dfrac{d\theta}{dt} = \dfrac{(4/5)^2}{4000}(600) = 0.096$ rad/s.

35.

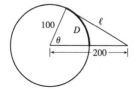

We are given that $\dfrac{dx}{dt} = 300$ km/h. By the Law of Cosines,

$y^2 = x^2 + 1^2 - 2(1)(x)\cos 120° = x^2 + 1 - 2x\left(-\tfrac{1}{2}\right) = x^2 + x + 1$, so

$2y\dfrac{dy}{dt} = 2x\dfrac{dx}{dt} + \dfrac{dx}{dt}$ $\Rightarrow$ $\dfrac{dy}{dt} = \dfrac{2x+1}{2y}\dfrac{dx}{dt}$. After 1 minute,

$x = \tfrac{300}{60} = 5$ km $\Rightarrow$ $y = \sqrt{5^2 + 5 + 1} = \sqrt{31}$ km $\Rightarrow$

$\dfrac{dy}{dt} = \dfrac{2(5)+1}{2\sqrt{31}}(300) = \dfrac{1650}{\sqrt{31}} \approx 296$ km/h.

37.

Let the distance between the runner and the friend be ℓ. Then by the Law of Cosines,

$\ell^2 = 200^2 + 100^2 - 2\cdot 200\cdot 100\cdot\cos\theta = 50{,}000 - 40{,}000\cos\theta$ $(\star)$.

Differentiating implicitly with respect to t, we obtain

$2\ell\dfrac{d\ell}{dt} = -40{,}000(-\sin\theta)\dfrac{d\theta}{dt}$. Now if D is the distance run when the angle is θ radians, then by the formula for the length of an arc on a circle, $s = r\theta$, we have $D = 100\theta$, so

$\theta = \dfrac{1}{100}D$ $\Rightarrow$ $\dfrac{d\theta}{dt} = \dfrac{1}{100}\dfrac{dD}{dt} = \dfrac{7}{100}$. To substitute into the expression for $\dfrac{d\ell}{dt}$, we must know $\sin\theta$ at the time

when $\ell = 200$, which we find from $(\star)$: $200^2 = 50{,}000 - 40{,}000\cos\theta$ $\Leftrightarrow$ $\cos\theta = \dfrac{1}{4}$ $\Rightarrow$

$\sin\theta = \sqrt{1 - \left(\tfrac{1}{4}\right)^2} = \dfrac{\sqrt{15}}{4}$. Substituting, we get $2(200)\dfrac{d\ell}{dt} = 40{,}000\dfrac{\sqrt{15}}{4}\left(\dfrac{7}{100}\right)$ $\Rightarrow$

$d\ell/dt = \tfrac{7\sqrt{15}}{4} \approx 6.78$ m/s. Whether the distance between them is increasing or decreasing depends on the direction in which the runner is running.

3.11 Linear Approximations and Differentials

1. As in Example 1, $T(0) = 185$, $T(10) = 172$, $T(20) = 160$, and

$$T'(20) \approx \frac{T(10) - T(20)}{10 - 20} = \frac{172 - 160}{-10} = -1.2 \,^\circ\text{F/min}.$$

$$T(30) \approx T(20) + T'(20)(30 - 20) \approx 160 - 1.2(10) = 148 \,^\circ\text{F}.$$

We would expect the temperature of the turkey to get closer to $75 \,^\circ\text{F}$

as time increases. Since the temperature decreased $13 \,^\circ\text{F}$ in the first

10 minutes and $12 \,^\circ\text{F}$ in the second 10 minutes, we can assume that the

slopes of the tangent line are increasing through negative values:

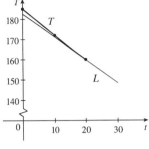

$-1.3, -1.2, \ldots$. Hence, the tangent lines are under the curve and $148 \,^\circ\text{F}$

is an underestimate. From the figure, we estimate the slope of the tangent line at $t = 20$ to be $\frac{184 - 147}{0 - 30} = -\frac{37}{30}$.

Then the linear approximation becomes $T(30) \approx T(20) + T'(20) \cdot 10 \approx 160 - \frac{37}{30}(10) = 147\frac{2}{3} \approx 147.7$.

3. Extend the tangent line at the point $(2030, 21)$ to the t-axis.

Answers will vary based on this approximation—we'll use

$t = 1900$ as our t-intercept. The linearization is then

$$P(t) \approx P(2030) + P'(2030)(t - 2030)$$

$$\approx 21 + \frac{21}{130}(t - 2030)$$

$$P(2040) = 21 + \frac{21}{130}(2040 - 2030) \approx 22.6\%$$

$$P(2050) = 21 + \frac{21}{130}(2050 - 2030) \approx 24.2\%$$

These predictions are probably too high since the tangent line lies above the graph at $t = 2030$.

5. $f(x) = x^3 \quad \Rightarrow \quad f'(x) = 3x^2$, so $f(1) = 1$ and $f'(1) = 3$. With $a = 1$, $L(x) = f(a) + f'(a)(x - a)$

becomes $L(x) = f(1) + f'(1)(x - 1) = 1 + 3(x - 1) = 3x - 2$.

7. $f(x) = \cos x \quad \Rightarrow \quad f'(x) = -\sin x$, so $f\left(\frac{\pi}{2}\right) = 0$ and $f'\left(\frac{\pi}{2}\right) = -1$. Thus,

$$L(x) = f\left(\frac{\pi}{2}\right) + f'\left(\frac{\pi}{2}\right)\left(x - \frac{\pi}{2}\right) = 0 - 1\left(x - \frac{\pi}{2}\right) = -x + \frac{\pi}{2}.$$

9. $f(x) = \sqrt{1 - x} \quad \Rightarrow \quad f'(x) = \dfrac{-1}{2\sqrt{1 - x}}$, so $f(0) = 1$ and

$f'(0) = -\frac{1}{2}$. Therefore,

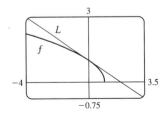

$$\sqrt{1 - x} = f(x) \approx f(0) + f'(0)(x - 0)$$

$$= 1 + \left(-\tfrac{1}{2}\right)(x - 0) = 1 - \tfrac{1}{2}x$$

So $\sqrt{0.9} = \sqrt{1 - 0.1} \approx 1 - \frac{1}{2}(0.1) = 0.95$ and

$\sqrt{0.99} = \sqrt{1 - 0.01} \approx 1 - \frac{1}{2}(0.01) = 0.995$.

11. $f(x) = \sqrt[3]{1-x} = (1-x)^{1/3} \;\Rightarrow\; f'(x) = -\frac{1}{3}(1-x)^{-2/3}$, so

$f(0) = 1$ and $f'(0) = -\frac{1}{3}$. Thus,

$f(x) \approx f(0) + f'(0)(x-0) = 1 - \frac{1}{3}x$. We need

$\sqrt[3]{1-x} - 0.1 < 1 - \frac{1}{3}x < \sqrt[3]{1-x} + 0.1$, which is true when

$-1.204 < x < 0.706$.

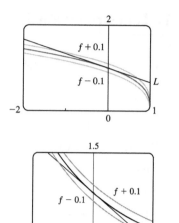

13. $f(x) = \dfrac{1}{(1+2x)^4} = (1+2x)^{-4} \;\Rightarrow\;$

$f'(x) = -4(1+2x)^{-5}(2) = \dfrac{-8}{(1+2x)^5}$, so $f(0) = 1$ and $f'(0) = -8$.

Thus, $f(x) \approx f(0) + f'(0)(x-0) = 1 + (-8)(x-0) = 1 - 8x$.

We need $1/(1+2x)^4 - 0.1 < 1 - 8x < 1/(1+2x)^4 + 0.1$, which is true

when $-0.045 < x < 0.055$.

15. If $y = f(x)$, then the differential dy is equal to $f'(x)\,dx$. $y = x^4 + 5x \;\Rightarrow\; dy = \left(4x^3 + 5\right) dx$.

17. $y = x \ln x \;\Rightarrow\; dy = \left(x \cdot \dfrac{1}{x} + \ln x \cdot 1\right) dx = (1 + \ln x)\,dx$

19. $y = \dfrac{u+1}{u-1} \;\Rightarrow\; dy = \dfrac{(u-1)(1) - (u+1)(1)}{(u-1)^2}\,du = \dfrac{-2}{(u-1)^2}\,du$

21. (a) $y = x^2 + 2x \;\Rightarrow\; dy = (2x+2)\,dx$

(b) When $x = 3$ and $dx = \frac{1}{2}$, $dy = [2(3)+2]\left(\frac{1}{2}\right) = 4$.

23. (a) $y = \sqrt{4+5x} \;\Rightarrow\; dy = \frac{1}{2}(4+5x)^{-1/2} \cdot 5\,dx = \dfrac{5}{2\sqrt{4+5x}}\,dx$

(b) When $x = 0$ and $dx = 0.04$, $dy = \dfrac{5}{2\sqrt{4}}(0.04) = \frac{5}{4} \cdot \frac{1}{25} = \frac{1}{20} = 0.05$.

25. (a) $y = \tan x \;\Rightarrow\; dy = \sec^2 x\,dx$

(b) When $x = \pi/4$ and $dx = -0.1$, $dy = [\sec(\pi/4)]^2 (-0.1) = \left(\sqrt{2}\right)^2 (-0.1) = -0.2$.

27. $y = x^2,\ x = 1,\ \Delta x = 0.5 \;\Rightarrow\;$

$\Delta y = (1.5)^2 - 1^2 = 1.25$.

$dy = 2x\,dx = 2(1)(0.5) = 1$

29. $y = 6 - x^2,\ x = -2,\ \Delta x = 0.4 \;\Rightarrow\;$

$\Delta y = \left(6 - (-1.6)^2\right) - \left(6 - (-2)^2\right) = 1.44$

$dy = -2x\,dx = -2(-2)(0.4) = 1.6$

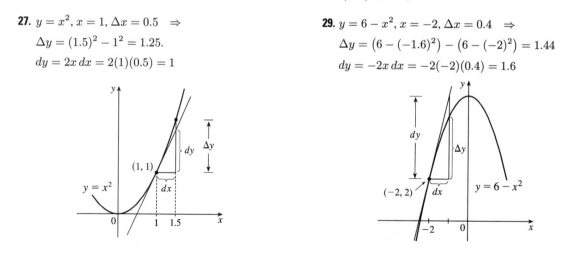

31. $y = f(x) = x^5 \Rightarrow dy = 5x^4\,dx$. When $x = 2$ and $dx = 0.001$, $dy = 5(2)^4(0.001) = 0.08$, so $(2.001)^5 = f(2.001) \approx f(2) + dy = 32 + 0.08 = 32.08$.

33. $y = f(x) = x^{2/3} \Rightarrow dy = \dfrac{2}{3\sqrt[3]{x}}\,dx$. When $x = 8$ and $dx = 0.06$, $dy = \dfrac{2}{3\sqrt[3]{8}}(0.06) = 0.02$, so $(8.06)^{2/3} = f(8.06) \approx f(8) + dy = 4 + 0.02 = 4.02$.

35. $y = f(x) = \tan x \Rightarrow dy = \sec^2 x\,dx$. When $x = 45°$ and $dx = -1°$,
$dy = \sec^2 45°(-\pi/180) = \left(\sqrt{2}\right)^2(-\pi/180) = -\pi/90$, so
$\tan 44° = f(44°) \approx f(45°) + dy = 1 - \pi/90 \approx 0.965$.

37. $y = f(x) = \sec x \Rightarrow f'(x) = \sec x \tan x$, so $f(0) = 1$ and $f'(0) = 1 \cdot 0 = 0$. The linear approximation of f at 0 is $f(0) + f'(0)(x - 0) = 1 + 0(x) = 1$. Since 0.08 is close to 0, approximating $\sec 0.08$ with 1 is reasonable.

39. $y = f(x) = \ln x \Rightarrow f'(x) = 1/x$, so $f(1) = 0$ and $f'(1) = 1$. The linear approximation of f at 1 is $f(1) + f'(1)(x - 1) = 0 + 1(x - 1) = x - 1$. Now $f(1.05) = \ln 1.05 \approx 1.05 - 1 = 0.05$, so the approximation is reasonable.

41. (a) If x is the edge length, then $V = x^3 \Rightarrow dV = 3x^2\,dx$. When $x = 30$ and $dx = 0.1$, $dV = 3(30)^2(0.1) = 270$, so the maximum possible error in computing the volume of the cube is about 270 cm^3. The relative error is calculated by dividing the change in V, ΔV, by V. We approximate ΔV with dV.

Relative error $= \dfrac{\Delta V}{V} \approx \dfrac{dV}{V} = \dfrac{3x^2\,dx}{x^3} = 3\dfrac{dx}{x} = 3\left(\dfrac{0.1}{30}\right) = 0.01$.

Percentage error = relative error $\times 100\% = 0.01 \times 100\% = 1\%$.

(b) $S = 6x^2 \Rightarrow dS = 12x\,dx$. When $x = 30$ and $dx = 0.1$, $dS = 12(30)(0.1) = 36$, so the maximum possible error in computing the surface area of the cube is about 36 cm^2.

Relative error $= \dfrac{\Delta S}{S} \approx \dfrac{dS}{S} = \dfrac{12x\,dx}{6x^2} = 2\dfrac{dx}{x} = 2\left(\dfrac{0.1}{30}\right) = 0.00\overline{6}$.

Percentage error = relative error $\times 100\% = 0.00\overline{6} \times 100\% = 0.\overline{6}\%$.

43. (a) For a sphere of radius r, the circumference is $C = 2\pi r$ and the surface area is $S = 4\pi r^2$, so $r = C/(2\pi) \Rightarrow$
$S = 4\pi(C/2\pi)^2 = C^2/\pi \Rightarrow dS = (2/\pi)C\,dC$. When $C = 84$ and $dC = 0.5$, $dS = \dfrac{2}{\pi}(84)(0.5) = \dfrac{84}{\pi}$,
so the maximum error is about $\dfrac{84}{\pi} \approx 27\text{ cm}^2$. Relative error $\approx \dfrac{dS}{S} = \dfrac{84/\pi}{84^2/\pi} = \dfrac{1}{84} \approx 0.012$

(b) $V = \dfrac{4}{3}\pi r^3 = \dfrac{4}{3}\pi\left(\dfrac{C}{2\pi}\right)^3 = \dfrac{C^3}{6\pi^2} \Rightarrow dV = \dfrac{1}{2\pi^2}C^2\,dC$. When $C = 84$ and $dC = 0.5$,
$dV = \dfrac{1}{2\pi^2}(84)^2(0.5) = \dfrac{1764}{\pi^2}$, so the maximum error is about $\dfrac{1764}{\pi^2} \approx 179\text{ cm}^3$. The relative error is
approximately $\dfrac{dV}{V} = \dfrac{1764/\pi^2}{(84)^3/(6\pi^2)} = \dfrac{1}{56} \approx 0.018$.

45. (a) $V = \pi r^2 h \Rightarrow \Delta V \approx dV = 2\pi rh\,dr = 2\pi rh\,\Delta r$

(b) The error is
$\Delta V - dV = [\pi(r + \Delta r)^2 h - \pi r^2 h] - 2\pi rh\,\Delta r = \pi r^2 h + 2\pi rh\,\Delta r + \pi(\Delta r)^2 h - \pi r^2 h - 2\pi rh\,\Delta r$
$= \pi(\Delta r)^2 h$

47. (a) $dc = \dfrac{dc}{dx}\, dx = 0\, dx = 0$

(b) $d(cu) = \dfrac{d}{dx}(cu)\, dx = c\,\dfrac{du}{dx}\, dx = c\, du$

(c) $d(u+v) = \dfrac{d}{dx}(u+v)\, dx = \left(\dfrac{du}{dx} + \dfrac{dv}{dx}\right) dx = \dfrac{du}{dx}\, dx + \dfrac{dv}{dx}\, dx = du + dv$

(d) $d(uv) = \dfrac{d}{dx}(uv)\, dx = \left(u\,\dfrac{dv}{dx} + v\,\dfrac{du}{dx}\right) dx = u\,\dfrac{dv}{dx}\, dx + v\,\dfrac{du}{dx}\, dx = u\, dv + v\, du$

(e) $d\left(\dfrac{u}{v}\right) = \dfrac{d}{dx}\left(\dfrac{u}{v}\right) dx = \dfrac{v\,\dfrac{du}{dx} - u\,\dfrac{dv}{dx}}{v^2}\, dx = \dfrac{v\,\dfrac{du}{dx}\, dx - u\,\dfrac{dv}{dx}\, dx}{v^2} = \dfrac{v\, du - u\, dv}{v^2}$

(f) $d\left(x^n\right) = \dfrac{d}{dx}\left(x^n\right) dx = nx^{n-1}\, dx$

49. (a) The graph shows that $f'(1) = 2$, so $L(x) = f(1) + f'(1)(x-1) = 5 + 2(x-1) = 2x + 3$.
$f(0.9) \approx L(0.9) = 4.8$ and $f(1.1) \approx L(1.1) = 5.2$.

(b) From the graph, we see that $f'(x)$ is positive and decreasing. This means that the slopes of the tangent lines are positive, but the tangents are becoming less steep. So the tangent lines lie *above* the curve. Thus, the estimates in part (a) are too large.

3 Review

———————————————— CONCEPT CHECK ————————————————

1. (a) The Power Rule: If n is any real number, then $\dfrac{d}{dx}(x^n) = nx^{n-1}$. The derivative of a variable base raised to a constant power is the power times the base raised to the power minus one.

(b) The Constant Multiple Rule: If c is a constant and f is a differentiable function, then $\dfrac{d}{dx}[cf(x)] = c\,\dfrac{d}{dx}f(x)$. The derivative of a constant times a function is the constant times the derivative of the function.

(c) The Sum Rule: If f and g are both differentiable, then $\dfrac{d}{dx}[f(x) + g(x)] = \dfrac{d}{dx}f(x) + \dfrac{d}{dx}g(x)$. The derivative of a sum of functions is the sum of the derivatives.

(d) The Difference Rule: If f and g are both differentiable, then $\dfrac{d}{dx}[f(x) - g(x)] = \dfrac{d}{dx}f(x) - \dfrac{d}{dx}g(x)$. The derivative of a difference of functions is the difference of the derivatives.

(e) The Product Rule: If f and g are both differentiable, then $\dfrac{d}{dx}[f(x)g(x)] = f(x)\,\dfrac{d}{dx}g(x) + g(x)\,\dfrac{d}{dx}f(x)$. The derivative of a product of two functions is the first function times the derivative of the second function plus the second function times the derivative of the first function.

(f) The Quotient Rule: If f and g are both differentiable, then $\dfrac{d}{dx}\left[\dfrac{f(x)}{g(x)}\right] = \dfrac{g(x)\,\dfrac{d}{dx}f(x) - f(x)\,\dfrac{d}{dx}g(x)}{[g(x)]^2}$. The derivative of a quotient of functions is the denominator times the derivative of the numerator minus the numerator times the derivative of the denominator, all divided by the square of the denominator.

(g) The Chain Rule: If f and g are both differentiable and $F = f \circ g$ is the composite function defined by $F(x) = f(g(x))$, then F is differentiable and F' is given by the product $F'(x) = f'(g(x))g'(x)$. The derivative of a composite function is the derivative of the outer function evaluated at the inner function times the derivative of the inner function.

2. (a) $y = x^n \implies y' = nx^{n-1}$

(b) $y = e^x \implies y' = e^x$

(c) $y = a^x \implies y' = a^x \ln a$

(d) $y = \ln x \implies y' = 1/x$

(e) $y = \log_a x \implies y' = 1/(x \ln a)$

(f) $y = \sin x \implies y' = \cos x$

(g) $y = \cos x \implies y' = -\sin x$

(h) $y = \tan x \implies y' = \sec^2 x$

(i) $y = \csc x \implies y' = -\csc x \cot x$

(j) $y = \sec x \implies y' = \sec x \tan x$

(k) $y = \cot x \implies y' = -\csc^2 x$

(l) $y = \sin^{-1} x \implies y' = 1/\sqrt{1 - x^2}$

(m) $y = \cos^{-1} x \implies y' = -1/\sqrt{1 - x^2}$

(n) $y = \tan^{-1} x \implies y' = 1/(1 + x^2)$

(o) $y = \sinh x \implies y' = \cosh x$

(p) $y = \cosh x \implies y' = \sinh x$

(q) $y = \tanh x \implies y' = \operatorname{sech}^2 x$

(r) $y = \sinh^{-1} x \implies y' = 1/\sqrt{1 + x^2}$

(s) $y = \cosh^{-1} x \implies y' = 1/\sqrt{x^2 - 1}$

(t) $y = \tanh^{-1} x \implies y' = 1/(1 - x^2)$

3. (a) e is the number such that $\displaystyle\lim_{h \to 0} \frac{e^h - 1}{h} = 1$.

(b) $e = \displaystyle\lim_{x \to 0} (1 + x)^{1/x}$

(c) The differentiation formula for $y = a^x$ $[y' = a^x \ln a]$ is simplest when $a = e$ because $\ln e = 1$.

(d) The differentiation formula for $y = \log_a x$ $[y' = 1/(x \ln a)]$ is simplest when $a = e$ because $\ln e = 1$.

4. (a) Implicit differentiation consists of differentiating both sides of an equation involving x and y with respect to x, and then solving the resulting equation for y'.

(b) Logarithmic differentiation consists of taking natural logarithms of both sides of an equation $y = f(x)$, simplifying, differentiating implicitly with respect to x, and then solving the resulting equation for y'.

5. The second derivative of a function f is the rate of change of the first derivative f'. The third derivative is the derivative (rate of change) of the second derivative. If f is the position function of an object, f' is its velocity function, f'' is its acceleration function, and f''' is its jerk function.

6. (a) The linearization L of f at $x = a$ is $L(x) = f(a) + f'(a)(x - a)$.

(b) If $y = f(x)$, then the differential dy is given by $dy = f'(x)\, dx$.

(c) See Figure 6 in Section 3.11.

————————————————— TRUE-FALSE QUIZ —————————————————

1. True. This is the Sum Rule.

3. True. This is the Chain Rule.

5. False. $\dfrac{d}{dx} f(\sqrt{x}) = \dfrac{f'(\sqrt{x})}{2\sqrt{x}}$ by the Chain Rule.

7. False. $\dfrac{d}{dx} 10^x = 10^x \ln 10$

9. True. $\dfrac{d}{dx} \left(\tan^2 x\right) = 2 \tan x \sec^2 x$, and $\dfrac{d}{dx} \left(\sec^2 x\right) = 2 \sec x \left(\sec x \tan x\right) = 2 \tan x \sec^2 x$.

11. True. $g(x) = x^5 \implies g'(x) = 5x^4 \implies g'(2) = 5(2)^4 = 80$, and by the definition of the derivative,

$$\lim_{x \to 2} \frac{g(x) - g(2)}{x - 2} = g'(2) = 80.$$

13. False. A tangent line to the parabola $y = x^2$ has slope $dy/dx = 2x$, so at $(-2, 4)$ the slope of the tangent is
$2(-2) = -4$ and an equation of the tangent line is $y - 4 = -4(x + 2)$. [The given equation,
$y - 4 = 2x(x + 2)$, is not even linear!]

―――――――――――――― EXERCISES ――――――――――――――

1. $y = (x^4 - 3x^2 + 5)^3$ ⟹

$y' = 3(x^4 - 3x^2 + 5)^2 \dfrac{d}{dx}(x^4 - 3x^2 + 5) = 3(x^4 - 3x^2 + 5)^2(4x^3 - 6x) = 6x(x^4 - 3x^2 + 5)^2(2x^2 - 3)$

3. $y = \sqrt{x} + \dfrac{1}{\sqrt[3]{x^4}} = x^{1/2} + x^{-4/3}$ ⟹ $y' = \frac{1}{2}x^{-1/2} - \frac{4}{3}x^{-7/3} = \dfrac{1}{2\sqrt{x}} - \dfrac{4}{3\sqrt[3]{x^7}}$

5. $y = 2x\sqrt{x^2 + 1}$ ⟹

$y' = 2x \cdot \frac{1}{2}(x^2 + 1)^{-1/2}(2x) + \sqrt{x^2 + 1}\,(2) = \dfrac{2x^2}{\sqrt{x^2 + 1}} + 2\sqrt{x^2 + 1} = \dfrac{2x^2 + 2(x^2 + 1)}{\sqrt{x^2 + 1}} = \dfrac{2(2x^2 + 1)}{\sqrt{x^2 + 1}}$

7. $y = e^{\sin 2\theta}$ ⟹ $y' = e^{\sin 2\theta}\dfrac{d}{d\theta}(\sin 2\theta) = e^{\sin 2\theta}(\cos 2\theta)(2) = 2\cos 2\theta\, e^{\sin 2\theta}$

9. $y = \dfrac{t}{1 - t^2}$ ⟹ $y' = \dfrac{(1 - t^2)(1) - t(-2t)}{(1 - t^2)^2} = \dfrac{1 - t^2 + 2t^2}{(1 - t^2)^2} = \dfrac{t^2 + 1}{(1 - t^2)^2}$

11. $y = xe^{-1/x}$ ⟹ $y' = xe^{-1/x}(1/x^2) + e^{-1/x} \cdot 1 = e^{-1/x}(1/x + 1)$

13. $y = \tan\sqrt{1 - x}$ ⟹ $y' = (\sec^2\sqrt{1 - x})\left(\dfrac{1}{2\sqrt{1 - x}}\right)(-1) = -\dfrac{\sec^2\sqrt{1 - x}}{2\sqrt{1 - x}}$

15. $\dfrac{d}{dx}(xy^4 + x^2y) = \dfrac{d}{dx}(x + 3y)$ ⟹ $x \cdot 4y^3 y' + y^4 \cdot 1 + x^2 \cdot y' + y \cdot 2x = 1 + 3y'$ ⟹

$y'(4xy^3 + x^2 - 3) = 1 - y^4 - 2xy$ ⟹ $y' = \dfrac{1 - y^4 - 2xy}{4xy^3 + x^2 - 3}$

17. $y = \dfrac{\sec 2\theta}{1 + \tan 2\theta}$ ⟹

$y' = \dfrac{(1 + \tan 2\theta)(\sec 2\theta \tan 2\theta \cdot 2) - (\sec 2\theta)(\sec^2 2\theta \cdot 2)}{(1 + \tan 2\theta)^2} = \dfrac{2\sec 2\theta\,[(1 + \tan 2\theta)\tan 2\theta - \sec^2 2\theta]}{(1 + \tan 2\theta)^2}$

$= \dfrac{2\sec 2\theta\,(\tan 2\theta + \tan^2 2\theta - \sec^2 2\theta)}{(1 + \tan 2\theta)^2} = \dfrac{2\sec 2\theta\,(\tan 2\theta - 1)}{(1 + \tan 2\theta)^2}$ $[1 + \tan^2 x = \sec^2 x]$

19. $y = e^{cx}(c\sin x - \cos x)$ ⟹

$y' = e^{cx}(c\cos x + \sin x) + ce^{cx}(c\sin x - \cos x)$

$= e^{cx}(c^2\sin x - c\cos x + c\cos x + \sin x) = e^{cx}(c^2\sin x + \sin x) = e^{cx}\sin x\,(c^2 + 1)$

21. $y = e^{e^x}$ ⟹ $y' = e^{e^x}\dfrac{d}{dx}(e^x) = e^{e^x}e^x = e^{x + e^x}$

23. $y = (1 - x^{-1})^{-1}$ ⟹

$y' = -1(1 - x^{-1})^{-2}[-(-1x^{-2})] = -(1 - 1/x)^{-2}x^{-2} = -((x - 1)/x)^{-2}x^{-2} = -(x - 1)^{-2}$

25. $\sin(xy) = x^2 - y$ $\Rightarrow$ $\cos(xy)(xy' + y \cdot 1) = 2x - y'$ $\Rightarrow$ $x\cos(xy)y' + y' = 2x - y\cos(xy)$ $\Rightarrow$

$y'[x\cos(xy) + 1] = 2x - y\cos(xy)$ $\Rightarrow$ $y' = \dfrac{2x - y\cos(xy)}{x\cos(xy) + 1}$

27. $y = \log_5(1 + 2x)$ $\Rightarrow$ $y' = \dfrac{1}{(1 + 2x)\ln 5}\dfrac{d}{dx}(1 + 2x) = \dfrac{2}{(1 + 2x)\ln 5}$

29. $y = \ln\sin x - \frac{1}{2}\sin^2 x$ $\Rightarrow$ $y' = \dfrac{1}{\sin x}\cdot\cos x - \frac{1}{2}\cdot 2\sin x\cdot\cos x = \cot x - \sin x\cos x$

31. $y = x\tan^{-1}(4x)$ $\Rightarrow$ $y' = x\cdot\dfrac{1}{1 + (4x)^2}\cdot 4 + \tan^{-1}(4x)\cdot 1 = \dfrac{4x}{1 + 16x^2} + \tan^{-1}(4x)$

33. $y = \ln|\sec 5x + \tan 5x|$ $\Rightarrow$

$y' = \dfrac{1}{\sec 5x + \tan 5x}(\sec 5x\tan 5x\cdot 5 + \sec^2 5x\cdot 5) = \dfrac{5\sec 5x(\tan 5x + \sec 5x)}{\sec 5x + \tan 5x} = 5\sec 5x$

35. $y = \cot(3x^2 + 5)$ $\Rightarrow$ $y' = -\csc^2(3x^2 + 5)(6x) = -6x\csc^2(3x^2 + 5)$

37. $y = \sin(\tan\sqrt{1 + x^3})$ $\Rightarrow$ $y' = \cos(\tan\sqrt{1 + x^3})(\sec^2\sqrt{1 + x^3})[3x^2/(2\sqrt{1 + x^3})]$

39. $y = \tan^2(\sin\theta) = [\tan(\sin\theta)]^2$ $\Rightarrow$ $y' = 2[\tan(\sin\theta)]\cdot\sec^2(\sin\theta)\cdot\cos\theta$

41. $y = \dfrac{\sqrt{x + 1}(2 - x)^5}{(x + 3)^7}$ $\Rightarrow$ $\ln y = \frac{1}{2}\ln(x + 1) + 5\ln(2 - x) - 7\ln(x + 3)$ $\Rightarrow$

$\dfrac{y'}{y} = \dfrac{1}{2(x + 1)} + \dfrac{-5}{2 - x} - \dfrac{7}{x + 3}$ $\Rightarrow$ $y' = \dfrac{\sqrt{x + 1}(2 - x)^5}{(x + 3)^7}\left[\dfrac{1}{2(x + 1)} - \dfrac{5}{2 - x} - \dfrac{7}{x + 3}\right]$ or

$y' = \dfrac{(2 - x)^4(3x^2 - 55x - 52)}{2\sqrt{x + 1}(x + 3)^8}$.

43. $y = x\sinh(x^2)$ $\Rightarrow$ $y' = x\cosh(x^2)\cdot 2x + \sinh(x^2)\cdot 1 = 2x^2\cosh(x^2) + \sinh(x^2)$

45. $y = \ln(\cosh 3x)$ $\Rightarrow$ $y' = (1/\cosh 3x)(\sinh 3x)(3) = 3\tanh 3x$

47. $y = \cosh^{-1}(\sinh x)$ $\Rightarrow$ $y' = \dfrac{1}{\sqrt{(\sinh x)^2 - 1}}\cdot\cosh x = \dfrac{\cosh x}{\sqrt{\sinh^2 x - 1}}$

49. $f(t) = \sqrt{4t + 1}$ $\Rightarrow$ $f'(t) = \frac{1}{2}(4t + 1)^{-1/2}\cdot 4 = 2(4t + 1)^{-1/2}$ $\Rightarrow$

$f''(t) = 2(-\frac{1}{2})(4t + 1)^{-3/2}\cdot 4 = -4/(4t + 1)^{3/2}$, so $f''(2) = -4/9^{3/2} = -\frac{4}{27}$.

51. $x^6 + y^6 = 1$ $\Rightarrow$ $6x^5 + 6y^5y' = 0$ $\Rightarrow$ $y' = -x^5/y^5$ $\Rightarrow$

$y'' = -\dfrac{y^5(5x^4) - x^5(5y^4y')}{(y^5)^2} = -\dfrac{5x^4y^4[y - x(-x^5/y^5)]}{y^{10}} = -\dfrac{5x^4[(y^6 + x^6)/y^5]}{y^6} = -\dfrac{5x^4}{y^{11}}$

53. We first show it is true for $n = 1$: $f(x) = xe^x$ $\Rightarrow$ $f'(x) = xe^x + e^x = (x + 1)e^x$. We now assume it is true

for $n = k$: $f^{(k)}(x) = (x + k)e^x$. With this assumption, we must show it is true for $n = k + 1$:

$f^{(k+1)}(x) = \dfrac{d}{dx}\left[f^{(k)}(x)\right] = \dfrac{d}{dx}[(x + k)e^x] = (x + k)e^x + e^x = [(x + k) + 1]e^x = [x + (k + 1)]e^x$.

Therefore, $f^{(n)}(x) = (x + n)e^x$ by mathematical induction.

55. $y = 4\sin^2 x \Rightarrow y' = 4 \cdot 2\sin x \cos x$. At $\left(\frac{\pi}{6}, 1\right)$, $y' = 8 \cdot \frac{1}{2} \cdot \frac{\sqrt{3}}{2} = 2\sqrt{3}$, so an equation of the tangent line is

$y - 1 = 2\sqrt{3}\left(x - \frac{\pi}{6}\right)$, or $y = 2\sqrt{3}\,x + 1 - \pi\sqrt{3}/3$.

57. $y = \sqrt{1 + 4\sin x} \Rightarrow y' = \frac{1}{2}(1 + 4\sin x)^{-1/2} \cdot 4\cos x = \dfrac{2\cos x}{\sqrt{1 + 4\sin x}}$. At $(0, 1)$, $y' = \dfrac{2}{\sqrt{1}} = 2$, so an

equation of the tangent line is $y - 1 = 2(x - 0)$, or $y = 2x + 1$.

59. $y = (2 + x)e^{-x} \Rightarrow y' = (2 + x)(-e^{-x}) + e^{-x} \cdot 1 = e^{-x}[-(2 + x) + 1] = e^{-x}(-x - 1)$. At $(0, 2)$,

$y' = 1(-1) = -1$, so an equation of the tangent line is $y - 2 = -1(x - 0)$, or $y = -x + 2$.

61. (a) $f(x) = x\sqrt{5 - x} \Rightarrow$

$$f'(x) = x\left[\frac{1}{2}(5 - x)^{-1/2}(-1)\right] + \sqrt{5 - x} = \frac{-x}{2\sqrt{5 - x}} + \sqrt{5 - x} \cdot \frac{2\sqrt{5 - x}}{2\sqrt{5 - x}}$$

$$= \frac{-x}{2\sqrt{5 - x}} + \frac{2(5 - x)}{2\sqrt{5 - x}} = \frac{-x + 10 - 2x}{2\sqrt{5 - x}} = \frac{10 - 3x}{2\sqrt{5 - x}}$$

(b) At $(1, 2)$: $f'(1) = \frac{7}{4}$. So an equation of the tangent line is $y - 2 = \frac{7}{4}(x - 1)$ or $y = \frac{7}{4}x + \frac{1}{4}$.

At $(4, 4)$: $f'(4) = -\frac{2}{2} = -1$. So an equation of the tangent line is $y - 4 = -1(x - 4)$ or $y = -x + 8$.

(c)

(d)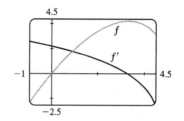

The graphs look reasonable, since f' is positive where f has tangents with positive slope, and f' is negative where f has tangents with negative slope.

63. $y = \sin x + \cos x \Rightarrow y' = \cos x - \sin x = 0 \Leftrightarrow \cos x = \sin x$ and $0 \leq x \leq 2\pi \Leftrightarrow x = \frac{\pi}{4}$ or $\frac{5\pi}{4}$, so the

points are $\left(\frac{\pi}{4}, \sqrt{2}\right)$ and $\left(\frac{5\pi}{4}, -\sqrt{2}\right)$.

65. $f(x) = (x - a)(x - b)(x - c) \Rightarrow f'(x) = (x - b)(x - c) + (x - a)(x - c) + (x - a)(x - b)$. So

$\dfrac{f'(x)}{f(x)} = \dfrac{(x - b)(x - c) + (x - a)(x - c) + (x - a)(x - b)}{(x - a)(x - b)(x - c)} = \dfrac{1}{x - a} + \dfrac{1}{x - b} + \dfrac{1}{x - c}$.

Or: $f(x) = (x - a)(x - b)(x - c) \Rightarrow \ln|f(x)| = \ln|x - a| + \ln|x - b| + \ln|x - c| \Rightarrow$

$\dfrac{f'(x)}{f(x)} = \dfrac{1}{x - a} + \dfrac{1}{x - b} + \dfrac{1}{x - c}$

67. (a) $h(x) = f(x)g(x) \Rightarrow h'(x) = f(x)g'(x) + g(x)f'(x) \Rightarrow$

$h'(2) = f(2)g'(2) + g(2)f'(2) = (3)(4) + (5)(-2) = 12 - 10 = 2$

(b) $F(x) = f(g(x)) \Rightarrow F'(x) = f'(g(x))g'(x) \Rightarrow F'(2) = f'(g(2))g'(2) = f'(5)(4) = 11 \cdot 4 = 44$

69. $f(x) = x^2 g(x) \Rightarrow f'(x) = x^2 g'(x) + g(x)(2x) = x\left[xg'(x) + 2g(x)\right]$

71. $f(x) = [g(x)]^2 \Rightarrow f'(x) = 2[g(x)]^1 \cdot g'(x) = 2g(x)g'(x)$

73. $f(x) = g(e^x) \quad \Rightarrow \quad f'(x) = g'(e^x)e^x$

75. $f(x) = \ln |g(x)| \quad \Rightarrow \quad f'(x) = \dfrac{1}{g(x)} g'(x) = \dfrac{g'(x)}{g(x)}$

77. $h(x) = \dfrac{f(x)g(x)}{f(x) + g(x)} \quad \Rightarrow$

$$h'(x) = \frac{[f(x) + g(x)]\left[f(x)g'(x) + g(x)f'(x)\right] - f(x)g(x)\left[f'(x) + g'(x)\right]}{[f(x) + g(x)]^2}$$

$$= \frac{[f(x)]^2 \, g'(x) + f(x)g(x)f'(x) + f(x)g(x)g'(x) + [g(x)]^2 \, f'(x) - f(x)g(x)f'(x) - f(x)g(x)g'(x)}{[f(x) + g(x)]^2}$$

$$= \frac{f'(x)[g(x)]^2 + g'(x)[f(x)]^2}{[f(x) + g(x)]^2}$$

79. Using the Chain Rule repeatedly, $h(x) = f(g(\sin 4x)) \quad \Rightarrow$

$$h'(x) = f'(g(\sin 4x)) \cdot \frac{d}{dx}\left(g(\sin 4x)\right) = f'(g(\sin 4x)) \cdot g'(\sin 4x) \cdot \frac{d}{dx}\left(\sin 4x\right)$$

$$= f'(g(\sin 4x))g'(\sin 4x)(\cos 4x)(4)$$

81. $y = [\ln(x + 4)]^2 \quad \Rightarrow \quad y' = 2[\ln(x + 4)]^1 \cdot \dfrac{1}{x + 4} \cdot 1 = 2\,\dfrac{\ln(x + 4)}{x + 4}$ and $y' = 0 \quad \Leftrightarrow \quad \ln(x + 4) = 0 \quad \Leftrightarrow$

$x + 4 = e^0 \quad \Rightarrow \quad x + 4 = 1 \quad \Leftrightarrow \quad x = -3$, so the tangent is horizontal at the point $(-3, 0)$.

83. $y = f(x) = ax^2 + bx + c \quad \Rightarrow \quad f'(x) = 2ax + b$. We know that $f'(-1) = 6$ and $f'(5) = -2$, so $-2a + b = 6$

and $10a + b = -2$. Subtracting the first equation from the second gives $12a = -8 \quad \Rightarrow \quad a = -\frac{2}{3}$. Substituting

$-\frac{2}{3}$ for a in the first equation gives $b = \frac{14}{3}$. Now $f(1) = 4 \quad \Rightarrow \quad 4 = a + b + c$, so $c = 4 + \frac{2}{3} - \frac{14}{3} = 0$ and

hence, $f(x) = -\frac{2}{3}x^2 + \frac{14}{3}x$.

85. $s(t) = Ae^{-ct}\cos(\omega t + \delta) \quad \Rightarrow$

$$v(t) = s'(t) = A\{e^{-ct}\left[-\omega\sin(\omega t + \delta)\right] + \cos(\omega t + \delta)(-ce^{-ct})\}$$
$$= -Ae^{-ct}\left[\omega\sin(\omega t + \delta) + c\cos(\omega t + \delta)\right] \quad \Rightarrow$$

$$a(t) = v'(t) = -A\{e^{-ct}[\omega^2\cos(\omega t + \delta) - c\omega\sin(\omega t + \delta)] + [\omega\sin(\omega t + \delta) + c\cos(\omega t + \delta)](-ce^{-ct})\}$$
$$= -Ae^{-ct}[\omega^2\cos(\omega t + \delta) - c\omega\sin(\omega t + \delta) - c\omega\sin(\omega t + \delta) - c^2\cos(\omega t + \delta)]$$
$$= -Ae^{-ct}[(\omega^2 - c^2)\cos(\omega t + \delta) - 2c\omega\sin(\omega t + \delta)]$$
$$= Ae^{-ct}[(c^2 - \omega^2)\cos(\omega t + \delta) + 2c\omega\sin(\omega t + \delta)]$$

87. (a) $y = t^3 - 12t + 3 \quad \Rightarrow \quad v(t) = y' = 3t^2 - 12 \quad \Rightarrow \quad a(t) = v'(t) = 6t$

(b) $v(t) = 3(t^2 - 4) > 0$ when $t > 2$, so it moves upward when $t > 2$ and downward when $0 \le t < 2$.

(c) Distance upward $= y(3) - y(2) = -6 - (-13) = 7$,

Distance downward $= y(0) - y(2) = 3 - (-13) = 16$. Total distance $= 7 + 16 = 23$.

89. The linear density ρ is the rate of change of mass m with respect to length x. $m = x(1 + \sqrt{x}) = x + x^{3/2}$ $\Rightarrow$

$\rho = dm/dx = 1 + \frac{3}{2}\sqrt{x}$, so the linear density when $x = 4$ is $1 + \frac{3}{2}\sqrt{4} = 4$ kg/m.

91. If $x =$ edge length, then $V = x^3$ $\Rightarrow$ $dV/dt = 3x^2\,dx/dt = 10$ $\Rightarrow$ $dx/dt = 10/(3x^2)$ and $S = 6x^2$ $\Rightarrow$

$dS/dt = (12x)\,dx/dt = 12x[10/(3x^2)] = 40/x$. When $x = 30$, $dS/dt = \frac{40}{30} = \frac{4}{3}$ cm^2/min.

93. Given $dh/dt = 5$ and $dx/dt = 15$, find dz/dt. $z^2 = x^2 + h^2$ $\Rightarrow$

$2z\dfrac{dz}{dt} = 2x\dfrac{dx}{dt} + 2h\dfrac{dh}{dt}$ $\Rightarrow$ $\dfrac{dz}{dt} = \dfrac{1}{z}(15x + 5h)$. When $t = 3$,

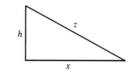

$h = 45 + 3(5) = 60$ and $x = 15(3) = 45$ $\Rightarrow$ $z = \sqrt{45^2 + 60^2} = 75$, so

$\dfrac{dz}{dt} = \frac{1}{75}[15(45) + 5(60)] = 13$ ft/s.

95. We are given $d\theta/dt = -0.25$ rad/h. $\tan\theta = 400/x$ $\Rightarrow$

$x = 400\cot\theta$ $\Rightarrow$ $\dfrac{dx}{dt} = -400\csc^2\theta\,\dfrac{d\theta}{dt}$. When $\theta = \frac{\pi}{6}$,

$\dfrac{dx}{dt} = -400(2)^2(-0.25) = 400$ ft/h.

97. (a) $f(x) = \sqrt[3]{1 + 3x} = (1 + 3x)^{1/3}$ $\Rightarrow$ $f'(x) = (1 + 3x)^{-2/3}$, so the linearization of f at $a = 0$ is

$L(x) = f(0) + f'(0)(x - 0) = 1^{1/3} + 1^{-2/3}x = 1 + x$. Thus, $\sqrt[3]{1 + 3x} \approx 1 + x$ $\Rightarrow$

$\sqrt[3]{1.03} = \sqrt[3]{1 + 3(0.01)} \approx 1 + (0.01) = 1.01$.

(b) The linear approximation is $\sqrt[3]{1 + 3x} \approx 1 + x$, so for the required

accuracy we want $\sqrt[3]{1 + 3x} - 0.1 < 1 + x < \sqrt[3]{1 + 3x} + 0.1$. From

the graph, it appears that this is true when $-0.23 < x < 0.40$.

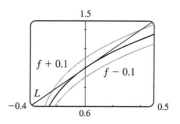

99. $A = x^2 + \frac{1}{2}\pi\left(\frac{1}{2}x\right)^2 = \left(1 + \frac{\pi}{8}\right)x^2$ $\Rightarrow$ $dA = \left(2 + \frac{\pi}{4}\right)x\,dx$.

When $x = 60$ and $dx = 0.1$, $dA = \left(2 + \frac{\pi}{4}\right)60(0.1) = 12 + \frac{3\pi}{2}$, so

the maximum error is approximately $12 + \frac{3\pi}{2} \approx 16.7$ cm^2.

101. $\displaystyle\lim_{h\to 0}\dfrac{\sqrt[4]{16 + h} - 2}{h} = \left[\dfrac{d}{dx}\sqrt[4]{x}\right]_{x = 16} = \frac{1}{4}x^{-3/4}\Big|_{x = 16} = \dfrac{1}{4\left(\sqrt[4]{16}\right)^3} = \dfrac{1}{32}$

103. $\displaystyle\lim_{x\to 0} \frac{\sqrt{1+\tan x} - \sqrt{1+\sin x}}{x^3} = \lim_{x\to 0} \frac{\left(\sqrt{1+\tan x} - \sqrt{1+\sin x}\right)\left(\sqrt{1+\tan x} + \sqrt{1+\sin x}\right)}{x^3\left(\sqrt{1+\tan x} + \sqrt{1+\sin x}\right)}$

$\displaystyle = \lim_{x\to 0} \frac{(1+\tan x) - (1+\sin x)}{x^3\left(\sqrt{1+\tan x} + \sqrt{1+\sin x}\right)} = \lim_{x\to 0} \frac{\sin x\,(1/\cos x - 1)}{x^3\left(\sqrt{1+\tan x} + \sqrt{1+\sin x}\right)} \cdot \frac{\cos x}{\cos x}$

$\displaystyle = \lim_{x\to 0} \frac{\sin x\,(1 - \cos x)}{x^3\left(\sqrt{1+\tan x} + \sqrt{1+\sin x}\right)\cos x} \cdot \frac{1+\cos x}{1+\cos x}$

$\displaystyle = \lim_{x\to 0} \frac{\sin x \cdot \sin^2 x}{x^3\left(\sqrt{1+\tan x} + \sqrt{1+\sin x}\right)\cos x\,(1+\cos x)}$

$\displaystyle = \left(\lim_{x\to 0} \frac{\sin x}{x}\right)^3 \lim_{x\to 0} \frac{1}{\left(\sqrt{1+\tan x} + \sqrt{1+\sin x}\right)\cos x\,(1+\cos x)}$

$\displaystyle = 1^3 \cdot \frac{1}{\left(\sqrt{1} + \sqrt{1}\right)\cdot 1 \cdot (1+1)} = \frac{1}{4}$

105. $\dfrac{d}{dx}\,[f(2x)] = x^2 \ \Rightarrow\ f'(2x)\cdot 2 = x^2 \ \Rightarrow\ f'(2x) = \frac{1}{2}x^2$. Let $t = 2x$. Then $f'(t) = \frac{1}{2}\left(\frac{1}{2}t\right)^2 = \frac{1}{8}t^2$,

so $f'(x) = \frac{1}{8}x^2$.

☐ PROBLEMS PLUS

1. Let a be the x-coordinate of Q. Since the derivative of $y = 1 - x^2$ is $y' = -2x$, the slope at Q is $-2a$. But since the triangle is equilateral, $\overline{AO}/\overline{OC} = \sqrt{3}/1$, so the slope at Q is $-\sqrt{3}$. Therefore, we must have that $-2a = -\sqrt{3}$

$\Rightarrow \quad a = \frac{\sqrt{3}}{2}$. Thus, the point Q has coordinates $\left(\frac{\sqrt{3}}{2}, 1 - \left(\frac{\sqrt{3}}{2}\right)^2\right) = \left(\frac{\sqrt{3}}{2}, \frac{1}{4}\right)$ and by symmetry, P has

coordinates $\left(-\frac{\sqrt{3}}{2}, \frac{1}{4}\right)$.

3. Let $y = \tan^{-1} x$. Then $\tan y = x$, so from the triangle we see that

$\sin(\tan^{-1} x) = \sin y = \dfrac{x}{\sqrt{1 + x^2}}$. Using this fact we have that

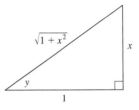

$\sin(\tan^{-1}(\sinh x)) = \dfrac{\sinh x}{\sqrt{1 + \sinh^2 x}} = \dfrac{\sinh x}{\cosh x} = \tanh x$. Hence,

$\sin^{-1}(\tanh x) = \sin^{-1}\left(\sin(\tan^{-1}(\sinh x))\right) = \tan^{-1}(\sinh x)$.

5. We use mathematical induction. Let S_n be the statement that $\dfrac{d^n}{dx^n}\left(\sin^4 x + \cos^4 x\right) = 4^{n-1}\cos(4x + n\pi/2)$.

S_1 is true because

$$\frac{d}{dx}\left(\sin^4 x + \cos^4 x\right) = 4\sin^3 x \cos x - 4\cos^3 x \sin x = 4\sin x \cos x \left(\sin^2 x - \cos^2 x\right)$$

$$= -4\sin x \cos x \cos 2x = -2\sin 2x \cos 2x = -\sin 4x = \sin(-4x)$$

$$= \cos\left(\frac{\pi}{2} - (-4x)\right) = \cos\left(\frac{\pi}{2} + 4x\right) = 4^{n-1}\cos\left(4x + n\frac{\pi}{2}\right) \text{ when } n = 1$$

Now assume S_k is true, that is, $\dfrac{d^k}{dx^k}\left(\sin^4 x + \cos^4 x\right) = 4^{k-1}\cos\left(4x + k\frac{\pi}{2}\right)$. Then

$$\frac{d^{k+1}}{dx^{k+1}}\left(\sin^4 x + \cos^4 x\right) = \frac{d}{dx}\left[\frac{d^k}{dx^k}\left(\sin^4 x + \cos^4 x\right)\right] = \frac{d}{dx}\left[4^{k-1}\cos\left(4x + k\frac{\pi}{2}\right)\right]$$

$$= -4^{k-1}\sin\left(4x + k\frac{\pi}{2}\right) \cdot \frac{d}{dx}\left(4x + k\frac{\pi}{2}\right) = -4^k\sin\left(4x + k\frac{\pi}{2}\right)$$

$$= 4^k\sin\left(-4x - k\frac{\pi}{2}\right) = 4^k\cos\left(\frac{\pi}{2} - \left(-4x - k\frac{\pi}{2}\right)\right)$$

$$= 4^k\cos\left(4x + (k+1)\frac{\pi}{2}\right)$$

which shows that S_{k+1} is true.

Therefore, $\dfrac{d^n}{dx^n}\left(\sin^4 x + \cos^4 x\right) = 4^{n-1}\cos\left(4x + n\frac{\pi}{2}\right)$ for every positive integer n, by mathematical induction.

Another proof: First write

$$\sin^4 x + \cos^4 x = \left(\sin^2 x + \cos^2 x\right)^2 - 2\sin^2 x \cos^2 x = 1 - \frac{1}{2}\sin^2 2x = 1 - \frac{1}{4}(1 - \cos 4x) = \frac{3}{4} + \frac{1}{4}\cos 4x.$$

Then we have $\dfrac{d^n}{dx^n}\left(\sin^4 x + \cos^4 x\right) = \dfrac{d^n}{dx^n}\left(\frac{3}{4} + \frac{1}{4}\cos 4x\right) = \frac{1}{4} \cdot 4^n\cos\left(4x + n\frac{\pi}{2}\right) = 4^{n-1}\cos\left(4x + n\frac{\pi}{2}\right)$.

7. We must find a value x_0 such that the normal lines to the parabola $y = x^2$ at $x = \pm x_0$ intersect at a point one unit

from the points $(\pm x_0, x_0^2)$. The normals to $y = x^2$ at $x = \pm x_0$ have slopes $-\dfrac{1}{\pm 2x_0}$ and pass through $(\pm x_0, x_0^2)$

respectively, so the normals have the equations $y - x_0^2 = -\dfrac{1}{2x_0}(x - x_0)$ and $y - x_0^2 = \dfrac{1}{2x_0}(x + x_0)$. The

common y-intercept is $x_0^2 + \frac{1}{2}$. We want to find the value of x_0 for which the distance from $\left(0, x_0^2 + \frac{1}{2}\right)$ to $\left(x_0, x_0^2\right)$

equals 1. The square of the distance is $(x_0 - 0)^2 + \left[x_0^2 - \left(x_0^2 + \frac{1}{2}\right)\right]^2 = x_0^2 + \frac{1}{4} = 1$ ⟺ $x_0 = \pm\frac{\sqrt{3}}{2}$. For these

values of x_0, the y-intercept is $x_0^2 + \frac{1}{2} = \frac{5}{4}$, so the center of the circle is at $\left(0, \frac{5}{4}\right)$.

Another solution: Let the center of the circle be $(0, a)$. Then the equation of the circle is $x^2 + (y - a)^2 = 1$.

Solving with the equation of the parabola, $y = x^2$, we get $x^2 + \left(x^2 - a\right)^2 = 1$ ⟺ $x^2 + x^4 - 2ax^2 + a^2 = 1$

⟺ $x^4 + (1 - 2a)x^2 + a^2 - 1 = 0$. The parabola and the circle will be tangent to each other when this quadratic

equation in x^2 has equal roots; that is, when the discriminant is 0. Thus, $(1 - 2a)^2 - 4\left(a^2 - 1\right) = 0$ ⟺

$1 - 4a + 4a^2 - 4a^2 + 4 = 0$ ⟺ $4a = 5$, so $a = \frac{5}{4}$. The center of the circle is $\left(0, \frac{5}{4}\right)$.

9. We can assume without loss of generality that $\theta = 0$ at time $t = 0$, so that $\theta = 12\pi t$ rad. [The angular velocity of

the wheel is 360 rpm $= 360 \cdot (2\pi \text{ rad})/(60 \text{ s}) = 12\pi$ rad/s.] Then the position of A as a function of time is

$A = (40 \cos\theta, 40 \sin\theta) = (40 \cos 12\pi t, 40 \sin 12\pi t)$, so $\sin\alpha = \dfrac{y}{1.2 \text{ m}} = \dfrac{40 \sin\theta}{120} = \dfrac{\sin\theta}{3} = \frac{1}{3}\sin 12\pi t$.

(a) Differentiating the expression for $\sin\alpha$, we get $\cos\alpha \cdot \dfrac{d\alpha}{dt} = \frac{1}{3} \cdot 12\pi \cdot \cos 12\pi t = 4\pi \cos\theta$.

When $\theta = \frac{\pi}{3}$, we have $\sin\alpha = \frac{1}{3}\sin\theta = \frac{\sqrt{3}}{6}$, so $\cos\alpha = \sqrt{1 - \left(\frac{\sqrt{3}}{6}\right)^2} = \sqrt{\frac{11}{12}}$ and

$\dfrac{d\alpha}{dt} = \dfrac{4\pi \cos\frac{\pi}{3}}{\cos\alpha} = \dfrac{2\pi}{\sqrt{11/12}} = \dfrac{4\pi\sqrt{3}}{\sqrt{11}} \approx 6.56$ rad/s.

(b) By the Law of Cosines, $|AP|^2 = |OA|^2 + |OP|^2 - 2|OA||OP|\cos\theta$ ⟹

$120^2 = 40^2 + |OP|^2 - 2 \cdot 40|OP|\cos\theta$ ⟹ $|OP|^2 - (80\cos\theta)|OP| - 12{,}800 = 0$ ⟹

$|OP| = \frac{1}{2}\left(80\cos\theta \pm \sqrt{6400\cos^2\theta + 51{,}200}\right) = 40\cos\theta \pm 40\sqrt{\cos^2\theta + 8}$

$= 40\left(\cos\theta + \sqrt{8 + \cos^2\theta}\right)$ cm (since $|OP| > 0$)

As a check, note that $|OP| = 160$ cm when $\theta = 0$ and $|OP| = 80\sqrt{2}$ cm when $\theta = \frac{\pi}{2}$.

(c) By part (b), the x-coordinate of P is given by $x = 40\left(\cos\theta + \sqrt{8 + \cos^2\theta}\right)$, so

$\dfrac{dx}{dt} = \dfrac{dx}{d\theta}\dfrac{d\theta}{dt} = 40\left(-\sin\theta - \dfrac{2\cos\theta\sin\theta}{2\sqrt{8 + \cos^2\theta}}\right) \cdot 12\pi = -480\pi\sin\theta\left(1 + \dfrac{\cos\theta}{\sqrt{8 + \cos^2\theta}}\right)$ cm/s.

In particular, $dx/dt = 0$ cm/s when $\theta = 0$ and $dx/dt = -480\pi$ cm/s when $\theta = \frac{\pi}{2}$.

11. Consider the statement that $\dfrac{d^n}{dx^n}(e^{ax}\sin bx) = r^n e^{ax}\sin(bx + n\theta)$. For $n = 1$,

$$\frac{d}{dx}(e^{ax}\sin bx) = ae^{ax}\sin bx + be^{ax}\cos bx, \text{ and}$$

$$re^{ax}\sin(bx + \theta) = re^{ax}[\sin bx\cos\theta + \cos bx\sin\theta] = re^{ax}\left(\frac{a}{r}\sin bx + \frac{b}{r}\cos bx\right)$$

$$= ae^{ax}\sin bx + be^{ax}\cos bx$$

since $\tan\theta = \dfrac{b}{a} \;\Rightarrow\; \sin\theta = \dfrac{b}{r}$ and $\cos\theta = \dfrac{a}{r}$.

So the statement is true for $n = 1$. Assume it is true for $n = k$. Then

$$\frac{d^{k+1}}{dx^{k+1}}(e^{ax}\sin bx) = \frac{d}{dx}\left[r^k e^{ax}\sin(bx + k\theta)\right] = r^k ae^{ax}\sin(bx + k\theta) + r^k e^{ax}b\cos(bx + k\theta)$$

$$= r^k e^{ax}[a\sin(bx + k\theta) + b\cos(bx + k\theta)]$$

But

$$\sin[bx + (k + 1)\theta] = \sin[(bx + k\theta) + \theta] = \sin(bx + k\theta)\cos\theta + \sin\theta\cos(bx + k\theta)$$

$$= \frac{a}{r}\sin(bx + k\theta) + \frac{b}{r}\cos(bx + k\theta)$$

Hence, $a\sin(bx + k\theta) + b\cos(bx + k\theta) = r\sin[bx + (k + 1)\theta]$. So

$$\frac{d^{k+1}}{dx^{k+1}}(e^{ax}\sin bx) = r^k e^{ax}[a\sin(bx + k\theta) + b\cos(bx + k\theta)] = r^k e^{ax}[r\sin(bx + (k + 1)\theta)]$$

$$= r^{k+1}e^{ax}[\sin(bx + (k + 1)\theta)]$$

Therefore, the statement is true for all n by mathematical induction.

13. It seems from the figure that as P approaches the point $(0, 2)$ from the right, $x_T \to \infty$ and $y_T \to 2^+$. As P approaches the point $(3, 0)$ from the left, it appears that $x_T \to 3^+$ and $y_T \to \infty$. So we guess that $x_T \in (3, \infty)$ and $y_T \in (2, \infty)$. It is more difficult to estimate the range of values for x_N and y_N. We might perhaps guess that $x_N \in (0, 3)$, and $y_N \in (-\infty, 0)$ or $(-2, 0)$.

In order to actually solve the problem, we implicitly differentiate the equation of the ellipse to find the equation

of the tangent line: $\dfrac{x^2}{9} + \dfrac{y^2}{4} = 1 \;\Rightarrow\; \dfrac{2x}{9} + \dfrac{2y}{4}y' = 0$, so $y' = -\dfrac{4}{9}\dfrac{x}{y}$. So at the point (x_0, y_0) on the ellipse, an

equation of the tangent line is $y - y_0 = -\dfrac{4}{9}\dfrac{x_0}{y_0}(x - x_0)$ or $4x_0 x + 9y_0 y = 4x_0^2 + 9y_0^2$. This can be written as

$\dfrac{x_0 x}{9} + \dfrac{y_0 y}{4} = \dfrac{x_0^2}{9} + \dfrac{y_0^2}{4} = 1$, because (x_0, y_0) lies on the ellipse. So an equation of the tangent line is

$\dfrac{x_0 x}{9} + \dfrac{y_0 y}{4} = 1$.

Therefore, the x-intercept x_T for the tangent line is given by $\dfrac{x_0 x_T}{9} = 1 \Leftrightarrow x_T = \dfrac{9}{x_0}$, and the y-intercept y_T

is given by $\dfrac{y_0 y_T}{4} = 1 \Leftrightarrow y_T = \dfrac{4}{y_0}$.

So as x_0 takes on all values in $(0, 3)$, x_T takes on all values in $(3, \infty)$, and as y_0 takes on all values in $(0, 2)$, y_T takes on all values in $(2, \infty)$. At the point (x_0, y_0) on the ellipse, the slope of the normal line is

$-\dfrac{1}{y'(x_0, y_0)} = \dfrac{9}{4}\dfrac{y_0}{x_0}$, and its equation is $y - y_0 = \dfrac{9}{4}\dfrac{y_0}{x_0}(x - x_0)$. So the x-intercept x_N for the normal line is

given by $0 - y_0 = \dfrac{9}{4}\dfrac{y_0}{x_0}(x_N - x_0) \Rightarrow x_N = -\dfrac{4x_0}{9} + x_0 = \dfrac{5x_0}{9}$, and the y-intercept y_N is given by

$y_N - y_0 = \dfrac{9}{4}\dfrac{y_0}{x_0}(0 - x_0) \Rightarrow y_N = -\dfrac{9y_0}{4} + y_0 = -\dfrac{5y_0}{4}$.

So as x_0 takes on all values in $(0, 3)$, x_N takes on all values in $\left(0, \tfrac{5}{3}\right)$, and as y_0 takes on all values in $(0, 2)$, y_N takes on all values in $\left(-\tfrac{5}{2}, 0\right)$.

15. (a)
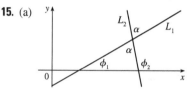
If the two lines L_1 and L_2 have slopes m_1 and m_2 and angles of inclination ϕ_1 and ϕ_2, then $m_1 = \tan\phi_1$ and $m_2 = \tan\phi_2$. The triangle in the figure shows that $\phi_1 + \alpha + (180° - \phi_2) = 180°$ and so $\alpha = \phi_2 - \phi_1$. Therefore, using the identity for $\tan(x - y)$, we have

$$\tan\alpha = \tan(\phi_2 - \phi_1) = \dfrac{\tan\phi_2 - \tan\phi_1}{1 + \tan\phi_2 \tan\phi_1} \text{ and so}$$

$$\tan\alpha = \dfrac{m_2 - m_1}{1 + m_1 m_2}.$$

(b) (i) The parabolas intersect when $x^2 = (x - 2)^2 \Rightarrow x = 1$. If $y = x^2$, then $y' = 2x$, so the slope of the tangent to $y = x^2$ at $(1, 1)$ is $m_1 = 2(1) = 2$. If $y = (x - 2)^2$, then $y' = 2(x - 2)$, so the slope of the tangent to $y = (x - 2)^2$ at $(1, 1)$ is $m_2 = 2(1 - 2) = -2$. Therefore,

$$\tan\alpha = \dfrac{m_2 - m_1}{1 + m_1 m_2} = \dfrac{-2 - 2}{1 + 2(-2)} = \tfrac{4}{3} \text{ and so } \alpha = \tan^{-1}\left(\tfrac{4}{3}\right) \approx 53° \text{ (or } 127°\text{)}.$$

(ii) $x^2 - y^2 = 3$ and $x^2 - 4x + y^2 + 3 = 0$ intersect when $x^2 - 4x + (x^2 - 3) + 3 = 0 \Leftrightarrow 2x(x - 2) = 0$ $\Rightarrow x = 0$ or 2, but 0 is extraneous. If $x = 2$, then $y = \pm 1$. If $x^2 - y^2 = 3$ then $2x - 2yy' = 0 \Rightarrow$ $y' = x/y$ and $x^2 - 4x + y^2 + 3 = 0 \Rightarrow 2x - 4 + 2yy' = 0 \Rightarrow y' = \dfrac{2 - x}{y}$. At $(2, 1)$ the slopes are

$m_1 = 2$ and $m_2 = 0$, so $\tan\alpha = \dfrac{0 - 2}{1 + 2 \cdot 0} = -2 \Rightarrow \alpha \approx 117°$. At $(2, -1)$ the slopes are $m_1 = -2$ and

$m_2 = 0$, so $\tan\alpha = \dfrac{0 - (-2)}{1 + (-2)(0)} = 2 \Rightarrow \alpha \approx 63°$ (or $117°$).

17. Since $\angle ROQ = \angle OQP = \theta$, the triangle QOR is isosceles, so
$|QR| = |RO| = x$. By the Law of Cosines, $x^2 = x^2 + r^2 - 2rx\cos\theta$. Hence,

$2rx\cos\theta = r^2$, so $x = \dfrac{r^2}{2r\cos\theta} = \dfrac{r}{2\cos\theta}$. Note that as $y \to 0^+$, $\theta \to 0^+$ (since

$\sin\theta = y/r$), and hence $x \to \dfrac{r}{2\cos 0} = \dfrac{r}{2}$. Thus, as P is taken closer and closer

to the x-axis, the point R approaches the midpoint of the radius AO.

19. $\lim\limits_{x \to 0} \dfrac{\sin(a + 2x) - 2\sin(a + x) + \sin a}{x^2}$

$= \lim\limits_{x \to 0} \dfrac{\sin a \cos 2x + \cos a \sin 2x - 2\sin a \cos x - 2\cos a \sin x + \sin a}{x^2}$

$= \lim\limits_{x \to 0} \dfrac{\sin a\,(\cos 2x - 2\cos x + 1) + \cos a\,(\sin 2x - 2\sin x)}{x^2}$

$= \lim\limits_{x \to 0} \dfrac{\sin a\,(2\cos^2 x - 1 - 2\cos x + 1) + \cos a\,(2\sin x \cos x - 2\sin x)}{x^2}$

$= \lim\limits_{x \to 0} \dfrac{\sin a\,(2\cos x)(\cos x - 1) + \cos a\,(2\sin x)(\cos x - 1)}{x^2}$

$= \lim\limits_{x \to 0} \dfrac{2(\cos x - 1)[\sin a \cos x + \cos a \sin x](\cos x + 1)}{x^2(\cos x + 1)}$

$= \lim\limits_{x \to 0} \dfrac{-2\sin^2 x\,[\sin(a + x)]}{x^2(\cos x + 1)} = -2\lim\limits_{x \to 0}\left(\dfrac{\sin x}{x}\right)^2 \cdot \dfrac{\sin(a + x)}{\cos x + 1} = -2(1)^2\dfrac{\sin(a + 0)}{\cos 0 + 1} = -\sin a$

21.

Let $f(x) = e^{2x}$ and $g(x) = k\sqrt{x}\ (k > 0)$. From the graphs of f and g, we see that f will intersect g exactly once when f and g share a tangent line. Thus, we must have $f = g$ and $f' = g'$ at $x = a$. $f(a) = g(a)\ \Rightarrow$

$e^{2a} = k\sqrt{a}$ **(1)** and $f'(a) = g'(a)\ \Rightarrow\ 2e^{2a} = \dfrac{k}{2\sqrt{a}}\ \Rightarrow$

$e^{2a} = \dfrac{k}{4\sqrt{a}}$. So we must have $k\sqrt{a} = \dfrac{k}{4\sqrt{a}}\ \Rightarrow\ (\sqrt{a})^2 = \dfrac{k}{4k}\ \Rightarrow$

$a = \frac{1}{4}$. From **(1)**, $e^{2(1/4)} = k\sqrt{1/4}\ \Rightarrow\ k = 2e^{1/2} = 2\sqrt{e} \approx 3.297$.

23. $y = \dfrac{x}{\sqrt{a^2 - 1}} - \dfrac{2}{\sqrt{a^2 - 1}}\arctan\dfrac{\sin x}{a + \sqrt{a^2 - 1} + \cos x}$. Let $k = a + \sqrt{a^2 - 1}$. Then

$y' = \dfrac{1}{\sqrt{a^2 - 1}} - \dfrac{2}{\sqrt{a^2 - 1}} \cdot \dfrac{1}{1 + \sin^2 x/(k + \cos x)^2} \cdot \dfrac{\cos x(k + \cos x) + \sin^2 x}{(k + \cos x)^2}$

$= \dfrac{1}{\sqrt{a^2 - 1}} - \dfrac{2}{\sqrt{a^2 - 1}} \cdot \dfrac{k\cos x + \cos^2 x + \sin^2 x}{(k + \cos x)^2 + \sin^2 x} = \dfrac{1}{\sqrt{a^2 - 1}} - \dfrac{2}{\sqrt{a^2 - 1}} \cdot \dfrac{k\cos x + 1}{k^2 + 2k\cos x + 1}$

$= \dfrac{k^2 + 2k\cos x + 1 - 2k\cos x - 2}{\sqrt{a^2 - 1}\,(k^2 + 2k\cos x + 1)} = \dfrac{k^2 - 1}{\sqrt{a^2 - 1}\,(k^2 + 2k\cos x + 1)}$

But $k^2 = 2a^2 + 2a\sqrt{a^2 - 1} - 1 = 2a(a + \sqrt{a^2 - 1}) - 1 = 2ak - 1$, so $k^2 + 1 = 2ak$, and

$k^2 - 1 = 2(ak - 1)$. So $y' = \dfrac{2(ak - 1)}{\sqrt{a^2 - 1}\,(2ak + 2k\cos x)} = \dfrac{ak - 1}{\sqrt{a^2 - 1}\,k\,(a + \cos x)}$. But

$ak - 1 = a^2 + a\sqrt{a^2 - 1} - 1 = k\sqrt{a^2 - 1}$, so $y' = 1/(a + \cos x)$.

25. $y = x^4 - 2x^2 - x\ \Rightarrow\ y' = 4x^3 - 4x - 1$. The equation of the tangent line at $x = a$ is

$y - (a^4 - 2a^2 - a) = (4a^3 - 4a - 1)(x - a)$ or $y = (4a^3 - 4a - 1)x + (-3a^4 + 2a^2)$ and similarly for $x = b$.

So if at $x = a$ and $x = b$ we have the same tangent line, then $4a^3 - 4a - 1 = 4b^3 - 4b - 1$ and

$-3a^4 + 2a^2 = -3b^4 + 2b^2$. The first equation gives $a^3 - b^3 = a - b\ \Rightarrow\ (a - b)(a^2 + ab + b^2) = (a - b)$.

Assuming $a \neq b$, we have $1 = a^2 + ab + b^2$. The second equation gives $3(a^4 - b^4) = 2(a^2 - b^2)\ \Rightarrow$

$3(a^2 - b^2)(a^2 + b^2) = 2(a^2 - b^2)$ which is true if $a = -b$. Substituting into $1 = a^2 + ab + b^2$ gives

$1 = a^2 - a^2 + a^2 \implies a = \pm 1$ so that $a = 1$ and $b = -1$ or vice versa. Thus, the points $(1, -2)$ and $(-1, 0)$ have a common tangent line.

As long as there are only two such points, we are done. So we show that these are in fact the only two such points. Suppose that $a^2 - b^2 \neq 0$. Then $3(a^2 - b^2)(a^2 + b^2) = 2(a^2 - b^2)$ gives $3(a^2 + b^2) = 2$ or $a^2 + b^2 = \frac{2}{3}$. Thus, $ab = (a^2 + ab + b^2) - (a^2 + b^2) = 1 - \frac{2}{3} = \frac{1}{3}$, so $b = \dfrac{1}{3a}$. Hence, $a^2 + \dfrac{1}{9a^2} = \dfrac{2}{3}$, so $9a^4 + 1 = 6a^2 \implies 0 = 9a^4 - 6a^2 + 1 = (3a^2 - 1)^2$. So $3a^2 - 1 = 0 \implies a^2 = \frac{1}{3} \implies b^2 = \dfrac{1}{9a^2} = \frac{1}{3} = a^2$, contradicting our assumption that $a^2 \neq b^2$.

27.

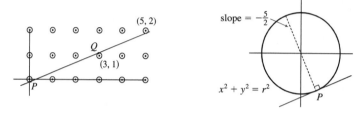

Because of the periodic nature of the lattice points, it suffices to consider the points in the 5×2 grid shown. We can see that the minimum value of r occurs when there is a line with slope $\frac{2}{5}$ which touches the circle centered at $(3, 1)$ and the circles centered at $(0, 0)$ and $(5, 2)$. To find P, the point at which the line is tangent to the circle at $(0, 0)$, we simultaneously solve $x^2 + y^2 = r^2$ and $y = -\frac{5}{2}x \implies x^2 + \frac{25}{4}x^2 = r^2 \implies x^2 = \frac{4}{29}r^2 \implies x = \frac{2}{\sqrt{29}}r$, $y = -\frac{5}{\sqrt{29}}r$. To find Q, we either use symmetry or solve $(x-3)^2 + (y-1)^2 = r^2$ and $y - 1 = -\frac{5}{2}(x - 3)$. As above, we get $x = 3 - \frac{2}{\sqrt{29}}r$, $y = 1 + \frac{5}{\sqrt{29}}r$. Now the slope of the line PQ is $\frac{2}{5}$, so

$$m_{PQ} = \frac{1 + \frac{5}{\sqrt{29}}r - \left(-\frac{5}{\sqrt{29}}r\right)}{3 - \frac{2}{\sqrt{29}}r - \frac{2}{\sqrt{29}}r} = \frac{1 + \frac{10}{\sqrt{29}}r}{3 - \frac{4}{\sqrt{29}}r} = \frac{\sqrt{29} + 10r}{3\sqrt{29} - 4r} = \frac{2}{5} \implies 5\sqrt{29} + 50r = 6\sqrt{29} - 8r \iff$$

$58r = \sqrt{29} \iff r = \frac{\sqrt{29}}{58}$. So the minimum value of r for which any line with slope $\frac{2}{5}$ intersects circles with radius r centered at the lattice points on the plane is $r = \frac{\sqrt{29}}{58} \approx 0.093$.

29.

By similar triangles, $\dfrac{r}{5} = \dfrac{h}{16} \implies r = \dfrac{5h}{16}$. The volume of the cone is $V = \frac{1}{3}\pi r^2 h = \frac{1}{3}\pi \left(\dfrac{5h}{16}\right)^2 h = \dfrac{25\pi}{768}h^3$, so $\dfrac{dV}{dt} = \dfrac{25\pi}{256}h^2 \dfrac{dh}{dt}$. Now the rate of change of the volume is also equal to the difference of what is being added $(2 \text{ cm}^3/\text{min})$ and what is oozing out $(k\pi r l$, where $\pi r l$ is the area of the cone and k is a proportionality constant). Thus, $\dfrac{dV}{dt} = 2 - k\pi r l$.

Equating the two expressions for $\dfrac{dV}{dt}$ and substituting $h = 10$, $\dfrac{dh}{dt} = -0.3$, $r = \dfrac{5(10)}{16} = \dfrac{25}{8}$, and $\dfrac{l}{\sqrt{281}} = \dfrac{10}{16}$

$\iff l = \frac{5}{8}\sqrt{281}$, we get $\frac{25\pi}{256}(10)^2(-0.3) = 2 - k\pi\frac{25}{8}\cdot\frac{5}{8}\sqrt{281} \iff \dfrac{125k\pi\sqrt{281}}{64} = 2 + \dfrac{750\pi}{256}$. Solving for k gives us $k = \dfrac{256 + 375\pi}{250\pi\sqrt{281}}$. To maintain a certain height, the rate of oozing, $k\pi r l$, must equal the rate of the liquid being poured in; that is, $\dfrac{dV}{dt} = 0$. $k\pi r l = \dfrac{256 + 375\pi}{250\pi\sqrt{281}} \cdot \pi \cdot \dfrac{25}{8} \cdot \dfrac{5\sqrt{281}}{8} = \dfrac{256 + 375\pi}{128} \approx 11.204 \text{ cm}^3/\text{min}$.

4 □ APPLICATIONS OF DIFFERENTIATION

4.1 Maximum and Minimum Values

1. A function f has an **absolute minimum** at $x = c$ if $f(c)$ is the smallest function value on the entire domain of f, whereas f has a **local minimum** at c if $f(c)$ is the smallest function value when x is near c.

3. Absolute maximum at b; absolute minimum at d; local maxima at b and e; local minima at d and s; neither a maximum nor a minimum at a, c, r, and t.

5. Absolute maximum value is $f(4) = 4$; absolute minimum value is $f(7) = 0$; local maximum values are $f(4) = 4$ and $f(6) = 3$; local minimum values are $f(2) = 1$ and $f(5) = 2$.

7. Absolute minimum at 2, absolute maximum at 3, local minimum at 4

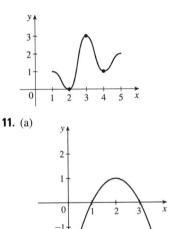

9. Absolute maximum at 5, absolute minimum at 2, local maximum at 3, local minima at 2 and 4

11. (a)

(b)

(c)

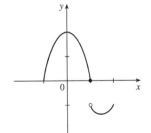

13. (a) *Note:* By the Extreme Value Theorem, f must *not* be continuous; because if it were, it would attain an absolute minimum.

(b)

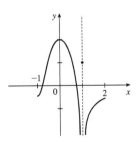

15. $f(x) = 8 - 3x$, $x \geq 1$. Absolute maximum $f(1) = 5$; no local maximum. No absolute or local minimum.

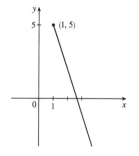

17. $f(x) = x^2$, $0 < x < 2$. No absolute or local maximum or minimum value.

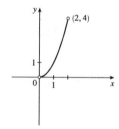

19. $f(x) = x^2$, $0 \leq x < 2$. Absolute minimum $f(0) = 0$; no local minimum. No absolute or local maximum.

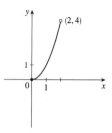

21. $f(x) = x^2$, $-3 \leq x \leq 2$. Absolute maximum $f(-3) = 9$. No local maximum. Absolute and local minimum $f(0) = 0$.

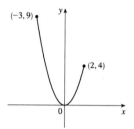

23. $f(t) = 1/t$, $0 < t < 1$. No maximum or minimum.

25. $f(\theta) = \sin \theta$, $-2\pi \leq \theta \leq 2\pi$. Absolute and local maxima $f\left(-\frac{3\pi}{2}\right) = f\left(\frac{\pi}{2}\right) = 1$. Absolute and local minima $f\left(-\frac{\pi}{2}\right) = f\left(\frac{3\pi}{2}\right) = -1$.

27. $f(x) = 1 - \sqrt{x}$. Absolute maximum $f(0) = 1$; no local maximum. No absolute or local minimum.

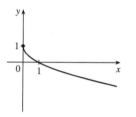

29. $f(x) = \begin{cases} 1 - x & \text{if } 0 \leq x < 2 \\ 2x - 4 & \text{if } 2 \leq x \leq 3 \end{cases}$

Absolute maximum $f(3) = 2$; no local maximum. No absolute or local minimum.

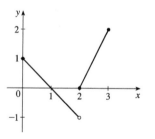

31. $f(x) = 5x^2 + 4x$ $\Rightarrow$ $f'(x) = 10x + 4.$ $f'(x) = 0$ $\Rightarrow$ $x = -\frac{2}{5}$, so $-\frac{2}{5}$ is the only critical number.

33. $f(x) = x^3 + 3x^2 - 24x$ $\Rightarrow$ $f'(x) = 3x^2 + 6x - 24 = 3(x^2 + 2x - 8).$

$f'(x) = 0$ $\Rightarrow$ $3(x+4)(x-2) = 0$ $\Rightarrow$ $x = -4, 2.$ These are the only critical numbers.

35. $s(t) = 3t^4 + 4t^3 - 6t^2$ $\Rightarrow$ $s'(t) = 12t^3 + 12t^2 - 12t.$ $s'(t) = 0$ $\Rightarrow$ $12t(t^2 + t - 1)$ $\Rightarrow$ $t = 0$ or

$t^2 + t - 1 = 0.$ Using the quadratic formula to solve the latter equation gives us

$t = \dfrac{-1 \pm \sqrt{1^2 - 4(1)(-1)}}{2(1)} = \dfrac{-1 \pm \sqrt{5}}{2} \approx 0.618, -1.618.$ The three critical numbers are $0, \dfrac{-1 \pm \sqrt{5}}{2}.$

37. $g(x) = |2x + 3| = \begin{cases} 2x + 3 & \text{if } 2x + 3 \ge 0 \\ -(2x + 3) & \text{if } 2x + 3 < 0 \end{cases}$ $\Rightarrow$ $g'(x) = \begin{cases} 2 & \text{if } x > -\frac{3}{2} \\ -2 & \text{if } x < -\frac{3}{2} \end{cases}$

$g'(x)$ is never 0, but $g'(x)$ does not exist for $x = -\frac{3}{2}$, so $-\frac{3}{2}$ is the only critical number.

39. $g(t) = 5t^{2/3} + t^{5/3}$ $\Rightarrow$ $g'(t) = \frac{10}{3}t^{-1/3} + \frac{5}{3}t^{2/3}.$ $g'(0)$ does not exist, so $t = 0$ is a critical number.

$g'(t) = \frac{5}{3}t^{-1/3}(2 + t) = 0$ $\Leftrightarrow$ $t = -2$, so $t = -2$ is also a critical number.

41. $F(x) = x^{4/5}(x - 4)^2$ $\Rightarrow$

$F'(x) = x^{4/5} \cdot 2(x - 4) + (x - 4)^2 \cdot \frac{4}{5}x^{-1/5} = \frac{1}{5}x^{-1/5}(x - 4)\left[5 \cdot x \cdot 2 + (x - 4) \cdot 4\right]$

$= \dfrac{(x - 4)(14x - 16)}{5x^{1/5}} = \dfrac{2(x - 4)(7x - 8)}{5x^{1/5}} = 0$ when $x = 4, \frac{8}{7};$ and $F'(0)$ does not exist.

Critical numbers are $0, \frac{8}{7}, 4.$

43. $f(\theta) = 2\cos\theta + \sin^2\theta$ $\Rightarrow$ $f'(\theta) = -2\sin\theta + 2\sin\theta\cos\theta.$ $f'(\theta) = 0$ $\Rightarrow$ $2\sin\theta(\cos\theta - 1) = 0$ $\Rightarrow$

$\sin\theta = 0$ or $\cos\theta = 1$ $\Rightarrow$ $\theta = n\pi$ (n an integer) or $\theta = 2n\pi.$ The solutions $\theta = n\pi$ include the solutions

$\theta = 2n\pi$, so the critical numbers are $\theta = n\pi.$

45. $f(x) = x\ln x$ $\Rightarrow$ $f'(x) = x(1/x) + (\ln x) \cdot 1 = \ln x + 1.$ $f'(x) = 0$ $\Leftrightarrow$ $\ln x = -1$ $\Leftrightarrow$

$x = e^{-1} = 1/e.$ Therefore, the only critical number is $x = 1/e.$

47. $f(x) = 3x^2 - 12x + 5$, $[0, 3].$ $f'(x) = 6x - 12 = 0$ $\Leftrightarrow$ $x = 2.$ Applying the Closed Interval Method, we

find that $f(0) = 5$, $f(2) = -7$, and $f(3) = -4.$ So $f(0) = 5$ is the absolute maximum value and $f(2) = -7$ is the

absolute minimum value.

49. $f(x) = 2x^3 - 3x^2 - 12x + 1$, $[-2, 3].$ $f'(x) = 6x^2 - 6x - 12 = 6(x^2 - x - 2) = 6(x - 2)(x + 1) = 0$ $\Leftrightarrow$

$x = 2, -1.$ $f(-2) = -3$, $f(-1) = 8$, $f(2) = -19$, and $f(3) = -8.$ So $f(-1) = 8$ is the absolute maximum

value and $f(2) = -19$ is the absolute minimum value.

51. $f(x) = x^4 - 2x^2 + 3$, $[-2, 3].$ $f'(x) = 4x^3 - 4x = 4x(x^2 - 1) = 4x(x + 1)(x - 1) = 0$ $\Leftrightarrow$ $x = -1, 0, 1.$

$f(-2) = 11$, $f(-1) = 2$, $f(0) = 3$, $f(1) = 2$, $f(3) = 66.$ So $f(3) = 66$ is the absolute maximum value and

$f(\pm 1) = 2$ is the absolute minimum value.

53. $f(x) = \dfrac{x}{x^2 + 1}$, $[0, 2]$. $f'(x) = \dfrac{(x^2 + 1) - x(2x)}{(x^2 + 1)^2} = \dfrac{1 - x^2}{(x^2 + 1)^2} = 0 \Leftrightarrow x = \pm 1$, but -1 is not in $[0, 2]$.

$f(0) = 0$, $f(1) = \frac{1}{2}$, $f(2) = \frac{2}{5}$. So $f(1) = \frac{1}{2}$ is the absolute maximum value and $f(0) = 0$ is the absolute minimum value.

55. $f(t) = t\sqrt{4 - t^2}$, $[-1, 2]$.

$f'(t) = t \cdot \frac{1}{2}(4 - t^2)^{-1/2}(-2t) + (4 - t^2)^{1/2} \cdot 1 = \dfrac{-t^2}{\sqrt{4 - t^2}} + \sqrt{4 - t^2} = \dfrac{-t^2 + (4 - t^2)}{\sqrt{4 - t^2}} = \dfrac{4 - 2t^2}{\sqrt{4 - t^2}}$.

$f'(t) = 0 \Rightarrow 4 - 2t^2 = 0 \Rightarrow t^2 = 2 \Rightarrow t = \pm\sqrt{2}$, but $t = -\sqrt{2}$ is not in the given interval, $[-1, 2]$.

$f'(t)$ does not exist if $4 - t^2 = 0 \Rightarrow t = \pm 2$, but -2 is not in the given interval. $f(-1) = -\sqrt{3}$, $f(\sqrt{2}) = 2$,

and $f(2) = 0$. So $f(\sqrt{2}) = 2$ is the absolute maximum value and $f(-1) = -\sqrt{3}$ is the absolute minimum value.

57. $f(x) = \sin x + \cos x$, $[0, \frac{\pi}{3}]$. $f'(x) = \cos x - \sin x = 0 \Leftrightarrow \sin x = \cos x \Rightarrow \dfrac{\sin x}{\cos x} = 1 \Rightarrow$

$\tan x = 1 \Rightarrow x = \frac{\pi}{4}$. $f(0) = 1$, $f(\frac{\pi}{4}) = \sqrt{2} \approx 1.41$, $f(\frac{\pi}{3}) = \frac{\sqrt{3}+1}{2} \approx 1.37$. So $f(\frac{\pi}{4}) = \sqrt{2}$ is the

absolute maximum value and $f(0) = 1$ is the absolute minimum value.

59. $f(x) = xe^{-x}$, $[0, 2]$. $f'(x) = x(-e^{-x}) + e^{-x} = e^{-x}(1 - x) = 0 \Leftrightarrow x = 1$.

$f(0) = 0$, $f(1) = e^{-1} = 1/e \approx 0.37$, $f(2) = 2/e^2 \approx 0.27$. So $f(1) = 1/e$ is the absolute maximum value and $f(0) = 0$ is the absolute minimum value.

61. $f(x) = x - 3\ln x$, $[1, 4]$. $f'(x) = 1 - \dfrac{3}{x} = \dfrac{x - 3}{x} = 0 \Leftrightarrow x = 3$. f' does not exist for $x = 0$, but 0 is not

in the domain of f. $f(1) = 1$, $f(3) = 3 - 3\ln 3 \approx -0.296$, $f(4) = 4 - 3\ln 4 \approx -0.159$. So $f(1) = 1$ is the

absolute maximum value and $f(3) = 3 - 3\ln 3 \approx -0.296$ is the absolute minimum value.

63. $f(x) = x^a(1 - x)^b$, $0 \le x \le 1$, $a > 0$, $b > 0$.

$f'(x) = x^a \cdot b(1 - x)^{b-1}(-1) + (1 - x)^b \cdot ax^{a-1} = x^{a-1}(1 - x)^{b-1}[x \cdot b(-1) + (1 - x) \cdot a]$

$\quad = x^{a-1}(1 - x)^{b-1}(a - ax - bx)$

At the endpoints, we have $f(0) = f(1) = 0$ [the minimum value of f]. In the interval $(0, 1)$, $f'(x) = 0 \Leftrightarrow$

$x = \dfrac{a}{a + b}$.

$f\left(\dfrac{a}{a + b}\right) = \left(\dfrac{a}{a + b}\right)^a\left(1 - \dfrac{a}{a + b}\right)^b = \dfrac{a^a}{(a + b)^a}\left(\dfrac{a + b - a}{a + b}\right)^b = \dfrac{a^a}{(a + b)^a} \cdot \dfrac{b^b}{(a + b)^b} = \dfrac{a^a b^b}{(a + b)^{a+b}}$.

So $f\left(\dfrac{a}{a + b}\right) = \dfrac{a^a b^b}{(a + b)^{a+b}}$ is the absolute maximum value.

65. (a)

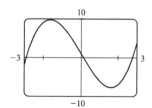

From the graph, it appears that the absolute maximum value is about $f(-1.63) = 9.71$, and the absolute minimum value is about $f(1.63) = -7.71$. These values make sense because the graph is symmetric about the point $(0, 1)$. ($y = x^3 - 8x$ is symmetric about the origin.)

(b) $f(x) = x^3 - 8x + 1 \implies f'(x) = 3x^2 - 8$. So $f'(x) = 0 \implies x = \pm\sqrt{\frac{8}{3}}$.

$$f\left(\pm\sqrt{\tfrac{8}{3}}\right) = \left(\pm\sqrt{\tfrac{8}{3}}\right)^3 - 8\left(\pm\sqrt{\tfrac{8}{3}}\right) + 1 = \pm\tfrac{8}{3}\sqrt{\tfrac{8}{3}} \mp 8\sqrt{\tfrac{8}{3}} + 1$$

$$= -\tfrac{16}{3}\sqrt{\tfrac{8}{3}} + 1 = 1 - \tfrac{32\sqrt{6}}{9} \text{ [minimum]} \quad \text{or} \quad \tfrac{16}{3}\sqrt{\tfrac{8}{3}} + 1 = 1 + \tfrac{32\sqrt{6}}{9} \text{ [maximum]}$$

(From the graph, we see that the extreme values do not occur at the endpoints.)

67. (a)

From the graph, it appears that the absolute maximum value is about $f(0.75) = 0.32$, and the absolute minimum value is $f(0) = f(1) = 0$; that is, at both endpoints.

(b) $f(x) = x\sqrt{x - x^2} \implies f'(x) = x \cdot \dfrac{1 - 2x}{2\sqrt{x - x^2}} + \sqrt{x - x^2} = \dfrac{(x - 2x^2) + (2x - 2x^2)}{2\sqrt{x - x^2}} = \dfrac{3x - 4x^2}{2\sqrt{x - x^2}}$.

So $f'(x) = 0 \implies 3x - 4x^2 = 0 \implies x(3 - 4x) = 0 \implies x = 0$ or $\frac{3}{4}$. $f(0) = f(1) = 0$ [minimum],

and $f\left(\frac{3}{4}\right) = \frac{3}{4}\sqrt{\frac{3}{4} - \left(\frac{3}{4}\right)^2} = \frac{3\sqrt{3}}{16}$ [maximum].

69. The density is defined as $\rho = \dfrac{\text{mass}}{\text{volume}} = \dfrac{1000}{V(T)}$ (in g/cm³). But a critical point of ρ will also be a critical point

of V [since $\dfrac{d\rho}{dT} = -1000V^{-2}\dfrac{dV}{dT}$ and V is never 0], and V is easier to differentiate than ρ.

$V(T) = 999.87 - 0.06426T + 0.0085043T^2 - 0.0000679T^3 \implies$

$V'(T) = -0.06426 + 0.0170086T - 0.0002037T^2$. Setting this equal to 0 and using the quadratic formula to

find T, we get $T = \dfrac{-0.0170086 \pm \sqrt{0.0170086^2 - 4 \cdot 0.0002037 \cdot 0.06426}}{2(-0.0002037)} \approx 3.9665°\text{C}$ or $79.5318°\text{C}$. Since

we are only interested in the region $0°\text{C} \le T \le 30°\text{C}$, we check the density ρ at the endpoints and at $3.9665°\text{C}$:

$\rho(0) \approx \dfrac{1000}{999.87} \approx 1.00013$; $\rho(30) \approx \dfrac{1000}{1003.7628} \approx 0.99625$; $\rho(3.9665) \approx \dfrac{1000}{999.7447} \approx 1.000255$. So water has

its maximum density at about $3.9665°\text{C}$.

71. We apply the Closed Interval Method to the continuous function

$I(t) = 0.00009045t^5 + 0.001438t^4 - 0.06561t^3 + 0.4598t^2 - 0.6270t + 99.33$ on $[0, 10]$. Its derivative is

$I'(t) = 0.00045225t^4 + 0.005752t^3 - 0.19683t^2 + 0.9196t - 0.6270$. Since I' exists for all t, the only critical

numbers of I occur when $I'(t) = 0$. We use a root-finder on a computer algebra system (or a graphing device) to

find that $I'(t) = 0$ when $t \approx -29.7186$, 0.8231, 5.1309, or 11.0459, but only the second and third roots lie in the

interval $[0, 10]$. The values of I at these critical numbers are $I(0.8231) \approx 99.09$ and $I(5.1309) \approx 100.67$. The

values of I at the endpoints of the interval are $I(0) = 99.33$ and $I(10) \approx 96.86$. Comparing these four numbers,

we see that food was most expensive at $t \approx 5.1309$ (corresponding roughly to August, 1989) and cheapest at

$t = 10$ (midyear 1994).

73. (a) $v(r) = k(r_0 - r)r^2 = kr_0r^2 - kr^3 \Rightarrow v'(r) = 2kr_0r - 3kr^2.$ $v'(r) = 0 \Rightarrow kr(2r_0 - 3r) = 0$

$\Rightarrow r = 0$ or $\frac{2}{3}r_0$ (but 0 is not in the interval). Evaluating v at $\frac{1}{2}r_0$, $\frac{2}{3}r_0$, and r_0, we get $v\left(\frac{1}{2}r_0\right) = \frac{1}{8}kr_0^3$,

$v\left(\frac{2}{3}r_0\right) = \frac{4}{27}kr_0^3$, and $v(r_0) = 0$. Since $\frac{4}{27} > \frac{1}{8}$, v attains its maximum value at $r = \frac{2}{3}r_0$. This supports the

statement in the text.

(b) From part (a), the maximum value of v is $\frac{4}{27}kr_0^3$.

(c)

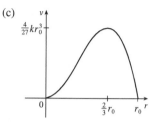

75. $f(x) = x^{101} + x^{51} + x + 1 \Rightarrow f'(x) = 101x^{100} + 51x^{50} + 1 \geq 1$ for all x, so $f'(x) = 0$ has no solution.

Thus, $f(x)$ has no critical number, so $f(x)$ can have no local maximum or minimum.

77. If f has a local minimum at c, then $g(x) = -f(x)$ has a local maximum at c, so $g'(c) = 0$ by the case of Fermat's

Theorem proved in the text. Thus, $f'(c) = -g'(c) = 0$.

4.2 The Mean Value Theorem

1. $f(x) = x^2 - 4x + 1$, $[0, 4]$. Since f is a polynomial, it is continuous and differentiable on $\mathbb{R}$, so it is continuous on

$[0, 4]$ and differentiable on $(0, 4)$. Also, $f(0) = 1 = f(4)$. $f'(c) = 0 \Leftrightarrow 2c - 4 = 0 \Leftrightarrow c = 2$, which is in

the open interval $(0, 4)$, so $c = 2$ satisfies the conclusion of Rolle's Theorem.

3. $f(x) = \sin 2\pi x$, $[-1, 1]$. f, being the composite of the sine function and the polynomial $2\pi x$, is continuous and

differentiable on $\mathbb{R}$, so it is continuous on $[-1, 1]$ and differentiable on $(-1, 1)$. Also, $f(-1) = 0 = f(1)$.

$f'(c) = 0 \Leftrightarrow 2\pi \cos 2\pi c = 0 \Leftrightarrow \cos 2\pi c = 0 \Leftrightarrow 2\pi c = \pm\frac{\pi}{2} + 2\pi n \Leftrightarrow c = \pm\frac{1}{4} + n$. If $n = 0$ or

± 1, then $c = \pm\frac{1}{4}, \pm\frac{3}{4}$ is in $(-1, 1)$.

5. $f(x) = 1 - x^{2/3}$. $f(-1) = 1 - (-1)^{2/3} = 1 - 1 = 0 = f(1)$. $f'(x) = -\frac{2}{3}x^{-1/3}$, so $f'(c) = 0$ has no

solution. This does not contradict Rolle's Theorem, since $f'(0)$ does not exist, and so f is not differentiable

on $(-1, 1)$.

7. $\dfrac{f(8) - f(0)}{8 - 0} = \dfrac{6 - 4}{8} = \dfrac{1}{4}$. The values of c which satisfy $f'(c) = \frac{1}{4}$ seem to be about $c = 0.8, 3.2, 4.4,$ and 6.1.

9. (a), (b) The equation of the secant line is

$$y - 5 = \frac{8.5 - 5}{8 - 1}(x - 1) \quad \Leftrightarrow$$

$$y = \tfrac{1}{2}x + \tfrac{9}{2}.$$

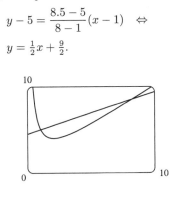

(c) $f(x) = x + 4/x \quad \Rightarrow \quad f'(x) = 1 - 4/x^2$.

So $f'(c) = \tfrac{1}{2} \quad \Rightarrow \quad c^2 = 8 \quad \Rightarrow \quad c = 2\sqrt{2}$, and

$f(c) = 2\sqrt{2} + \frac{4}{2\sqrt{2}} = 3\sqrt{2}$. Thus, an equation of the

tangent line is $y - 3\sqrt{2} = \tfrac{1}{2}\left(x - 2\sqrt{2}\right) \quad \Leftrightarrow$

$$y = \tfrac{1}{2}x + 2\sqrt{2}.$$

11. $f(x) = 3x^2 + 2x + 5$, $[-1, 1]$. f is continuous on $[-1, 1]$ and differentiable on $(-1, 1)$ since polynomials are

continuous and differentiable on $\mathbb{R}$. $\quad f'(c) = \dfrac{f(b) - f(a)}{b - a} \quad \Leftrightarrow \quad 6c + 2 = \dfrac{f(1) - f(-1)}{1 - (-1)} = \dfrac{10 - 6}{2} = 2 \quad \Leftrightarrow$

$6c = 0 \quad \Leftrightarrow \quad c = 0$, which is in $(-1, 1)$.

13. $f(x) = e^{-2x}$, $[0, 3]$. f is continuous and differentiable on $\mathbb{R}$, so it is continuous on $[0, 3]$ and differentiable on

$(0, 3)$. $\quad f'(c) = \dfrac{f(b) - f(a)}{b - a} \quad \Leftrightarrow \quad -2e^{-2c} = \dfrac{e^{-6} - e^0}{3 - 0} \quad \Leftrightarrow \quad e^{-2c} = \dfrac{1 - e^{-6}}{6} \quad \Leftrightarrow \quad -2c = \ln\left(\dfrac{1 - e^{-6}}{6}\right)$

$\Leftrightarrow \quad c = -\dfrac{1}{2}\ln\left(\dfrac{1 - e^{-6}}{6}\right) \approx 0.897$, which is in $(0, 3)$.

15. $f(x) = |x - 1|$. $f(3) - f(0) = |3 - 1| - |0 - 1| = 1$. Since $f'(c) = -1$ if $c < 1$ and $f'(c) = 1$ if $c > 1$,

$f'(c)(3 - 0) = \pm 3$ and so is never equal to 1. This does not contradict the Mean Value Theorem since $f'(1)$ does

not exist.

17. Let $f(x) = 1 + 2x + x^3 + 4x^5$. Then $f(-1) = -6 < 0$ and $f(0) = 1 > 0$. Since f is a polynomial, it is

continuous, so the Intermediate Value Theorem says that there is a number c between -1 and 0 such that $f(c) = 0$.

Thus, the given equation has a real root. Suppose the equation has distinct real roots a and b with $a < b$. Then

$f(a) = f(b) = 0$. Since f is a polynomial, it is differentiable on (a, b) and continuous on $[a, b]$. By Rolle's

Theorem, there is a number r in (a, b) such that $f'(r) = 0$. But $f'(x) = 2 + 3x^2 + 20x^4 \geq 2$ for all x, so $f'(x)$

can never be 0. This contradiction shows that the equation can't have two distinct real roots. Hence, it has exactly

one real root.

19. Let $f(x) = x^3 - 15x + c$ for x in $[-2, 2]$. If f has two real roots a and b in $[-2, 2]$, with $a < b$, then

$f(a) = f(b) = 0$. Since the polynomial f is continuous on $[a, b]$ and differentiable on (a, b), Rolle's Theorem

implies that there is a number r in (a, b) such that $f'(r) = 0$. Now $f'(r) = 3r^2 - 15$. Since r is in (a, b), which is

contained in $[-2, 2]$, we have $|r| < 2$, so $r^2 < 4$. It follows that $3r^2 - 15 < 3 \cdot 4 - 15 = -3 < 0$. This contradicts

$f'(r) = 0$, so the given equation can't have two real roots in $[-2, 2]$. Hence, it has at most one real root in $[-2, 2]$.

21. (a) Suppose that a cubic polynomial $P(x)$ has roots $a_1 < a_2 < a_3 < a_4$, so $P(a_1) = P(a_2) = P(a_3) = P(a_4)$.

By Rolle's Theorem there are numbers c_1, c_2, c_3 with $a_1 < c_1 < a_2$, $a_2 < c_2 < a_3$ and $a_3 < c_3 < a_4$ and

$P'(c_1) = P'(c_2) = P'(c_3) = 0$. Thus, the second-degree polynomial $P'(x)$ has three distinct real roots, which

is impossible.

(b) We prove by induction that a polynomial of degree n has at most n real roots. This is certainly true for $n = 1$.

Suppose that the result is true for all polynomials of degree n and let $P(x)$ be a polynomial of degree $n + 1$.

Suppose that $P(x)$ has more than $n + 1$ real roots, say $a_1 < a_2 < a_3 < \cdots < a_{n+1} < a_{n+2}$. Then

$P(a_1) = P(a_2) = \cdots = P(a_{n+2}) = 0$. By Rolle's Theorem there are real numbers $c_1, \ldots, c_{n+1}$ with

$a_1 < c_1 < a_2, \ldots, a_{n+1} < c_{n+1} < a_{n+2}$ and $P'(c_1) = \cdots = P'(c_{n+1}) = 0$. Thus, the nth degree

polynomial $P'(x)$ has at least $n + 1$ roots. This contradiction shows that $P(x)$ has at most $n + 1$ real roots.

23. By the Mean Value Theorem, $f(4) - f(1) = f'(c)(4 - 1)$ for some $c \in (1, 4)$. But for every $c \in (1, 4)$ we have

$f'(c) \geq 2$. Putting $f'(c) \geq 2$ into the above equation and substituting $f(1) = 10$, we get

$f(4) = f(1) + f'(c)(4 - 1) = 10 + 3f'(c) \geq 10 + 3 \cdot 2 = 16$. So the smallest possible value of $f(4)$ is 16.

25. Suppose that such a function f exists. By the Mean Value Theorem there is a number $0 < c < 2$ with

$f'(c) = \dfrac{f(2) - f(0)}{2 - 0} = \dfrac{5}{2}$. But this is impossible since $f'(x) \leq 2 < \frac{5}{2}$ for all x, so no such function can exist.

27. We use Exercise 26 with $f(x) = \sqrt{1 + x}$, $g(x) = 1 + \frac{1}{2}x$, and $a = 0$. Notice that $f(0) = 1 = g(0)$ and

$f'(x) = \dfrac{1}{2\sqrt{1 + x}} < \dfrac{1}{2} = g'(x)$ for $x > 0$. So by Exercise 26, $f(b) < g(b)$ $\Rightarrow$ $\sqrt{1 + b} < 1 + \frac{1}{2}b$ for $b > 0$.

Another method: Apply the Mean Value Theorem directly to either $f(x) = 1 + \frac{1}{2}x - \sqrt{1 + x}$ or $g(x) = \sqrt{1 + x}$

on $[0, b]$.

29. Let $f(x) = \sin x$ and let $b < a$. Then $f(x)$ is continuous on $[b, a]$ and differentiable on (b, a). By the Mean Value

Theorem, there is a number $c \in (b, a)$ with $\sin a - \sin b = f(a) - f(b) = f'(c)(a - b) = (\cos c)(a - b)$. Thus,

$|\sin a - \sin b| \leq |\cos c| \, |b - a| \leq |a - b|$. If $a < b$, then $|\sin a - \sin b| = |\sin b - \sin a| \leq |b - a| = |a - b|$. If

$a = b$, both sides of the inequality are 0.

31. For $x > 0$, $f(x) = g(x)$, so $f'(x) = g'(x)$. For $x < 0$, $f'(x) = (1/x)' = -1/x^2$ and

$g'(x) = (1 + 1/x)' = -1/x^2$, so again $f'(x) = g'(x)$. However, the domain of $g(x)$ is not an interval [it is

$(-\infty, 0) \cup (0, \infty)$] so we cannot conclude that $f - g$ is constant (in fact it is not).

33. Let $f(x) = \arcsin\left(\dfrac{x - 1}{x + 1}\right) - 2\arctan\sqrt{x} + \dfrac{\pi}{2}$. Note that the domain of f is $[0, \infty)$. Thus,

$f'(x) = \dfrac{1}{\sqrt{1 - \left(\dfrac{x - 1}{x + 1}\right)^2}} \dfrac{(x + 1) - (x - 1)}{(x + 1)^2} - \dfrac{2}{1 + x} \cdot \dfrac{1}{2\sqrt{x}} = \dfrac{1}{\sqrt{x}\,(x + 1)} - \dfrac{1}{\sqrt{x}\,(x + 1)} = 0$. Then

$f(x) = C$ on $(0, \infty)$ by Theorem 5. By continuity of f, $f(x) = C$ on $[0, \infty)$. To find C, we let $x = 0$ $\Rightarrow$
$\arcsin(-1) - 2\arctan(0) + \frac{\pi}{2} = C$ $\Rightarrow$ $-\frac{\pi}{2} - 0 + \frac{\pi}{2} = 0 = C$. Thus, $f(x) = 0$ $\Rightarrow$
$\arcsin\left(\dfrac{x-1}{x+1}\right) = 2\arctan\sqrt{x} - \frac{\pi}{2}$.

35. Let $g(t)$ and $h(t)$ be the position functions of the two runners and let $f(t) = g(t) - h(t)$. By hypothesis,
$f(0) = g(0) - h(0) = 0$ and $f(b) = g(b) - h(b) = 0$, where b is the finishing time. Then by the Mean Value
Theorem, there is a time c, with $0 < c < b$, such that $f'(c) = \dfrac{f(b) - f(0)}{b - 0}$. But $f(b) = f(0) = 0$, so $f'(c) = 0$.
Since $f'(c) = g'(c) - h'(c) = 0$, we have $g'(c) = h'(c)$. So at time c, both runners have the same speed
$g'(c) = h'(c)$.

4.3 How Derivatives Affect the Shape of a Graph

1. (a) f is increasing on $(0, 6)$ and $(8, 9)$.

(b) f is decreasing on $(6, 8)$.

(c) f is concave upward on $(2, 4)$ and $(7, 9)$.

(d) f is concave downward on $(0, 2)$ and $(4, 7)$.

(e) The points of inflection are $(2, 3)$, $(4, 4.5)$ and $(7, 4)$ (where the concavity changes).

3. (a) Use the Increasing/Decreasing (I/D) Test.

(b) Use the Concavity Test.

(c) At any value of x where the concavity changes, we have an inflection point at $(x, f(x))$.

5. (a) Since $f'(x) > 0$ on $(1, 5)$, f is increasing on this interval. Since $f'(x) < 0$ on $(0, 1)$ and $(5, 6)$, f is decreasing
on these intervals.

(b) Since $f'(x) = 0$ at $x = 1$ and f' changes from negative to positive there, f changes from decreasing to
increasing and has a local minimum at $x = 1$. Since $f'(x) = 0$ at $x = 5$ and f' changes from positive to
negative there, f changes from increasing to decreasing and has a local maximum at $x = 5$.

7. There is an inflection point at $x = 1$ because $f''(x)$ changes from negative to positive there, and so the graph of f
changes from concave downward to concave upward. There is an inflection point at $x = 7$ because $f''(x)$ changes
from positive to negative there, and so the graph of f changes from concave upward to concave downward.

9. The function must be always decreasing and concave downward.

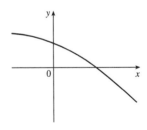

11. (a) $f(x) = x^3 - 12x + 1 \implies f'(x) = 3x^2 - 12 = 3(x+2)(x-2)$.

We don't need to include "3" in the chart to determine the sign of $f'(x)$.

Interval	$x+2$	$x-2$	$f'(x)$	f
$x < -2$	$-$	$-$	$+$	increasing on $(-\infty, -2)$
$-2 < x < 2$	$+$	$-$	$-$	decreasing on $(-2, 2)$
$x > 2$	$+$	$+$	$+$	increasing on $(2, \infty)$

So f is increasing on $(-\infty, -2)$ and $(2, \infty)$ and f is decreasing on $(-2, 2)$.

(b) f changes from increasing to decreasing at $x = -2$ and from decreasing to increasing at $x = 2$. Thus, $f(-2) = 17$ is a local maximum value and $f(2) = -15$ is a local minimum value.

(c) $f''(x) = 6x$. $f''(x) > 0 \iff x > 0$ and $f''(x) < 0 \iff x < 0$. Thus, f is concave upward on $(0, \infty)$ and concave downward on $(-\infty, 0)$. There is an inflection point where the concavity changes, at $(0, f(0)) = (0, 1)$.

13. (a) $f(x) = x^4 - 2x^2 + 3 \implies f'(x) = 4x^3 - 4x = 4x(x^2 - 1) = 4x(x+1)(x-1)$.

Interval	$x+1$	x	$x-1$	$f'(x)$	f
$x < -1$	$-$	$-$	$-$	$-$	decreasing on $(-\infty, -1)$
$-1 < x < 0$	$+$	$-$	$-$	$+$	increasing on $(-1, 0)$
$0 < x < 1$	$+$	$+$	$-$	$-$	decreasing on $(0, 1)$
$x > 1$	$+$	$+$	$+$	$+$	increasing on $(1, \infty)$

So f is increasing on $(-1, 0)$ and $(1, \infty)$ and f is decreasing on $(-\infty, -1)$ and $(0, 1)$.

(b) f changes from increasing to decreasing at $x = 0$ and from decreasing to increasing at $x = -1$ and $x = 1$. Thus, $f(0) = 3$ is a local maximum value and $f(\pm 1) = 2$ are local minimum values.

(c) $f''(x) = 12x^2 - 4 = 12(x^2 - \frac{1}{3}) = 12(x + 1/\sqrt{3})(x - 1/\sqrt{3})$. $f''(x) > 0 \iff x < -1/\sqrt{3}$ or $x > 1/\sqrt{3}$ and $f''(x) < 0 \iff -1/\sqrt{3} < x < 1/\sqrt{3}$. Thus, f is concave upward on $(-\infty, -\sqrt{3}/3)$ and $(\sqrt{3}/3, \infty)$ and concave downward on $(-\sqrt{3}/3, \sqrt{3}/3)$. There are inflection points at $(\pm\sqrt{3}/3, \frac{22}{9})$.

15. (a) $f(x) = x - 2\sin x$ on $(0, 3\pi) \implies f'(x) = 1 - 2\cos x$. $f'(x) > 0 \iff 1 - 2\cos x > 0 \iff \cos x < \frac{1}{2}$ $\iff \frac{\pi}{3} < x < \frac{5\pi}{3}$ or $\frac{7\pi}{3} < x < 3\pi$. $f'(x) < 0 \iff \cos x > \frac{1}{2} \iff 0 < x < \frac{\pi}{3}$ or $\frac{5\pi}{3} < x < \frac{7\pi}{3}$. So f is increasing on $(\frac{\pi}{3}, \frac{5\pi}{3})$ and $(\frac{7\pi}{3}, 3\pi)$, and f is decreasing on $(0, \frac{\pi}{3})$ and $(\frac{5\pi}{3}, \frac{7\pi}{3})$.

(b) f changes from increasing to decreasing at $x = \frac{5\pi}{3}$, and from decreasing to increasing at $x = \frac{\pi}{3}$ and at $x = \frac{7\pi}{3}$. Thus, $f(\frac{5\pi}{3}) = \frac{5\pi}{3} + \sqrt{3} \approx 6.97$ is a local maximum value and $f(\frac{\pi}{3}) = \frac{\pi}{3} - \sqrt{3} \approx -0.68$ and $f(\frac{7\pi}{3}) = \frac{7\pi}{3} - \sqrt{3} \approx 5.60$ are local minimum values.

(c) $f''(x) = 2\sin x > 0 \iff 0 < x < \pi$ and $2\pi < x < 3\pi$, $f''(x) < 0 \iff \pi < x < 2\pi$. Thus, f is concave upward on $(0, \pi)$ and $(2\pi, 3\pi)$, and f is concave downward on $(\pi, 2\pi)$. There are inflection points at (π, π) and $(2\pi, 2\pi)$.

17. (a) $y = f(x) = xe^x$ $\Rightarrow$ $f'(x) = xe^x + e^x = e^x(x+1)$. So $f'(x) > 0$ $\Leftrightarrow$ $x + 1 > 0$ $\Leftrightarrow$ $x > -1$.

Thus, f is increasing on $(-1, \infty)$ and decreasing on $(-\infty, -1)$.

(b) f changes from decreasing to increasing at its only critical number, $x = -1$. Thus, $f(-1) = -e^{-1}$ is a local minimum value.

(c) $f'(x) = e^x(x+1)$ $\Rightarrow$ $f''(x) = e^x(1) + (x+1)e^x = e^x(x+2)$. So $f''(x) > 0$ $\Leftrightarrow$ $x + 2 > 0$ $\Leftrightarrow$ $x > -2$. Thus, f is concave upward on $(-2, \infty)$ and concave downward on $(-\infty, -2)$. Since the concavity changes direction at $x = -2$, the point $(-2, -2e^{-2})$ is an inflection point.

19. (a) $y = f(x) = \dfrac{\ln x}{\sqrt{x}}$. (Note that f is only defined for $x > 0$.)

$$f'(x) = \frac{\sqrt{x}\,(1/x) - \ln x\left(\frac{1}{2}x^{-1/2}\right)}{x} = \frac{\dfrac{1}{\sqrt{x}} - \dfrac{\ln x}{2\sqrt{x}}}{x} \cdot \frac{2\sqrt{x}}{2\sqrt{x}} = \frac{2 - \ln x}{2x^{3/2}} > 0 \quad \Leftrightarrow$$

$2 - \ln x > 0$ $\Leftrightarrow$ $\ln x < 2$ $\Leftrightarrow$ $x < e^2$. Therefore f is increasing on $(0, e^2)$ and decreasing on (e^2, ∞).

(b) f changes from increasing to decreasing at $x = e^2$, so $f(e^2) = \dfrac{\ln e^2}{\sqrt{e^2}} = \dfrac{2}{e}$ is a local maximum value.

(c) $f''(x) = \dfrac{2x^{3/2}(-1/x) - (2 - \ln x)(3x^{1/2})}{(2x^{3/2})^2} = \dfrac{-2x^{1/2} + 3x^{1/2}(\ln x - 2)}{4x^3}$

$= \dfrac{x^{1/2}(-2 + 3\ln x - 6)}{4x^3} = \dfrac{3\ln x - 8}{4x^{5/2}}$

$f''(x) = 0$ $\Leftrightarrow$ $\ln x = \frac{8}{3}$ $\Leftrightarrow$ $x = e^{8/3}$. $f''(x) > 0$ $\Leftrightarrow$ $x > e^{8/3}$, so f is concave upward on $(e^{8/3}, \infty)$ and concave downward on $(0, e^{8/3})$. There is an inflection point at $\left(e^{8/3}, \frac{8}{3}e^{-4/3}\right) \approx (14.39, 0.70)$.

21. $f(x) = x^5 - 5x + 3$ $\Rightarrow$ $f'(x) = 5x^4 - 5 = 5(x^2 + 1)(x + 1)(x - 1)$.

First Derivative Test: $f'(x) < 0$ $\Rightarrow$ $-1 < x < 1$ and $f'(x) > 0$ $\Rightarrow$ $x > 1$ or $x < -1$. Since f' changes from positive to negative at $x = -1$, $f(-1) = 7$ is a local maximum value; and since f' changes from negative to positive at $x = 1$, $f(1) = -1$ is a local minimum value.

Second Derivative Test: $f''(x) = 20x^3$. $f'(x) = 0$ $\Leftrightarrow$ $x = \pm 1$. $f''(-1) = -20 < 0$ $\Rightarrow$ $f(-1) = 7$ is a local maximum value. $f''(1) = 20 > 0$ $\Rightarrow$ $f(1) = -1$ is a local minimum value.

Preference: For this function, the two tests are equally easy.

23. $f(x) = x + \sqrt{1 - x}$ $\Rightarrow$ $f'(x) = 1 + \frac{1}{2}(1 - x)^{-1/2}(-1) = 1 - \dfrac{1}{2\sqrt{1 - x}}$. Note that f is defined for

$1 - x \geq 0$; that is, for $x \leq 1$. $f'(x) = 0$ $\Rightarrow$ $2\sqrt{1 - x} = 1$ $\Rightarrow$ $\sqrt{1 - x} = \frac{1}{2}$ $\Rightarrow$ $1 - x = \frac{1}{4}$ $\Rightarrow$

$x = \frac{3}{4}$. f' does not exist at $x = 1$, but we can't have a local maximum or minimum at an endpoint.

First Derivative Test: $f'(x) > 0$ $\Rightarrow$ $x < \frac{3}{4}$ and $f'(x) < 0$ $\Rightarrow$ $\frac{3}{4} < x < 1$. Since f' changes from positive to negative at $x = \frac{3}{4}$, $f\left(\frac{3}{4}\right) = \frac{5}{4}$ is a local maximum value.

Second Derivative Test: $f''(x) = -\frac{1}{2}\left(-\frac{1}{2}\right)(1 - x)^{-3/2}(-1) = -\dfrac{1}{4\left(\sqrt{1 - x}\right)^3}$. $f''\left(\frac{3}{4}\right) = -2 < 0$ $\Rightarrow$

$f\left(\frac{3}{4}\right) = \frac{5}{4}$ is a local maximum value.

Preference: The First Derivative Test may be slightly easier to apply in this case.

25. (a) By the Second Derivative Test, if $f'(2) = 0$ and $f''(2) = -5 < 0$, f has a local maximum at $x = 2$.

(b) If $f'(6) = 0$, we know that f has a horizontal tangent at $x = 6$. Knowing that $f''(6) = 0$ does not provide any additional information since the Second Derivative Test fails. For example, the first and second derivatives of $y = (x-6)^4$, $y = -(x-6)^4$, and $y = (x-6)^3$ all equal zero for $x = 6$, but the first has a local minimum at $x = 6$, the second has a local maximum at $x = 6$, and the third has an inflection point at $x = 6$.

27. $f'(0) = f'(2) = f'(4) = 0$ ⇔ horizontal tangents at $x = 0, 2, 4$. $f'(x) > 0$ if $x < 0$ or $2 < x < 4$ ⇒ f is increasing on $(-\infty, 0)$ and $(2, 4)$. $f'(x) < 0$ if $0 < x < 2$ or $x > 4$ ⇒ f is decreasing on $(0, 2)$ and $(4, \infty)$. $f''(x) > 0$ if $1 < x < 3$ ⇒ f is concave upward on $(1, 3)$. $f''(x) < 0$ if $x < 1$ or $x > 3$ ⇒ f is concave downward on $(-\infty, 1)$ and $(3, \infty)$. There are inflection points when $x = 1$ and 3.

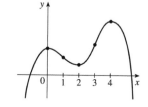

29. $f'(x) > 0$ if $|x| < 2$ ⇒ f is increasing on $(-2, 2)$. $f'(x) < 0$ if $|x| > 2$ ⇒ f is decreasing on $(-\infty, -2)$ and $(2, \infty)$. $f'(-2) = 0$ ⇒ horizontal tangent at $x = -2$. $\lim_{x \to 2} |f'(x)| = \infty$ ⇒ there is a vertical asymptote or vertical tangent (cusp) at $x = 2$. $f''(x) > 0$ if $x \neq 2$ ⇒ f is concave upward on $(-\infty, 2)$ and $(2, \infty)$.

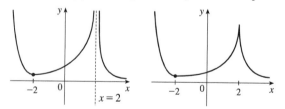

31. (a) f is increasing where f' is positive, that is, on $(0, 2)$, $(4, 6)$, and $(8, \infty)$; and decreasing where f' is negative, that is, on $(2, 4)$ and $(6, 8)$.

(b) f has local maxima where f' changes from positive to negative, at $x = 2$ and at $x = 6$, and local minima where f' changes from negative to positive, at $x = 4$ and at $x = 8$.

(c) f is concave upward (CU) where f' is increasing, that is, on $(3, 6)$ and $(6, \infty)$, and concave downward (CD) where f' is decreasing, that is, on $(0, 3)$.

(d) There is a point of inflection where f changes from being CD to being CU, that is, at $x = 3$.

(e)

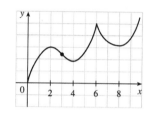

33. (a) $f(x) = 2x^3 - 3x^2 - 12x$ ⇒ $f'(x) = 6x^2 - 6x - 12 = 6(x^2 - x - 2) = 6(x-2)(x+1)$. $f'(x) > 0$ ⇔ $x < -1$ or $x > 2$ and $f'(x) < 0$ ⇔ $-1 < x < 2$. So f is increasing on $(-\infty, -1)$ and $(2, \infty)$, and f is decreasing on $(-1, 2)$.

(b) Since f changes from increasing to decreasing at $x = -1$, $f(-1) = 7$ is a local maximum value. Since f changes from decreasing to increasing at $x = 2$, $f(2) = -20$ is a local minimum value.

(c) $f''(x) = 6(2x - 1) \Rightarrow f''(x) > 0$ on $\left(\frac{1}{2}, \infty\right)$ and $f''(x) < 0$ on $\left(-\infty, \frac{1}{2}\right)$. So f is concave upward on $\left(\frac{1}{2}, \infty\right)$ and concave downward on $\left(-\infty, \frac{1}{2}\right)$. There is a change in concavity at $x = \frac{1}{2}$, and we have an inflection point at $\left(\frac{1}{2}, -\frac{13}{2}\right)$.

(d)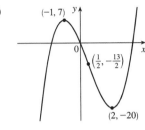

35. (a) $f(x) = x^4 - 6x^2 \Rightarrow f'(x) = 4x^3 - 12x = 4x(x^2 - 3) = 0$ when $x = 0, \pm\sqrt{3}$.

Interval	$4x$	$x^2 - 3$	$f'(x)$	f
$x < -\sqrt{3}$	$-$	$+$	$-$	decreasing on $\left(-\infty, -\sqrt{3}\right)$
$-\sqrt{3} < x < 0$	$-$	$-$	$+$	increasing on $\left(-\sqrt{3}, 0\right)$
$0 < x < \sqrt{3}$	$+$	$-$	$-$	decreasing on $\left(0, \sqrt{3}\right)$
$x > \sqrt{3}$	$+$	$+$	$+$	increasing on $\left(\sqrt{3}, \infty\right)$

(b) Local minimum values $f\left(\pm\sqrt{3}\right) = -9$, local maximum value $f(0) = 0$

(c) $f''(x) = 12x^2 - 12 = 12(x^2 - 1) > 0 \Leftrightarrow x^2 > 1 \Leftrightarrow |x| > 1 \Leftrightarrow x > 1$ or $x < -1$, so f is CU on $(-\infty, -1)$, $(1, \infty)$ and CD on $(-1, 1)$. Inflection points at $(\pm 1, -5)$

(d)

37. (a) $h(x) = 3x^5 - 5x^3 + 3 \Rightarrow h'(x) = 15x^4 - 15x^2 = 15x^2(x^2 - 1) = 0$ when $x = 0, \pm 1$. Since $15x^2$ is nonnegative, $h'(x) > 0 \Leftrightarrow x^2 > 1 \Leftrightarrow |x| > 1 \Leftrightarrow x > 1$ or $x < -1$, so h is increasing on $(-\infty, -1)$ and $(1, \infty)$ and decreasing on $(-1, 1)$, with a horizontal tangent at $x = 0$.

(b) Local maximum value $h(-1) = 5$, local minimum value $h(1) = 1$

(c) $h''(x) = 60x^3 - 30x = 30x(2x^2 - 1)$

$\qquad = 60x\left(x + \frac{1}{\sqrt{2}}\right)\left(x - \frac{1}{\sqrt{2}}\right) \Rightarrow$

$h''(x) > 0$ when $x > \frac{1}{\sqrt{2}}$ or $-\frac{1}{\sqrt{2}} < x < 0$, so h is CU on

$\left(-\frac{1}{\sqrt{2}}, 0\right)$ and $\left(\frac{1}{\sqrt{2}}, \infty\right)$ and CD on $\left(-\infty, -\frac{1}{\sqrt{2}}\right)$ and $\left(0, \frac{1}{\sqrt{2}}\right)$.

Inflection points at $(0, 3)$ and $\left(\pm\frac{1}{\sqrt{2}}, 3 \mp \frac{7}{8}\sqrt{2}\right)$ [about

$(-0.71, 4.24)$ and $(0.71, 1.76)$].

(d)

39. (a) $A(x) = x\sqrt{x+3}$ $\Rightarrow$

$$A'(x) = x \cdot \tfrac{1}{2}(x+3)^{-1/2} + \sqrt{x+3} \cdot 1 = \frac{x}{2\sqrt{x+3}} + \sqrt{x+3} = \frac{x + 2(x+3)}{2\sqrt{x+3}} = \frac{3x+6}{2\sqrt{x+3}}.$$

The domain of A is $[-3, \infty)$. $A'(x) > 0$ for $x > -2$ and $A'(x) < 0$ for $-3 < x < -2$, so A is increasing on

$(-2, \infty)$ and decreasing on $(-3, -2)$.

(b) $A(-2) = -2$ is a local minimum value.

(d)

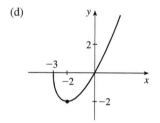

(c) $A''(x) = \dfrac{2\sqrt{x+3} \cdot 3 - (3x+6) \cdot \dfrac{1}{\sqrt{x+3}}}{\left(2\sqrt{x+3}\right)^2}$

$= \dfrac{6(x+3) - (3x+6)}{4(x+3)^{3/2}} = \dfrac{3x+12}{4(x+3)^{3/2}} = \dfrac{3(x+4)}{4(x+3)^{3/2}}$

$A''(x) > 0$ for all $x > -3$, so A is concave upward on $(-3, \infty)$.

There is no inflection point.

41. (a) $C(x) = x^{1/3}(x+4) = x^{4/3} + 4x^{1/3}$ $\Rightarrow$ $C'(x) = \tfrac{4}{3}x^{1/3} + \tfrac{4}{3}x^{-2/3} = \tfrac{4}{3}x^{-2/3}(x+1) = \dfrac{4(x+1)}{3\sqrt[3]{x^2}}$.

$C'(x) > 0$ if $-1 < x < 0$ or $x > 0$ and $C'(x) < 0$ for $x < -1$, so C is increasing on $(-1, \infty)$ and C is

decreasing on $(-\infty, -1)$.

(b) $C(-1) = -3$ is a local minimum value.

(d)

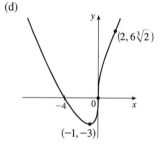

(c) $C''(x) = \tfrac{4}{9}x^{-2/3} - \tfrac{8}{9}x^{-5/3} = \tfrac{4}{9}x^{-5/3}(x-2) = \dfrac{4(x-2)}{9\sqrt[3]{x^5}}$.

$C''(x) < 0$ for $0 < x < 2$ and $C''(x) > 0$ for $x < 0$ and $x > 2$, so

C is concave downward on $(0, 2)$ and concave upward on $(-\infty, 0)$

and $(2, \infty)$. There are inflection points at $(0, 0)$ and

$\left(2, 6\sqrt[3]{2}\right) \approx (2, 7.56)$.

43. (a) $f(\theta) = 2\cos\theta - \cos 2\theta$, $0 \le \theta \le 2\pi$.

$f'(\theta) = -2\sin\theta + 2\sin 2\theta = -2\sin\theta + 2(2\sin\theta\cos\theta) = 2\sin\theta(2\cos\theta - 1)$.

Interval	$\sin\theta$	$2\cos\theta - 1$	$f'(\theta)$	f
$0 < \theta < \tfrac{\pi}{3}$	$+$	$+$	$+$	increasing on $\left(0, \tfrac{\pi}{3}\right)$
$\tfrac{\pi}{3} < \theta < \pi$	$+$	$-$	$-$	decreasing on $\left(\tfrac{\pi}{3}, \pi\right)$
$\pi < \theta < \tfrac{5\pi}{3}$	$-$	$-$	$+$	increasing on $\left(\pi, \tfrac{5\pi}{3}\right)$
$\tfrac{5\pi}{3} < \theta < 2\pi$	$-$	$+$	$-$	decreasing on $\left(\tfrac{5\pi}{3}, 2\pi\right)$

(b) $f\left(\tfrac{\pi}{3}\right) = \tfrac{3}{2}$ and $f\left(\tfrac{5\pi}{3}\right) = \tfrac{3}{2}$ are local maximum values and $f(\pi) = -3$ is a local minimum value.

(c) $f'(\theta) = -2\sin\theta + 2\sin 2\theta \;\Rightarrow$

$\quad f''(\theta) = -2\cos\theta + 4\cos 2\theta = -2\cos\theta + 4(2\cos^2\theta - 1)$

$\quad\quad = 2(4\cos^2\theta - \cos\theta - 2)$

$\quad f''(\theta) = 0 \;\Leftrightarrow\; \cos\theta = \dfrac{1 \pm \sqrt{33}}{8} \;\Leftrightarrow\; \theta = \cos^{-1}\!\left(\dfrac{1 \pm \sqrt{33}}{8}\right)$

$\quad\Leftrightarrow\; \theta = \cos^{-1}\!\left(\dfrac{1 + \sqrt{33}}{8}\right) \approx 0.5678,$

$\quad 2\pi - \cos^{-1}\!\left(\dfrac{1+\sqrt{33}}{8}\right) \approx 5.7154,\; \cos^{-1}\!\left(\dfrac{1-\sqrt{33}}{8}\right) \approx 2.2057,\text{ or }2\pi - \cos^{-1}\!\left(\dfrac{1-\sqrt{33}}{8}\right) \approx 4.0775.$

(d)

Denote these four values of θ by $\theta_1, \theta_4, \theta_2,$ and θ_3, respectively. Then f is CU on $(0, \theta_1)$, CD on (θ_1, θ_2), CU on (θ_2, θ_3), CD on (θ_3, θ_4), and CU on $(\theta_4, 2\pi)$. To find the *exact* y-coordinate for $\theta = \theta_1$, we have

$$f(\theta_1) = 2\cos\theta_1 - \cos 2\theta_1 = 2\cos\theta_1 - (2\cos^2\theta_1 - 1) = 2\left(\dfrac{1+\sqrt{33}}{8}\right) - 2\left(\dfrac{1+\sqrt{33}}{8}\right)^2 + 1$$

$$= \tfrac{1}{4} + \tfrac{1}{4}\sqrt{33} - \tfrac{1}{32} - \tfrac{1}{16}\sqrt{33} - \tfrac{33}{32} + 1 = \tfrac{3}{16} + \tfrac{3}{16}\sqrt{33} = \tfrac{3}{16}\left(1 + \sqrt{33}\right) = y_1 \approx 1.26.$$

Similarly, $f(\theta_2) = \tfrac{3}{16}\left(1 - \sqrt{33}\right) = y_2 \approx -0.89$. So f has inflection points at $(\theta_1, y_1), (\theta_2, y_2), (\theta_3, y_2)$, and (θ_4, y_1).

45. $f(x) = \dfrac{x^2}{x^2 - 1} = \dfrac{x^2}{(x+1)(x-1)}$ has domain $(-\infty, -1) \cup (-1, 1) \cup (1, \infty)$.

(a) $\displaystyle\lim_{x\to\pm\infty} f(x) = \lim_{x\to\pm\infty} \dfrac{x^2/x^2}{(x^2-1)/x^2} = \lim_{x\to\pm\infty} \dfrac{1}{1-1/x^2} = \dfrac{1}{1-0} = 1$, so $y = 1$ is a HA.

$\quad\displaystyle\lim_{x\to-1^-}\dfrac{x^2}{x^2-1} = \infty$ since $x^2 \to 1$ and $(x^2 - 1) \to 0^+$ as $x \to -1^-$, so $x = -1$ is a VA.

$\quad\displaystyle\lim_{x\to 1^+}\dfrac{x^2}{x^2-1} = \infty$ since $x^2 \to 1$ and $(x^2 - 1) \to 0^+$ as $x \to 1^+$, so $x = 1$ is a VA.

(b) $f(x) = \dfrac{x^2}{x^2-1} \;\Rightarrow\; f'(x) = \dfrac{(x^2-1)(2x) - x^2(2x)}{(x^2-1)^2} = \dfrac{2x[(x^2-1) - x^2]}{(x^2-1)^2} = \dfrac{-2x}{(x^2-1)^2}.$ Since

$(x^2 - 1)^2$ is positive for all x in the domain of f, the sign of the derivative is determined by the sign of $-2x$. Thus, $f'(x) > 0$ if $x < 0$ $(x \ne -1)$ and $f'(x) < 0$ if $x > 0$ $(x \ne 1)$. So f is increasing on $(-\infty, -1)$ and $(-1, 0)$, and f is decreasing on $(0, 1)$ and $(1, \infty)$.

(c) $f'(x) = 0 \;\Rightarrow\; x = 0$ and $f(0) = 0$ is a local maximum value.

(d) $f''(x) = \dfrac{(x^2-1)^2(-2) - (-2x)\cdot 2(x^2-1)(2x)}{[(x^2-1)^2]^2}$

$\quad = \dfrac{2(x^2-1)[-(x^2-1) + 4x^2]}{(x^2-1)^4} = \dfrac{2(3x^2+1)}{(x^2-1)^3}.$

The sign of $f''(x)$ is determined by the denominator; that is, $f''(x) > 0$ if $|x| > 1$ and $f''(x) < 0$ if $|x| < 1$. Thus, f is CU on $(-\infty, -1)$ and $(1, \infty)$, and f is CD on $(-1, 1)$. There are no inflection points.

(e)

47. (a) $\lim\limits_{x \to -\infty} \left(\sqrt{x^2+1} - x\right) = \infty$ and

$$\lim_{x \to \infty} \left(\sqrt{x^2+1} - x\right) = \lim_{x \to \infty} \left(\sqrt{x^2+1} - x\right) \frac{\sqrt{x^2+1}+x}{\sqrt{x^2+1}+x} = \lim_{x \to \infty} \frac{1}{\sqrt{x^2+1}+x} = 0, \text{ so } y = 0 \text{ is a HA.}$$

(b) $f(x) = \sqrt{x^2+1} - x \;\Rightarrow\; f'(x) = \dfrac{x}{\sqrt{x^2+1}} - 1$. Since $\dfrac{x}{\sqrt{x^2+1}} < 1$ for all x, $f'(x) < 0$, so f is

decreasing on $\mathbb{R}$.

(c) No minimum or maximum

(e)

(d) $f''(x) = \dfrac{\left(x^2+1\right)^{1/2}(1) - x \cdot \frac{1}{2}\left(x^2+1\right)^{-1/2}(2x)}{\left(\sqrt{x^2+1}\right)^2}$

$\qquad = \dfrac{\left(x^2+1\right)^{1/2} - \dfrac{x^2}{\left(x^2+1\right)^{1/2}}}{x^2+1} = \dfrac{\left(x^2+1\right) - x^2}{\left(x^2+1\right)^{3/2}}$

$\qquad = \dfrac{1}{\left(x^2+1\right)^{3/2}} > 0$, so f is CU on $\mathbb{R}$. No IP

49. $f(x) = \ln(1 - \ln x)$ is defined when $x > 0$ (so that $\ln x$ is defined) and $1 - \ln x > 0$ [so that $\ln(1 - \ln x)$ is

defined]. The second condition is equivalent to $1 > \ln x \;\Leftrightarrow\; x < e$, so f has domain $(0, e)$.

(a) As $x \to 0^+$, $\ln x \to -\infty$, so $1 - \ln x \to \infty$ and $f(x) \to \infty$. As $x \to e^-$, $\ln x \to 1^-$, so $1 - \ln x \to 0^+$ and

$f(x) \to -\infty$. Thus, $x = 0$ and $x = e$ are vertical asymptotes. There is no horizontal asymptote.

(b) $f'(x) = \dfrac{1}{1 - \ln x}\left(-\dfrac{1}{x}\right) = -\dfrac{1}{x(1 - \ln x)} < 0$ on $(0, e)$. Thus, f is decreasing on its domain, $(0, e)$.

(c) $f'(x) \neq 0$ on $(0, e)$, so f has no local maximum or minimum value.

(d) $f''(x) = -\dfrac{-[x(1 - \ln x)]'}{[x(1 - \ln x)]^2} = \dfrac{x(-1/x) + (1 - \ln x)}{x^2(1 - \ln x)^2}$

(e)

$\qquad = -\dfrac{\ln x}{x^2(1 - \ln x)^2}$

so $f''(x) > 0 \;\Leftrightarrow\; \ln x < 0 \;\Leftrightarrow\; 0 < x < 1$. Thus, f is CU on

$(0, 1)$ and CD on $(1, e)$. There is an inflection point at $(1, 0)$.

51. (a) $\lim\limits_{x \to \pm\infty} e^{-1/(x+1)} = 1$ since $-1/(x+1) \to 0$, so $y = 1$ is a HA. $\lim\limits_{x \to -1^+} e^{-1/(x+1)} = 0$ since

$-1/(x+1) \to -\infty$, $\lim\limits_{x \to -1^-} e^{-1/(x+1)} = \infty$ since $-1/(x+1) \to \infty$, so $x = -1$ is a VA.

(b) $f(x) = e^{-1/(x+1)} \;\Rightarrow\; f'(x) = e^{-1/(x+1)}\left[-(-1)\dfrac{1}{(x+1)^2}\right]$ [Reciprocal Rule] $= e^{-1/(x+1)}/(x+1)^2$

$\;\Rightarrow\; f'(x) > 0$ for all x except -1, so f is increasing on $(-\infty, -1)$ and $(-1, \infty)$.

(c) No local maximum or minimum

(d) $f''(x) = \dfrac{(x+1)^2 e^{-1/(x+1)}\left[1/(x+1)^2\right] - e^{-1/(x+1)}\left[2(x+1)\right]}{\left[(x+1)^2\right]^2}$

$= \dfrac{e^{-1/(x+1)}\left[1 - (2x+2)\right]}{(x+1)^4} = -\dfrac{e^{-1/(x+1)}(2x+1)}{(x+1)^4} \quad \Rightarrow$

$f''(x) > 0 \quad \Leftrightarrow \quad 2x+1 < 0 \quad \Leftrightarrow \quad x < -\frac{1}{2}$, so f is CU on

$(-\infty, -1)$ and $\left(-1, -\frac{1}{2}\right)$, and CD on $\left(-\frac{1}{2}, \infty\right)$. f has an IP at

$\left(-\frac{1}{2}, e^{-2}\right)$.

(e)

53. (a)

From the graph, we get an estimate of $f(1) \approx 1.41$ as a local

maximum value, and no local minimum value.

$f(x) = \dfrac{x+1}{\sqrt{x^2+1}} \quad \Rightarrow \quad f'(x) = \dfrac{1-x}{(x^2+1)^{3/2}}.$

$f'(x) = 0 \quad \Leftrightarrow \quad x = 1.$ $f(1) = \frac{2}{\sqrt{2}} = \sqrt{2}$ is the exact value.

(b) From the graph in part (a), f increases most rapidly somewhere between $x = -\frac{1}{2}$ and $x = -\frac{1}{4}$. To find the

exact value, we need to find the maximum value of f', which we can do by finding the critical numbers of f'.

$f''(x) = \dfrac{2x^2 - 3x - 1}{(x^2+1)^{5/2}} = 0 \quad \Leftrightarrow \quad x = \dfrac{3 \pm \sqrt{17}}{4}.$ $x = \dfrac{3+\sqrt{17}}{4}$ corresponds to the *minimum* value of f'.

The maximum value of f' is at $\left(\dfrac{3-\sqrt{17}}{4}, \sqrt{\dfrac{7}{6} - \dfrac{\sqrt{17}}{6}}\right) \approx (-0.28, 0.69).$

55. $f(x) = \cos x + \frac{1}{2}\cos 2x \quad \Rightarrow \quad f'(x) = -\sin x - \sin 2x \quad \Rightarrow \quad f''(x) = -\cos x - 2\cos 2x$

(a)

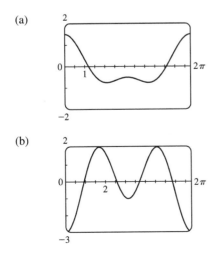

From the graph of f, it seems that f is CD on $(0,1)$, CU on $(1, 2.5)$,

CD on $(2.5, 3.7)$, CU on $(3.7, 5.3)$, and CD on $(5.3, 2\pi)$. The points

of inflection appear to be at $(1, 0.4)$, $(2.5, -0.6)$, $(3.7, -0.6)$, and

$(5.3, 0.4)$.

(b)

From the graph of f'' (and zooming in near the zeros), it seems that f

is CD on $(0, 0.94)$, CU on $(0.94, 2.57)$, CD on $(2.57, 3.71)$, CU on

$(3.71, 5.35)$, and CD on $(5.35, 2\pi)$. Refined estimates of the

inflection points are $(0.94, 0.44)$, $(2.57, -0.63)$, $(3.71, -0.63)$, and

$(5.35, 0.44)$.

57. In Maple, we define f and then use the command

plot (diff (diff (f,x) ,x) ,x=-3..3);. In Mathematica, we

define f and then use Plot [Dt [Dt [f,x] ,x] , {x,-3,3}].We

see that $f'' > 0$ for $x > 0.1$ and $f'' < 0$ for $x < 0.1$. So f is concave

up on $(0.1, \infty)$ and concave down on $(-\infty, 0.1)$.

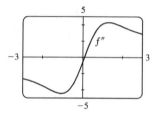

59. Most students learn more in the third hour of studying than in the eighth hour, so $K(3) - K(2)$ is larger than

$K(8) - K(7)$. In other words, as you begin studying for a test, the rate of knowledge gain is large and then starts to

taper off, so $K'(t)$ decreases and the graph of K is concave downward.

61.

From the graph, we estimate that the most rapid increase in the

percentage of households in the United States with at least one VCR

occurs at about $t = 8$. To maximize the first derivative, we need to

determine the values for which the second derivative is 0. We'll use

$V(t) = \dfrac{a}{1 + be^{ct}}$, and substitute $a = 85$, $b = 53$, and $c = -0.5$ later.

$$V'(t) = -\frac{a(bce^{ct})}{(1 + be^{ct})^2} \quad \text{[by the Reciprocal Rule]} \quad \text{and}$$

$$V''(t) = -abc \cdot \frac{(1 + be^{ct})^2 \cdot ce^{ct} - e^{ct} \cdot 2(1 + be^{ct}) \cdot bce^{ct}}{\left[(1 + be^{ct})^2\right]^2}$$

$$= \frac{-abc \cdot ce^{ct}(1 + be^{ct})[(1 + be^{ct}) - 2be^{ct}]}{(1 + be^{ct})^4} = \frac{-abc^2e^{ct}(1 - be^{ct})}{(1 + be^{ct})^3}$$

So $V''(t) = 0 \iff 1 = be^{ct} \iff e^{ct} = 1/b$. Now graph $y = e^{-0.5t}$ and $y = \frac{1}{53}$. These graphs intersect at

$t \approx 7.94$ years, which corresponds to roughly midyear 1988. [Alternatively, we could use the rootfinder on a

calculator to solve $e^{-0.5t} = \frac{1}{53}$. Or, if you have already studied logarithms, you can solve $e^{ct} = 1/b$ as follows:

$ct = \ln(1/b) \iff t = (1/c) \ln(1/b) = -2 \ln \frac{1}{53} \approx 7.94$ years.]

63. $f(x) = ax^3 + bx^2 + cx + d \Rightarrow f'(x) = 3ax^2 + 2bx + c$. We are

given that $f(1) = 0$ and $f(-2) = 3$, so $f(1) = a + b + c + d = 0$ and

$f(-2) = -8a + 4b - 2c + d = 3$. Also $f'(1) = 3a + 2b + c = 0$ and

$f'(-2) = 12a - 4b + c = 0$ by Fermat's Theorem. Solving these four

equations, we get $a = \frac{2}{9}, b = \frac{1}{3}, c = -\frac{4}{3}, d = \frac{7}{9}$, so the function is

$f(x) = \frac{1}{9}(2x^3 + 3x^2 - 12x + 7)$.

65. Suppose that f is differentiable on an interval I and $f'(x) > 0$ for all x in I except $x = c$. To show that f is increasing on I, let x_1, x_2 be two numbers in I with $x_1 < x_2$.

> *Case 1* $x_1 < x_2 < c$. Let J be the interval $\{x \in I \mid x < c\}$. By applying the Increasing/Decreasing Test to f on J, we see that f is increasing on J, so $f(x_1) < f(x_2)$.
>
> *Case 2* $c < x_1 < x_2$. Apply the Increasing/Decreasing Test to f on $K = \{x \in I \mid x > c\}$.
>
> *Case 3* $x_1 < x_2 = c$. Apply the proof of the Increasing/Decreasing Test, using the Mean Value Theorem (MVT) on the interval $[x_1, x_2]$ and noting that the MVT does not require f to be differentiable at the endpoints of $[x_1, x_2]$.
>
> *Case 4* $c = x_1 < x_2$. Same proof as in Case 3.
>
> *Case 5* $x_1 < c < x_2$. By Cases 3 and 4, f is increasing on $[x_1, c]$ and on $[c, x_2]$, so $f(x_1) < f(c) < f(x_2)$.

In all cases, we have shown that $f(x_1) < f(x_2)$. Since x_1, x_2 were any numbers in I with $x_1 < x_2$, we have shown that f is increasing on I.

67. (a) Since f and g are positive, increasing, and CU on I with f'' and g'' never equal to 0, we have $f > 0$,

$f' \geq 0$, $f'' > 0$, $g > 0$, $g' \geq 0$, $g'' > 0$ on I. Then $(fg)' = f'g + fg'$ $\Rightarrow$

$(fg)'' = f''g + 2f'g' + fg'' \geq f''g + fg'' > 0$ on I $\Rightarrow$ fg is CU on I.

(b) In part (a), if f and g are both decreasing instead of increasing, then $f' \leq 0$ and $g' \leq 0$ on I, so we still have

$2f'g' \geq 0$ on I. Thus, $(fg)'' = f''g + 2f'g' + fg'' \geq f''g + fg'' > 0$ on I $\Rightarrow$ fg is CU on I as in part (a).

(c) Suppose f is increasing and g is decreasing [with f and g positive and CU]. Then $f' \geq 0$ and $g' \leq 0$ on I,

so $2f'g' \leq 0$ on I and the argument in parts (a) and (b) fails.

> **Example 1.** $I = (0, \infty)$, $f(x) = x^3$, $g(x) = 1/x$. Then $(fg)(x) = x^2$, so $(fg)'(x) = 2x$ and
> $(fg)''(x) = 2 > 0$ on I. Thus, fg is CU on I.

> **Example 2.** $I = (0, \infty)$, $f(x) = 4x\sqrt{x}$, $g(x) = 1/x$. Then $(fg)(x) = 4\sqrt{x}$, so $(fg)'(x) = 2/\sqrt{x}$ and
> $(fg)''(x) = -1/\sqrt{x^3} < 0$ on I. Thus, fg is CD on I.

> **Example 3.** $I = (0, \infty)$, $f(x) = x^2$, $g(x) = 1/x$. Thus, $(fg)(x) = x$, so fg is linear on I.

69. $f(x) = \tan x - x$ $\Rightarrow$ $f'(x) = \sec^2 x - 1 > 0$ for $0 < x < \frac{\pi}{2}$ since $\sec^2 x > 1$ for $0 < x < \frac{\pi}{2}$. So f is

increasing on $\left(0, \frac{\pi}{2}\right)$. Thus, $f(x) > f(0) = 0$ for $0 < x < \frac{\pi}{2}$ $\Rightarrow$ $\tan x - x > 0$ $\Rightarrow$ $\tan x > x$ for

$0 < x < \frac{\pi}{2}$.

71. Let the cubic function be $f(x) = ax^3 + bx^2 + cx + d \Rightarrow f'(x) = 3ax^2 + 2bx + c \Rightarrow f''(x) = 6ax + 2b$.

So f is CU when $6ax + 2b > 0 \Leftrightarrow x > -b/(3a)$, CD when $x < -b/(3a)$, and so the only point of inflection

occurs when $x = -b/(3a)$. If the graph has three x-intercepts x_1, x_2 and x_3, then the expression for $f(x)$ must

factor as $f(x) = a(x - x_1)(x - x_2)(x - x_3)$. Multiplying these factors together gives us

$f(x) = a\left[x^3 - (x_1 + x_2 + x_3)x^2 + (x_1x_2 + x_1x_3 + x_2x_3)x - x_1x_2x_3\right]$. Equating the coefficients of the

x^2-terms for the two forms of f gives us $b = -a(x_1 + x_2 + x_3)$. Hence, the x-coordinate of the point of inflection

is $-\dfrac{b}{3a} = -\dfrac{-a(x_1 + x_2 + x_3)}{3a} = \dfrac{x_1 + x_2 + x_3}{3}$.

73. By hypothesis $g = f'$ is differentiable on an open interval containing c. Since $(c, f(c))$ is a point of inflection, the

concavity changes at $x = c$, so $f''(x)$ changes signs at $x = c$. Hence, by the First Derivative Test, f' has a local

extremum at $x = c$. Thus, by Fermat's Theorem $f''(c) = 0$.

75. Using the fact that $|x| = \sqrt{x^2}$, we have that $g(x) = x\sqrt{x^2} \Rightarrow g'(x) = \sqrt{x^2} + \sqrt{x^2} = 2\sqrt{x^2} = 2|x| \Rightarrow$

$g''(x) = 2x(x^2)^{-1/2} = \dfrac{2x}{|x|} < 0$ for $x < 0$ and $g''(x) > 0$ for $x > 0$, so $(0, 0)$ is an inflection point. But $g''(0)$

does not exist.

4.4 Indeterminate Forms and L'Hospital's Rule

The use of l'Hospital's Rule is indicated by an **H** above the equal sign: $\overset{H}{=}$.

1. (a) $\lim\limits_{x \to a} \dfrac{f(x)}{g(x)}$ is an indeterminate form of type $\dfrac{0}{0}$.

(b) $\lim\limits_{x \to a} \dfrac{f(x)}{p(x)} = 0$ because the numerator approaches 0 while the denominator becomes large.

(c) $\lim\limits_{x \to a} \dfrac{h(x)}{p(x)} = 0$ because the numerator approaches a finite number while the denominator becomes large.

(d) If $\lim\limits_{x \to a} p(x) = \infty$ and $f(x) \to 0$ through positive values, then $\lim\limits_{x \to a} \dfrac{p(x)}{f(x)} = \infty$. [For example, take $a = 0$,

$p(x) = 1/x^2$, and $f(x) = x^2$.] If $f(x) \to 0$ through negative values, then $\lim\limits_{x \to a} \dfrac{p(x)}{f(x)} = -\infty$. [For example,

take $a = 0$, $p(x) = 1/x^2$, and $f(x) = -x^2$.] If $f(x) \to 0$ through both positive and negative values, then the

limit might not exist. [For example, take $a = 0$, $p(x) = 1/x^2$, and $f(x) = x$.]·

(e) $\lim\limits_{x \to a} \dfrac{p(x)}{q(x)}$ is an indeterminate form of type $\dfrac{\infty}{\infty}$.

3. (a) When x is near a, $f(x)$ is near 0 and $p(x)$ is large, so $f(x) - p(x)$ is large negative. Thus,

$\lim\limits_{x \to a} [f(x) - p(x)] = -\infty$.

(b) $\lim\limits_{x \to a} [p(x) - q(x)]$ is an indeterminate form of type $\infty - \infty$.

(c) When x is near a, $p(x)$ and $q(x)$ are both large, so $p(x) + q(x)$ is large. Thus, $\lim\limits_{x \to a} [p(x) + q(x)] = \infty$.

5. This limit has the form $\frac{0}{0}$. We can simply factor the numerator to evaluate this limit.

$$\lim_{x \to -1} \frac{x^2 - 1}{x + 1} = \lim_{x \to -1} \frac{(x + 1)(x - 1)}{x + 1} = \lim_{x \to -1} (x - 1) = -2$$

7. This limit has the form $\frac{0}{0}$. $\lim_{x \to 1} \frac{x^9 - 1}{x^5 - 1} \overset{\text{H}}{=} \lim_{x \to 1} \frac{9x^8}{5x^4} = \frac{9}{5} \lim_{x \to 1} x^4 = \frac{9}{5}(1) = \frac{9}{5}$

9. This limit has the form $\frac{0}{0}$. $\lim_{x \to (\pi/2)^+} \frac{\cos x}{1 - \sin x} \overset{\text{H}}{=} \lim_{x \to (\pi/2)^+} \frac{-\sin x}{-\cos x} = \lim_{x \to (\pi/2)^+} \tan x = -\infty.$

11. This limit has the form $\frac{0}{0}$. $\lim_{t \to 0} \frac{e^t - 1}{t^3} \overset{\text{H}}{=} \lim_{t \to 0} \frac{e^t}{3t^2} = \infty$ since $e^t \to 1$ and $3t^2 \to 0^+$ as $t \to 0$.

13. This limit has the form $\frac{0}{0}$. $\lim_{x \to 0} \frac{\tan px}{\tan qx} \overset{\text{H}}{=} \lim_{x \to 0} \frac{p \sec^2 px}{q \sec^2 qx} = \frac{p(1)^2}{q(1)^2} = \frac{p}{q}$

15. This limit has the form $\frac{\infty}{\infty}$. $\lim_{x \to \infty} \frac{\ln x}{x} \overset{\text{H}}{=} \lim_{x \to \infty} \frac{1/x}{1} = 0$

17. $\lim_{x \to 0^+} [(\ln x)/x] = -\infty$ since $\ln x \to -\infty$ as $x \to 0^+$ and dividing by small values of x just increases the magnitude of the quotient $(\ln x)/x$. L'Hospital's Rule does not apply.

19. This limit has the form $\frac{0}{0}$. $\lim_{t \to 0} \frac{5^t - 3^t}{t} \overset{\text{H}}{=} \lim_{t \to 0} \frac{5^t \ln 5 - 3^t \ln 3}{1} = \ln 5 - \ln 3 = \ln \frac{5}{3}$

21. This limit has the form $\frac{0}{0}$. $\lim_{x \to 0} \frac{e^x - 1 - x}{x^2} \overset{\text{H}}{=} \lim_{x \to 0} \frac{e^x - 1}{2x} \overset{\text{H}}{=} \lim_{x \to 0} \frac{e^x}{2} = \frac{1}{2}$

23. This limit has the form $\frac{\infty}{\infty}$. $\lim_{x \to \infty} \frac{e^x}{x^3} \overset{\text{H}}{=} \lim_{x \to \infty} \frac{e^x}{3x^2} \overset{\text{H}}{=} \lim_{x \to \infty} \frac{e^x}{6x} \overset{\text{H}}{=} \lim_{x \to \infty} \frac{e^x}{6} = \infty$

25. This limit has the form $\frac{0}{0}$. $\lim_{x \to 0} \frac{\sin^{-1} x}{x} \overset{\text{H}}{=} \lim_{x \to 0} \frac{1/\sqrt{1 - x^2}}{1} = \lim_{x \to 0} \frac{1}{\sqrt{1 - x^2}} = \frac{1}{1} = 1$

27. This limit has the form $\frac{0}{0}$. $\lim_{x \to 0} \frac{1 - \cos x}{x^2} \overset{\text{H}}{=} \lim_{x \to 0} \frac{\sin x}{2x} \overset{\text{H}}{=} \lim_{x \to 0} \frac{\cos x}{2} = \frac{1}{2}$

29. $\lim_{x \to 0} \frac{x + \sin x}{x + \cos x} = \frac{0 + 0}{0 + 1} = \frac{0}{1} = 0$. L'Hospital's Rule does not apply.

31. This limit has the form $\frac{\infty}{\infty}$. $\lim_{x \to \infty} \frac{x}{\ln(1 + 2e^x)} \overset{\text{H}}{=} \lim_{x \to \infty} \frac{1}{\dfrac{1}{1 + 2e^x} \cdot 2e^x} = \lim_{x \to \infty} \frac{1 + 2e^x}{2e^x} \overset{\text{H}}{=} \lim_{x \to \infty} \frac{2e^x}{2e^x} = 1$

33. This limit has the form $\frac{0}{0}$. $\lim_{x \to 1} \frac{1 - x + \ln x}{1 + \cos \pi x} \overset{\text{H}}{=} \lim_{x \to 1} \frac{-1 + 1/x}{-\pi \sin \pi x} \overset{\text{H}}{=} \lim_{x \to 1} \frac{-1/x^2}{-\pi^2 \cos \pi x} = \frac{-1}{-\pi^2 (-1)} = -\frac{1}{\pi^2}$

35. This limit has the form $\frac{0}{0}$. $\lim_{x \to 1} \frac{x^a - ax + a - 1}{(x - 1)^2} \overset{\text{H}}{=} \lim_{x \to 1} \frac{ax^{a-1} - a}{2(x - 1)} \overset{\text{H}}{=} \lim_{x \to 1} \frac{a(a - 1)x^{a-2}}{2} = \frac{a(a - 1)}{2}$

37. This limit has the form $0 \cdot (-\infty)$. We need to write this product as a quotient, but keep in mind that we will have to differentiate both the numerator and the denominator. If we differentiate $\dfrac{1}{\ln x}$, we get a complicated expression that results in a more difficult limit. Instead we write the quotient as $\dfrac{\ln x}{x^{-1/2}}$.

$$\lim_{x \to 0^+} \sqrt{x} \ln x = \lim_{x \to 0^+} \frac{\ln x}{x^{-1/2}} \overset{\text{H}}{=} \lim_{x \to 0^+} \frac{1/x}{-\frac{1}{2}x^{-3/2}} \cdot \frac{-2x^{3/2}}{-2x^{3/2}} = \lim_{x \to 0^+} (-2\sqrt{x}) = 0$$

39. This limit has the form $\infty \cdot 0$. We'll change it to the form $\frac{0}{0}$.

$$\lim_{x \to 0} \cot 2x \sin 6x = \lim_{x \to 0} \frac{\sin 6x}{\tan 2x} \overset{H}{=} \lim_{x \to 0} \frac{6 \cos 6x}{2 \sec^2 2x} = \frac{6(1)}{2(1)^2} = 3$$

41. This limit has the form $\infty \cdot 0$. $\lim_{x \to \infty} x^3 e^{-x^2} = \lim_{x \to \infty} \frac{x^3}{e^{x^2}} \overset{H}{=} \lim_{x \to \infty} \frac{3x^2}{2xe^{x^2}} = \lim_{x \to \infty} \frac{3x}{2e^{x^2}} \overset{H}{=} \lim_{x \to \infty} \frac{3}{4xe^{x^2}} = 0$

43. This limit has the form $0 \cdot (-\infty)$.

$$\lim_{x \to 1^+} \ln x \tan(\pi x/2) = \lim_{x \to 1^+} \frac{\ln x}{\cot(\pi x/2)} \overset{H}{=} \lim_{x \to 1^+} \frac{1/x}{(-\pi/2)\csc^2(\pi x/2)} = \frac{1}{(-\pi/2)(1)^2} = -\frac{2}{\pi}$$

45. $\lim_{x \to 0} \left(\frac{1}{x} - \csc x \right) = \lim_{x \to 0} \left(\frac{1}{x} - \frac{1}{\sin x} \right) = \lim_{x \to 0} \frac{\sin x - x}{x \sin x}$

$$\overset{H}{=} \lim_{x \to 0} \frac{\cos x - 1}{x \cos x + \sin x} \overset{H}{=} \lim_{x \to 0} \frac{-\sin x}{2 \cos x - x \sin x} = \frac{0}{2} = 0$$

47. We will multiply and divide by the conjugate of the expression to change the form of the expression.

$$\lim_{x \to \infty} \left(\sqrt{x^2 + x} - x \right) = \lim_{x \to \infty} \left(\frac{\sqrt{x^2 + x} - x}{1} \cdot \frac{\sqrt{x^2 + x} + x}{\sqrt{x^2 + x} + x} \right) = \lim_{x \to \infty} \frac{(x^2 + x) - x^2}{\sqrt{x^2 + x} + x}$$

$$= \lim_{x \to \infty} \frac{x}{\sqrt{x^2 + x} + x} = \lim_{x \to \infty} \frac{1}{\sqrt{1 + 1/x} + 1} = \frac{1}{\sqrt{1} + 1} = \frac{1}{2}.$$

As an alternate solution, write $\sqrt{x^2 + x} - x$ as $\sqrt{x^2 + x} - \sqrt{x^2}$, factor out $\sqrt{x^2}$, rewrite as

$(\sqrt{1 + 1/x} - 1)/(1/x)$, and apply l'Hospital's Rule.

49. The limit has the form $\infty - \infty$ and we will change the form to a product by factoring out x.

$$\lim_{x \to \infty} (x - \ln x) = \lim_{x \to \infty} x \left(1 - \frac{\ln x}{x} \right) = \infty \text{ since } \lim_{x \to \infty} \frac{\ln x}{x} \overset{H}{=} \lim_{x \to \infty} \frac{1/x}{1} = 0.$$

51. $y = x^{x^2} \quad \Rightarrow \quad \ln y = x^2 \ln x$, so

$$\lim_{x \to 0^+} \ln y = \lim_{x \to 0^+} x^2 \ln x = \lim_{x \to 0^+} \frac{\ln x}{1/x^2} \overset{H}{=} \lim_{x \to 0^+} \frac{1/x}{-2/x^3} = \lim_{x \to 0^+} \left(-\frac{1}{2}x^2 \right) = 0 \quad \Rightarrow$$

$$\lim_{x \to 0^+} x^{x^2} = \lim_{x \to 0^+} e^{\ln y} = e^0 = 1.$$

53. $y = (1 - 2x)^{1/x} \quad \Rightarrow \quad \ln y = \frac{1}{x} \ln(1 - 2x)$, so $\lim_{x \to 0} \ln y = \lim_{x \to 0} \frac{\ln(1 - 2x)}{x} \overset{H}{=} \lim_{x \to 0} \frac{-2/(1 - 2x)}{1} = -2 \quad \Rightarrow$

$\lim_{x \to 0} (1 - 2x)^{1/x} = \lim_{x \to 0} e^{\ln y} = e^{-2}.$

55. $y = \left(1 + \frac{3}{x} + \frac{5}{x^2} \right)^x \quad \Rightarrow \quad \ln y = x \ln \left(1 + \frac{3}{x} + \frac{5}{x^2} \right) \quad \Rightarrow$

$$\lim_{x \to \infty} \ln y = \lim_{x \to \infty} \frac{\ln \left(1 + \dfrac{3}{x} + \dfrac{5}{x^2} \right)}{1/x} \overset{H}{=} \lim_{x \to \infty} \frac{\left(-\dfrac{3}{x^2} - \dfrac{10}{x^3} \right) \Big/ \left(1 + \dfrac{3}{x} + \dfrac{5}{x^2} \right)}{-1/x^2} = \lim_{x \to \infty} \frac{3 + \dfrac{10}{x}}{1 + \dfrac{3}{x} + \dfrac{5}{x^2}} = 3,$$

so $\lim_{x \to \infty} \left(1 + \frac{3}{x} + \frac{5}{x^2} \right)^x = \lim_{x \to \infty} e^{\ln y} = e^3.$

57. $y = x^{1/x} \quad \Rightarrow \quad \ln y = (1/x) \ln x \quad \Rightarrow \quad \lim_{x \to \infty} \ln y = \lim_{x \to \infty} \frac{\ln x}{x} \overset{H}{=} \lim_{x \to \infty} \frac{1/x}{1} = 0 \quad \Rightarrow$

$\lim_{x \to \infty} x^{1/x} = \lim_{x \to \infty} e^{\ln y} = e^0 = 1$

59. $y = \left(\dfrac{x}{x+1}\right)^x \quad \Rightarrow \quad \ln y = x \ln\left(\dfrac{x}{x+1}\right) \quad \Rightarrow$

$$\lim_{x\to\infty} \ln y = \lim_{x\to\infty} x \ln\left(\dfrac{x}{x+1}\right) = \lim_{x\to\infty} \dfrac{\ln x - \ln(x+1)}{1/x} \overset{\text{H}}{=} \lim_{x\to\infty} \dfrac{1/x - 1/(x+1)}{-1/x^2}$$

$$= \lim_{x\to\infty} \left(-x + \dfrac{x^2}{x+1}\right) = \lim_{x\to\infty} \dfrac{-x}{x+1} = -1$$

so $\displaystyle\lim_{x\to\infty} \left(\dfrac{x}{x+1}\right)^x = \lim_{x\to\infty} e^{\ln y} = e^{-1}$

Or: $\displaystyle\lim_{x\to\infty} \left(\dfrac{x}{x+1}\right)^x = \lim_{x\to\infty} \left[\left(\dfrac{x+1}{x}\right)^{-1}\right]^x = \left[\lim_{x\to\infty}\left(1 + \dfrac{1}{x}\right)^x\right]^{-1} = e^{-1}$

61. $y = (\cos x)^{1/x^2} \quad \Rightarrow \quad \ln y = \dfrac{1}{x^2} \ln \cos x \quad \Rightarrow$

$$\lim_{x\to 0+} \ln y = \lim_{x\to 0+} \dfrac{\ln \cos x}{x^2} \overset{\text{H}}{=} \lim_{x\to 0+} \dfrac{-\tan x}{2x} \overset{\text{H}}{=} \lim_{x\to 0+} \dfrac{-\sec^2 x}{2} = -\dfrac{1}{2} \quad \Rightarrow$$

$$\lim_{x\to 0+} (\cos x)^{1/x^2} = \lim_{x\to 0+} e^{\ln y} = e^{-1/2} = 1/\sqrt{e}$$

63.

From the graph, it appears that $\displaystyle\lim_{x\to\infty} x\left[\ln(x+5) - \ln x\right] = 5$.

To prove this, we first note that

$$\ln(x+5) - \ln x = \ln\dfrac{x+5}{x} = \ln\left(1 + \dfrac{5}{x}\right) \to \ln 1 = 0 \text{ as } x \to \infty. \text{ Thus,}$$

$$\lim_{x\to\infty} x\left[\ln(x+5) - \ln x\right] = \lim_{x\to\infty} \dfrac{\ln(x+5) - \ln x}{1/x} \overset{\text{H}}{=} \lim_{x\to\infty} \dfrac{\dfrac{1}{x+5} - \dfrac{1}{x}}{-1/x^2}$$

$$= \lim_{x\to\infty} \left[\dfrac{x - (x+5)}{x(x+5)} \cdot \dfrac{-x^2}{1}\right] = \lim_{x\to\infty} \dfrac{5x^2}{x^2 + 5x} = 5$$

65.

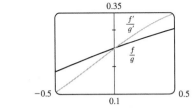

From the graph, it appears that

$$\lim_{x\to 0} \dfrac{f(x)}{g(x)} = \lim_{x\to 0} \dfrac{f'(x)}{g'(x)} = 0.25. \text{ We calculate}$$

$$\lim_{x\to 0} \dfrac{f(x)}{g(x)} = \lim_{x\to 0} \dfrac{e^x - 1}{x^3 + 4x} \overset{\text{H}}{=} \lim_{x\to 0} \dfrac{e^x}{3x^2 + 4} = \dfrac{1}{4}.$$

67. $\displaystyle\lim_{x\to\infty} \dfrac{e^x}{x^n} \overset{\text{H}}{=} \lim_{x\to\infty} \dfrac{e^x}{nx^{n-1}} \overset{\text{H}}{=} \lim_{x\to\infty} \dfrac{e^x}{n(n-1)x^{n-2}} \overset{\text{H}}{=} \cdots \overset{\text{H}}{=} \lim_{x\to\infty} \dfrac{e^x}{n!} = \infty$

69. First we will find $\displaystyle\lim_{n\to\infty} \left(1 + \dfrac{i}{n}\right)^{nt}$, which is of the form 1^∞. $y = \left(1 + \dfrac{i}{n}\right)^{nt} \quad \Rightarrow \quad \ln y = nt \ln\left(1 + \dfrac{i}{n}\right)$, so

$$\lim_{n\to\infty} \ln y = \lim_{n\to\infty} nt \ln\left(1 + \dfrac{i}{n}\right) = t \lim_{n\to\infty} \dfrac{\ln(1 + i/n)}{1/n} \overset{\text{H}}{=} t \lim_{n\to\infty} \dfrac{(-i/n^2)}{(1 + i/n)(-1/n^2)} = t \lim_{n\to\infty} \dfrac{i}{1 + i/n} = ti$$

$\Rightarrow \quad \displaystyle\lim_{n\to\infty} y = e^{it}.$ Thus, as $n \to \infty$, $A = A_0 \left(1 + \dfrac{i}{n}\right)^{nt} \to A_0 e^{it}$.

71. We see that both numerator and denominator approach 0, so we can use l'Hospital's Rule:

$$\lim_{x \to a} \frac{\sqrt{2a^3x - x^4} - a\sqrt[3]{aax}}{a - \sqrt[4]{ax^3}} \overset{\text{H}}{=} \lim_{x \to a} \frac{\frac{1}{2}\left(2a^3x - x^4\right)^{-1/2}\left(2a^3 - 4x^3\right) - a\left(\frac{1}{3}\right)(aax)^{-2/3}a^2}{-\frac{1}{4}(ax^3)^{-3/4}(3ax^2)}$$

$$= \frac{\frac{1}{2}\left(2a^3a - a^4\right)^{-1/2}\left(2a^3 - 4a^3\right) - \frac{1}{3}a^3\left(a^2a\right)^{-2/3}}{-\frac{1}{4}(aa^3)^{-3/4}(3aa^2)}$$

$$= \frac{\left(a^4\right)^{-1/2}\left(-a^3\right) - \frac{1}{3}a^3\left(a^3\right)^{-2/3}}{-\frac{3}{4}a^3\left(a^4\right)^{-3/4}} = \frac{-a - \frac{1}{3}a}{-\frac{3}{4}} = \frac{4}{3}\left(\frac{4}{3}a\right) = \frac{16}{9}a$$

73. Since $f(2) = 0$, the given limit has the form $\frac{0}{0}$.

$$\lim_{x \to 0} \frac{f(2 + 3x) + f(2 + 5x)}{x} \overset{\text{H}}{=} \lim_{x \to 0} \frac{f'(2 + 3x) \cdot 3 + f'(2 + 5x) \cdot 5}{1} = f'(2) \cdot 3 + f'(2) \cdot 5 = 8f'(2) = 8 \cdot 7 = 56$$

75. Since $\lim_{h \to 0}[f(x + h) - f(x - h)] = f(x) - f(x) = 0$ (f is differentiable and hence continuous) and $\lim_{h \to 0} 2h = 0$,

we use l'Hospital's Rule:

$$\lim_{h \to 0} \frac{f(x + h) - f(x - h)}{2h} \overset{\text{H}}{=} \lim_{h \to 0} \frac{f'(x + h)(1) - f'(x - h)(-1)}{2} = \frac{f'(x) + f'(x)}{2} = \frac{2f'(x)}{2} = f'(x)$$

$\dfrac{f(x + h) - f(x - h)}{2h}$ is the slope of the secant line

between $(x - h, f(x - h))$ and $(x + h, f(x + h))$. As

$h \to 0$, this line gets closer to the tangent line and its slope

approaches $f'(x)$.

77. (a) We show that $\lim_{x \to 0} \dfrac{f(x)}{x^n} = 0$ for every integer $n \geq 0$. Let $y = \dfrac{1}{x^2}$. Then

$$\lim_{x \to 0} \frac{f(x)}{x^{2n}} = \lim_{x \to 0} \frac{e^{-1/x^2}}{(x^2)^n} = \lim_{y \to \infty} \frac{y^n}{e^y} \overset{\text{H}}{=} \lim_{y \to \infty} \frac{ny^{n-1}}{e^y} \overset{\text{H}}{=} \cdots \overset{\text{H}}{=} \lim_{y \to \infty} \frac{n!}{e^y} = 0 \;\Rightarrow$$

$$\lim_{x \to 0} \frac{f(x)}{x^n} = \lim_{x \to 0} x^n \frac{f(x)}{x^{2n}} = \lim_{x \to 0} x^n \lim_{x \to 0} \frac{f(x)}{x^{2n}} = 0. \text{ Thus, } f'(0) = \lim_{x \to 0} \frac{f(x) - f(0)}{x - 0} = \lim_{x \to 0} \frac{f(x)}{x} = 0.$$

(b) Using the Chain Rule and the Quotient Rule we see that $f^{(n)}(x)$ exists for $x \neq 0$. In fact, we prove by induction

that for each $n \geq 0$, there is a polynomial p_n and a non-negative integer k_n with $f^{(n)}(x) = p_n(x)f(x)/x^{k_n}$ for

$x \neq 0$. This is true for $n = 0$; suppose it is true for the nth derivative. Then $f'(x) = f(x)(2/x^3)$, so

$$f^{(n+1)}(x) = \left[x^{k_n}\left[p_n'(x)f(x) + p_n(x)f'(x)\right] - k_n x^{k_n - 1}p_n(x)f(x)\right]x^{-2k_n}$$

$$= \left[x^{k_n}p_n'(x) + p_n(x)(2/x^3) - k_n x^{k_n - 1}p_n(x)\right]f(x)x^{-2k_n}$$

$$= \left[x^{k_n + 3}p_n'(x) + 2p_n(x) - k_n x^{k_n + 2}p_n(x)\right]f(x)x^{-(2k_n + 3)}$$

which has the desired form.

Now we show by induction that $f^{(n)}(0) = 0$ for all n. By part (a), $f'(0) = 0$. Suppose that $f^{(n)}(0) = 0$. Then

$$f^{(n+1)}(0) = \lim_{x \to 0} \frac{f^{(n)}(x) - f^{(n)}(0)}{x - 0} = \lim_{x \to 0} \frac{f^{(n)}(x)}{x} = \lim_{x \to 0} \frac{p_n(x)f(x)/x^{k_n}}{x} = \lim_{x \to 0} \frac{p_n(x)f(x)}{x^{k_n + 1}}$$

$$= \lim_{x \to 0} p_n(x) \lim_{x \to 0} \frac{f(x)}{x^{k_n + 1}} = p_n(0) \cdot 0 = 0$$

4.5 Summary of Curve Sketching

1. $y = f(x) = x^3 + x = x(x^2 + 1)$ **A.** f is a polynomial, so $D = \mathbb{R}$.

B. x-intercept $= 0$, y-intercept $= f(0) = 0$ **C.** $f(-x) = -f(x)$, so f is odd; the curve is symmetric about the origin.   **D.** f is a polynomial, so there is no asymptote.   **E.** $f'(x) = 3x^2 + 1 > 0$, so f is increasing on $(-\infty, \infty)$.   **F.** There is no critical number and hence, no local maximum or minimum value.   **G.** $f''(x) = 6x > 0$ on $(0, \infty)$ and $f''(x) < 0$ on $(-\infty, 0)$, so f is CU on $(0, \infty)$ and CD on $(-\infty, 0)$. Since the concavity changes at $x = 0$, there is an inflection point at $(0, 0)$.

H.

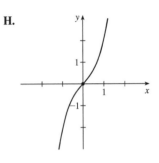

3. $y = f(x) = 2 - 15x + 9x^2 - x^3 = -(x-2)(x^2 - 7x + 1)$ **A.** $D = \mathbb{R}$ **B.** y-intercept: $f(0) = 2$;

x-intercepts: $f(x) = 0 \implies x = 2$ or (by the quadratic formula) $x = \frac{7 \pm \sqrt{45}}{2} \approx 0.15, 6.85$

C. No symmetry   **D.** No asymptote

E. $f'(x) = -15 + 18x - 3x^2 = -3(x^2 - 6x + 5)$
$$= -3(x - 1)(x - 5) > 0 \iff 1 < x < 5$$
so f is increasing on $(1, 5)$ and decreasing on $(-\infty, 1)$ and $(5, \infty)$.

F. Local maximum value $f(5) = 27$, local minimum value $f(1) = -5$

G. $f''(x) = 18 - 6x = -6(x - 3) > 0 \iff x < 3$, so f is CU on $(-\infty, 3)$ and CD on $(3, \infty)$. IP at $(3, 11)$

H.

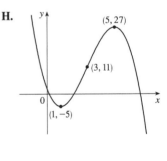

5. $y = f(x) = x^4 + 4x^3 = x^3(x + 4)$ **A.** $D = \mathbb{R}$ **B.** y-intercept: $f(0) = 0$; x-intercepts: $f(x) = 0 \iff x = -4, 0$ **C.** No symmetry

D. No asymptote   **E.** $f'(x) = 4x^3 + 12x^2 = 4x^2(x + 3) > 0 \iff$ $x > -3$, so f is increasing on $(-3, \infty)$ and decreasing on $(-\infty, -3)$.

F. Local minimum value $f(-3) = -27$, no local maximum

G. $f''(x) = 12x^2 + 24x = 12x(x + 2) < 0 \iff -2 < x < 0$, so f is CD on $(-2, 0)$ and CU on $(-\infty, -2)$ and $(0, \infty)$.

IP at $(0, 0)$ and $(-2, -16)$

H.

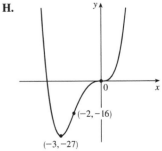

7. $y = f(x) = 2x^5 - 5x^2 + 1$ **A.** $D = \mathbb{R}$ **B.** y-intercept: $f(0) = 1$ **C.** No symmetry **D.** No asymptote

E. $f'(x) = 10x^4 - 10x = 10x(x^3 - 1) = 10x(x - 1)(x^2 + x + 1)$, so $f'(x) < 0 \Leftrightarrow 0 < x < 1$ and

$f'(x) > 0 \Leftrightarrow x < 0$ or $x > 1$. Thus, f is increasing on $(-\infty, 0)$ and $(1, \infty)$ and decreasing on $(0, 1)$.

F. Local maximum value $f(0) = 1$, local minimum value $f(1) = -2$ **H.**

G. $f''(x) = 40x^3 - 10 = 10(4x^3 - 1)$ so $f''(x) = 0 \Leftrightarrow x = 1/\sqrt[3]{4}$.

$f''(x) > 0 \Leftrightarrow x > 1/\sqrt[3]{4}$ and $f''(x) < 0 \Leftrightarrow x < 1/\sqrt[3]{4}$,

so f is CD on $\left(-\infty, 1/\sqrt[3]{4}\right)$ and CU on $\left(1/\sqrt[3]{4}, \infty\right)$.

IP at $\left(\dfrac{1}{\sqrt[3]{4}}, 1 - \dfrac{9}{2\left(\sqrt[3]{4}\right)^2}\right) \approx (0.630, -0.786)$

9. $y = f(x) = x/(x - 1)$ **A.** $D = \{x \mid x \neq 1\} = (-\infty, 1) \cup (1, \infty)$ **B.** x-intercept $= 0$,

y-intercept $= f(0) = 0$ **C.** No symmetry **D.** $\lim\limits_{x \to \pm\infty} \dfrac{x}{x - 1} = 1$, so $y = 1$ is a HA.

$\lim\limits_{x \to 1^-} \dfrac{x}{x - 1} = -\infty$, $\lim\limits_{x \to 1^+} \dfrac{x}{x - 1} = \infty$, so $x = 1$ is a VA.   **H.**

E. $f'(x) = \dfrac{(x - 1) - x}{(x - 1)^2} = \dfrac{-1}{(x - 1)^2} < 0$ for $x \neq 1$, so f is decreasing

on $(-\infty, 1)$ and $(1, \infty)$.   **F.** No extreme values

G. $f''(x) = \dfrac{2}{(x - 1)^3} > 0 \Leftrightarrow x > 1$, so f is CU on $(1, \infty)$ and CD

on $(-\infty, 1)$. No IP

11. $y = f(x) = 1/(x^2 - 9)$ **A.** $D = \{x \mid x \neq \pm 3\} = (-\infty, -3) \cup (-3, 3) \cup (3, \infty)$

B. y-intercept $= f(0) = -\frac{1}{9}$, no x-intercept **C.** $f(-x) = f(x) \Rightarrow f$ is even; the curve is symmetric about

the y-axis. **D.** $\lim\limits_{x \to \pm\infty} \dfrac{1}{x^2 - 9} = 0$, so $y = 0$ is a HA. $\lim\limits_{x \to 3^-} \dfrac{1}{x^2 - 9} = -\infty$, $\lim\limits_{x \to 3^+} \dfrac{1}{x^2 - 9} = \infty$,

$\lim\limits_{x \to -3^-} \dfrac{1}{x^2 - 9} = \infty$, $\lim\limits_{x \to -3^+} \dfrac{1}{x^2 - 9} = -\infty$, so $x = 3$ and $x = -3$ are VA.

E. $f'(x) = -\dfrac{2x}{(x^2 - 9)^2} > 0 \Leftrightarrow x < 0$ $(x \neq -3)$ so f is increasing   **H.**

on $(-\infty, -3)$ and $(-3, 0)$ and decreasing on $(0, 3)$ and $(3, \infty)$.

F. Local maximum value $f(0) = -\frac{1}{9}$.

G. $y'' = \dfrac{-2(x^2 - 9)^2 + (2x)2(x^2 - 9)(2x)}{(x^2 - 9)^4} = \dfrac{6(x^2 + 3)}{(x^2 - 9)^3} > 0 \Leftrightarrow$

$x^2 > 9 \Leftrightarrow x > 3$ or $x < -3$, so f is CU on $(-\infty, -3)$ and $(3, \infty)$

and CD on $(-3, 3)$. No IP

13. $y = f(x) = x/(x^2 + 9)$ **A.** $D = \mathbb{R}$ **B.** y-intercept: $f(0) = 0$; x-intercept: $f(x) = 0 \iff x = 0$

C. $f(-x) = -f(x)$, so f is odd and the curve is symmetric about the origin.   **D.** $\lim\limits_{x \to \pm\infty} \left[x/(x^2 + 9) \right] = 0$, so

$y = 0$ is a HA; no VA   **E.** $f'(x) = \dfrac{(x^2 + 9)(1) - x(2x)}{(x^2 + 9)^2} = \dfrac{9 - x^2}{(x^2 + 9)^2} = \dfrac{(3 + x)(3 - x)}{(x^2 + 9)^2} > 0 \iff$

$-3 < x < 3$, so f is increasing on $(-3, 3)$ and decreasing on $(-\infty, -3)$ and $(3, \infty)$.

F. Local minimum value $f(-3) = -\frac{1}{6}$, local maximum value $f(3) = \frac{1}{6}$

G. $f''(x) = \dfrac{(x^2 + 9)^2 (-2x) - (9 - x^2) \cdot 2(x^2 + 9)(2x)}{\left[(x^2 + 9)^2 \right]^2}$

$= \dfrac{(2x)(x^2 + 9)\left[-(x^2 + 9) - 2(9 - x^2) \right]}{(x^2 + 9)^4}$

$= \dfrac{2x(x^2 - 27)}{(x^2 + 9)^3} = 0 \iff x = 0, \pm\sqrt{27} = \pm 3\sqrt{3}$

H.

$(3, \frac{1}{6})$

$(-3, -\frac{1}{6})$

$f''(x) > 0 \iff -3\sqrt{3} < x < 0$ or $x > 3\sqrt{3}$, so f is CU on

$(-3\sqrt{3}, 0)$ and $(3\sqrt{3}, \infty)$, and CD on $(-\infty, -3\sqrt{3})$ and $(0, 3\sqrt{3})$.

There are three inflection points: $(0, 0)$ and $\left(\pm 3\sqrt{3}, \pm\frac{1}{12}\sqrt{3} \right)$.

15. $y = f(x) = \dfrac{x - 1}{x^2}$ **A.** $D = \{x \mid x \neq 0\} = (-\infty, 0) \cup (0, \infty)$ **B.** No y-intercept; x-intercept: $f(x) = 0 \iff$

$x = 1$ **C.** No symmetry   **D.** $\lim\limits_{x \to \pm\infty} \dfrac{x - 1}{x^2} = 0$, so $y = 0$ is a HA. $\lim\limits_{x \to 0} \dfrac{x - 1}{x^2} = -\infty$, so $x = 0$ is a VA.

E. $f'(x) = \dfrac{x^2 \cdot 1 - (x - 1) \cdot 2x}{(x^2)^2} = \dfrac{-x^2 + 2x}{x^4} = \dfrac{-(x - 2)}{x^3}$, so $f'(x) > 0 \iff 0 < x < 2$ and

$f'(x) < 0 \iff x < 0$ or $x > 2$. Thus, f is increasing on $(0, 2)$ and

decreasing on $(-\infty, 0)$ and $(2, \infty)$.

F. No local minimum, local maximum value $f(2) = \frac{1}{4}$.

G. $f''(x) = \dfrac{x^3 \cdot (-1) - [-(x - 2)] \cdot 3x^2}{(x^3)^2} = \dfrac{2x^3 - 6x^2}{x^6} = \dfrac{2(x - 3)}{x^4}$.

$f''(x)$ is negative on $(-\infty, 0)$ and $(0, 3)$ and positive on $(3, \infty)$, so f is

CD on $(-\infty, 0)$ and $(0, 3)$ and CU on $(3, \infty)$. IP at $\left(3, \frac{2}{9} \right)$

H.

$\left(2, \frac{1}{4}\right) \left(3, \frac{2}{9}\right)$

17. $y = f(x) = \dfrac{x^2}{x^2 + 3} = \dfrac{(x^2 + 3) - 3}{x^2 + 3} = 1 - \dfrac{3}{x^2 + 3}$ **A.** $D = \mathbb{R}$ **B.** y-intercept: $f(0) = 0$; x-intercepts:

$f(x) = 0 \iff x = 0$ **C.** $f(-x) = f(x)$, so f is even; the graph is symmetric about the y-axis.

D. $\lim\limits_{x \to \pm\infty} \dfrac{x^2}{x^2 + 3} = 1$, so $y = 1$ is a HA. No VA.   **E.** Using the Reciprocal Rule,

$f'(x) = -3 \cdot \dfrac{-2x}{(x^2+3)^2} = \dfrac{6x}{(x^2+3)^2}$. $f'(x) > 0 \Leftrightarrow x > 0$ and $f'(x) < 0 \Leftrightarrow x < 0$, so f is decreasing

on $(-\infty, 0)$ and increasing on $(0, \infty)$. **F.** Local minimum value $f(0) = 0$, no local maximum.

G. $f''(x) = \dfrac{(x^2+3)^2 \cdot 6 - 6x \cdot 2(x^2+3) \cdot 2x}{[(x^2+3)^2]^2}$

$= \dfrac{6(x^2+3)[(x^2+3) - 4x^2]}{(x^2+3)^4}$

$= \dfrac{6(3-3x^2)}{(x^2+3)^3} = \dfrac{-18(x+1)(x-1)}{(x^2+3)^3}$

H.

$f''(x)$ is negative on $(-\infty, -1)$ and $(1, \infty)$ and positive on $(-1, 1)$, so f is CD on $(-\infty, -1)$ and $(1, \infty)$ and

CU on $(-1, 1)$. IP at $\left(\pm 1, \frac{1}{4}\right)$

19. $y = f(x) = x\sqrt{5-x}$ **A.** The domain is $\{x \mid 5 - x \geq 0\} = (-\infty, 5]$ **B.** y-intercept: $f(0) = 0$;

x-intercepts: $f(x) = 0 \Leftrightarrow x = 0, 5$ **C.** No symmetry **D.** No asymptote

E. $f'(x) = x \cdot \frac{1}{2}(5-x)^{-1/2}(-1) + (5-x)^{1/2} \cdot 1 = \frac{1}{2}(5-x)^{-1/2}[-x + 2(5-x)] = \dfrac{10-3x}{2\sqrt{5-x}} > 0 \Leftrightarrow$

$x < \frac{10}{3}$, so f is increasing on $\left(-\infty, \frac{10}{3}\right)$ and decreasing on $\left(\frac{10}{3}, 5\right)$.

F. Local maximum value $f\left(\frac{10}{3}\right) = \frac{10}{9}\sqrt{15} \approx 4.3$; no local minimum

G. $f''(x) = \dfrac{2(5-x)^{1/2}(-3) - (10-3x) \cdot 2\left(\frac{1}{2}\right)(5-x)^{-1/2}(-1)}{\left(2\sqrt{5-x}\right)^2}$

$= \dfrac{(5-x)^{-1/2}[-6(5-x) + (10-3x)]}{4(5-x)} = \dfrac{3x-20}{4(5-x)^{3/2}}$

$f''(x) < 0$ for $x < 5$, so f is CD on $(-\infty, 5)$. No IP

H.

21. $y = f(x) = \sqrt{x^2+1} - x$ **A.** $D = \mathbb{R}$ **B.** No x-intercept, y-intercept $= 1$ **C.** No symmetry

D. $\displaystyle\lim_{x \to -\infty} \left(\sqrt{x^2+1} - x\right) = \infty$ and

$\displaystyle\lim_{x \to \infty} \left(\sqrt{x^2+1} - x\right) = \lim_{x \to \infty} \left(\sqrt{x^2+1} - x\right) \dfrac{\sqrt{x^2+1} + x}{\sqrt{x^2+1} + x} = \lim_{x \to \infty} \dfrac{1}{\sqrt{x^2+1} + x} = 0$,

so $y = 0$ is a HA. **E.** $f'(x) = \dfrac{x}{\sqrt{x^2+1}} - 1 = \dfrac{x - \sqrt{x^2+1}}{\sqrt{x^2+1}} \Rightarrow$

H.

$f'(x) < 0$, so f is decreasing on $\mathbb{R}$. **F.** No extreme values

G. $f''(x) = \dfrac{1}{(x^2+1)^{3/2}} > 0$, so f is CU on $\mathbb{R}$. No IP

23. $y = f(x) = x/\sqrt{x^2 + 1}$ **A.** $D = \mathbb{R}$ **B.** y-intercept: $f(0) = 0$; x-intercepts: $f(x) = 0 \;\Rightarrow\; x = 0$

C. $f(-x) = -f(x)$, so f is odd; the graph is symmetric about the origin.

D. $\displaystyle\lim_{x\to\infty} f(x) = \lim_{x\to\infty} \frac{x}{\sqrt{x^2+1}} = \lim_{x\to\infty} \frac{x/x}{\sqrt{x^2+1}/x} = \lim_{x\to\infty} \frac{x/x}{\sqrt{x^2+1}/\sqrt{x^2}} = \lim_{x\to\infty} \frac{1}{\sqrt{1+1/x^2}}$

$$= \frac{1}{\sqrt{1+0}} = 1$$

and

$$\lim_{x\to-\infty} f(x) = \lim_{x\to-\infty} \frac{x}{\sqrt{x^2+1}} = \lim_{x\to-\infty} \frac{x/x}{\sqrt{x^2+1}/x} = \lim_{x\to-\infty} \frac{x/x}{\sqrt{x^2+1}\big/\left(-\sqrt{x^2}\right)}$$

$$= \lim_{x\to-\infty} \frac{1}{-\sqrt{1+1/x^2}} = \frac{1}{-\sqrt{1+0}} = -1$$

so $y = \pm 1$ are HA. No VA.

E. $f'(x) = \dfrac{\sqrt{x^2+1} - x \cdot \dfrac{2x}{2\sqrt{x^2+1}}}{[(x^2+1)^{1/2}]^2} = \dfrac{x^2+1-x^2}{(x^2+1)^{3/2}} = \dfrac{1}{(x^2+1)^{3/2}} > 0$ for all x, so f is increasing on $\mathbb{R}$.

F. No extreme values

G. $f''(x) = -\frac{3}{2}(x^2+1)^{-5/2} \cdot 2x = \dfrac{-3x}{(x^2+1)^{5/2}}$, so $f''(x) > 0$ for

$x < 0$ and $f''(x) < 0$ for $x > 0$. Thus, f is CU on $(-\infty, 0)$ and

CD on $(0, \infty)$. IP at $(0,0)$

H.

25. $y = f(x) = \sqrt{1-x^2}/x$ **A.** $D = \left\{ x \;\middle|\; |x| \le 1, x \ne 0 \right\} = [-1, 0) \cup (0, 1]$ **B.** x-intercepts ± 1, no y-intercept

C. $f(-x) = -f(x)$, so the curve is symmetric about $(0,0)$. **D.** $\displaystyle\lim_{x\to 0^+} \frac{\sqrt{1-x^2}}{x} = \infty$, $\displaystyle\lim_{x\to 0^-} \frac{\sqrt{1-x^2}}{x} = -\infty$,

so $x = 0$ is a VA. **E.** $f'(x) = \dfrac{\left(-x^2/\sqrt{1-x^2}\right) - \sqrt{1-x^2}}{x^2} = -\dfrac{1}{x^2\sqrt{1-x^2}} < 0$, so f is decreasing on $(-1, 0)$

and $(0, 1)$. **F.** No extreme values

G. $f''(x) = \dfrac{2-3x^2}{x^3(1-x^2)^{3/2}} > 0 \;\Leftrightarrow\; -1 < x < -\sqrt{\frac{2}{3}}$ or

H.

$0 < x < \sqrt{\frac{2}{3}}$, so f is CU on $\left(-1, -\sqrt{\frac{2}{3}}\right)$ and $\left(0, \sqrt{\frac{2}{3}}\right)$ and CD on

$\left(-\sqrt{\frac{2}{3}}, 0\right)$ and $\left(\sqrt{\frac{2}{3}}, 1\right)$. IP at $\left(\pm\sqrt{\frac{2}{3}}, \pm\frac{1}{\sqrt{2}}\right)$

27. $y = f(x) = x - 3x^{1/3}$ **A.** $D = \mathbb{R}$ **B.** y-intercept: $f(0) = 0$; x-intercepts: $f(x) = 0 \implies x = 3x^{1/3} \implies$

$x^3 = 27x \implies x^3 - 27x = 0 \implies x(x^2 - 27) = 0 \implies x = 0, \pm 3\sqrt{3}$ **C.** $f(-x) = -f(x)$, so f is odd;

the graph is symmetric about the origin. **D.** No asymptote **E.** $f'(x) = 1 - x^{-2/3} = 1 - \dfrac{1}{x^{2/3}} = \dfrac{x^{2/3} - 1}{x^{2/3}}$.

$f'(x) > 0$ when $|x| > 1$ and $f'(x) < 0$ when $0 < |x| < 1$, so f is increasing on $(-\infty, -1)$ and $(1, \infty)$, and

decreasing on $(-1, 0)$ and $(0, 1)$ [hence decreasing on $(-1, 1)$ since f is **H.**

continuous on $(-1, 1)$]. **F.** Local maximum value $f(-1) = 2$, local

minimum value $f(1) = -2$ **G.** $f''(x) = \frac{2}{3}x^{-5/3} < 0$ when $x < 0$

and $f''(x) > 0$ when $x > 0$, so f is CD on $(-\infty, 0)$ and CU on $(0, \infty)$.

IP at $(0, 0)$

29. $y = f(x) = x + \sqrt{|x|}$ **A.** $D = \mathbb{R}$ **B.** x-intercepts $0, -1$; y-intercept 0 **C.** No symmetry

D. $\lim\limits_{x \to \infty} \left(x + \sqrt{|x|} \right) = \infty$, $\lim\limits_{x \to -\infty} \left(x + \sqrt{|x|} \right) = -\infty$. No asymptote **E.** For $x > 0$, $f(x) = x + \sqrt{x} \implies$

$f'(x) = 1 + \dfrac{1}{2\sqrt{x}} > 0$, so f increases on $(0, \infty)$. For $x < 0$, $f(x) = x + \sqrt{-x} \implies f'(x) = 1 - \dfrac{1}{2\sqrt{-x}} > 0$

$\iff 2\sqrt{-x} > 1 \iff -x > \frac{1}{4} \iff x < -\frac{1}{4}$, so f increases on $\left(-\infty, -\frac{1}{4}\right)$ and decreases on $\left(-\frac{1}{4}, 0\right)$.

F. Local maximum value $f\left(-\frac{1}{4}\right) = \frac{1}{4}$, local minimum value $f(0) = 0$ **H.**

G. For $x > 0$, $f''(x) = -\frac{1}{4}x^{-3/2} \implies f''(x) < 0$, so f is CD on

$(0, \infty)$. For $x < 0$, $f''(x) = -\frac{1}{4}(-x)^{-3/2} \implies f''(x) < 0$, so f is

CD on $(-\infty, 0)$. No IP

31. $y = f(x) = 3\sin x - \sin^3 x$ **A.** $D = \mathbb{R}$ **B.** y-intercept: $f(0) = 0$; x-intercepts: $f(x) = 0 \implies$

$\sin x \,(3 - \sin^2 x) = 0 \implies \sin x = 0$ [since $\sin^2 x \leq 1 < 3$] $\implies x = n\pi$, n an integer.

C. $f(-x) = -f(x)$, so f is odd; the graph (shown for $-2\pi \leq x \leq 2\pi$) is symmetric about the origin and periodic

with period 2π. **D.** No asymptote **E.** $f'(x) = 3\cos x - 3\sin^2 x \cos x = 3\cos x\,(1 - \sin^2 x) = 3\cos^3 x$.

$f'(x) > 0 \iff \cos x > 0 \iff x \in \left(2n\pi - \frac{\pi}{2}, 2n\pi + \frac{\pi}{2}\right)$ for each integer n, and $f'(x) < 0 \iff$

$\cos x < 0 \iff x \in \left(2n\pi + \frac{\pi}{2}, 2n\pi + \frac{3\pi}{2}\right)$ for each integer n. Thus, f is increasing on $\left(2n\pi - \frac{\pi}{2}, 2n\pi + \frac{\pi}{2}\right)$

for each integer n, and f is decreasing on $\left(2n\pi + \frac{\pi}{2}, 2n\pi + \frac{3\pi}{2}\right)$ for each integer n.

F. f has local maximum values $f\left(2n\pi + \frac{\pi}{2}\right) = 2$ and local minimum values $f\left(2n\pi + \frac{3\pi}{2}\right) = -2$.

G. $f''(x) = -9 \sin x \cos^2 x = -9 \sin x \left(1 - \sin^2 x\right) = -9 \sin x \left(1 - \sin x\right)(1 + \sin x)$. $f''(x) < 0 \iff$

$\sin x > 0$ and $\sin x \neq \pm 1 \iff x \in \left(2n\pi, 2n\pi + \frac{\pi}{2}\right) \cup \left(2n\pi + \frac{\pi}{2}, 2n\pi + \pi\right)$ for some integer n.

$f''(x) > 0 \iff \sin x < 0$ and $\sin x \neq \pm 1 \iff x \in \left((2n-1)\pi, (2n-1)\pi + \frac{\pi}{2}\right) \cup \left((2n-1)\pi + \frac{\pi}{2}, 2n\pi\right)$

for some integer n. Thus, f is CD on the intervals $\left(2n\pi, \left(2n + \frac{1}{2}\right)\pi\right)$ and

$\left(\left(2n + \frac{1}{2}\right)\pi, (2n + 1)\pi\right)$ [hence CD on the intervals $\left(2n\pi, (2n + 1)\pi\right)$]

for each integer n, and f is CU on the intervals $\left((2n - 1)\pi, \left(2n - \frac{1}{2}\right)\pi\right)$

and $\left(\left(2n - \frac{1}{2}\right)\pi, 2n\pi\right)$ [hence CU on the intervals $((2n - 1)\pi, 2n\pi)$] for

each integer n. f has inflection points at $(n\pi, 0)$ for each integer n.

H.

33. $y = f(x) = x \tan x, -\frac{\pi}{2} < x < \frac{\pi}{2}$ **A.** $D = \left(-\frac{\pi}{2}, \frac{\pi}{2}\right)$ **B.** Intercepts are 0 **C.** $f(-x) = f(x)$, so the curve is

symmetric about the y-axis. **D.** $\displaystyle \lim_{x \to (\pi/2)^-} x \tan x = \infty$ and $\displaystyle \lim_{x \to -(\pi/2)^+} x \tan x = \infty$, so $x = \frac{\pi}{2}$ and

$x = -\frac{\pi}{2}$ are VA. **E.** $f'(x) = \tan x + x \sec^2 x > 0 \iff 0 < x < \frac{\pi}{2}$,

so f increases on $\left(0, \frac{\pi}{2}\right)$ and decreases on $\left(-\frac{\pi}{2}, 0\right)$.

F. Absolute and local minimum value $f(0) = 0$.

G. $y'' = 2 \sec^2 x + 2x \tan x \sec^2 x > 0$ for $-\frac{\pi}{2} < x < \frac{\pi}{2}$, so f is CU

on $\left(-\frac{\pi}{2}, \frac{\pi}{2}\right)$. No IP

H.

35. $y = f(x) = \frac{1}{2}x - \sin x, 0 < x < 3\pi$ **A.** $D = (0, 3\pi)$ **B.** No y-intercept. The x-intercept, approximately 1.9,

can be found using Newton's Method. **C.** No symmetry **D.** No asymptote **E.** $f'(x) = \frac{1}{2} - \cos x > 0 \iff$

$\cos x < \frac{1}{2} \iff \frac{\pi}{3} < x < \frac{5\pi}{3}$ or $\frac{7\pi}{3} < x < 3\pi$, so f is increasing on $\left(\frac{\pi}{3}, \frac{5\pi}{3}\right)$ and $\left(\frac{7\pi}{3}, 3\pi\right)$ and decreasing

on $\left(0, \frac{\pi}{3}\right)$ and $\left(\frac{5\pi}{3}, \frac{7\pi}{3}\right)$. **F.** Local minimum value $f\left(\frac{\pi}{3}\right) = \frac{\pi}{6} - \frac{\sqrt{3}}{2}$,

local maximum value $f\left(\frac{5\pi}{3}\right) = \frac{5\pi}{6} + \frac{\sqrt{3}}{2}$, local minimum value

$f\left(\frac{7\pi}{3}\right) = \frac{7\pi}{6} - \frac{\sqrt{3}}{2}$ **G.** $f''(x) = \sin x > 0 \iff 0 < x < \pi$ or

$2\pi < x < 3\pi$, so f is CU on $(0, \pi)$ and $(2\pi, 3\pi)$ and CD on $(\pi, 2\pi)$.

IPs at $\left(\pi, \frac{\pi}{2}\right)$ and $(2\pi, \pi)$.

H.

37. $y = f(x) = \sin 2x - 2\sin x$ **A.** $D = \mathbb{R}$ **B.** y-intercept $= f(0) = 0$. $y = 0 \Leftrightarrow$

$2\sin x = \sin 2x = 2\sin x \cos x \Leftrightarrow \sin x = 0$ or $\cos x = 1 \Leftrightarrow x = n\pi$ (x-intercepts)

C. $f(-x) = -f(x)$, so the curve is symmetric about $(0,0)$.

Note: f is periodic with period 2π, so we determine E–G for $-\pi \le x \le \pi$. **D.** No asymptotes

E. $f'(x) = 2\cos 2x - 2\cos x = 2(2\cos^2 x - 1 - \cos x) = 2(2\cos x + 1)(\cos x - 1) > 0 \Leftrightarrow \cos x < -\frac{1}{2}$

$\Leftrightarrow -\pi < x < -\frac{2\pi}{3}$ or $\frac{2\pi}{3} < x < \pi$, so f is increasing on $\left(-\pi, -\frac{2\pi}{3}\right)$, $\left(\frac{2\pi}{3}, \pi\right)$ and decreasing on $\left(-\frac{2\pi}{3}, \frac{2\pi}{3}\right)$.

F. Local maximum value $f\left(-\frac{2\pi}{3}\right) = \frac{3\sqrt{3}}{2}$,

local minimum value $f\left(\frac{2\pi}{3}\right) = -\frac{3\sqrt{3}}{2}$

G. $f''(x) = -4\sin 2x + 2\sin x = 2\sin x\,(1 - 4\cos x) = 0$ when

$x = 0$, $\pm\pi$ or $\cos x = \frac{1}{4}$. If $\alpha = \cos^{-1}\frac{1}{4}$, then f is CU on $(-\alpha, 0)$ and

(α, π) and CD on $(-\pi, -\alpha)$ and $(0, \alpha)$.

IPs at $(0,0)$, $(\pm\pi, 0)$, $\left(\alpha, -\frac{3\sqrt{15}}{8}\right)$, $\left(-\alpha, \frac{3\sqrt{15}}{8}\right)$.

H.

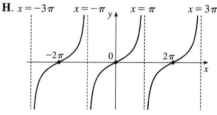

39. $y = f(x) = \dfrac{\sin x}{1 + \cos x}$ $\left[\overset{\text{when}}{\underset{\cos x \neq 1}{=}} \dfrac{\sin x}{1 + \cos x} \cdot \dfrac{1 - \cos x}{1 - \cos x} = \dfrac{\sin x\,(1 - \cos x)}{\sin^2 x} = \dfrac{1 - \cos x}{\sin x} = \csc x - \cot x\right]$

A. The domain of f is the set of all real numbers except odd integer multiples of π. **B.** y-intercept: $f(0) = 0$;

x-intercepts: $x = n\pi$, n an even integer. **C.** $f(-x) = -f(x)$, so f is an odd function; the graph is symmetric

about the origin and has period 2π. **D.** When n is an odd integer, $\displaystyle\lim_{x \to (n\pi)^-} f(x) = \infty$ and $\displaystyle\lim_{x \to (n\pi)^+} f(x) = -\infty$,

so $x = n\pi$ is a VA for each odd integer n. No HA.

E. $f'(x) = \dfrac{(1 + \cos x) \cdot \cos x - \sin x(-\sin x)}{(1 + \cos x)^2} = \dfrac{1 + \cos x}{(1 + \cos x)^2} = \dfrac{1}{1 + \cos x}$. $f'(x) > 0$ for all x except odd

multiples of π, so f is increasing on $((2k - 1)\pi, (2k + 1)\pi)$ for each integer k. **F.** No extreme values

G. $f''(x) = \dfrac{\sin x}{(1 + \cos x)^2} > 0 \Rightarrow \sin x > 0 \Rightarrow$

$x \in (2k\pi, (2k + 1)\pi)$ and $f''(x) < 0$ on $((2k - 1)\pi, 2k\pi)$

for each integer k. f is CU on $(2k\pi, (2k + 1)\pi)$ and CD on

$((2k - 1)\pi, 2k\pi)$ for each integer k. f has IPs at $(2k\pi, 0)$

for each integer k.

H.

41. $y = 1/(1 + e^{-x})$ **A.** $D = \mathbb{R}$ **B.** No x-intercept; y-intercept $= f(0) = \frac{1}{2}$.   **C.** No symmetry

D. $\lim_{x \to \infty} 1/(1 + e^{-x}) = \frac{1}{1+0} = 1$ and $\lim_{x \to -\infty} 1/(1 + e^{-x}) = 0$ (since $\lim_{x \to -\infty} e^{-x} = \infty$), so f has horizontal

asymptotes $y = 0$ and $y = 1$.   **E.** $f'(x) = -(1 + e^{-x})^{-2}(-e^{-x}) = e^{-x}/(1 + e^{-x})^2$. This is positive for all x,

so f is increasing on $\mathbb{R}$.   **F.** No extreme values

G. $f''(x) = \dfrac{(1 + e^{-x})^2(-e^{-x}) - e^{-x}(2)(1 + e^{-x})(-e^{-x})}{(1 + e^{-x})^4}$

$= \dfrac{e^{-x}(e^{-x} - 1)}{(1 + e^{-x})^3}$

H.

The second factor in the numerator is negative for $x > 0$ and positive for

$x < 0$, and the other factors are always positive, so f is CU on $(-\infty, 0)$

and CD on $(0, \infty)$. f has an inflection point at $(0, \frac{1}{2})$.

43. $y = f(x) = x \ln x$ **A.** $D = (0, \infty)$ **B.** x-intercept when $\ln x = 0$ $\Leftrightarrow$ $x = 1$, no y-intercept

C. No symmetry   **D.** $\lim_{x \to \infty} x \ln x = \infty$,

$\lim_{x \to 0^+} x \ln x = \lim_{x \to 0^+} \dfrac{\ln x}{1/x} \overset{\text{H}}{=} \lim_{x \to 0^+} \dfrac{1/x}{-1/x^2} = \lim_{x \to 0^+} (-x) = 0$, no

H.

asymptote.   **E.** $f'(x) = \ln x + 1 = 0$ when $\ln x = -1$ $\Leftrightarrow$ $x = e^{-1}$.

$f'(x) > 0$ $\Leftrightarrow$ $\ln x > -1$ $\Leftrightarrow$ $x > e^{-1}$, so f is increasing on

$(1/e, \infty)$ and decreasing on $(0, 1/e)$.   **F.** $f(1/e) = -1/e$ is an absolute

and local minimum value.   **G.** $f''(x) = 1/x > 0$, so f is CU on $(0, \infty)$.

No IP

45. $y = f(x) = xe^{-x}$ **A.** $D = \mathbb{R}$ **B.** Intercepts are 0   **C.** No symmetry   **H.**

D. $\lim_{x \to \infty} xe^{-x} = \lim_{x \to \infty} \dfrac{x}{e^x} \overset{\text{H}}{=} \lim_{x \to \infty} \dfrac{1}{e^x} = 0$, so $y = 0$ is a HA.

$\lim_{x \to -\infty} xe^{-x} = -\infty$ **E.** $f'(x) = e^{-x} - xe^{-x} = e^{-x}(1 - x) > 0$ $\Leftrightarrow$

$x < 1$, so f is increasing on $(-\infty, 1)$ and decreasing on $(1, \infty)$.

F. Absolute and local maximum value $f(1) = 1/e$.

G. $f''(x) = e^{-x}(x - 2) > 0$ $\Leftrightarrow$ $x > 2$, so f is CU on $(2, \infty)$ and CD on $(-\infty, 2)$. IP at $(2, 2/e^2)$

47. $y = f(x) = \ln(\sin x)$

A. $D = \{x \text{ in } \mathbb{R} \mid \sin x > 0\} = \bigcup_{n=-\infty}^{\infty} (2n\pi, (2n+1)\pi)$

$$= \cdots \cup (-4\pi, -3\pi) \cup (-2\pi, -\pi) \cup (0, \pi) \cup (2\pi, 3\pi) \cup \cdots$$

B. No y-intercept; x-intercepts: $f(x) = 0 \iff \ln(\sin x) = 0 \iff \sin x = e^0 = 1 \iff x = 2n\pi + \frac{\pi}{2}$ for

each integer n. **C.** f is periodic with period 2π. **D.** $\lim\limits_{x\to(2n\pi)^+} f(x) = -\infty$ and $\lim\limits_{x\to[(2n+1)\pi]^-} f(x) = -\infty$, so

the lines $x = n\pi$ are VAs for all integers n. **E.** $f'(x) = \frac{\cos x}{\sin x} = \cot x$, so $f'(x) > 0$ when $2n\pi < x < 2n\pi + \frac{\pi}{2}$

for each integer n, and $f'(x) < 0$ when $2n\pi + \frac{\pi}{2} < x < (2n+1)\pi$. Thus, f is increasing on $\left(2n\pi, 2n\pi + \frac{\pi}{2}\right)$ and

decreasing on $\left(2n\pi + \frac{\pi}{2}, (2n+1)\pi\right)$ for each integer n. **F.** Local maximum values $f\left(2n\pi + \frac{\pi}{2}\right) = 0$, no local

minimum. **G.** $f''(x) = -\csc^2 x < 0$, so f is CD on $(2n\pi, (2n+1)\pi)$ for each integer n. No IP

H.

49. $y = f(x) = xe^{-x^2}$ **A.** $D = \mathbb{R}$ **B.** Intercepts are 0 **C.** $f(-x) = -f(x)$, so the curve is symmetric

about the origin. **D.** $\lim\limits_{x\to\pm\infty} xe^{-x^2} = \lim\limits_{x\to\pm\infty} \frac{x}{e^{x^2}} \overset{\text{H}}{=} \lim\limits_{x\to\pm\infty} \frac{1}{2xe^{x^2}} = 0$, so $y = 0$ is a HA.

E. $f'(x) = e^{-x^2} - 2x^2 e^{-x^2} = e^{-x^2}(1 - 2x^2) > 0 \iff x^2 < \frac{1}{2} \iff |x| < \frac{1}{\sqrt{2}}$, so f is increasing on

$\left(-\frac{1}{\sqrt{2}}, \frac{1}{\sqrt{2}}\right)$ and decreasing on $\left(-\infty, -\frac{1}{\sqrt{2}}\right)$ and $\left(\frac{1}{\sqrt{2}}, \infty\right)$. **F.** Local maximum value $f\left(\frac{1}{\sqrt{2}}\right) = 1/\sqrt{2e}$, local

minimum value $f\left(-\frac{1}{\sqrt{2}}\right) = -1/\sqrt{2e}$ **G.** $f''(x) = -2xe^{-x^2}(1 - 2x^2) - 4xe^{-x^2} = 2xe^{-x^2}(2x^2 - 3) > 0$

$\iff x > \sqrt{\frac{3}{2}}$ or $-\sqrt{\frac{3}{2}} < x < 0$, so f is CU on $\left(\sqrt{\frac{3}{2}}, \infty\right)$ **H.**

and $\left(-\sqrt{\frac{3}{2}}, 0\right)$ and CD on $\left(-\infty, -\sqrt{\frac{3}{2}}\right)$ and $\left(0, \sqrt{\frac{3}{2}}\right)$.

IP are $(0, 0)$ and $\left(\pm\sqrt{\frac{3}{2}}, \pm\sqrt{\frac{3}{2}}e^{-3/2}\right)$.

51. $y = f(x) = e^{3x} + e^{-2x}$ **A.** $D = \mathbb{R}$ **B.** y-intercept $= f(0) = 2$; **H.**

no x-intercept **C.** No symmetry **D.** No asymptotes

E. $f'(x) = 3e^{3x} - 2e^{-2x}$, so $f'(x) > 0 \iff 3e^{3x} > 2e^{-2x}$

[multiply by e^{2x}] $\iff e^{5x} > \frac{2}{3} \iff 5x > \ln\frac{2}{3} \iff$

$x > \frac{1}{5}\ln\frac{2}{3} \approx -0.081$. Similarly, $f'(x) < 0 \iff x < \frac{1}{5}\ln\frac{2}{3}$.

f is decreasing on $\left(-\infty, \frac{1}{5}\ln\frac{2}{3}\right)$ and increasing on $\left(\frac{1}{5}\ln\frac{2}{3}, \infty\right)$.

F. Local minimum value $f\left(\frac{1}{5}\ln\frac{2}{3}\right) = \left(\frac{2}{3}\right)^{3/5} + \left(\frac{2}{3}\right)^{-2/5} \approx 1.96$; no local maximum.

G. $f''(x) = 9e^{3x} + 4e^{-2x}$, so $f''(x) > 0$ for all x, and f is CU on $(-\infty, \infty)$. No IP

53. $y = -\dfrac{W}{24EI}x^4 + \dfrac{WL}{12EI}x^3 - \dfrac{WL^2}{24EI}x^2 = -\dfrac{W}{24EI}x^2\left(x^2 - 2Lx + L^2\right)$

$= \dfrac{-W}{24EI}x^2(x-L)^2 = cx^2(x-L)^2$

where $c = -\dfrac{W}{24EI}$ is a negative constant and $0 \le x \le L$. We sketch

$f(x) = cx^2(x-L)^2$ for $c = -1$. $f(0) = f(L) = 0$.

$f'(x) = cx^2[2(x-L)] + (x-L)^2(2cx) = 2cx(x-L)[x + (x-L)] = 2cx(x-L)(2x-L)$. So for

$0 < x < L$, $f'(x) > 0 \Leftrightarrow x(x-L)(2x-L) < 0$ (since $c < 0$) $\Leftrightarrow L/2 < x < L$ and $f'(x) < 0 \Leftrightarrow$

$0 < x < L/2$. So f is increasing on $(L/2, L)$ and decreasing on $(0, L/2)$, and there is a local and absolute

minimum at $(L/2, f(L/2)) = (L/2, cL^4/16)$.

$f'(x) = 2c[x(x-L)(2x-L)] \Rightarrow$

$f''(x) = 2c[1(x-L)(2x-L) + x(1)(2x-L) + x(x-L)(2)] = 2c(6x^2 - 6Lx + L^2) = 0 \Leftrightarrow$

$x = \dfrac{6L \pm \sqrt{12L^2}}{12} = \tfrac{1}{2}L \pm \tfrac{\sqrt{3}}{6}L$, and these are the x-coordinates of the two inflection points.

55. $y = \dfrac{x^2+1}{x+1}$. Long division gives us:

$$
\begin{array}{r}
x - 1 \\
x+1{\overline{\smash{\big)}\,x^2 \phantom{{}+x} +1}} \\
\underline{x^2 + x} \\
-x+1 \\
\underline{-x-1} \\
2
\end{array}
$$

Thus, $y = f(x) = \dfrac{x^2+1}{x+1} = x - 1 + \dfrac{2}{x+1}$ and $f(x) - (x-1) = \dfrac{2}{x+1} = \dfrac{\frac{2}{x}}{1 + \frac{1}{x}}$ [for $x \ne 0$] $\to 0$

as $x \to \pm\infty$. So the line $y = x - 1$ is a slant asymptote (SA).

57. $y = \dfrac{4x^3 - 2x^2 + 5}{2x^2 + x - 3}$. Long division gives us:

$$
\begin{array}{r}
2x - 2 \\
2x^2 + x - 3{\overline{\smash{\big)}\,4x^3 - 2x^2 \phantom{{}-6x} + 5}} \\
\underline{4x^3 + 2x^2 - 6x} \\
-4x^2 + 6x + 5 \\
\underline{-4x^2 - 2x + 6} \\
8x - 1
\end{array}
$$

Thus, $y = f(x) = \dfrac{4x^3 - 2x^2 + 5}{2x^2 + x - 3} = 2x - 2 + \dfrac{8x - 1}{2x^2 + x - 3}$ and

$f(x) - (2x - 2) = \dfrac{8x - 1}{2x^2 + x - 3} = \dfrac{\frac{8}{x} - \frac{1}{x^2}}{2 + \frac{1}{x} - \frac{3}{x^2}}$ [for $x \ne 0$] $\to 0$ as $x \to \pm\infty$. So the line $y = 2x - 2$ is

a SA.

59. $y = f(x) = \dfrac{-2x^2 + 5x - 1}{2x - 1} = -x + 2 + \dfrac{1}{2x - 1}$ **A.** $D = \{x \in \mathbb{R} \mid x \neq \frac{1}{2}\} = \left(-\infty, \frac{1}{2}\right) \cup \left(\frac{1}{2}, \infty\right)$

B. y-intercept: $f(0) = 1$; x-intercepts: $f(x) = 0 \;\Rightarrow\; -2x^2 + 5x - 1 = 0 \;\Rightarrow\; x = \dfrac{-5 \pm \sqrt{17}}{-4} \;\Rightarrow$

$x \approx 0.22, 2.28$.   **C.** No symmetry

D. $\displaystyle\lim_{x \to (1/2)^-} f(x) = -\infty$ and $\displaystyle\lim_{x \to (1/2)^+} f(x) = \infty$, so $x = \frac{1}{2}$ is a VA.

$\displaystyle\lim_{x \to \pm\infty} [f(x) - (-x + 2)] = \lim_{x \to \pm\infty} \dfrac{1}{2x - 1} = 0$, so the line $y = -x + 2$ is a SA.

E. $f'(x) = -1 - \dfrac{2}{(2x - 1)^2} < 0$ for $x \neq \frac{1}{2}$, so f is decreasing

H.

on $\left(-\infty, \frac{1}{2}\right)$ and $\left(\frac{1}{2}, \infty\right)$.   **F.** No extreme values

G. $f'(x) = -1 - 2(2x - 1)^{-2} \;\Rightarrow$

$f''(x) = -2(-2)(2x - 1)^{-3}(2) = \dfrac{8}{(2x - 1)^3}$, so $f''(x) > 0$ when

$x > \frac{1}{2}$ and $f''(x) < 0$ when $x < \frac{1}{2}$. Thus, f is CU on $\left(\frac{1}{2}, \infty\right)$ and CD

on $\left(-\infty, \frac{1}{2}\right)$. No IP

61. $y = f(x) = \left(x^2 + 4\right)/x = x + 4/x$ **A.** $D = \{x \mid x \neq 0\} = (-\infty, 0) \cup (0, \infty)$ **B.** No intercept

C. $f(-x) = -f(x) \;\Rightarrow$ symmetry about the origin   **D.** $\displaystyle\lim_{x \to \infty} (x + 4/x) = \infty$ but $f(x) - x = 4/x \to 0$ as

$x \to \pm\infty$, so $y = x$ is a slant asymptote. $\displaystyle\lim_{x \to 0^+} (x + 4/x) = \infty$ and

H.

$\displaystyle\lim_{x \to 0^-} (x + 4/x) = -\infty$, so $x = 0$ is a VA.   **E.** $f'(x) = 1 - 4/x^2 > 0$

$\Leftrightarrow\; x^2 > 4 \;\Leftrightarrow\; x > 2$ or $x < -2$, so f is increasing on $(-\infty, -2)$

and $(2, \infty)$ and decreasing on $(-2, 0)$ and $(0, 2)$.

F. Local maximum value $f(-2) = -4$, local minimum value $f(2) = 4$

G. $f''(x) = 8/x^3 > 0 \;\Leftrightarrow\; x > 0$ so f is CU on $(0, \infty)$ and CD

on $(-\infty, 0)$. No IP

63. $y = f(x) = \dfrac{2x^3 + x^2 + 1}{x^2 + 1} = 2x + 1 + \dfrac{-2x}{x^2 + 1}$ **A.** $D = \mathbb{R}$ **B.** y-intercept: $f(0) = 1$; x-intercept: $f(x) = 0$

$\Rightarrow\; 0 = 2x^3 + x^2 + 1 = (x + 1)(2x^2 - x + 1) \;\Rightarrow\; x = -1$ **C.** No symmetry   **D.** No VA

$\displaystyle\lim_{x \to \pm\infty} [f(x) - (2x + 1)] = \lim_{x \to \pm\infty} \dfrac{-2x}{x^2 + 1} = \lim_{x \to \pm\infty} \dfrac{-2/x}{1 + 1/x^2} = 0$, so the line $y = 2x + 1$ is a slant asymptote.

E. $f'(x) = 2 + \dfrac{(x^2 + 1)(-2) - (-2x)(2x)}{(x^2 + 1)^2} = \dfrac{2(x^4 + 2x^2 + 1) - 2x^2 - 2 + 4x^2}{(x^2 + 1)^2}$

$= \dfrac{2x^4 + 6x^2}{(x^2 + 1)^2} = \dfrac{2x^2(x^2 + 3)}{(x^2 + 1)^2}$

so $f'(x) > 0$ if $x \neq 0$. Thus, f is increasing on $(-\infty, 0)$ and $(0, \infty)$. Since f is continuous at 0, f is increasing

on $\mathbb{R}$. **F.** No extreme values

G. $f''(x) = \dfrac{(x^2+1)^2 \cdot (8x^3+12x) - (2x^4+6x^2) \cdot 2(x^2+1)(2x)}{[(x^2+1)^2]^2}$ **H.**

$$= \dfrac{4x(x^2+1)[(x^2+1)(2x^2+3) - 2x^4 - 6x^2]}{(x^2+1)^4}$$

$$= \dfrac{4x(-x^2+3)}{(x^2+1)^3}$$

so $f''(x) > 0$ for $x < -\sqrt{3}$ and $0 < x < \sqrt{3}$, and $f''(x) < 0$ for

$-\sqrt{3} < x < 0$ and $x > \sqrt{3}$. f is CU on $(-\infty, -\sqrt{3}\,)$ and $(0, \sqrt{3}\,)$,

and CD on $(-\sqrt{3}, 0)$ and $(\sqrt{3}, \infty)$. There are three IPs: $(0, 1)$, $\left(-\sqrt{3}, -\frac{3}{2}\sqrt{3}+1\right) \approx (-1.73, -1.60)$, and

$\left(\sqrt{3}, \frac{3}{2}\sqrt{3}+1\right) \approx (1.73, 3.60)$.

65. $y = f(x) = x - \tan^{-1} x$, $f'(x) = 1 - \dfrac{1}{1+x^2} = \dfrac{1+x^2-1}{1+x^2} = \dfrac{x^2}{1+x^2}$,

$f''(x) = \dfrac{(1+x^2)(2x) - x^2(2x)}{(1+x^2)^2} = \dfrac{2x(1+x^2-x^2)}{(1+x^2)^2} = \dfrac{2x}{(1+x^2)^2}$.

$\displaystyle\lim_{x \to \infty} \left[f(x) - \left(x - \tfrac{\pi}{2}\right) \right] = \lim_{x \to \infty} \left(\tfrac{\pi}{2} - \tan^{-1} x\right) = \tfrac{\pi}{2} - \tfrac{\pi}{2} = 0$, so $y = x - \tfrac{\pi}{2}$ is a SA. Also,

$\displaystyle\lim_{x \to -\infty} \left[f(x) - \left(x + \tfrac{\pi}{2}\right) \right] = \lim_{x \to -\infty} \left(-\tfrac{\pi}{2} - \tan^{-1} x\right)$

$\qquad\qquad\qquad\qquad\quad = -\tfrac{\pi}{2} - \left(-\tfrac{\pi}{2}\right) = 0$

so $y = x + \tfrac{\pi}{2}$ is also a SA. $f'(x) \geq 0$ for all x, with equality $\Leftrightarrow$

$x = 0$, so f is increasing on $\mathbb{R}$. $f''(x)$ has the same sign as x, so f is CD

on $(-\infty, 0)$ and CU on $(0, \infty)$. $f(-x) = -f(x)$, so f is an odd function;

its graph is symmetric about the origin. f has no local extreme values. Its

only IP is at $(0, 0)$.

67. $\dfrac{x^2}{a^2} - \dfrac{y^2}{b^2} = 1 \;\Rightarrow\; y = \pm\dfrac{b}{a}\sqrt{x^2-a^2}$. Now

$\displaystyle\lim_{x \to \infty} \left[\dfrac{b}{a}\sqrt{x^2-a^2} - \dfrac{b}{a}x \right] = \dfrac{b}{a} \cdot \lim_{x \to \infty} \left(\sqrt{x^2-a^2} - x\right) \dfrac{\sqrt{x^2-a^2}+x}{\sqrt{x^2-a^2}+x} = \dfrac{b}{a} \cdot \lim_{x \to \infty} \dfrac{-a^2}{\sqrt{x^2-a^2}+x} = 0$,

which shows that $y = \dfrac{b}{a}x$ is a slant asymptote. Similarly,

$\displaystyle\lim_{x \to \infty} \left[-\dfrac{b}{a}\sqrt{x^2-a^2} - \left(-\dfrac{b}{a}x\right) \right] = -\dfrac{b}{a} \cdot \lim_{x \to \infty} \dfrac{-a^2}{\sqrt{x^2-a^2}+x} = 0$, so $y = -\dfrac{b}{a}x$ is a slant asymptote.

69. $\lim\limits_{x \to \pm\infty} \left[f(x) - x^3 \right] = \lim\limits_{x \to \pm\infty} \dfrac{x^4 + 1}{x} - \dfrac{x^4}{x} = \lim\limits_{x \to \pm\infty} \dfrac{1}{x} = 0$, so the graph of f is asymptotic to that of $y = x^3$.

A. $D = \{x \mid x \neq 0\}$ **B.** No intercept   **C.** f is symmetric about the origin.   **D.** $\lim\limits_{x \to 0^-} \left(x^3 + \dfrac{1}{x} \right) = -\infty$ and

$\lim\limits_{x \to 0^+} \left(x^3 + \dfrac{1}{x} \right) = \infty$, so $x = 0$ is a vertical asymptote, and as shown above, the graph of f is asymptotic to

that of $y = x^3$.   **E.** $f'(x) = 3x^2 - 1/x^2 > 0 \;\Leftrightarrow\; x^4 > \frac{1}{3} \;\Leftrightarrow\;$ **H.**

$|x| > \frac{1}{\sqrt[4]{3}}$, so f is increasing on $\left(-\infty, -\frac{1}{\sqrt[4]{3}} \right)$ and $\left(\frac{1}{\sqrt[4]{3}}, \infty \right)$ and

decreasing on $\left(-\frac{1}{\sqrt[4]{3}}, 0 \right)$ and $\left(0, \frac{1}{\sqrt[4]{3}} \right)$.   **F.** Local maximum value

$f\left(-\frac{1}{\sqrt[4]{3}} \right) = -4 \cdot 3^{-5/4}$, local minimum value $f\left(\frac{1}{\sqrt[4]{3}} \right) = 4 \cdot 3^{-5/4}$

G. $f''(x) = 6x + 2/x^3 > 0 \;\Leftrightarrow\; x > 0$, so f is CU on $(0, \infty)$ and
CD on $(-\infty, 0)$. No IP

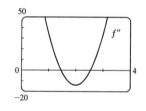

4.6 Graphing with Calculus and Calculators

1. $f(x) = 4x^4 - 32x^3 + 89x^2 - 95x + 29 \;\Rightarrow\; f'(x) = 16x^3 - 96x^2 + 178x - 95 \;\Rightarrow$

$f''(x) = 48x^2 - 192x + 178.$ $f(x) = 0 \;\Leftrightarrow\; x \approx 0.5, \, 1.60;$ $f'(x) = 0 \;\Leftrightarrow\; x \approx 0.92, \, 2.5, \, 2.58$ and

$f''(x) = 0 \;\Leftrightarrow\; x \approx 1.46, \, 2.54.$

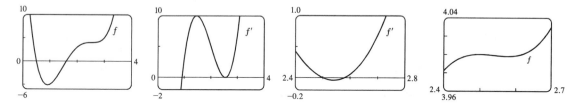

From the graphs of f', we estimate that $f' < 0$ and that f is decreasing on $(-\infty, 0.92)$ and $(2.5, 2.58)$, and that

$f' > 0$ and f is increasing on $(0.92, 2.5)$ and $(2.58, \infty)$ with local minimum values $f(0.92) \approx -5.12$ and

$f(2.58) \approx 3.998$ and local maximum value $f(2.5) = 4$. The graphs of f' make it clear that f has a maximum and a

minimum near $x = 2.5$, shown more clearly in the fourth graph.

From the graph of f'', we estimate that $f'' > 0$ and that f is CU

on $(-\infty, 1.46)$ and $(2.54, \infty)$, and that $f'' < 0$ and f is CD on

$(1.46, 2.54)$. There are inflection points at about $(1.46, -1.40)$ and

$(2.54, 3.999)$.

3. $f(x) = \sqrt[3]{x^2 - 3x - 5}$ $\Rightarrow$ $f'(x) = \dfrac{1}{3}\dfrac{2x - 3}{(x^2 - 3x - 5)^{2/3}}$ $\Rightarrow$ $f''(x) = -\dfrac{2}{9}\dfrac{x^2 - 3x + 24}{(x^2 - 3x - 5)^{5/3}}$

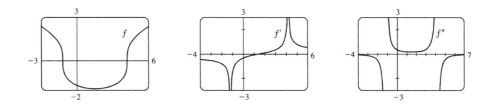

Note: With some CAS's, including Maple, it is necessary to define $f(x) = \dfrac{x^2 - 3x - 5}{|x^2 - 3x - 5|}\,|x^2 - 3x - 5|^{1/3}$,

since the CAS does not compute real cube roots of negative numbers. We estimate from the graph of f' that f is

increasing on $(1.5, \infty)$, and decreasing on $(-\infty, 1.5)$. f has no maximum. Minimum value: $f(1.5) \approx -1.9$.

From the graph of f'', we estimate that f is CU on $(-1.2, 4.2)$ and CD on $(-\infty, -1.2)$ and $(4.2, \infty)$. IP at

$(-1.2, 0)$ and $(4.2, 0)$.

5. $f(x) = \dfrac{x}{x^3 - x^2 - 4x + 1}$ $\Rightarrow$ $f'(x) = \dfrac{-2x^3 + x^2 + 1}{(x^3 - x^2 - 4x + 1)^2}$ $\Rightarrow$

$$f''(x) = \frac{2(3x^5 - 3x^4 + 5x^3 - 6x^2 + 3x + 4)}{(x^3 - x^2 - 4x + 1)^3}$$

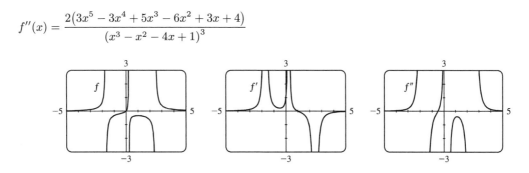

We estimate from the graph of f that $y = 0$ is a horizontal asymptote, and that there are vertical asymptotes at

$x = -1.7$, $x = 0.24$, and $x = 2.46$. From the graph of f', we estimate that f is increasing on $(-\infty, -1.7)$,

$(-1.7, 0.24)$, and $(0.24, 1)$, and that f is decreasing on $(1, 2.46)$ and $(2.46, \infty)$. There is a local maximum value

at $f(1) = -\frac{1}{3}$. From the graph of f'', we estimate that f is CU on $(-\infty, -1.7)$, $(-0.506, 0.24)$, and $(2.46, \infty)$,

and that f is CD on $(-1.7, -0.506)$ and $(0.24, 2.46)$. There is an inflection point at $(-0.506, -0.192)$.

7. $f(x) = x^2 - 4x + 7\cos x$, $-4 \le x \le 4$. $f'(x) = 2x - 4 - 7\sin x$ $\Rightarrow$ $f''(x) = 2 - 7\cos x$.
$f(x) = 0$ $\Leftrightarrow$ $x \approx 1.10$; $f'(x) = 0$ $\Leftrightarrow$ $x \approx -1.49, -1.07$, or 2.89; $f''(x) = 0$ $\Leftrightarrow$
$x = \pm\cos^{-1}\left(\frac{2}{7}\right) \approx \pm 1.28$.

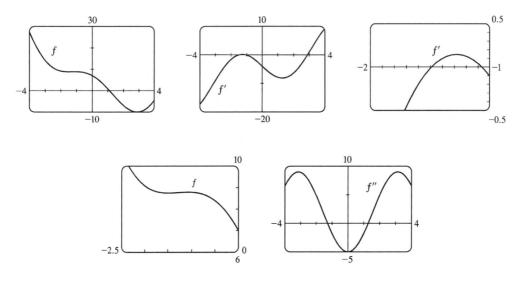

From the graphs of f', we estimate that f is decreasing ($f' < 0$) on $(-4, -1.49)$, increasing on $(-1.49, -1.07)$,
decreasing on $(-1.07, 2.89)$, and increasing on $(2.89, 4)$, with local minimum values $f(-1.49) \approx 8.75$ and
$f(2.89) \approx -9.99$ and local maximum value $f(-1.07) \approx 8.79$ (notice the second graph of f). From the graph
of f'', we estimate that f is CU ($f'' > 0$) on $(-4, -1.28)$, CD on $(-1.28, 1.28)$, and CU on $(1.28, 4)$. There are
inflection points at about $(-1.28, 8.77)$ and $(1.28, -1.48)$.

9. $f(x) = 8x^3 - 3x^2 - 10$ $\Rightarrow$ $f'(x) = 24x^2 - 6x$ $\Rightarrow$ $f''(x) = 48x - 6$

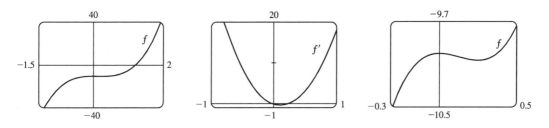

From the graphs, it appears that $f(x) = 8x^3 - 3x^2 - 10$ increases on $(-\infty, 0)$ and $(0.25, \infty)$ and decreases on
$(0, 0.25)$; that f has a local maximum value of $f(0) = -10.0$ and a local minimum value of $f(0.25) \approx -10.1$; that
f is CU on $(0.1, \infty)$ and CD on $(-\infty, 0.1)$; and that f has an IP at $(0.1, -10)$. To find the exact values, note that
$f'(x) = 24x^2 - 6x = 6x(4x - 1)$, which is positive ($f$ is increasing) for $(-\infty, 0)$ and $\left(\frac{1}{4}, \infty\right)$, and negative
(f is decreasing) on $\left(0, \frac{1}{4}\right)$. By the FDT, f has a local maximum at $x = 0$: $f(0) = -10$; and f has a local
minimum at $\frac{1}{4}$: $f\left(\frac{1}{4}\right) = \frac{1}{8} - \frac{3}{16} - 10 = -\frac{161}{16}$. $f''(x) = 48x - 6 = 6(8x - 1)$, which is positive ($f$ is CU) on
$\left(\frac{1}{8}, \infty\right)$ and negative (f is CD) on $\left(-\infty, \frac{1}{8}\right)$. f has an IP at $\left(\frac{1}{8}, f\left(\frac{1}{8}\right)\right) = \left(\frac{1}{8}, -\frac{321}{32}\right)$.

11. From the graph, it appears that f increases on $(-2.1, 2.1)$ and decreases

on $(-3, -2.1)$ and $(2.1, 3)$; that f has a local maximum of $f(2.1) \approx 4.5$

and a local minimum of $f(-2.1) \approx -4.5$; that f is CU on $(-3.0, 0)$ and

CD on $(0, 3.0)$, and that f has an IP at $(0, 0)$. $f(x) = x\sqrt{9 - x^2}$ $\Rightarrow$

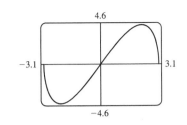

$f'(x) = \dfrac{-x^2}{\sqrt{9 - x^2}} + \sqrt{9 - x^2} = \dfrac{9 - 2x^2}{\sqrt{9 - x^2}}$, which is positive

(f is increasing) on $\left(\dfrac{-3\sqrt{2}}{2}, \dfrac{3\sqrt{2}}{2}\right)$ and negative (f is decreasing) on $\left(-3, \dfrac{-3\sqrt{2}}{2}\right)$ and $\left(\dfrac{3\sqrt{2}}{2}, 3\right)$. By the FDT,

f has a local maximum value of $f\left(\dfrac{3\sqrt{2}}{2}\right) = \dfrac{3\sqrt{2}}{2}\sqrt{9 - \left(\dfrac{3\sqrt{2}}{2}\right)^2} = \dfrac{9}{2}$; and f has a local minimum value of

$f\left(\dfrac{-3\sqrt{2}}{2}\right) = -\dfrac{9}{2}$ (since f is an odd function). $f'(x) = \dfrac{-x^2}{\sqrt{9 - x^2}} + \sqrt{9 - x^2}$ $\Rightarrow$

$f''(x) = \dfrac{\sqrt{9 - x^2}\,(-2x) + x^2\left(\frac{1}{2}\right)\left(9 - x^2\right)^{-1/2}(-2x)}{9 - x^2} - x\left(9 - x^2\right)^{-1/2} = \dfrac{-2x - x^3\left(9 - x^2\right)^{-1} - x}{\sqrt{9 - x^2}}$

$= \dfrac{-3x}{\sqrt{9 - x^2}} - \dfrac{x^3}{\left(9 - x^2\right)^{3/2}} = \dfrac{x\left(2x^2 - 27\right)}{\left(9 - x^2\right)^{3/2}}$

which is positive (f is CU) on $(-3, 0)$ and negative (f is CD) on $(0, 3)$. f has an IP at $(0, 0)$.

13. (a) $f(x) = x^2 \ln x$. The domain of f is $(0, \infty)$.

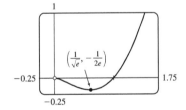

(b) $\lim\limits_{x \to 0^+} x^2 \ln x = \lim\limits_{x \to 0^+} \dfrac{\ln x}{1/x^2} \overset{\text{H}}{=} \lim\limits_{x \to 0^+} \dfrac{1/x}{-2/x^3} = \lim\limits_{x \to 0^+} \left(-\dfrac{x^2}{2}\right) = 0$. There is a hole at $(0, 0)$.

(c) It appears that there is an IP at about $(0.2, -0.06)$ and a local minimum at $(0.6, -0.18)$. $f(x) = x^2 \ln x$ $\Rightarrow$

$f'(x) = x^2(1/x) + (\ln x)(2x) = x(2\ln x + 1) > 0$ $\Leftrightarrow$ $\ln x > -\dfrac{1}{2}$ $\Leftrightarrow$ $x > e^{-1/2}$, so f is increasing on

$(1/\sqrt{e}, \infty)$, decreasing on $(0, 1/\sqrt{e})$. By the FDT, $f(1/\sqrt{e}) = -1/(2e)$ is a local minimum value. This point

is approximately $(0.6065, -0.1839)$, which agrees with our estimate.

$f''(x) = x(2/x) + (2\ln x + 1) = 2\ln x + 3 > 0$ $\Leftrightarrow$ $\ln x > -\dfrac{3}{2}$ $\Leftrightarrow$ $x > e^{-3/2}$, so f is CU on

$\left(e^{-3/2}, \infty\right)$ and CD on $\left(0, e^{-3/2}\right)$. IP is $\left(e^{-3/2}, -3/(2e^3)\right) \approx (0.2231, -0.0747)$.

15.

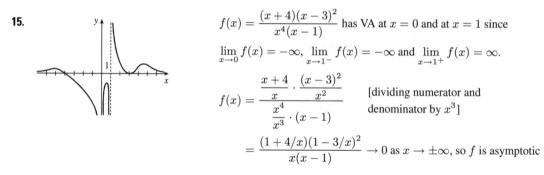

$f(x) = \dfrac{(x+4)(x-3)^2}{x^4(x-1)}$ has VA at $x = 0$ and at $x = 1$ since

$$\lim_{x \to 0} f(x) = -\infty, \ \lim_{x \to 1^-} f(x) = -\infty \text{ and } \lim_{x \to 1^+} f(x) = \infty.$$

$$f(x) = \dfrac{\dfrac{x+4}{x} \cdot \dfrac{(x-3)^2}{x^2}}{\dfrac{x^4}{x^3} \cdot (x-1)} \qquad \text{[dividing numerator and denominator by } x^3\text{]}$$

$$= \dfrac{(1 + 4/x)(1 - 3/x)^2}{x(x-1)} \to 0 \text{ as } x \to \pm\infty, \text{ so } f \text{ is asymptotic}$$

to the x-axis. Since f is undefined at $x = 0$, it has no y-intercept. $f(x) = 0 \ \Rightarrow \ (x+4)(x-3)^2 = 0 \ \Rightarrow$
$x = -4$ or $x = 3$, so f has x-intercepts -4 and 3. Note, however, that the graph of f is only tangent to the x-axis
and does not cross it at $x = 3$, since f is positive as $x \to 3^-$ and as $x \to 3^+$.

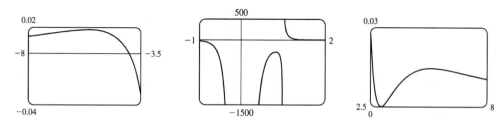

From these graphs, it appears that f has three maximum values and one minimum value. The maximum values are
approximately $f(-5.6) = 0.0182$, $f(0.82) = -281.5$ and $f(5.2) = 0.0145$ and we know (since the graph is
tangent to the x-axis at $x = 3$) that the minimum value is $f(3) = 0$.

17. $f(x) = \dfrac{x^2(x+1)^3}{(x-2)^2(x-4)^4} \ \Rightarrow \ f'(x) = -\dfrac{x(x+1)^2(x^3 + 18x^2 - 44x - 16)}{(x-2)^3(x-4)^5}$ (from CAS).

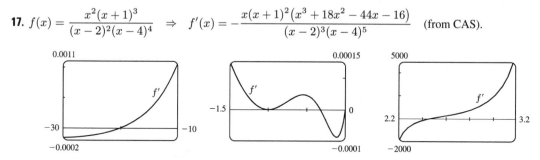

From the graphs of f', it seems that the critical points which indicate extrema occur at $x \approx -20$, -0.3, and 2.5,
as estimated in Example 3. (There is another critical point at $x = -1$, but the sign of f' does not change there.)
We differentiate again, obtaining $f''(x) = 2 \dfrac{(x+1)(x^6 + 36x^5 + 6x^4 - 628x^3 + 684x^2 + 672x + 64)}{(x-2)^4(x-4)^6}$.

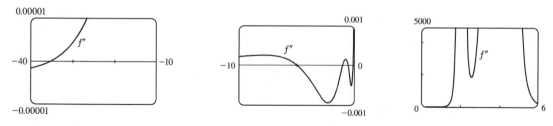

From the graphs of f'', it appears that f is CU on $(-35.3, -5.0)$, $(-1, -0.5)$, $(-0.1, 2)$, $(2, 4)$ and $(4, \infty)$ and CD on $(-\infty, -35.3)$, $(-5.0, -1)$ and $(-0.5, -0.1)$. We check back on the graphs of f to find the y-coordinates of the inflection points, and find that these points are approximately $(-35.3, -0.015)$, $(-5.0, -0.005)$, $(-1, 0)$, $(-0.5, 0.00001)$, and $(-0.1, 0.0000066)$.

19. $y = f(x) = \dfrac{\sin^2 x}{\sqrt{x^2 + 1}}$ with $0 \le x \le 3\pi$. From a CAS, $y' = \dfrac{\sin x \left[2(x^2 + 1)\cos x - x\sin x\right]}{(x^2 + 1)^{3/2}}$ and

$y'' = \dfrac{(4x^4 + 6x^2 + 5)\cos^2 x - 4x(x^2 + 1)\sin x \cos x - 2x^4 - 2x^2 - 3}{(x^2 + 1)^{5/2}}$.

From the graph of f' and the formula for y', we determine that $y' = 0$ when $x = \pi$, 2π, 3π, or $x \approx 1.3$, 4.6, or 7.8. So f is increasing on $(0, 1.3)$, $(\pi, 4.6)$, and $(2\pi, 7.8)$. f is decreasing on $(1.3, \pi)$, $(4.6, 2\pi)$, and $(7.8, 3\pi)$. Local maximum values: $f(1.3) \approx 0.6$, $f(4.6) \approx 0.21$, and $f(7.8) \approx 0.13$. Local minimum values: $f(\pi) = f(2\pi) = 0$. From the graph of f'', we see that $y'' = 0 \Leftrightarrow x \approx 0.6$, 2.1, 3.8, 5.4, 7.0, or 8.6. So f is CU on $(0, 0.6)$, $(2.1, 3.8)$, $(5.4, 7.0)$, and $(8.6, 3\pi)$. f is CD on $(0.6, 2.1)$, $(3.8, 5.4)$, and $(7.0, 8.6)$. There are IP at $(0.6, 0.25)$, $(2.1, 0.31)$, $(3.8, 0.10)$, $(5.4, 0.11)$, $(7.0, 0.061)$, and $(8.6, 0.065)$.

21. $y = f(x) = \dfrac{1 - e^{1/x}}{1 + e^{1/x}}$. From a CAS, $y' = \dfrac{2e^{1/x}}{x^2(1 + e^{1/x})^2}$ and $y'' = \dfrac{-2e^{1/x}(1 - e^{1/x} + 2x + 2xe^{1/x})}{x^4(1 + e^{1/x})^3}$.

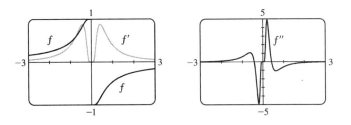

f is an odd function defined on $(-\infty, 0) \cup (0, \infty)$. Its graph has no x- or y-intercepts. Since $\displaystyle \lim_{x \to \pm\infty} f(x) = 0$, the x-axis is a HA. $f'(x) > 0$ for $x \ne 0$, so f is increasing on $(-\infty, 0)$ and $(0, \infty)$. It has no local extreme values. $f''(x) = 0$ for $x \approx \pm 0.417$, so f is CU on $(-\infty, -0.417)$, CD on $(-0.417, 0)$, CU on $(0, 0.417)$, and CD on $(0.417, \infty)$. f has IPs at $(-0.417, 0.834)$ and $(0.417, -0.834)$.

23. (a) $f(x) = x^{1/x}$

(b) Recall that $a^b = e^{b \ln a}$. $\lim\limits_{x \to 0^+} x^{1/x} = \lim\limits_{x \to 0^+} e^{(1/x) \ln x}$. As $x \to 0^+$,

$\dfrac{\ln x}{x} \to -\infty$, so $x^{1/x} = e^{(1/x) \ln x} \to 0$. This indicates that there is

a hole at $(0, 0)$. As $x \to \infty$, we have the indeterminate form ∞^0.

$\lim\limits_{x \to \infty} x^{1/x} = \lim\limits_{x \to \infty} e^{(1/x) \ln x}$, but $\lim\limits_{x \to \infty} \dfrac{\ln x}{x} \overset{\text{H}}{=} \lim\limits_{x \to \infty} \dfrac{1/x}{1} = 0$, so

$\lim\limits_{x \to \infty} x^{1/x} = e^0 = 1$. This indicates that $y = 1$ is a HA.

(c) Estimated maximum: $(2.72, 1.45)$. No estimated minimum. We use logarithmic differentiation to find any

critical numbers. $y = x^{1/x} \quad \Rightarrow \quad \ln y = \dfrac{1}{x} \ln x \quad \Rightarrow \quad \dfrac{y'}{y} = \dfrac{1}{x} \cdot \dfrac{1}{x} + (\ln x)\left(-\dfrac{1}{x^2}\right) \quad \Rightarrow$

$y' = x^{1/x}\left(\dfrac{1 - \ln x}{x^2}\right) = 0 \quad \Rightarrow \quad \ln x = 1 \quad \Rightarrow \quad x = e$. For $0 < x < e$, $y' > 0$ and for $x > e$, $y' < 0$, so

$f(e) = e^{1/e}$ is a local maximum value. This point is approximately $(2.7183, 1.4447)$, which agrees with our

estimate.

(d)

From the graph, we see that $f''(x) = 0$ at $x \approx 0.58$ and $x \approx 4.37$.

Since f'' changes sign at these values, they are x-coordinates of

inflection points.

25.

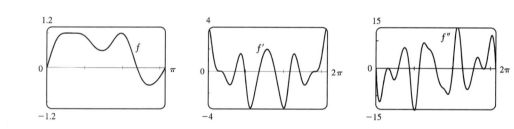

From the graph of $f(x) = \sin(x + \sin 3x)$ in the viewing rectangle $[0, \pi]$ by $[-1.2, 1.2]$, it looks like f has two

maxima and two minima. If we calculate and graph $f'(x) = [\cos(x + \sin 3x)](1 + 3\cos 3x)$ on $[0, 2\pi]$,

we see that the graph of f' appears to be almost tangent to the x-axis at about $x = 0.7$. The graph of

$f'' = -[\sin(x + \sin 3x)](1 + 3\cos 3x)^2 + \cos(x + \sin 3x)(-9 \sin 3x)$ is even more interesting near this x-value:

it seems to just touch the x-axis.

If we zoom in on this place on the graph of f'', we see that f'' actually does cross the axis twice near $x = 0.65$,

indicating a change in concavity for a very short interval. If we look at the graph of f' on the same interval, we see

that it changes sign three times near $x = 0.65$, indicating that what we had thought was a broad extremum at about

$x = 0.7$ actually consists of three extrema (two maxima and a minimum). These maximum values are roughly

$f(0.59) = 1$ and $f(0.68) = 1$, and the minimum value is roughly $f(0.64) = 0.99996$. There are also a maximum

value of about $f(1.96) = 1$ and minimum values of about $f(1.46) = 0.49$ and $f(2.73) = -0.51$. The points of

inflection on $(0, \pi)$ are about $(0.61, 0.99998)$, $(0.66, 0.99998)$, $(1.17, 0.72)$, $(1.75, 0.77)$, and $(2.28, 0.34)$. On

$(\pi, 2\pi)$, they are about $(4.01, -0.34)$, $(4.54, -0.77)$, $(5.11, -0.72)$, $(5.62, -0.99998)$, and $(5.67, -0.99998)$.

There are also IP at $(0, 0)$ and $(\pi, 0)$. Note that the function is odd and periodic with period 2π, and it is also

rotationally symmetric about all points of the form $((2n + 1)\pi, 0)$, n an integer.

27. $f(x) = x^4 + cx^2 = x^2(x^2 + c)$. Note that f is an even function. For $c \geq 0$, the only x-intercept is the point $(0, 0)$.

We calculate $f'(x) = 4x^3 + 2cx = 4x(x^2 + \frac{1}{2}c)$ $\Rightarrow$ $f''(x) = 12x^2 + 2c$. If $c \geq 0$, $x = 0$ is the only critical

point and there is no inflection point. As we can see from the examples, there is no change in the basic shape of the

graph for $c \geq 0$; it merely becomes steeper as c increases. For $c = 0$, the graph is the simple curve

$y = x^4$. For $c < 0$, there are x-intercepts at 0 and at $\pm\sqrt{-c}$. Also,

there is a maximum at $(0, 0)$, and there are minima at

$\left(\pm\sqrt{-\frac{1}{2}c}, -\frac{1}{4}c^2\right)$. As $c \to -\infty$, the x-coordinates of these minima

get larger in absolute value, and the minimum points move

downward. There are inflection points at $\left(\pm\sqrt{-\frac{1}{6}c}, -\frac{5}{36}c^2\right)$, which

also move away from the origin as $c \to -\infty$.

29.

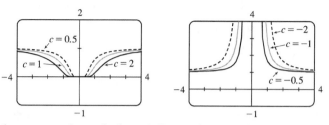

$c = 0$ is a transitional value — we get the graph of $y = 1$. For $c > 0$, we see that there is a HA at $y = 1$, and that the graph spreads out as c increases. At first glance there appears to be a minimum at $(0, 0)$, but $f(0)$ is undefined, so there is no minimum or maximum. For $c < 0$, we still have the HA at $y = 1$, but the range is $(1, \infty)$ rather than $(0, 1)$. We also have a VA at $x = 0$. $f(x) = e^{-c/x^2}$ $\Rightarrow$ $f'(x) = e^{-c/x^2}\left(\dfrac{2c}{x^3}\right)$ $\Rightarrow$

$f''(x) = \dfrac{2c(2c - 3x^2)}{x^6 e^{c/x^2}}$. $f'(x) \neq 0$ and $f'(x)$ exists for all $x \neq 0$ (and 0 is not in the domain of f), so there are no maxima or minima. $f''(x) = 0$ $\Rightarrow$ $x = \pm\sqrt{2c/3}$, so if $c > 0$, the inflection points spread out as c increases, and if $c < 0$, there are no IP. For $c > 0$, there are IP at $\left(\pm\sqrt{2c/3}, e^{-3/2}\right)$. Note that the y-coordinate of the IP is constant.

31. Note that $c = 0$ is a transitional value at which the graph consists of the x-axis. Also, we can see that if we substitute $-c$ for c, the function $f(x) = \dfrac{cx}{1 + c^2 x^2}$ will be reflected in the x-axis, so we investigate only positive values of c (except $c = -1$, as a demonstration of this reflective property). Also, f is an odd function. $\displaystyle\lim_{x \to \pm\infty} f(x) = 0$, so $y = 0$ is a horizontal asymptote for all c. We calculate

$f'(x) = \dfrac{(1 + c^2 x^2)c - cx(2c^2 x)}{(1 + c^2 x^2)^2} = -\dfrac{c(c^2 x^2 - 1)}{(1 + c^2 x^2)^2}$. $f'(x) = 0$ $\Leftrightarrow$ $c^2 x^2 - 1 = 0$ $\Leftrightarrow$ $x = \pm 1/c$. So there

is an absolute maximum value of $f(1/c) = \frac{1}{2}$ and an absolute minimum value of $f(-1/c) = -\frac{1}{2}$. These extrema have the same value regardless of c, but the maximum points move closer to the y-axis as c increases.

$f''(x) = \dfrac{(-2c^3 x)(1 + c^2 x^2)^2 - (-c^3 x^2 + c)[2(1 + c^2 x^2)(2c^2 x)]}{(1 + c^2 x^2)^4}$

$= \dfrac{(-2c^3 x)(1 + c^2 x^2) + (c^3 x^2 - c)(4c^2 x)}{(1 + c^2 x^2)^3} = \dfrac{2c^3 x(c^2 x^2 - 3)}{(1 + c^2 x^2)^3}$

$f''(x) = 0$ $\Leftrightarrow$ $x = 0$ or $\pm\sqrt{3}/c$, so there are inflection points at $(0, 0)$ and at $\left(\pm\sqrt{3}/c, \pm\sqrt{3}/4\right)$.

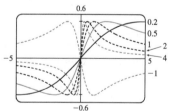

Again, the y-coordinate of the inflection points does not depend on c, but as c increases, both inflection points approach the y-axis.

33. $f(x) = cx + \sin x$ $\Rightarrow$ $f'(x) = c + \cos x$ $\Rightarrow$ $f''(x) = -\sin x$

$f(-x) = -f(x)$, so f is an odd function and its graph is symmetric with respect to the origin.

$f(x) = 0$ $\Leftrightarrow$ $\sin x = -cx$, so 0 is always an x-intercept.

$f'(x) = 0$ $\Leftrightarrow$ $\cos x = -c$, so there is no critical number when $|c| > 1$. If $|c| \leq 1$, then there are infinitely many critical numbers. If x_1 is the unique solution of $\cos x = -c$ in the interval $[0, \pi]$, then the critical numbers are $2n\pi \pm x_1$, where n ranges over the integers. (Special cases: When $c = 1$, $x_1 = 0$; when $c = 0$, $x = \frac{\pi}{2}$; and when $c = -1$, $x_1 = \pi$.)

$f''(x) < 0 \quad \Leftrightarrow \quad \sin x > 0$, so f is CD on intervals of the form $(2n\pi, (2n+1)\pi)$. f is CU on intervals of the form $((2n-1)\pi, 2n\pi)$. The inflection points of f are the points $(2n\pi, 2n\pi c)$, where n is an integer.

If $c \geq 1$, then $f'(x) \geq 0$ for all x, so f is increasing and has no extremum. If $c \leq -1$, then $f'(x) \leq 0$ for all x, so f is decreasing and has no extremum. If $|c| < 1$, then $f'(x) > 0 \quad \Leftrightarrow \quad \cos x > -c \quad \Leftrightarrow \quad x$ is in an interval of the form $(2n\pi - x_1, 2n\pi + x_1)$ for some integer n. These are the intervals on which f is increasing. Similarly, we find that f is decreasing on the intervals of the form $(2n\pi + x_1, 2(n+1)\pi - x_1)$. Thus, f has local maxima at the points $2n\pi + x_1$, where f has the values $c(2n\pi + x_1) + \sin x_1 = c(2n\pi + x_1) + \sqrt{1 - c^2}$, and f has local minima at the points $2n\pi - x_1$, where we have $f(2n\pi - x_1) = c(2n\pi - x_1) - \sin x_1 = c(2n\pi - x_1) - \sqrt{1 - c^2}$.

The transitional values of c are -1 and 1. The inflection points move vertically, but not horizontally, when c changes. When $|c| \geq 1$, there is no extremum. For $|c| < 1$, the maxima are spaced 2π apart horizontally, as are the minima. The horizontal spacing between maxima and adjacent minima is regular (and equals π) when $c = 0$, but the horizontal space between a local maximum and the nearest local minimum shrinks as $|c|$ approaches 1.

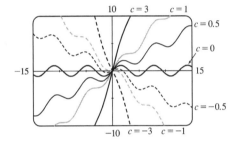

35. If $c < 0$, then $\displaystyle\lim_{x \to -\infty} f(x) = \lim_{x \to -\infty} \frac{x}{e^{cx}} \overset{\text{H}}{=} \lim_{x \to -\infty} \frac{1}{ce^{cx}} = 0$, and $\displaystyle\lim_{x \to \infty} f(x) = \infty$.

If $c > 0$, then $\displaystyle\lim_{x \to -\infty} f(x) = -\infty$, and $\displaystyle\lim_{x \to \infty} f(x) \overset{\text{H}}{=} \lim_{x \to \infty} \frac{1}{ce^{cx}} = 0$.

If $c = 0$, then $f(x) = x$, so $\displaystyle\lim_{x \to \pm\infty} f(x) = \pm\infty$ respectively.

So we see that $c = 0$ is a transitional value. We now exclude the case $c = 0$, since we know how the function behaves in that case. To find the maxima and minima of f, we differentiate: $f(x) = xe^{-cx} \quad \Rightarrow$ $f'(x) = x(-ce^{-cx}) + e^{-cx} = (1 - cx)e^{-cx}$. This is 0 when $1 - cx = 0 \quad \Leftrightarrow \quad x = 1/c$. If $c < 0$ then this represents a minimum value of $f(1/c) = 1/(ce)$, since $f'(x)$ changes from negative to positive at $x = 1/c$; and if $c > 0$, it represents a maximum value. As $|c|$ increases, the maximum or minimum point gets closer to the origin. To find the inflection points, we differentiate again: $f'(x) = e^{-cx}(1 - cx) \quad \Rightarrow$ $f''(x) = e^{-cx}(-c) + (1 - cx)(-ce^{-cx}) = (cx - 2)ce^{-cx}$. This changes sign when $cx - 2 = 0 \quad \Leftrightarrow \quad x = 2/c$. So as $|c|$ increases, the points of inflection get closer to the origin.

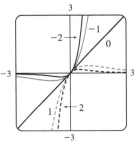

37. (a) $f(x) = cx^4 - 2x^2 + 1$. For $c = 0$, $f(x) = -2x^2 + 1$, a parabola whose vertex, $(0, 1)$, is the absolute

maximum. For $c > 0$, $f(x) = cx^4 - 2x^2 + 1$ opens upward with two minimum points. As $c \to 0$, the minimum

points spread apart and move downward; they are below the x-axis for $0 < c < 1$ and above for $c > 1$. For

$c < 0$, the graph opens downward, and has an absolute maximum at $x = 0$ and no local minimum.

(b) $f'(x) = 4cx^3 - 4x = 4cx(x^2 - 1/c)$ $(c \neq 0)$. If $c \leq 0$, 0 is the

only critical number. $f''(x) = 12cx^2 - 4$, so $f''(0) = -4$ and there

is a local maximum at $(0, f(0)) = (0, 1)$, which lies on $y = 1 - x^2$.

If $c > 0$, the critical numbers are 0 and $\pm 1/\sqrt{c}$. As before, there is a

local maximum at $(0, f(0)) = (0, 1)$, which lies on $y = 1 - x^2$.

$f''(\pm 1/\sqrt{c}) = 12 - 4 = 8 > 0$, so there is a local minimum at

$x = \pm 1/\sqrt{c}$. Here $f(\pm 1/\sqrt{c}) = c(1/c^2) - 2/c + 1 = -1/c + 1$.

But $(\pm 1/\sqrt{c}, -1/c + 1)$ lies on $y = 1 - x^2$ since

$1 - (\pm 1/\sqrt{c})^2 = 1 - 1/c$.

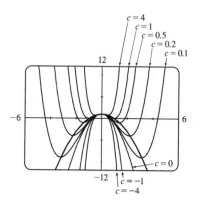

4.7 Optimization Problems

1. (a)

First Number	Second Number	Product
1	22	22
2	21	42
3	20	60
4	19	76
5	18	90
6	17	102
7	16	112
8	15	120
9	14	126
10	13	130
11	12	132

We needn't consider pairs where the first number

is larger than the second, since we can just

interchange the numbers in such cases. The

answer appears to be 11 and 12, but we have

considered only integers in the table.

(b) Call the two numbers x and y. Then $x + y = 23$, so $y = 23 - x$. Call the product P. Then

$P = xy = x(23 - x) = 23x - x^2$, so we wish to maximize the function $P(x) = 23x - x^2$. Since

$P'(x) = 23 - 2x$, we see that $P'(x) = 0 \iff x = \frac{23}{2} = 11.5$. Thus, the maximum value of P is

$P(11.5) = (11.5)^2 = 132.25$ and it occurs when $x = y = 11.5$.

Or: Note that $P''(x) = -2 < 0$ for all x, so P is everywhere concave downward and the local maximum at

$x = 11.5$ must be an absolute maximum.

3. The two numbers are x and $\dfrac{100}{x}$, where $x > 0$. Minimize $f(x) = x + \dfrac{100}{x}$. $f'(x) = 1 - \dfrac{100}{x^2} = \dfrac{x^2 - 100}{x^2}$.

The critical number is $x = 10$. Since $f'(x) < 0$ for $0 < x < 10$ and $f'(x) > 0$ for $x > 10$, there is an absolute minimum at $x = 10$. The numbers are 10 and 10.

5. If the rectangle has dimensions x and y, then its perimeter is $2x + 2y = 100$ m, so $y = 50 - x$. Thus, the area is $A = xy = x(50 - x)$. We wish to maximize the function $A(x) = x(50 - x) = 50x - x^2$, where $0 < x < 50$. Since $A'(x) = 50 - 2x = -2(x - 25)$, $A'(x) > 0$ for $0 < x < 25$ and $A'(x) < 0$ for $25 < x < 50$. Thus, A has an absolute maximum at $x = 25$, and $A(25) = 25^2 = 625$ m². The dimensions of the rectangle that maximize its area are $x = y = 25$ m. (The rectangle is a square.)

7. (a)

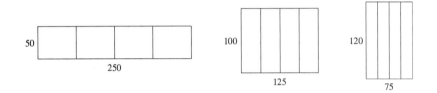

The areas of the three figures are 12,500, 12,500, and 9000 ft². There appears to be a maximum area of at least 12,500 ft².

(b) Let x denote the length of each of two sides and three dividers.

Let y denote the length of the other two sides.

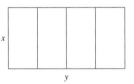

(c) Area $A =$ length × width $= y \cdot x$

(d) Length of fencing $= 750 \quad \Rightarrow \quad 5x + 2y = 750$

(e) $5x + 2y = 750 \quad \Rightarrow \quad y = 375 - \frac{5}{2}x \quad \Rightarrow \quad A(x) = \left(375 - \frac{5}{2}x\right)x = 375x - \frac{5}{2}x^2$

(f) $A'(x) = 375 - 5x = 0 \quad \Rightarrow \quad x = 75$. Since $A''(x) = -5 < 0$ there is an absolute maximum when $x = 75$. Then $y = \frac{375}{2} = 187.5$. The largest area is $75\left(\frac{375}{2}\right) = 14{,}062.5$ ft². These values of x and y are between the values in the first and second figures in part (a). Our original estimate was low.

9.

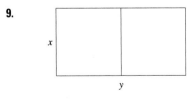

$xy = 1.5 \times 10^6$, so $y = 1.5 \times 10^6/x$. Minimize the amount of fencing, which is $3x + 2y = 3x + 2(1.5 \times 10^6/x) = 3x + 3 \times 10^6/x = F(x)$. $F'(x) = 3 - 3 \times 10^6/x^2 = 3(x^2 - 10^6)/x^2$. The critical number is $x = 10^3$ and $F'(x) < 0$ for $0 < x < 10^3$ and $F'(x) > 0$ if $x > 10^3$, so the absolute minimum occurs when $x = 10^3$ and $y = 1.5 \times 10^3$.

The field should be 1000 feet by 1500 feet with the middle fence parallel to the short side of the field.

11. Let b be the length of the base of the box and h the height. The surface area is $1200 = b^2 + 4hb$ $\Rightarrow$

$h = (1200 - b^2)/(4b)$. The volume is $V = b^2 h = b^2 (1200 - b^2)/4b = 300b - b^3/4$ $\Rightarrow$ $V'(b) = 300 - \frac{3}{4}b^2$.

$V'(b) = 0$ $\Rightarrow$ $300 = \frac{3}{4}b^2$ $\Rightarrow$ $b^2 = 400$ $\Rightarrow$ $b = \sqrt{400} = 20$. Since $V'(b) > 0$ for $0 < b < 20$ and

$V'(b) < 0$ for $b > 20$, there is an absolute maximum when $b = 20$ by the First Derivative Test for Absolute

Extreme Values (see page 334). If $b = 20$, then $h = (1200 - 20^2)/(4 \cdot 20) = 10$, so the largest possible volume

is $b^2 h = (20)^2 (10) = 4000$ cm^3.

13.

$10 = (2w)(w)h = 2w^2 h$, so $h = 5/w^2$. The cost is

$$C(w) = 10(2w^2) + 6[2(2wh) + 2hw] + 6(2w^2)$$

$$= 32w^2 + 36wh = 32w^2 + 180/w$$

$$C'(w) = 64w - 180/w^2 = 4(16w^3 - 45)/w^2 \quad \Rightarrow \quad w = \sqrt[3]{\tfrac{45}{16}} \text{ is the}$$

critical number. $C'(w) < 0$ for $0 < w < \sqrt[3]{\frac{45}{16}}$ and $C'(w) > 0$ for $w > \sqrt[3]{\frac{45}{16}}$. The minimum cost is

$C\left(\sqrt[3]{\frac{45}{16}}\right) = 32(2.8125)^{2/3} + 180/\sqrt{2.8125} \approx \$191.28.$

15. The distance from a point (x, y) on the line $y = 4x + 7$ to the origin is $\sqrt{(x-0)^2 + (y-0)^2} = \sqrt{x^2 + y^2}$.

However, it is easier to work with the *square* of the distance; that is,

$D(x) = \left(\sqrt{x^2 + y^2}\right)^2 = x^2 + y^2 = x^2 + (4x + 7)^2$. Because the distance is positive, its minimum value will

occur at the same point as the minimum value of D.

$D'(x) = 2x + 2(4x + 7)(4) = 34x + 56$, so $D'(x) = 0$ $\Leftrightarrow$ $x = -\frac{28}{17}$.

$D''(x) = 34 > 0$, so D is concave upward for all x. Thus, D has an absolute minimum at $x = -\frac{28}{17}$. The point

closest to the origin is $(x, y) = \left(-\frac{28}{17}, 4\left(-\frac{28}{17}\right) + 7\right) = \left(-\frac{28}{17}, \frac{7}{17}\right)$.

17.

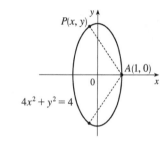

From the figure, we see that there are two points that are farthest away

from $A(1, 0)$. The distance d from A to an arbitrary point $P(x, y)$ on the

ellipse is $d = \sqrt{(x-1)^2 + (y-0)^2}$ and the square of the distance is

$S = d^2 = x^2 - 2x + 1 + y^2 = x^2 - 2x + 1 + (4 - 4x^2) = -3x^2 - 2x + 5.$

$S' = -6x - 2$ and $S' = 0$ $\Rightarrow$ $x = -\frac{1}{3}$. Now $S'' = -6 < 0$, so we

know that S has a maximum at $x = -\frac{1}{3}$. Since $-1 \le x \le 1$, $S(-1) = 4$,

$S\left(-\frac{1}{3}\right) = \frac{16}{3}$, and $S(1) = 0$, we see that the maximum distance is $\sqrt{\frac{16}{3}}$. The corresponding y-values are

$y = \pm\sqrt{4 - 4\left(-\frac{1}{3}\right)^2} = \pm\sqrt{\frac{32}{9}} = \pm\frac{4}{3}\sqrt{2} \approx \pm 1.89$. The points are $\left(-\frac{1}{3}, \pm\frac{4}{3}\sqrt{2}\right)$.

19.

The area of the rectangle is $(2x)(2y) = 4xy$. Also $r^2 = x^2 + y^2$ so $y = \sqrt{r^2 - x^2}$, so the area is $A(x) = 4x\sqrt{r^2 - x^2}$. Now

$$A'(x) = 4\left(\sqrt{r^2 - x^2} - \frac{x^2}{\sqrt{r^2 - x^2}}\right) = 4\frac{r^2 - 2x^2}{\sqrt{r^2 - x^2}}.$$ The critical

number is $x = \frac{1}{\sqrt{2}}r$. Clearly this gives a maximum.

$$y = \sqrt{r^2 - \left(\frac{1}{\sqrt{2}}r\right)^2} = \sqrt{\tfrac{1}{2}r^2} = \tfrac{1}{\sqrt{2}}r = x,$$ which tells us that the

rectangle is a square. The dimensions are $2x = \sqrt{2}\,r$ and $2y = \sqrt{2}\,r$.

21.

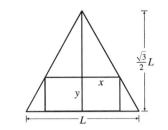

The height h of the equilateral triangle with sides of length L is $\frac{\sqrt{3}}{2}L$,

since $h^2 + (L/2)^2 = L^2 \;\Rightarrow\; h^2 = L^2 - \tfrac{1}{4}L^2 = \tfrac{3}{4}L^2 \;\Rightarrow$

$h = \frac{\sqrt{3}}{2}L$. Using similar triangles, $\dfrac{\frac{\sqrt{3}}{2}L - y}{x} = \dfrac{\frac{\sqrt{3}}{2}L}{L/2} = \sqrt{3} \;\Rightarrow$

$$\sqrt{3}\,x = \frac{\sqrt{3}}{2}L - y \;\Rightarrow\; y = \frac{\sqrt{3}}{2}L - \sqrt{3}\,x \;\Rightarrow\; y = \frac{\sqrt{3}}{2}(L - 2x).$$

The area of the inscribed rectangle is $A(x) = (2x)y = \sqrt{3}\,x(L - 2x) = \sqrt{3}\,Lx - 2\sqrt{3}\,x^2$, where $0 \le x \le L/2$.

Now $0 = A'(x) = \sqrt{3}\,L - 4\sqrt{3}\,x \;\Rightarrow\; x = \sqrt{3}\,L/(4\sqrt{3}) = L/4$. Since $A(0) = A(L/2) = 0$, the maximum

occurs when $x = L/4$, and $y = \frac{\sqrt{3}}{2}L - \frac{\sqrt{3}}{4}L = \frac{\sqrt{3}}{4}L$, so the dimensions are $L/2$ and $\frac{\sqrt{3}}{4}L$.

23.

The area of the triangle is

$$A(x) = \tfrac{1}{2}(2t)(r + x) = t(r + x) = \sqrt{r^2 - x^2}(r + x).$$ Then

$$0 = A'(x) = r\frac{-2x}{2\sqrt{r^2 - x^2}} + \sqrt{r^2 - x^2} + x\frac{-2x}{2\sqrt{r^2 - x^2}}$$

$$= -\frac{x^2 + rx}{\sqrt{r^2 - x^2}} + \sqrt{r^2 - x^2} \;\Rightarrow$$

$\dfrac{x^2 + rx}{\sqrt{r^2 - x^2}} = \sqrt{r^2 - x^2} \;\Rightarrow\; x^2 + rx = r^2 - x^2 \;\Rightarrow\; 0 = 2x^2 + rx - r^2 = (2x - r)(x + r) \;\Rightarrow$

$x = \tfrac{1}{2}r$ or $x = -r$. Now $A(r) = 0 = A(-r) \;\Rightarrow$ the maximum occurs where $x = \tfrac{1}{2}r$, so the triangle has height

$r + \tfrac{1}{2}r = \tfrac{3}{2}r$ and base $2\sqrt{r^2 - \left(\tfrac{1}{2}r\right)^2} = 2\sqrt{\tfrac{3}{4}r^2} = \sqrt{3}\,r$.

25.

The cylinder has volume $V = \pi y^2(2x)$. Also $x^2 + y^2 = r^2 \;\Rightarrow$ $y^2 = r^2 - x^2$, so $V(x) = \pi(r^2 - x^2)(2x) = 2\pi(r^2 x - x^3)$, where $0 \le x \le r$. $V'(x) = 2\pi(r^2 - 3x^2) = 0 \;\Rightarrow\; x = r/\sqrt{3}$. Now $V(0) = V(r) = 0$, so there is a maximum when $x = r/\sqrt{3}$ and $V(r/\sqrt{3}) = \pi(r^2 - r^2/3)(2r/\sqrt{3}) = 4\pi r^3/(3\sqrt{3})$.

27.

The cylinder has surface area

2(area of the base) + (lateral surface area)

$= 2\pi(\text{radius})^2 + 2\pi(\text{radius})(\text{height}) = 2\pi y^2 + 2\pi y(2x)$.

Now $x^2 + y^2 = r^2 \;\Rightarrow\; y^2 = r^2 - x^2 \;\Rightarrow\; y = \sqrt{r^2 - x^2}$, so the

surface area is

$$S(x) = 2\pi\left(r^2 - x^2\right) + 4\pi x\,\sqrt{r^2 - x^2}, \quad 0 \le x \le r$$

$$= 2\pi r^2 - 2\pi x^2 + 4\pi\left(x\,\sqrt{r^2 - x^2}\right)$$

Thus, $S'(x) = 0 - 4\pi x + 4\pi\left[x \cdot \frac{1}{2}\left(r^2 - x^2\right)^{-1/2}(-2x) + \left(r^2 - x^2\right)^{1/2} \cdot 1\right]$

$$= 4\pi\left[-x - \frac{x^2}{\sqrt{r^2 - x^2}} + \sqrt{r^2 - x^2}\right] = 4\pi \cdot \frac{-x\sqrt{r^2 - x^2} - x^2 + r^2 - x^2}{\sqrt{r^2 - x^2}}$$

$S'(x) = 0 \;\Rightarrow\; x\sqrt{r^2 - x^2} = r^2 - 2x^2 \;\;(\star) \;\Rightarrow\; \left(x\sqrt{r^2 - x^2}\right)^2 = \left(r^2 - 2x^2\right)^2 \;\Rightarrow$

$x^2\left(r^2 - x^2\right) = r^4 - 4r^2 x^2 + 4x^4 \;\Rightarrow\; r^2 x^2 - x^4 = r^4 - 4r^2 x^2 + 4x^4 \;\Rightarrow\; 5x^4 - 5r^2 x^2 + r^4 = 0$.

This is a quadratic equation in x^2. By the quadratic formula, $x^2 = \frac{5 \pm \sqrt{5}}{10}r^2$, but we reject the root with the + sign

since it doesn't satisfy $(\star)$. [The right side is negative and the left side is positive.] So $x = \sqrt{\frac{5 - \sqrt{5}}{10}}\, r$. Since

$S(0) = S(r) = 0$, the maximum surface area occurs at the critical number and $x^2 = \frac{5 - \sqrt{5}}{10}r^2 \;\Rightarrow$

$y^2 = r^2 - \frac{5 - \sqrt{5}}{10}r^2 = \frac{5 + \sqrt{5}}{10}r^2 \;\Rightarrow$ the surface area is

$2\pi\left(\frac{5 + \sqrt{5}}{10}\right)r^2 + 4\pi\sqrt{\frac{5 - \sqrt{5}}{10}}\sqrt{\frac{5 + \sqrt{5}}{10}}r^2 = \pi r^2\left[2 \cdot \frac{5 + \sqrt{5}}{10} + 4\frac{\sqrt{(5 - \sqrt{5})(5 + \sqrt{5})}}{10}\right] = \pi r^2\left[\frac{5 + \sqrt{5}}{5} + \frac{2\sqrt{20}}{5}\right] =$

$\pi r^2\left[\frac{5 + \sqrt{5} + 2 \cdot 2\sqrt{5}}{5}\right] = \pi r^2\left[\frac{5 + 5\sqrt{5}}{5}\right] = \pi r^2\left(1 + \sqrt{5}\right)$.

29.

$xy = 384 \;\Rightarrow\; y = 384/x$. Total area is

$A(x) = (8 + x)(12 + 384/x) = 12(40 + x + 256/x)$, so

$A'(x) = 12\left(1 - 256/x^2\right) = 0 \;\Rightarrow\; x = 16$. There is an absolute

minimum when $x = 16$ since $A'(x) < 0$ for $0 < x < 16$ and $A'(x) > 0$

for $x > 16$. When $x = 16$, $y = 384/16 = 24$, so the dimensions are

24 cm and 36 cm.

31.

Let x be the length of the wire used for the square. The total area is

$$A(x) = \left(\frac{x}{4}\right)^2 + \frac{1}{2}\left(\frac{10 - x}{3}\right)\frac{\sqrt{3}}{2}\left(\frac{10 - x}{3}\right)$$

$$= \frac{1}{16}x^2 + \frac{\sqrt{3}}{36}(10 - x)^2, \quad 0 \le x \le 10$$

$A'(x) = \frac{1}{8}x - \frac{\sqrt{3}}{18}(10 - x) = 0 \;\Leftrightarrow\; \frac{9}{72}x + \frac{4\sqrt{3}}{72}x - \frac{40\sqrt{3}}{72} = 0 \;\Leftrightarrow\; x = \frac{40\sqrt{3}}{9 + 4\sqrt{3}}$. Now

$A(0) = \left(\frac{\sqrt{3}}{36}\right)100 \approx 4.81$, $A(10) = \frac{100}{16} = 6.25$ and $A\left(\frac{40\sqrt{3}}{9 + 4\sqrt{3}}\right) \approx 2.72$, so

(a) The maximum area occurs when $x = 10$ m, and all the wire is used for the square.

(b) The minimum area occurs when $x = \frac{40\sqrt{3}}{9 + 4\sqrt{3}} \approx 4.35$ m.

33.

The volume is $V = \pi r^2 h$ and the surface area is

$$S(r) = \pi r^2 + 2\pi r h = \pi r^2 + 2\pi r \left(\frac{V}{\pi r^2}\right) = \pi r^2 + \frac{2V}{r}.$$

$$S'(r) = 2\pi r - \frac{2V}{r^2} = 0 \;\Rightarrow\; 2\pi r^3 = 2V \;\Rightarrow\; r = \sqrt[3]{\frac{V}{\pi}} \text{ cm}.$$

This gives an absolute minimum since $S'(r) < 0$ for $0 < r < \sqrt[3]{\frac{V}{\pi}}$ and $S'(r) > 0$ for $r > \sqrt[3]{\frac{V}{\pi}}$. When

$$r = \sqrt[3]{\frac{V}{\pi}}, h = \frac{V}{\pi r^2} = \frac{V}{\pi (V/\pi)^{2/3}} = \sqrt[3]{\frac{V}{\pi}} \text{ cm}.$$

35.

$h^2 + r^2 = R^2 \;\Rightarrow\; V = \frac{\pi}{3}r^2 h = \frac{\pi}{3}(R^2 - h^2)h = \frac{\pi}{3}(R^2 h - h^3)$.

$V'(h) = \frac{\pi}{3}(R^2 - 3h^2) = 0$ when $h = \frac{1}{\sqrt{3}}R$. This gives an absolute

maximum, since $V'(h) > 0$ for $0 < h < \frac{1}{\sqrt{3}}R$ and $V'(h) < 0$ for

$h > \frac{1}{\sqrt{3}}R$. The maximum volume is

$$V\left(\frac{1}{\sqrt{3}}R\right) = \frac{\pi}{3}\left(\frac{1}{\sqrt{3}}R^3 - \frac{1}{3\sqrt{3}}R^3\right) = \frac{2}{9\sqrt{3}}\pi R^3.$$

37.

By similar triangles, $\dfrac{H}{R} = \dfrac{H-h}{r}$ **(1)**. The volume of the inner cone is

$V = \frac{1}{3}\pi r^2 h$, so we'll solve **(1)** for h. $\dfrac{Hr}{R} = H - h \;\Rightarrow\;$

$$h = H - \frac{Hr}{R} = \frac{HR - Hr}{R} = \frac{H}{R}(R - r) \quad \textbf{(2)}.$$

Thus, $V(r) = \dfrac{\pi}{3}r^2 \cdot \dfrac{H}{R}(R - r) = \dfrac{\pi H}{3R}(Rr^2 - r^3) \;\Rightarrow\;$

$$V'(r) = \frac{\pi H}{3R}(2Rr - 3r^2) = \frac{\pi H}{3R}r(2R - 3r).$$

$V'(r) = 0 \;\Rightarrow\; r = 0$ or $2R = 3r \;\Rightarrow\; r = \frac{2}{3}R$ and from **(2)**, $h = \dfrac{H}{R}\left(R - \frac{2}{3}R\right) = \dfrac{H}{R}\left(\frac{1}{3}R\right) = \frac{1}{3}H$.

$V'(r)$ changes from positive to negative at $r = \frac{2}{3}R$, so the inner cone has a maximum volume of

$V = \frac{1}{3}\pi r^2 h = \frac{1}{3}\pi\left(\frac{2}{3}R\right)^2\left(\frac{1}{3}H\right) = \frac{4}{27} \cdot \frac{1}{3}\pi R^2 H$, which is approximately 15% of the volume of the larger cone.

39. $S = 6sh - \frac{3}{2}s^2 \cot\theta + 3s^2\frac{\sqrt{3}}{2}\csc\theta$

(a) $\dfrac{dS}{d\theta} = \frac{3}{2}s^2\csc^2\theta - 3s^2\frac{\sqrt{3}}{2}\csc\theta\cot\theta$ or $\frac{3}{2}s^2\csc\theta\left(\csc\theta - \sqrt{3}\cot\theta\right)$.

(b) $\dfrac{dS}{d\theta} = 0$ when $\csc\theta - \sqrt{3}\cot\theta = 0 \;\Rightarrow\; \dfrac{1}{\sin\theta} - \sqrt{3}\dfrac{\cos\theta}{\sin\theta} = 0 \;\Rightarrow\; \cos\theta = \frac{1}{\sqrt{3}}$. The First Derivative

Test shows that the minimum surface area occurs when $\theta = \cos^{-1}\left(\frac{1}{\sqrt{3}}\right) \approx 55°$.

(c)

If $\cos\theta = \frac{1}{\sqrt{3}}$, then $\cot\theta = \frac{1}{\sqrt{2}}$ and $\csc\theta = \frac{\sqrt{3}}{\sqrt{2}}$, so the surface area is

$$S = 6sh - \frac{3}{2}s^2\frac{1}{\sqrt{2}} + 3s^2\frac{\sqrt{3}}{2}\frac{\sqrt{3}}{\sqrt{2}} = 6sh - \frac{3}{2\sqrt{2}}s^2 + \frac{9}{2\sqrt{2}}s^2$$

$$= 6sh + \frac{6}{2\sqrt{2}}s^2 = 6s\left(h + \frac{1}{2\sqrt{2}}s\right)$$

41. Here $T(x) = \dfrac{\sqrt{x^2 + 25}}{6} + \dfrac{5 - x}{8}, 0 \le x \le 5 \;\Rightarrow\; T'(x) = \dfrac{x}{6\sqrt{x^2 + 25}} - \dfrac{1}{8} = 0 \;\Leftrightarrow\; 8x = 6\sqrt{x^2 + 25}$

$\Leftrightarrow\; 16x^2 = 9(x^2 + 25) \;\Leftrightarrow\; x = \frac{15}{\sqrt{7}}$. But $\frac{15}{\sqrt{7}} > 5$, so T has no critical number. Since $T(0) \approx 1.46$ and

$T(5) \approx 1.18$, he should row directly to B.

43.

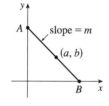

The total illumination is $I(x) = \dfrac{3k}{x^2} + \dfrac{k}{(10 - x)^2}, 0 < x < 10$. Then

$I'(x) = \dfrac{-6k}{x^3} + \dfrac{2k}{(10 - x)^3} = 0 \;\Rightarrow\; 6k(10 - x)^3 = 2kx^3 \;\Rightarrow$

$3(10 - x)^3 = x^3 \;\Rightarrow\; \sqrt[3]{3}(10 - x) = x \;\Rightarrow\; 10\sqrt[3]{3} - \sqrt[3]{3}\,x = x$

$\Rightarrow\; 10\sqrt[3]{3} = x + \sqrt[3]{3}\,x \;\Rightarrow\; 10\sqrt[3]{3} = (1 + \sqrt[3]{3})x \;\Rightarrow$

$x = \dfrac{10\sqrt[3]{3}}{1 + \sqrt[3]{3}} \approx 5.9$ ft. This gives a minimum since $I''(x) > 0$ for

$0 < x < 10$.

45.

Every line segment in the first quadrant passing through (a, b) with endpoints on the x- and y-axes satisfies an equation of the form $y - b = m(x - a)$, where $m < 0$. By setting $x = 0$ and then $y = 0$, we find its endpoints, $A(0, b - am)$ and $B\left(a - \frac{b}{m}, 0\right)$. The distance d from A to B is given by $d = \sqrt{[(a - \frac{b}{m}) - 0]^2 + [0 - (b - am)]^2}$.

It follows that the square of the length of the line segment, as a function of m, is given by

$$S(m) = \left(a - \frac{b}{m}\right)^2 + (am - b)^2 = a^2 - \frac{2ab}{m} + \frac{b^2}{m^2} + a^2m^2 - 2abm + b^2. \text{ Thus,}$$

$$S'(m) = \frac{2ab}{m^2} - \frac{2b^2}{m^3} + 2a^2m - 2ab = \frac{2}{m^3}(abm - b^2 + a^2m^4 - abm^3)$$

$$= \frac{2}{m^3}[b(am - b) + am^3(am - b)] = \frac{2}{m^3}(am - b)(b + am^3)$$

Thus, $S'(m) = 0 \;\Leftrightarrow\; m = b/a$ or $m = -\sqrt[3]{\frac{b}{a}}$. Since $b/a > 0$ and $m < 0$, m must equal $-\sqrt[3]{\frac{b}{a}}$. Since

$\dfrac{2}{m^3} < 0$, we see that $S'(m) < 0$ for $m < -\sqrt[3]{\frac{b}{a}}$ and $S'(m) > 0$ for $m > -\sqrt[3]{\frac{b}{a}}$. Thus, S has its absolute

minimum value when $m = -\sqrt[3]{\frac{b}{a}}$. That value is

$$S\left(-\sqrt[3]{\tfrac{b}{a}}\right) = \left(a + b\sqrt[3]{\tfrac{a}{b}}\right)^2 + \left(-a\sqrt[3]{\tfrac{b}{a}} - b\right)^2 = \left(a + \sqrt[3]{ab^2}\right)^2 + \left(\sqrt[3]{a^2b} + b\right)^2$$

$$= a^2 + 2a^{4/3}b^{2/3} + a^{2/3}b^{4/3} + a^{4/3}b^{2/3} + 2a^{2/3}b^{4/3} + b^2 = a^2 + 3a^{4/3}b^{2/3} + 3a^{2/3}b^{4/3} + b^2$$

The last expression is of the form $x^3 + 3x^2y + 3xy^2 + y^3 \quad [= (x + y)^3]$ with $x = a^{2/3}$ and $y = b^{2/3}$,

so we can write it as $(a^{2/3} + b^{2/3})^3$ and the shortest such line segment has length $\sqrt{S} = (a^{2/3} + b^{2/3})^{3/2}$.

47.

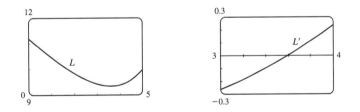

Here $s^2 = h^2 + b^2/4$, so $h^2 = s^2 - b^2/4$. The area is $A = \frac{1}{2}b\sqrt{s^2 - b^2/4}$.

Let the perimeter be p, so $2s + b = p$ or $s = (p - b)/2$ $\Rightarrow$

$A(b) = \frac{1}{2}b\sqrt{(p-b)^2/4 - b^2/4} = b\sqrt{p^2 - 2pb}/4$. Now

$A'(b) = \dfrac{\sqrt{p^2 - 2pb}}{4} - \dfrac{bp/4}{\sqrt{p^2 - 2pb}} = \dfrac{-3pb + p^2}{4\sqrt{p^2 - 2pb}}$. Therefore, $A'(b) = 0$ $\Rightarrow$

$-3pb + p^2 = 0$ $\Rightarrow$ $b = p/3$. Since $A'(b) > 0$ for $b < p/3$ and $A'(b) < 0$ for $b > p/3$, there is an absolute

maximum when $b = p/3$. But then $2s + p/3 = p$, so $s = p/3$ $\Rightarrow$ $s = b$ $\Rightarrow$ the triangle is equilateral.

49. Note that $|AD| = |AP| + |PD|$ $\Rightarrow$ $5 = x + |PD|$ $\Rightarrow$ $|PD| = 5 - x$. Using the Pythagorean Theorem

for $\triangle PDB$ and $\triangle PDC$ gives us

$$L(x) = |AP| + |BP| + |CP| = x + \sqrt{(5-x)^2 + 2^2} + \sqrt{(5-x)^2 + 3^2}$$

$$= x + \sqrt{x^2 - 10x + 29} + \sqrt{x^2 - 10x + 34}$$

$$\Rightarrow \quad L'(x) = 1 + \frac{x-5}{\sqrt{x^2 - 10x + 29}} + \frac{x-5}{\sqrt{x^2 - 10x + 34}}.$$

From the graphs of L and L', it seems that the minimum value of L is about $L(3.59) = 9.35$ m.

51.

The total time is

$$T(x) = (\text{time from } A \text{ to } C) + (\text{time from } C \text{ to } B)$$

$$= \frac{\sqrt{a^2 + x^2}}{v_1} + \frac{\sqrt{b^2 + (d-x)^2}}{v_2}, \quad 0 < x < d$$

$$T'(x) = \frac{x}{v_1\sqrt{a^2 + x^2}} - \frac{d-x}{v_2\sqrt{b^2 + (d-x)^2}} = \frac{\sin\theta_1}{v_1} - \frac{\sin\theta_2}{v_2}$$

The minimum occurs when $T'(x) = 0$ $\Rightarrow$ $\dfrac{\sin\theta_1}{v_1} = \dfrac{\sin\theta_2}{v_2}$.

[*Note:* $T''(x) > 0$]

53.

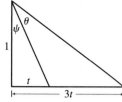

$y^2 = x^2 + z^2$, but triangles CDE and BCA are similar, so

$z/8 = x/(4\sqrt{x-4}) \Rightarrow z = 2x/\sqrt{x-4}$. Thus, we minimize

$f(x) = y^2 = x^2 + 4x^2/(x-4) = x^3/(x-4), \quad 4 < x \le 8$.

$f'(x) = \dfrac{(x-4)(3x^2) - x^3}{(x-4)^2} = \dfrac{x^2[3(x-4)-x]}{(x-4)^2} = \dfrac{2x^2(x-6)}{(x-4)^2} = 0$

when $x = 6$. $f'(x) < 0$ when $x < 6$, $f'(x) > 0$ when $x > 6$, so the

minimum occurs when $x = 6$ in.

55.

It suffices to maximize $\tan\theta$. Now

$\dfrac{3t}{1} = \tan(\psi + \theta) = \dfrac{\tan\psi + \tan\theta}{1 - \tan\psi\,\tan\theta} = \dfrac{t + \tan\theta}{1 - t\tan\theta}$.

So $3t(1 - t\tan\theta) = t + \tan\theta \Rightarrow 2t = (1 + 3t^2)\tan\theta \Rightarrow$

$\tan\theta = \dfrac{2t}{1 + 3t^2}$. Let $f(t) = \tan\theta = \dfrac{2t}{1 + 3t^2} \Rightarrow$

$f'(t) = \dfrac{2(1 + 3t^2) - 2t(6t)}{(1 + 3t^2)^2} = \dfrac{2(1 - 3t^2)}{(1 + 3t^2)^2} = 0 \Leftrightarrow$

$1 - 3t^2 = 0 \Leftrightarrow t = \frac{1}{\sqrt{3}}$ since $t \ge 0$.

Now $f'(t) > 0$ for $0 \le t < \frac{1}{\sqrt{3}}$ and $f'(t) < 0$ for $t > \frac{1}{\sqrt{3}}$, so f has an absolute maximum when $t = \frac{1}{\sqrt{3}}$

and $\tan\theta = \dfrac{2(1/\sqrt{3})}{1 + 3(1/\sqrt{3})^2} = \frac{1}{\sqrt{3}} \Rightarrow \theta = \frac{\pi}{6}$. Substituting for t and θ in $3t = \tan(\psi + \theta)$ gives us

$\sqrt{3} = \tan(\psi + \frac{\pi}{6}) \Rightarrow \psi = \frac{\pi}{6}$.

57.

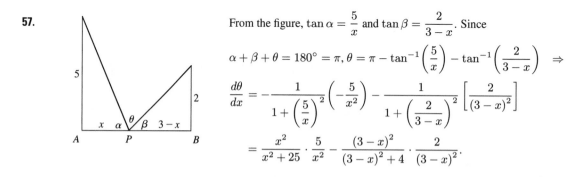

From the figure, $\tan\alpha = \dfrac{5}{x}$ and $\tan\beta = \dfrac{2}{3-x}$. Since

$\alpha + \beta + \theta = 180° = \pi,\ \theta = \pi - \tan^{-1}\left(\dfrac{5}{x}\right) - \tan^{-1}\left(\dfrac{2}{3-x}\right) \Rightarrow$

$\dfrac{d\theta}{dx} = -\dfrac{1}{1 + \left(\dfrac{5}{x}\right)^2}\left(-\dfrac{5}{x^2}\right) - \dfrac{1}{1 + \left(\dfrac{2}{3-x}\right)^2}\left[\dfrac{2}{(3-x)^2}\right]$

$= \dfrac{x^2}{x^2 + 25} \cdot \dfrac{5}{x^2} - \dfrac{(3-x)^2}{(3-x)^2 + 4} \cdot \dfrac{2}{(3-x)^2}$.

Now $\dfrac{d\theta}{dx} = 0 \Rightarrow \dfrac{5}{x^2 + 25} = \dfrac{2}{x^2 - 6x + 13} \Rightarrow 2x^2 + 50 = 5x^2 - 30x + 65 \Rightarrow$

$3x^2 - 30x + 15 = 0 \Rightarrow x^2 - 10x + 5 = 0 \Rightarrow x = 5 \pm 2\sqrt{5}$. We reject the root with the $+$ sign,

since it is larger than 3. $d\theta/dx > 0$ for $x < 5 - 2\sqrt{5}$ and $d\theta/dx < 0$ for $x > 5 - 2\sqrt{5}$, so θ is maximized

when $|AP| = x = 5 - 2\sqrt{5} \approx 0.53$.

59.

In the small triangle with sides a and c and hypotenuse W, $\sin\theta = \dfrac{a}{W}$ and $\cos\theta = \dfrac{c}{W}$. In the triangle with sides b and d and hypotenuse L,

$\sin\theta = \dfrac{d}{L}$ and $\cos\theta = \dfrac{b}{L}$. Thus, $a = W\sin\theta$, $c = W\cos\theta$, $d = L\sin\theta$, and $b = L\cos\theta$, so the area of the circumscribed rectangle is

$$A(\theta) = (a+b)(c+d) = (W\sin\theta + L\cos\theta)(W\cos\theta + L\sin\theta)$$

$$= W^2\sin\theta\,\cos\theta + WL\sin^2\theta + LW\cos^2\theta + L^2\sin\theta\,\cos\theta$$

$$= LW\sin^2\theta + LW\cos^2\theta + \left(L^2 + W^2\right)\sin\theta\,\cos\theta$$

$$= LW\left(\sin^2\theta + \cos^2\theta\right) + \left(L^2 + W^2\right)\cdot\tfrac{1}{2}\cdot 2\sin\theta\,\cos\theta$$

$$= LW + \tfrac{1}{2}\left(L^2 + W^2\right)\sin 2\theta, \quad 0 \le \theta \le \tfrac{\pi}{2}$$

This expression shows, without calculus, that the maximum value of $A(\theta)$ occurs when $\sin 2\theta = 1 \;\Leftrightarrow\; 2\theta = \tfrac{\pi}{2}$ $\Rightarrow\; \theta = \tfrac{\pi}{4}$. So the maximum area is $A\left(\tfrac{\pi}{4}\right) = LW + \tfrac{1}{2}\left(L^2 + W^2\right) = \tfrac{1}{2}\left(L^2 + 2LW + W^2\right) = \tfrac{1}{2}(L + W)^2$.

61. (a) If $k = $ energy/km over land, then energy/km over water $= 1.4k$. So the total energy is

$$E = 1.4k\sqrt{25 + x^2} + k(13 - x),\ 0 \le x \le 13,$$

and so $\dfrac{dE}{dx} = \dfrac{1.4kx}{(25 + x^2)^{1/2}} - k.$

Set $\dfrac{dE}{dx} = 0$: $1.4kx = k\left(25 + x^2\right)^{1/2}\;\Rightarrow\;1.96x^2 = x^2 + 25\;\Rightarrow\;0.96x^2 = 25\;\Rightarrow\;x = \dfrac{5}{\sqrt{0.96}} \approx 5.1.$

Testing against the value of E at the endpoints: $E(0) = 1.4k(5) + 13k = 20k$, $E(5.1) \approx 17.9k$, $E(13) \approx 19.5k$. Thus, to minimize energy, the bird should fly to a point about 5.1 km from B.

(b) If W/L is large, the bird would fly to a point C that is closer to B than to D to minimize the energy used flying over water. If W/L is small, the bird would fly to a point C that is closer to D than to B to minimize the distance of the flight. $E = W\sqrt{25 + x^2} + L(13 - x)\;\Rightarrow\;\dfrac{dE}{dx} = \dfrac{Wx}{\sqrt{25 + x^2}} - L = 0$ when

$\dfrac{W}{L} = \dfrac{\sqrt{25 + x^2}}{x}$. By the same sort of argument as in part (a), this ratio will give the minimal expenditure of energy if the bird heads for the point x km from B.

(c) For flight direct to D, $x = 13$, so from part (b), $W/L = \dfrac{\sqrt{25 + 13^2}}{13} \approx 1.07$. There is no value of W/L for which the bird should fly directly to B. But note that $\lim\limits_{x \to 0^+}(W/L) = \infty$, so if the point at which E is a minimum is close to B, then W/L is large.

(d) Assuming that the birds instinctively choose the path that minimizes the energy expenditure, we can use the equation for $dE/dx = 0$ from part (a) with $1.4k = c$, $x = 4$, and $k = 1$: $(c)(4) = 1\cdot\left(25 + 4^2\right)^{1/2}\;\Rightarrow\;c = \sqrt{41}/4 \approx 1.6.$

4.8 Applications to Business and Economics

1. (a) $C(0)$ represents the fixed costs of production, such as rent, utilities, machinery etc., which are incurred even when nothing is produced.

(b) The inflection point is the point at which $C''(x)$ changes from negative to positive; that is, the marginal cost $C'(x)$ changes from decreasing to increasing. Thus, the marginal cost is minimized at the inflection point.

(c) The marginal cost function is $C'(x)$. We graph it as in Example 1 in Section 2.9.

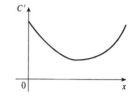

3. $c(x) = 21.4 - 0.002x$ and $c(x) = C(x)/x \Rightarrow C(x) = 21.4x - 0.002x^2$. $C'(x) = 21.4 - 0.004x$ and $C'(1000) = 17.4$. This means that the cost of producing the 1001st unit is about $17.40.

5. (a) The cost function is $C(x) = 40{,}000 + 300x + x^2$, so the cost at a production level of 1000 is $C(1000) = \$1{,}340{,}000$. The average cost function is $c(x) = \dfrac{C(x)}{x} = \dfrac{40{,}000}{x} + 300 + x$ and $c(1000) = \$1340/\text{unit}$. The marginal cost function is $C'(x) = 300 + 2x$ and $C'(1000) = \$2300/\text{unit}$.

(b) See the box preceding Example 1. We must have $C'(x) = c(x) \Leftrightarrow 300 + 2x = \dfrac{40{,}000}{x} + 300 + x \Leftrightarrow$

$x = \dfrac{40{,}000}{x} \Rightarrow x^2 = 40{,}000 \Rightarrow x = \sqrt{40{,}000} = 200$. This gives a minimum value of the average cost function $c(x)$ since $c''(x) = \dfrac{80{,}000}{x^3} > 0$.

(c) The minimum average cost is $c(200) = \$700/\text{unit}$.

7. (a) $C(x) = 16{,}000 + 200x + 4x^{3/2}$, $C(1000) = 16{,}000 + 200{,}000 + 40{,}000\sqrt{10} \approx 216{,}000 + 126{,}491$, so $C(1000) \approx \$342{,}491$. $c(x) = C(x)/x = \dfrac{16{,}000}{x} + 200 + 4x^{1/2}$, $c(1000) \approx \$342.49/\text{unit}$. $C'(x) = 200 + 6x^{1/2}$, $C'(1000) = 200 + 60\sqrt{10} \approx \$389.74/\text{unit}$.

(b) We must have $C'(x) = c(x) \Leftrightarrow 200 + 6x^{1/2} = \dfrac{16{,}000}{x} + 200 + 4x^{1/2} \Leftrightarrow 2x^{3/2} = 16{,}000 \Leftrightarrow$

$x = (8{,}000)^{2/3} = 400$ units. To check that this is a minimum, we calculate $c'(x) = \dfrac{-16{,}000}{x^2} + \dfrac{2}{\sqrt{x}} = \dfrac{2}{x^2}(x^{3/2} - 8000)$. This is negative for $x < (8000)^{2/3} = 400$, zero at $x = 400$, and positive for $x > 400$, so c is decreasing on $(0, 400)$ and increasing on $(400, \infty)$. Thus, c has an absolute minimum at $x = 400$. [*Note:* $c''(x)$ is *not* positive for all $x > 0$.]

(c) The minimum average cost is $c(400) = 40 + 200 + 80 = \$320/\text{unit}$.

9. (a) $C(x) = 3700 + 5x - 0.04x^2 + 0.0003x^3 \Rightarrow C'(x) = 5 - 0.08x + 0.0009x^2$ (marginal cost).

$c(x) = \dfrac{C(x)}{x} = \dfrac{3700}{x} + 5 - 0.04x + 0.0003x^2$ (average cost).

(b)

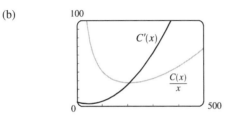

The graphs intersect at $(208.51, 27.45)$, so the production level that minimizes average cost is about 209 units.

(c) $c'(x) = -\dfrac{3700}{x^2} - 0.04 + 0.0006x = 0 \quad\Rightarrow\quad 3700 + 0.04x^2 - 0.0006x^3 = 0 \quad\Rightarrow\quad x_1 \approx 208.51.$

$c(x_1) \approx \$27.45/\text{unit}.$

(d) The marginal cost is given by $C'(x)$, so to find its minimum value we'll find the derivative of C'; that is, C''.

$C''(x) = -0.08 + 0.0018x = 0 \quad\Rightarrow\quad x_1 = \frac{800}{18} = 44.\overline{44}.\quad C'(x_1) = \$3.22/\text{unit}.$

$C'''(x) = 0.0018 > 0$ for all x, so this is the minimum marginal cost. C''' is the second derivative of C'.

11. $C(x) = 680 + 4x + 0.01x^2,\ p(x) = 12 \quad\Rightarrow\quad R(x) = xp(x) = 12x.$ If the profit is maximum, then

$R'(x) = C'(x) \quad\Rightarrow\quad 12 = 4 + 0.02x \quad\Rightarrow\quad 0.02x = 8 \quad\Rightarrow\quad x = 400.$ The profit is maximized if $P''(x) < 0$, but since $P''(x) = R''(x) - C''(x)$, we can just check the condition $R''(x) < C''(x)$. Now $R''(x) = 0 < 0.02 = C''(x)$, so $x = 400$ gives a maximum.

13. $C(x) = 1450 + 36x - x^2 + 0.001x^3,\ p(x) = 60 - 0.01x.$ Then $R(x) = xp(x) = 60x - 0.01x^2.$ If the profit is maximum, then $R'(x) = C'(x) \quad\Leftrightarrow\quad 60 - 0.02x = 36 - 2x + 0.003x^2 \quad\Rightarrow\quad 0.003x^2 - 1.98x - 24 = 0.$ By

the quadratic formula, $x = \dfrac{1.98 \pm \sqrt{(-1.98)^2 + 4(0.003)(24)}}{2(0.003)} = \dfrac{1.98 \pm \sqrt{4.2084}}{0.006}.$ Since $x > 0,$

$x \approx (1.98 + 2.05)/0.006 \approx 672.$ Now $R''(x) = -0.02$ and $C''(x) = -2 + 0.006x \quad\Rightarrow\quad C''(672) = 2.032$

$\Rightarrow\quad R''(672) < C''(672) \quad\Rightarrow\quad$ there is a maximum at $x = 672.$

15. $C(x) = 0.001x^3 - 0.3x^2 + 6x + 900.$ The marginal cost is $C'(x) = 0.003x^2 - 0.6x + 6.$

$C'(x)$ is increasing when $C''(x) > 0 \quad\Leftrightarrow\quad 0.006x - 0.6 > 0 \quad\Leftrightarrow\quad x > 0.6/0.006 = 100.$ So $C'(x)$ starts to increase when $x = 100.$

17. (a) $C(x) = 1200 + 12x - 0.1x^2 + 0.0005x^3.$

$R(x) = xp(x) = 29x - 0.00021x^2.$

Since the profit is maximized when $R'(x) = C'(x),$

we examine the curves R and C in the figure, looking for x-values at which the slopes of the tangent lines are equal. It appears that $x = 200$ is a good estimate.

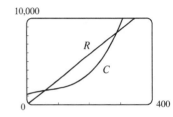

(b) $R'(x) = C'(x) \quad\Rightarrow\quad 29 - 0.00042x = 12 - 0.2x + 0.0015x^2 \quad\Rightarrow\quad 0.0015x^2 - 0.19958x - 17 = 0 \quad\Rightarrow$

$x \approx 192.06$ (for $x > 0$). As in Exercise 11, $R''(x) < C''(x) \quad\Rightarrow\quad -0.00042 < -0.2 + 0.003x \quad\Leftrightarrow$

$0.003x > 0.19958 \quad\Leftrightarrow\quad x > 66.5.$ Our value of 192 is in this range, so we have a maximum profit when we produce 192 yards of fabric.

19. (a) We are given that the demand function p is linear and $p(27,000) = 10$, $p(33,000) = 8$, so the slope is

$\frac{10 - 8}{27,000 - 33,000} = -\frac{1}{3000}$ and an equation of the line is $y - 10 = \left(-\frac{1}{3000}\right)(x - 27,000) \Rightarrow$

$y = p(x) = -\frac{1}{3000}x + 19 = 19 - (x/3000)$.

(b) The revenue is $R(x) = xp(x) = 19x - (x^2/3000) \Rightarrow R'(x) = 19 - (x/1500) = 0$ when $x = 28,500$.

Since $R''(x) = -1/1500 < 0$, the maximum revenue occurs when $x = 28,500 \Rightarrow$ the price is

$p(28,500) = \$9.50$.

21. (a) As in Example 3, we see that the demand function p is linear. We are given that $p(1000) = 450$ and deduce that

$p(1100) = 440$, since a \$10 reduction in price increases sales by 100 per week. The slope for p is

$\frac{440 - 450}{1100 - 1000} = -\frac{1}{10}$, so an equation is $p - 450 = -\frac{1}{10}(x - 1000)$ or $p(x) = -\frac{1}{10}x + 550$.

(b) $R(x) = xp(x) = -\frac{1}{10}x^2 + 550x$. $R'(x) = -\frac{1}{5}x + 550 = 0$ when $x = 5(550) = 2750$.

$p(2750) = 275$, so the rebate should be $450 - 275 = \$175$.

(c) $C(x) = 68,000 + 150x \Rightarrow$

$P(x) = R(x) - C(x) = -\frac{1}{10}x^2 + 550x - 68,000 - 150x = -\frac{1}{10}x^2 + 400x - 68,000$,

$P'(x) = -\frac{1}{5}x + 400 = 0$ when $x = 2000$. $p(2000) = 350$. Therefore, the rebate to maximize profits should

be $450 - 350 = \$100$.

23. If the reorder quantity is x, then the manager places $\dfrac{800}{x}$ orders per year. Storage costs for the year are

$\frac{1}{2}x \cdot 4 = 2x$ dollars. Handling costs are \$100 per delivery, for a total of $\dfrac{800}{x} \cdot 100 = \dfrac{80,000}{x}$ dollars. The total

costs for the year are $C(x) = 2x + \dfrac{80,000}{x}$. To minimize $C(x)$, we calculate

$C'(x) = 2 - \dfrac{80,000}{x^2} = \dfrac{2}{x^2}(x^2 - 40,000)$. This is negative when $x < 200$, zero when $x = 200$, and positive when

$x > 200$, so C is decreasing on $(0, 200)$ and increasing on $(200, \infty)$, reaching its absolute minimum when

$x = 200$. Thus, the optimal reorder quantity is 200 cases. The manager will place 4 orders per year for a total cost

of $C(200) = \$800$.

4.9 Newton's Method

1. (a)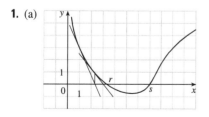

The tangent line at $x = 1$ intersects the x-axis at

$x \approx 2.3$, so $x_2 \approx 2.3$. The tangent line at

$x = 2.3$ intersects the x-axis at $x \approx 3$,

so $x_3 \approx 3.0$.

(b) $x_1 = 5$ would *not* be a better first approximation than $x_1 = 1$ since the tangent line is nearly horizontal. In fact,

the second approximation for $x_1 = 5$ appears to be to the left of $x = 1$.

3. Since $x_1 = 3$ and $y = 5x - 4$ is tangent to $y = f(x)$ at $x = 3$, we simply need to find where the tangent line

intersects the x-axis. $y = 0 \Rightarrow 5x_2 - 4 = 0 \Rightarrow x_2 = \frac{4}{5}$.

5. $f(x) = x^3 + 2x - 4 \Rightarrow f'(x) = 3x^2 + 2$, so $x_{n+1} = x_n - \dfrac{x_n^3 + 2x_n - 4}{3x_n^2 + 2}$. Now $x_1 = 1 \Rightarrow$

$$x_2 = 1 - \frac{1 + 2 - 4}{3 \cdot 1^2 + 2} = 1 - \frac{-1}{5} = 1.2 \Rightarrow x_3 = 1.2 - \frac{(1.2)^3 + 2(1.2) - 4}{3(1.2)^2 + 2} \approx 1.1797.$$

7. $f(x) = x^4 - 20 \Rightarrow f'(x) = 4x^3$, so $x_{n+1} = x_n - \dfrac{f(x_n)}{f'(x_n)} = x_n - \dfrac{x_n^4 - 20}{4x_n^3}$.

Now $x_1 = 2 \Rightarrow x_2 = 2 - \dfrac{2^4 - 20}{4(2)^3} = 2.125 \Rightarrow x_3 = 2.125 - \dfrac{(2.125)^4 - 20}{4(2.125)^3} \approx 2.1148$.

9. $f(x) = x^3 + x + 3 \Rightarrow f'(x) = 3x^2 + 1$, so

$x_{n+1} = x_n - \dfrac{x_n^3 + x_n + 3}{3x_n^2 + 1}$. Now $x_1 = -1 \Rightarrow$

$x_2 = -1 - \dfrac{(-1)^3 + (-1) + 3}{3(-1)^2 + 1} = -1 - \dfrac{-1 - 1 + 3}{3 + 1} = -1 - \dfrac{1}{4} = -1.25$.

Newton's method follows the tangent line at $(-1, 1)$ down to its

intersection with the x-axis at $(-1.25, 0)$, giving the second

approximation $x_2 = -1.25$.

11. To approximate $x = \sqrt[3]{30}$ (so that $x^3 = 30$), we can take $f(x) = x^3 - 30$. So $f'(x) = 3x^2$, and thus,

$x_{n+1} = x_n - \dfrac{x_n^3 - 30}{3x_n^2}$. Since $\sqrt[3]{27} = 3$ and 27 is close to 30, we'll use $x_1 = 3$. We need to find approximations

until they agree to eight decimal places. $x_1 = 3 \Rightarrow x_2 \approx 3.11111111$, $x_3 \approx 3.10723734$,

$x_4 \approx 3.10723251 \approx x_5$. So $\sqrt[3]{30} \approx 3.10723251$, to eight decimal places.

Here is a quick and easy method for finding the iterations for Newton's method on a programmable calculator.
(The screens shown are from the TI-83 Plus, but the method is similar on other calculators.) Assign $f(x) = x^3 - 30$
to Y_1, and $f'(x) = 3x^2$ to Y_2. Now store $x_1 = 3$ in X and then enter $X - Y_1/Y_2 \to X$ to get $x_2 = 3.\overline{1}$. By
successively pressing the ENTER key, you get the approximations $x_3, x_4, \ldots$.

In Derive, load the utility file SOLVE. Enter NEWTON(x^3-30,x,3) and then APPROXIMATE to get
[3, 3.11111111, 3.10723733, 3.10723250, 3.10723250]. You can request a specific iteration by adding a fourth
argument. For example, NEWTON(x^3-30,x,3,2) gives [3, 3.11111111, 3.10723733].

In Maple, make the assignments $f := x \to x^3 - 30;$, $g := x \to x - f(x)/D(f)(x);$, and $x := 3.;$.
Repeatedly execute the command $x := g(x);$ to generate successive approximations.

In Mathematica, make the assignments $f[x\_] := x^3 - 30$, $g[x\_] := x - f[x]/f'[x]$, and $x = 3$.
Repeatedly execute the command $x = g[x]$ to generate successive approximations.

13. $f(x) = 2x^3 - 6x^2 + 3x + 1$ $\Rightarrow$ $f'(x) = 6x^2 - 12x + 3$ $\Rightarrow$ $x_{n+1} = x_n - \dfrac{2x_n^3 - 6x_n^2 + 3x_n + 1}{6x_n^2 - 12x_n + 3}$. We

need to find approximations until they agree to six decimal places. $x_1 = 2.5$ $\Rightarrow$ $x_2 \approx 2.285714$,

$x_3 \approx 2.228824$, $x_4 \approx 2.224765$, $x_5 \approx 2.224745 \approx x_6$. So the root is 2.224745, to six decimal places.

15. $\sin x = x^2$, so $f(x) = \sin x - x^2$ $\Rightarrow$ $f'(x) = \cos x - 2x$ $\Rightarrow$

$x_{n+1} = x_n - \dfrac{\sin x_n - x_n^2}{\cos x_n - 2x_n}$. From the figure, the positive root of

$\sin x = x^2$ is near 1. $x_1 = 1$ $\Rightarrow$ $x_2 \approx 0.891396$, $x_3 \approx 0.876985$,

$x_4 \approx 0.876726 \approx x_5$. So the positive root is 0.876726, to six decimal

places.

17.

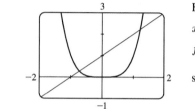

From the graph, we see that there appear to be points of intersection near

$x = -0.7$ and $x = 1.2$. Solving $x^4 = 1 + x$ is the same as solving

$f(x) = x^4 - x - 1 = 0$. $f(x) = x^4 - x - 1$ $\Rightarrow$ $f'(x) = 4x^3 - 1$,

so $x_{n+1} = x_n - \dfrac{x_n^4 - x_n - 1}{4x_n^3 - 1}$.

$x_1 = -0.7$	$x_1 = 1.2$
$x_2 \approx -0.725253$	$x_2 \approx 1.221380$
$x_3 \approx -0.724493$	$x_3 \approx 1.220745$
$x_4 \approx -0.724492 \approx x_5$	$x_4 \approx 1.220744 \approx x_5$

To six decimal places, the roots of the equation are -0.724492 and 1.220744.

19.

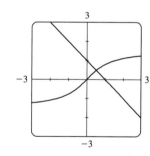

From the graph, there appears to be a point of intersection near $x = 0.5$.

Solving $\tan^{-1} x = 1 - x$ is the same as solving

$f(x) = \tan^{-1} x + x - 1 = 0$. $f(x) = \tan^{-1} x + x - 1$ $\Rightarrow$

$f'(x) = \dfrac{1}{1 + x^2} + 1$, so $x_{n+1} = x_n - \dfrac{\tan^{-1} x_n + x_n - 1}{1/(1 + x_n^2) + 1}$.

$x_1 = 0.5$ $\Rightarrow$ $x_2 \approx 0.520196$, $x_3 \approx 0.520269 \approx x_4$. So to six decimal

places, the root of the equation is 0.520269.

21. From the graph, there appears to be a point of intersection near $x = 0.6$.

Solving $\cos x = \sqrt{x}$ is the same as solving $f(x) = \cos x - \sqrt{x} = 0$.

$f(x) = \cos x - \sqrt{x}$ $\Rightarrow$ $f'(x) = -\sin x - 1/(2\sqrt{x})$, so

$x_{n+1} = x_n - \dfrac{\cos x_n - \sqrt{x_n}}{-\sin x_n - 1/(2\sqrt{x})}$. Now $x_1 = 0.6$ $\Rightarrow$

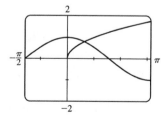

$x_2 \approx 0.641928$, $x_3 \approx 0.641714 \approx x_4$. To six decimal places, the root of

the equation is 0.641714.

23.

$f(x) = x^5 - x^4 - 5x^3 - x^2 + 4x + 3 \Rightarrow$
$f'(x) = 5x^4 - 4x^3 - 15x^2 - 2x + 4 \Rightarrow$
$$x_{n+1} = x_n - \frac{x_n^5 - x_n^4 - 5x_n^3 - x_n^2 + 4x_n + 3}{5x_n^4 - 4x_n^3 - 15x_n^2 - 2x_n + 4}. \text{ From the graph of } f,$$
there appear to be roots near -1.4, 1.1, and 2.7.

$x_1 = -1.4$	$x_1 = 1.1$	$x_1 = 2.7$
$x_2 \approx -1.39210970$	$x_2 \approx 1.07780402$	$x_2 \approx 2.72046250$
$x_3 \approx -1.39194698$	$x_3 \approx 1.07739442$	$x_3 \approx 2.71987870$
$x_4 \approx -1.39194691 \approx x_5$	$x_4 \approx 1.07739428 \approx x_5$	$x_4 \approx 2.71987822 \approx x_5$

To eight decimal places, the roots of the equation are -1.39194691, 1.07739428, and 2.71987822.

25.

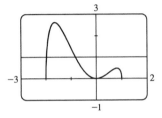

From the graph, $y = x^2\sqrt{2 - x - x^2}$ and $y = 1$ intersect twice, at
$x \approx -2$ and at $x \approx -1$. $f(x) = x^2\sqrt{2 - x - x^2} - 1 \Rightarrow$
$f'(x) = x^2 \cdot \frac{1}{2}(2 - x - x^2)^{-1/2}(-1 - 2x) + (2 - x - x^2)^{1/2} \cdot 2x$
$\quad = \frac{1}{2}x(2 - x - x^2)^{-1/2}\left[x(-1 - 2x) + 4(2 - x - x^2)\right]$
$\quad = \frac{x(8 - 5x - 6x^2)}{2\sqrt{(2 + x)(1 - x)}},$

so $x_{n+1} = x_n - \dfrac{x_n^2\sqrt{2 - x_n - x_n^2} - 1}{\dfrac{x_n(8 - 5x_n - 6x_n^2)}{2\sqrt{(2 + x_n)(1 - x_n)}}}$. Trying $x_1 = -2$ won't work because $f'(-2)$ is undefined, so we'll

try $x_1 = -1.95$.

$x_1 = -1.95$	$x_1 = -0.8$
$x_2 \approx -1.98580357$	$x_2 \approx -0.82674444$
$x_3 \approx -1.97899778$	$x_3 \approx -0.82646236$
$x_4 \approx -1.97807848$	$x_4 \approx -0.82646233 \approx x_5$
$x_5 \approx -1.97806682$	
$x_6 \approx -1.97806681 \approx x_7$	

To eight decimal places, the roots of the equation are -1.97806681 and -0.82646233.

27.

From the graph, we see that $y = e^{-x^2}$ and $y = x^2 - x$ intersect
twice. Good first approximations are $x = -0.5$ and $x = 1.1$.
$f(x) = e^{-x^2} - x^2 + x \Rightarrow f'(x) = -2xe^{-x^2} - 2x + 1$, so
$$x_{n+1} = x_n - \frac{e^{-x_n^2} - x_n^2 + x_n}{-2x_ne^{-x_n^2} - 2x_n + 1}.$$

$x_1 = -0.5$	$x_1 = 1.1$
$x_2 \approx -0.51036446$	$x_2 \approx 1.20139754$
$x_3 \approx -0.51031156 \approx x_4$	$x_3 \approx 1.19844118$
	$x_4 \approx 1.19843871 \approx x_5$

To eight decimal places, the roots of the equation are -0.51031156 and 1.19843871.

29. (a) $f(x) = x^2 - a \Rightarrow f'(x) = 2x$, so Newton's method gives

$$x_{n+1} = x_n - \frac{x_n^2 - a}{2x_n} = x_n - \frac{1}{2}x_n + \frac{a}{2x_n} = \frac{1}{2}x_n + \frac{a}{2x_n} = \frac{1}{2}\left(x_n + \frac{a}{x_n}\right).$$

(b) Using (a) with $a = 1000$ and $x_1 = \sqrt{900} = 30$, we get $x_2 \approx 31.666667$, $x_3 \approx 31.622807$, and $x_4 \approx 31.622777 \approx x_5$. So $\sqrt{1000} \approx 31.622777$.

31. $f(x) = x^3 - 3x + 6 \Rightarrow f'(x) = 3x^2 - 3$. If $x_1 = 1$, then $f'(x_1) = 0$ and the tangent line used for approximating x_2 is horizontal. Attempting to find x_2 results in trying to divide by zero.

33. For $f(x) = x^{1/3}$, $f'(x) = \frac{1}{3}x^{-2/3}$ and

$$x_{n+1} = x_n - \frac{f(x_n)}{f'(x_n)} = x_n - \frac{x_n^{1/3}}{\frac{1}{3}x_n^{-2/3}} = x_n - 3x_n = -2x_n.$$

Therefore, each successive approximation becomes twice as large as the previous one in absolute value, so the sequence of approximations fails to converge to the root, which is 0. In the figure, we have $x_1 = 0.5$, $x_2 = -2(0.5) = -1$, and $x_3 = -2(-1) = 2$.

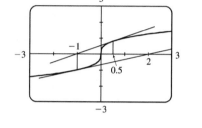

35. (a) $f(x) = 3x^4 - 28x^3 + 6x^2 + 24x \Rightarrow f'(x) = 12x^3 - 84x^2 + 12x + 24 \Rightarrow$

$f''(x) = 36x^2 - 168x + 12$. Now to solve $f'(x) = 0$, try $x_1 = \frac{1}{2} \Rightarrow x_2 = x_1 - \frac{f'(x_1)}{f''(x_1)} = \frac{2}{3} \Rightarrow$

$x_3 \approx 0.6455 \Rightarrow x_4 \approx 0.6452 \Rightarrow x_5 \approx 0.6452$. Now try $x_1 = 6 \Rightarrow x_2 = 7.12 \Rightarrow$

$x_3 \approx 6.8353 \Rightarrow x_4 \approx 6.8102 \Rightarrow x_5 \approx 6.8100$. Finally try $x_1 = -0.5 \Rightarrow x_2 \approx -0.4571 \Rightarrow$

$x_3 \approx -0.4552 \Rightarrow x_4 \approx -0.4552$. Therefore, $x = -0.455, 6.810$ and 0.645 are all critical numbers correct to three decimal places.

(b) $f(-1) = 13$, $f(7) = -1939$, $f(6.810) \approx -1949.07$, $f(-0.455) \approx -6.912$, $f(0.645) \approx 10.982$. Therefore, $f(6.810) \approx -1949.07$ is the absolute minimum correct to two decimal places.

37.

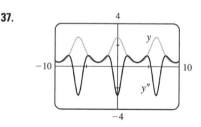

From the figure, we see that $y = f(x) = e^{\cos x}$ is periodic with period 2π. To find the x-coordinates of the IP, we only need to approximate the zeros of y'' on $[0, \pi]$. $f'(x) = -e^{\cos x} \sin x \Rightarrow$

$f''(x) = e^{\cos x}(\sin^2 x - \cos x)$. Since $e^{\cos x} \neq 0$, we will use Newton's method with $g(x) = \sin^2 x - \cos x$, $g'(x) = 2\sin x \cos x + \sin x$, and $x_1 = 1$. $x_2 \approx 0.904173$, $x_3 \approx 0.904557 \approx x_4$. Thus, $(0.904557, 1.855277)$ is the IP.

39.

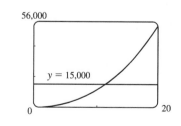

The volume of the silo, in terms of its radius, is

$$V(r) = \pi r^2(30) + \frac{1}{2}\left(\frac{4}{3}\pi r^3\right) = 30\pi r^2 + \frac{2}{3}\pi r^3.$$

From a graph of V, we see that $V(r) = 15{,}000$ at $r \approx 11$ ft. Now we use Newton's method to solve the equation $V(r) - 15{,}000 = 0$.

$\frac{dV}{dr} = 60\pi r + 2\pi r^2$, so $r_{n+1} = r_n - \dfrac{30\pi r_n^2 + \frac{2}{3}\pi r_n^3 - 15{,}000}{60\pi r_n + 2\pi r_n^2}$. Taking

$r_1 = 11$, we get $r_2 \approx 11.2853$, $r_3 \approx 11.2807 \approx r_4$. So in order for the silo to hold $15{,}000$ ft^3 of grain, its radius must be about 11.2807 ft.

41. In this case, $A = 18,000$, $R = 375$, and $n = 5(12) = 60$. So the formula $A = \dfrac{R}{i}\left[1 - (1+i)^{-n}\right]$ becomes

$$18,000 = \frac{375}{x}\left[1 - (1+x)^{-60}\right] \quad \Leftrightarrow \quad 48x = 1 - (1+x)^{-60} \quad \text{[multiply each term by } (1+x)^{60}] \quad \Leftrightarrow$$

$48x(1+x)^{60} - (1+x)^{60} + 1 = 0$. Let the LHS be called $f(x)$, so that

$$f'(x) = 48x(60)(1+x)^{59} + 48(1+x)^{60} - 60(1+x)^{59}$$

$$= 12(1+x)^{59}\left[4x(60) + 4(1+x) - 5\right] = 12(1+x)^{59}(244x - 1)$$

$$x_{n+1} = x_n - \frac{48x_n(1+x_n)^{60} - (1+x_n)^{60} + 1}{12(1+x_n)^{59}(244x_n - 1)}.$$ An interest rate of 1% per month seems like a reasonable

estimate for $x = i$. So let $x_1 = 1\% = 0.01$, and we get $x_2 \approx 0.0082202$, $x_3 \approx 0.0076802$, $x_4 \approx 0.0076291$,
$x_5 \approx 0.0076286 \approx x_6$. Thus, the dealer is charging a monthly interest rate of 0.76286% (or 9.55% per year,
compounded monthly).

4.10 Antiderivatives

1. $f(x) = 6x^2 - 8x + 3 \quad \Rightarrow \quad F(x) = 6\dfrac{x^{2+1}}{2+1} - 8\dfrac{x^{1+1}}{1+1} + 3x + C = 2x^3 - 4x^2 + 3x + C$

   Check: $F'(x) = 2 \cdot 3x^2 - 4 \cdot 2x + 3 + 0 = 6x^2 - 8x + 3 = f(x)$

3. $f(x) = 1 - x^3 + 5x^5 - 3x^7 \quad \Rightarrow \quad F(x) = x - \dfrac{x^{3+1}}{3+1} + 5\dfrac{x^{5+1}}{5+1} - 3\dfrac{x^{7+1}}{7+1} + C = x - \frac{1}{4}x^4 + \frac{5}{6}x^6 - \frac{3}{8}x^8 + C$

5. $f(x) = 5x^{1/4} - 7x^{3/4} \quad \Rightarrow \quad F(x) = 5\dfrac{x^{1/4+1}}{\frac{1}{4}+1} - 7\dfrac{x^{3/4+1}}{\frac{3}{4}+1} + C = 5\dfrac{x^{5/4}}{5/4} - 7\dfrac{x^{7/4}}{7/4} + C = 4x^{5/4} - 4x^{7/4} + C$

7. $f(x) = 6\sqrt{x} - \sqrt[6]{x} = 6x^{1/2} - x^{1/6} \quad \Rightarrow$

   $F(x) = 6\dfrac{x^{1/2+1}}{\frac{1}{2}+1} - \dfrac{x^{1/6+1}}{\frac{1}{6}+1} + C = 6\dfrac{x^{3/2}}{3/2} - \dfrac{x^{7/6}}{7/6} + C = 4x^{3/2} - \frac{6}{7}x^{7/6} + C$

9. $f(x) = \dfrac{10}{x^9} = 10x^{-9}$ has domain $(-\infty, 0) \cup (0, \infty)$, so $F(x) = \begin{cases} \dfrac{10x^{-8}}{-8} + C_1 = -\dfrac{5}{4x^8} + C_1 & \text{if } x < 0 \\[2mm] -\dfrac{5}{4x^8} + C_2 & \text{if } x > 0 \end{cases}$

   See Example 1 for a similar problem.

11. $f(u) = \dfrac{u^4 + 3\sqrt{u}}{u^2} = \dfrac{u^4}{u^2} + \dfrac{3u^{1/2}}{u^2} = u^2 + 3u^{-3/2} \quad \Rightarrow$

   $F(u) = \dfrac{u^3}{3} + 3\dfrac{u^{-3/2+1}}{-3/2+1} + C = \dfrac{1}{3}u^3 + 3\dfrac{u^{-1/2}}{-1/2} + C = \dfrac{1}{3}u^3 - \dfrac{6}{\sqrt{u}} + C$

13. $g(\theta) = \cos\theta - 5\sin\theta \quad \Rightarrow \quad G(\theta) = \sin\theta - 5(-\cos\theta) + C = \sin\theta + 5\cos\theta + C$

15. $f(x) = 2x + 5(1 - x^2)^{-1/2} = 2x + \dfrac{5}{\sqrt{1 - x^2}} \quad \Rightarrow \quad F(x) = x^2 + 5\sin^{-1}x + C$

17. $f(x) = 5x^4 - 2x^5 \Rightarrow F(x) = 5 \cdot \dfrac{x^5}{5} - 2 \cdot \dfrac{x^6}{6} + C = x^5 - \frac{1}{3}x^6 + C.$

$F(0) = 4 \Rightarrow 0^5 - \frac{1}{3} \cdot 0^6 + C = 4 \Rightarrow C = 4$, so

$F(x) = x^5 - \frac{1}{3}x^6 + 4$. The graph confirms our answer since $f(x) = 0$

when F has a local maximum, f is positive when F is increasing, and f is

negative when F is decreasing.

19. $f''(x) = 6x + 12x^2 \Rightarrow f'(x) = 6 \cdot \dfrac{x^2}{2} + 12 \cdot \dfrac{x^3}{3} + C = 3x^2 + 4x^3 + C \Rightarrow$

$f(x) = 3 \cdot \dfrac{x^3}{3} + 4 \cdot \dfrac{x^4}{4} + Cx + D = x^3 + x^4 + Cx + D$ [C and D are just arbitrary constants]

21. $f''(x) = 1 + x^{4/5} \Rightarrow f'(x) = x + \frac{5}{9}x^{9/5} + C \Rightarrow$

$f(x) = \frac{1}{2}x^2 + \frac{5}{9} \cdot \frac{5}{14}x^{14/5} + Cx + D = \frac{1}{2}x^2 + \frac{25}{126}x^{14/5} + Cx + D$

23. $f'''(t) = e^t \Rightarrow f''(t) = e^t + C \Rightarrow f'(t) = e^t + Ct + D \Rightarrow f(t) = e^t + \frac{1}{2}Ct^2 + Dt + E$

25. $f'(x) = 1 - 6x \Rightarrow f(x) = x - 3x^2 + C.$ $f(0) = C$ and $f(0) = 8 \Rightarrow C = 8$, so $f(x) = x - 3x^2 + 8.$

27. $f'(x) = \sqrt{x}(6 + 5x) = 6x^{1/2} + 5x^{3/2} \Rightarrow f(x) = 4x^{3/2} + 2x^{5/2} + C.$

$f(1) = 6 + C$ and $f(1) = 10 \Rightarrow C = 4$, so $f(x) = 4x^{3/2} + 2x^{5/2} + 4.$

29. $f'(t) = 2\cos t + \sec^2 t \Rightarrow f(t) = 2\sin t + \tan t + C$ because $-\pi/2 < t < \pi/2.$

$f\left(\frac{\pi}{3}\right) = 2(\sqrt{3}/2) + \sqrt{3} + C = 2\sqrt{3} + C$ and $f\left(\frac{\pi}{3}\right) = 4 \Rightarrow C = 4 - 2\sqrt{3}$, so

$f(t) = 2\sin t + \tan t + 4 - 2\sqrt{3}.$

31. $f'(x) = 2/x \Rightarrow f(x) = 2\ln|x| + C = 2\ln(-x) + C$ (since $x < 0$). Now

$f(-1) = 2\ln 1 + C = 2(0) + C = 7 \Rightarrow C = 7$. Therefore, $f(x) = 2\ln(-x) + 7, x < 0.$

33. $f''(x) = 24x^2 + 2x + 10 \Rightarrow f'(x) = 8x^3 + x^2 + 10x + C.$ $f'(1) = 8 + 1 + 10 + C$ and $f'(1) = -3 \Rightarrow$

$19 + C = -3 \Rightarrow C = -22$, so $f'(x) = 8x^3 + x^2 + 10x - 22$ and hence,

$f(x) = 2x^4 + \frac{1}{3}x^3 + 5x^2 - 22x + D.$ $f(1) = 2 + \frac{1}{3} + 5 - 22 + D$ and $f(1) = 5 \Rightarrow D = 22 - \frac{7}{3} = \frac{59}{3},$

so $f(x) = 2x^4 + \frac{1}{3}x^3 + 5x^2 - 22x + \frac{59}{3}.$

35. $f''(\theta) = \sin\theta + \cos\theta \Rightarrow f'(\theta) = -\cos\theta + \sin\theta + C.$ $f'(0) = -1 + C$ and $f'(0) = 4 \Rightarrow C = 5$, so

$f'(\theta) = -\cos\theta + \sin\theta + 5$ and hence, $f(\theta) = -\sin\theta - \cos\theta + 5\theta + D.$ $f(0) = -1 + D$ and $f(0) = 3 \Rightarrow$

$D = 4$, so $f(\theta) = -\sin\theta - \cos\theta + 5\theta + 4.$

37. $f''(x) = 2 - 12x \Rightarrow f'(x) = 2x - 6x^2 + C \Rightarrow f(x) = x^2 - 2x^3 + Cx + D.$

$f(0) = D$ and $f(0) = 9 \Rightarrow D = 9.$ $f(2) = 4 - 16 + 2C + 9 = 2C - 3$ and $f(2) = 15 \Rightarrow 2C = 18 \Rightarrow$

$C = 9$, so $f(x) = x^2 - 2x^3 + 9x + 9.$

39. $f''(x) = 2 + \cos x$ ⇒ $f'(x) = 2x + \sin x + C$ ⇒ $f(x) = x^2 - \cos x + Cx + D$. $f(0) = -1 + D$ and
$f(0) = -1$ ⇒ $D = 0$. $f\left(\frac{\pi}{2}\right) = \pi^2/4 + \left(\frac{\pi}{2}\right)C$ and $f\left(\frac{\pi}{2}\right) = 0$ ⇒ $\left(\frac{\pi}{2}\right)C = -\pi^2/4$ ⇒ $C = -\frac{\pi}{2}$, so
$f(x) = x^2 - \cos x - \left(\frac{\pi}{2}\right)x$.

41. $f''(x) = x^{-2}$, $x > 0$ ⇒ $f'(x) = -1/x + C$ ⇒ $f(x) = -\ln|x| + Cx + D = -\ln x + Cx + D$
(since $x > 0$). $f(1) = 0$ ⇒ $C + D = 0$ and $f(2) = 0$ ⇒ $-\ln 2 + 2C + D = 0$ ⇒
$-\ln 2 + 2C - C = 0$ [since $D = -C$] ⇒ $-\ln 2 + C = 0$ ⇒ $C = \ln 2$ and $D = -\ln 2$.
So $f(x) = -\ln x + (\ln 2)x - \ln 2$.

43. Given $f'(x) = 2x + 1$, we have $f(x) = x^2 + x + C$. Since f passes through $(1, 6)$,
$f(1) = 6$ ⇒ $1^2 + 1 + C = 6$ ⇒ $C = 4$. Therefore, $f(x) = x^2 + x + 4$ and $f(2) = 2^2 + 2 + 4 = 10$.

45. b is the antiderivative of f. For small x, f is negative, so the graph of its antiderivative must be decreasing. But
both a and c are increasing for small x, so only b can be f's antiderivative. Also, f is positive where b is increasing,
which supports our conclusion.

47. The graph of F will have a minimum at 0 and a maximum at 2, since $f = F'$ goes from negative to positive at
$x = 0$, and from positive to negative at $x = 2$.

49.

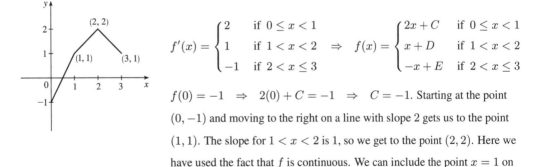

$f'(x) = \begin{cases} 2 & \text{if } 0 \le x < 1 \\ 1 & \text{if } 1 < x < 2 \\ -1 & \text{if } 2 < x \le 3 \end{cases}$ ⇒ $f(x) = \begin{cases} 2x + C & \text{if } 0 \le x < 1 \\ x + D & \text{if } 1 < x < 2 \\ -x + E & \text{if } 2 < x \le 3 \end{cases}$

$f(0) = -1$ ⇒ $2(0) + C = -1$ ⇒ $C = -1$. Starting at the point
$(0, -1)$ and moving to the right on a line with slope 2 gets us to the point
$(1, 1)$. The slope for $1 < x < 2$ is 1, so we get to the point $(2, 2)$. Here we
have used the fact that f is continuous. We can include the point $x = 1$ on
either the first or the second part of f. The line connecting $(1, 1)$ to $(2, 2)$ is $y = x$, so $D = 0$. The slope for
$2 < x \le 3$ is -1, so we get to $(3, 1)$. $f(3) = 1$ ⇒ $-3 + E = 1$ ⇒ $E = 4$. Thus,

$$f(x) = \begin{cases} 2x - 1 & \text{if } 0 \le x \le 1 \\ x & \text{if } 1 < x < 2 \\ -x + 4 & \text{if } 2 \le x \le 3 \end{cases}$$

Note that $f'(x)$ does not exist at $x = 1$ or at $x = 2$.

51. $f(x) = \sin(x^2)$, $0 \le x \le 4$

53.

55.

x	$f(x)$
0	1
0.5	0.959
1.0	0.841
1.5	0.665
2.0	0.455
2.5	0.239
3.0	0.047

x	$f(x)$
3.5	-0.100
4.0	-0.189
4.5	-0.217
5.0	-0.192
5.5	-0.128
6.0	-0.047

We compute slopes [values of $f(x) = (\sin x)/x$ for $0 < x < 2\pi$] as in the table [$\lim_{x \to 0^+} f(x) = 1$] and draw a direction field as in Example 6. Then we use the direction field to graph F starting at $(0, 0)$.

57.

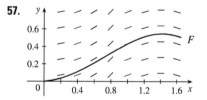

Remember that the given table values of f are the slopes of F at any x. For example, at $x = 1.4$, the slope of F is $f(1.4) = 0$.

59. $v(t) = s'(t) = \sin t - \cos t$ $\Rightarrow$ $s(t) = -\cos t - \sin t + C$. $s(0) = -1 + C$ and $s(0) = 0$ $\Rightarrow$ $C = 1$, so $s(t) = -\cos t - \sin t + 1$.

61. $a(t) = v'(t) = t - 2$ $\Rightarrow$ $v(t) = \frac{1}{2}t^2 - 2t + C$. $v(0) = C$ and $v(0) = 3$ $\Rightarrow$ $C = 3$, so $v(t) = \frac{1}{2}t^2 - 2t + 3$ and $s(t) = \frac{1}{6}t^3 - t^2 + 3t + D$. $s(0) = D$ and $s(0) = 1$ $\Rightarrow$ $D = 1$, and $s(t) = \frac{1}{6}t^3 - t^2 + 3t + 1$.

63. $a(t) = v'(t) = 10\sin t + 3\cos t \quad \Rightarrow \quad v(t) = -10\cos t + 3\sin t + C \quad \Rightarrow$

$s(t) = -10\sin t - 3\cos t + Ct + D.$ $s(0) = -3 + D = 0$ and $s(2\pi) = -3 + 2\pi C + D = 12 \quad \Rightarrow \quad D = 3$ and

$C = \frac{6}{\pi}.$ Thus, $s(t) = -10\sin t - 3\cos t + \frac{6}{\pi}t + 3.$

65. (a) We first observe that since the stone is dropped 450 m above the ground, $v(0) = 0$ and $s(0) = 450.$

$v'(t) = a(t) = -9.8 \quad \Rightarrow \quad v(t) = -9.8t + C.$ Now $v(0) = 0 \quad \Rightarrow \quad C = 0,$ so $v(t) = -9.8t \quad \Rightarrow$

$s(t) = -4.9t^2 + D.$ Last, $s(0) = 450 \quad \Rightarrow \quad D = 450 \quad \Rightarrow \quad s(t) = 450 - 4.9t^2.$

(b) The stone reaches the ground when $s(t) = 0.$ $450 - 4.9t^2 = 0 \quad \Rightarrow \quad t^2 = 450/4.9 \quad \Rightarrow$

$t_1 = \sqrt{450/4.9} \approx 9.58$ s.

(c) The velocity with which the stone strikes the ground is $v(t_1) = -9.8\sqrt{450/4.9} \approx -93.9$ m/s.

(d) This is just reworking parts (a) and (b) with $v(0) = -5.$ Using $v(t) = -9.8t + C,$ $v(0) = -5 \quad \Rightarrow$

$0 + C = -5 \quad \Rightarrow \quad v(t) = -9.8t - 5.$ So $s(t) = -4.9t^2 - 5t + D$ and $s(0) = 450 \quad \Rightarrow \quad D = 450 \quad \Rightarrow$

$s(t) = -4.9t^2 - 5t + 450.$ Solving $s(t) = 0$ by using the quadratic formula gives us

$t = \left(5 \pm \sqrt{8845}\right)/(-9.8) \quad \Rightarrow \quad t_1 \approx 9.09$ s.

67. By Exercise 66 with $a = -9.8,$ $s(t) = -4.9t^2 + v_0 t + s_0$ and $v(t) = s'(t) = -9.8t + v_0.$ So

$[v(t)]^2 = (-9.8t + v_0)^2 = (9.8)^2 t^2 - 19.6v_0 t + v_0^2 = v_0^2 + 96.04t^2 - 19.6v_0 t = v_0^2 - 19.6\left(-4.9t^2 + v_0 t\right).$

But $-4.9t^2 + v_0 t$ is just $s(t)$ without the s_0 term; that is, $s(t) - s_0.$ Thus, $[v(t)]^2 = v_0^2 - 19.6\left[s(t) - s_0\right].$

69. Using Exercise 66 with $a = -32,$ $v_0 = 0,$ and $s_0 = h$ (the height of the cliff), we know that the height at time t is

$s(t) = -16t^2 + h.$ $v(t) = s'(t) = -32t$ and $v(t) = -120 \quad \Rightarrow \quad -32t = -120 \quad \Rightarrow \quad t = 3.75,$ so

$0 = s(3.75) = -16(3.75)^2 + h \quad \Rightarrow \quad h = 16(3.75)^2 = 225$ ft.

71. Marginal cost $= 1.92 - 0.002x = C'(x) \quad \Rightarrow \quad C(x) = 1.92x - 0.001x^2 + K.$ But

$C(1) = 1.92 - 0.001 + K = 562 \quad \Rightarrow \quad K = 560.081.$ Therefore, $C(x) = 1.92x - 0.001x^2 + 560.081 \quad \Rightarrow$

$C(100) = 742.081,$ so the cost of producing 100 items is \$742.08.

73. Taking the upward direction to be positive we have that for $0 \le t \le 10$ (using the subscript 1 to refer to

$0 \le t \le 10),$ $a_1(t) = -(9 - 0.9t) = v_1'(t) \quad \Rightarrow \quad v_1(t) = -9t + 0.45t^2 + v_0,$ but $v_1(0) = v_0 = -10 \quad \Rightarrow$

$v_1(t) = -9t + 0.45t^2 - 10 = s_1'(t) \quad \Rightarrow \quad s_1(t) = -\frac{9}{2}t^2 + 0.15t^3 - 10t + s_0.$ But $s_1(0) = 500 = s_0 \quad \Rightarrow$

$s_1(t) = -\frac{9}{2}t^2 + 0.15t^3 - 10t + 500.$ $s_1(10) = -450 + 150 - 100 + 500 = 100,$ so it takes more

than 10 seconds for the raindrop to fall. Now for $t > 10,$ $a(t) = 0 = v'(t) \quad \Rightarrow$

$v(t) = \text{constant} = v_1(10) = -9(10) + 0.45(10)^2 - 10 = -55 \quad \Rightarrow \quad v(t) = -55.$ At 55 ft/s, it will take

$100/55 \approx 1.8$ s to fall the last 100 ft. Hence, the total time is $10 + \frac{100}{55} = \frac{130}{11} \approx 11.8$ s.

75. $a(t) = k,$ the initial velocity is 30 mi/h $= 30 \cdot \frac{5280}{3600} = 44$ ft/s, and the final velocity (after 5 seconds) is

50 mi/h $= 50 \cdot \frac{5280}{3600} = \frac{220}{3}$ ft/s. So $v(t) = kt + C$ and $v(0) = 44 \quad \Rightarrow \quad C = 44.$ Thus, $v(t) = kt + 44 \quad \Rightarrow$

$v(5) = 5k + 44.$ But $v(5) = \frac{220}{3},$ so $5k + 44 = \frac{220}{3} \quad \Rightarrow \quad 5k = \frac{88}{3} \quad \Rightarrow \quad k = \frac{88}{15} \approx 5.87$ ft/s$^2.$

77. Let the acceleration be $a(t) = k$ km/h². We have $v(0) = 100$ km/h and we can take the initial position $s(0)$ to

be 0. We want the time t_f for which $v(t) = 0$ to satisfy $s(t) < 0.08$ km. In general, $v'(t) = a(t) = k$, so

$v(t) = kt + C$, where $C = v(0) = 100$. Now $s'(t) = v(t) = kt + 100$, so $s(t) = \frac{1}{2}kt^2 + 100t + D$, where

$D = s(0) = 0$. Thus, $s(t) = \frac{1}{2}kt^2 + 100t$. Since $v(t_f) = 0$, we have $kt_f + 100 = 0$ or $t_f = -100/k$, so

$$s(t_f) = \frac{1}{2}k\left(-\frac{100}{k}\right)^2 + 100\left(-\frac{100}{k}\right) = 10{,}000\left(\frac{1}{2k} - \frac{1}{k}\right) = -\frac{5{,}000}{k}. \text{ The condition } s(t_f) \text{ must satisfy is}$$

$-\dfrac{5{,}000}{k} < 0.08 \quad \Rightarrow \quad -\dfrac{5{,}000}{0.08} > k \quad [k \text{ is negative}] \quad \Rightarrow \quad k < -62{,}500$ km/h², or equivalently,

$k < -\frac{3125}{648} \approx -4.82$ m/s².

79. (a) First note that 90 mi/h $= 90 \times \frac{5280}{3600}$ ft/s $= 132$ ft/s. Then $a(t) = 4$ ft/s² $\Rightarrow v(t) = 4t + C$, but $v(0) = 0$

$\Rightarrow \quad C = 0$. Now $4t = 132$ when $t = \frac{132}{4} = 33$ s, so it takes 33 s to reach 132 ft/s. Therefore, taking

$s(0) = 0$, we have $s(t) = 2t^2, 0 \le t \le 33$. So $s(33) = 2178$ ft. 15 minutes $= 15(60) = 900$ s, so for

$33 < t \le 933$ we have $v(t) = 132$ ft/s $\Rightarrow s(933) = 132(900) + 2178 = 120{,}978$ ft $= 22.9125$ mi.

(b) As in part (a), the train accelerates for 33 s and travels 2178 ft while doing so. Similarly, it decelerates for 33 s

and travels 2178 ft at the end of its trip. During the remaining $900 - 66 = 834$ s it travels at 132 ft/s, so

the distance traveled is $132 \cdot 834 = 110{,}088$ ft. Thus, the total distance is

$2178 + 110{,}088 + 2178 = 114{,}444$ ft $= 21.675$ mi.

(c) 45 mi $= 45(5280) = 237{,}600$ ft. Subtract $2(2178)$ to take care of the speeding up and slowing down, and we

have $233{,}244$ ft at 132 ft/s for a trip of $233{,}244/132 = 1767$ s at 90 mi/h. The total time is

$1767 + 2(33) = 1833$ s $= 30$ min 33 s $= 30.55$ min.

(d) $37.5(60) = 2250$ s. $2250 - 2(33) = 2184$ s at maximum speed. $2184(132) + 2(2178) = 292{,}644$ total feet

or $292{,}644/5280 = 55.425$ mi.

4 Review

──────────────── CONCEPT CHECK ────────────────

1. A function f has an **absolute maximum** at $x = c$ if $f(c)$ is the largest function value on the entire domain of f,
whereas f has a **local maximum** at c if $f(c)$ is the largest function value when x is near c. See Figure 4 in
Section 4.1.

2. (a) See Theorem 4.1.3.

(b) See the Closed Interval Method before Example 8 in Section 4.1.

3. (a) See Theorem 4.1.4.

(b) See Definition 4.1.6.

4. (a) See Rolle's Theorem at the beginning of Section 4.2.

(b) See the Mean Value Theorem in Section 4.2. Geometric interpretation—there is some point P on the graph of a function f [on the interval (a, b)] where the tangent line is parallel to the secant line that connects $(a, f(a))$ and $(b, f(b))$.

5. (a) See the I/D Test before Example 1 in Section 4.3.

(b) See the Concavity Test before Example 4 in Section 4.3.

6. (a) See the First Derivative Test after Example 1 in Section 4.3.

(b) See the Second Derivative Test before Example 6 in Section 4.3.

(c) See the note before Example 7 in Section 4.3.

7. (a) See l'Hospital's Rule and the three notes that follow it in Section 4.4.

(b) Write fg as $\dfrac{f}{1/g}$ or $\dfrac{g}{1/f}$.

(c) Convert the difference into a quotient using a common denominator, rationalizing, factoring, or some other method.

(d) Convert the power to a product by taking the natural logarithm of both sides of $y = f^g$ or by writing f^g as $e^{g \ln f}$.

8. Without calculus you could get misleading graphs that fail to show the most interesting features of a function. See the discussion following Figure 3 in Section 4.5 and the first paragraph in Section 4.6.

9. (a) See Figure 3 in Section 4.9.

(b) $x_2 = x_1 - \dfrac{f(x_1)}{f'(x_1)}$

(c) $x_{n+1} = x_n - \dfrac{f(x_n)}{f'(x_n)}$

(d) Newton's method is likely to fail or to work very slowly when $f'(x_1)$ is close to 0.

10. (a) See the definition at the beginning of Section 4.10.

(b) If F_1 and F_2 are both antiderivatives of f on an interval I, then they differ by a constant.

──────────────── TRUE-FALSE QUIZ ────────────────

1. False.   For example, take $f(x) = x^3$, then $f'(x) = 3x^2$ and $f'(0) = 0$, but $f(0) = 0$ is not a maximum or minimum; $(0, 0)$ is an inflection point.

3. False.   For example, $f(x) = x$ is continuous on $(0, 1)$ but attains neither a maximum nor a minimum value on $(0, 1)$. Don't confuse this with f being continuous on the *closed* interval $[a, b]$, which would make the statement true.

5. True.   This is an example of part (b) of the I/D Test.

7. False. $f'(x) = g'(x)$ $\Rightarrow$ $f(x) = g(x) + C$. For example, if $f(x) = x + 2$ and $g(x) = x + 1$, then $f'(x) = g'(x) = 1$, but $f(x) \neq g(x)$.

9. True. The graph of one such function is sketched.

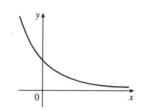

11. True. Let $x_1 < x_2$ where $x_1, x_2 \in I$. Then $f(x_1) < f(x_2)$ and $g(x_1) < g(x_2)$ (since f and g are increasing on I), so $(f+g)(x_1) = f(x_1) + g(x_1) < f(x_2) + g(x_2) = (f+g)(x_2)$.

13. False. Take $f(x) = x$ and $g(x) = x - 1$. Then both f and g are increasing on $(0,1)$. But $f(x)g(x) = x(x-1)$ is not increasing on $(0,1)$.

15. True. Let $x_1, x_2 \in I$ and $x_1 < x_2$. Then $f(x_1) < f(x_2)$ (f is increasing) $\Rightarrow$

$$\frac{1}{f(x_1)} > \frac{1}{f(x_2)} \ (f \text{ is positive}) \ \Rightarrow \ g(x_1) > g(x_2) \ \Rightarrow \ g(x) = 1/f(x) \text{ is decreasing on } I.$$

17. True. By the Mean Value Theorem, there exists a number c in $(0, 1)$ such that
$$f(1) - f(0) = f'(c)(1 - 0) = f'(c). \text{ Since } f'(c) \text{ is nonzero, } f(1) - f(0) \neq 0, \text{ so } f(1) \neq f(0).$$

———————————— EXERCISES ————————————

1. $f(x) = 10 + 27x - x^3, \ 0 \le x \le 4.$ $f'(x) = 27 - 3x^2 = -3(x^2 - 9) = -3(x+3)(x-3) = 0$ only when $x = 3$ (since -3 is not in the domain). $f'(x) > 0$ for $x < 3$ and $f'(x) < 0$ for $x > 3$, so $f(3) = 64$ is a local maximum value. Checking the endpoints, we find $f(0) = 10$ and $f(4) = 54$. Thus, $f(0) = 10$ is the absolute minimum value and $f(3) = 64$ is the absolute maximum value.

3. $f(x) = \dfrac{x}{x^2 + x + 1}, \ -2 \le x \le 0.$ $f'(x) = \dfrac{(x^2 + x + 1)(1) - x(2x + 1)}{(x^2 + x + 1)^2} = \dfrac{1 - x^2}{(x^2 + x + 1)^2} = 0 \ \Leftrightarrow$

$x = -1$ (since 1 is not in the domain). $f'(x) < 0$ for $-2 < x < -1$ and $f'(x) > 0$ for $-1 < x < 0$, so $f(-1) = -1$ is a local and absolute minimum value. $f(-2) = -\frac{2}{3}$ and $f(0) = 0$, so $f(0) = 0$ is an absolute maximum value.

5. $f(x) = x + \sin 2x, \ [0, \pi].$ $f'(x) = 1 + 2\cos 2x = 0 \ \Leftrightarrow \ \cos 2x = -\frac{1}{2} \ \Leftrightarrow \ 2x = \frac{2\pi}{3} \text{ or } \frac{4\pi}{3} \ \Leftrightarrow \ x = \frac{\pi}{3}$

or $\frac{2\pi}{3}$. $f''(x) = -4\sin 2x$, so $f''\left(\frac{\pi}{3}\right) = -4\sin\frac{2\pi}{3} = -2\sqrt{3} < 0$ and $f''\left(\frac{2\pi}{3}\right) = -4\sin\frac{4\pi}{3} = 2\sqrt{3} > 0$, so $f\left(\frac{\pi}{3}\right) = \frac{\pi}{3} + \frac{\sqrt{3}}{2} \approx 1.91$ is a local maximum value and $f\left(\frac{2\pi}{3}\right) = \frac{2\pi}{3} - \frac{\sqrt{3}}{2} \approx 1.23$ is a local minimum value. Also $f(0) = 0$ and $f(\pi) = \pi$, so $f(0) = 0$ is the absolute minimum value and $f(\pi) = \pi$ is the absolute maximum value.

7. $\lim\limits_{x \to 0} \dfrac{\tan \pi x}{\ln(1 + x)} \overset{\text{H}}{=} \lim\limits_{x \to 0} \dfrac{\pi \sec^2 \pi x}{1/(1 + x)} = \dfrac{\pi \cdot 1^2}{1/1} = \pi$

9. $\lim\limits_{x \to 0} \dfrac{e^{4x} - 1 - 4x}{x^2} \overset{\text{H}}{=} \lim\limits_{x \to 0} \dfrac{4e^{4x} - 4}{2x} \overset{\text{H}}{=} \lim\limits_{x \to 0} \dfrac{16e^{4x}}{2} = \lim\limits_{x \to 0} 8e^{4x} = 8 \cdot 1 = 8$

11. $\lim\limits_{x \to \infty} x^3 e^{-x} = \lim\limits_{x \to \infty} \dfrac{x^3}{e^x} \overset{\text{H}}{=} \lim\limits_{x \to \infty} \dfrac{3x^2}{e^x} \overset{\text{H}}{=} \lim\limits_{x \to \infty} \dfrac{6x}{e^x} \overset{\text{H}}{=} \lim\limits_{x \to \infty} \dfrac{6}{e^x} = 0$

13. $\displaystyle\lim_{x \to 1^+} \left(\frac{x}{x-1} - \frac{1}{\ln x}\right) = \lim_{x \to 1^+} \left(\frac{x\ln x - x + 1}{(x-1)\ln x}\right) \overset{\text{H}}{=} \lim_{x \to 1^+} \frac{x \cdot (1/x) + \ln x - 1}{(x-1) \cdot (1/x) + \ln x}$

$\displaystyle = \lim_{x \to 1^+} \frac{\ln x}{1 - 1/x + \ln x} \overset{\text{H}}{=} \lim_{x \to 1^+} \frac{1/x}{1/x^2 + 1/x} = \frac{1}{1+1} = \frac{1}{2}$

15. $f(0) = 0,\ f'(-2) = f'(1) = f'(9) = 0,\ \lim\limits_{x \to \infty} f(x) = 0,$

$\lim\limits_{x \to 6} f(x) = -\infty,\ f'(x) < 0$ on $(-\infty, -2),\ (1,6),$ and $(9, \infty),$

$f'(x) > 0$ on $(-2, 1)$ and $(6, 9),\ f''(x) > 0$ on $(-\infty, 0)$

and $(12, \infty),\ f''(x) < 0$ on $(0, 6)$ and $(6, 12)$

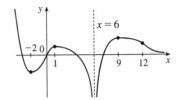

17. f is odd, $f'(x) < 0$ for $0 < x < 2,\quad f'(x) > 0$ for $x > 2,$

$f''(x) > 0$ for $0 < x < 3,\quad f''(x) < 0$ for $x > 3,$

$\lim_{x \to \infty} f(x) = -2$

19. $y = f(x) = 2 - 2x - x^3$ **A.** $D = \mathbb{R}$ **B.** y-intercept: $f(0) = 2.$

The x-intercept (approximately 0.770917) can be found using Newton's

Method. **C.** No symmetry **D.** No asymptote

E. $f'(x) = -2 - 3x^2 = -(3x^2 + 2) < 0$, so f is decreasing on $\mathbb{R}$.

F. No extreme value **G.** $f''(x) = -6x < 0$ on $(0, \infty)$ and $f''(x) > 0$

on $(-\infty, 0)$, so f is CD on $(0, \infty)$ and CU on $(-\infty, 0)$.

There is an IP at $(0, 2)$.

H.

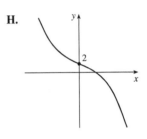

21. $y = f(x) = x^4 - 3x^3 + 3x^2 - x = x(x-1)^3$ **A.** $D = \mathbb{R}$ **B.** y-intercept: $f(0) = 0$; x-intercepts: $f(x) = 0$

$\Leftrightarrow\ x = 0$ or $x = 1$ **C.** No symmetry **D.** f is a polynomial function and hence, it has no asymptote.

E. $f'(x) = 4x^3 - 9x^2 + 6x - 1$. Since the sum of the coefficients is 0, 1 is a root of f', so

$f'(x) = (x-1)(4x^2 - 5x + 1) = (x-1)^2(4x-1)$. $f'(x) < 0\ \Rightarrow\ x < \frac{1}{4}$, so f is decreasing on $\left(-\infty, \frac{1}{4}\right)$

and f is increasing on $\left(\frac{1}{4}, \infty\right)$. **F.** $f'(x)$ does not change sign at $x = 1$,

so there is not a local extremum there. $f\left(\frac{1}{4}\right) = -\frac{27}{256}$ is a local minimum

value. **G.** $f''(x) = 12x^2 - 18x + 6 = 6(2x-1)(x-1)$.

$f''(x) = 0\ \Leftrightarrow\ x = \frac{1}{2}$ or 1. $f''(x) < 0\ \Leftrightarrow\ \frac{1}{2} < x < 1\ \Rightarrow$

f is CD on $\left(\frac{1}{2}, 1\right)$ and CU on $\left(-\infty, \frac{1}{2}\right)$ and $(1, \infty)$. There are inflection

points at $\left(\frac{1}{2}, -\frac{1}{16}\right)$ and $(1, 0)$.

H.

23. $y = f(x) = \dfrac{1}{x(x-3)^2}$ **A.** $D = \{x \mid x \neq 0, 3\} = (-\infty, 0) \cup (0, 3) \cup (3, \infty)$ **B.** No intercepts.

C. No symmetry.   **D.** $\displaystyle\lim_{x \to \pm\infty} \dfrac{1}{x(x-3)^2} = 0$, so $y = 0$ is a HA. $\displaystyle\lim_{x \to 0^+} \dfrac{1}{x(x-3)^2} = \infty$,

$\displaystyle\lim_{x \to 0^-} \dfrac{1}{x(x-3)^2} = -\infty$, $\displaystyle\lim_{x \to 3} \dfrac{1}{x(x-3)^2} = \infty$, so $x = 0$ and $x = 3$ are VA.

E. $f'(x) = -\dfrac{(x-3)^2 + 2x(x-3)}{x^2(x-3)^4} = \dfrac{3(1-x)}{x^2(x-3)^3}$ $\Rightarrow$

$f'(x) > 0$ ⇔ $1 < x < 3$, so f is increasing on $(1, 3)$ and decreasing   **H.**

on $(-\infty, 0)$, $(0, 1)$, and $(3, \infty)$.   **F.** Local minimum value $f(1) = \frac{1}{4}$

G. $f''(x) = \dfrac{6(2x^2 - 4x + 3)}{x^3(x-3)^4}$. Note that $2x^2 - 4x + 3 > 0$ for all x

since it has negative discriminant. So $f''(x) > 0$ ⇔ $x > 0$ ⇒ f is

CU on $(0, 3)$ and $(3, \infty)$ and CD on $(-\infty, 0)$. No IP

25. $y = f(x) = \dfrac{x^2}{x+8} = x - 8 + \dfrac{64}{x+8}$ **A.** $D = \{x \mid x \neq -8\}$ **B.** Intercepts are 0   **C.** No symmetry

D. $\displaystyle\lim_{x \to \infty} \dfrac{x^2}{x+8} = \infty$, but $f(x) - (x-8) = \dfrac{64}{x+8} \to 0$ as $x \to \infty$, so $y = x - 8$ is a slant asymptote.

$\displaystyle\lim_{x \to -8^+} \dfrac{x^2}{x+8} = \infty$ and $\displaystyle\lim_{x \to -8^-} \dfrac{x^2}{x+8} = -\infty$, so $x = -8$ is a VA.   **H.**

E. $f'(x) = 1 - \dfrac{64}{(x+8)^2} = \dfrac{x(x+16)}{(x+8)^2} > 0$ ⇔ $x > 0$ or $x < -16$,

so f is increasing on $(-\infty, -16)$ and $(0, \infty)$ and decreasing on

$(-16, -8)$ and $(-8, 0)$.   **F.** Local maximum value $f(-16) = -32$,

local minimum value $f(0) = 0$ **G.** $f''(x) = 128/(x+8)^3 > 0$ ⇔

$x > -8$, so f is CU on $(-8, \infty)$ and CD on $(-\infty, -8)$. No IP

27. $y = f(x) = x\sqrt{2+x}$ **A.** $D = [-2, \infty)$ **B.** y-intercept: $f(0) = 0$; x-intercepts: -2 and 0 **C.** No

symmetry   **D.** No asymptote   **E.** $f'(x) = \dfrac{x}{2\sqrt{2+x}} + \sqrt{2+x} = \dfrac{1}{2\sqrt{2+x}}[x + 2(2+x)] = \dfrac{3x+4}{2\sqrt{2+x}} = 0$

when $x = -\frac{4}{3}$, so f is decreasing on $\left(-2, -\frac{4}{3}\right)$ and increasing on $\left(-\frac{4}{3}, \infty\right)$.   **F.** Local minimum value

$f\left(-\frac{4}{3}\right) = -\frac{4}{3}\sqrt{\frac{2}{3}} = -\frac{4\sqrt{6}}{9} \approx -1.09$, no local maximum   **H.**

G. $f''(x) = \dfrac{2\sqrt{2+x} \cdot 3 - (3x+4)\dfrac{1}{\sqrt{2+x}}}{4(2+x)}$

$= \dfrac{6(2+x) - (3x+4)}{4(2+x)^{3/2}} = \dfrac{3x+8}{4(2+x)^{3/2}}$

$f''(x) > 0$ for $x > -2$, so f is CU on $(-2, \infty)$. No IP

29. $y = f(x) = \sin^2 x - 2 \cos x$ **A.** $D = \mathbb{R}$ **B.** y-intercept: $f(0) = -2$ **C.** $f(-x) = f(x)$, so

f is symmetric with respect to the y-axis. f has period 2π. **D.** No asymptote

E. $y' = 2 \sin x \cos x + 2 \sin x = 2 \sin x (\cos x + 1)$. $y' = 0$ $\Leftrightarrow$ $\sin x = 0$ or $\cos x = -1$ $\Leftrightarrow$ $x = n\pi$ or

$x = (2n+1)\pi$. $y' > 0$ when $\sin x > 0$, since $\cos x + 1 \geq 0$ for all x. Therefore, $y' > 0$ (and so f is increasing)

on $(2n\pi, (2n+1)\pi)$; $y' < 0$ (and so f is decreasing) on $((2n-1)\pi, 2n\pi)$. **F.** Local maximum values are

$f((2n+1)\pi) = 2$; local minimum values are $f(2n\pi) = -2$. **G.** $y' = \sin 2x + 2 \sin x$ $\Rightarrow$

$$y'' = 2 \cos 2x + 2 \cos x = 2(2 \cos^2 x - 1) + 2 \cos x = 4 \cos^2 x + 2 \cos x - 2$$

$$= 2(2 \cos^2 x + \cos x - 1) = 2(2 \cos x - 1)(\cos x + 1)$$

$y'' = 0$ $\Leftrightarrow$ $\cos x = \frac{1}{2}$ or -1 $\Leftrightarrow$ $x = 2n\pi \pm \frac{\pi}{3}$ or $x = (2n+1)\pi$. $y'' > 0$ (and so f is CU) on

$\left(2n\pi - \frac{\pi}{3}, 2n\pi + \frac{\pi}{3}\right)$; $y'' \leq 0$ (and so f is CD) on $\left(2n\pi + \frac{\pi}{3}, 2n\pi + \frac{5\pi}{3}\right)$. There are inflection points

at $\left(2n\pi \pm \frac{\pi}{3}, -\frac{1}{4}\right)$.

H.

31. $y = f(x) = \sin^{-1}(1/x)$ **A.** $D = \{x \mid -1 \leq 1/x \leq 1\} = (-\infty, -1] \cup [1, \infty)$. **B.** No intercept

C. $f(-x) = -f(x)$, symmetric about the origin **D.** $\lim\limits_{x \to \pm\infty} \sin^{-1}(1/x) = \sin^{-1}(0) = 0$, so $y = 0$ is a HA.

E. $f'(x) = \dfrac{1}{\sqrt{1 - (1/x)^2}}\left(-\dfrac{1}{x^2}\right) = \dfrac{-1}{\sqrt{x^4 - x^2}} < 0$, so f is decreasing on $(-\infty, -1)$ and $(1, \infty)$.

F. No local extreme value, but $f(1) = \frac{\pi}{2}$ is the absolute maximum value

and $f(-1) = -\frac{\pi}{2}$ is the absolute minimum value.

G. $f''(x) = \dfrac{4x^3 - 2x}{2(x^4 - x^2)^{3/2}} = \dfrac{x(2x^2 - 1)}{(x^4 - x^2)^{3/2}} > 0$ for $x > 1$ and

$f''(x) < 0$ for $x < -1$, so f is CU on $(1, \infty)$ and CD on $(-\infty, -1)$.

No IP

H.

33. $y = f(x) = e^x + e^{-3x}$ **A.** $D = \mathbb{R}$ **B.** y-intercept 2; no x-intercept **C.** No symmetry

D. $\lim\limits_{x \to \pm\infty}\left(e^x + e^{-3x}\right) = \infty$, no asymptote **E.** $y = f(x) = e^x + e^{-3x}$ $\Rightarrow$

$f'(x) = e^x - 3e^{-3x} = e^{-3x}(e^{4x} - 3) > 0$ $\Leftrightarrow$ $e^{4x} > 3$ $\Leftrightarrow$

$4x > \ln 3$ $\Leftrightarrow$ $x > \frac{1}{4} \ln 3 \approx 0.27$, so f is increasing on $\left(\frac{1}{4} \ln 3, \infty\right)$

and decreasing on $\left(-\infty, \frac{1}{4} \ln 3\right)$.

F. Absolute minimum value $f\left(\frac{1}{4} \ln 3\right) = 3^{1/4} + 3^{-3/4} \approx 1.75$.

G. $f''(x) = e^x + 9e^{-3x} > 0$, so f is CU on $(-\infty, \infty)$. No IP

H.

35. $f(x) = \dfrac{x^2 - 1}{x^3}$ $\Rightarrow$ $f'(x) = \dfrac{x^3(2x) - (x^2 - 1)3x^2}{x^6} = \dfrac{3 - x^2}{x^4}$ $\Rightarrow$

$f''(x) = \dfrac{x^4(-2x) - (3 - x^2)4x^3}{x^8} = \dfrac{2x^2 - 12}{x^5}$

Estimates: From the graphs of f' and f'', it appears that f is increasing on $(-1.73, 0)$ and $(0, 1.73)$ and decreasing

on $(-\infty, -1.73)$ and $(1.73, \infty)$; f has a local maximum of about $f(1.73) = 0.38$ and a local minimum of about

$f(-1.7) = -0.38$; f is CU on $(-2.45, 0)$ and $(2.45, \infty)$, and CD on $(-\infty, -2.45)$ and $(0, 2.45)$; and f has

inflection points at about $(-2.45, -0.34)$ and $(2.45, 0.34)$.

Exact: Now $f'(x) = \dfrac{3 - x^2}{x^4}$ is positive for $0 < x^2 < 3$, that is, f is increasing on $(-\sqrt{3}, 0)$ and $(0, \sqrt{3})$; and

$f'(x)$ is negative (and so f is decreasing) on $(-\infty, -\sqrt{3})$ and $(\sqrt{3}, \infty)$. $f'(x) = 0$ when $x = \pm\sqrt{3}$. f' goes

from positive to negative at $x = \sqrt{3}$, so f has a local maximum of $f(\sqrt{3}) = \dfrac{(\sqrt{3})^2 - 1}{(\sqrt{3})^3} = \dfrac{2\sqrt{3}}{9}$; and since f is odd,

we know that maxima on the interval $(0, \infty)$ correspond to minima on $(-\infty, 0)$, so f has a local minimum of

$f(-\sqrt{3}) = -\dfrac{2\sqrt{3}}{9}$. Also, $f''(x) = \dfrac{2x^2 - 12}{x^5}$ is positive (so f is CU) on $(-\sqrt{6}, 0)$ and $(\sqrt{6}, \infty)$, and negative

(so f is CD) on $(-\infty, -\sqrt{6})$ and $(0, \sqrt{6})$. There are IP at $\left(\sqrt{6}, \dfrac{5\sqrt{6}}{36}\right)$ and $\left(-\sqrt{6}, -\dfrac{5\sqrt{6}}{36}\right)$.

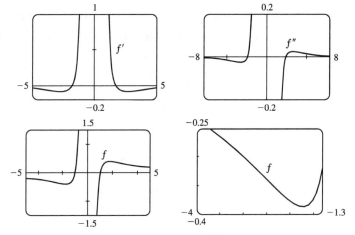

37. $f(x) = 3x^6 - 5x^5 + x^4 - 5x^3 - 2x^2 + 2$ $\Rightarrow$ $f'(x) = 18x^5 - 25x^4 + 4x^3 - 15x^2 - 4x$ $\Rightarrow$

$f''(x) = 90x^4 - 100x^3 + 12x^2 - 30x - 4$

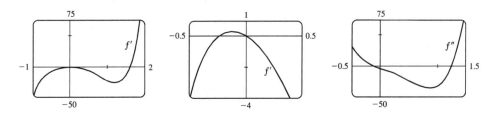

From the graphs of f' and f'', it appears that f is increasing on $(-0.23, 0)$ and $(1.62, \infty)$ and decreasing on $(-\infty, -0.23)$ and $(0, 1.62)$; f has a local maximum of about $f(0) = 2$ and local minima

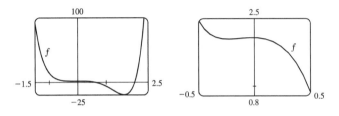

of about $f(-0.23) = 1.96$ and $f(1.62) = -19.2$; f is CU on $(-\infty, -0.12)$ and $(1.24, \infty)$ and CD on $(-0.12, 1.24)$; and f has inflection points at about $(-0.12, 1.98)$ and $(1.24, -12.1)$.

39.

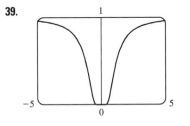

From the graph, we estimate the points of inflection to be about

$(\pm 0.82, 0.22)$. $f(x) = e^{-1/x^2}$ $\Rightarrow$ $f'(x) = 2x^{-3}e^{-1/x^2}$ $\Rightarrow$

$f''(x) = 2\left[x^{-3}\left(2x^{-3}\right)e^{-1/x^2} + e^{-1/x^2}\left(-3x^{-4}\right)\right]$

$= 2x^{-6}e^{-1/x^2}\left(2 - 3x^2\right)$.

This is 0 when $2 - 3x^2 = 0$ $\Leftrightarrow$ $x = \pm\sqrt{\frac{2}{3}}$, so the inflection points

are $\left(\pm\sqrt{\frac{2}{3}}, e^{-3/2}\right)$.

41. $f(x) = \arctan(\cos(3\arcsin x))$. We use a CAS to compute f' and f'', and to graph f, f', and f'':

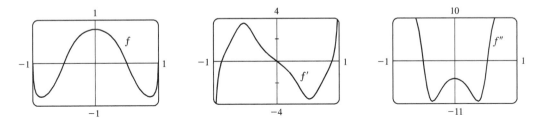

From the graph of f', it appears that the only maximum occurs at $x = 0$ and there are minima at $x = \pm 0.87$.

From the graph of f'', it appears that there are inflection points at $x = \pm 0.52$.

43. The family of functions $f(x) = \ln(\sin x + C)$ all have the same period and all have maximum values at $x = \frac{\pi}{2} + 2\pi n$. Since the domain of ln is $(0, \infty)$, f has a graph only if $\sin x + C > 0$ somewhere. Since $-1 \le \sin x \le 1$, this happens if $C > -1$, that is, f has no graph if $C \le -1$. Similarly, if $C > 1$, then $\sin x + C > 0$ and f is continuous on $(-\infty, \infty)$. As C increases, the graph of f is shifted vertically upward and flattens out.

If $-1 < C \le 1$, f is defined where $\sin x + C > 0$ $\Leftrightarrow$ $\sin x > -C$ $\Leftrightarrow$ $\sin^{-1}(-C) < x < \pi - \sin^{-1}(-C)$.

Since the period is 2π, the domain of f is $\left(2n\pi + \sin^{-1}(-C), (2n+1)\pi - \sin^{-1}(-C)\right)$, n an integer.

45. $f(x) = x^{101} + x^{51} + x - 1 = 0$. Since f is continuous and $f(0) = -1$ and $f(1) = 2$, the equation has at least one

root in $(0, 1)$, by the Intermediate Value Theorem. Suppose the equation has two roots, a and b, with $a < b$.

Then $f(a) = 0 = f(b)$, so by the Mean Value Theorem, there is a number x in (a, b) such that

$$f'(x) = \frac{f(b) - f(a)}{b - a} = \frac{0}{b - a} = 0, \text{ so } f' \text{ has a root in } (a, b). \text{ But this is impossible since}$$

$f'(x) = 101x^{100} + 51x^{50} + 1 \geq 1$ for all x.

47. Since f is continuous on $[32, 33]$ and differentiable on $(32, 33)$, then by the Mean Value Theorem there exists a

number c in $(32, 33)$ such that $f'(c) = \frac{1}{5}c^{-4/5} = \dfrac{\sqrt[5]{33} - \sqrt[5]{32}}{33 - 32} = \sqrt[5]{33} - 2$, but $\frac{1}{5}c^{-4/5} > 0 \Rightarrow \sqrt[5]{33} - 2 > 0$

$\Rightarrow \sqrt[5]{33} > 2$. Also f' is decreasing, so that $f'(c) < f'(32) = \frac{1}{5}(32)^{-4/5} = 0.0125 \Rightarrow$

$0.0125 > f'(c) = \sqrt[5]{33} - 2 \Rightarrow \sqrt[5]{33} < 2.0125$. Therefore, $2 < \sqrt[5]{33} < 2.0125$.

49. (a) $g(x) = f(x^2) \Rightarrow g'(x) = 2xf'(x^2)$ by the Chain Rule. Since $f'(x) > 0$ for all $x \neq 0$, we must have

$f'(x^2) > 0$ for $x \neq 0$, so $g'(x) = 0 \Leftrightarrow x = 0$. Now $g'(x)$ changes sign (from negative to positive) at

$x = 0$, since one of its factors, $f'(x^2)$, is positive for all x, and its other factor, $2x$, changes from negative to

positive at this point, so by the First Derivative Test, f has a local and absolute minimum at $x = 0$.

(b) $g'(x) = 2xf'(x^2) \Rightarrow g''(x) = 2[xf''(x^2)(2x) + f'(x^2)] = 4x^2f''(x^2) + 2f'(x^2)$ by the Product Rule

and the Chain Rule. But $x^2 > 0$ for all $x \neq 0$, $f''(x^2) > 0$ (since f is CU for $x > 0$), and $f'(x^2) > 0$ for all

$x \neq 0$, so since all of its factors are positive, $g''(x) > 0$ for $x \neq 0$. Whether $g''(0)$ is positive or 0 doesn't

matter (since the sign of g'' does not change there); g is concave upward on $\mathbb{R}$.

51. If $B = 0$, the line is vertical and the distance from $x = -\dfrac{C}{A}$ to (x_1, y_1) is $\left| x_1 + \dfrac{C}{A} \right| = \dfrac{|Ax_1 + By_1 + C|}{\sqrt{A^2 + B^2}}$, so

assume $B \neq 0$. The square of the distance from (x_1, y_1) to the line is $f(x) = (x - x_1)^2 + (y - y_1)^2$ where

$Ax + By + C = 0$, so we minimize $f(x) = (x - x_1)^2 + \left(-\dfrac{A}{B}x - \dfrac{C}{B} - y_1 \right)^2 \Rightarrow$

$f'(x) = 2(x - x_1) + 2\left(-\dfrac{A}{B}x - \dfrac{C}{B} - y_1 \right)\left(-\dfrac{A}{B} \right)$. $f'(x) = 0 \Rightarrow x = \dfrac{B^2x_1 - ABy_1 - AC}{A^2 + B^2}$ and this gives

a minimum since $f''(x) = 2\left(1 + \dfrac{A^2}{B^2} \right) > 0$. Substituting this value of x into $f(x)$ and simplifying gives

$f(x) = \dfrac{(Ax_1 + By_1 + C)^2}{A^2 + B^2}$, so the minimum distance is $\sqrt{f(x)} = \dfrac{|Ax_1 + By_1 + C|}{\sqrt{A^2 + B^2}}$.

53.

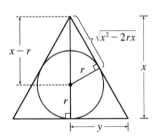

By similar triangles, $\dfrac{y}{x} = \dfrac{r}{\sqrt{x^2 - 2rx}}$, so the area of the triangle is

$$A(x) = \tfrac{1}{2}(2y)x = xy = \dfrac{rx^2}{\sqrt{x^2 - 2rx}} \quad \Rightarrow$$

$$A'(x) = \dfrac{2rx\sqrt{x^2 - 2rx} - rx^2(x - r)/\sqrt{x^2 - 2rx}}{x^2 - 2rx}$$

$$= \dfrac{rx^2(x - 3r)}{(x^2 - 2rx)^{3/2}} = 0 \text{ when } x = 3r.$$

$A'(x) < 0$ when $2r < x < 3r$, $A'(x) > 0$ when $x > 3r$. So $x = 3r$ gives a minimum and $A(3r) = r(9r^2)/(\sqrt{3}\,r) = 3\sqrt{3}\,r^2$.

55.

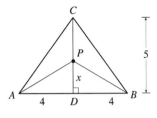

We minimize

$$L(x) = |PA| + |PB| + |PC| = 2\sqrt{x^2 + 16} + (5 - x),$$

$0 \le x \le 5$. $L'(x) = 2x/\sqrt{x^2 + 16} - 1 = 0 \quad \Leftrightarrow$

$$2x = \sqrt{x^2 + 16} \quad \Leftrightarrow \quad 4x^2 = x^2 + 16 \quad \Leftrightarrow \quad x = \tfrac{4}{\sqrt{3}}.$$

$L(0) = 13$, $L\left(\tfrac{4}{\sqrt{3}}\right) \approx 11.9$, $L(5) \approx 12.8$, so the minimum

occurs when $x = \tfrac{4}{\sqrt{3}} \approx 2.3$.

57. $v = K\sqrt{\dfrac{L}{C} + \dfrac{C}{L}} \quad \Rightarrow \quad \dfrac{dv}{dL} = \dfrac{K}{2\sqrt{(L/C) + (C/L)}}\left(\dfrac{1}{C} - \dfrac{C}{L^2}\right) = 0 \quad \Leftrightarrow \quad \dfrac{1}{C} = \dfrac{C}{L^2} \quad \Leftrightarrow \quad L^2 = C^2 \quad \Leftrightarrow$

$L = C$. This gives the minimum velocity since $v' < 0$ for $0 < L < C$ and $v' > 0$ for $L > C$.

59. Let x denote the number of $1 decreases in ticket price. Then the ticket price is $12 - 1(x)$, and the average

attendance is $11,000 + 1000(x)$. Now the revenue per game is

$$R(x) = (\text{price per person}) \times (\text{number of people per game})$$

$$= (12 - x)(11,000 + 1000x) = -1000x^2 + 1000x + 132,000$$

for $0 \le x \le 4$ (since the seating capacity is 15,000) $\quad \Rightarrow \quad R'(x) = -2000x + 1000 = 0 \quad \Leftrightarrow \quad x = 0.5$.

This is a maximum since $R''(x) = -2000 < 0$ for all x. Now we must check the value of

$R(x) = (12 - x)(11,000 + 1000x)$ at $x = 0.5$ and at the endpoints of the domain to see which value of x

gives the maximum value of R. $R(0) = (12)(11,000) = 132,000$, $R(0.5) = (11.5)(11,500) = 132,250$, and

$R(4) = (8)(15,000) = 120,000$. Thus, the maximum revenue of $132,250 per game occurs when the average

attendance is 11,500 and the ticket price is $11.50.

61. $f(x) = x^5 - x^4 + 3x^2 - 3x - 2 \;\Rightarrow\; f'(x) = 5x^4 - 4x^3 + 6x - 3$, so

$$x_{n+1} = x_n - \frac{x_n^5 - x_n^4 + 3x_n^2 - 3x_n - 2}{5x_n^4 - 4x_n^3 + 6x_n - 3}. \text{ Now } x_1 = 1 \;\Rightarrow\; x_2 = 1.5 \;\Rightarrow\; x_3 \approx 1.343860 \;\Rightarrow$$

$x_4 \approx 1.300320 \;\Rightarrow\; x_5 \approx 1.297396 \;\Rightarrow\; x_6 \approx 1.297383 \approx x_7$, so the root in $[1, 2]$ is 1.297383, to six

decimal places.

63. $f(t) = \cos t + t - t^2 \;\Rightarrow\; f'(t) = -\sin t + 1 - 2t.$ $f'(t)$ exists

for all t, so to find the maximum of f, we can examine the zeros of f'.

From the graph of f', we see that a good choice for t_1 is $t_1 = 0.3$.

Use $g(t) = -\sin t + 1 - 2t$ and $g'(t) = -\cos t - 2$ to obtain

$t_2 \approx 0.33535293$, $t_3 \approx 0.33541803 \approx t_4$. Since

$f''(t) = -\cos t - 2 < 0$ for all t, $f(0.33541803) \approx 1.16718557$ is

the absolute maximum.

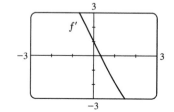

65. $f'(x) = \sqrt{x^5} - 4/\sqrt[5]{x} = x^{5/2} - 4x^{-1/5} \;\Rightarrow\; f(x) = \frac{2}{7}x^{7/2} - 4\left(\frac{5}{4}x^{4/5}\right) + C = \frac{2}{7}x^{7/2} - 5x^{4/5} + C$

67. $f'(x) = e^x - (2/\sqrt{x}) = e^x - 2x^{-1/2} \;\Rightarrow$

$$f(x) = e^x - 2\frac{x^{-1/2+1}}{-1/2+1} + C = e^x - 2\frac{x^{1/2}}{1/2} + C = e^x - 4\sqrt{x} + C$$

69. $f'(t) = 2t - 3\sin t \;\Rightarrow\; f(t) = t^2 + 3\cos t + C.$

$f(0) = 3 + C$ and $f(0) = 5 \;\Rightarrow\; C = 2$, so $f(t) = t^2 + 3\cos t + 2$.

71. $f''(x) = 1 - 6x + 48x^2 \;\Rightarrow\; f'(x) = x - 3x^2 + 16x^3 + C.$ $f'(0) = C$ and $f'(0) = 2 \;\Rightarrow\; C = 2$, so
$f'(x) = x - 3x^2 + 16x^3 + 2$ and hence, $f(x) = \frac{1}{2}x^2 - x^3 + 4x^4 + 2x + D.$ $f(0) = D$ and $f(0) = 1 \;\Rightarrow$
$D = 1$, so $f(x) = \frac{1}{2}x^2 - x^3 + 4x^4 + 2x + 1.$

73. (a) Since f is 0 just to the left of the y-axis,
we must have a minimum of F at the
same place since we are increasing
through $(0, 0)$ on F. There must be a
local maximum to the left of $x = -3$,
since f changes from positive to negative
there.

(b) $f(x) = 0.1e^x + \sin x \;\Rightarrow\; F(x) = 0.1e^x - \cos x + C.$
$F(0) = 0 \;\Rightarrow\; 0.1 - 1 + C = 0 \;\Rightarrow\; C = 0.9$, so
$F(x) = 0.1e^x - \cos x + 0.9.$

(c)

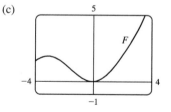

75. Choosing the positive direction to be upward, we have $a(t) = -9.8 \Rightarrow v(t) = -9.8t + v_0$, but $v(0) = 0 = v_0$

$\Rightarrow v(t) = -9.8t = s'(t) \Rightarrow s(t) = -4.9t^2 + s_0$, but $s(0) = s_0 = 500 \Rightarrow s(t) = -4.9t^2 + 500$. When

$s = 0$, $-4.9t^2 + 500 = 0 \Rightarrow t_1 = \sqrt{\frac{500}{4.9}} \approx 10.1 \Rightarrow v(t_1) = -9.8\sqrt{\frac{500}{4.9}} \approx -98.995$ m/s. Since the

canister has been designed to withstand an impact velocity of 100 m/s, the canister will *not burst*.

77. (a)

The cross-sectional area of the rectangular beam is

$A = 2x \cdot 2y = 4xy = 4x\sqrt{100 - x^2}, 0 \le x \le 10$, so

$\dfrac{dA}{dx} = 4x\left(\tfrac{1}{2}\right)\left(100 - x^2\right)^{-1/2}(-2x) + \left(100 - x^2\right)^{1/2} \cdot 4$

$= \dfrac{-4x^2}{\left(100 - x^2\right)^{1/2}} + 4\left(100 - x^2\right)^{1/2}$

$= \dfrac{4\left[-x^2 + \left(100 - x^2\right)\right]}{\left(100 - x^2\right)^{1/2}}.$

$\dfrac{dA}{dx} = 0$ when $-x^2 + \left(100 - x^2\right) = 0 \Rightarrow x^2 = 50 \Rightarrow x = \sqrt{50} \approx 7.07 \Rightarrow$

$y = \sqrt{100 - \left(\sqrt{50}\right)^2} = \sqrt{50}$. Since $A(0) = A(10) = 0$, the rectangle of maximum area is a square.

(b)

The cross-sectional area of each rectangular plank (shaded in the figure) is

$A = 2x\left(y - \sqrt{50}\right) = 2x\left[\sqrt{100 - x^2} - \sqrt{50}\right], 0 \le x \le \sqrt{50}$, so

$\dfrac{dA}{dx} = 2\left(\sqrt{100 - x^2} - \sqrt{50}\right) + 2x\left(\tfrac{1}{2}\right)\left(100 - x^2\right)^{-1/2}(-2x)$

$= 2\left(100 - x^2\right)^{1/2} - 2\sqrt{50} - \dfrac{2x^2}{\left(100 - x^2\right)^{1/2}}$

Set $\dfrac{dA}{dx} = 0$: $\left(100 - x^2\right) - \sqrt{50}\left(100 - x^2\right)^{1/2} - x^2 = 0 \Rightarrow 100 - 2x^2 = \sqrt{50}\left(100 - x^2\right)^{1/2} \Rightarrow$

$10{,}000 - 400x^2 + 4x^4 = 50\left(100 - x^2\right) \Rightarrow 4x^4 - 350x^2 + 5000 = 0 \Rightarrow$

$2x^4 - 175x^2 + 2500 = 0 \Rightarrow x^2 = \dfrac{175 \pm \sqrt{10{,}625}}{4} \approx 69.52$ or $17.98 \Rightarrow x \approx 8.34$ or 4.24.

But $8.34 > \sqrt{50}$, so $x_1 \approx 4.24 \Rightarrow y - \sqrt{50} = \sqrt{100 - x_1^2} - \sqrt{50} \approx 1.99$. Each plank should have

dimensions about $8\frac{1}{2}$ inches by 2 inches.

(c) From the figure in part (a), the width is $2x$ and the depth is $2y$, so the strength is

$S = k(2x)(2y)^2 = 8kxy^2 = 8kx\left(100 - x^2\right) = 800kx - 8kx^3, 0 \le x \le 10$. $dS/dx = 800k - 24kx^2 = 0$

when $24kx^2 = 800k \Rightarrow x^2 = \frac{100}{3} \Rightarrow x = \frac{10}{\sqrt{3}} \Rightarrow y = \sqrt{\frac{200}{3}} = \frac{10\sqrt{2}}{\sqrt{3}} = \sqrt{2}\,x$. Since

$S(0) = S(10) = 0$, the maximum strength occurs when $x = \frac{10}{\sqrt{3}}$. The dimensions should be

$\frac{20}{\sqrt{3}} \approx 11.55$ inches by $\frac{20\sqrt{2}}{\sqrt{3}} \approx 16.33$ inches.

79. (a) $I = \dfrac{k\cos\theta}{d^2} = \dfrac{k(h/d)}{d^2} = k\dfrac{h}{d^3} = k\dfrac{h}{\left(\sqrt{40^2 + h^2}\,\right)^3} = k\dfrac{h}{(1600 + h^2)^{3/2}}$ ⇒

$$\dfrac{dI}{dh} = k\dfrac{(1600 + h^2)^{3/2} - h\frac{3}{2}(1600 + h^2)^{1/2} \cdot 2h}{\left[(1600 + h^2)^{3/2}\right]^2} = \dfrac{k(1600 + h^2)^{1/2}(1600 + h^2 - 3h^2)}{(1600 + h^2)^3}$$

$$= \dfrac{k(1600 - 2h^2)}{(1600 + h^2)^{5/2}} \quad [k \text{ is the constant of proportionality}]$$

Set $dI/dh = 0$: $1600 - 2h^2 = 0$ ⇒ $h^2 = 800$ ⇒ $h = \sqrt{800} = 20\sqrt{2}$. By the First Derivative Test,

I has a local maximum at $h = 20\sqrt{2} \approx 28$ ft.

(b)

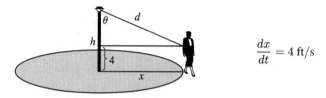

$\dfrac{dx}{dt} = 4 \text{ ft/s}$

$I = \dfrac{k\cos\theta}{d^2} = \dfrac{k[(h-4)/d]}{d^2} = \dfrac{k(h-4)}{d^3} = \dfrac{k(h-4)}{[(h-4)^2 + x^2]^{3/2}} = k(h-4)[(h-4)^2 + x^2]^{-3/2}$

$\dfrac{dI}{dt} = \dfrac{dI}{dx} \cdot \dfrac{dx}{dt} = k(h-4)\left(-\frac{3}{2}\right)[(h-4)^2 + x^2]^{-5/2} \cdot 2x \cdot \dfrac{dx}{dt}$

$\qquad = k(h-4)(-3x)[(h-4)^2 + x^2]^{-5/2} \cdot 4 = \dfrac{-12xk(h-4)}{[(h-4)^2 + x^2]^{5/2}}$

$\dfrac{dI}{dt}\bigg|_{x=40} = -\dfrac{480k(h-4)}{[(h-4)^2 + 1600]^{5/2}}$

81. We first show that $\dfrac{x}{1 + x^2} < \tan^{-1}x$ for $x > 0$. Let $f(x) = \tan^{-1}x - \dfrac{x}{1+x^2}$. Then

$f'(x) = \dfrac{1}{1+x^2} - \dfrac{1(1+x^2) - x(2x)}{(1+x^2)^2} = \dfrac{(1+x^2) - (1-x^2)}{(1+x^2)^2} = \dfrac{2x^2}{(1+x^2)^2} > 0$ for $x > 0$. So $f(x)$ is

increasing on $(0, \infty)$. Hence, $0 < x$ ⇒ $0 = f(0) < f(x) = \tan^{-1}x - \dfrac{x}{1+x^2}$. So $\dfrac{x}{1+x^2} < \tan^{-1}x$

for $0 < x$. We next show that $\tan^{-1}x < x$ for $x > 0$. Let $h(x) = x - \tan^{-1}x$. Then

$h'(x) = 1 - \dfrac{1}{1+x^2} = \dfrac{x^2}{1+x^2} > 0$. Hence, $h(x)$ is increasing on $(0, \infty)$. So for $0 < x$,

$0 = h(0) < h(x) = x - \tan^{-1}x$. Hence, $\tan^{-1}x < x$ for $x > 0$, and we conclude that $\dfrac{x}{1+x^2} < \tan^{-1}x < x$

for $x > 0$.

☐ PROBLEMS PLUS

1. Let $y = f(x) = e^{-x^2}$. The area of the rectangle under the curve from $-x$ to x is $A(x) = 2xe^{-x^2}$ where $x \geq 0$.

We maximize $A(x)$: $A'(x) = 2e^{-x^2} - 4x^2e^{-x^2} = 2e^{-x^2}\left(1 - 2x^2\right) = 0 \Rightarrow x = \frac{1}{\sqrt{2}}$. This gives a maximum

since $A'(x) > 0$ for $0 \leq x < \frac{1}{\sqrt{2}}$ and $A'(x) < 0$ for $x > \frac{1}{\sqrt{2}}$. We next determine the points of inflection of $f(x)$.

Notice that $f'(x) = -2xe^{-x^2} = -A(x)$. So $f''(x) = -A'(x)$ and hence, $f''(x) < 0$ for $-\frac{1}{\sqrt{2}} < x < \frac{1}{\sqrt{2}}$ and

$f''(x) > 0$ for $x < -\frac{1}{\sqrt{2}}$ and $x > \frac{1}{\sqrt{2}}$. So $f(x)$ changes concavity at $x = \pm\frac{1}{\sqrt{2}}$, and the two vertices of the

rectangle of largest area are at the inflection points.

3. First, we recognize some symmetry in the inequality: $\dfrac{e^{x+y}}{xy} \geq e^2 \Leftrightarrow \dfrac{e^x}{x} \cdot \dfrac{e^y}{y} \geq e \cdot e$. This suggests that we

need to show that $\dfrac{e^x}{x} \geq e$ for $x > 0$. If we can do this, then the inequality $\dfrac{e^y}{y} \geq e$ is true, and the given inequality

follows. $f(x) = \dfrac{e^x}{x} \Rightarrow f'(x) = \dfrac{xe^x - e^x}{x^2} = \dfrac{e^x(x-1)}{x^2} = 0 \Rightarrow x = 1$. By the First Derivative Test, we

have a minimum of $f(1) = e$, so $e^x/x \geq e$ for all x.

5. First we show that $x(1-x) \leq \frac{1}{4}$ for all x. Let $f(x) = x(1-x) = x - x^2$. Then $f'(x) = 1 - 2x$. This is 0 when

$x = \frac{1}{2}$ and $f'(x) > 0$ for $x < \frac{1}{2}$, $f'(x) < 0$ for $x > \frac{1}{2}$, so the absolute maximum of f is $f\left(\frac{1}{2}\right) = \frac{1}{4}$. Thus,

$x(1-x) \leq \frac{1}{4}$ for all x.

Now suppose that the given assertion is false, that is, $a(1-b) > \frac{1}{4}$ and $b(1-a) > \frac{1}{4}$. Multiply these

inequalities: $a(1-b)b(1-a) > \frac{1}{16} \Rightarrow [a(1-a)][b(1-b)] > \frac{1}{16}$. But we know that $a(1-a) \leq \frac{1}{4}$ and

$b(1-b) \leq \frac{1}{4} \Rightarrow [a(1-a)][b(1-b)] \leq \frac{1}{16}$. Thus, we have a contradiction, so the given assertion is proved.

7. Differentiating $x^2 + xy + y^2 = 12$ implicitly with respect to x gives $2x + y + x\dfrac{dy}{dx} + 2y\dfrac{dy}{dx} = 0$, so

$\dfrac{dy}{dx} = -\dfrac{2x+y}{x+2y}$. At a highest or lowest point, $\dfrac{dy}{dx} = 0 \Leftrightarrow y = -2x$. Substituting $-2x$ for y in the original

equation gives $x^2 + x(-2x) + (-2x)^2 = 12$, so $3x^2 = 12$ and $x = \pm2$. If $x = 2$, then $y = -2x = -4$, and if

$x = -2$ then $y = 4$. Thus, the highest and lowest points are $(-2, 4)$ and $(2, -4)$.

9. $A = (x_1, x_1^2)$ and $B = (x_2, x_2^2)$, where x_1 and x_2 are the solutions of the quadratic equation $x^2 = mx + b$. Let

$P = (x, x^2)$ and set $A_1 = (x_1, 0)$, $B_1 = (x_2, 0)$, and $P_1 = (x, 0)$. Let $f(x)$ denote the area of triangle PAB.

Then $f(x)$ can be expressed in terms of the areas of three trapezoids as follows:

$$f(x) = \text{area}\,(A_1 ABB_1) - \text{area}\,(A_1 APP_1) - \text{area}\,(B_1 BPP_1)$$

$$= \tfrac{1}{2}\left(x_1^2 + x_2^2\right)(x_2 - x_1) - \tfrac{1}{2}\left(x_1^2 + x^2\right)(x - x_1) - \tfrac{1}{2}\left(x^2 + x_2^2\right)(x_2 - x)$$

After expanding and canceling terms, we get

$$f(x) = \tfrac{1}{2}\left(x_2 x_1^2 - x_1 x_2^2 - x x_1^2 + x_1 x^2 - x_2 x^2 + x x_2^2\right) = \tfrac{1}{2}\left[x_1^2(x_2 - x) + x_2^2(x - x_1) + x^2(x_1 - x_2)\right]$$

$$f'(x) = \tfrac{1}{2}\left[-x_1^2 + x_2^2 + 2x(x_1 - x_2)\right]. \quad f''(x) = \tfrac{1}{2}\left[2(x_1 - x_2)\right] = x_1 - x_2 < 0 \text{ since } x_2 > x_1.$$

$$f'(x) = 0 \;\Rightarrow\; 2x(x_1 - x_2) = x_1^2 - x_2^2 \;\Rightarrow\; x_P = \tfrac{1}{2}(x_1 + x_2).$$

$$f(x_P) = \tfrac{1}{2}\left(x_1^2\left[\tfrac{1}{2}(x_2 - x_1)\right] + x_2^2\left[\tfrac{1}{2}(x_2 - x_1)\right] + \tfrac{1}{4}(x_1 + x_2)^2(x_1 - x_2)\right)$$

$$= \tfrac{1}{2}\left[\tfrac{1}{2}(x_2 - x_1)(x_1^2 + x_2^2) - \tfrac{1}{4}(x_2 - x_1)(x_1 + x_2)^2\right]$$

$$= \tfrac{1}{8}(x_2 - x_1)\left[2(x_1^2 + x_2^2) - (x_1^2 + 2x_1 x_2 + x_2^2)\right]$$

$$= \tfrac{1}{8}(x_2 - x_1)(x_1^2 - 2x_1 x_2 + x_2^2) = \tfrac{1}{8}(x_2 - x_1)(x_1 - x_2)^2 = \tfrac{1}{8}(x_2 - x_1)(x_2 - x_1)^2$$

$$= \tfrac{1}{8}(x_2 - x_1)^3$$

To put this in terms of m and b, we solve the system $y = x_1^2$ and $y = mx_1 + b$, giving us $x_1^2 - mx_1 - b = 0 \;\Rightarrow\;$

$x_1 = \tfrac{1}{2}\left(m - \sqrt{m^2 + 4b}\right)$. Similarly, $x_2 = \tfrac{1}{2}\left(m + \sqrt{m^2 + 4b}\right)$. The area is then

$\tfrac{1}{8}(x_2 - x_1)^3 = \tfrac{1}{8}\left(\sqrt{m^2 + 4b}\right)^3$, and is attained at the point $P(x_P, x_P^2) = P\left(\tfrac{1}{2}m, \tfrac{1}{4}m^2\right)$.

Note: Another way to get an expression for $f(x)$ is to use the formula for an area of a triangle in terms of the

coordinates of the vertices: $f(x) = \tfrac{1}{2}\left[\left(x_2 x_1^2 - x_1 x_2^2\right) + \left(x_1 x^2 - x x_1^2\right) + \left(x x_2^2 - x_2 x^2\right)\right]$.

11. $f(x) = \left(a^2 + a - 6\right)\cos 2x + (a - 2)x + \cos 1 \;\Rightarrow\; f'(x) = -\left(a^2 + a - 6\right)\sin 2x\,(2) + (a - 2)$.

The derivative exists for all x, so the only possible critical points will occur where $f'(x) = 0 \;\Leftrightarrow$

$2(a - 2)(a + 3)\sin 2x = a - 2 \;\Leftrightarrow\;$ either $a = 2$ or $2(a + 3)\sin 2x = 1$, with the latter implying that

$\sin 2x = \dfrac{1}{2(a + 3)}$. Since the range of $\sin 2x$ is $[-1, 1]$, this equation has no solution whenever either

$\dfrac{1}{2(a + 3)} < -1$ or $\dfrac{1}{2(a + 3)} > 1$. Solving these inequalities, we get $-\tfrac{7}{2} < a < -\tfrac{5}{2}$.

13. (a) Let $y = |AD|$, $x = |AB|$, and $1/x = |AC|$, so that $|AB| \cdot |AC| = 1$.

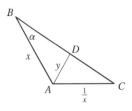

We compute the area $\mathcal{A}$ of $\triangle ABC$ in two ways. First,

$\mathcal{A} = \frac{1}{2}|AB||AC|\sin\frac{2\pi}{3} = \frac{1}{2} \cdot 1 \cdot \frac{\sqrt{3}}{2} = \frac{\sqrt{3}}{4}$. Second,

$\mathcal{A} = (\text{area of } \triangle ABD) + (\text{area of } \triangle ACD)$

$\quad = \frac{1}{2}|AB||AD|\sin\frac{\pi}{3} + \frac{1}{2}|AD||AC|\sin\frac{\pi}{3}$

$\quad = \frac{1}{2}xy\frac{\sqrt{3}}{2} + \frac{1}{2}y(1/x)\frac{\sqrt{3}}{2} = \frac{\sqrt{3}}{4}y(x + 1/x)$

Equating the two expressions for the area, we get $\frac{\sqrt{3}}{4}y\left(x + \frac{1}{x}\right) = \frac{\sqrt{3}}{4} \Leftrightarrow y = \frac{1}{x + 1/x} = \frac{x}{x^2 + 1}$, $x > 0$.

Another method: Use the Law of Sines on the triangles ABD and ABC. In $\triangle ABD$, we have

$\angle A + \angle B + \angle D = 180° \Leftrightarrow 60° + \alpha + \angle D = 180° \Leftrightarrow \angle D = 120° - \alpha$. Thus,

$\frac{x}{y} = \frac{\sin(120° - \alpha)}{\sin\alpha} = \frac{\sin 120° \cos\alpha - \cos 120° \sin\alpha}{\sin\alpha} = \frac{\frac{\sqrt{3}}{2}\cos\alpha + \frac{1}{2}\sin\alpha}{\sin\alpha} \Rightarrow \frac{x}{y} = \frac{\sqrt{3}}{2}\cot\alpha + \frac{1}{2}$,

and by a similar argument with $\triangle ABC$, $\frac{\sqrt{3}}{2}\cot\alpha = x^2 + \frac{1}{2}$. Eliminating $\cot\alpha$ gives $\frac{x}{y} = \left(x^2 + \frac{1}{2}\right) + \frac{1}{2} \Rightarrow$

$y = \frac{x}{x^2 + 1}$, $x > 0$.

(b) We differentiate our expression for y with respect to x to find the maximum:

$\frac{dy}{dx} = \frac{(x^2 + 1) - x(2x)}{(x^2 + 1)^2} = \frac{1 - x^2}{(x^2 + 1)^2} = 0$ when $x = 1$. This indicates a maximum by the First Derivative

Test, since $y'(x) > 0$ for $0 < x < 1$ and $y'(x) < 0$ for $x > 1$, so the maximum value of y is $y(1) = \frac{1}{2}$.

15. Suppose that the curve $y = a^x$ intersects the line $y = x$. Then $a^{x_0} = x_0$ for some $x_0 > 0$, and hence $a = x_0^{1/x_0}$.

We find the maximum value of $g(x) = x^{1/x}$, > 0, because if a is larger than the maximum

value of this function, then the curve $y = a^x$ does not intersect the line $y = x$.

$g'(x) = e^{(1/x)\ln x}\left(-\frac{1}{x^2}\ln x + \frac{1}{x} \cdot \frac{1}{x}\right) = x^{1/x}\left(\frac{1}{x^2}\right)(1 - \ln x)$. This is 0 only where $x = e$, and for $0 < x < e$,

$f'(x) > 0$, while for $x > e$, $f'(x) < 0$, so g has an absolute maximum of $g(e) = e^{1/e}$. So if $y = a^x$ intersects

$y = x$, we must have $0 < a \le e^{1/e}$. Conversely, suppose that $0 < a \le e^{1/e}$. Then $a^e \le e$, so the graph of $y = a^x$

lies below or touches the graph of $y = x$ at $x = e$. Also $a^0 = 1 > 0$, so the graph of $y = a^x$ lies above that of

$y = x$ at $x = 0$. Therefore, by the Intermediate Value Theorem, the graphs of $y = a^x$ and $y = x$ must intersect

somewhere between $x = 0$ and $x = e$.

17. Note that $f(0) = 0$, so for $x \neq 0$, $\left| \dfrac{f(x) - f(0)}{x - 0} \right| = \left| \dfrac{f(x)}{x} \right| = \dfrac{|f(x)|}{|x|} \leq \dfrac{|\sin x|}{|x|} = \dfrac{\sin x}{x}$.

Therefore, $|f'(0)| = \left| \lim\limits_{x \to 0} \dfrac{f(x) - f(0)}{x - 0} \right| = \lim\limits_{x \to 0} \left| \dfrac{f(x) - f(0)}{x - 0} \right| \leq \lim\limits_{x \to 0} \dfrac{\sin x}{x} = 1$. But

$f'(x) = a_1 \cos x + 2a_2 \cos 2x + \cdots + na_n \cos nx$, so $|f'(0)| = |a_1 + 2a_2 + \cdots + na_n| \leq 1$.

Another solution: We are given that $\left| \sum_{k=1}^{n} a_k \sin kx \right| \leq |\sin x|$. So for x close to 0, and $x \neq 0$, we have

$\left| \sum_{k=1}^{n} a_k \dfrac{\sin kx}{\sin x} \right| \leq 1 \quad \Rightarrow \quad \lim\limits_{x \to 0} \left| \sum_{k=1}^{n} a_k \dfrac{\sin kx}{\sin x} \right| \leq 1 \quad \Rightarrow \quad \left| \sum_{k=1}^{n} a_k \lim\limits_{x \to 0} \dfrac{\sin kx}{\sin x} \right| \leq 1$. But by l'Hospital's Rule,

$\lim\limits_{x \to 0} \dfrac{\sin kx}{\sin x} = \lim\limits_{x \to 0} \dfrac{k \cos kx}{\cos x} = k$, so $\left| \sum_{k=1}^{n} ka_k \right| \leq 1$.

19. (a) Distance = rate × time, so time = distance/rate. $T_1 = \dfrac{D}{c_1}$,

$T_2 = \dfrac{2|PR|}{c_1} + \dfrac{|RS|}{c_2} = \dfrac{2h \sec \theta}{c_1} + \dfrac{D - 2h \tan \theta}{c_2}$, $T_3 = \dfrac{2\sqrt{h^2 + D^2/4}}{c_1} = \dfrac{\sqrt{4h^2 + D^2}}{c_1}$.

(b) $\dfrac{dT_2}{d\theta} = \dfrac{2h}{c_1} \cdot \sec \theta \tan \theta - \dfrac{2h}{c_2} \sec^2 \theta = 0$ when $2h \sec \theta \left(\dfrac{1}{c_1} \tan \theta - \dfrac{1}{c_2} \sec \theta \right) = 0 \quad \Rightarrow$

$\dfrac{1}{c_1} \dfrac{\sin \theta}{\cos \theta} - \dfrac{1}{c_2} \dfrac{1}{\cos \theta} = 0 \quad \Rightarrow \quad \dfrac{\sin \theta}{c_1 \cos \theta} = \dfrac{1}{c_2 \cos \theta} \quad \Rightarrow \quad \sin \theta = \dfrac{c_1}{c_2}$. The First Derivative Test shows that

this gives a minimum.

(c) Using part (a) with $D = 1$ and $T_1 = 0.26$, we have $T_1 = \dfrac{D}{c_1} \quad \Rightarrow$

$c_1 = \dfrac{1}{0.26} \approx 3.85$ km/s. $T_3 = \dfrac{\sqrt{4h^2 + D^2}}{c_1} \quad \Rightarrow \quad 4h^2 + D^2 = T_3^2 c_1^2 \quad \Rightarrow$

$h = \frac{1}{2} \sqrt{T_3^2 c_1^2 - D^2} = \frac{1}{2} \sqrt{(0.34)^2 (1/0.26)^2 - 1^2} \approx 0.42$ km. To find c_2, we use $\sin \theta = \dfrac{c_1}{c_2}$ from part (b)

and $T_2 = \dfrac{2h \sec \theta}{c_1} + \dfrac{D - 2h \tan \theta}{c_2}$ from part (a). From the figure,

$\sin \theta = \dfrac{c_1}{c_2} \quad \Rightarrow \quad \sec \theta = \dfrac{c_2}{\sqrt{c_2^2 - c_1^2}}$ and $\tan \theta = \dfrac{c_1}{\sqrt{c_2^2 - c_1^2}}$, so

$T_2 = \dfrac{2hc_2}{c_1 \sqrt{c_2^2 - c_1^2}} + \dfrac{D\sqrt{c_2^2 - c_1^2} - 2hc_1}{c_2 \sqrt{c_2^2 - c_1^2}}$.

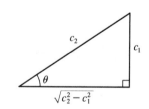

Using the values for T_2 [given as 0.32], h, c_1, and D, we can graph

$Y_1 = T_2$ and $Y_2 = \dfrac{2hc_2}{c_1 \sqrt{c_2^2 - c_1^2}} + \dfrac{D\sqrt{c_2^2 - c_1^2} - 2hc_1}{c_2 \sqrt{c_2^2 - c_1^2}}$ and find their intersection points. Doing so gives us

$c_2 \approx 4.10$ and 7.66, but if $c_2 = 4.10$, then $\theta = \arcsin(c_1/c_2) \approx 69.6°$, which implies that point S is to the left

of point R in the diagram. So $c_2 = 7.66$ km/s.

21.

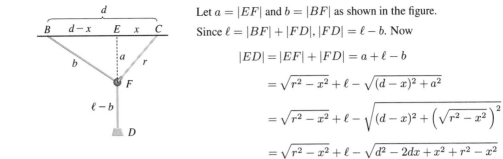

Let $a = |EF|$ and $b = |BF|$ as shown in the figure. Since $\ell = |BF| + |FD|$, $|FD| = \ell - b$. Now

$$|ED| = |EF| + |FD| = a + \ell - b$$

$$= \sqrt{r^2 - x^2} + \ell - \sqrt{(d-x)^2 + a^2}$$

$$= \sqrt{r^2 - x^2} + \ell - \sqrt{(d-x)^2 + \left(\sqrt{r^2 - x^2}\right)^2}$$

$$= \sqrt{r^2 - x^2} + \ell - \sqrt{d^2 - 2dx + x^2 + r^2 - x^2}$$

Let $f(x) = \sqrt{r^2 - x^2} + \ell - \sqrt{d^2 + r^2 - 2dx}$.

$$f'(x) = \tfrac{1}{2}\left(r^2 - x^2\right)^{-1/2}(-2x) - \tfrac{1}{2}\left(d^2 + r^2 - 2dx\right)^{-1/2}(-2d) = \frac{-x}{\sqrt{r^2 - x^2}} + \frac{d}{\sqrt{d^2 + r^2 - 2dx}}.$$

$f'(x) = 0 \ \Rightarrow\ \dfrac{x}{\sqrt{r^2 - x^2}} = \dfrac{d}{\sqrt{d^2 + r^2 - 2dx}} \ \Rightarrow\ \dfrac{x^2}{r^2 - x^2} = \dfrac{d^2}{d^2 + r^2 - 2dx} \ \Rightarrow$

$d^2 x^2 + r^2 x^2 - 2dx^3 = d^2 r^2 - d^2 x^2 \ \Rightarrow\ 0 = 2dx^3 - 2d^2 x^2 - r^2 x^2 + d^2 r^2 \ \Rightarrow$

$0 = 2dx^2(x - d) - r^2\left(x^2 - d^2\right) \ \Rightarrow\ 0 = 2dx^2(x - d) - r^2(x + d)(x - d) \ \Rightarrow$

$0 = (x - d)\left[2dx^2 - r^2(x + d)\right]$

But $d > r > x$, so $x \neq d$. Thus, we solve $2dx^2 - r^2 x - dr^2 = 0$ for x:

$$x = \frac{-(-r^2) \pm \sqrt{(-r^2)^2 - 4(2d)(-dr^2)}}{2(2d)} = \frac{r^2 \pm \sqrt{r^4 + 8d^2 r^2}}{4d}.$$ Because $\sqrt{r^4 + 8d^2 r^2} > r^2$, the "negative" can be discarded. Thus,

$$x = \frac{r^2 + \sqrt{r^2}\sqrt{r^2 + 8d^2}}{4d} = \frac{r^2 + r\sqrt{r^2 + 8d^2}}{4d} \quad (r > 0)$$

$$= \frac{r}{4d}\left(r + \sqrt{r^2 + 8d^2}\right)$$

The maximum value of $|ED|$ occurs at this value of x.

23. $V = \tfrac{4}{3}\pi r^3 \ \Leftrightarrow\ \dfrac{dV}{dt} = 4\pi r^2 \dfrac{dr}{dt}$. But $\dfrac{dV}{dt}$ is proportional to the surface area, so $\dfrac{dV}{dt} = k \cdot 4\pi r^2$ for some

constant k. Therefore, $4\pi r^2 \dfrac{dr}{dt} = k \cdot 4\pi r^2 \ \Leftrightarrow\ \dfrac{dr}{dt} = k = $ constant. An antiderivative of k with respect to t is

kt, so $r = kt + C$. When $t = 0$, the radius r must equal the original radius r_0, so $C = r_0$, and $r = kt + r_0$. To find

k we use the fact that when $t = 3$, $r = 3k + r_0$ and $V = \tfrac{1}{2}V_0 \ \Rightarrow\ \tfrac{4}{3}\pi(3k + r_0)^3 = \tfrac{1}{2} \cdot \tfrac{4}{3}\pi r_0^3 \ \Rightarrow$

$(3k + r_0)^3 = \tfrac{1}{2}r_0^3 \ \Rightarrow\ 3k + r_0 = \dfrac{1}{\sqrt[3]{2}}r_0 \ \Rightarrow\ k = \tfrac{1}{3}r_0\left(\dfrac{1}{\sqrt[3]{2}} - 1\right)$. Since $r = kt + r_0$,

$r = \tfrac{1}{3}r_0\left(\dfrac{1}{\sqrt[3]{2}} - 1\right)t + r_0$. When the snowball has melted completely we have $r = 0 \ \Rightarrow$

$\tfrac{1}{3}r_0\left(\dfrac{1}{\sqrt[3]{2}} - 1\right)t + r_0 = 0$ which gives $t = \dfrac{3\sqrt[3]{2}}{\sqrt[3]{2} - 1}$. Hence, it takes $\dfrac{3\sqrt[3]{2}}{\sqrt[3]{2} - 1} - 3 = \dfrac{3}{\sqrt[3]{2} - 1} \approx 11$ h 33 min

longer.

5 □ INTEGRALS

5.1 Areas and Distances

1. (a) Since f is *increasing*, we can obtain a *lower* estimate by using *left* endpoints. We are instructed to use five rectangles, so $n = 5$.

$$L_5 = \sum_{i=1}^{5} f(x_{i-1}) \, \Delta x \qquad [\Delta x = \tfrac{b-a}{n} = \tfrac{10-0}{5} = 2]$$

$$= f(x_0) \cdot 2 + f(x_1) \cdot 2 + f(x_2) \cdot 2 + f(x_3) \cdot 2 + f(x_4) \cdot 2$$

$$= 2 \left[f(0) + f(2) + f(4) + f(6) + f(8) \right]$$

$$\approx 2(1 + 3 + 4.3 + 5.4 + 6.3) = 2(20) = 40$$

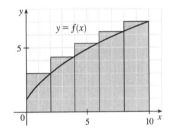

Since f is *increasing*, we can obtain an *upper* estimate by using *right* endpoints.

$$R_5 = \sum_{i=1}^{5} f(x_i) \, \Delta x$$

$$= 2 \left[f(x_1) + f(x_2) + f(x_3) + f(x_4) + f(x_5) \right]$$

$$= 2 \left[f(2) + f(4) + f(6) + f(8) + f(10) \right]$$

$$\approx 2(3 + 4.3 + 5.4 + 6.3 + 7) = 2(26) = 52$$

Comparing R_5 to L_5, we see that we have added the area of the rightmost upper rectangle, $f(10) \cdot 2$, to the sum and subtracted the area of the leftmost lower rectangle, $f(0) \cdot 2$, from the sum.

(b) $L_{10} = \sum_{i=1}^{10} f(x_{i-1}) \, \Delta x \qquad [\Delta x = \tfrac{10-0}{10} = 1]$

$$= 1 \left[f(x_0) + f(x_1) + \cdots + f(x_9) \right]$$

$$= f(0) + f(1) + \cdots + f(9)$$

$$\approx 1 + 2.1 + 3 + 3.7 + 4.3 + 4.9 + 5.4 + 5.8 + 6.3 + 6.7$$

$$= 43.2$$

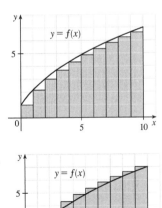

$R_{10} = \sum_{i=1}^{10} f(x_i) \, \Delta x = f(1) + f(2) + \cdots + f(10)$

$$= L_{10} + 1 \cdot f(10) - 1 \cdot f(0) \quad \begin{bmatrix} \text{add rightmost upper rectangle,} \\ \text{subtract leftmost lower rectangle} \end{bmatrix}$$

$$= 43.2 + 7 - 1 = 49.2$$

3. (a) $R_4 = \sum\limits_{i=1}^{4} f(x_i)\,\Delta x \quad [\Delta x = \frac{5-1}{4} = 1]$

$= f(x_1) \cdot 1 + f(x_2) \cdot 1 + f(x_3) \cdot 1 + f(x_4) \cdot 1$

$= f(2) + f(3) + f(4) + f(5)$

$= \frac{1}{2} + \frac{1}{3} + \frac{1}{4} + \frac{1}{5} = \frac{77}{60} = 1.28\overline{3}$

Since f is *decreasing* on $[1,5]$, an *underestimate* is obtained by using the *right* endpoint approximation, R_4.

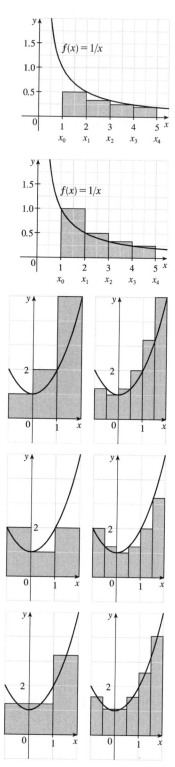

(b) $L_4 = \sum\limits_{i=1}^{4} f(x_{i-1})\,\Delta x$

$= f(1) + f(2) + f(3) + f(4)$

$= 1 + \frac{1}{2} + \frac{1}{3} + \frac{1}{4} = \frac{25}{12} = 2.08\overline{3}$

L_4 is an overestimate. Alternatively, we could just add the area of the leftmost upper rectangle and subtract the area of the rightmost lower rectangle; that is, $L_4 = R_4 + f(1) \cdot 1 - f(5) \cdot 1$.

5. (a) $f(x) = 1 + x^2$ and $\Delta x = \dfrac{2-(-1)}{3} = 1 \ \Rightarrow$

$R_3 = 1 \cdot f(0) + 1 \cdot f(1) + 1 \cdot f(2) = 1 \cdot 1 + 1 \cdot 2 + 1 \cdot 5 = 8.$

$\Delta x = \dfrac{2-(-1)}{6} = 0.5 \ \Rightarrow$

$R_6 = 0.5[f(-0.5) + f(0) + f(0.5) + f(1) + f(1.5) + f(2)]$

$= 0.5(1.25 + 1 + 1.25 + 2 + 3.25 + 5)$

$= 0.5(13.75) = 6.875$

(b) $L_3 = 1 \cdot f(-1) + 1 \cdot f(0) + 1 \cdot f(1) = 1 \cdot 2 + 1 \cdot 1 + 1 \cdot 2 = 5$

$L_6 = 0.5[f(-1) + f(-0.5) + f(0) + f(0.5) + f(1) + f(1.5)]$

$= 0.5(2 + 1.25 + 1 + 1.25 + 2 + 3.25)$

$= 0.5(10.75) = 5.375$

(c) $M_3 = 1 \cdot f(-0.5) + 1 \cdot f(0.5) + 1 \cdot f(1.5)$

$= 1 \cdot 1.25 + 1 \cdot 1.25 + 1 \cdot 3.25 = 5.75$

$M_6 = 0.5[f(-0.75) + f(-0.25) + f(0.25)$

$\qquad + f(0.75) + f(1.25) + f(1.75)]$

$= 0.5(1.5625 + 1.0625 + 1.0625 + 1.5625 + 2.5625 + 4.0625)$

$= 0.5(11.875) = 5.9375$

(d) M_6 appears to be the best estimate.

7. Here is one possible algorithm (ordered sequence of operations) for calculating the sums:

1 Let SUM = 0, X_MIN = 0, X_MAX = π, N = 10 (or 30 or 50, depending on which sum we are calculating), DELTA_X = (X_MAX − X_MIN)/N, and RIGHT_ENDPOINT = X_MIN + DELTA_X.

2 Repeat steps 2a, 2b in sequence until RIGHT_ENDPOINT > X_MAX.

2a Add sin (RIGHT_ENDPOINT) to SUM.

2b Add DELTA_X to RIGHT_ENDPOINT.

At the end of this procedure, (DELTA_X) · (SUM) is equal to the answer we are looking for. We find that

$$R_{10} = \frac{\pi}{10} \sum_{i=1}^{10} \sin\left(\frac{i\pi}{10}\right) \approx 1.9835, \; R_{30} = \frac{\pi}{30} \sum_{i=1}^{30} \sin\left(\frac{i\pi}{30}\right) \approx 1.9982, \text{ and } R_{50} = \frac{\pi}{50} \sum_{i=1}^{50} \sin\left(\frac{i\pi}{50}\right) \approx 1.9993.$$

It appears that the exact area is 2.

Shown below is program SUMRIGHT and its output from a TI-83 Plus calculator. To generalize the program, we have input (rather than assigned) values for Xmin, Xmax, and N. Also, the function, sin x, is assigned to Y_1, enabling us to evaluate any right sum merely by changing Y_1 and running the program.

9. In Maple, we have to perform a number of steps before getting a numerical answer. After loading the student package [command: `with(student);`] we use the command
`left_sum:=leftsum(x^(1/2),x=1..4,10 [or 30, or 50]);` which gives us the expression in summation notation. To get a numerical approximation to the sum, we use `evalf(left_sum);`. Mathematica does not have a special command for these sums, so we must type them in manually. For example, the first left sum is given by `(3/10)*Sum[Sqrt[1+3(i-1)/10],{i,1,10}]`, and we use the `N` command on the resulting output to get a numerical approximation.

In Derive, we use the `LEFT_RIEMANN` command to get the left sums, but must define the right sums ourselves. (We can define a new function using `LEFT_RIEMANN` with k ranging from 1 to n instead of from 0 to $n − 1$.)

(a) With $f(x) = \sqrt{x}$, $1 \le x \le 4$, the left sums are of the form $L_n = \frac{3}{n} \sum_{i=1}^{n} \sqrt{1 + \frac{3(i-1)}{n}}$. Specifically,

$L_{10} \approx 4.5148$, $L_{30} \approx 4.6165$, and $L_{50} \approx 4.6366$. The right sums are of the form $R_n = \frac{3}{n} \sum_{i=1}^{n} \sqrt{1 + \frac{3i}{n}}$.

Specifically, $R_{10} \approx 4.8148$, $R_{30} \approx 4.7165$, and $R_{50} \approx 4.6966$.

(b) In Maple, we use the `leftbox` and `rightbox` commands (with the same arguments as `leftsum` and `rightsum` above) to generate the graphs.

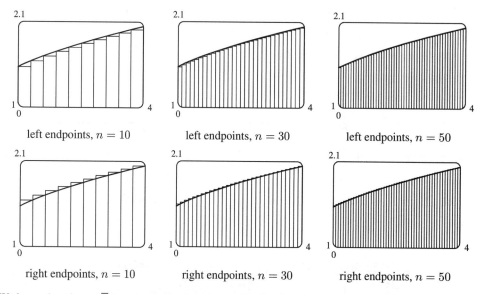

left endpoints, $n = 10$	left endpoints, $n = 30$	left endpoints, $n = 50$
right endpoints, $n = 10$	right endpoints, $n = 30$	right endpoints, $n = 50$

(c) We know that since $\sqrt{x}$ is an increasing function on $(1, 4)$, all of the left sums are smaller than the actual area, and all of the right sums are larger than the actual area. Since the left sum with $n = 50$ is about $4.637 > 4.6$ and the right sum with $n = 50$ is about $4.697 < 4.7$, we conclude that $4.6 < L_{50} <$ exact area $< R_{50} < 4.7$, so the exact area is between 4.6 and 4.7.

11. Since v is an increasing function, L_6 will give us a lower estimate and R_6 will give us an upper estimate.

$$L_6 = (0 \text{ ft/s})(0.5 \text{ s}) + (6.2)(0.5) + (10.8)(0.5) + (14.9)(0.5) + (18.1)(0.5) + (19.4)(0.5)$$
$$= 0.5(69.4) = 34.7 \text{ ft}$$

$$R_6 = 0.5(6.2 + 10.8 + 14.9 + 18.1 + 19.4 + 20.2) = 0.5(89.6) = 44.8 \text{ ft}$$

13. Lower estimate for oil leakage: $R_5 = (7.6 + 6.8 + 6.2 + 5.7 + 5.3)(2) = (31.6)(2) = 63.2 \text{ L}$.

Upper estimate for oil leakage: $L_5 = (8.7 + 7.6 + 6.8 + 6.2 + 5.7)(2) = (35)(2) = 70 \text{ L}$.

15. For a decreasing function, using left endpoints gives us an overestimate and using right endpoints results in an underestimate. We will use M_6 to get an estimate. $\Delta t = 1$, so

$$M_6 = 1[v(0.5) + v(1.5) + v(2.5) + v(3.5) + v(4.5) + v(5.5)]$$

$$\approx 55 + 40 + 28 + 18 + 10 + 4 = 155 \text{ ft}$$

For a very rough check on the above calculation, we can draw a line from $(0, 70)$ to $(6, 0)$ and calculate the area of the triangle: $\frac{1}{2}(70)(6) = 210$. This is clearly an overestimate, so our midpoint estimate of 155 is reasonable.

17. $f(x) = \sqrt[4]{x}$, $1 \le x \le 16$. $\Delta x = (16-1)/n = 15/n$ and $x_i = 1 + i\,\Delta x = 1 + 15i/n$.

$$A = \lim_{n \to \infty} R_n = \lim_{n \to \infty} \sum_{i=1}^{n} f(x_i)\,\Delta x = \lim_{n \to \infty} \sum_{i=1}^{n} \sqrt[4]{1 + \tfrac{15i}{n}} \cdot \tfrac{15}{n}.$$

19. $f(x) = x\cos x$, $0 \le x \le \frac{\pi}{2}$. $\Delta x = (\frac{\pi}{2} - 0)/n = \frac{\pi}{2}/n$ and $x_i = 0 + i\,\Delta x = \frac{\pi}{2}i/n$.

$$A = \lim_{n \to \infty} R_n = \lim_{n \to \infty} \sum_{i=1}^{n} f(x_i)\,\Delta x = \lim_{n \to \infty} \sum_{i=1}^{n} \frac{i\pi}{2n} \cos\!\left(\frac{i\pi}{2n}\right) \cdot \frac{\pi}{2n}.$$

21. $\displaystyle\lim_{n \to \infty} \sum_{i=1}^{n} \frac{\pi}{4n}\tan\frac{i\pi}{4n}$ can be interpreted as the area of the region lying under the graph of $y = \tan x$ on the interval

$\left[0, \frac{\pi}{4}\right]$, since for $y = \tan x$ on $\left[0, \frac{\pi}{4}\right]$ with $\Delta x = \dfrac{\pi/4 - 0}{n} = \dfrac{\pi}{4n}$, $x_i = 0 + i\,\Delta x = \dfrac{i\pi}{4n}$, and $x_i^* = x_i$, the

expression for the area is $A = \displaystyle\lim_{n \to \infty} \sum_{i=1}^{n} f\left(x_i^*\right)\Delta x = \lim_{n \to \infty} \sum_{i=1}^{n} \tan\!\left(\frac{i\pi}{4n}\right) \frac{\pi}{4n}$. Note that this answer is not unique,

since the expression for the area is the same for the function $y = \tan(x - k\pi)$ on the interval $\left[k\pi, k\pi + \frac{\pi}{4}\right]$, where

k is any integer.

23. (a) $y = f(x) = x^5$. $\Delta x = \dfrac{2-0}{n} = \dfrac{2}{n}$ and $x_i = 0 + i\,\Delta x = \dfrac{2i}{n}$.

$$A = \lim_{n \to \infty} R_n = \lim_{n \to \infty} \sum_{i=1}^{n} f(x_i)\,\Delta x = \lim_{n \to \infty} \sum_{i=1}^{n} \left(\frac{2i}{n}\right)^5 \cdot \frac{2}{n} = \lim_{n \to \infty} \sum_{i=1}^{n} \frac{32i^5}{n^5} \cdot \frac{2}{n} = \lim_{n \to \infty} \frac{64}{n^6} \sum_{i=1}^{n} i^5.$$

(b) $\displaystyle\sum_{i=1}^{n} i^5 \overset{\text{CAS}}{=} \frac{n^2(n+1)^2(2n^2 + 2n - 1)}{12}$

(c) $\displaystyle\lim_{n \to \infty} \frac{64}{n^6} \cdot \frac{n^2(n+1)^2(2n^2 + 2n - 1)}{12} = \frac{64}{12}\lim_{n \to \infty}\frac{(n^2 + 2n + 1)(2n^2 + 2n - 1)}{n^2 \cdot n^2}$

$$= \frac{16}{3}\lim_{n \to \infty}\left(1 + \frac{2}{n} + \frac{1}{n^2}\right)\left(2 + \frac{2}{n} - \frac{1}{n^2}\right) = \frac{16}{3} \cdot 1 \cdot 2 = \frac{32}{3}$$

25. $y = f(x) = \cos x$. $\Delta x = \dfrac{b-0}{n} = \dfrac{b}{n}$ and $x_i = 0 + i\,\Delta x = \dfrac{bi}{n}$.

$$A = \lim_{n \to \infty} R_n = \lim_{n \to \infty} \sum_{i=1}^{n} f(x_i)\,\Delta x = \lim_{n \to \infty} \sum_{i=1}^{n} \cos\!\left(\frac{bi}{n}\right) \cdot \frac{b}{n} \overset{\text{CAS}}{=} \lim_{n \to \infty}\left[\frac{b\sin\!\left(b\left(\frac{1}{2n}+1\right)\right)}{2n\sin\!\left(\frac{b}{2n}\right)} - \frac{b}{2n}\right] \overset{\text{CAS}}{=} \sin b$$

If $b = \frac{\pi}{2}$, then $A = \sin\frac{\pi}{2} = 1$.

5.2 The Definite Integral

1. $R_4 = \sum\limits_{i=1}^{4} f(x_i)\,\Delta x$ $[x_i^* = x_i$ is a right endpoint and $\Delta x = 0.5]$

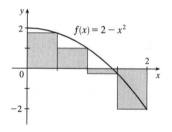

$= 0.5\,[f(0.5) + f(1) + f(1.5) + f(2)]$ $[f(x) = 2 - x^2]$

$= 0.5\,[1.75 + 1 + (-0.25) + (-2)]$

$= 0.5(0.5) = 0.25$

The Riemann sum represents the sum of the areas of the two rectangles above the x-axis minus the sum of the areas of the two rectangles below the x-axis; that is, the *net area* of the rectangles with respect to the x-axis.

3. $M_5 = \sum\limits_{i=1}^{5} f(\overline{x}_i)\,\Delta x$ $[x_i^* = \overline{x}_i = \frac{1}{2}(x_{i-1} + x_i)$ is a midpoint and $\Delta x = 1]$

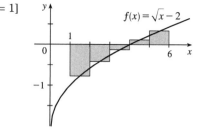

$= 1\,[f(1.5) + f(2.5) + f(3.5)$

$\qquad\qquad + f(4.5) + f(5.5)]$ $[f(x) = \sqrt{x} - 2]$

≈ -0.856759

The Riemann sum represents the sum of the areas of the two rectangles above the x-axis minus the sum of the areas of the three rectangles below the x-axis.

5. $\Delta x = (b - a)/n = (8 - 0)/4 = 8/4 = 2$.

  (a) Using the right endpoints to approximate $\int_0^8 f(x)\,dx$, we have

$$\sum_{i=1}^{4} f(x_i)\,\Delta x = 2[f(2) + f(4) + f(6) + f(8)] \approx 2[1 + 2 + (-2) + 1] = 4.$$

  (b) Using the left endpoints to approximate $\int_0^8 f(x)\,dx$, we have

$$\sum_{i=1}^{4} f(x_{i-1})\,\Delta x = 2[f(0) + f(2) + f(4) + f(6)] \approx 2[2 + 1 + 2 + (-2)] = 6.$$

  (c) Using the midpoint of each subinterval to approximate $\int_0^8 f(x)\,dx$, we have

$$\sum_{i=1}^{4} f(\overline{x}_i)\,\Delta x = 2[f(1) + f(3) + f(5) + f(7)] \approx 2[3 + 2 + 1 + (-1)] = 10.$$

7. Since f is increasing, $L_5 \leq \int_0^{25} f(x)\,dx \leq R_5$.

$$\text{Lower estimate} = L_5 = \sum_{i=1}^{5} f(x_{i-1})\,\Delta x = 5[f(0) + f(5) + f(10) + f(15) + f(20)]$$

$$= 5(-42 - 37 - 25 - 6 + 15) = 5(-95) = -475$$

$$\text{Upper estimate} = R_5 = \sum_{i=1}^{5} f(x_i)\,\Delta x = 5[f(5) + f(10) + f(15) + f(20) + f(25)]$$

$$= 5(-37 - 25 - 6 + 15 + 36) = 5(-17) = -85$$

9. $\Delta x = (10 - 2)/4 = 2$, so the endpoints are 2, 4, 6, 8, and 10, and the midpoints are 3, 5, 7, and 9. The Midpoint

Rule gives $\int_2^{10} \sqrt{x^3 + 1}\,dx \approx \sum\limits_{i=1}^{4} f(\overline{x}_i)\,\Delta x = 2\left(\sqrt{3^3 + 1} + \sqrt{5^3 + 1} + \sqrt{7^3 + 1} + \sqrt{9^3 + 1}\right) \approx 124.1644$.

11. $\Delta x = (1 - 0)/5 = 0.2$, so the endpoints are 0, 0.2, 0.4, 0.6, 0.8, and 1, and the midpoints are 0.1, 0.3, 0.5, 0.7, and 0.9. The Midpoint Rule gives

$$\int_0^1 \sin(x^2)\,dx \approx \sum_{i=1}^{5} f(\overline{x}_i)\,\Delta x = 0.2\big[\sin(0.1)^2 + \sin(0.3)^2 + \sin(0.5)^2 + \sin(0.7)^2 + \sin(0.9)^2\big] \approx 0.3084.$$

13. In Maple, we use the command `with(student);` to load the sum and box commands, then
`m:=middlesum(sin(x^2),x=0..1,5);` which gives us the sum in summation notation, then
`M:=evalf(m);` which gives $M_5 \approx 0.30843908$, confirming the result of Exercise 11. The command
`middlebox(sin(x^2),x=0..1,5)` generates the graph. Repeating for $n = 10$ and $n = 20$ gives
$M_{10} \approx 0.30981629$ and $M_{20} \approx 0.31015563$.

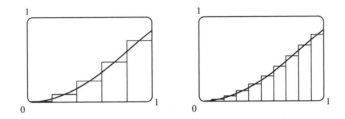

15. We'll create the table of values to approximate $\int_0^\pi \sin x\,dx$ by using the program in the solution to Exercise 5.1.7 with $Y_1 = \sin x$, Xmin $= 0$, Xmax $= \pi$, and $n = 5, 10, 50,$ and 100.

The values of R_n appear to be approaching 2.

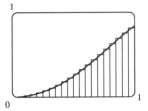

n	R_n
5	1.933766
10	1.983524
50	1.999342
100	1.999836

17. On $[0, \pi]$, $\displaystyle\lim_{n\to\infty} \sum_{i=1}^{n} x_i \sin x_i\,\Delta x = \int_0^\pi x \sin x\,dx$.

19. On $[1, 8]$, $\displaystyle\lim_{n\to\infty} \sum_{i=1}^{n} \sqrt{2x_i^* + (x_i^*)^2}\,\Delta x = \int_1^8 \sqrt{2x + x^2}\,dx$.

21. Note that $\Delta x = \dfrac{5 - (-1)}{n} = \dfrac{6}{n}$ and $x_i = -1 + i\,\Delta x = -1 + \dfrac{6i}{n}$.

$$\int_{-1}^{5} (1 + 3x)\,dx = \lim_{n\to\infty} \sum_{i=1}^{n} f(x_i)\,\Delta x = \lim_{n\to\infty} \sum_{i=1}^{n} \left[1 + 3\left(-1 + \frac{6i}{n}\right)\right]\frac{6}{n}$$

$$= \lim_{n\to\infty} \frac{6}{n} \sum_{i=1}^{n} \left[-2 + \frac{18i}{n}\right] = \lim_{n\to\infty} \frac{6}{n}\left[\sum_{i=1}^{n}(-2) + \sum_{i=1}^{n}\frac{18i}{n}\right]$$

$$= \lim_{n\to\infty} \frac{6}{n}\left[-2n + \frac{18}{n}\sum_{i=1}^{n} i\right] = \lim_{n\to\infty} \frac{6}{n}\left[-2n + \frac{18}{n}\cdot\frac{n(n+1)}{2}\right]$$

$$= \lim_{n\to\infty}\left[-12 + \frac{108}{n^2}\cdot\frac{n(n+1)}{2}\right] = \lim_{n\to\infty}\left[-12 + 54\frac{n+1}{n}\right]$$

$$= \lim_{n\to\infty}\left[-12 + 54\left(1 + \frac{1}{n}\right)\right] = -12 + 54\cdot 1 = 42$$

23. Note that $\Delta x = \dfrac{2-0}{n} = \dfrac{2}{n}$ and $x_i = 0 + i\,\Delta x = \dfrac{2i}{n}$.

$$\int_0^2 (2 - x^2)\,dx = \lim_{n \to \infty} \sum_{i=1}^n f(x_i)\,\Delta x = \lim_{n \to \infty} \sum_{i=1}^n \left(2 - \frac{4i^2}{n^2}\right)\left(\frac{2}{n}\right)$$

$$= \lim_{n \to \infty} \frac{2}{n}\left[\sum_{i=1}^n 2 - \frac{4}{n^2}\sum_{i=1}^n i^2\right] = \lim_{n \to \infty} \frac{2}{n}\left(2n - \frac{4}{n^2}\sum_{i=1}^n i^2\right)$$

$$= \lim_{n \to \infty}\left[4 - \frac{8}{n^3}\cdot\frac{n(n+1)(2n+1)}{6}\right] = \lim_{n \to \infty}\left(4 - \frac{4}{3}\cdot\frac{n+1}{n}\cdot\frac{2n+1}{n}\right)$$

$$= \lim_{n \to \infty}\left[4 - \frac{4}{3}\left(1 + \frac{1}{n}\right)\left(2 + \frac{1}{n}\right)\right] = 4 - \tfrac{4}{3}\cdot 1 \cdot 2 = \tfrac{4}{3}$$

25. Note that $\Delta x = \dfrac{2-1}{n} = \dfrac{1}{n}$ and $x_i = 1 + i\,\Delta x = 1 + i(1/n) = 1 + i/n$.

$$\int_1^2 x^3\,dx = \lim_{n \to \infty}\sum_{i=1}^n f(x_i)\,\Delta x = \lim_{n \to \infty}\sum_{i=1}^n\left(1 + \frac{i}{n}\right)^3\left(\frac{1}{n}\right) = \lim_{n \to \infty}\frac{1}{n}\sum_{i=1}^n\left(\frac{n+i}{n}\right)^3$$

$$= \lim_{n \to \infty}\frac{1}{n^4}\sum_{i=1}^n (n^3 + 3n^2 i + 3ni^2 + i^3) = \lim_{n \to \infty}\frac{1}{n^4}\left[\sum_{i=1}^n n^3 + \sum_{i=1}^n 3n^2 i + \sum_{i=1}^n 3ni^2 + \sum_{i=1}^n i^3\right]$$

$$= \lim_{n \to \infty}\frac{1}{n^4}\left[n\cdot n^3 + 3n^2\sum_{i=1}^n i + 3n\sum_{i=1}^n i^2 + \sum_{i=1}^n i^3\right]$$

$$= \lim_{n \to \infty}\left[1 + \frac{3}{n^2}\cdot\frac{n(n+1)}{2} + \frac{3}{n^3}\cdot\frac{n(n+1)(2n+1)}{6} + \frac{1}{n^4}\cdot\frac{n^2(n+1)^2}{4}\right]$$

$$= \lim_{n \to \infty}\left[1 + \frac{3}{2}\cdot\frac{n+1}{n} + \frac{1}{2}\cdot\frac{n+1}{n}\cdot\frac{2n+1}{n} + \frac{1}{4}\cdot\frac{(n+1)^2}{n^2}\right]$$

$$= \lim_{n \to \infty}\left[1 + \frac{3}{2}\left(1 + \frac{1}{n}\right) + \frac{1}{2}\left(1 + \frac{1}{n}\right)\left(2 + \frac{1}{n}\right) + \frac{1}{4}\left(1 + \frac{1}{n}\right)^2\right] = 1 + \frac{3}{2} + \frac{1}{2}\cdot 2 + \frac{1}{4} = 3.75$$

27. $\displaystyle\int_a^b x\,dx = \lim_{n \to \infty}\frac{b-a}{n}\sum_{i=1}^n\left[a + \frac{b-a}{n}i\right] = \lim_{n \to \infty}\left[\frac{a(b-a)}{n}\sum_{i=1}^n 1 + \frac{(b-a)^2}{n^2}\sum_{i=1}^n i\right]$

$$= \lim_{n \to \infty}\left[\frac{a(b-a)}{n}n + \frac{(b-a)^2}{n^2}\cdot\frac{n(n+1)}{2}\right] = a(b-a) + \lim_{n \to \infty}\frac{(b-a)^2}{2}\left(1 + \frac{1}{n}\right)$$

$$= a(b-a) + \tfrac{1}{2}(b-a)^2 = (b-a)\left(a + \tfrac{1}{2}b - \tfrac{1}{2}a\right) = (b-a)\tfrac{1}{2}(b+a) = \tfrac{1}{2}(b^2 - a^2)$$

29. $f(x) = \dfrac{x}{1 + x^5}$, $a = 2$, $b = 6$, and $\Delta x = \dfrac{6-2}{n} = \dfrac{4}{n}$. Using Equation 3, we get $x_i^* = x_i = 2 + i\,\Delta x = 2 + \dfrac{4i}{n}$,

so $\displaystyle\int_2^6 \frac{x}{1+x^5}\,dx = \lim_{n \to \infty} R_n = \lim_{n \to \infty}\sum_{i=1}^n \frac{2 + \dfrac{4i}{n}}{1 + \left(2 + \dfrac{4i}{n}\right)^5}\cdot\frac{4}{n}$.

31. $\Delta x = (\pi - 0)/n = \pi/n$ and $x_i^* = x_i = \pi i/n$.

$$\int_0^\pi \sin 5x\,dx = \lim_{n \to \infty}\sum_{i=1}^n (\sin 5x_i)\left(\frac{\pi}{n}\right) = \lim_{n \to \infty}\sum_{i=1}^n\left(\sin\frac{5\pi i}{n}\right)\frac{\pi}{n} \overset{\text{CAS}}{=} \pi\lim_{n \to \infty}\frac{1}{n}\cot\left(\frac{5\pi}{2n}\right) \overset{\text{CAS}}{=} \pi\left(\frac{2}{5\pi}\right) = \frac{2}{5}$$

33. (a) Think of $\int_0^2 f(x)\,dx$ as the area of a trapezoid with bases 1 and 3 and height 2. The area of a trapezoid is

$A = \frac{1}{2}(b+B)h$, so $\int_0^2 f(x)\,dx = \frac{1}{2}(1+3)2 = 4$.

(b) $\int_0^5 f(x)\,dx = \int_0^2 f(x)\,dx + \int_2^3 f(x)\,dx + \int_3^5 f(x)\,dx$

 trapezoid   rectangle   triangle

$= \frac{1}{2}(1+3)2 +$ $3 \cdot 1$ $+$ $\frac{1}{2} \cdot 2 \cdot 3$ $= 4+3+3 = 10$

(c) $\int_5^7 f(x)\,dx$ is the negative of the area of the triangle with base 2 and height 3. $\int_5^7 f(x)\,dx = -\frac{1}{2} \cdot 2 \cdot 3 = -3$.

(d) $\int_7^9 f(x)\,dx$ is the negative of the area of a trapezoid with bases 3 and 2 and height 2, so it equals

$-\frac{1}{2}(B+b)h = -\frac{1}{2}(3+2)2 = -5$. Thus,

$\int_0^9 f(x)\,dx = \int_0^5 f(x)\,dx + \int_5^7 f(x)\,dx + \int_7^9 f(x)\,dx = 10 + (-3) + (-5) = 2$.

35. $\int_0^3 \left(\frac{1}{2}x - 1\right) dx$ can be interpreted as the area of the triangle above the x-axis

minus the area of the triangle below the x-axis; that is,

$\frac{1}{2}(1)\left(\frac{1}{2}\right) - \frac{1}{2}(2)(1) = \frac{1}{4} - 1 = -\frac{3}{4}$.

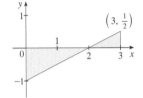

37. $\int_{-3}^0 \left(1 + \sqrt{9 - x^2}\right) dx$ can be interpreted as the area under the graph of

$f(x) = 1 + \sqrt{9 - x^2}$ between $x = -3$ and $x = 0$. This is equal to one-quarter

the area of the circle with radius 3, plus the area of the rectangle, so

$\int_{-3}^0 \left(1 + \sqrt{9 - x^2}\right) dx = \frac{1}{4}\pi \cdot 3^2 + 1 \cdot 3 = 3 + \frac{9}{4}\pi$.

39. $\int_{-1}^2 |x|\,dx$ can be interpreted as the sum of the areas of the two shaded

triangles; that is, $\frac{1}{2}(1)(1) + \frac{1}{2}(2)(2) = \frac{1}{2} + \frac{4}{2} = \frac{5}{2}$.

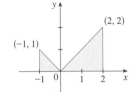

41. $\int_9^4 \sqrt{t}\,dt = -\int_4^9 \sqrt{t}\,dt$ [because we reversed the limits of integration]

$\phantom{\int_9^4 \sqrt{t}\,dt}= -\int_4^9 \sqrt{x}\,dx$ [we can use any letter without changing the value of the integral]

$\phantom{\int_9^4 \sqrt{t}\,dt}= -\frac{38}{3}$

43. $\int_0^1 \left(5 - 6x^2\right) dx = \int_0^1 5\,dx - 6\int_0^1 x^2\,dx = 5(1 - 0) - 6\left(\frac{1}{3}\right) = 5 - 2 = 3$

45. $\int_1^3 e^{x+2}\,dx = \int_1^3 e^x \cdot e^2\,dx = e^2 \int_1^3 e^x\,dx = e^2\left(e^3 - e\right) = e^5 - e^3$

47. $\int_{-2}^2 f(x)\,dx + \int_2^5 f(x)\,dx - \int_{-2}^{-1} f(x)\,dx = \int_{-2}^5 f(x)\,dx + \int_{-1}^{-2} f(x)\,dx$ [by Property 5 and reversing limits]

$\phantom{\int_{-2}^2 f(x)\,dx + \int_2^5 f(x)\,dx}= \int_{-1}^5 f(x)\,dx$ [Property 5]

49. $\int_0^9 [2f(x) + 3g(x)]\,dx = 2\int_0^9 f(x)\,dx + 3\int_0^9 g(x)\,dx = 2(37) + 3(16) = 122$

51. $0 \le \sin x < 1$ on $\left[0, \frac{\pi}{4}\right]$, so $\sin^3 x \le \sin^2 x$ on $\left[0, \frac{\pi}{4}\right]$. Hence, $\int_0^{\pi/4} \sin^3 x \, dx \le \int_0^{\pi/4} \sin^2 x \, dx$ (Property 7).

53. If $-1 \le x \le 1$, then $0 \le x^2 \le 1$ and $1 \le 1 + x^2 \le 2$, so $1 \le \sqrt{1+x^2} \le \sqrt{2}$ and
$1[1 - (-1)] \le \int_{-1}^1 \sqrt{1+x^2} \, dx \le \sqrt{2}\,[1 - (-1)]$ [Property 8]; that is, $2 \le \int_{-1}^1 \sqrt{1+x^2} \, dx \le 2\sqrt{2}$.

55. If $1 \le x \le 2$, then $\frac{1}{2} \le \frac{1}{x} \le 1$, so $\frac{1}{2}(2-1) \le \int_1^2 \frac{1}{x} \, dx \le 1(2-1)$ or $\frac{1}{2} \le \int_1^2 \frac{1}{x} \, dx \le 1$.

57. If $\frac{\pi}{4} \le x \le \frac{\pi}{3}$, then $1 \le \tan x \le \sqrt{3}$, so $1\left(\frac{\pi}{3} - \frac{\pi}{4}\right) \le \int_{\pi/4}^{\pi/3} \tan x \, dx \le \sqrt{3}\left(\frac{\pi}{3} - \frac{\pi}{4}\right)$ or
$\frac{\pi}{12} \le \int_{\pi/4}^{\pi/3} \tan x \, dx \le \frac{\pi}{12}\sqrt{3}$.

59. The only critical number of $f(x) = xe^{-x}$ on $[0, 2]$ is $x = 1$. Since $f(0) = 0$, $f(1) = e^{-1} \approx 0.368$, and
$f(2) = 2e^{-2} \approx 0.271$, we know that the absolute minimum value of f on $[0, 2]$ is 0, and the absolute maximum
is e^{-1}. By Property 8, $0 \le xe^{-x} \le e^{-1}$ for $0 \le x \le 2$ $\Rightarrow$ $0(2-0) \le \int_0^2 xe^{-x} \, dx \le e^{-1}(2-0)$ $\Rightarrow$
$0 \le \int_0^2 xe^{-x} \, dx \le 2/e$.

61. $\sqrt{x^4 + 1} \ge \sqrt{x^4} = x^2$, so $\int_1^3 \sqrt{x^4 + 1} \, dx \ge \int_1^3 x^2 \, dx = \frac{1}{3}(3^3 - 1^3) = \frac{26}{3}$.

63. Using a regular partition and right endpoints as in the proof of Property 2, we calculate
$$\int_a^b cf(x) \, dx = \lim_{n \to \infty} \sum_{i=1}^n cf(x_i) \, \Delta x_i = \lim_{n \to \infty} c \sum_{i=1}^n f(x_i) \, \Delta x_i = c \lim_{n \to \infty} \sum_{i=1}^n f(x_i) \, \Delta x_i = c \int_a^b f(x) \, dx.$$

65. Since $-|f(x)| \le f(x) \le |f(x)|$, it follows from Property 7 that
$$-\int_a^b |f(x)| \, dx \le \int_a^b f(x) \, dx \le \int_a^b |f(x)| \, dx \quad \Rightarrow \quad \left|\int_a^b f(x) \, dx\right| \le \int_a^b |f(x)| \, dx$$

Note that the definite integral is a real number, and so the following property applies: $-a \le b \le a$ $\Rightarrow$ $|b| \le a$
for all real numbers b and nonnegative numbers a.

67. $\lim_{n \to \infty} \sum_{i=1}^n \frac{i^4}{n^5} = \lim_{n \to \infty} \frac{1}{n} \sum_{i=1}^n \left(\frac{i}{n}\right)^4 = \int_0^1 x^4 \, dx$

69. Choose $x_i = 1 + \frac{i}{n}$ and $x_i^* = \sqrt{x_{i-1}x_i} = \sqrt{\left(1 + \frac{i-1}{n}\right)\left(1 + \frac{i}{n}\right)}$. Then

$$\int_1^2 x^{-2} \, dx = \lim_{n \to \infty} \frac{1}{n} \sum_{i=1}^n \frac{1}{\left(1 + \frac{i-1}{n}\right)\left(1 + \frac{i}{n}\right)}$$

$$= \lim_{n \to \infty} n \sum_{i=1}^n \frac{1}{(n+i-1)(n+i)}$$

$$= \lim_{n \to \infty} n \sum_{i=1}^n \left(\frac{1}{n+i-1} - \frac{1}{n+i}\right) \qquad \text{[by the hint]}$$

$$= \lim_{n \to \infty} n \left(\sum_{i=0}^{n-1} \frac{1}{n+i} - \sum_{i=1}^n \frac{1}{n+i}\right)$$

$$= \lim_{n \to \infty} n \left(\left[\frac{1}{n} + \frac{1}{n+1} + \cdots + \frac{1}{2n-1}\right] - \left[\frac{1}{n+1} + \cdots + \frac{1}{2n-1} + \frac{1}{2n}\right]\right)$$

$$= \lim_{n \to \infty} n \left(\frac{1}{n} - \frac{1}{2n}\right) = \lim_{n \to \infty} \left(1 - \frac{1}{2}\right) = \frac{1}{2}$$

5.3 The Fundamental Theorem of Calculus

1. The precise version of this statement is given by the Fundamental Theorem of Calculus. See the statement of this theorem and the paragraph that follows it at the end of Section 5.3.

3. (a) $g(x) = \int_0^x f(t)\, dt$.

$g(0) = \int_0^0 f(t)\, dt = 0$

$g(1) = \int_0^1 f(t)\, dt = 1 \cdot 2 = 2$ [rectangle],

$g(2) = \int_0^2 f(t)\, dt = \int_0^1 f(t)\, dt + \int_1^2 f(t)\, dt = g(1) + \int_1^2 f(t)\, dt$

$\quad = 2 + 1 \cdot 2 + \frac{1}{2} \cdot 1 \cdot 2 = 5$ [rectangle plus triangle],

$g(3) = \int_0^3 f(t)\, dt = g(2) + \int_2^3 f(t)\, dt = 5 + \frac{1}{2} \cdot 1 \cdot 4 = 7,$

$g(6) = g(3) + \int_3^6 f(t)\, dt$ [the integral is negative since f lies under the x-axis]

$\quad = 7 + \left[-\left(\frac{1}{2} \cdot 2 \cdot 2 + 1 \cdot 2\right)\right] = 7 - 4 = 3$

(d)

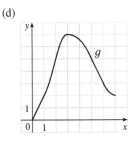

(b) g is increasing on $(0, 3)$ because as x increases from 0 to 3, we keep adding more area.

(c) g has a maximum value when we start subtracting area; that is, at $x = 3$.

5.

(a) By FTC1 with $f(t) = t^2$ and $a = 1$, $g(x) = \int_1^x t^2\, dt \Rightarrow$

$g'(x) = f(x) = x^2$.

(b) Using FTC2, $g(x) = \int_1^x t^2\, dt = \left[\frac{1}{3}t^3\right]_1^x = \frac{1}{3}x^3 - \frac{1}{3} \Rightarrow$

$g'(x) = x^2$.

7. $f(t) = \sqrt{1 + 2t}$ and $g(x) = \int_0^x \sqrt{1 + 2t}\, dt$, so by FTC1, $g'(x) = f(x) = \sqrt{1 + 2x}$.

9. $f(t) = t^2 \sin t$ and $g(y) = \int_2^y t^2 \sin t\, dt$, so by FTC1, $g'(y) = f(y) = y^2 \sin y$.

11. $F(x) = \int_x^2 \cos(t^2)\, dt = -\int_2^x \cos(t^2)\, dt \Rightarrow F'(x) = -\cos(x^2)$

13. Let $u = \dfrac{1}{x}$. Then $\dfrac{du}{dx} = -\dfrac{1}{x^2}$. Also, $\dfrac{dh}{dx} = \dfrac{dh}{du}\dfrac{du}{dx}$, so

$h'(x) = \dfrac{d}{dx}\int_2^{1/x} \arctan t\, dt = \dfrac{d}{du}\int_2^u \arctan t\, dt \cdot \dfrac{du}{dx} = \arctan u \dfrac{du}{dx} = -\dfrac{\arctan(1/x)}{x^2}.$

15. Let $u = \sqrt{x}$. Then $\dfrac{du}{dx} = \dfrac{1}{2\sqrt{x}}$. Also, $\dfrac{dy}{dx} = \dfrac{dy}{du}\dfrac{du}{dx}$, so

$y' = \dfrac{d}{dx}\int_3^{\sqrt{x}} \dfrac{\cos t}{t}\, dt = \dfrac{d}{du}\int_3^u \dfrac{\cos t}{t}\, dt \cdot \dfrac{du}{dx} = \dfrac{\cos u}{u} \cdot \dfrac{1}{2\sqrt{x}} = \dfrac{\cos\sqrt{x}}{\sqrt{x}} \cdot \dfrac{1}{2\sqrt{x}} = \dfrac{\cos\sqrt{x}}{2x}.$

17. Let $w = 1 - 3x$. Then $\dfrac{dw}{dx} = -3$. Also, $\dfrac{dy}{dx} = \dfrac{dy}{dw}\dfrac{dw}{dx}$, so

$y' = \dfrac{d}{dx}\int_{1-3x}^1 \dfrac{u^3}{1 + u^2}\, du = \dfrac{d}{dw}\int_w^1 \dfrac{u^3}{1 + u^2}\, du \cdot \dfrac{dw}{dx}$

$\quad = -\dfrac{d}{dw}\int_1^w \dfrac{u^3}{1 + u^2}\, du \cdot \dfrac{dw}{dx} = -\dfrac{w^3}{1 + w^2}(-3) = \dfrac{3(1 - 3x)^3}{1 + (1 - 3x)^2}$

19. $\displaystyle\int_{-1}^3 x^5\, dx = \left[\dfrac{x^6}{6}\right]_{-1}^3 = \dfrac{3^6}{6} - \dfrac{(-1)^6}{6} = \dfrac{729 - 1}{6} = \dfrac{364}{3}$

21. $\int_2^8 (4x+3)\,dx = \left[\frac{4}{2}x^2 + 3x\right]_2^8 = \left(2\cdot 8^2 + 3\cdot 8\right) - \left(2\cdot 2^2 + 3\cdot 2\right) = 152 - 14 = 138$

23. $\int_0^1 x^{4/5}\,dx = \left[\frac{5}{9}x^{9/5}\right]_0^1 = \frac{5}{9} - 0 = \frac{5}{9}$

25. $\int_1^2 \frac{3}{t^4}\,dt = 3\int_1^2 t^{-4}\,dt = 3\left[\frac{t^{-3}}{-3}\right]_1^2 = \frac{3}{-3}\left[\frac{1}{t^3}\right]_1^2 = -1\left(\frac{1}{8} - 1\right) = \frac{7}{8}$

27. $\int_{-5}^5 \frac{2}{x^3}\,dx$ does not exist because the function $f(x) = \dfrac{2}{x^3}$ has an infinite discontinuity at $x = 0$; that is, f is

discontinuous on the interval $[-5, 5]$.

29. $\int_0^2 x(2+x^5)\,dx = \int_0^2 (2x+x^6)\,dx = \left[x^2 + \frac{1}{7}x^7\right]_0^2 = \left(4 + \frac{128}{7}\right) - (0+0) = \frac{156}{7}$

31. $\int_0^{\pi/4} \sec^2 t\,dt = \left[\tan t\right]_0^{\pi/4} = \tan\frac{\pi}{4} - \tan 0 = 1 - 0 = 1$

33. $\int_\pi^{2\pi} \csc^2\theta\,d\theta$ does not exist because the function $f(\theta) = \csc^2\theta$ has infinite discontinuities at $\theta = \pi$ and $\theta = 2\pi$; that is, f is discontinuous on the interval $[\pi, 2\pi]$.

35. $\int_1^9 \frac{1}{2x}\,dx = \frac{1}{2}\int_1^9 \frac{1}{x}\,dx = \frac{1}{2}\Big[\ln|x|\Big]_1^9 = \frac{1}{2}(\ln 9 - \ln 1) = \frac{1}{2}\ln 9 - 0 = \ln 9^{1/2} = \ln 3$

37. $\int_{1/2}^{\sqrt{3}/2} \frac{6}{\sqrt{1-t^2}}\,dt = 6\int_{1/2}^{\sqrt{3}/2} \frac{1}{\sqrt{1-t^2}}\,dt = 6\left[\sin^{-1} t\right]_{1/2}^{\sqrt{3}/2} = 6\left[\sin^{-1}\left(\frac{\sqrt{3}}{2}\right) - \sin^{-1}\left(\frac{1}{2}\right)\right]$

$\qquad = 6\left(\frac{\pi}{3} - \frac{\pi}{6}\right) = 6\left(\frac{\pi}{6}\right) = \pi$

39. $\int_{-1}^1 e^{u+1}\,du = \left[e^{u+1}\right]_{-1}^1 = e^2 - e^0 = e^2 - 1$ [or start with $e^{u+1} = e^u e^1$]

41. $\int_0^2 f(x)\,dx = \int_0^1 x^4\,dx + \int_1^2 x^5\,dx = \left[\frac{1}{5}x^5\right]_0^1 + \left[\frac{1}{6}x^6\right]_1^2 = \left(\frac{1}{5} - 0\right) + \left(\frac{64}{6} - \frac{1}{6}\right) = 10.7$

43. From the graph, it appears that the area is about 60. The actual area is

$\int_0^{27} x^{1/3}\,dx = \left[\frac{3}{4}x^{4/3}\right]_0^{27} = \frac{3}{4}\cdot 81 - 0 = \frac{243}{4} = 60.75$. This is $\frac{3}{4}$ of the

area of the viewing rectangle.

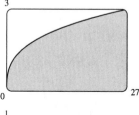

45. It appears that the area under the graph is about $\frac{2}{3}$ of the area of the

viewing rectangle, or about $\frac{2}{3}\pi \approx 2.1$. The actual area is

$\int_0^\pi \sin x\,dx = [-\cos x]_0^\pi = (-\cos\pi) - (-\cos 0) = -(-1) + 1 = 2$.

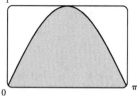

47. $\int_{-1}^2 x^3\,dx = \left[\frac{1}{4}x^4\right]_{-1}^2 = 4 - \frac{1}{4} = \frac{15}{4} = 3.75$

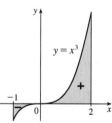

49. $g(x) = \int_{2x}^{3x} \frac{u^2 - 1}{u^2 + 1}\, du = \int_{2x}^{0} \frac{u^2 - 1}{u^2 + 1}\, du + \int_{0}^{3x} \frac{u^2 - 1}{u^2 + 1}\, du = -\int_{0}^{2x} \frac{u^2 - 1}{u^2 + 1}\, du + \int_{0}^{3x} \frac{u^2 - 1}{u^2 + 1}\, du \;\Rightarrow$

$g'(x) = -\frac{(2x)^2 - 1}{(2x)^2 + 1} \cdot \frac{d}{dx}(2x) + \frac{(3x)^2 - 1}{(3x)^2 + 1} \cdot \frac{d}{dx}(3x) = -2 \cdot \frac{4x^2 - 1}{4x^2 + 1} + 3 \cdot \frac{9x^2 - 1}{9x^2 + 1}$

51. $y = \int_{\sqrt{x}}^{x^3} \sqrt{t}\sin t\, dt = \int_{\sqrt{x}}^{1} \sqrt{t}\sin t\, dt + \int_{1}^{x^3} \sqrt{t}\sin t\, dt = -\int_{1}^{\sqrt{x}} \sqrt{t}\sin t\, dt + \int_{1}^{x^3} \sqrt{t}\sin t\, dt \;\Rightarrow$

$y' = -\sqrt[4]{x}\,(\sin\sqrt{x}) \cdot \frac{d}{dx}(\sqrt{x}) + x^{3/2}\sin(x^3) \cdot \frac{d}{dx}(x^3) = -\frac{\sqrt[4]{x}\sin\sqrt{x}}{2\sqrt{x}} + x^{3/2}\sin(x^3)(3x^2)$

$= 3x^{7/2}\sin(x^3) - \frac{\sin\sqrt{x}}{2\sqrt[4]{x}}$

53. $F(x) = \int_{1}^{x} f(t)\, dt \;\Rightarrow\; F'(x) = f(x) = \int_{1}^{x^2} \frac{\sqrt{1 + u^4}}{u}\, du \left[\text{since } f(t) = \int_{1}^{t^2} \frac{\sqrt{1 + u^4}}{u}\, du \right] \;\Rightarrow$

$F''(x) = f'(x) = \frac{\sqrt{1 + (x^2)^4}}{x^2} \cdot \frac{d}{dx}(x^2) = \frac{\sqrt{1 + x^8}}{x^2} \cdot 2x = \frac{2\sqrt{1 + x^8}}{x}$. So $F''(2) = \sqrt{1 + 2^8} = \sqrt{257}$.

55. By FTC2, $\int_{1}^{4} f'(x)\, dx = f(4) - f(1)$, so $17 = f(4) - 12 \;\Rightarrow\; f(4) = 17 + 12 = 29$.

57. (a) The Fresnel function $S(x) = \int_{0}^{x} \sin(\frac{\pi}{2}t^2)\, dt$ has local maximum values where $0 = S'(x) = \sin(\frac{\pi}{2}x^2)$ and S' changes from positive to negative. For $x > 0$, this happens when $\frac{\pi}{2}x^2 = (2n - 1)\pi$ [odd multiples of π] $\Leftrightarrow$ $x^2 = 2(2n - 1)$ $\Leftrightarrow$ $x = \sqrt{4n - 2}$, n any positive integer. For $x < 0$, S' changes from positive to negative where $\frac{\pi}{2}x^2 = 2n\pi$ [even multiples of π] $\Leftrightarrow$ $x^2 = 4n$ $\Leftrightarrow$ $x = -2\sqrt{n}$. S' does not change sign at $x = 0$.

(b) S is concave upward on those intervals where $S''(x) > 0$. Differentiating our expression for $S'(x)$, we get $S''(x) = \cos(\frac{\pi}{2}x^2)(2\frac{\pi}{2}x) = \pi x \cos(\frac{\pi}{2}x^2)$. For $x > 0$, $S''(x) > 0$ where $\cos(\frac{\pi}{2}x^2) > 0$ $\Leftrightarrow$ $0 < \frac{\pi}{2}x^2 < \frac{\pi}{2}$ or $(2n - \frac{1}{2})\pi < \frac{\pi}{2}x^2 < (2n + \frac{1}{2})\pi$, n any integer $\Leftrightarrow$ $0 < x < 1$ or $\sqrt{4n - 1} < x < \sqrt{4n + 1}$, n any positive integer. For $x < 0$, $S''(x) > 0$ where $\cos(\frac{\pi}{2}x^2) < 0$ $\Leftrightarrow$ $(2n - \frac{3}{2})\pi < \frac{\pi}{2}x^2 < (2n - \frac{1}{2})\pi$, n any integer $\Leftrightarrow$ $4n - 3 < x^2 < 4n - 1$ $\Leftrightarrow$ $\sqrt{4n - 3} < |x| < \sqrt{4n - 1}$ $\Rightarrow$ $\sqrt{4n - 3} < -x < \sqrt{4n - 1}$ $\Rightarrow$ $-\sqrt{4n - 3} > x > -\sqrt{4n - 1}$, so the intervals of upward concavity for $x < 0$ are $(-\sqrt{4n - 1}, -\sqrt{4n - 3})$, n any positive integer. To summarize: S is concave upward on the intervals $(0, 1)$, $(-\sqrt{3}, -1)$, $(\sqrt{3}, \sqrt{5})$, $(-\sqrt{7}, -\sqrt{5})$, $(\sqrt{7}, 3)$, ...

(c) In Maple, we use `plot({int(sin(Pi*t^2/2),t=0..x),0.2},x=0..2);`. Note that Maple recognizes the Fresnel function, calling it `FresnelS(x)`. In Mathematica, we use `Plot[{Integrate[Sin[Pi*t^2/2],{t,0,x}],0.2},{x,0,2}]`. In Derive, we load the utility file FRESNEL and plot FRESNEL_SIN(x). From the graphs, we see that $\int_{0}^{x} \sin(\frac{\pi}{2}t^2)\, dt = 0.2$ at $x \approx 0.74$.

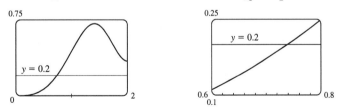

59. (a) By FTC1, $g'(x) = f(x)$. So $g'(x) = f(x) = 0$ at $x = 1, 3, 5, 7$, and 9. g has local maxima at $x = 1$ and 5 (since $f = g'$ changes from positive to negative there) and local minima at $x = 3$ and 7. There is no local maximum or minimum at $x = 9$, since f is not defined for $x > 9$.

(b) We can see from the graph that $\left| \int_0^1 f\,dt \right| < \left| \int_1^3 f\,dt \right| < \left| \int_3^5 f\,dt \right| < \left| \int_5^7 f\,dt \right| < \left| \int_7^9 f\,dt \right|$.

So $g(1) = \left| \int_0^1 f\,dt \right|$, $g(5) = \int_0^5 f\,dt = g(1) - \left| \int_1^3 f\,dt \right| + \left| \int_3^5 f\,dt \right|$, and

$g(9) = \int_0^9 f\,dt = g(5) - \left| \int_5^7 f\,dt \right| + \left| \int_7^9 f\,dt \right|$. Thus, $g(1) < g(5) < g(9)$, and so the absolute maximum of $g(x)$ occurs at $x = 9$.

(c) g is concave downward on those intervals where $g'' < 0$. But $g'(x) = f(x)$, so $g''(x) = f'(x)$, which is negative on (approximately) $\left(\frac{1}{2}, 2\right)$, $(4, 6)$ and $(8, 9)$. So g is concave downward on these intervals.

(d)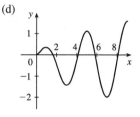

61. $\displaystyle \lim_{n\to\infty} \sum_{i=1}^{n} \frac{i^3}{n^4} = \lim_{n\to\infty} \frac{1-0}{n} \sum_{i=1}^{n} \left(\frac{i}{n}\right)^3 = \int_0^1 x^3\,dx = \left[\frac{x^4}{4}\right]_0^1 = \frac{1}{4}$

63. Suppose $h < 0$. Since f is continuous on $[x + h, x]$, the Extreme Value Theorem says that there are numbers u and v in $[x + h, x]$ such that $f(u) = m$ and $f(v) = M$, where m and M are the absolute minimum and maximum values of f on $[x + h, x]$. By Property 8 of integrals, $m(-h) \le \int_{x+h}^x f(t)\,dt \le M(-h)$; that is,

$f(u)(-h) \le -\int_x^{x+h} f(t)\,dt \le f(v)(-h)$. Since $-h > 0$, we can divide this inequality by $-h$:

$f(u) \le \dfrac{1}{h} \displaystyle\int_x^{x+h} f(t)\,dt \le f(v)$. By Equation 2, $\dfrac{g(x+h) - g(x)}{h} = \dfrac{1}{h}\displaystyle\int_x^{x+h} f(t)\,dt$ for $h \ne 0$, and hence

$f(u) \le \dfrac{g(x+h) - g(x)}{h} \le f(v)$, which is Equation 3 in the case where $h < 0$.

65. (a) Let $f(x) = \sqrt{x} \;\Rightarrow\; f'(x) = 1/(2\sqrt{x}) > 0$ for $x > 0 \;\Rightarrow\; f$ is increasing on $(0, \infty)$. If $x \ge 0$, then $x^3 \ge 0$, so $1 + x^3 \ge 1$ and since f is increasing, this means that $f(1 + x^3) \ge f(1) \;\Rightarrow\; \sqrt{1 + x^3} \ge 1$ for $x \ge 0$. Next let $g(t) = t^2 - t \;\Rightarrow\; g'(t) = 2t - 1 \;\Rightarrow\; g'(t) > 0$ when $t \ge 1$. Thus, g is increasing on $(1, \infty)$. And since $g(1) = 0$, $g(t) \ge 0$ when $t \ge 1$. Now let $t = \sqrt{1 + x^3}$, where $x \ge 0$. $\sqrt{1 + x^3} \ge 1$ (from above) $\;\Rightarrow\; t \ge 1 \;\Rightarrow\; g(t) \ge 0 \;\Rightarrow\; (1 + x^3) - \sqrt{1 + x^3} \ge 0$ for $x \ge 0$. Therefore, $1 \le \sqrt{1 + x^3} \le 1 + x^3$ for $x \ge 0$.

(b) From part (a) and Property 7: $\int_0^1 1\,dx \le \int_0^1 \sqrt{1 + x^3}\,dx \le \int_0^1 (1 + x^3)\,dx \;\Leftrightarrow\;$ $[x]_0^1 \le \int_0^1 \sqrt{1 + x^3}\,dx \le \left[x + \frac{1}{4}x^4\right]_0^1 \;\Leftrightarrow\; 1 \le \int_0^1 \sqrt{1 + x^3}\,dx \le 1 + \frac{1}{4} = 1.25$.

67. Using FTC1, we differentiate both sides of $6 + \displaystyle\int_a^x \frac{f(t)}{t^2}\,dt = 2\sqrt{x}$ to get $\dfrac{f(x)}{x^2} = 2\dfrac{1}{2\sqrt{x}} \;\Rightarrow\; f(x) = x^{3/2}$.

To find a, we substitute $x = a$ in the original equation to obtain $6 + \displaystyle\int_a^a \frac{f(t)}{t^2}\,dt = 2\sqrt{a} \;\Rightarrow\; 6 + 0 = 2\sqrt{a} \;\Rightarrow\;$ $3 = \sqrt{a} \;\Rightarrow\; a = 9$.

69. (a) Let $F(t) = \int_0^t f(s)\, ds$. Then, by FTC1, $F'(t) = f(t) = $ rate of depreciation, so $F(t)$ represents the loss in value over the interval $[0, t]$.

(b) $C(t) = \dfrac{1}{t}\left[A + \displaystyle\int_0^t f(s)\, ds\right] = \dfrac{A + F(t)}{t}$ represents the average expenditure per unit of t during the interval $[0, t]$, assuming that there has been only one overhaul during that time period. The company wants to minimize average expenditure.

(c) $C(t) = \dfrac{1}{t}\left[A + \displaystyle\int_0^t f(s)\, ds\right]$. Using FTC1, we have $C'(t) = -\dfrac{1}{t^2}\left[A + \displaystyle\int_0^t f(s)\, ds\right] + \dfrac{1}{t}f(t)$.

$C'(t) = 0 \;\Rightarrow\; t\,f(t) = A + \displaystyle\int_0^t f(s)\, ds \;\Rightarrow\; f(t) = \dfrac{1}{t}\left[A + \displaystyle\int_0^t f(s)\, ds\right] = C(t)$.

5.4 Indefinite Integrals and the Net Change Theorem

1. $\dfrac{d}{dx}\left[\sqrt{x^2 + 1} + C\right] = \dfrac{d}{dx}\left[(x^2 + 1)^{1/2} + C\right] = \frac{1}{2}(x^2 + 1)^{-1/2} \cdot 2x = \dfrac{x}{\sqrt{x^2 + 1}}$

3. $\dfrac{d}{dx}\left[\dfrac{x}{a^2\sqrt{a^2 - x^2}} + C\right] = \dfrac{1}{a^2}\,\dfrac{\sqrt{a^2 - x^2} - x(-x)/\sqrt{a^2 - x^2}}{a^2 - x^2} = \dfrac{1}{a^2}\,\dfrac{(a^2 - x^2) + x^2}{(a^2 - x^2)^{3/2}} = \dfrac{1}{\sqrt{(a^2 - x^2)^3}}$

5. $\displaystyle\int x^{-3/4}\, dx = \dfrac{x^{-3/4 + 1}}{-3/4 + 1} + C = \dfrac{x^{1/4}}{1/4} + C = 4x^{1/4} + C$

7. $\displaystyle\int (x^3 + 6x + 1)\, dx = \dfrac{x^4}{4} + 6\dfrac{x^2}{2} + x + C = \frac{1}{4}x^4 + 3x^2 + x + C$

9. $\displaystyle\int (1 - t)(2 + t^2)\, dt = \int (2 - 2t + t^2 - t^3)\, dt = 2t - 2\dfrac{t^2}{2} + \dfrac{t^3}{3} - \dfrac{t^4}{4} + C = 2t - t^2 + \frac{1}{3}t^3 - \frac{1}{4}t^4 + C$

11. $\displaystyle\int (2 - \sqrt{x})^2\, dx = \int (4 - 4\sqrt{x} + x)\, dx = 4x - 4\dfrac{x^{3/2}}{3/2} + \dfrac{x^2}{2} + C = 4x - \frac{8}{3}x^{3/2} + \frac{1}{2}x^2 + C$

13. $\displaystyle\int \dfrac{\sin x}{1 - \sin^2 x}\, dx = \int \dfrac{\sin x}{\cos^2 x}\, dx = \int \dfrac{1}{\cos x} \cdot \dfrac{\sin x}{\cos x}\, dx = \int \sec x \tan x\, dx = \sec x + C$

15. $\int x\sqrt{x}\, dx = \int x^{3/2}\, dx = \frac{2}{5}x^{5/2} + C$.

The members of the family in the figure correspond to $C = 5, 3, 0, -2,$ and -4.

17. $\int_0^2 (6x^2 - 4x + 5)\, dx = \left[6 \cdot \frac{1}{3}x^3 - 4 \cdot \frac{1}{2}x^2 + 5x\right]_0^2 = \left[2x^3 - 2x^2 + 5x\right]_0^2 = (16 - 8 + 10) - 0 = 18$

19. $\int_{-1}^0 (2x - e^x)\, dx = \left[x^2 - e^x\right]_{-1}^0 = (0 - 1) - (1 - e^{-1}) = -2 + 1/e$

21. $\int_{-2}^{2}(3u+1)^2\,du = \int_{-2}^{2}\left(9u^2+6u+1\right)\,du = \left[9\cdot\frac{1}{3}u^3+6\cdot\frac{1}{2}u^2+u\right]_{-2}^{2} = \left[3u^3+3u^2+u\right]_{-2}^{2}$
$$= (24+12+2)-(-24+12-2) = 38-(-14) = 52$$

23. $\int_{1}^{4}\sqrt{t}\,(1+t)\,dt = \int_{1}^{4}(t^{1/2}+t^{3/2})\,dt = \left[\frac{2}{3}t^{3/2}+\frac{2}{5}t^{5/2}\right]_{1}^{4} = \left(\frac{16}{3}+\frac{64}{5}\right)-\left(\frac{2}{3}+\frac{2}{5}\right) = \frac{14}{3}+\frac{62}{5} = \frac{256}{15}$

25. $\int_{-2}^{-1}\left(4y^3+\frac{2}{y^3}\right)dy = \left[4\cdot\frac{1}{4}y^4+2\cdot\frac{1}{-2}y^{-2}\right]_{-2}^{-1} = \left[y^4-\frac{1}{y^2}\right]_{-2}^{-1} = (1-1)-\left(16-\frac{1}{4}\right) = -\frac{63}{4}$

27. $\int_{0}^{1}x\left(\sqrt[3]{x}+\sqrt[4]{x}\right)dx = \int_{0}^{1}(x^{4/3}+x^{5/4})\,dx = \left[\frac{3}{7}x^{7/3}+\frac{4}{9}x^{9/4}\right]_{0}^{1} = \left(\frac{3}{7}+\frac{4}{9}\right)-0 = \frac{55}{63}$

29. $\int_{1}^{4}\sqrt{5/x}\,dx = \sqrt{5}\int_{1}^{4}x^{-1/2}\,dx = \sqrt{5}\left[2\sqrt{x}\right]_{1}^{4} = \sqrt{5}\,(2\cdot2-2\cdot1) = 2\sqrt{5}$

31. $\int_{0}^{\pi}(4\sin\theta-3\cos\theta)\,d\theta = \left[-4\cos\theta-3\sin\theta\right]_{0}^{\pi} = (4-0)-(-4-0) = 8$

33. $\int_{0}^{\pi/4}\frac{1+\cos^2\theta}{\cos^2\theta}\,d\theta = \int_{0}^{\pi/4}\left(\frac{1}{\cos^2\theta}+\frac{\cos^2\theta}{\cos^2\theta}\right)d\theta = \int_{0}^{\pi/4}(\sec^2\theta+1)\,d\theta$
$$= \left[\tan\theta+\theta\right]_{0}^{\pi/4} = \left(\tan\frac{\pi}{4}+\frac{\pi}{4}\right)-(0+0) = 1+\frac{\pi}{4}$$

35. $\int_{1}^{64}\frac{1+\sqrt[3]{x}}{\sqrt{x}}\,dx = \int_{1}^{64}\left(\frac{1}{x^{1/2}}+\frac{x^{1/3}}{x^{1/2}}\right)dx = \int_{1}^{64}\left(x^{-1/2}+x^{(1/3)-(1/2)}\right)dx = \int_{1}^{64}\left(x^{-1/2}+x^{-1/6}\right)dx$
$$= \left[2x^{1/2}+\frac{6}{5}x^{5/6}\right]_{1}^{64} = \left(16+\frac{192}{5}\right)-\left(2+\frac{6}{5}\right) = 14+\frac{186}{5} = \frac{256}{5}$$

37. $\int_{1}^{e}\frac{x^2+x+1}{x}\,dx = \int_{1}^{e}\left(x+1+\frac{1}{x}\right)dx = \left[\frac{1}{2}x^2+x+\ln|x|\right]_{1}^{e}$
$$= \left(\frac{1}{2}e^2+e+\ln e\right)-\left(\frac{1}{2}+1+\ln 1\right) = \frac{1}{2}e^2+e-\frac{1}{2}$$

39. $\int_{-1}^{2}(x-2|x|)\,dx = \int_{-1}^{0}[x-2(-x)]\,dx + \int_{0}^{2}[x-2(x)]\,dx = \int_{-1}^{0}3x\,dx + \int_{0}^{2}(-x)\,dx = 3\left[\frac{1}{2}x^2\right]_{-1}^{0} - \left[\frac{1}{2}x^2\right]_{0}^{2}$
$$= 3\left(0-\frac{1}{2}\right)-(2-0) = -\frac{7}{2} = -3.5$$

41. The graph shows that $y = x+x^2-x^4$ has x-intercepts at $x = 0$ and at $x = a \approx 1.32$. So the area of the region that lies under the curve and above the x-axis is
$$\int_{0}^{a}\left(x+x^2-x^4\right)dx = \left[\frac{1}{2}x^2+\frac{1}{3}x^3-\frac{1}{5}x^5\right]_{0}^{a}$$
$$= \left(\frac{1}{2}a^2+\frac{1}{3}a^3-\frac{1}{5}a^5\right)-0$$
$$\approx 0.84$$

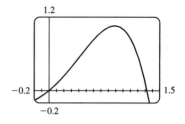

43. $A = \int_{0}^{2}(2y-y^2)\,dy = \left[y^2-\frac{1}{3}y^3\right]_{0}^{2} = \left(4-\frac{8}{3}\right)-0 = \frac{4}{3}$

45. If $w'(t)$ is the rate of change of weight in pounds per year, then $w(t)$ represents the weight in pounds of the child at age t. We know from the Net Change Theorem that $\int_{5}^{10}w'(t)\,dt = w(10)-w(5)$, so the integral represents the increase in the child's weight (in pounds) between the ages of 5 and 10.

47. Since $r(t)$ is the rate at which oil leaks, we can write $r(t) = -V'(t)$, where $V(t)$ is the volume of oil at time t. [Note that the minus sign is needed because V is decreasing, so $V'(t)$ is negative, but $r(t)$ is positive.] Thus, by the Net Change Theorem, $\int_{0}^{120}r(t)\,dt = -\int_{0}^{120}V'(t)\,dt = -[V(120)-V(0)] = V(0)-V(120)$, which is the number of gallons of oil that leaked from the tank in the first two hours (120 minutes).

49. By the Net Change Theorem, $\int_{1000}^{5000} R'(x)\,dx = R(5000) - R(1000)$, so it represents the increase in revenue when production is increased from 1000 units to 5000 units.

51. In general, the unit of measurement for $\int_a^b f(x)\,dx$ is the product of the unit for $f(x)$ and the unit for x. Since $f(x)$ is measured in newtons and x is measured in meters, the units for $\int_0^{100} f(x)\,dx$ are newton-meters. (A newton-meter is abbreviated N-m and is called a joule.)

53. (a) displacement $= \int_0^3 (3t-5)\,dt = \left[\frac{3}{2}t^2 - 5t\right]_0^3 = \frac{27}{2} - 15 = -\frac{3}{2}$ m

(b) distance traveled $= \int_0^3 |3t-5|\,dt = \int_0^{5/3} (5-3t)\,dt + \int_{5/3}^3 (3t-5)\,dt$

$$= \left[5t - \tfrac{3}{2}t^2\right]_0^{5/3} + \left[\tfrac{3}{2}t^2 - 5t\right]_{5/3}^3 = \tfrac{25}{3} - \tfrac{3}{2}\cdot\tfrac{25}{9} + \tfrac{27}{2} - 15 - \left(\tfrac{3}{2}\cdot\tfrac{25}{9} - \tfrac{25}{3}\right) = \tfrac{41}{6}\ \text{m}$$

55. (a) $v'(t) = a(t) = t + 4 \ \Rightarrow\ v(t) = \frac{1}{2}t^2 + 4t + C \ \Rightarrow\ v(0) = C = 5 \ \Rightarrow\ v(t) = \frac{1}{2}t^2 + 4t + 5$ m/s

(b) distance traveled $= \int_0^{10} |v(t)|\,dt = \int_0^{10} \left|\frac{1}{2}t^2 + 4t + 5\right|\,dt = \int_0^{10} \left(\frac{1}{2}t^2 + 4t + 5\right)\,dt$

$$= \left[\tfrac{1}{6}t^3 + 2t^2 + 5t\right]_0^{10} = \tfrac{500}{3} + 200 + 50 = 416\tfrac{2}{3}\ \text{m}$$

57. Since $m'(x) = \rho(x)$, $m = \int_0^4 \rho(x)\,dx = \int_0^4 (9 + 2\sqrt{x})\,dx = \left[9x + \tfrac{4}{3}x^{3/2}\right]_0^4 = 36 + \tfrac{32}{3} - 0 = \tfrac{140}{3} = 46\tfrac{2}{3}$ kg.

59. Let s be the position of the car. We know from Equation 2 that $s(100) - s(0) = \int_0^{100} v(t)\,dt$. We use the Midpoint Rule for $0 \le t \le 100$ with $n = 5$. Note that the length of each of the five time intervals is 20 seconds $= \frac{20}{3600}$ hour $= \frac{1}{180}$ hour. So the distance traveled is

$$\int_0^{100} v(t)\,dt \approx \tfrac{1}{180}[v(10) + v(30) + v(50) + v(70) + v(90)]$$

$$= \tfrac{1}{180}(38 + 58 + 51 + 53 + 47)$$

$$= \tfrac{247}{180} \approx 1.4\ \text{miles}$$

61. From the Net Change Theorem, the increase in cost if the production level is raised from 2000 yards to 4000 yards is $C(4000) - C(2000) = \int_{2000}^{4000} C'(x)\,dx$.

$$\int_{2000}^{4000} C'(x)\,dx = \int_{2000}^{4000} (3 - 0.01x + 0.000006x^2)\,dx$$

$$= \left[3x - 0.005x^2 + 0.000002x^3\right]_{2000}^{4000} = 60{,}000 - 2{,}000 = \$58{,}000$$

63. (a) We can find the area between the Lorenz curve and the line $y = x$ by subtracting the area under $y = L(x)$ from the area under $y = x$. Thus,

$$\text{coefficient of inequality} = \frac{\text{area between Lorenz curve and line } y = x}{\text{area under line } y = x} = \frac{\int_0^1 [x - L(x)]\,dx}{\int_0^1 x\,dx}$$

$$= \frac{\int_0^1 [x - L(x)]\,dx}{[x^2/2]_0^1} = \frac{\int_0^1 [x - L(x)]\,dx}{1/2} = 2\int_0^1 [x - L(x)]\,dx$$

(b) $L(x) = \frac{5}{12}x^2 + \frac{7}{12}x \ \Rightarrow\ L(50\%) = L\left(\frac{1}{2}\right) = \frac{5}{48} + \frac{7}{24} = \frac{19}{48} = 0.3958\overline{3}$, so the bottom 50% of the households receive at most about 40% of the income. Using the result in part (a),

$$\text{coefficient of inequality} = 2\int_0^1 [x - L(x)]\,dx = 2\int_0^1 \left(x - \tfrac{5}{12}x^2 - \tfrac{7}{12}x\right)\,dx$$

$$= 2\int_0^1 \left(\tfrac{5}{12}x - \tfrac{5}{12}x^2\right)\,dx = 2\int_0^1 \tfrac{5}{12}\left(x - x^2\right)\,dx$$

$$= \tfrac{5}{6}\left[\tfrac{1}{2}x^2 - \tfrac{1}{3}x^3\right]_0^1 = \tfrac{5}{6}\left(\tfrac{1}{2} - \tfrac{1}{3}\right) = \tfrac{5}{6}\left(\tfrac{1}{6}\right) = \tfrac{5}{36}$$

5.5 The Substitution Rule

1. Let $u = 3x$. Then $du = 3\,dx$, so $dx = \frac{1}{3}du$. Thus,

$\int \cos 3x\,dx = \int \cos u\left(\frac{1}{3}\,du\right) = \frac{1}{3}\int \cos u\,du = \frac{1}{3}\sin u + C = \frac{1}{3}\sin 3x + C$. Don't forget that it is often very easy to check an indefinite integration by differentiating your answer. In this case,

$\dfrac{d}{dx}\left(\frac{1}{3}\sin 3x + C\right) = \frac{1}{3}(\cos 3x)\cdot 3 = \cos 3x$, the desired result.

3. Let $u = x^3 + 1$. Then $du = 3x^2\,dx$ and $x^2\,dx = \frac{1}{3}\,du$, so

$\int x^2\sqrt{x^3 + 1}\,dx = \int \sqrt{u}\left(\frac{1}{3}\,du\right) = \dfrac{1}{3}\dfrac{u^{3/2}}{3/2} + C = \frac{1}{3}\cdot\frac{2}{3}u^{3/2} + C = \frac{2}{9}(x^3 + 1)^{3/2} + C.$

5. Let $u = 1 + 2x$. Then $du = 2\,dx$ and $dx = \frac{1}{2}\,du$, so

$\displaystyle\int \frac{4}{(1 + 2x)^3}\,dx = 4\int u^{-3}\left(\frac{1}{2}\,du\right) = 2\frac{u^{-2}}{-2} + C = -\frac{1}{u^2} + C = -\frac{1}{(1 + 2x)^2} + C.$

7. Let $u = x^2 + 3$. Then $du = 2x\,dx$, so $\int 2x(x^2 + 3)^4\,dx = \int u^4\,du = \frac{1}{5}u^5 + C = \frac{1}{5}(x^2 + 3)^5 + C.$

9. Let $u = 3x - 2$. Then $du = 3\,dx$ and $dx = \frac{1}{3}\,du$, so

$\int (3x - 2)^{20}\,dx = \int u^{20}\left(\frac{1}{3}\,du\right) = \frac{1}{3}\cdot\frac{1}{21}u^{21} + C = \frac{1}{63}(3x - 2)^{21} + C.$

11. Let $u = 1 + x + 2x^2$. Then $du = (1 + 4x)\,dx$, so

$\displaystyle\int \frac{1 + 4x}{\sqrt{1 + x + 2x^2}}\,dx = \int \frac{du}{\sqrt{u}} = \int u^{-1/2}\,du = \frac{u^{1/2}}{1/2} + C = 2\sqrt{1 + x + 2x^2} + C.$

13. Let $u = 5 - 3x$. Then $du = -3\,dx$ and $dx = -\frac{1}{3}\,du$, so

$\displaystyle\int \frac{dx}{5 - 3x} = \int \frac{1}{u}\left(-\frac{1}{3}\,du\right) = -\frac{1}{3}\ln|u| + C = -\frac{1}{3}\ln|5 - 3x| + C.$

15. Let $u = 2y + 1$. Then $du = 2\,dy$ and $dy = \frac{1}{2}\,du$, so

$\displaystyle\int \frac{3}{(2y + 1)^5}\,dy = \int 3u^{-5}\left(\frac{1}{2}\,du\right) = \frac{3}{2}\cdot\frac{1}{-4}u^{-4} + C = \frac{-3}{8(2y + 1)^4} + C.$

17. Let $u = 4 - t$. Then $du = -dt$ and $dt = -du$, so

$\int \sqrt{4 - t}\,dt = \int u^{1/2}\,(-du) = -\frac{2}{3}u^{3/2} + C = -\frac{2}{3}(4 - t)^{3/2} + C.$

19. Let $u = \pi t$. Then $du = \pi\,dt$ and $dt = \frac{1}{\pi}\,du$, so

$\int \sin \pi t\,dt = \int \sin u\left(\frac{1}{\pi}\,du\right) = \frac{1}{\pi}(-\cos u) + C = -\frac{1}{\pi}\cos \pi t + C.$

21. Let $u = \ln x$. Then $du = \dfrac{dx}{x}$, so $\displaystyle\int \frac{(\ln x)^2}{x}\,dx = \int u^2\,du = \frac{1}{3}u^3 + C = \frac{1}{3}(\ln x)^3 + C.$

23. Let $u = \sqrt{t}$. Then $du = \dfrac{dt}{2\sqrt{t}}$ and $\dfrac{1}{\sqrt{t}}\,dt = 2\,du$, so

$\displaystyle\int \frac{\cos \sqrt{t}}{\sqrt{t}}\,dt = \int \cos u\,(2\,du) = 2\sin u + C = 2\sin\sqrt{t} + \dot{C}.$

25. Let $u = \sin\theta$. Then $du = \cos\theta\,d\theta$, so $\int \cos\theta \sin^6\theta\,d\theta = \int u^6\,du = \frac{1}{7}u^7 + C = \frac{1}{7}\sin^7\theta + C.$

27. Let $u = 1 + e^x$. Then $du = e^x\,dx$, so $\int e^x\sqrt{1 + e^x}\,dx = \int \sqrt{u}\,du = \frac{2}{3}u^{3/2} + C = \frac{2}{3}(1 + e^x)^{3/2} + C.$

Or: Let $u = \sqrt{1 + e^x}$. Then $u^2 = 1 + e^x$ and $2u\,du = e^x\,dx$, so

$\int e^x\sqrt{1 + e^x}\,dx = \int u\cdot 2u\,du = \frac{2}{3}u^3 + C = \frac{2}{3}(1 + e^x)^{3/2} + C.$

29. Let $u = 1 + z^3$. Then $du = 3z^2\,dz$ and $z^2\,dz = \frac{1}{3}\,du$, so

$$\int \frac{z^2}{\sqrt[3]{1+z^3}}\,dz = \int u^{-1/3}\left(\tfrac{1}{3}\,du\right) = \tfrac{1}{3}\cdot\tfrac{3}{2}u^{2/3} + C = \tfrac{1}{2}(1+z^3)^{2/3} + C.$$

31. Let $u = \ln x$. Then $du = \dfrac{dx}{x}$, so $\displaystyle\int \frac{dx}{x\ln x} = \int \frac{du}{u} = \ln|u| + C = \ln|\ln x| + C.$

33. Let $u = \cot x$. Then $du = -\csc^2 x\,dx$ and $\csc^2 x\,dx = -du$, so

$$\int \sqrt{\cot x}\,\csc^2 x\,dx = \int \sqrt{u}\,(-du) = -\frac{u^{3/2}}{3/2} + C = -\tfrac{2}{3}(\cot x)^{3/2} + C.$$

35. $\displaystyle\int \cot x\,dx = \int \frac{\cos x}{\sin x}\,dx.$ Let $u = \sin x$. Then $du = \cos x\,dx$, so

$\int \cot x\,dx = \int \frac{1}{u}\,du = \ln|u| + C = \ln|\sin x| + C.$

37. Let $u = \sec x$. Then $du = \sec x\tan x\,dx$, so

$\int \sec^3 x\tan x\,dx = \int \sec^2 x\,(\sec x\tan x)\,dx = \int u^2\,du = \tfrac{1}{3}u^3 + C = \tfrac{1}{3}\sec^3 x + C.$

39. Let $u = b + cx^{a+1}$. Then $du = (a+1)cx^a\,dx$, so

$$\int x^a\sqrt{b+cx^{a+1}}\,dx = \int u^{1/2}\frac{1}{(a+1)c}\,du = \frac{1}{(a+1)c}\left(\tfrac{2}{3}u^{3/2}\right) + C = \frac{2}{3c(a+1)}\left(b+cx^{a+1}\right)^{3/2} + C.$$

41. Let $u = 1 + x^2$. Then $du = 2x\,dx$, so

$$\int \frac{1+x}{1+x^2}\,dx = \int \frac{1}{1+x^2}\,dx + \int \frac{x}{1+x^2}\,dx = \tan^{-1}x + \int \frac{\tfrac{1}{2}du}{u} = \tan^{-1}x + \tfrac{1}{2}\ln|u| + C$$
$$= \tan^{-1}x + \tfrac{1}{2}\ln\left|1+x^2\right| + C = \tan^{-1}x + \tfrac{1}{2}\ln\left(1+x^2\right) + C \ \ [\text{since } 1+x^2 > 0].$$

43. Let $u = x + 2$. Then $du = dx$, so

$$\int \frac{x}{\sqrt[4]{x+2}}\,dx = \int \frac{u-2}{\sqrt[4]{u}}\,du = \int \left(u^{3/4} - 2u^{-1/4}\right)du = \tfrac{4}{7}u^{7/4} - 2\cdot\tfrac{4}{3}u^{3/4} + C$$
$$= \tfrac{4}{7}(x+2)^{7/4} - \tfrac{8}{3}(x+2)^{3/4} + C$$

In Exercises 45–48, let $f(x)$ denote the integrand and $F(x)$ its antiderivative (with $C = 0$).

45. $f(x) = \dfrac{3x-1}{(3x^2-2x+1)^4}.$

$u = 3x^2 - 2x + 1 \quad\Rightarrow\quad du = (6x-2)\,dx = 2(3x-1)\,dx$, so

$$\int \frac{3x-1}{(3x^2-2x+1)^4}\,dx = \int \frac{1}{u^4}\left(\frac{1}{2}\,du\right) = \frac{1}{2}\int u^{-4}\,du$$

$$= -\frac{1}{6}u^{-3} + C = -\frac{1}{6\,(3x^2-2x+1)^3} + C$$

Notice that at $x = \frac{1}{3}$, f changes from negative to positive, and F has a local minimum.

47. $f(x) = \sin^3 x\cos x. \ \ u = \sin x \quad\Rightarrow\quad du = \cos x\,dx$, so

$$\int \sin^3 x\cos x\,dx = \int u^3\,du = \tfrac{1}{4}u^4 + C = \tfrac{1}{4}\sin^4 x + C$$

Note that at $x = \frac{\pi}{2}$, f changes from positive to negative and F has a local maximum. Also, both f and F are periodic with period π, so at $x = 0$ and at $x = \pi$, f changes from negative to positive and F has local minima.

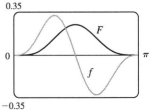

49. Let $u = x - 1$, so $du = dx$. When $x = 0$, $u = -1$; when $x = 2$, $u = 1$. Thus, $\int_0^2 (x - 1)^{25}\,dx = \int_{-1}^1 u^{25}\,du = 0$ by Theorem 7(b), since $f(u) = u^{25}$ is an odd function.

51. Let $u = 1 + 2x^3$, so $du = 6x^2\,dx$. When $x = 0$, $u = 1$; when $x = 1$, $u = 3$. Thus,

$$\int_0^1 x^2 \left(1 + 2x^3\right)^5 dx = \int_1^3 u^5 \left(\tfrac{1}{6}\,du\right) = \tfrac{1}{6}\left[\tfrac{1}{6}u^6\right]_1^3 = \tfrac{1}{36}(3^6 - 1^6) = \tfrac{1}{36}(729 - 1) = \tfrac{728}{36} = \tfrac{182}{9}$$

53. Let $u = t/4$, so $du = \tfrac{1}{4}\,dt$. When $t = 0$, $u = 0$; when $t = \pi$, $u = \pi/4$. Thus,

$$\int_0^\pi \sec^2(t/4)\,dt = \int_0^{\pi/4} \sec^2 u\,(4\,du) = 4\big[\tan u\big]_0^{\pi/4} = 4\left(\tan\tfrac{\pi}{4} - \tan 0\right) = 4(1 - 0) = 4.$$

55. $\int_{-\pi/6}^{\pi/6} \tan^3\theta\,d\theta = 0$ by Theorem 7(b), since $f(\theta) = \tan^3\theta$ is an odd function.

57. Let $u = 1/x$, so $du = -1/x^2\,dx$. When $x = 1$, $u = 1$; when $x = 2$, $u = \tfrac{1}{2}$. Thus,

$$\int_1^2 \frac{e^{1/x}}{x^2}\,dx = \int_1^{1/2} e^u\,(-du) = -\big[e^u\big]_1^{1/2} = -(e^{1/2} - e) = e - \sqrt{e}.$$

59. Let $u = \cos\theta$, so $du = -\sin\theta\,d\theta$. When $\theta = 0$, $u = 1$; when $\theta = \tfrac{\pi}{3}$, $u = \tfrac{1}{2}$. Thus,

$$\int_0^{\pi/3} \frac{\sin\theta}{\cos^2\theta}\,d\theta = \int_1^{1/2} \frac{-du}{u^2} = \int_{1/2}^1 u^{-2}\,du = \left[-\frac{1}{u}\right]_{1/2}^1 = -1 - (-2) = 1.$$

61. Let $u = 1 + 2x$, so $du = 2\,dx$. When $x = 0$, $u = 1$; when $x = 13$, $u = 27$. Thus,

$$\int_0^{13} \frac{dx}{\sqrt[3]{(1 + 2x)^2}} = \int_1^{27} u^{-2/3}\left(\tfrac{1}{2}\,du\right) = \left[\tfrac{1}{2}\cdot 3u^{1/3}\right]_1^{27} = \tfrac{3}{2}(3 - 1) = 3.$$

63. Let $u = x - 1$, so $u + 1 = x$ and $du = dx$. When $x = 1$, $u = 0$; when $x = 2$, $u = 1$. Thus,

$$\int_1^2 x\sqrt{x - 1}\,dx = \int_0^1 (u + 1)\sqrt{u}\,du = \int_0^1 (u^{3/2} + u^{1/2})\,du = \left[\tfrac{2}{5}u^{5/2} + \tfrac{2}{3}u^{3/2}\right]_0^1 = \tfrac{2}{5} + \tfrac{2}{3} = \tfrac{16}{15}.$$

65. Let $u = \ln x$, so $du = \dfrac{dx}{x}$. When $x = e$, $u = 1$; when $x = e^4$; $u = 4$. Thus,

$$\int_e^{e^4} \frac{dx}{x\sqrt{\ln x}} = \int_1^4 u^{-1/2}\,du = 2\big[u^{1/2}\big]_1^4 = 2(2 - 1) = 2.$$

67. $\int_0^4 \dfrac{dx}{(x - 2)^3}$ does not exist since $f(x) = \dfrac{1}{(x - 2)^3}$ has an infinite discontinuity at $x = 2$.

69. Let $u = x^2 + a^2$, so $du = 2x\,dx$ and $x\,dx = \tfrac{1}{2}\,du$. When $x = 0$, $u = a^2$; when $x = a$, $u = 2a^2$. Thus,

$$\int_0^a x\sqrt{x^2 + a^2}\,dx = \int_{a^2}^{2a^2} u^{1/2}\left(\tfrac{1}{2}\,du\right) = \tfrac{1}{2}\left[\tfrac{2}{3}u^{3/2}\right]_{a^2}^{2a^2} = \left[\tfrac{1}{3}u^{3/2}\right]_{a^2}^{2a^2}$$
$$= \tfrac{1}{3}\left[(2a^2)^{3/2} - (a^2)^{3/2}\right] = \tfrac{1}{3}\left(2\sqrt{2} - 1\right)a^3$$

71. From the graph, it appears that the area under the curve is about

$1 + \left(\text{a little more than }\tfrac{1}{2}\cdot 1\cdot 0.7\right)$, or about 1.4. The exact area is given by

$A = \int_0^1 \sqrt{2x + 1}\,dx$. Let $u = 2x + 1$, so $du = 2\,dx$. The limits change to

$2\cdot 0 + 1 = 1$ and $2\cdot 1 + 1 = 3$, and

$A = \int_1^3 \sqrt{u}\left(\tfrac{1}{2}\,du\right) = \tfrac{1}{2}\left[\tfrac{2}{3}u^{3/2}\right]_1^3 = \tfrac{1}{3}\left(3\sqrt{3} - 1\right) = \sqrt{3} - \tfrac{1}{3} \approx 1.399.$

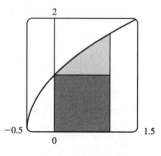

73. First write the integral as a sum of two integrals:

$I = \int_{-2}^{2}(x+3)\sqrt{4-x^2}\,dx = I_1 + I_2 = \int_{-2}^{2} x\sqrt{4-x^2}\,dx + \int_{-2}^{2} 3\sqrt{4-x^2}\,dx$. $I_1 = 0$ by Theorem 7(b), since

$f(x) = x\sqrt{4-x^2}$ is an odd function and we are integrating from $x = -2$ to $x = 2$. We interpret I_2 as three times

the area of a semicircle with radius 2, so $I = 0 + 3 \cdot \frac{1}{2}(\pi \cdot 2^2) = 6\pi$.

75. First Figure Let $u = \sqrt{x}$, so $x = u^2$ and $dx = 2u\,du$. When $x = 0$, $u = 0$; when $x = 1$, $u = 1$. Thus,

$$A_1 = \int_0^1 e^{\sqrt{x}}\,dx = \int_0^1 e^u(2u\,du) = 2\int_0^1 ue^u\,du.$$

Second Figure $A_2 = \int_0^1 2xe^x\,dx = 2\int_0^1 ue^u\,du.$

Third Figure Let $u = \sin x$, so $du = \cos x\,dx$. When $x = 0$, $u = 0$; when $x = \frac{\pi}{2}$, $u = 1$. Thus,

$$A_3 = \int_0^{\pi/2} e^{\sin x}\sin 2x\,dx = \int_0^{\pi/2} e^{\sin x}(2\sin x \cos x)\,dx = \int_0^1 e^u(2u\,du) = 2\int_0^1 ue^u\,du.$$

Since $A_1 = A_2 = A_3$, all three areas are equal.

77. The volume of inhaled air in the lungs at time t is

$$V(t) = \int_0^t f(u)\,du = \int_0^t \tfrac{1}{2}\sin\left(\tfrac{2\pi}{5}u\right)du = \int_0^{2\pi t/5} \tfrac{1}{2}\sin v\left(\tfrac{5}{2\pi}\,dv\right) \quad \text{[substitute } v = \tfrac{2\pi}{5}u,\ dv = \tfrac{2\pi}{5}\,du\text{]}$$

$$= \tfrac{5}{4\pi}\left[-\cos v\right]_0^{2\pi t/5} = \tfrac{5}{4\pi}\left[-\cos\left(\tfrac{2\pi}{5}t\right)+1\right] = \tfrac{5}{4\pi}\left[1 - \cos\left(\tfrac{2\pi}{5}t\right)\right] \text{ liters}$$

79. Let $u = 2x$. Then $du = 2\,dx$, so $\int_0^2 f(2x)\,dx = \int_0^4 f(u)\left(\tfrac{1}{2}\,du\right) = \tfrac{1}{2}\int_0^4 f(u)\,du = \tfrac{1}{2}(10) = 5.$

81. Let $u = -x$. Then $du = -dx$, so

$$\int_a^b f(-x)\,dx = \int_{-a}^{-b} f(u)(-du) = \int_{-b}^{-a} f(u)\,du = \int_{-b}^{-a} f(x)\,dx$$

From the diagram, we see that the equality follows from the fact that we
are reflecting the graph of f, and the limits of integration, about the
y-axis.

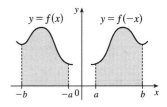

83. Let $u = 1 - x$. Then $x = 1 - u$ and $dx = -du$, so

$$\int_0^1 x^a(1-x)^b\,dx = \int_1^0 (1-u)^a u^b(-du) = \int_0^1 u^b(1-u)^a\,du = \int_0^1 x^b(1-x)^a\,dx.$$

85. $\dfrac{x\sin x}{1+\cos^2 x} = x \cdot \dfrac{\sin x}{2-\sin^2 x} = xf(\sin x)$, where $f(t) = \dfrac{t}{2-t^2}$. By Exercise 84,

$$\int_0^\pi \frac{x\sin x}{1+\cos^2 x}\,dx = \int_0^\pi xf(\sin x)\,dx = \frac{\pi}{2}\int_0^\pi f(\sin x)\,dx = \frac{\pi}{2}\int_0^\pi \frac{\sin x}{1+\cos^2 x}\,dx$$

Let $u = \cos x$. Then $du = -\sin x\,dx$. When $x = \pi$, $u = -1$ and when $x = 0$, $u = 1$. So

$$\frac{\pi}{2}\int_0^\pi \frac{\sin x}{1+\cos^2 x}\,dx = -\frac{\pi}{2}\int_1^{-1} \frac{du}{1+u^2} = \frac{\pi}{2}\int_{-1}^1 \frac{du}{1+u^2} = \frac{\pi}{2}\left[\tan^{-1} u\right]_{-1}^1$$

$$= \frac{\pi}{2}\left[\tan^{-1} 1 - \tan^{-1}(-1)\right] = \frac{\pi}{2}\left[\frac{\pi}{4} - \left(-\frac{\pi}{4}\right)\right] = \frac{\pi^2}{4}$$

5.6 The Logarithm Defined as an Integral

1. (a)

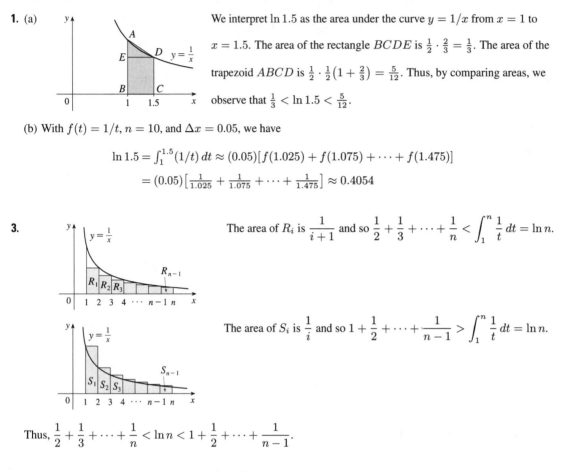

We interpret $\ln 1.5$ as the area under the curve $y = 1/x$ from $x = 1$ to $x = 1.5$. The area of the rectangle $BCDE$ is $\frac{1}{2} \cdot \frac{2}{3} = \frac{1}{3}$. The area of the trapezoid $ABCD$ is $\frac{1}{2} \cdot \frac{1}{2}\left(1 + \frac{2}{3}\right) = \frac{5}{12}$. Thus, by comparing areas, we observe that $\frac{1}{3} < \ln 1.5 < \frac{5}{12}$.

(b) With $f(t) = 1/t$, $n = 10$, and $\Delta x = 0.05$, we have

$$\ln 1.5 = \int_1^{1.5}(1/t)\,dt \approx (0.05)[f(1.025) + f(1.075) + \cdots + f(1.475)]$$
$$= (0.05)\left[\tfrac{1}{1.025} + \tfrac{1}{1.075} + \cdots + \tfrac{1}{1.475}\right] \approx 0.4054$$

3.

The area of R_i is $\dfrac{1}{i+1}$ and so $\dfrac{1}{2} + \dfrac{1}{3} + \cdots + \dfrac{1}{n} < \displaystyle\int_1^n \dfrac{1}{t}\,dt = \ln n$.

The area of S_i is $\dfrac{1}{i}$ and so $1 + \dfrac{1}{2} + \cdots + \dfrac{1}{n-1} > \displaystyle\int_1^n \dfrac{1}{t}\,dt = \ln n$.

Thus, $\dfrac{1}{2} + \dfrac{1}{3} + \cdots + \dfrac{1}{n} < \ln n < 1 + \dfrac{1}{2} + \cdots + \dfrac{1}{n-1}$.

5. If $f(x) = \ln(x^r)$, then $f'(x) = (1/x^r)(rx^{r-1}) = r/x$. But if $g(x) = r\ln x$, then $g'(x) = r/x$. So f and g must differ by a constant: $\ln(x^r) = r\ln x + C$. Put $x = 1$: $\ln(1^r) = r\ln 1 + C \;\Rightarrow\; C = 0$, so $\ln(x^r) = r\ln x$.

7. Using the third law of logarithms and Equation 10, we have $\ln e^{rx} = rx = r\ln e^x = \ln(e^x)^r$. Since $\ln$ is a one-to-one function, it follows that $e^{rx} = (e^x)^r$.

9. Using Definition 13, the first law of logarithms, and the first law of exponents for e^x, we have

$$(ab)^x = e^{x\ln(ab)} = e^{x(\ln a + \ln b)} = e^{x\ln a + x\ln b} = e^{x\ln a}e^{x\ln b} = a^x b^x.$$

5 Review

───────────────────────────────── CONCEPT CHECK ─────────────────────────────────

1. (a) $\sum_{i=1}^{n} f(x_i^*) \Delta x$ is an expression for a Riemann sum of a function f.

   x_i^* is a point in the ith subinterval $[x_{i-1}, x_i]$ and Δx is the length of the subintervals.

   (b) See Figure 1 in Section 5.2.

   (c) In Section 5.2, see Figure 3 and the paragraph beside it.

2. (a) See Definition 5.2.2.

   (b) See Figure 2 in Section 5.2.

   (c) In Section 5.2, see Figure 4 and the paragraph above it.

3. See the Fundamental Theorem of Calculus after Example 8 in Section 5.3.

4. (a) See the Net Change Theorem after Example 5 in Section 5.4.

   (b) $\int_{t_1}^{t_2} r(t)\, dt$ represents the change in the amount of water in the reservoir between time t_1 and time t_2.

5. (a) $\int_{60}^{120} v(t)\, dt$ represents the change in position of the particle from $t = 60$ to $t = 120$ seconds.

   (b) $\int_{60}^{120} |v(t)|\, dt$ represents the total distance traveled by the particle from $t = 60$ to 120 seconds.

   (c) $\int_{60}^{120} a(t)\, dt$ represents the change in the velocity of the particle from $t = 60$ to $t = 120$ seconds.

6. (a) $\int f(x)\, dx$ is the family of functions $\{F \mid F' = f\}$. Any two such functions differ by a constant.

   (b) The connection is given by the Evaluation Theorem: $\int_a^b f(x)\, dx = \left[\int f(x)\, dx \right]_a^b$ if f is continuous.

7. The precise version of this statement is given by the Fundamental Theorem of Calculus. See the statement of this theorem and the paragraph that follows it at the end of Section 5.3.

8. See the Substitution Rule (5.5.4). This says that it is permissible to operate with the dx after an integral sign as if it were a differential.

───────────────────────────────── TRUE-FALSE QUIZ ─────────────────────────────────

1. True by Property 2 of the Integral in Section 5.2.

3. True by Property 3 of the Integral in Section 5.2.

5. False.   For example, let $f(x) = x^2$. Then $\int_0^1 \sqrt{x^2}\, dx = \int_0^1 x\, dx = \frac{1}{2}$, but $\sqrt{\int_0^1 x^2\, dx} = \sqrt{\frac{1}{3}} = \frac{1}{\sqrt{3}}$.

7. True by Comparison Property 7 of the Integral in Section 5.2.

9. True.   The integrand is an odd function that is continuous on $[-1, 1]$, so the result follows from Theorem 5.5.6(b).

11. False.   The function $f(x) = 1/x^4$ is not bounded on the interval $[-2, 1]$. It has an infinite discontinuity at $x = 0$, so it is not integrable on the interval. (If the integral were to exist, a positive value would be expected, by Comparison Property 6 of Integrals.)

13. False.   For example, the function $y = |x|$ is continuous on $\mathbb{R}$, but has no derivative at $x = 0$.

―――――――――――――――――――――――― EXERCISES ――――――――――――――――

1. (a)

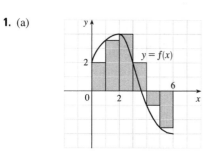

$$L_6 = \sum_{i=1}^{6} f(x_{i-1}) \, \Delta x \quad [\Delta x = \tfrac{6-0}{6} = 1]$$

$$= f(x_0) \cdot 1 + f(x_1) \cdot 1 + f(x_2) \cdot 1$$
$$+ f(x_3) \cdot 1 + f(x_4) \cdot 1 + f(x_5) \cdot 1$$
$$\approx 2 + 3.5 + 4 + 2 + (-1) + (-2.5) = 8$$

The Riemann sum represents the sum of the areas of the four rectangles above the x-axis minus the sum of the areas of the two rectangles below the x-axis.

(b)

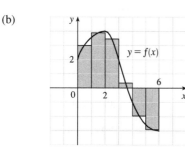

$$M_6 = \sum_{i=1}^{6} f(\overline{x}_i) \, \Delta x \quad [\Delta x = \tfrac{6-0}{6} = 1]$$

$$= f(\overline{x}_1) \cdot 1 + f(\overline{x}_2) \cdot 1 + f(\overline{x}_3) \cdot 1$$
$$+ f(\overline{x}_4) \cdot 1 + f(\overline{x}_5) \cdot 1 + f(\overline{x}_6) \cdot 1$$
$$= f(0.5) + f(1.5) + f(2.5) + f(3.5) + f(4.5) + f(5.5)$$
$$\approx 3 + 3.9 + 3.4 + 0.3 + (-2) + (-2.9) = 5.7$$

The Riemann sum represents the sum of the areas of the four rectangles above the x-axis minus the sum of the areas of the two rectangles below the x-axis.

3. $\int_0^1 \left(x + \sqrt{1-x^2}\right) dx = \int_0^1 x \, dx + \int_0^1 \sqrt{1-x^2} \, dx = I_1 + I_2$.

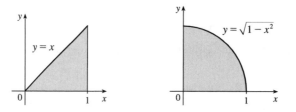

I_1 can be interpreted as the area of the triangle shown in the figure and I_2 can be interpreted as the area of the quarter-circle. Area $= \tfrac{1}{2}(1)(1) + \tfrac{1}{4}(\pi)(1)^2 = \tfrac{1}{2} + \tfrac{\pi}{4}$.

5. $\int_0^6 f(x) \, dx = \int_0^4 f(x) \, dx + \int_4^6 f(x) \, dx \quad \Rightarrow \quad 10 = 7 + \int_4^6 f(x) \, dx \quad \Rightarrow \quad \int_4^6 f(x) \, dx = 10 - 7 = 3$

7. First note that either a or b must be the graph of $\int_0^x f(t) \, dt$, since $\int_0^0 f(t) \, dt = 0$, and $c(0) \neq 0$. Now notice that $b > 0$ when c is increasing, and that $c > 0$ when a is increasing. It follows that c is the graph of $f(x)$, b is the graph of $f'(x)$, and a is the graph of $\int_0^x f(t) \, dt$.

9. $\int_1^2 \left(8x^3 + 3x^2\right) dx = \left[8 \cdot \tfrac{1}{4}x^4 + 3 \cdot \tfrac{1}{3}x^3\right]_1^2 = \left[2x^4 + x^3\right]_1^2 = \left(2 \cdot 2^4 + 2^3\right) - (2 + 1) = 40 - 3 = 37$

11. $\int_0^1 \left(1 - x^9\right) dx = \left[x - \tfrac{1}{10}x^{10}\right]_0^1 = \left(1 - \tfrac{1}{10}\right) - 0 = \tfrac{9}{10}$

13. $\int_1^9 \dfrac{\sqrt{u} - 2u^2}{u} \, du = \int_1^9 (u^{-1/2} - 2u) \, du = \left[2u^{1/2} - u^2 \right]_1^9 = (6 - 81) - (2 - 1) = -76$

15. Let $u = y^2 + 1$, so $du = 2y \, dy$ and $y \, dy = \frac{1}{2} \, du$. When $y = 0$, $u = 1$; when $y = 1$, $u = 2$. Thus,

$\int_0^1 y(y^2 + 1)^5 \, dy = \int_1^2 u^5 \left(\frac{1}{2} \, du \right) = \frac{1}{2} \left[\frac{1}{6} u^6 \right]_1^2 = \frac{1}{12} (64 - 1) = \frac{63}{12} = \frac{21}{4}.$

17. $\int_1^5 \dfrac{dt}{(t - 4)^2}$ does not exist because the function $f(t) = \dfrac{1}{(t - 4)^2}$ has an infinite discontinuity at $t = 4$;

that is, f is discontinuous on the interval $[1, 5]$.

19. Let $u = v^3$, so $du = 3v^2 \, dv$. When $v = 0$, $u = 0$; when $v = 1$, $u = 1$. Thus,

$\int_0^1 v^2 \cos(v^3) \, dv = \int_0^1 \cos u \left(\frac{1}{3} \, du \right) = \frac{1}{3} \left[\sin u \right]_0^1 = \frac{1}{3} (\sin 1 - 0) = \frac{1}{3} \sin 1.$

21. $\int_0^1 e^{\pi t} \, dt = \left[\frac{1}{\pi} e^{\pi t} \right]_0^1 = \frac{1}{\pi} (e^{\pi} - 1)$

23. $\int_2^4 \dfrac{1 + x - x^2}{x^2} \, dx = \int_2^4 \left(\dfrac{1}{x^2} + \dfrac{x}{x^2} - \dfrac{x^2}{x^2} \right) dx = \int_2^4 \left(x^{-2} + \dfrac{1}{x} - 1 \right) dx = \left[-\dfrac{1}{x} + \ln|x| - x \right]_2^4$

$\qquad = \left(-\frac{1}{4} + \ln 4 - 4 \right) - \left(-\frac{1}{2} + \ln 2 - 2 \right) = \ln 2 - \frac{7}{4}$

25. Let $u = x^2 + 4x$. Then $du = (2x + 4) \, dx = 2(x + 2) \, dx$, so

$\int \dfrac{x + 2}{\sqrt{x^2 + 4x}} \, dx = \int u^{-1/2} \left(\frac{1}{2} \, du \right) = \frac{1}{2} \cdot 2u^{1/2} + C = \sqrt{u} + C = \sqrt{x^2 + 4x} + C.$

27. Let $u = \sin \pi t$. Then $du = \pi \cos \pi t \, dt$, so $\int \sin \pi t \cos \pi t \, dt = \int u \left(\frac{1}{\pi} \, du \right) = \frac{1}{\pi} \cdot \frac{1}{2} u^2 + C = \frac{1}{2\pi} (\sin \pi t)^2 + C.$

29. Let $u = \sqrt{x}$. Then $du = \dfrac{dx}{2\sqrt{x}}$, so $\int \dfrac{e^{\sqrt{x}}}{\sqrt{x}} \, dx = 2 \int e^u \, du = 2e^u + C = 2e^{\sqrt{x}} + C.$

31. Let $u = \ln(\cos x)$. Then $du = \dfrac{-\sin x}{\cos x} \, dx = -\tan x \, dx$, so

$\int \tan x \ln(\cos x) \, dx = -\int u \, du = -\frac{1}{2} u^2 + C = -\frac{1}{2} [\ln(\cos x)]^2 + C.$

33. Let $u = 1 + x^4$. Then $du = 4x^3 \, dx$, so $\int \dfrac{x^3}{1 + x^4} \, dx = \frac{1}{4} \int \dfrac{1}{u} \, du = \frac{1}{4} \ln|u| + C = \frac{1}{4} \ln(1 + x^4) + C.$

35. Let $u = 1 + \sec \theta$. Then $du = \sec \theta \tan \theta \, d\theta$, so

$\int \dfrac{\sec \theta \tan \theta}{1 + \sec \theta} \, d\theta = \int \dfrac{1}{1 + \sec \theta} (\sec \theta \tan \theta \, d\theta) = \int \dfrac{1}{u} \, du = \ln|u| + C = \ln|1 + \sec \theta| + C.$

37. Since $x^2 - 4 < 0$ for $0 \le x < 2$ and $x^2 - 4 > 0$ for $2 < x \le 3$, we have $|x^2 - 4| = -(x^2 - 4) = 4 - x^2$ for

$0 \le x < 2$ and $|x^2 - 4| = x^2 - 4$ for $2 < x \le 3$. Thus,

$$\int_0^3 |x^2 - 4| \, dx = \int_0^2 (4 - x^2) \, dx + \int_2^3 (x^2 - 4) \, dx = \left[4x - \dfrac{x^3}{3} \right]_0^2 + \left[\dfrac{x^3}{3} - 4x \right]_2^3$$

$$= \left(8 - \frac{8}{3} \right) - 0 + (9 - 12) - \left(\frac{8}{3} - 8 \right) = \frac{16}{3} - 3 + \frac{16}{3} = \frac{32}{3} - \frac{9}{3} = \frac{23}{3}$$

In Exercises 39 and 40, let $f(x)$ denote the integrand and $F(x)$ its antiderivative (with $C = 0$).

39. Let $u = 1 + \sin x$. Then $du = \cos x \, dx$, so

$$\int \frac{\cos x \, dx}{\sqrt{1 + \sin x}} = \int u^{-1/2} \, du = 2u^{1/2} + C = 2\sqrt{1 + \sin x} + C.$$

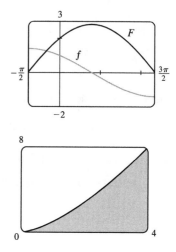

41. From the graph, it appears that the area under the curve $y = x\sqrt{x}$ between $x = 0$ and $x = 4$ is somewhat less than half the area of an 8×4 rectangle, so perhaps about 13 or 14. To find the exact value, we evaluate

$$\int_0^4 x\sqrt{x} \, dx = \int_0^4 x^{3/2} \, dx = \left[\tfrac{2}{5} x^{5/2}\right]_0^4 = \tfrac{2}{5}(4)^{5/2} = \tfrac{64}{5} = 12.8.$$

43. By FTC1, $F(x) = \int_1^x \sqrt{1 + t^4} \, dt \Rightarrow F'(x) = \sqrt{1 + x^4}$.

45. $g(x) = \displaystyle\int_0^{x^3} \frac{t \, dt}{\sqrt{1 + t^3}}$. Let $y = g(u)$ and $u = x^3$.

Then $g'(x) = \dfrac{dy}{dx} = \dfrac{dy}{du}\dfrac{du}{dx} = \dfrac{u}{\sqrt{1 + u^3}} 3x^2 = \dfrac{x^3}{\sqrt{1 + x^9}} 3x^2 = \dfrac{3x^5}{\sqrt{1 + x^9}}$.

47. $y = \displaystyle\int_{\sqrt{x}}^x \frac{e^t}{t} \, dt = \int_{\sqrt{x}}^1 \frac{e^t}{t} \, dt + \int_1^x \frac{e^t}{t} \, dt = -\int_1^{\sqrt{x}} \frac{e^t}{t} \, dt + \int_1^x \frac{e^t}{t} \, dt \Rightarrow$

$\dfrac{dy}{dx} = -\dfrac{d}{dx}\left(\displaystyle\int_1^{\sqrt{x}} \frac{e^t}{t} \, dt\right) + \dfrac{d}{dx}\left(\displaystyle\int_1^x \frac{e^t}{t} \, dt\right)$. Let $u = \sqrt{x}$. Then

$$\frac{d}{dx}\int_1^{\sqrt{x}} \frac{e^t}{t} \, dt = \frac{d}{dx}\int_1^u \frac{e^t}{t} \, dt = \frac{d}{du}\left(\int_1^u \frac{e^t}{t} \, dt\right)\frac{du}{dx} = \frac{e^u}{u}\cdot\frac{1}{2\sqrt{x}} = \frac{e^{\sqrt{x}}}{\sqrt{x}}\cdot\frac{1}{2\sqrt{x}} = \frac{e^{\sqrt{x}}}{2x},$$

so $\dfrac{dy}{dx} = -\dfrac{e^{\sqrt{x}}}{2x} + \dfrac{e^x}{x}$.

49. If $1 \le x \le 3$, then $\sqrt{1^2 + 3} \le \sqrt{x^2 + 3} \le \sqrt{3^2 + 3} \Rightarrow 2 \le \sqrt{x^2 + 3} \le 2\sqrt{3}$, so $2(3 - 1) \le \int_1^3 \sqrt{x^2 + 3} \, dx \le 2\sqrt{3}(3 - 1)$; that is, $4 \le \int_1^3 \sqrt{x^2 + 3} \, dx \le 4\sqrt{3}$.

51. $0 \le x \le 1 \Rightarrow 0 \le \cos x \le 1 \Rightarrow x^2 \cos x \le x^2 \Rightarrow$
$\int_0^1 x^2 \cos x \, dx \le \int_0^1 x^2 \, dx = \tfrac{1}{3}\left[x^3\right]_0^1 = \tfrac{1}{3}$ [Property 7].

53. $\cos x \le 1 \Rightarrow e^x \cos x \le e^x \Rightarrow \int_0^1 e^x \cos x \, dx \le \int_0^1 e^x \, dx = \left[e^x\right]_0^1 = e - 1$

55. Let $f(x) = \sqrt{1+x^3}$ on $[0, 1]$. The Midpoint Rule with $n = 5$ gives

$$\int_0^1 \sqrt{1+x^3}\, dx \approx \tfrac{1}{5}[f(0.1) + f(0.3) + f(0.5) + f(0.7) + f(0.9)]$$

$$= \tfrac{1}{5}\left[\sqrt{1+(0.1)^3} + \sqrt{1+(0.3)^3} + \cdots + \sqrt{1+(0.9)^3}\right] \approx 1.110$$

57. Note that $r(t) = b'(t)$, where $b(t) =$ the number of barrels of oil consumed up to time t. So, by the Net Change Theorem, $\int_0^3 r(t)\, dt = b(3) - b(0)$ represents the number of barrels of oil consumed from Jan. 1, 2000, through Jan. 1, 2003.

59. We use the Midpoint Rule with $n = 6$ and $\Delta t = \frac{24-0}{6} = 4$. The increase in the bee population was

$$\int_0^{24} r(t)\, dt \approx M_6 = 4[r(2) + r(6) + r(10) + r(14) + r(18) + r(22)]$$

$$\approx 4[50 + 1000 + 7000 + 8550 + 1350 + 150]$$

$$= 4(18{,}100) = 72{,}400$$

61. By the Fundamental Theorem of Calculus, we know that $F(x) = \int_a^x t^2 \sin(t^2)\, dt$ is an antiderivative of $f(x) = x^2 \sin(x^2)$. This integral cannot be expressed in any simpler form. Since $\int_a^a f\, dt = 0$ for any a, we can take $a = 1$, and then $F(1) = 0$, as required. So $F(x) = \int_1^x t^2 \sin(t^2)\, dt$ is the desired function.

63. Area under the curve $y = \sinh cx$ between $x = 0$ and $x = 1$ is

equal to $1 \Rightarrow \int_0^1 \sinh cx\, dx = 1 \Rightarrow \tfrac{1}{c}[\cosh cx]_0^1 = 1 \Rightarrow$

$\tfrac{1}{c}(\cosh c - 1) = 1 \Rightarrow \cosh c - 1 = c \Rightarrow \cosh c = c + 1$.

From the graph, we get $c = 0$ and $c \approx 1.6161$, but $c = 0$ isn't a

solution for this problem since the curve $y = \sinh cx$ becomes $y = 0$

and the area under it is 0. Thus, $c \approx 1.6161$.

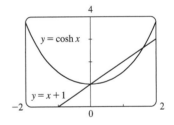

65. Using FTC1, we differentiate both sides of the given equation, $\int_0^x f(t)\, dt = xe^{2x} + \int_0^x e^{-t} f(t)\, dt$, and get

$$f(x) = e^{2x} + 2xe^{2x} + e^{-x} f(x) \Rightarrow f(x)(1 - e^{-x}) = e^{2x} + 2xe^{2x} \Rightarrow f(x) = \frac{e^{2x}(1+2x)}{1-e^{-x}}.$$

67. Let $u = f(x)$ and $du = f'(x)\, dx$. So $2\int_a^b f(x)f'(x)\, dx = 2\int_{f(a)}^{f(b)} u\, du = \left[u^2\right]_{f(a)}^{f(b)} = [f(b)]^2 - [f(a)]^2$.

69. Let $u = 1 - x$. Then $du = -dx$, so $\int_0^1 f(1-x)\, dx = \int_1^0 f(u)(-du) = \int_0^1 f(u)\, du = \int_0^1 f(x)\, dx$.

1. Differentiating both sides of the equation $x \sin \pi x = \int_0^{x^2} f(t)\,dt$ (using FTC1 and the Chain Rule for the right side)

gives $\sin \pi x + \pi x \cos \pi x = 2x f(x^2)$. Letting $x = 2$ so that $f(x^2) = f(4)$, we obtain

$\sin 2\pi + 2\pi \cos 2\pi = 4f(4)$, so $f(4) = \frac{1}{4}(0 + 2\pi \cdot 1) = \frac{\pi}{2}$.

3. For $1 \le x \le 2$, we have $x^4 \le 2^4 = 16$, so $1 + x^4 \le 17$ and $\dfrac{1}{1+x^4} \ge \dfrac{1}{17}$. Thus,

$$\int_1^2 \frac{1}{1+x^4}\,dx \ge \int_1^2 \frac{1}{17}\,dx = \frac{1}{17}.$$ Also $1 + x^4 > x^4$ for $1 \le x \le 2$, so $\dfrac{1}{1+x^4} < \dfrac{1}{x^4}$ and

$$\int_1^2 \frac{1}{1+x^4}\,dx < \int_1^2 x^{-4}\,dx = \left[\frac{x^{-3}}{-3}\right]_1^2 = -\frac{1}{24} + \frac{1}{3} = \frac{7}{24}.$$ Thus, we have the estimate

$$\frac{1}{17} \le \int_1^2 \frac{1}{1+x^4}\,dx \le \frac{7}{24}.$$

5. $f(x) = \displaystyle\int_0^{g(x)} \frac{1}{\sqrt{1+t^3}}\,dt$, where $g(x) = \int_0^{\cos x} \left[1 + \sin(t^2)\right]dt$. Using FTC1 and the Chain Rule (twice) we

have $f'(x) = \dfrac{1}{\sqrt{1+[g(x)]^3}}g'(x) = \dfrac{1}{\sqrt{1+[g(x)]^3}}\left[1 + \sin(\cos^2 x)\right](-\sin x)$. Now

$g\!\left(\frac{\pi}{2}\right) = \int_0^0 \left[1 + \sin(t^2)\right]dt = 0$, so $f'\!\left(\frac{\pi}{2}\right) = \frac{1}{\sqrt{1+0}}(1 + \sin 0)(-1) = 1 \cdot 1 \cdot (-1) = -1$.

7. By l'Hospital's Rule and the Fundamental Theorem, using the notation $\exp(y) = e^y$,

$$\lim_{x\to 0} \frac{\int_0^x (1 - \tan 2t)^{1/t}\,dt}{x} \overset{\text{H}}{=} \lim_{x\to 0} \frac{(1 - \tan 2x)^{1/x}}{1} = \exp\left(\lim_{x\to 0} \frac{\ln(1 - \tan 2x)}{x}\right)$$

$$\overset{\text{H}}{=} \exp\left(\lim_{x\to 0} \frac{-2\sec^2 2x}{1 - \tan 2x}\right) = \exp\left(\frac{-2 \cdot 1^2}{1 - 0}\right) = e^{-2}.$$

9. $f(x) = 2 + x - x^2 = (-x + 2)(x + 1) = 0 \iff x = 2$ or $x = -1$. $f(x) \ge 0$ for $x \in [-1, 2]$ and $f(x) < 0$

everywhere else. The integral $\int_a^b (2 + x - x^2)\,dx$ has a maximum on the interval where the integrand is positive,

which is $[-1, 2]$. So $a = -1$, $b = 2$. (Any larger interval gives a smaller integral since $f(x) < 0$ outside $[-1, 2]$.

Any smaller interval also gives a smaller integral since $f(x) \ge 0$ in $[-1, 2]$.)

11. (a) We can split the integral $\int_0^n [\![x]\!]\,dx$ into the sum $\displaystyle\sum_{i=1}^n \left[\int_{i-1}^i [\![x]\!]\,dx\right]$. But on each of the intervals $[i - 1, i)$ of

integration, $[\![x]\!]$ is a constant function, namely $i - 1$. So the ith integral in the sum is equal to

$(i - 1)[i - (i - 1)] = (i - 1)$. So the original integral is equal to $\displaystyle\sum_{i=1}^n (i - 1) = \sum_{i=1}^{n-1} i = \frac{(n-1)n}{2}$.

(b) We can write $\int_a^b [\![x]\!] \, dx = \int_0^b [\![x]\!] \, dx - \int_0^a [\![x]\!] \, dx$.

Now $\int_0^b [\![x]\!] \, dx = \int_0^{[\![b]\!]} [\![x]\!] \, dx + \int_{[\![b]\!]}^b [\![x]\!] \, dx$. The first of these integrals is equal to $\frac{1}{2}([\![b]\!] - 1) [\![b]\!]$, by part (a),

and since $[\![x]\!] = [\![b]\!]$ on $[[\![b]\!], b]$, the second integral is just $[\![b]\!] (b - [\![b]\!])$. So

$\int_0^b [\![x]\!] \, dx = \frac{1}{2}([\![b]\!] - 1) [\![b]\!] + [\![b]\!] (b - [\![b]\!]) = \frac{1}{2} [\![b]\!] (2b - [\![b]\!] - 1)$ and similarly

$\int_0^a [\![x]\!] \, dx = \frac{1}{2} [\![a]\!] (2a - [\![a]\!] - 1)$. Therefore, $\int_a^b [\![x]\!] \, dx = \frac{1}{2} [\![b]\!] (2b - [\![b]\!] - 1) - \frac{1}{2} [\![a]\!] (2a - [\![a]\!] - 1)$.

13. Differentiating the equation $\int_0^x f(t) \, dt = [f(x)]^2$ using FTC1 gives $f(x) = 2f(x)f'(x) \ \Rightarrow$

$f(x)[2f'(x) - 1] = 0$, so $f(x) = 0$ or $f'(x) = \frac{1}{2}$. $f'(x) = \frac{1}{2} \ \Rightarrow \ f(x) = \frac{1}{2}x + C$. To find C we substitute into

the original equation to get $\int_0^x \left(\frac{1}{2}t + C\right) dt = \left(\frac{1}{2}x + C\right)^2 \ \Leftrightarrow \ \frac{1}{4}x^2 + Cx = \frac{1}{4}x^2 + Cx + C^2$. It follows that

$C = 0$, so $f(x) = \frac{1}{2}x$. Therefore, $f(x) = 0$ or $f(x) = \frac{1}{2}x$.

15. Note that $\dfrac{d}{dx} \left(\int_0^x \left[\int_0^u f(t) \, dt \right] du \right) = \int_0^x f(t) \, dt$ by FTC1, while

$$\frac{d}{dx} \left[\int_0^x f(u)(x - u) \, du \right] = \frac{d}{dx} \left[x \int_0^x f(u) \, du \right] - \frac{d}{dx} \left[\int_0^x f(u)u \, du \right]$$

$$= \int_0^x f(u) \, du + xf(x) - f(x)x = \int_0^x f(u) \, du$$

Hence, $\int_0^x f(u)(x - u) \, du = \int_0^x \left[\int_0^u f(t) \, dt \right] du + C$. Setting $x = 0$ gives $C = 0$.

17. $\displaystyle\lim_{n\to\infty} \left(\frac{1}{\sqrt{n}\sqrt{n+1}} + \frac{1}{\sqrt{n}\sqrt{n+2}} + \cdots + \frac{1}{\sqrt{n}\sqrt{n+n}} \right)$

$\displaystyle = \lim_{n\to\infty} \frac{1}{n} \left(\sqrt{\frac{n}{n+1}} + \sqrt{\frac{n}{n+2}} + \cdots + \sqrt{\frac{n}{n+n}} \right)$

$\displaystyle = \lim_{n\to\infty} \frac{1}{n} \left(\frac{1}{\sqrt{1+1/n}} + \frac{1}{\sqrt{1+2/n}} + \cdots + \frac{1}{\sqrt{1+1}} \right)$

$\displaystyle = \lim_{n\to\infty} \frac{1}{n} \sum_{i=1}^{n} f\left(\frac{i}{n}\right) \qquad \left[\text{where } f(x) = \frac{1}{\sqrt{1+x}}\right]$

$\displaystyle = \int_0^1 \frac{1}{\sqrt{1+x}} \, dx = \left[2\sqrt{1+x}\right]_0^1 = 2\left(\sqrt{2} - 1\right)$

19. The shaded region has area $\int_0^1 f(x) \, dx = \frac{1}{3}$. The integral $\int_0^1 f^{-1}(y) \, dy$

gives the area of the unshaded region, which we know to be $1 - \frac{1}{3} = \frac{2}{3}$.

So $\int_0^1 f^{-1}(y) \, dy = \frac{2}{3}$.

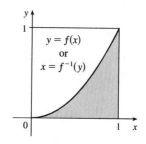

6 □ APPLICATIONS OF INTEGRATION

6.1 Areas between Curves

1. $A = \int_{x=0}^{x=4} (y_T - y_B)\, dx = \int_0^4 \left[(5x - x^2) - x \right] dx = \int_0^4 \left(4x - x^2 \right) dx$

$= \left[2x^2 - \frac{1}{3}x^3 \right]_0^4 = \left(32 - \frac{64}{3} \right) - (0) = \frac{32}{3}$

3. $A = \int_{y=-1}^{y=1} (x_R - x_L)\, dy = \int_{-1}^1 \left[e^y - (y^2 - 2) \right] dy$

$= \int_{-1}^1 \left(e^y - y^2 + 2 \right) dy = \left[e^y - \frac{1}{3}y^3 + 2y \right]_{-1}^1 = \left(e^1 - \frac{1}{3} + 2 \right) - \left(e^{-1} + \frac{1}{3} - 2 \right) = e - \frac{1}{e} + \frac{10}{3}$

5. $A = \int_{-1}^2 \left[(9 - x^2) - (x + 1) \right] dx$

$= \int_{-1}^2 \left(8 - x - x^2 \right) dx$

$= \left[8x - \frac{x^2}{2} - \frac{x^3}{3} \right]_{-1}^2$

$= \left(16 - 2 - \frac{8}{3} \right) - \left(-8 - \frac{1}{2} + \frac{1}{3} \right)$

$= 22 - 3 + \frac{1}{2} = \frac{39}{2}$

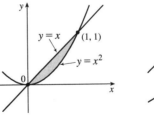

7. The curves intersect when $x = x^2 \;\Rightarrow\; x^2 - x = 0 \;\Leftrightarrow\; x(x - 1) = 0 \;\Leftrightarrow\; x = 0, 1$.

$A = \int_0^1 \left(x - x^2 \right) dx$

$= \left[\frac{1}{2}x^2 - \frac{1}{3}x^3 \right]_0^1$

$= \frac{1}{2} - \frac{1}{3}$

$= \frac{1}{6}$

9. $A = \int_1^2 \left(\frac{1}{x} - \frac{1}{x^2} \right) dx = \left[\ln x + \frac{1}{x} \right]_1^2$

$= \left(\ln 2 + \frac{1}{2} \right) - (\ln 1 + 1)$

$= \ln 2 - \frac{1}{2} \approx 0.19$

11. $A = \int_0^1 \left(\sqrt{x} - x^2 \right) dx$

$= \left[\frac{2}{3} x^{3/2} - \frac{1}{3} x^3 \right]_0^1$

$= \frac{2}{3} - \frac{1}{3}$

$= \frac{1}{3}$

13. $12 - x^2 = x^2 - 6 \quad \Leftrightarrow \quad 2x^2 = 18 \quad \Leftrightarrow \quad x^2 = 9 \quad \Leftrightarrow \quad x = \pm 3$, so

$A = \int_{-3}^3 \left[(12 - x^2) - (x^2 - 6) \right] dx = 2 \int_0^3 \left(18 - 2x^2 \right) dx \quad$ [by symmetry]

$= 2 \left[18x - \frac{2}{3} x^3 \right]_0^3 = 2 \left[(54 - 18) - 0 \right] = 2(36) = 72$

15. $\frac{1}{2} x = \sqrt{x} \quad \Rightarrow \quad \frac{1}{4} x^2 = x \quad \Rightarrow \quad x^2 - 4x = 0 \quad \Rightarrow \quad x(x - 4) = 0 \quad \Rightarrow \quad x = 0$ or 4, so

$A = \int_0^4 \left(\sqrt{x} - \frac{1}{2} x \right) dx + \int_4^9 \left(\frac{1}{2} x - \sqrt{x} \right) dx = \left[\frac{2}{3} x^{3/2} - \frac{1}{4} x^2 \right]_0^4 + \left[\frac{1}{4} x^2 - \frac{2}{3} x^{3/2} \right]_4^9$

$= \left[\left(\frac{16}{3} - 4 \right) - 0 \right] + \left[\left(\frac{81}{4} - 18 \right) - \left(4 - \frac{16}{3} \right) \right] = \frac{81}{4} + \frac{32}{3} - 26 = \frac{59}{12}$

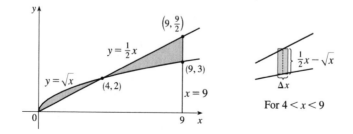

17. $2y^2 = 1 - y$ ⇔ $2y^2 + y - 1 = 0$ ⇔ $(2y - 1)(y + 1) = 0$ ⇔ $y = \frac{1}{2}$ or -1, so $x = \frac{1}{2}$ or 2 and

$A = \int_{-1}^{1/2} \left[(1 - y) - 2y^2 \right] dy = \int_{-1}^{1/2} \left(1 - y - 2y^2 \right) dy = \left[y - \frac{1}{2}y^2 - \frac{2}{3}y^3 \right]_{-1}^{1/2}$

$= \left(\frac{1}{2} - \frac{1}{8} - \frac{1}{12} \right) - \left(-1 - \frac{1}{2} + \frac{2}{3} \right) = \frac{7}{24} - \left(-\frac{5}{6} \right) = \frac{7}{24} + \frac{20}{24} = \frac{27}{24} = \frac{9}{8}$

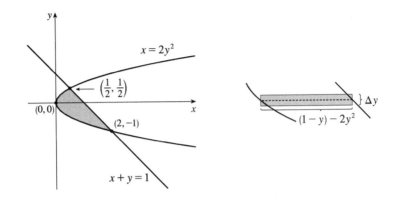

19. The curves intersect when $1 - y^2 = y^2 - 1$ ⇔ $2 = 2y^2$ ⇔ $y^2 = 1$ ⇔ $y = \pm 1$.

$A = \int_{-1}^{1} \left[(1 - y^2) - (y^2 - 1) \right] dy$

$= \int_{-1}^{1} 2(1 - y^2)\, dy$

$= 2 \cdot 2 \int_{0}^{1} (1 - y^2)\, dy$

$= 4 \left[y - \frac{1}{3}y^3 \right]_0^1 = 4 \left(1 - \frac{1}{3} \right) = \frac{8}{3}$

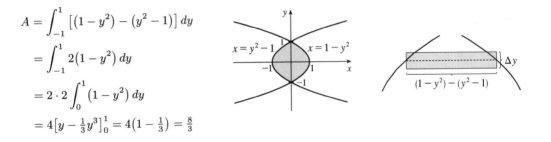

21. Notice that $\cos x = \sin 2x = 2 \sin x \cos x$ ⇔

$2 \sin x \cos x - \cos x = 0$ ⇔ $\cos x (2 \sin x - 1) = 0$ ⇔

$2 \sin x = 1$ or $\cos x = 0$ ⇔ $x = \frac{\pi}{6}$ or $\frac{\pi}{2}$.

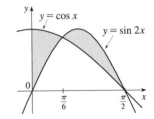

$A = \int_0^{\pi/6} (\cos x - \sin 2x)\, dx + \int_{\pi/6}^{\pi/2} (\sin 2x - \cos x)\, dx$

$= \left[\sin x + \frac{1}{2} \cos 2x \right]_0^{\pi/6} + \left[-\frac{1}{2} \cos 2x - \sin x \right]_{\pi/6}^{\pi/2}$

$= \frac{1}{2} + \frac{1}{2} \cdot \frac{1}{2} - \left(0 + \frac{1}{2} \cdot 1 \right) + \left(\frac{1}{2} - 1 \right) - \left(-\frac{1}{2} \cdot \frac{1}{2} - \frac{1}{2} \right) = \frac{1}{2}$

23. From the graph, we see that the curves intersect at $x = 0$, $x = \frac{\pi}{2}$, and $x = \pi$. By symmetry,

$$A = \int_0^\pi \left| \cos x - \left(1 - \frac{2x}{\pi}\right) \right| dx = 2 \int_0^{\pi/2} \left[\cos x - \left(1 - \frac{2x}{\pi}\right) \right] dx = 2 \int_0^{\pi/2} \left(\cos x - 1 + \frac{2x}{\pi} \right) dx$$

$$= 2 \left[\sin x - x + \tfrac{1}{\pi} x^2 \right]_0^{\pi/2} = 2 \left[\left(1 - \tfrac{\pi}{2} + \tfrac{1}{\pi} \cdot \tfrac{\pi^2}{4} \right) - 0 \right] = 2 \left(1 - \tfrac{\pi}{2} + \tfrac{\pi}{4} \right) = 2 - \tfrac{\pi}{2}$$

25. The curves intersect when $x^2 = \dfrac{2}{x^2 + 1}$ ⇔ $x^4 + x^2 = 2$ ⇔ $x^4 + x^2 - 2 = 0$ ⇔

$(x^2 + 2)(x^2 - 1) = 0$ ⇔ $x^2 = 1$ ⇔ $x = \pm 1$.

$$A = \int_{-1}^1 \left(\frac{2}{x^2 + 1} - x^2 \right) dx$$

$$= 2 \int_0^1 \left(\frac{2}{x^2 + 1} - x^2 \right) dx$$

$$= 2 \left[2 \tan^{-1} x - \tfrac{1}{3} x^3 \right]_0^1 = 2 \left(2 \cdot \tfrac{\pi}{4} - \tfrac{1}{3} \right)$$

$$= \pi - \tfrac{2}{3} \approx 2.47$$

27. An equation of the line through $(0, 0)$ and $(2, 1)$ is $y = \frac{1}{2} x$; through $(0, 0)$

and $(-1, 6)$ is $y = -6x$; through $(2, 1)$ and $(-1, 6)$ is $y = -\frac{5}{3}x + \frac{13}{3}$.

$$A = \int_{-1}^0 \left[\left(-\tfrac{5}{3}x + \tfrac{13}{3} \right) - (-6x) \right] dx + \int_0^2 \left[\left(-\tfrac{5}{3}x + \tfrac{13}{3} \right) - \tfrac{1}{2}x \right] dx$$

$$= \int_{-1}^0 \left(\tfrac{13}{3}x + \tfrac{13}{3} \right) dx + \int_0^2 \left(-\tfrac{13}{6}x + \tfrac{13}{3} \right) dx$$

$$= \tfrac{13}{3} \int_{-1}^0 (x + 1)\, dx + \tfrac{13}{3} \int_0^2 \left(-\tfrac{1}{2}x + 1 \right) dx$$

$$= \tfrac{13}{3} \left[\tfrac{1}{2}x^2 + x \right]_{-1}^0 + \tfrac{13}{3} \left[-\tfrac{1}{4}x^2 + x \right]_0^2$$

$$= \tfrac{13}{3} \left[0 - \left(\tfrac{1}{2} - 1 \right) \right] + \tfrac{13}{3} \left[(-1 + 2) - 0 \right] = \tfrac{13}{3} \cdot \tfrac{1}{2} + \tfrac{13}{3} \cdot 1 = \tfrac{13}{2}$$

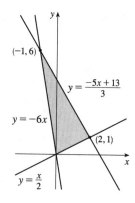

29. $A = \int_{-1}^{1} \left| x^3 - x \right| dx$

$= 2 \int_{0}^{1} \left(x - x^3 \right) dx$ [by symmetry]

$= 2 \left[\frac{1}{2} x^2 - \frac{1}{4} x^4 \right]_{0}^{1}$

$= 2 \left(\frac{1}{2} - \frac{1}{4} \right) = \frac{1}{2}$

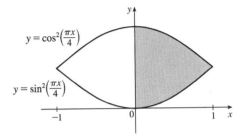

31. Let $f(x) = \cos^2 \left(\frac{\pi x}{4} \right) - \sin^2 \left(\frac{\pi x}{4} \right)$ and $\Delta x = \frac{1 - 0}{4}$.

The shaded area is given by

$A = \int_{0}^{1} f(x) \, dx \approx M_4$

$\qquad = \frac{1}{4} \left[f \left(\frac{1}{8} \right) + f \left(\frac{3}{8} \right) + f \left(\frac{5}{8} \right) + f \left(\frac{7}{8} \right) \right]$

$\qquad \approx 0.6407$

33.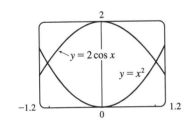

From the graph, we see that the curves intersect at $x = \pm a \approx \pm 1.02$, with $2 \cos x > x^2$ on $(-a, a)$. So the area of the region bounded by the curves is

$A = \int_{-a}^{a} \left(2 \cos x - x^2 \right) dx = 2 \int_{0}^{a} \left(2 \cos x - x^2 \right) dx$

$\qquad = 2 \left[2 \sin x - \frac{1}{3} x^3 \right]_{0}^{a} \approx 2.70$

35.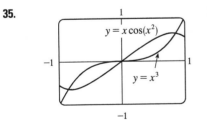

From the graph, we see that the curves intersect at $x = \pm a \approx \pm 0.86$. So the area of the region bounded by the curves is

$A = 2 \int_{0}^{a} \left[x \cos(x^2) - x^3 \right] dx = 2 \left[\frac{1}{2} \sin(x^2) - \frac{1}{4} x^4 \right]_{0}^{a}$

$\qquad \approx 0.40$

37. As the figure illustrates, the curves $y = x$ and

$y = x^5 - 6x^3 + 4x$ enclose a four-part region symmetric

about the origin (since $x^5 - 6x^3 + 4x$ and x are odd

functions of x). The curves intersect at values of x where

$x^5 - 6x^3 + 4x = x$; that is, where $x(x^4 - 6x^2 + 3) = 0$.

That happens at $x = 0$ and where

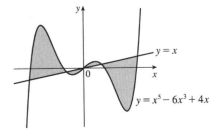

$$x^2 = \frac{6 \pm \sqrt{36 - 12}}{2} = 3 \pm \sqrt{6}; \text{ that is, at } x = -\sqrt{3 + \sqrt{6}}, -\sqrt{3 - \sqrt{6}}, 0, \sqrt{3 - \sqrt{6}}, \text{ and } \sqrt{3 + \sqrt{6}}. \text{ The exact}$$

area is

$$2 \int_0^{\sqrt{3+\sqrt{6}}} \left| (x^5 - 6x^3 + 4x) - x \right| dx = 2 \int_0^{\sqrt{3+\sqrt{6}}} \left| x^5 - 6x^3 + 3x \right| dx$$

$$= 2 \int_0^{\sqrt{3-\sqrt{6}}} (x^5 - 6x^3 + 3x) \, dx + 2 \int_{\sqrt{3-\sqrt{6}}}^{\sqrt{3+\sqrt{6}}} (-x^5 + 6x^3 - 3x) \, dx$$

$$\overset{\text{CAS}}{=} 12\sqrt{6} - 9$$

39. 1 second $= \frac{1}{3600}$ hour, so 10 s $= \frac{1}{360}$ h. With the given data, we can take $n = 5$ to use the Midpoint Rule.

$\Delta t = \frac{1/360 - 0}{5} = \frac{1}{1800}$, so

$$\text{distance}_{\text{Kelly}} - \text{distance}_{\text{Chris}} = \int_0^{1/360} v_K \, dt - \int_0^{1/360} v_C \, dt = \int_0^{1/360} (v_K - v_C) \, dt$$

$$\approx M_5 = \frac{1}{1800} \left[(v_K - v_C)(1) + (v_K - v_C)(3) + (v_K - v_C)(5) \right.$$

$$\left. + (v_K - v_C)(7) + (v_K - v_C)(9) \right]$$

$$= \frac{1}{1800} \left[(22 - 20) + (52 - 46) + (71 - 62) + (86 - 75) + (98 - 86) \right]$$

$$= \frac{1}{1800} (2 + 6 + 9 + 11 + 12) = \frac{1}{1800} (40) = \frac{1}{45} \text{ mile, or } 117\frac{1}{3} \text{ feet}$$

41. We know that the area under curve A between $t = 0$ and $t = x$ is $\int_0^x v_A(t) \, dt = s_A(x)$, where $v_A(t)$ is the velocity of car A and s_A is its displacement. Similarly, the area under curve B between $t = 0$ and $t = x$ is $\int_0^x v_B(t) \, dt = s_B(x)$.

(a) After one minute, the area under curve A is greater than the area under curve B. So car A is ahead after one minute.

(b) The area of the shaded region has numerical value $s_A(1) - s_B(1)$, which is the distance by which A is ahead of B after 1 minute.

(c) After two minutes, car B is traveling faster than car A and has gained some ground, but the area under curve A from $t = 0$ to $t = 2$ is still greater than the corresponding area for curve B, so car A is still ahead.

(d) From the graph, it appears that the area between curves A and B for $0 \le t \le 1$ (when car A is going faster), which corresponds to the distance by which car A is ahead, seems to be about 3 squares. Therefore, the cars will be side by side at the time x where the area between the curves for $1 \le t \le x$ (when car B is going faster) is the same as the area for $0 \le t \le 1$. From the graph, it appears that this time is $x \approx 2.2$. So the cars are side by side when $t \approx 2.2$ minutes.

43.

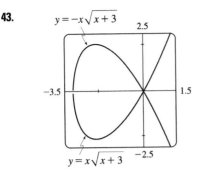

$y = -x\sqrt{x+3}$

$y = x\sqrt{x+3}$

To graph this function, we must first express it as a combination of explicit functions of y; namely, $y = \pm x\sqrt{x+3}$. We can see from the graph that the loop extends from $x = -3$ to $x = 0$, and that by symmetry, the area we seek is just twice the area under the top half of the curve on this interval, the equation of the top half being $y = -x\sqrt{x+3}$. So the area is $A = 2\int_{-3}^{0}\left(-x\sqrt{x+3}\right)dx$. We substitute $u = x+3$, so $du = dx$ and the limits change to 0 and 3, and we get

$$A = -2\int_{0}^{3}\left[(u-3)\sqrt{u}\,\right]du = -2\int_{0}^{3}\left(u^{3/2} - 3u^{1/2}\right)du$$

$$= -2\left[\tfrac{2}{5}u^{5/2} - 2u^{3/2}\right]_{0}^{3} = -2\left[\tfrac{2}{5}\left(3^{2}\sqrt{3}\right) - 2\left(3\sqrt{3}\right)\right] = \tfrac{24}{5}\sqrt{3}$$

45.

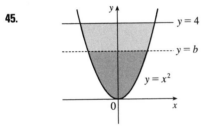

$y = 4$

$y = b$

$y = x^2$

By the symmetry of the problem, we consider only the first quadrant, where $y = x^2 \Rightarrow x = \sqrt{y}$. We are looking for a number b such that $\int_{0}^{b}\sqrt{y}\,dy = \int_{b}^{4}\sqrt{y}\,dy \Rightarrow \tfrac{2}{3}\left[y^{3/2}\right]_{0}^{b} = \tfrac{2}{3}\left[y^{3/2}\right]_{b}^{4} \Rightarrow$

$b^{3/2} = 4^{3/2} - b^{3/2} \Rightarrow 2b^{3/2} = 8 \Rightarrow b^{3/2} = 4 \Rightarrow$

$b = 4^{2/3} \approx 2.52$.

47. We first assume that $c > 0$, since c can be replaced by $-c$ in both equations without changing the graphs, and if $c = 0$ the curves do not enclose a region. We see from the graph that the enclosed area A lies between $x = -c$ and $x = c$, and by symmetry, it is equal to four times the area in the first quadrant.

The enclosed area is

$$A = 4\int_{0}^{c}\left(c^2 - x^2\right)dx = 4\left[c^2 x - \tfrac{1}{3}x^3\right]_{0}^{c}$$

$$= 4\left(c^3 - \tfrac{1}{3}c^3\right) = 4\left(\tfrac{2}{3}c^3\right) = \tfrac{8}{3}c^3$$

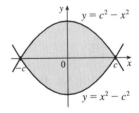

$y = c^2 - x^2$

$y = x^2 - c^2$

So $A = 576 \Leftrightarrow \tfrac{8}{3}c^3 = 576 \Leftrightarrow c^3 = 216 \Leftrightarrow c = \sqrt[3]{216} = 6$.

Note that $c = -6$ is another solution, since the graphs are the same.

49. The curve and the line will determine a region when they intersect at two or more points. So we solve the equation

$x/(x^2 + 1) = mx \Rightarrow x = x\left(mx^2 + m\right) \Rightarrow$

$x\left(mx^2 + m\right) - x = 0 \Rightarrow x\left(mx^2 + m - 1\right) = 0 \Rightarrow$

$x = 0$ or $mx^2 + m - 1 = 0 \Rightarrow x = 0$ or $x^2 = \dfrac{1 - m}{m} \Rightarrow$

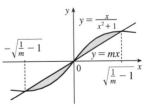

$y = \dfrac{x}{x^2 + 1}$

$y = mx$

$-\sqrt{\dfrac{1}{m} - 1}$

$\sqrt{\dfrac{1}{m} - 1}$

$x = 0$ or $x = \pm\sqrt{\dfrac{1}{m} - 1}$. Note that if $m = 1$, this has only the solution $x = 0$, and no region is determined. But if

$1/m - 1 > 0 \;\Leftrightarrow\; 1/m > 1 \;\Leftrightarrow\; 0 < m < 1$, then there are two solutions. [Another way of seeing this is to

observe that the slope of the tangent to $y = x/(x^2 + 1)$ at the origin is $y' = 1$ and therefore we must have

$0 < m < 1$.] Note that we cannot just integrate between the positive and negative roots, since the curve and the line

cross at the origin. Since mx and $x/(x^2 + 1)$ are both odd functions, the total area is twice the area between the

curves on the interval $\left[0, \sqrt{1/m - 1}\right]$. So the total area enclosed is

$$2\int_0^{\sqrt{1/m-1}} \left[\frac{x}{x^2 + 1} - mx\right] dx = 2\left[\tfrac{1}{2}\ln(x^2 + 1) - \tfrac{1}{2}mx^2\right]_0^{\sqrt{1/m-1}}$$

$$= [\ln(1/m - 1 + 1) - m(1/m - 1)] - (\ln 1 - 0)$$

$$= \ln(1/m) - 1 + m = m - \ln m - 1$$

6.2 Volumes

1. A cross-section is circular with radius x^2, so its area is $A(x) = \pi(x^2)^2$.

$$V = \int_0^1 A(x)\,dx = \int_0^1 \pi(x^2)^2\,dx = \pi\int_0^1 x^4\,dx = \pi\left[\tfrac{1}{5}x^5\right]_0^1 = \tfrac{\pi}{5}$$

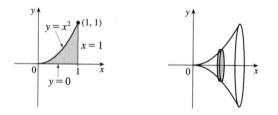

3. A cross-section is a disk with radius $1/x$, so its area is $A(x) = \pi(1/x)^2$.

$$V = \int_1^2 A(x)\,dx = \int_1^2 \pi\left(\frac{1}{x}\right)^2\,dx = \pi\int_1^2 \frac{1}{x^2}\,dx = \pi\left[-\frac{1}{x}\right]_1^2 = \pi\left[-\tfrac{1}{2} - (-1)\right] = \tfrac{\pi}{2}$$

5. A cross-section is a disk with radius $\sqrt{y}$, so its area is $A(y) = \pi\left(\sqrt{y}\right)^2$.

$$V = \int_0^4 A(y)\,dy = \int_0^4 \pi\left(\sqrt{y}\right)^2 dy = \pi \int_0^4 y\,dy = \pi\left[\tfrac{1}{2}y^2\right]_0^4 = 8\pi$$

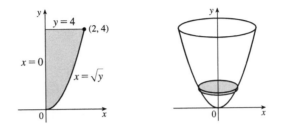

7. A cross-section is a washer (annulus) with inner radius x^2 and outer radius $\sqrt{x}$, so its area is
$A(x) = \pi\left(\sqrt{x}\right)^2 - \pi\left(x^2\right)^2 = \pi\left(x - x^4\right)$.

$$V = \int_0^1 A(x)\,dx = \pi\int_0^1 \left(x - x^4\right)dx = \pi\left[\tfrac{1}{2}x^2 - \tfrac{1}{5}x^5\right]_0^1 = \pi\left(\tfrac{1}{2} - \tfrac{1}{5}\right) = \tfrac{3\pi}{10}$$

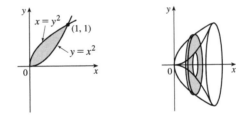

9. A cross-section is a washer with inner radius y^2 and outer radius $2y$, so its area is
$A(y) = \pi(2y)^2 - \pi\left(y^2\right)^2 = \pi\left(4y^2 - y^4\right)$.

$$V = \int_0^2 A(y)\,dy = \pi\int_0^2 \left(4y^2 - y^4\right)dy = \pi\left[\tfrac{4}{3}y^3 - \tfrac{1}{5}y^5\right]_0^2 = \pi\left(\tfrac{32}{3} - \tfrac{32}{5}\right) = \tfrac{64\pi}{15}$$

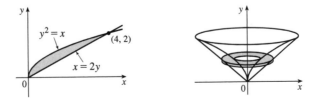

11. A cross-section is a washer with inner radius $1 - \sqrt{x}$ and outer radius $1 - x$, so its area is

$$A(x) = \pi(1 - x)^2 - \pi(1 - \sqrt{x})^2 = \pi\left[(1 - 2x + x^2) - (1 - 2\sqrt{x} + x)\right] = \pi(-3x + x^2 + 2\sqrt{x}).$$

$$V = \int_0^1 A(x)\,dx = \pi \int_0^1 (-3x + x^2 + 2\sqrt{x})\,dx$$

$$= \pi\left[-\tfrac{3}{2}x^2 + \tfrac{1}{3}x^3 + \tfrac{4}{3}x^{3/2}\right]_0^1 = \pi\left(-\tfrac{3}{2} + \tfrac{5}{3}\right) = \tfrac{\pi}{6}$$

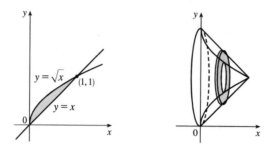

13. A cross-section is an annulus with inner radius $2 - 1$ and outer radius $2 - x^4$, so its area is

$$A(x) = \pi(2 - x^4)^2 - \pi(2 - 1)^2 = \pi(3 - 4x^4 + x^8).$$

$$V = \int_{-1}^1 A(x)\,dx = 2\int_0^1 A(x)\,dx = 2\pi \int_0^1 (3 - 4x^4 + x^8)\,dx = 2\pi\left[3x - \tfrac{4}{5}x^5 + \tfrac{1}{9}x^9\right]_0^1$$

$$= 2\pi\left(3 - \tfrac{4}{5} + \tfrac{1}{9}\right) = \tfrac{208}{45}\pi$$

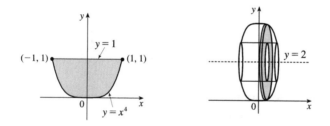

15. $V = \int_{-1}^1 \pi(1 - y^2)^2\,dy = 2\int_0^1 \pi(1 - y^2)^2\,dy = 2\pi \int_0^1 (1 - 2y^2 + y^4)\,dy$

$$= 2\pi\left[y - \tfrac{2}{3}y^3 + \tfrac{1}{5}y^5\right]_0^1 = 2\pi \cdot \tfrac{8}{15} = \tfrac{16}{15}\pi$$

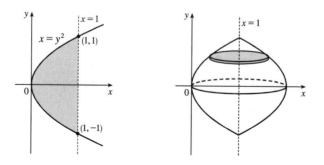

17. $y = x^2 \;\Rightarrow\; x = \sqrt{y}$ for $x \geq 0$. The outer radius is the distance from $x = -1$ to $x = \sqrt{y}$ and the inner radius is

the distance from $x = -1$ to $x = y^2$.

$$V = \int_0^1 \pi \left\{ \left[\sqrt{y} - (-1)\right]^2 - \left[y^2 - (-1)\right]^2 \right\} dy = \pi \int_0^1 \left[(\sqrt{y}+1)^2 - (y^2+1)^2\right] dy$$

$$= \pi \int_0^1 \left(y + 2\sqrt{y} + 1 - y^4 - 2y^2 - 1\right) dy = \pi \int_0^1 \left(y + 2\sqrt{y} - y^4 - 2y^2\right) dy$$

$$= \pi \left[\tfrac{1}{2}y^2 + \tfrac{4}{3}y^{3/2} - \tfrac{1}{5}y^5 - \tfrac{2}{3}y^3\right]_0^1 = \pi\left(\tfrac{1}{2} + \tfrac{4}{3} - \tfrac{1}{5} - \tfrac{2}{3}\right) = \tfrac{29}{30}\pi$$

19. $\mathcal{R}_1$ about OA (the line $y = 0$): $V = \int_0^1 A(x)\, dx = \int_0^1 \pi(x^3)^2\, dx = \pi \int_0^1 x^6\, dx = \pi \left[\tfrac{1}{7}x^7\right]_0^1 = \tfrac{\pi}{7}$

21. $\mathcal{R}_1$ about AB (the line $x = 1$):

$$V = \int_0^1 A(y)\, dy = \int_0^1 \pi\left(1 - \sqrt[3]{y}\right)^2 dy = \pi \int_0^1 \left(1 - 2y^{1/3} + y^{2/3}\right) dy$$

$$= \pi\left[y - \tfrac{3}{2}y^{4/3} + \tfrac{3}{5}y^{5/3}\right]_0^1 = \pi\left(1 - \tfrac{3}{2} + \tfrac{3}{5}\right) = \tfrac{\pi}{10}$$

23. $\mathcal{R}_2$ about OA (the line $y = 0$):

$$V = \int_0^1 A(x)\, dx = \int_0^1 \left[\pi(1)^2 - \pi(\sqrt{x})^2\right] dx = \pi \int_0^1 (1 - x)\, dx = \pi\left[x - \tfrac{1}{2}x^2\right]_0^1 = \pi\left(1 - \tfrac{1}{2}\right) = \tfrac{\pi}{2}$$

25. $\mathcal{R}_2$ about AB (the line $x = 1$):

$$V = \int_0^1 A(y)\, dy = \int_0^1 \left[\pi(1)^2 - \pi(1 - y^2)^2\right] dy = \pi \int_0^1 \left[1 - (1 - 2y^2 + y^4)\right] dy$$

$$= \pi \int_0^1 (2y^2 - y^4)\, dy = \pi\left[\tfrac{2}{3}y^3 - \tfrac{1}{5}y^5\right]_0^1 = \pi\left(\tfrac{2}{3} - \tfrac{1}{5}\right) = \tfrac{7\pi}{15}$$

27. $\mathcal{R}_3$ about OA (the line $y = 0$):

$$V = \int_0^1 A(x)\, dx = \int_0^1 \left[\pi(\sqrt{x})^2 - \pi(x^3)^2\right] dx = \pi \int_0^1 (x - x^6)\, dx = \pi\left[\tfrac{1}{2}x^2 - \tfrac{1}{7}x^7\right]_0^1 = \pi\left(\tfrac{1}{2} - \tfrac{1}{7}\right) = \tfrac{5\pi}{14}.$$

Note: Let $\mathcal{R} = \mathcal{R}_1 + \mathcal{R}_2 + \mathcal{R}_3$. If we rotate $\mathcal{R}$ about any of the segments OA, OC, AB, or BC, we obtain a right

circular cylinder of height 1 and radius 1. Its volume is $\pi r^2 h = \pi(1)^2 \cdot 1 = \pi$. As a check for Exercises 19, 23,

and 27, we can add the answers, and that sum must equal π. Thus, $\tfrac{\pi}{7} + \tfrac{\pi}{2} + \tfrac{5\pi}{14} = \left(\tfrac{2 + 7 + 5}{14}\right)\pi = \pi$.

29. $\mathcal{R}_3$ about AB (the line $x = 1$):

$$V = \int_0^1 A(y)\, dy = \int_0^1 \left[\pi(1 - y^2)^2 - \pi\left(1 - \sqrt[3]{y}\right)^2\right] dy = \pi \int_0^1 \left[(1 - 2y^2 + y^4) - (1 - 2y^{1/3} + y^{2/3})\right] dy$$

$$= \pi \int_0^1 \left(-2y^2 + y^4 + 2y^{1/3} - y^{2/3}\right) dy = \pi\left[-\tfrac{2}{3}y^3 + \tfrac{1}{5}y^5 + \tfrac{3}{2}y^{4/3} - \tfrac{3}{5}y^{5/3}\right]_0^1$$

$$= \pi\left(-\tfrac{2}{3} + \tfrac{1}{5} + \tfrac{3}{2} - \tfrac{3}{5}\right) = \tfrac{13\pi}{30}$$

Note: See the note in Exercise 27. For Exercises 21, 25, and 29, we have $\tfrac{\pi}{10} + \tfrac{7\pi}{15} + \tfrac{13\pi}{30} = \left(\tfrac{3 + 14 + 13}{30}\right)\pi = \pi$.

31. $V = \pi \displaystyle\int_0^{\pi/4} (1 - \tan^3 x)^2 \, dx$

33. $V = \pi \displaystyle\int_0^{\pi} \left[(1-0)^2 - (1 - \sin x)^2\right] dx$

$= \pi \displaystyle\int_0^{\pi} \left[1^2 - (1 - \sin x)^2\right] dx$

35. $V = \pi \displaystyle\int_{-\sqrt{8}}^{\sqrt{8}} \left\{ [3 - (-2)]^2 - \left[\sqrt{y^2 + 1} - (-2)\right]^2 \right\} dy$

$= \pi \displaystyle\int_{-2\sqrt{2}}^{2\sqrt{2}} \left[5^2 - \left(\sqrt{1 + y^2} + 2\right)^2 \right] dy$

37.

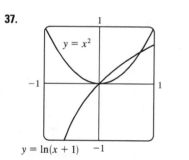

$y = x^2$ and $y = \ln(x+1)$ intersect at $x = 0$ and at $x = a \approx 0.747$.

$$V = \pi \int_0^a \left\{ [\ln(x+1)]^2 - \left(x^2\right)^2 \right\} dx \approx 0.132$$

39. $V = \pi \displaystyle\int_0^{\pi} \left\{ \left[\sin^2 x - (-1)\right]^2 - [0 - (-1)]^2 \right\} dx$

$\overset{\text{CAS}}{=} \dfrac{11}{8}\pi^2$

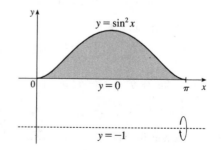

41. $\pi \int_0^{\pi/2} \cos^2 x \, dx$ describes the volume of the solid obtained by rotating the region

$\mathcal{R} = \{(x, y) \mid 0 \leq x \leq \frac{\pi}{2}, 0 \leq y \leq \cos x\}$ of the xy-plane about the x-axis.

43. $\pi \int_0^1 (y^4 - y^8) dy = \pi \int_0^1 \left[(y^2)^2 - (y^4)^2 \right] dy$ describes the volume of the solid obtained by rotating the region

$\mathcal{R} = \{(x, y) \mid 0 \leq y \leq 1, y^4 \leq x \leq y^2\}$ of the xy-plane about the y-axis.

45. There are 10 subintervals over the 15-cm length, so we'll use $n = 10/2 = 5$ for the Midpoint Rule.

$$V = \int_0^{15} A(x) \, dx \approx M_5 = \frac{15-0}{5}[A(1.5) + A(4.5) + A(7.5) + A(10.5) + A(13.5)]$$
$$= 3(18 + 79 + 106 + 128 + 39) = 3 \cdot 370 = 1110 \text{ cm}^3$$

47. We'll form a right circular cone with height h and base radius r by

revolving the line $y = \frac{r}{h}x$ about the x-axis.

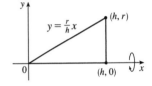

$$V = \pi \int_0^h \left(\frac{r}{h} x \right)^2 dx = \pi \int_0^h \frac{r^2}{h^2} x^2 dx = \pi \frac{r^2}{h^2} \left[\frac{1}{3} x^3 \right]_0^h$$
$$= \pi \frac{r^2}{h^2} \left(\frac{1}{3} h^3 \right) = \frac{1}{3} \pi r^2 h$$

Another solution: Revolve $x = -\dfrac{r}{h} y + r$ about the y-axis.

$$V = \pi \int_0^h \left(-\frac{r}{h} y + r \right)^2 dy \overset{*}{=} \pi \int_0^h \left[\frac{r^2}{h^2} y^2 - \frac{2r^2}{h} y + r^2 \right] dy$$
$$= \pi \left[\frac{r^2}{3h^2} y^3 - \frac{r^2}{h} y^2 + r^2 y \right]_0^h = \pi \left(\frac{1}{3} r^2 h - r^2 h + r^2 h \right) = \frac{1}{3} \pi r^2 h$$

* Or use substitution with $u = r - \dfrac{r}{h} y$ and $du = -\dfrac{r}{h} dy$ to get

$$\pi \int_r^0 u^2 \left(-\frac{h}{r} du \right) = -\pi \frac{h}{r} \left[\frac{1}{3} u^3 \right]_r^0 = -\pi \frac{h}{r} \left(-\frac{1}{3} r^3 \right) = \frac{1}{3} \pi r^2 h.$$

49. $x^2 + y^2 = r^2 \iff x^2 = r^2 - y^2$

$$V = \pi \int_{r-h}^r (r^2 - y^2) \, dy = \pi \left[r^2 y - \frac{y^3}{3} \right]_{r-h}^r$$
$$= \pi \left\{ \left[r^3 - \frac{r^3}{3} \right] - \left[r^2 (r - h) - \frac{(r - h)^3}{3} \right] \right\}$$
$$= \pi \left\{ \frac{2}{3} r^3 - \frac{1}{3}(r - h)\left[3r^2 - (r - h)^2 \right] \right\}$$
$$= \frac{1}{3} \pi \left\{ 2r^3 - (r - h)\left[3r^2 - (r^2 - 2rh + h^2) \right] \right\}$$
$$= \frac{1}{3} \pi \left\{ 2r^3 - (r - h)\left[2r^2 + 2rh - h^2 \right] \right\}$$
$$= \frac{1}{3} \pi \left(2r^3 - 2r^3 - 2r^2 h + rh^2 + 2r^2 h + 2rh^2 - h^3 \right)$$
$$= \frac{1}{3} \pi \left(3rh^2 - h^3 \right) = \frac{1}{3} \pi h^2 (3r - h), \text{ or, equivalently, } \pi h^2 \left(r - \frac{h}{3} \right)$$

51. For a cross-section at height y, we see from similar triangles that $\dfrac{a/2}{b/2} = \dfrac{h-y}{h}$, so $a = b\left(1 - \dfrac{y}{h}\right)$.

Similarly, for cross-sections having $2b$ as their base and β replacing a, $\beta = 2b\left(1 - \dfrac{y}{h}\right)$. So

$$
\begin{aligned}
V &= \int_0^h A(y)\,dy = \int_0^h \left[b\left(1 - \frac{y}{h}\right)\right]\left[2b\left(1 - \frac{y}{h}\right)\right]dy \\
&= \int_0^h 2b^2\left(1 - \frac{y}{h}\right)^2 dy = 2b^2 \int_0^h \left(1 - \frac{2y}{h} + \frac{y^2}{h^2}\right)dy \\
&= 2b^2\left[y - \frac{y^2}{h} + \frac{y^3}{3h^2}\right]_0^h = 2b^2\left[h - h + \tfrac{1}{3}h\right] \\
&= \tfrac{2}{3}b^2 h \quad [= \tfrac{1}{3}Bh \text{ where } B \text{ is the area of the base, as with any pyramid.}]
\end{aligned}
$$

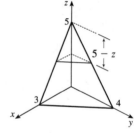

53. A cross-section at height z is a triangle similar to the base, so we'll multiply the legs of the base triangle, 3 and 4, by a proportionality factor of $(5 - z)/5$. Thus, the triangle at height z has area

$$
A(z) = \frac{1}{2}\cdot 3\left(\frac{5-z}{5}\right)\cdot 4\left(\frac{5-z}{5}\right) = 6\left(1 - \frac{z}{5}\right)^2, \text{ so}
$$

$$
\begin{aligned}
V &= \int_0^5 A(z)\,dz = 6\int_0^5 (1 - z/5)^2\,dz \\
&= 6\int_1^0 u^2(-5\,du) \quad [u = 1 - z/5,\ du = -\tfrac{1}{5}dz] \\
&= -30\left[\tfrac{1}{3}u^3\right]_1^0 = -30\left(-\tfrac{1}{3}\right) = 10 \text{ cm}^3
\end{aligned}
$$

55. If l is a leg of the isosceles right triangle and $2y$ is the hypotenuse,

then $l^2 + l^2 = (2y)^2 \ \Rightarrow\ 2l^2 = 4y^2 \ \Rightarrow\ l^2 = 2y^2$.

$$
V = \int_{-2}^2 A(x)\,dx = 2\int_0^2 A(x)\,dx = 2\int_0^2 \tfrac{1}{2}(l)(l)\,dx = 2\int_0^2 y^2\,dx
$$

$$
= 2\int_0^2 \tfrac{1}{4}(36 - 9x^2)\,dx = \tfrac{9}{2}\int_0^2 (4 - x^2)\,dx
$$

$$
= \tfrac{9}{2}\left[4x - \tfrac{1}{3}x^3\right]_0^2 = \tfrac{9}{2}\left(8 - \tfrac{8}{3}\right) = 24
$$

typical cross-section of length
$2y = \sqrt{36 - 9x^2}$

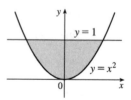

57. The cross-section of the base corresponding to the coordinate y has length

$2x = 2\sqrt{y}$. The square has area $A(y) = \left(2\sqrt{y}\right)^2 = 4y$, so

$$
V = \int_0^1 A(y)\,dy = \int_0^1 4y\,dy = \left[2y^2\right]_0^1 = 2.
$$

59. A typical cross-section perpendicular to the y-axis in the base has length
$\ell(y) = 3 - \frac{3}{2}y$. This length is the leg of an isosceles right triangle, so

$$A(y) = \frac{1}{2}\left[\ell(y)\right]^2 \quad \left[\frac{1}{2}bh \text{ with base} = \text{height}\right]$$

$$= \frac{1}{2}\left[3\left(1 - \frac{1}{2}y\right)\right]^2 = \frac{9}{2}\left(1 - \frac{1}{2}y\right)^2$$

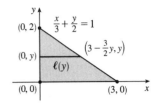

Thus,

$$V = \int_0^2 A(y)\,dy = \frac{9}{2}\int_1^0 u^2(-2\,du) \quad \left[u = 1 - \frac{1}{2}y,\ du = -\frac{1}{2}\,dy\right]$$

$$= -9\left[\frac{1}{3}u^3\right]_1^0 = -9\left(-\frac{1}{3}\right) = 3$$

61. (a) The torus is obtained by rotating the circle $(x - R)^2 + y^2 = r^2$
about the y-axis. Solving for x, we see that the right half of the
circle is given by $x = R + \sqrt{r^2 - y^2} = f(y)$ and the left half by
$x = R - \sqrt{r^2 - y^2} = g(y)$. So

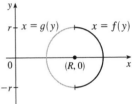

$$V = \pi \int_{-r}^{r} \left\{ [f(y)]^2 - [g(y)]^2 \right\} dy$$

$$= 2\pi \int_0^r \left[\left(R^2 + 2R\sqrt{r^2 - y^2} + r^2 - y^2 \right) - \left(R^2 - 2R\sqrt{r^2 - y^2} + r^2 - y^2 \right) \right] dy$$

$$= 2\pi \int_0^r 4R\sqrt{r^2 - y^2}\,dy = 8\pi R \int_0^r \sqrt{r^2 - y^2}\,dy$$

(b) Observe that the integral represents a quarter of the area of a circle with radius r, so
$$8\pi R \int_0^r \sqrt{r^2 - y^2}\,dy = 8\pi R \cdot \frac{1}{4}\pi r^2 = 2\pi^2 r^2 R.$$

63. (a) Volume$(S_1) = \int_0^h A(z)\,dz =$ Volume(S_2) since the cross-sectional area $A(z)$ at height z is the same for both
solids.

(b) By Cavalieri's Principle, the volume of the cylinder in the figure is the same as that of a right circular cylinder
with radius r and height h, that is, $\pi r^2 h$.

65. The volume is obtained by rotating the area common to two circles of
radius r, as shown. The volume of the right half is

$$V_{\text{right}} = \pi \int_0^{r/2} y^2\,dx = \pi \int_0^{r/2} \left[r^2 - \left(\frac{1}{2}r + x \right)^2 \right] dx$$

$$= \pi\left[r^2 x - \frac{1}{3}\left(\frac{1}{2}r + x \right)^3 \right]_0^{r/2} = \pi\left[\left(\frac{1}{2}r^3 - \frac{1}{3}r^3 \right) - \left(0 - \frac{1}{24}r^3 \right) \right] = \frac{5}{24}\pi r^3$$

So by symmetry, the total volume is twice this, or $\frac{5}{12}\pi r^3$.

Another solution: We observe that the volume is the twice the volume of a cap of a sphere, so we can use the

formula from Exercise 49 with $h = \frac{1}{2}r$: $V = 2 \cdot \frac{1}{3}\pi h^2(3r - h) = \frac{2}{3}\pi\left(\frac{1}{2}r\right)^2\left(3r - \frac{1}{2}r\right) = \frac{5}{12}\pi r^3.$

67. Take the x-axis to be the axis of the cylindrical hole of radius r.
A quarter of the cross-section through y, perpendicular to the
y-axis, is the rectangle shown. Using the Pythagorean Theorem
twice, we see that the dimensions of this rectangle are

$x = \sqrt{R^2 - y^2}$ and $z = \sqrt{r^2 - y^2}$, so

$\frac{1}{4}A(y) = xz = \sqrt{r^2 - y^2}\sqrt{R^2 - y^2}$, and

$$V = \int_{-r}^{r} A(y)\,dy = \int_{-r}^{r} 4\sqrt{r^2 - y^2}\sqrt{R^2 - y^2}\,dy$$

$$= 8\int_{0}^{r}\sqrt{r^2 - y^2}\sqrt{R^2 - y^2}\,dy$$

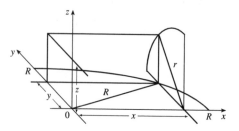

69. (a) The radius of the barrel is the same at each end by symmetry, since the
function $y = R - cx^2$ is even. Since the barrel is obtained by rotating
the graph of the function y about the x-axis, this radius is equal to the
value of y at $x = \frac{1}{2}h$, which is $R - c\left(\frac{1}{2}h\right)^2 = R - d = r$.

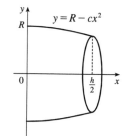

(b) The barrel is symmetric about the y-axis, so its volume is twice the volume of that part of the barrel for $x > 0$.
Also, the barrel is a volume of rotation, so

$$V = 2\int_{0}^{h/2}\pi y^2\,dx = 2\pi\int_{0}^{h/2}\left(R - cx^2\right)^2 dx = 2\pi\left[R^2 x - \frac{2}{3}Rcx^3 + \frac{1}{5}c^2x^5\right]_{0}^{h/2}$$

$$= 2\pi\left(\frac{1}{2}R^2 h - \frac{1}{12}Rch^3 + \frac{1}{160}c^2h^5\right)$$

Trying to make this look more like the expression we want, we rewrite it as

$V = \frac{1}{3}\pi h\left[2R^2 + \left(R^2 - \frac{1}{2}Rch^2 + \frac{3}{80}c^2h^4\right)\right]$. But

$R^2 - \frac{1}{2}Rch^2 + \frac{3}{80}c^2h^4 = \left(R - \frac{1}{4}ch^2\right)^2 - \frac{1}{40}c^2h^4 = (R - d)^2 - \frac{2}{5}\left(\frac{1}{4}ch^2\right)^2 = r^2 - \frac{2}{5}d^2$.

Substituting this back into V, we see that $V = \frac{1}{3}\pi h\left(2R^2 + r^2 - \frac{2}{5}d^2\right)$, as required.

6.3 Volumes by Cylindrical Shells

1.

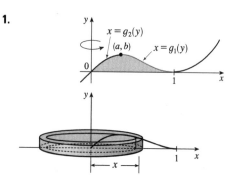

If we were to use the "washer" method, we would first have
to locate the local maximum point (a, b) of $y = x(x - 1)^2$
using the methods of Chapter 4. Then we would have to
solve the equation $y = x(x - 1)^2$ for x in terms of y to
obtain the functions $x = g_1(y)$ and $x = g_2(y)$ shown in the
first figure. This step would be difficult because it involves
the cubic formula. Finally we would find the volume using

$$V = \pi\int_{0}^{b}\left\{[g_1(y)]^2 - [g_2(y)]^2\right\}dy.$$

Using shells, we find that a typical approximating shell has radius x, so its circumference is $2\pi x$. Its height is y, that is, $x(x-1)^2$. So the total volume is

$$V = \int_0^1 2\pi x \left[x(x-1)^2\right] dx = 2\pi \int_0^1 \left(x^4 - 2x^3 + x^2\right) dx = 2\pi \left[\frac{x^5}{5} - 2\frac{x^4}{4} + \frac{x^3}{3}\right]_0^1 = \frac{\pi}{15}$$

3. $V = \displaystyle\int_1^2 2\pi x \cdot \frac{1}{x}\,dx = 2\pi \int_1^2 1\,dx$

$\qquad = 2\pi\,[x]_1^2 = 2\pi(2-1) = 2\pi$

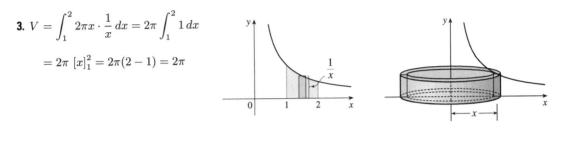

5. $V = \int_0^1 2\pi x e^{-x^2}\,dx$. Let $u = x^2$.

Thus, $du = 2x\,dx$, so

$V = \pi \int_0^1 e^{-u}\,du = \pi\left[-e^{-u}\right]_0^1$

$\qquad = \pi(1 - 1/e)$

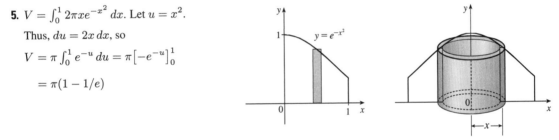

7. The curves intersect when $4(x-2)^2 = x^2 - 4x + 7 \;\Leftrightarrow\; 4x^2 - 16x + 16 = x^2 - 4x + 7 \;\Leftrightarrow$

$\quad 3x^2 - 12x + 9 = 0 \;\Leftrightarrow\; 3(x^2 - 4x + 3) = 0 \;\Leftrightarrow\; 3(x-1)(x-3) = 0$, so $x = 1$ or 3.

$\quad V = 2\pi \int_1^3 \left\{x\left[(x^2 - 4x + 7) - 4(x-2)^2\right]\right\} dx = 2\pi \int_1^3 \left[x(x^2 - 4x + 7 - 4x^2 + 16x - 16)\right] dx$

$\quad = 2\pi \int_1^3 \left[x(-3x^2 + 12x - 9)\right] dx = 2\pi(-3) \int_1^3 (x^3 - 4x^2 + 3x)\,dx = -6\pi\left[\frac{1}{4}x^4 - \frac{4}{3}x^3 + \frac{3}{2}x^2\right]_1^3$

$\quad = -6\pi\left[\left(\frac{81}{4} - 36 + \frac{27}{2}\right) - \left(\frac{1}{4} - \frac{4}{3} + \frac{3}{2}\right)\right] = -6\pi\left(20 - 36 + 12 + \frac{4}{3}\right) = -6\pi\left(-\frac{8}{3}\right) = 16\pi$

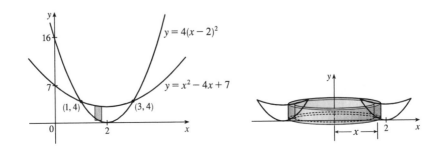

9. $V = \int_1^2 2\pi y (1 + y^2)\, dy = 2\pi \int_1^2 (y + y^3)\, dy = 2\pi \left[\frac{1}{2}y^2 + \frac{1}{4}y^4\right]_1^2$

$= 2\pi \left[(2 + 4) - \left(\frac{1}{2} + \frac{1}{4}\right)\right] = 2\pi \left(\frac{21}{4}\right) = \frac{21\pi}{2}$

11. $V = 2\pi \int_0^8 \left[y\left(\sqrt[3]{y} - 0\right)\right] dy$

$= 2\pi \int_0^8 y^{4/3}\, dy = 2\pi \left[\frac{3}{7}y^{7/3}\right]_0^8$

$= \frac{6\pi}{7}(8^{7/3}) = \frac{6\pi}{7}(2^7) = \frac{768\pi}{7}$

13. The curves intersect when $4x^2 = 6 - 2x \iff 2x^2 + x - 3 = 0 \iff (2x + 3)(x - 1) = 0 \iff x = -\frac{3}{2}$ or 1.

Solving the equations for x gives us $y = 4x^2 \implies x = \pm\frac{1}{2}\sqrt{y}$ and $2x + y = 6 \implies x = -\frac{1}{2}y + 3$.

$V = 2\pi \int_0^4 \left\{y\left[\left(\frac{1}{2}\sqrt{y}\right) - \left(-\frac{1}{2}\sqrt{y}\right)\right]\right\} dy + 2\pi \int_4^9 \left\{y\left[\left(-\frac{1}{2}y + 3\right) - \left(-\frac{1}{2}\sqrt{y}\right)\right]\right\} dy$

$= 2\pi \int_0^4 \left(y\sqrt{y}\right) dy + 2\pi \int_4^9 \left(-\frac{1}{2}y^2 + 3y + \frac{1}{2}y^{3/2}\right) dy = 2\pi \left[\frac{2}{5}y^{5/2}\right]_0^4 + 2\pi \left[-\frac{1}{6}y^3 + \frac{3}{2}y^2 + \frac{1}{5}y^{5/2}\right]_4^9$

$= 2\pi\left(\frac{2}{5} \cdot 32\right) + 2\pi\left[\left(-\frac{243}{2} + \frac{243}{2} + \frac{243}{5}\right) - \left(-\frac{32}{3} + 24 + \frac{32}{5}\right)\right]$

$= \frac{128}{5}\pi + 2\pi\left(\frac{433}{15}\right) = \frac{1250}{15}\pi = \frac{250}{3}\pi$

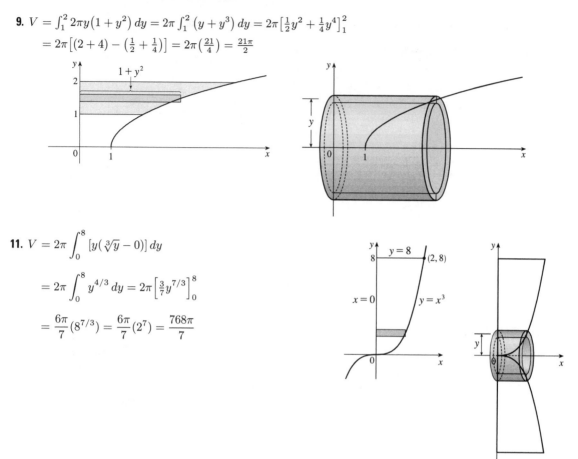

15. $V = \int_1^2 2\pi(x-1)x^2\,dx = 2\pi\left[\frac{1}{4}x^4 - \frac{1}{3}x^3\right]_1^2$

$= 2\pi\left[\left(4 - \frac{8}{3}\right) - \left(\frac{1}{4} - \frac{1}{3}\right)\right] = \frac{17}{6}\pi$

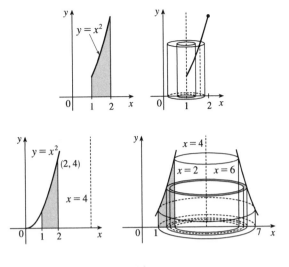

17. $V = \int_1^2 2\pi(4-x)x^2\,dx = 2\pi\left[\frac{4}{3}x^3 - \frac{1}{4}x^4\right]_1^2$

$= 2\pi\left[\left(\frac{32}{3} - 4\right) - \left(\frac{4}{3} - \frac{1}{4}\right)\right] = \frac{67}{6}\pi$

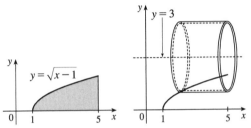

19. $V = \int_0^2 2\pi(3-y)(5-x)\,dy$

$= \int_0^2 2\pi(3-y)\left(5 - y^2 - 1\right)dy$

$= \int_0^2 2\pi\left(12 - 4y - 3y^2 + y^3\right)dy$

$= 2\pi\left[12y - 2y^2 - y^3 + \frac{1}{4}y^4\right]_0^2$

$= 2\pi(24 - 8 - 8 + 4) = 24\pi$

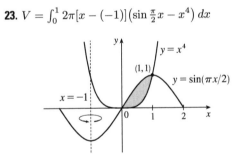

21. $V = \int_1^2 2\pi x \ln x\,dx$

23. $V = \int_0^1 2\pi[x-(-1)]\left(\sin\frac{\pi}{2}x - x^4\right)dx$

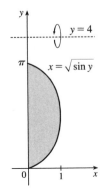

25. $V = \int_0^\pi 2\pi(4-y)\sqrt{\sin y}\,dy$

27. $\Delta x = \dfrac{\pi/4 - 0}{4} = \dfrac{\pi}{16}$.

$V = \int_0^{\pi/4} 2\pi x \tan x \, dx \approx 2\pi \cdot \frac{\pi}{16} \left(\frac{\pi}{32} \tan \frac{\pi}{32} + \frac{3\pi}{32} \tan \frac{3\pi}{32} + \frac{5\pi}{32} \tan \frac{5\pi}{32} + \frac{7\pi}{32} \tan \frac{7\pi}{32} \right) \approx 1.142$

29. $\int_0^3 2\pi x^5 \, dx = 2\pi \int_0^3 x(x^4) \, dx$. The solid is obtained by rotating the region $0 \le y \le x^4, 0 \le x \le 3$ about the y-axis using cylindrical shells.

31. $\int_0^1 2\pi(3 - y)(1 - y^2) \, dy$. The solid is obtained by rotating the region bounded by (i) $x = 1 - y^2$, $x = 0$, and $y = 0$ or (ii) $x = y^2$, $x = 1$, and $y = 0$ about the line $y = 3$ using cylindrical shells.

33.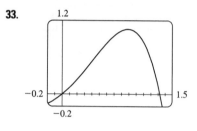

From the graph, the curves intersect at $x = 0$ and at $x = a \approx 1.32$, with $x + x^2 - x^4 > 0$ on the interval $(0, a)$. So the volume of the solid obtained by rotating the region about the y-axis is

$V = 2\pi \int_0^a \left[x(x + x^2 - x^4) \right] dx = 2\pi \int_0^a (x^2 + x^3 - x^5) \, dx$

$= 2\pi \left[\frac{1}{3}x^3 + \frac{1}{4}x^4 - \frac{1}{6}x^6 \right]_0^a \approx 4.05$

35. $V = 2\pi \displaystyle\int_0^{\pi/2} \left[\left(\frac{\pi}{2} - x \right)\left(\sin^2 x - \sin^4 x \right) \right] dx$

$\stackrel{\text{CAS}}{=} \frac{1}{32}\pi^3$

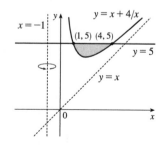

37. Use disks:

$V = \int_{-2}^1 \pi\left(x^2 + x - 2\right)^2 dx = \pi \int_{-2}^1 \left(x^4 + 2x^3 - 3x^2 - 4x + 4\right) dx$

$= \pi\left[\frac{1}{5}x^5 + \frac{1}{2}x^4 - x^3 - 2x^2 + 4x\right]_{-2}^1 = \pi\left[\left(\frac{1}{5} + \frac{1}{2} - 1 - 2 + 4\right) - \left(-\frac{32}{5} + 8 + 8 - 8 - 8\right)\right]$

$= \pi\left(\frac{33}{5} + \frac{3}{2}\right) = \frac{81}{10}\pi$

39. Use shells:

$V = \int_1^4 2\pi[x - (-1)][5 - (x + 4/x)] \, dx$

$= 2\pi \int_1^4 (x + 1)(5 - x - 4/x) \, dx$

$= 2\pi \int_1^4 \left(5x - x^2 - 4 + 5 - x - 4/x\right) dx$

$= 2\pi \int_1^4 \left(-x^2 + 4x + 1 - 4/x\right) dx = 2\pi\left[-\frac{1}{3}x^3 + 2x^2 + x - 4\ln x\right]_1^4$

$= 2\pi\left[\left(-\frac{64}{3} + 32 + 4 - 4\ln 4\right) - \left(-\frac{1}{3} + 2 + 1 - 0\right)\right]$

$= 2\pi(12 - 4\ln 4) = 8\pi(3 - \ln 4)$

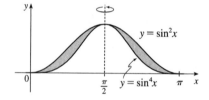

41. Use disks: $V = \pi \int_0^2 \left[\sqrt{1 - (y-1)^2}\right]^2 dy = \pi \int_0^2 (2y - y^2)\, dy = \pi\left[y^2 - \tfrac{1}{3}y^3\right]_0^2 = \pi\left(4 - \tfrac{8}{3}\right) = \tfrac{4}{3}\pi$

43. $V = 2\int_0^r 2\pi x \sqrt{r^2 - x^2}\, dx = -2\pi \int_0^r (r^2 - x^2)^{1/2}(-2x)\, dx = \left[-2\pi \cdot \tfrac{2}{3}(r^2 - x^2)^{3/2}\right]_0^r$

$= -\tfrac{4}{3}\pi(0 - r^3) = \tfrac{4}{3}\pi r^3$

45. $V = 2\pi \int_0^r x\left(-\dfrac{h}{r}x + h\right) dx = 2\pi h \int_0^r \left(-\dfrac{x^2}{r} + x\right) dx = 2\pi h\left[-\dfrac{x^3}{3r} + \dfrac{x^2}{2}\right]_0^r = 2\pi h\,\dfrac{r^2}{6} = \dfrac{\pi r^2 h}{3}$

6.4 Work

1. By Equation 2, $W = Fd = (900)(8) = 7200$ J.

3. $W = \displaystyle\int_a^b f(x)\, dx = \int_0^9 \dfrac{10}{(1+x)^2}\, dx = 10\int_1^{10} \dfrac{1}{u^2}\, du \qquad [u = 1+x,\ du = dx]$

$= 10\left[-\dfrac{1}{u}\right]_1^{10} = 10\left(-\tfrac{1}{10} + 1\right) = 9$ ft-lb

5. The force function is given by $F(x)$ (in newtons) and the work (in joules) is the area under the curve, given by

$\int_0^8 F(x)\, dx = \int_0^4 F(x)\, dx + \int_4^8 F(x)\, dx = \tfrac{1}{2}(4)(30) + (4)(30) = 180$ J.

7. $10 = f(x) = kx = \tfrac{1}{3}k$ [4 inches $= \tfrac{1}{3}$ foot], so $k = 30$ lb/ft and $f(x) = 30x$. Now 6 inches $= \tfrac{1}{2}$ foot, so

$W = \int_0^{1/2} 30x\, dx = \left[15x^2\right]_0^{1/2} = \tfrac{15}{4}$ ft-lb.

9. If $\int_0^{0.12} kx\, dx = 2$ J, then $2 = \left[\tfrac{1}{2}kx^2\right]_0^{0.12} = \tfrac{1}{2}k(0.0144) = 0.0072k$ and $k = \dfrac{2}{0.0072} = \dfrac{2500}{9} \approx 277.78$ N/m.

Thus, the work needed to stretch the spring from 35 cm to 40 cm is

$\int_{0.05}^{0.10} \tfrac{2500}{9}x\, dx = \left[\tfrac{1250}{9}x^2\right]_{1/20}^{1/10} = \tfrac{1250}{9}\left(\tfrac{1}{100} - \tfrac{1}{400}\right) = \tfrac{25}{24} \approx 1.04$ J.

11. $f(x) = kx$, so $30 = \dfrac{2500}{9}x$ and $x = \dfrac{270}{2500}$ m $= 10.8$ cm

In Exercises 13–20, n is the number of subintervals of length Δx, and x_i^* is a sample point in the ith subinterval $[x_{i-1}, x_i]$.

13. (a) The portion of the rope from x ft to $(x + \Delta x)$ ft below the top of the building weighs $\tfrac{1}{2}\Delta x$ lb and must be

lifted x_i^* ft, so its contribution to the total work is $\tfrac{1}{2}x_i^*\,\Delta x$ ft-lb. The total work is

$$W = \lim_{n\to\infty}\sum_{i=1}^n \tfrac{1}{2}x_i^*\,\Delta x = \int_0^{50} \tfrac{1}{2}x\, dx = \left[\tfrac{1}{4}x^2\right]_0^{50} = \tfrac{2500}{4} = 625 \text{ ft-lb}$$

Notice that the exact height of the building does not matter (as long as it is more than 50 ft).

(b) When half the rope is pulled to the top of the building, the work to lift the top half of the rope is

$W_1 = \int_0^{25} \tfrac{1}{2}x\, dx = \left[\tfrac{1}{4}x^2\right]_0^{25} = \tfrac{625}{4}$ ft-lb. The bottom half of the rope is lifted 25 ft and the work needed to

accomplish that is $W_2 = \int_{25}^{50} \tfrac{1}{2} \cdot 25\, dx = \tfrac{25}{2}[x]_{25}^{50} = \tfrac{625}{2}$ ft-lb. The total work done in pulling half the rope to

the top of the building is $W = W_1 + W_2 = \tfrac{625}{2} + \tfrac{625}{4} = \tfrac{3}{4} \cdot 625 = \tfrac{1875}{4}$ ft-lb.

15. The work needed to lift the cable is $\displaystyle\lim_{n\to\infty}\sum_{i=1}^n 2x_i^*\,\Delta x = \int_0^{500} 2x\, dx = \left[x^2\right]_0^{500} = 250{,}000$ ft-lb. The work

needed to lift the coal is 800 lb $\cdot\ 500$ ft $= 400{,}000$ ft-lb. Thus, the total work required is

$250{,}000 + 400{,}000 = 650{,}000$ ft-lb.

17. At a height of x meters ($0 \le x \le 12$), the mass of the rope is $(0.8 \text{ kg/m})(12 - x \text{ m}) = (9.6 - 0.8x)$ kg and the mass of the water is $\left(\frac{36}{12} \text{ kg/m}\right)(12 - x \text{ m}) = (36 - 3x)$ kg. The mass of the bucket is 10 kg, so the total mass is $(9.6 - 0.8x) + (36 - 3x) + 10 = (55.6 - 3.8x)$ kg, and hence, the total force is $9.8(55.6 - 3.8x)$ N.

The work needed to lift the bucket Δx m through the ith subinterval of $[0, 12]$ is $9.8(55.6 - 3.8x_i^*)\Delta x$, so the total work is

$$W = \lim_{n \to \infty} \sum_{i=1}^{n} 9.8(55.6 - 3.8x_i^*)\, \Delta x = \int_0^{12}(9.8)(55.6 - 3.8x)\, dx = 9.8\left[55.6x - 1.9x^2\right]_0^{12}$$
$$= 9.8(393.6) \approx 3857 \text{ J}$$

19. A "slice" of water Δx m thick and lying at a depth of x_i^* m (where $0 \le x_i^* \le \frac{1}{2}$) has volume $(2 \times 1 \times \Delta x)$ m³, a mass of $2000\,\Delta x$ kg, weighs about $(9.8)(2000\,\Delta x) = 19{,}600\,\Delta x$ N, and thus requires about $19{,}600 x_i^*\,\Delta x$ J of work for its removal. So $W = \lim_{n \to \infty} \sum_{i=1}^{n} 19{,}600 x_i^*\,\Delta x = \int_0^{1/2} 19{,}600x\, dx = \left[9800x^2\right]_0^{1/2} = 2450 \text{ J}.$

21. A rectangular "slice" of water Δx m thick and lying x ft above the bottom has width x ft and volume $8x\,\Delta x$ m³. It weighs about $(9.8 \times 1000)(8x\,\Delta x)$ N, and must be lifted $(5 - x)$ m by the pump, so the work needed is about $(9.8 \times 10^3)(5 - x)(8x\,\Delta x)$ J. The total work required is

$$W \approx \int_0^3 (9.8 \times 10^3)(5 - x)8x\, dx = (9.8 \times 10^3)\int_0^3(40x - 8x^2)\, dx = (9.8 \times 10^3)\left[20x^2 - \tfrac{8}{3}x^3\right]_0^3$$
$$= (9.8 \times 10^3)(180 - 72) = (9.8 \times 10^3)(108) = 1058.4 \times 10^3 \approx 1.06 \times 10^6 \text{ J}$$

23. Measure depth x downward from the flat top of the tank, so that $0 \le x \le 2$ ft. Then $\Delta W = (62.5)(2\sqrt{4 - x^2})(8\,\Delta x)(x + 1)$ ft-lb, so

$$W \approx (62.5)(16)\int_0^2(x + 1)\sqrt{4 - x^2}\, dx = 1000\left(\int_0^2 x\sqrt{4 - x^2}\, dx + \int_0^2 \sqrt{4 - x^2}\, dx\right)$$
$$= 1000\left[\int_0^4 u^{1/2}\left(\tfrac{1}{2}\, du\right) + \tfrac{1}{4}\pi(2^2)\right] \qquad \text{[Put } u = 4 - x^2, \text{ so } du = -2x\, dx]$$
$$= 1000\left(\left[\tfrac{1}{2} \cdot \tfrac{2}{3}u^{3/2}\right]_0^4 + \pi\right) = 1000\left(\tfrac{8}{3} + \pi\right) \approx 5.8 \times 10^3 \text{ ft-lb}$$

Note: The second integral represents the area of a quarter-circle of radius 2.

25. If only 4.7×10^5 J of work is done, then only the water above a certain level (call it h) will be pumped out. So we use the same formula as in Exercise 21, except that the work is fixed, and we are trying to find the lower limit of integration: $4.7 \times 10^5 \approx \int_h^3 (9.8 \times 10^3)(5 - x)8x\, dx = (9.8 \times 10^3)\left[20x^2 - \tfrac{8}{3}x^3\right]_h^3 \Leftrightarrow$

$\frac{4.7}{9.8} \times 10^2 \approx 48 = (20 \cdot 3^2 - \tfrac{8}{3} \cdot 3^3) - (20h^2 - \tfrac{8}{3}h^3) \Leftrightarrow$

$2h^3 - 15h^2 + 45 = 0$. To find the solution of this equation, we plot $2h^3 - 15h^2 + 45$ between $h = 0$ and $h = 3$. We see that the equation is satisfied for $h \approx 2.0$. So the depth of water remaining in the tank is about 2.0 m.

27. $V = \pi r^2 x$, so V is a function of x and P can also be regarded as a function of x. If $V_1 = \pi r^2 x_1$ and $V_2 = \pi r^2 x_2$, then

$$W = \int_{x_1}^{x_2} F(x)\, dx = \int_{x_1}^{x_2} \pi r^2 P(V(x))\, dx$$

$$= \int_{x_1}^{x_2} P(V(x))\, dV(x) \qquad \text{[Let } V(x) = \pi r^2 x, \text{ so } dV(x) = \pi r^2\, dx.\text{]}$$

$$= \int_{V_1}^{V_2} P(V)\, dV \quad \text{by the Substitution Rule.}$$

29. $W = \int_a^b F(r)\, dr = \int_a^b G\dfrac{m_1 m_2}{r^2}\, dr = G m_1 m_2 \left[\dfrac{-1}{r}\right]_a^b = G m_1 m_2 \left(\dfrac{1}{a} - \dfrac{1}{b}\right)$

6.5 Average Value of a Function

1. $f_{ave} = \dfrac{1}{b-a} \int_a^b f(x)\, dx = \dfrac{1}{1-(-1)} \int_{-1}^{1} x^2\, dx = \dfrac{1}{2} \cdot 2 \int_0^1 x^2\, dx = \left[\dfrac{1}{3}x^3\right]_0^1 = \dfrac{1}{3}$

3. $g_{ave} = \dfrac{1}{\frac{\pi}{2} - 0} \int_0^{\pi/2} \cos x\, dx = \dfrac{2}{\pi}[\sin x]_0^{\pi/2} = \dfrac{2}{\pi}(1 - 0) = \dfrac{2}{\pi}$

5. $f_{ave} = \dfrac{1}{5-0} \int_0^5 t e^{-t^2}\, dt = \dfrac{1}{5} \int_0^{-25} e^u \left(-\dfrac{1}{2}\, du\right)$ $[u = -t^2,\, du = -2t\, dt,\, t\, dt = -\dfrac{1}{2}\, du]$

$= -\dfrac{1}{10}[e^u]_0^{-25} = -\dfrac{1}{10}\left(e^{-25} - 1\right) = \dfrac{1}{10}\left(1 - e^{-25}\right)$

7. $h_{ave} = \dfrac{1}{\pi - 0} \int_0^\pi \cos^4 x \sin x\, dx = \dfrac{1}{\pi} \int_1^{-1} u^4(-du)$ $[u = \cos x,\, du = -\sin x\, dx]$

$= \dfrac{1}{\pi} \int_{-1}^1 u^4\, du = \dfrac{1}{\pi} \cdot 2 \int_0^1 u^4\, du = \dfrac{2}{\pi}\left[\dfrac{1}{5}u^5\right]_0^1 = \dfrac{2}{5\pi}$

9. (a) $f_{ave} = \dfrac{1}{5-2} \int_2^5 (x-3)^2\, dx = \dfrac{1}{3}\left[\dfrac{1}{3}(x-3)^3\right]_2^5$

$= \dfrac{1}{9}\left[2^3 - (-1)^3\right] = \dfrac{1}{9}(8+1) = 1$

(b) $f(c) = f_{ave} \iff (c-3)^2 = 1 \iff c - 3 = \pm 1$
$\iff c = 2 \text{ or } 4$

(c)

11. (a) $f_{ave} = \dfrac{1}{\pi - 0} \int_0^\pi (2\sin x - \sin 2x)\, dx$

$= \dfrac{1}{\pi}\left[-2\cos x + \dfrac{1}{2}\cos 2x\right]_0^\pi$

$= \dfrac{1}{\pi}\left[(2 + \dfrac{1}{2}) - (-2 + \dfrac{1}{2})\right] = \dfrac{4}{\pi}$

(b) $f(c) = f_{ave} \iff 2\sin c - \sin 2c = \dfrac{4}{\pi} \iff$
$c_1 \approx 1.238 \text{ or } c_2 \approx 2.808$

(c)

13. f is continuous on $[1, 3]$, so by the Mean Value Theorem for Integrals there exists a number c in $[1, 3]$ such that
$\int_1^3 f(x)\, dx = f(c)(3 - 1) \quad \Rightarrow \quad 8 = 2f(c)$; that is, there is a number c such that $f(c) = \frac{8}{2} = 4$.

15. $f_{\text{ave}} = \dfrac{1}{50 - 20} \displaystyle\int_{20}^{50} f(x)\, dx \approx \frac{1}{30} M_3 = \frac{1}{30} \cdot \frac{50 - 20}{3} [f(25) + f(35) + f(45)]$

$\qquad = \frac{1}{3}(38 + 29 + 48) = \frac{115}{3} = 38\frac{1}{3}$

17. Let $t = 0$ and $t = 12$ correspond to 9 A.M. and 9 P.M., respectively.

$T_{\text{ave}} = \frac{1}{12 - 0} \displaystyle\int_0^{12} \left[50 + 14 \sin \frac{1}{12}\pi t \right] dt = \frac{1}{12} \left[50t - 14 \cdot \frac{12}{\pi} \cos \frac{1}{12}\pi t \right]_0^{12}$

$\qquad = \frac{1}{12} \left[50 \cdot 12 + 14 \cdot \frac{12}{\pi} + 14 \cdot \frac{12}{\pi} \right] = \left(50 + \frac{28}{\pi} \right)°\text{F} \approx 59°\text{F}$

19. $\rho_{\text{ave}} = \dfrac{1}{8} \displaystyle\int_0^8 \frac{12}{\sqrt{x + 1}}\, dx = \frac{3}{2} \int_0^8 (x + 1)^{-1/2}\, dx = \left[3\sqrt{x + 1} \right]_0^8 = 9 - 3 = 6\ \text{kg/m}$

21. $V_{\text{ave}} = \dfrac{1}{5} \displaystyle\int_0^5 V(t)\, dt = \frac{1}{5} \int_0^5 \frac{5}{4\pi} \left[1 - \cos\left(\frac{2}{5}\pi t\right) \right] dt = \frac{1}{4\pi} \int_0^5 \left[1 - \cos\left(\frac{2}{5}\pi t\right) \right] dt$

$\qquad = \frac{1}{4\pi} \left[t - \frac{5}{2\pi} \sin\left(\frac{2}{5}\pi t\right) \right]_0^5 = \frac{1}{4\pi} [(5 - 0) - 0] = \frac{5}{4\pi} \approx 0.4\ \text{L}$

23. Let $F(x) = \int_a^x f(t)\, dt$ for x in $[a, b]$. Then F is continuous on $[a, b]$ and differentiable on (a, b), so by the Mean Value Theorem there is a number c in (a, b) such that $F(b) - F(a) = F'(c)(b - a)$. But $F'(x) = f(x)$ by the Fundamental Theorem of Calculus. Therefore, $\int_a^b f(t)\, dt - 0 = f(c)(b - a)$.

6 Review

——————————————————— CONCEPT CHECK ———————————————————

1. (a) See Section 6.1, Figure 2 and Equations 6.1.1 and 6.1.2.

(b) Instead of using "top minus bottom" and integrating from left to right, we use "right minus left" and integrate from bottom to top. See Figures 11 and 12 in Section 6.1.

2. The numerical value of the area represents the number of meters by which Sue is ahead of Kathy after 1 minute.

3. (a) See the discussion in Section 6.2, near Figures 2 and 3, ending in the Definition of Volume.

(b) See the discussion between Examples 5 and 6 in Section 6.2. If the cross-section is a disk, find the radius in terms of x or y and use $A = \pi(\text{radius})^2$. If the cross-section is a washer, find the inner radius r_{in} and outer radius r_{out} and use $A = \pi\left(r_{\text{out}}^2\right) - \pi\left(r_{\text{in}}^2\right)$.

4. (a) $V = 2\pi r h\, \Delta r = (\text{circumference}) (\text{height}) (\text{thickness})$

(b) For a typical shell, find the circumference and height in terms of x or y and calculate
$V = \int_a^b (\text{circumference}) (\text{height}) (dx \text{ or } dy)$, where a and b are the limits on x or y.

(c) Sometimes slicing produces washers or disks whose radii are difficult (or impossible) to find explicitly. On other occasions, the cylindrical shell method leads to an easier integral than slicing does.

5. $\int_0^6 f(x)\, dx$ represents the amount of work done. Its units are newton-meters, or joules.

6. (a) The average value of a function f on an interval $[a, b]$ is $f_{\text{ave}} = \dfrac{1}{b - a} \displaystyle\int_a^b f(x)\, dx$.

(b) The Mean Value Theorem for Integrals says that there is a number c at which the value of f is exactly equal to the average value of the function, that is, $f(c) = f_{\text{ave}}$. For a geometric interpretation of the Mean Value Theorem for Integrals, see Figure 2 in Section 6.5 and the discussion that accompanies it.

─────────────── EXERCISES ───────────────

1. $0 = x^2 - x - 6 = (x - 3)(x + 2)$ ⇔ $x = 3$ or -2. So

$$A = \int_{-2}^{3} \left[0 - \left(x^2 - x - 6\right)\right] dx = \int_{-2}^{3} \left(-x^2 + x + 6\right) dx$$

$$= \left[-\tfrac{1}{3}x^3 + \tfrac{1}{2}x^2 + 6x\right]_{-2}^{3}$$

$$= \left(-9 + \tfrac{9}{2} + 18\right) - \left(\tfrac{8}{3} + 2 - 12\right)$$

$$= \tfrac{125}{6}$$

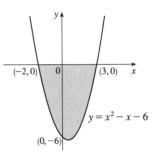

3. $A = \int_0^1 \left[\left(e^x - 1\right) - \left(x^2 - x\right)\right] dx$

$\quad = \int_0^1 \left(e^x - 1 - x^2 + x\right) dx = \left[e^x - x - \tfrac{1}{3}x^3 + \tfrac{1}{2}x^2\right]_0^1$

$\quad = \left(e - 1 - \tfrac{1}{3} + \tfrac{1}{2}\right) - (1 - 0 - 0 + 0) = e - \tfrac{11}{6}$

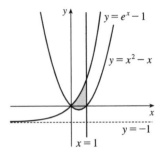

5. $A = \displaystyle\int_0^2 \left[\sin\left(\tfrac{\pi x}{2}\right) - \left(x^2 - 2x\right)\right] dx$

$\quad = \left[-\tfrac{2}{\pi}\cos\left(\tfrac{\pi x}{2}\right) - \tfrac{1}{3}x^3 + x^2\right]_0^2$

$\quad = \left(\tfrac{2}{\pi} - \tfrac{8}{3} + 4\right) - \left(-\tfrac{2}{\pi} - 0 + 0\right) = \tfrac{4}{3} + \tfrac{4}{\pi}$

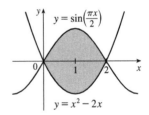

7. Using washers with inner radius x^2 and outer radius $2x$, we have

$$V = \pi \int_0^2 \left[(2x)^2 - \left(x^2\right)^2\right] dx = \pi \int_0^2 \left(4x^2 - x^4\right) dx$$

$$= \pi\left[\tfrac{4}{3}x^3 - \tfrac{1}{5}x^5\right]_0^2 = \pi\left(\tfrac{32}{3} - \tfrac{32}{5}\right) = 32\pi \cdot \tfrac{2}{15}$$

$$= \tfrac{64\pi}{15}$$

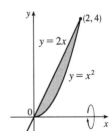

9. $V = \pi \displaystyle\int_{-3}^{3} \left\{\left[(9 - y^2) - (-1)\right]^2 - \left[0 - (-1)\right]^2\right\} dy$

$\quad = 2\pi \displaystyle\int_0^3 \left[(10 - y^2)^2 - 1\right] dy$

$\quad = 2\pi \displaystyle\int_0^3 \left(100 - 20y^2 + y^4 - 1\right) dy$

$\quad = 2\pi \displaystyle\int_0^3 \left(99 - 20y^2 + y^4\right) dy = 2\pi\left[99y - \tfrac{20}{3}y^3 + \tfrac{1}{5}y^5\right]_0^3$

$\quad = 2\pi\left(297 - 180 + \tfrac{243}{5}\right) = \tfrac{1656\pi}{5}$

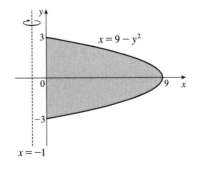

11. The graph of $x^2 - y^2 = a^2$ is a hyperbola with right and
left branches. Solving for y gives us $y^2 = x^2 - a^2$ $\Rightarrow$
$y = \pm\sqrt{x^2 - a^2}$. We'll use shells and the height of each
shell is $\sqrt{x^2 - a^2} - \left(-\sqrt{x^2 - a^2}\right) = 2\sqrt{x^2 - a^2}$.

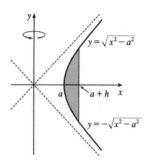

  The volume is $V = \int_a^{a+h} 2\pi x \cdot 2\sqrt{x^2 - a^2}\, dx$. To evaluate, let $u = x^2 - a^2$, so $du = 2x\, dx$ and $x\, dx = \frac{1}{2}\, du$.
When $x = a$, $u = 0$, and when $x = a + h$,
$u = (a + h)^2 - a^2 = a^2 + 2ah + h^2 - a^2 = 2ah + h^2$.

Thus, $V = 4\pi \int_0^{2ah+h^2} \sqrt{u}\left(\frac{1}{2}\, du\right) = 2\pi\left[\frac{2}{3}u^{3/2}\right]_0^{2ah+h^2} = \frac{4}{3}\pi\left(2ah + h^2\right)^{3/2}$.

13. $V = \int_0^1 \pi\left[\left(1 - x^3\right)^2 - \left(1 - x^2\right)^2\right] dx$

15. (a) A cross-section is a washer with inner radius x^2 and outer radius x.

  $V = \int_0^1 \pi\left[(x)^2 - \left(x^2\right)^2\right] dx = \int_0^1 \pi\left(x^2 - x^4\right) dx = \pi\left[\frac{1}{3}x^3 - \frac{1}{5}x^5\right]_0^1 = \pi\left[\frac{1}{3} - \frac{1}{5}\right] = \frac{2\pi}{15}$

 (b) A cross-section is a washer with inner radius y and outer radius $\sqrt{y}$.

  $V = \int_0^1 \pi\left[\left(\sqrt{y}\right)^2 - y^2\right] dy = \int_0^1 \pi\left(y - y^2\right) dy = \pi\left[\frac{1}{2}y^2 - \frac{1}{3}y^3\right]_0^1 = \pi\left[\frac{1}{2} - \frac{1}{3}\right] = \frac{\pi}{6}$

 (c) A cross-section is a washer with inner radius $2 - x$ and outer radius $2 - x^2$.

  $V = \int_0^1 \pi\left[\left(2 - x^2\right)^2 - (2 - x)^2\right] dx = \int_0^1 \pi\left(x^4 - 5x^2 + 4x\right) dx = \pi\left[\frac{1}{5}x^5 - \frac{5}{3}x^3 + 2x^2\right]_0^1$

  $= \pi\left[\frac{1}{5} - \frac{5}{3} + 2\right] = \frac{8\pi}{15}$

17. (a) Using the Midpoint Rule on $[0, 1]$ with $f(x) = \tan\left(x^2\right)$ and $n = 4$, we estimate

  $A = \int_0^1 \tan\left(x^2\right) dx \approx \frac{1}{4}\left[\tan\left(\left(\frac{1}{8}\right)^2\right) + \tan\left(\left(\frac{3}{8}\right)^2\right) + \tan\left(\left(\frac{5}{8}\right)^2\right) + \tan\left(\left(\frac{7}{8}\right)^2\right)\right] \approx \frac{1}{4}(1.53) \approx 0.38$

 (b) Using the Midpoint Rule on $[0, 1]$ with $f(x) = \pi\tan^2\left(x^2\right)$ (for disks) and $n = 4$, we estimate

  $V = \int_0^1 f(x)\, dx \approx \frac{1}{4}\pi\left[\tan^2\left(\left(\frac{1}{8}\right)^2\right) + \tan^2\left(\left(\frac{3}{8}\right)^2\right) + \tan^2\left(\left(\frac{5}{8}\right)^2\right) + \tan^2\left(\left(\frac{7}{8}\right)^2\right)\right] \approx \frac{\pi}{4}(1.114) \approx 0.87$

19. The solid is obtained by rotating the region $\mathcal{R} = \left\{(x, y) \mid 0 \le x \le \frac{\pi}{2}, 0 \le y \le \cos x\right\}$ about the y-axis.

21. The solid is obtained by rotating the region $\mathcal{R} = \left\{(x, y) \mid 0 \le y \le 2, 0 \le x \le 4 - y^2\right\}$ about the x-axis.

23. Take the base to be the disk $x^2 + y^2 \le 9$. Then $V = \int_{-3}^3 A(x)\, dx$, where $A(x_0)$ is the area of the isosceles right
triangle whose hypotenuse lies along the line $x = x_0$ in the xy-plane. The length of the hypotenuse is $2\sqrt{9 - x^2}$
and the length of each leg is $\sqrt{2}\sqrt{9 - x^2}$. $A(x) = \frac{1}{2}\left(\sqrt{2}\sqrt{9 - x^2}\right)^2 = 9 - x^2$, so
$V = 2\int_0^3 A(x)\, dx = 2\int_0^3 \left(9 - x^2\right) dx = 2\left[9x - \frac{1}{3}x^3\right]_0^3 = 2(27 - 9) = 36$.

25. Equilateral triangles with sides measuring $\frac{1}{4}x$ meters have height $\frac{1}{4}x \sin 60° = \frac{\sqrt{3}}{8}x$. Therefore,

$$A(x) = \frac{1}{2} \cdot \frac{1}{4}x \cdot \frac{\sqrt{3}}{8}x = \frac{\sqrt{3}}{64}x^2. \quad V = \int_0^{20} A(x)\,dx = \frac{\sqrt{3}}{64}\int_0^{20} x^2\,dx = \frac{\sqrt{3}}{64}\left[\frac{1}{3}x^3\right]_0^{20} = \frac{8000\sqrt{3}}{64\cdot 3} = \frac{125\sqrt{3}}{3}\ \text{m}^3.$$

27. $f(x) = kx \implies 30\,\text{N} = k(15 - 12)\,\text{cm} \implies k = 10\,\text{N/cm} = 1000\,\text{N/m}.\ 20\,\text{cm} - 12\,\text{cm} = 0.08\,\text{m} \implies$

$$W = \int_0^{0.08} kx\,dx = 1000\int_0^{0.08} x\,dx = 500\left[x^2\right]_0^{0.08} = 500(0.08)^2 = 3.2\ \text{N-m} = 3.2\ \text{J}.$$

29. (a) The parabola has equation $y = ax^2$ with vertex at the origin and passing

through $(4, 4)$. $4 = a \cdot 4^2 \implies a = \frac{1}{4} \implies y = \frac{1}{4}x^2 \implies x^2 = 4y$

$\implies x = 2\sqrt{y}$. Each circular disk has radius $2\sqrt{y}$ and is moved $4 - y$ ft.

$$W = \int_0^4 \pi\left(2\sqrt{y}\right)^2 62.5\,(4 - y)\,dy = 250\pi\int_0^4 y(4 - y)\,dy$$

$$= 250\pi\left[2y^2 - \tfrac{1}{3}y^3\right]_0^4 = 250\pi\left(32 - \tfrac{64}{3}\right) = \frac{8000\pi}{3} \approx 8378\ \text{ft-lb}$$

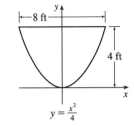

(b) In part (a) we knew the final water level (0) but not the amount of work

done. Here we use the same equation, except with the work fixed, and the

lower limit of integration (that is, the final water level — call it h)

unknown: $W = 4000 \iff 250\pi\left[2y^2 - \tfrac{1}{3}y^3\right]_h^4 = 4000 \iff$

$\frac{16}{\pi} = \left[\left(32 - \tfrac{64}{3}\right) - \left(2h^2 - \tfrac{1}{3}h^3\right)\right] \iff h^3 - 6h^2 + 32 - \frac{48}{\pi} = 0.$

We graph the function $f(h) = h^3 - 6h^2 + 32 - \frac{48}{\pi}$ on the interval $[0, 4]$ to see where it is 0. From the graph,

$f(h) = 0$ for $h \approx 2.1$. So the depth of water remaining is about 2.1 ft.

31. $\displaystyle\lim_{h\to 0} f_{\text{ave}} = \lim_{h\to 0} \frac{1}{(x + h) - x}\int_x^{x+h} f(t)\,dt = \lim_{h\to 0} \frac{F(x + h) - F(x)}{h}$, where $F(x) = \int_a^x f(t)\,dt$. But we

recognize this limit as being $F'(x)$ by the definition of a derivative. Therefore, $\displaystyle\lim_{h\to 0} f_{\text{ave}} = F'(x) = f(x)$

by FTC1.

□ PROBLEMS PLUS

1. (a) The area under the graph of f from 0 to t is equal to $\int_0^t f(x)\,dx$, so the requirement is that $\int_0^t f(x)\,dx = t^3$ for all t. We differentiate both sides of this equation with respect to t (with the help of FTC1) to get $f(t) = 3t^2$. This function is positive and continuous, as required.

(b) The volume generated from $x = 0$ to $x = b$ is $\int_0^b \pi[f(x)]^2\,dx$. Hence, we are given that $b^2 = \int_0^b \pi[f(x)]^2\,dx$ for all $b > 0$. Differentiating both sides of this equation with respect to b using the Fundamental Theorem of Calculus gives $2b = \pi[f(b)]^2 \quad \Rightarrow \quad f(b) = \sqrt{2b/\pi}$, since f is positive. Therefore, $f(x) = \sqrt{2x/\pi}$.

3. Let a and b be the x-coordinates of the points where the line intersects the curve. From the figure, $R_1 = R_2 \quad \Rightarrow$

$$\int_0^a \left[c - (8x - 27x^3)\right]dx = \int_a^b \left[(8x - 27x^3) - c\right]dx$$

$$\left[cx - 4x^2 + \tfrac{27}{4}x^4\right]_0^a = \left[4x^2 - \tfrac{27}{4}x^4 - cx\right]_a^b$$

$$ac - 4a^2 + \tfrac{27}{4}a^4 = \left(4b^2 - \tfrac{27}{4}b^4 - bc\right) - \left(4a^2 - \tfrac{27}{4}a^4 - ac\right)$$

$$0 = 4b^2 - \tfrac{27}{4}b^4 - bc = 4b^2 - \tfrac{27}{4}b^4 - b\left(8b - 27b^3\right)$$

$$= 4b^2 - \tfrac{27}{4}b^4 - 8b^2 + 27b^4 = \tfrac{81}{4}b^4 - 4b^2$$

$$= b^2\left(\tfrac{81}{4}b^2 - 4\right)$$

So for $b > 0$, $b^2 = \tfrac{16}{81} \quad \Rightarrow \quad b = \tfrac{4}{9}$. Thus, $c = 8b - 27b^3 = 8\left(\tfrac{4}{9}\right) - 27\left(\tfrac{64}{729}\right) = \tfrac{32}{9} - \tfrac{64}{27} = \tfrac{32}{27}$.

$y = 8x - 27x^3$
R_2
$y = c$
(a,c) (b,c)
R_1

5. (a) $V = \pi h^2(r - h/3) = \tfrac{1}{3}\pi h^2(3r - h)$. See the solution to Exercise 6.2.49.

(b) The smaller segment has height $h = 1 - x$ and so by part (a) its volume is

$V = \tfrac{1}{3}\pi(1 - x)^2\left[3(1) - (1 - x)\right] = \tfrac{1}{3}\pi(x - 1)^2(x + 2)$. This volume must be $\tfrac{1}{3}$ of the total volume of the sphere, which is $\tfrac{4}{3}\pi(1)^3$. So $\tfrac{1}{3}\pi(x - 1)^2(x + 2) = \tfrac{1}{3}\left(\tfrac{4}{3}\pi\right) \quad \Rightarrow \quad (x^2 - 2x + 1)(x + 2) = \tfrac{4}{3} \quad \Rightarrow$

$x^3 - 3x + 2 = \tfrac{4}{3} \quad \Rightarrow \quad 3x^3 - 9x + 2 = 0$. Using Newton's method with $f(x) = 3x^3 - 9x + 2$,

$f'(x) = 9x^2 - 9$, we get $x_{n+1} = x_n - \dfrac{3x_n^3 - 9x_n + 2}{9x_n^2 - 9}$. Taking $x_1 = 0$, we get $x_2 \approx 0.2222$, and

$x_3 \approx 0.2261 \approx x_4$, so, correct to four decimal places, $x \approx 0.2261$.

(c) With $r = 0.5$ and $s = 0.75$, the equation $x^3 - 3rx^2 + 4r^3s = 0$ becomes $x^3 - 3(0.5)x^2 + 4(0.5)^3(0.75) = 0$

$\Rightarrow \quad x^3 - \tfrac{3}{2}x^2 + 4\left(\tfrac{1}{8}\right)\tfrac{3}{4} = 0 \quad \Rightarrow \quad 8x^3 - 12x^2 + 3 = 0$. We use Newton's method with

$f(x) = 8x^3 - 12x^2 + 3$, $f'(x) = 24x^2 - 24x$, so $x_{n+1} = x_n - \dfrac{8x_n^3 - 12x_n^2 + 3}{24x_n^2 - 24x_n}$. Take $x_1 = 0.5$. Then

$x_2 \approx 0.6667$, and $x_3 \approx 0.6736 \approx x_4$. So to four decimal places the depth is 0.6736 m.

(d) (i) From part (a) with $r = 5$ in., the volume of water in the bowl is

$$V = \tfrac{1}{3}\pi h^2(3r - h) = \tfrac{1}{3}\pi h^2(15 - h) = 5\pi h^2 - \tfrac{1}{3}\pi h^3.$$ We are given that $\dfrac{dV}{dt} = 0.2 \text{ m}^3/\text{s}$ and we want to

find $\dfrac{dh}{dt}$ when $h = 3$. Now $\dfrac{dV}{dt} = 10\pi h \dfrac{dh}{dt} - \pi h^2 \dfrac{dh}{dt}$, so $\dfrac{dh}{dt} = \dfrac{0.2}{\pi(10h - h^2)}$. When $h = 3$, we have

$\dfrac{dh}{dt} = \dfrac{0.2}{\pi(10 \cdot 3 - 3^2)} = \dfrac{1}{105\pi} \approx 0.003 \text{ in/s}$.

(ii) From part (a), the volume of water required to fill the bowl from the instant that the water is 4 in. deep is

$V = \tfrac{1}{2} \cdot \tfrac{4}{3}\pi(5)^3 - \tfrac{1}{3}\pi(4)^2(15 - 4) = \tfrac{2}{3} \cdot 125\pi - \tfrac{16}{3} \cdot 11\pi = \tfrac{74}{3}\pi$. To find the time required to fill the

bowl we divide this volume by the rate: Time $= \dfrac{74\pi/3}{0.2} = \dfrac{370\pi}{3} \approx 387 \text{ s} \approx 6.5 \text{ min}$

7. We are given that the rate of change of the volume of water is $\dfrac{dV}{dt} = -kA(x)$, where k is some positive constant

and $A(x)$ is the area of the surface when the water has depth x. Now we are concerned with the rate of change of

the depth of the water with respect to time, that is, $\dfrac{dx}{dt}$. But by the Chain Rule, $\dfrac{dV}{dt} = \dfrac{dV}{dx}\dfrac{dx}{dt}$, so the first equation

can be written $\dfrac{dV}{dx}\dfrac{dx}{dt} = -kA(x)$ $(\star)$. Also, we know that the total volume of water up to a depth x is

$V(x) = \int_0^x A(s)\,ds$, where $A(s)$ is the area of a cross-section of the water at a depth s. Differentiating this

equation with respect to x, we get $dV/dx = A(x)$. Substituting this into equation $\star$, we get

$A(x)(dx/dt) = -kA(x)$ $\Rightarrow$ $dx/dt = -k$, a constant.

9. We must find expressions for the areas A and B, and then set them equal and see what this says about the curve C.

If $P = (a, 2a^2)$, then area A is just $\int_0^a (2x^2 - x^2)\,dx = \int_0^a x^2\,dx = \tfrac{1}{3}a^3$. To find area B, we use y as the variable

of integration. So we find the equation of the middle curve as a function of y: $y = 2x^2$ $\Leftrightarrow$ $x = \sqrt{y/2}$,

since we are concerned with the first quadrant only. We can express area B as

$\int_0^{2a^2} \left[\sqrt{y/2} - C(y)\right] dy = \left[\tfrac{4}{3}(y/2)^{3/2}\right]_0^{2a^2} - \int_0^{2a^2} C(y)\,dy = \tfrac{4}{3}a^3 - \int_0^{2a^2} C(y)\,dy$, where $C(y)$ is the function

with graph C. Setting $A = B$, we get $\tfrac{1}{3}a^3 = \tfrac{4}{3}a^3 - \int_0^{2a^2} C(y)\,dy$ $\Leftrightarrow$ $\int_0^{2a^2} C(y)\,dy = a^3$. Now we

differentiate this equation with respect to a using the Chain Rule and the Fundamental Theorem:

$C(2a^2)(4a) = 3a^2$ $\Rightarrow$ $C(y) = \tfrac{3}{4}\sqrt{y/2}$, where $y = 2a^2$. Now we can solve for y: $x = \tfrac{3}{4}\sqrt{y/2}$ $\Rightarrow$

$x^2 = \tfrac{9}{16}(y/2)$ $\Rightarrow$ $y = \tfrac{32}{9}x^2$.

11. (a) Stacking disks along the y-axis gives us $V = \int_0^h \pi \left[f(y)\right]^2 \, dy$.

(b) Using the Chain Rule, $\dfrac{dV}{dt} = \dfrac{dV}{dh} \cdot \dfrac{dh}{dt} = \pi \left[f(h)\right]^2 \dfrac{dh}{dt}$.

(c) $kA\sqrt{h} = \pi[f(h)]^2 \dfrac{dh}{dt}$. Set $\dfrac{dh}{dt} = C$: $\pi[f(h)]^2 \, C = kA\sqrt{h} \;\Rightarrow\; [f(h)]^2 = \dfrac{kA}{\pi C}\sqrt{h} \;\Rightarrow\;$

$f(h) = \sqrt{\dfrac{kA}{\pi C}}\, h^{1/4}$; that is, $f(y) = \sqrt{\dfrac{kA}{\pi C}}\, y^{1/4}$. The advantage of having $\dfrac{dh}{dt} = C$ is that the markings on the

container are equally spaced.

13. We assume that P lies in the region of positive x. Since $y = x^3$ is an odd

function, this assumption will not affect the result of the calculation. Let

$P = (a, a^3)$. The slope of the tangent to the curve $y = x^3$ at P is $3a^2$, and so

the equation of the tangent is $y - a^3 = 3a^2(x - a) \;\Leftrightarrow\; y = 3a^2 x - 2a^3$.

We solve this simultaneously with $y = x^3$ to find the other point of intersection:

$x^3 = 3a^2 x - 2a^3 \;\Leftrightarrow\; (x - a)^2(x + 2a) = 0$. So $Q = (-2a, -8a^3)$ is

the other point of intersection. The equation of the tangent at Q is

$y - (-8a^3) = 12a^2[x - (-2a)] \;\Leftrightarrow\; y = 12a^2 x + 16a^3$. By symmetry,

this tangent will intersect the curve again at $x = -2(-2a) = 4a$. The curve lies above the first tangent, and below

the second, so we are looking for a relationship between $A = \int_{-2a}^{a} \left[x^3 - \left(3a^2 x - 2a^3\right)\right] dx$ and

$B = \int_{-2a}^{4a} \left[\left(12a^2 x + 16a^3\right) - x^3\right] dx$. We calculate $A = \left[\frac{1}{4}x^4 - \frac{3}{2}a^2 x^2 + 2a^3 x\right]_{-2a}^{a} = \frac{3}{4}a^4 - \left(-6a^4\right) = \frac{27}{4}a^4$,

and $B = \left[6a^2 x^2 + 16a^3 x - \frac{1}{4}x^4\right]_{-2a}^{4a} = 96a^4 - \left(-12a^4\right) = 108a^4$. We see that $B = 16A = 2^4 A$. This is

because our calculation of area B was essentially the same as that of area A, with a replaced by $-2a$, so if we

replace a with $-2a$ in our expression for A, we get $\frac{27}{4}(-2a)^4 = 108a^4 = B$.

7 □ TECHNIQUES OF INTEGRATION

7.1 Integration by Parts

1. Let $u = \ln x$, $dv = x\,dx$ $\Rightarrow$ $du = dx/x$, $v = \frac{1}{2}x^2$. Then by Equation 2, $\int u\,dv = uv - \int v\,du$,

$$\int x\ln x\,dx = \frac{1}{2}x^2\ln x - \int \frac{1}{2}x^2(dx/x) = \frac{1}{2}x^2\ln x - \frac{1}{2}\int x\,dx = \frac{1}{2}x^2\ln x - \frac{1}{2}\cdot\frac{1}{2}x^2 + C$$

$$= \frac{1}{2}x^2\ln x - \frac{1}{4}x^2 + C$$

3. Let $u = x$, $dv = \cos 5x\,dx$ $\Rightarrow$ $du = dx$, $v = \frac{1}{5}\sin 5x$. Then by Equation 2,

$\int x\cos 5x\,dx = \frac{1}{5}x\sin 5x - \int \frac{1}{5}\sin 5x\,dx = \frac{1}{5}x\sin 5x + \frac{1}{25}\cos 5x + C$.

5. Let $u = r$, $dv = e^{r/2}\,dr$ $\Rightarrow$ $du = dr$, $v = 2e^{r/2}$. Then

$\int re^{r/2}\,dr = 2re^{r/2} - \int 2e^{r/2}\,dr = 2re^{r/2} - 4e^{r/2} + C$.

7. Let $u = x^2$, $dv = \sin \pi x\,dx$ $\Rightarrow$ $du = 2x\,dx$ and $v = -\frac{1}{\pi}\cos \pi x$. Then

$I = \int x^2\sin \pi x\,dx = -\frac{1}{\pi}x^2\cos \pi x + \frac{2}{\pi}\int x\cos \pi x\,dx$ (∗).

Next let $U = x$, $dV = \cos \pi x\,dx$ $\Rightarrow$ $dU = dx$, $V = \frac{1}{\pi}\sin \pi x$, so

$\int x\cos \pi x\,dx = \frac{1}{\pi}x\sin \pi x - \frac{1}{\pi}\int \sin \pi x\,dx = \frac{1}{\pi}x\sin \pi x + \frac{1}{\pi^2}\cos \pi x + C_1$. Substituting for $\int x\cos \pi x\,dx$ in (∗),

we get $I = -\frac{1}{\pi}x^2\cos \pi x + \frac{2}{\pi}\left(\frac{1}{\pi}x\sin \pi x + \frac{1}{\pi^2}\cos \pi x + C_1\right) = -\frac{1}{\pi}x^2\cos \pi x + \frac{2}{\pi^2}x\sin \pi x + \frac{2}{\pi^3}\cos \pi x + C$,

where $C = \frac{2}{\pi}C_1$.

9. Let $u = \ln(2x + 1)$, $dv = dx$ $\Rightarrow$ $du = \dfrac{2}{2x + 1}\,dx$, $v = x$. Then

$$\int \ln(2x + 1)\,dx = x\ln(2x + 1) - \int \frac{2x}{2x + 1}\,dx = x\ln(2x + 1) - \int \frac{(2x + 1) - 1}{2x + 1}\,dx$$

$$= x\ln(2x + 1) - \int \left(1 - \frac{1}{2x + 1}\right)dx = x\ln(2x + 1) - x + \frac{1}{2}\ln(2x + 1) + C$$

$$= \frac{1}{2}(2x + 1)\ln(2x + 1) - x + C$$

11. Let $u = \arctan 4t$, $dv = dt$ $\Rightarrow$ $du = \dfrac{4}{1 + (4t)^2}\,dt = \dfrac{4}{1 + 16t^2}\,dt$, $v = t$. Then

$$\int \arctan 4t\,dt = t\arctan 4t - \int \frac{4t}{1 + 16t^2}\,dt = t\arctan 4t - \frac{1}{8}\int \frac{32t}{1 + 16t^2}\,dt$$

$$= t\arctan 4t - \frac{1}{8}\ln(1 + 16t^2) + C$$

13. First let $u = (\ln x)^2$, $dv = dx$ $\Rightarrow$ $du = 2\ln x \cdot \frac{1}{x}\,dx$, $v = x$. Then by Equation 2,

$I = \int (\ln x)^2\,dx = x(\ln x)^2 - 2\int x\ln x \cdot \frac{1}{x}\,dx = x(\ln x)^2 - 2\int \ln x\,dx$. Next let $U = \ln x$, $dV = dx$ $\Rightarrow$

$dU = 1/x\,dx$, $V = x$ to get $\int \ln x\,dx = x\ln x - \int x \cdot (1/x)\,dx = x\ln x - \int dx = x\ln x - x + C_1$. Thus,

$I = x(\ln x)^2 - 2(x\ln x - x + C_1) = x(\ln x)^2 - 2x\ln x + 2x + C$, where $C = -2C_1$.

15. First let $u = \sin 3\theta$, $dv = e^{2\theta}\,d\theta$ $\Rightarrow$ $du = 3\cos 3\theta\,d\theta$, $v = \frac{1}{2}e^{2\theta}$. Then

$I = \int e^{2\theta}\sin 3\theta\,d\theta = \frac{1}{2}e^{2\theta}\sin 3\theta - \frac{3}{2}\int e^{2\theta}\cos 3\theta\,d\theta$. Next let $U = \cos 3\theta$,

$dV = e^{2\theta}\,d\theta$ $\Rightarrow$ $dU = -3\sin 3\theta\,d\theta$, $V = \frac{1}{2}e^{2\theta}$ to get

$\int e^{2\theta}\cos 3\theta\,d\theta = \frac{1}{2}e^{2\theta}\cos 3\theta + \frac{3}{2}\int e^{2\theta}\sin 3\theta\,d\theta$. Substituting in the previous formula gives

$I = \frac{1}{2}e^{2\theta}\sin 3\theta - \frac{3}{4}e^{2\theta}\cos 3\theta - \frac{9}{4}\int e^{2\theta}\sin 3\theta\, d\theta = \frac{1}{2}e^{2\theta}\sin 3\theta - \frac{3}{4}e^{2\theta}\cos 3\theta - \frac{9}{4}I \quad\Rightarrow$

$\frac{13}{4}I = \frac{1}{2}e^{2\theta}\sin 3\theta - \frac{3}{4}e^{2\theta}\cos 3\theta + C_1$. Hence, $I = \frac{1}{13}e^{2\theta}(2\sin 3\theta - 3\cos 3\theta) + C$, where $C = \frac{4}{13}C_1$.

17. Let $u = y$, $dv = \sinh y\, dy \quad\Rightarrow\quad du = dy$, $v = \cosh y$. Then
$\int y\sinh y\, dy = y\cosh y - \int \cosh y\, dy = y\cosh y - \sinh y + C$.

19. Let $u = t$, $dv = \sin 3t\, dt \quad\Rightarrow\quad du = dt$, $v = -\frac{1}{3}\cos 3t$. Then
$\int_0^\pi t\sin 3t\, dt = \left[-\frac{1}{3}t\cos 3t\right]_0^\pi + \frac{1}{3}\int_0^\pi \cos 3t\, dt = \left(\frac{1}{3}\pi - 0\right) + \frac{1}{9}\left[\sin 3t\right]_0^\pi = \frac{\pi}{3}$.

21. Let $u = \ln x$, $dv = x^{-2}\, dx \quad\Rightarrow\quad du = \dfrac{1}{x}\, dx$, $v = -x^{-1}$. By (6),

$\displaystyle\int_1^2 \frac{\ln x}{x^2}\, dx = \left[-\frac{\ln x}{x}\right]_1^2 + \int_1^2 x^{-2}\, dx = -\frac{1}{2}\ln 2 + \ln 1 + \left[-\frac{1}{x}\right]_1^2 = -\frac{1}{2}\ln 2 + 0 - \frac{1}{2} + 1 = \frac{1}{2} - \frac{1}{2}\ln 2$.

23. Let $u = y$, $dv = \dfrac{dy}{e^{2y}} = e^{-2y}dy \quad\Rightarrow\quad du = dy$, $v = -\dfrac{1}{2}e^{-2y}$. Then

$\displaystyle\int_0^1 \frac{y}{e^{2y}}\, dy = \left[-\frac{1}{2}ye^{-2y}\right]_0^1 + \frac{1}{2}\int_0^1 e^{-2y}dy = \left(-\frac{1}{2}e^{-2} + 0\right) - \frac{1}{4}\left[e^{-2y}\right]_0^1 = -\frac{1}{2}e^{-2} - \frac{1}{4}e^{-2} + \frac{1}{4} = \frac{1}{4} - \frac{3}{4}e^{-2}$.

25. Let $u = \cos^{-1} x$, $dv = dx \quad\Rightarrow\quad du = -\dfrac{dx}{\sqrt{1-x^2}}$, $v = x$. Then

$\displaystyle I = \int_0^{1/2} \cos^{-1} x\, dx = \left[x\cos^{-1} x\right]_0^{1/2} + \int_0^{1/2} \frac{x\, dx}{\sqrt{1-x^2}} = \frac{1}{2}\cdot\frac{\pi}{3} + \int_1^{3/4} t^{-1/2}\left[-\frac{1}{2}\, dt\right]$, where $t = 1 - x^2$

$\Rightarrow\quad dt = -2x\, dx$. Thus, $I = \frac{\pi}{6} + \frac{1}{2}\int_{3/4}^1 t^{-1/2}\, dt = \frac{\pi}{6} + \left[\sqrt{t}\right]_{3/4}^1 = \frac{\pi}{6} + 1 - \frac{\sqrt{3}}{2} = \frac{1}{6}\left(\pi + 6 - 3\sqrt{3}\right)$.

27. Let $u = \ln(\sin x)$, $dv = \cos x\, dx \quad\Rightarrow\quad du = \dfrac{\cos x}{\sin x}\, dx$, $v = \sin x$. Then
$I = \int \cos x\ln(\sin x)\, dx = \sin x\ln(\sin x) - \int \cos x\, dx = \sin x\ln(\sin x) - \sin x + C$.
Another method: Substitute $t = \sin x$, so $dt = \cos x\, dx$. Then $I = \int \ln t\, dt = t\ln t - t + C$ (see Example 2) and
so $I = \sin x(\ln\sin x - 1) + C$.

29. Let $w = \ln x \quad\Rightarrow\quad dw = dx/x$. Then $x = e^w$ and $dx = e^w\, dw$, so

$\int \cos(\ln x)\, dx = \int e^w\cos w\, dw = \frac{1}{2}e^w(\sin w + \cos w) + C$ [by the method of Example 4]

$= \frac{1}{2}x[\sin(\ln x) + \cos(\ln x)] + C$

31. Let $u = (\ln x)^2$, $dv = x^4\, dx \quad\Rightarrow\quad du = 2\dfrac{\ln x}{x}\, dx$, $v = \dfrac{x^5}{5}$. By (6),

$\displaystyle\int_1^2 x^4(\ln x)^2\, dx = \left[\frac{x^5}{5}(\ln x)^2\right]_1^2 - 2\int_1^2 \frac{x^4}{5}\ln x\, dx = \frac{32}{5}(\ln 2)^2 - 0 - 2\int_1^2 \frac{x^4}{5}\ln x\, dx$.

Let $U = \ln x$, $dV = \dfrac{x^4}{5}\, dx \quad\Rightarrow\quad dU = \dfrac{1}{x}\, dx$, $V = \dfrac{x^5}{25}$.

Then $\displaystyle\int_1^2 \frac{x^4}{5}\ln x\, dx = \left[\frac{x^5}{25}\ln x\right]_1^2 - \int_1^2 \frac{x^4}{25}\, dx = \frac{32}{25}\ln 2 - 0 - \left[\frac{x^5}{125}\right]_1^2 = \frac{32}{25}\ln 2 - \left(\frac{32}{125} - \frac{1}{125}\right)$.

So $\int_1^2 x^4(\ln x)^2\, dx = \frac{32}{5}(\ln 2)^2 - 2\left(\frac{32}{25}\ln 2 - \frac{31}{125}\right) = \frac{32}{5}(\ln 2)^2 - \frac{64}{25}\ln 2 + \frac{62}{125}$.

33. Let $w = \sqrt{x}$, so that $x = w^2$ and $dx = 2w\, dw$. Thus, $\int \sin\sqrt{x}\, dx = \int 2w\sin w\, dw$. Now use parts with $u = 2w$,
$dv = \sin w\, dw$, $du = 2\, dw$, $v = -\cos w$ to get

$\int 2w\sin w\, dw = -2w\cos w + \int 2\cos w\, dw = -2w\cos w + 2\sin w + C$

$= -2\sqrt{x}\cos\sqrt{x} + 2\sin\sqrt{x} + C = 2\left(\sin\sqrt{x} - \sqrt{x}\cos\sqrt{x}\right) + C$

35. Let $x = \theta^2$, so that $dx = 2\theta\,d\theta$. Thus, $\int_{\sqrt{\pi/2}}^{\sqrt{\pi}} \theta^3 \cos(\theta^2)\,d\theta = \int_{\sqrt{\pi/2}}^{\sqrt{\pi}} \theta^2 \cos(\theta^2) \cdot \frac{1}{2}(2\theta\,d\theta) = \frac{1}{2}\int_{\pi/2}^{\pi} x \cos x\,dx$.

Now use parts with $u = x$, $dv = \cos x\,dx$, $du = dx$, $v = \sin x$ to get

$$\frac{1}{2}\int_{\pi/2}^{\pi} x \cos x\,dx = \frac{1}{2}\left([x \sin x]_{\pi/2}^{\pi} - \int_{\pi/2}^{\pi} \sin x\,dx\right) = \frac{1}{2}[x \sin x + \cos x]_{\pi/2}^{\pi}$$

$$= \frac{1}{2}(\pi \sin \pi + \cos \pi) - \frac{1}{2}\left(\frac{\pi}{2}\sin\frac{\pi}{2} + \cos\frac{\pi}{2}\right) = \frac{1}{2}(\pi \cdot 0 - 1) - \frac{1}{2}\left(\frac{\pi}{2}\cdot 1 + 0\right) = -\frac{1}{2} - \frac{\pi}{4}$$

In Exercises 37–40, let $f(x)$ denote the integrand and $F(x)$ its antiderivative (with $C = 0$).

37. Let $u = x$, $dv = \cos \pi x\,dx$ $\Rightarrow$ $du = dx$, $v = (\sin \pi x)/\pi$. Then

$$\int x \cos \pi x\,dx = x \cdot \frac{\sin \pi x}{\pi} - \int \frac{\sin \pi x}{\pi}\,dx = \frac{x \sin \pi x}{\pi} + \frac{\cos \pi x}{\pi^2} + C.$$

We see from the graph that this is reasonable, since F has extreme values where f is 0.

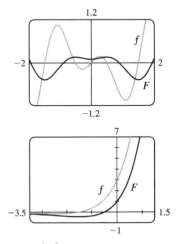

39. Let $u = 2x + 3$, $dv = e^x\,dx$ $\Rightarrow$ $du = 2\,dx$, $v = e^x$. Then
$\int(2x+3)e^x\,dx = (2x+3)e^x - 2\int e^x\,dx = (2x+3)e^x - 2e^x + C = (2x+1)\,e^x + C$. We see from the graph that this is reasonable, since F has a minimum where f changes from negative to positive.

41. (a) Take $n = 2$ in Example 6 to get $\int \sin^2 x\,dx = -\frac{1}{2}\cos x \sin x + \frac{1}{2}\int 1\,dx = \frac{x}{2} - \frac{\sin 2x}{4} + C$.

(b) $\int \sin^4 x\,dx = -\frac{1}{4}\cos x \sin^3 x + \frac{3}{4}\int \sin^2 x\,dx = -\frac{1}{4}\cos x \sin^3 x + \frac{3}{8}x - \frac{3}{16}\sin 2x + C$.

43. (a) From Example 6, $\displaystyle\int \sin^n x\,dx = -\frac{1}{n}\cos x \sin^{n-1} x + \frac{n-1}{n}\int \sin^{n-2} x\,dx$. Using (6),

$$\int_0^{\pi/2} \sin^n x\,dx = \left[-\frac{\cos x \sin^{n-1} x}{n}\right]_0^{\pi/2} + \frac{n-1}{n}\int_0^{\pi/2} \sin^{n-2} x\,dx$$

$$= (0 - 0) + \frac{n-1}{n}\int_0^{\pi/2} \sin^{n-2} x\,dx = \frac{n-1}{n}\int_0^{\pi/2} \sin^{n-2} x\,dx$$

(b) Using $n = 3$ in part (a), we have $\int_0^{\pi/2} \sin^3 x\,dx = \frac{2}{3}\int_0^{\pi/2} \sin x\,dx = \left[-\frac{2}{3}\cos x\right]_0^{\pi/2} = \frac{2}{3}$.

Using $n = 5$ in part (a), we have $\int_0^{\pi/2} \sin^5 x\,dx = \frac{4}{5}\int_0^{\pi/2} \sin^3 x\,dx = \frac{4}{5}\cdot\frac{2}{3} = \frac{8}{15}$.

(c) The formula holds for $n = 1$ (that is, $2n + 1 = 3$) by (b). Assume it holds for some $k \geq 1$. Then

$$\int_0^{\pi/2} \sin^{2k+1} x\,dx = \frac{2 \cdot 4 \cdot 6 \cdot \cdots \cdot (2k)}{3 \cdot 5 \cdot 7 \cdot \cdots \cdot (2k+1)}.$$ By Example 6,

$$\int_0^{\pi/2} \sin^{2k+3} x\,dx = \frac{2k+2}{2k+3}\int_0^{\pi/2} \sin^{2k+1} x\,dx = \frac{2k+2}{2k+3}\cdot\frac{2 \cdot 4 \cdot 6 \cdot \cdots \cdot (2k)}{3 \cdot 5 \cdot 7 \cdot \cdots \cdot (2k+1)}$$

$$= \frac{2 \cdot 4 \cdot 6 \cdot \cdots \cdot (2k)[2\,(k+1)]}{3 \cdot 5 \cdot 7 \cdot \cdots \cdot (2k+1)[2\,(k+1)+1]},$$

so the formula holds for $n = k + 1$. By induction, the formula holds for all $n \geq 1$.

45. Let $u = (\ln x)^n$, $dv = dx$ $\Rightarrow$ $du = n(\ln x)^{n-1}(dx/x)$, $v = x$. By Equation 2,
$\int (\ln x)^n\,dx = x(\ln x)^n - \int nx(\ln x)^{n-1}(dx/x) = x(\ln x)^n - n\int (\ln x)^{n-1}\,dx$.

47. Let $u = \left(x^2 + a^2\right)^n$, $dv = dx$ $\Rightarrow$ $du = n\left(x^2 + a^2\right)^{n-1} 2x\,dx$, $v = x$. Then

$$\int \left(x^2 + a^2\right)^n dx = x\left(x^2 + a^2\right)^n - 2n \int x^2 \left(x^2 + a^2\right)^{n-1} dx$$

$$= x\left(x^2 + a^2\right)^n - 2n \left[\int \left(x^2 + a^2\right)^n dx - a^2 \int \left(x^2 + a^2\right)^{n-1} dx\right] \quad [\text{since } x^2 = \left(x^2 + a^2\right) - a^2]$$

$$\Rightarrow (2n+1) \int \left(x^2 + a^2\right)^n dx = x\left(x^2 + a^2\right)^n + 2na^2 \int \left(x^2 + a^2\right)^{n-1} dx, \text{ and}$$

$$\int \left(x^2 + a^2\right)^n dx = \frac{x\left(x^2 + a^2\right)^n}{2n+1} + \frac{2na^2}{2n+1} \int \left(x^2 + a^2\right)^{n-1} dx \quad [\text{provided } 2n + 1 \neq 0].$$

49. Take $n = 3$ in Exercise 45 to get

$$\int (\ln x)^3 dx = x(\ln x)^3 - 3\int (\ln x)^2 dx = x(\ln x)^3 - 3x(\ln x)^2 + 6x \ln x - 6x + C \quad [\text{by Exercise 13}].$$

Or: Instead of using Exercise 13, apply Exercise 45 again with $n = 2$.

51. Area $= \int_0^5 xe^{-0.4x}dx$. Let $u = x$, $dv = e^{-0.4x}dx$ $\Rightarrow$

$du = dx$, $v = -2.5e^{-0.4x}$. Then

area $= \left[-2.5xe^{-0.4x}\right]_0^5 + 2.5\int_0^5 e^{-0.4x}dx$

$\quad = -12.5e^{-2} + 0 + 2.5\left[-2.5e^{-0.4x}\right]_0^5$

$\quad = -12.5e^{-2} - 6.25(e^{-2} - 1) = 6.25 - 18.75e^{-2} \quad$ or $\quad \frac{25}{4} - \frac{75}{4}e^{-2}$

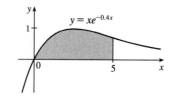

53. The curves $y = x \sin x$ and $y = (x-2)^2$ intersect at $a \approx 1.04748$ and
$b \approx 2.87307$, so

area $= \int_a^b \left[x \sin x - (x-2)^2\right] dx$

$\quad = \left[-x \cos x + \sin x - \frac{1}{3}(x-2)^3\right]_a^b \quad [\text{by Example 1}]$

$\quad \approx 2.81358 - 0.63075 = 2.18283$

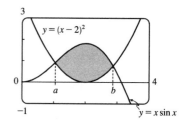

55. $V = \int_0^1 2\pi x \cos(\pi x/2)dx$. Let $u = x$, $dv = \cos(\pi x/2)dx$ $\Rightarrow$ $du = dx$, $v = \frac{2}{\pi}\sin(\pi x/2)$.

$$V = 2\pi\left[\frac{2}{\pi}x \sin\left(\frac{\pi x}{2}\right)\right]_0^1 - 2\pi \cdot \frac{2}{\pi}\int_0^1 \sin\left(\frac{\pi x}{2}\right) dx = 2\pi\left(\frac{2}{\pi} - 0\right) - 4\left[-\frac{2}{\pi}\cos\left(\frac{\pi x}{2}\right)\right]_0^1$$

$\quad = 4 + \frac{8}{\pi}(0 - 1) = 4 - \frac{8}{\pi}$.

57. Volume $= \int_{-1}^0 2\pi(1 - x)e^{-x} dx$. Let $u = 1 - x$, $dv = e^{-x} dx$ $\Rightarrow$ $du = -dx$, $v = -e^{-x}$.

$V = 2\pi\left[(1 - x)(-e^{-x})\right]_{-1}^0 - 2\pi\int_{-1}^0 e^{-x} dx = 2\pi\left[(x - 1)(e^{-x}) + e^{-x}\right]_{-1}^0$

$\quad = 2\pi\left[xe^{-x}\right]_{-1}^0 = 2\pi(0 + e) = 2\pi e$

59. The average value of $f(x) = x^2 \ln x$ on the interval $[1, 3]$ is $f_{\text{ave}} = \frac{1}{3 - 1}\int_1^3 x^2 \ln x\,dx = \frac{1}{2}I$.

Let $u = \ln x$, $dv = x^2 dx$ $\Rightarrow$ $du = (1/x)\,dx$, $v = \frac{1}{3}x^3$. So

$I = \left[\frac{1}{3}x^3 \ln x\right]_1^3 - \int_1^3 \frac{1}{3}x^2 dx = (9\ln 3 - 0) - \left[\frac{1}{9}x^3\right]_1^3 = 9\ln 3 - \left(3 - \frac{1}{9}\right) = 9\ln 3 - \frac{26}{9}$.

Thus, $f_{\text{ave}} = \frac{1}{2}I = \frac{1}{2}\left(9\ln 3 - \frac{26}{9}\right) = \frac{9}{2}\ln 3 - \frac{13}{9}$.

61. Since $v(t) > 0$ for all t, the desired distance is $s(t) = \int_0^t v(w)dw = \int_0^t w^2 e^{-w}\,dw$.

First let $u = w^2$, $dv = e^{-w}\,dw$ $\Rightarrow$ $du = 2w\,dw$, $v = -e^{-w}$. Then $s(t) = \left[-w^2 e^{-w}\right]_0^t + 2\int_0^t we^{-w}\,dw$.

Next let $U = w$, $dV = e^{-w}\,dw$ $\Rightarrow$ $dU = dw$, $V = -e^{-w}$. Then

$$s(t) = -t^2 e^{-t} + 2\left(\left[-we^{-w}\right]_0^t + \int_0^t e^{-w}\,dw\right) = -t^2 e^{-t} + 2\left(-te^{-t} + 0 + \left[-e^{-w}\right]_0^t\right)$$

$$= -t^2 e^{-t} + 2(-te^{-t} - e^{-t} + 1) = -t^2 e^{-t} - 2te^{-t} - 2e^{-t} + 2$$

$$= 2 - e^{-t}(t^2 + 2t + 2) \text{ meters}$$

63. For $I = \int_1^4 xf''(x)\,dx$, let $u = x$, $dv = f''(x)\,dx$ $\Rightarrow$ $du = dx$, $v = f'(x)$. Then

$I = [xf'(x)]_1^4 - \int_1^4 f'(x)\,dx = 4f'(4) - 1 \cdot f'(1) - [f(4) - f(1)] = 4 \cdot 3 - 1 \cdot 5 - (7 - 2) = 12 - 5 - 5 = 2$.

We used the fact that f'' is continuous to guarantee that I exists.

65. Using the formula for volumes of rotation and the figure, we see that

Volume $= \int_0^d \pi b^2\,dy - \int_0^c \pi a^2\,dy - \int_c^d \pi[g(y)]^2\,dy = \pi b^2 d - \pi a^2 c - \int_c^d \pi[g(y)]^2\,dy$. Let $y = f(x)$, which

gives $dy = f'(x)\,dx$ and $g(y) = x$, so that $V = \pi b^2 d - \pi a^2 c - \pi \int_a^b x^2 f'(x)\,dx$. Now integrate

by parts with $u = x^2$, and $dv = f'(x)\,dx$ $\Rightarrow$ $du = 2x\,dx$, $v = f(x)$, and

$\int_a^b x^2 f'(x)\,dx = \left[x^2 f(x)\right]_a^b - \int_a^b 2x f(x)\,dx = b^2 f(b) - a^2 f(a) - \int_a^b 2x f(x)\,dx$, but $f(a) = c$ and $f(b) = d$

$\Rightarrow$ $V = \pi b^2 d - \pi a^2 c - \pi\left[b^2 d - a^2 c - \int_a^b 2xf(x)\,dx\right] = \int_a^b 2\pi xf(x)\,dx$.

7.2 Trigonometric Integrals

The symbols $\overset{s}{=}$ and $\overset{c}{=}$ indicate the use of the substitutions $\{u = \sin x, du = \cos x\,dx\}$ and $\{u = \cos x, du = -\sin x\,dx\}$, respectively.

1. $\int \sin^3 x \cos^2 x\,dx = \int \sin^2 x \cos^2 x \sin x\,dx = \int (1 - \cos^2 x)\cos^2 x \sin x\,dx \overset{c}{=} \int (1 - u^2)u^2(-du)$

$\quad = \int (u^2 - 1)u^2\,du = \int (u^4 - u^2)du = \tfrac{1}{5}u^5 - \tfrac{1}{3}u^3 + C = \tfrac{1}{5}\cos^5 x - \tfrac{1}{3}\cos^3 x + C$

3. $\int_{\pi/2}^{3\pi/4} \sin^5 x \cos^3 x\,dx = \int_{\pi/2}^{3\pi/4} \sin^5 x \cos^2 x \cos x\,dx = \int_{\pi/2}^{3\pi/4} \sin^5 x (1 - \sin^2 x)\cos x\,dx$

$\quad \overset{s}{=} \int_1^{\sqrt{2}/2} u^5(1 - u^2)du = \int_1^{\sqrt{2}/2} (u^5 - u^7)du = \left[\tfrac{1}{6}u^6 - \tfrac{1}{8}u^8\right]_1^{\sqrt{2}/2}$

$\quad = \left(\tfrac{1/8}{6} - \tfrac{1/16}{8}\right) - \left(\tfrac{1}{6} - \tfrac{1}{8}\right) = -\tfrac{11}{384}$

5. $\int \cos^5 x \sin^4 x\,dx = \int \cos^4 x \sin^4 x \cos x\,dx = \int (1 - \sin^2 x)^2 \sin^4 x \cos x\,dx \overset{s}{=} \int (1 - u^2)^2 u^4\,du$

$\quad = \int (1 - 2u^2 + u^4)u^4\,du = \int (u^4 - 2u^6 + u^8)\,du = \tfrac{1}{5}u^5 - \tfrac{2}{7}u^7 + \tfrac{1}{9}u^9 + C$

$\quad = \tfrac{1}{5}\sin^5 x - \tfrac{2}{7}\sin^7 x + \tfrac{1}{9}\sin^9 x + C$

7. $\int_0^{\pi/2} \cos^2 \theta\,d\theta = \int_0^{\pi/2} \tfrac{1}{2}(1 + \cos 2\theta)\,d\theta$ [half-angle identity]

$\quad = \tfrac{1}{2}\left[\theta + \tfrac{1}{2}\sin 2\theta\right]_0^{\pi/2} = \tfrac{1}{2}\left[\left(\tfrac{\pi}{2} + 0\right) - (0 + 0)\right] = \tfrac{\pi}{4}$

9. $\int_0^\pi \sin^4(3t)\,dt = \int_0^\pi \left[\sin^2(3t)\right]^2\,dt = \int_0^\pi \left[\tfrac{1}{2}(1 - \cos 6t)\right]^2\,dt = \tfrac{1}{4}\int_0^\pi (1 - 2\cos 6t + \cos^2 6t)\,dt$

$\quad = \tfrac{1}{4}\int_0^\pi \left[1 - 2\cos 6t + \tfrac{1}{2}(1 + \cos 12t)\right]\,dt = \tfrac{1}{4}\int_0^\pi \left(\tfrac{3}{2} - 2\cos 6t + \tfrac{1}{2}\cos 12t\right)\,dt$

$\quad = \tfrac{1}{4}\left[\tfrac{3}{2}t - \tfrac{1}{3}\sin 6t + \tfrac{1}{24}\sin 12t\right]_0^\pi = \tfrac{1}{4}\left[\left(\tfrac{3\pi}{2} - 0 + 0\right) - (0 - 0 + 0)\right] = \tfrac{3\pi}{8}$

11. $\int (1 + \cos \theta)^2 \, d\theta = \int (1 + 2 \cos \theta + \cos^2 \theta) \, d\theta = \theta + 2 \sin \theta + \frac{1}{2} \int (1 + \cos 2\theta) \, d\theta$

$\qquad = \theta + 2 \sin \theta + \frac{1}{2}\theta + \frac{1}{4} \sin 2\theta + C = \frac{3}{2}\theta + 2 \sin \theta + \frac{1}{4} \sin 2\theta + C$

13. $\int_0^{\pi/4} \sin^4 x \cos^2 x \, dx = \int_0^{\pi/4} \sin^2 x \, (\sin x \cos x)^2 \, dx = \int_0^{\pi/4} \frac{1}{2}(1 - \cos 2x)\left(\frac{1}{2} \sin 2x\right)^2 \, dx$

$\qquad = \frac{1}{8} \int_0^{\pi/4} (1 - \cos 2x) \sin^2 2x \, dx = \frac{1}{8} \int_0^{\pi/4} \sin^2 2x \, dx - \frac{1}{8} \int_0^{\pi/4} \sin^2 2x \cos 2x \, dx$

$\qquad = \frac{1}{16} \int_0^{\pi/4} (1 - \cos 4x) \, dx - \frac{1}{16}\left[\frac{1}{3} \sin^3 2x\right]_0^{\pi/4} = \frac{1}{16}\left[x - \frac{1}{4} \sin 4x - \frac{1}{3} \sin^3 2x\right]_0^{\pi/4}$

$\qquad = \frac{1}{16}\left(\frac{\pi}{4} - 0 - \frac{1}{3}\right) = \frac{1}{192}(3\pi - 4)$

15. $\int \sin^3 x \sqrt{\cos x} \, dx = \int (1 - \cos^2 x) \sqrt{\cos x} \sin x \, dx \overset{c}{=} \int (1 - u^2) \, u^{1/2} \, (-du) = \int \left(u^{5/2} - u^{1/2}\right) du$

$\qquad = \frac{2}{7} u^{7/2} - \frac{2}{3} u^{3/2} + C = \frac{2}{7} (\cos x)^{7/2} - \frac{2}{3} (\cos x)^{3/2} + C$

$\qquad = \left(\frac{2}{7} \cos^3 x - \frac{2}{3} \cos x\right) \sqrt{\cos x} + C$

17. $\displaystyle\int \cos^2 x \tan^3 x \, dx = \int \frac{\sin^3 x}{\cos x} \, dx \overset{c}{=} \int \frac{(1 - u^2)(-du)}{u} = \int \left[\frac{-1}{u} + u\right] du$

$\qquad = -\ln|u| + \frac{1}{2} u^2 + C = \frac{1}{2} \cos^2 x - \ln|\cos x| + C$

19. $\displaystyle\int \frac{1 - \sin x}{\cos x} \, dx = \int (\sec x - \tan x) \, dx = \ln|\sec x + \tan x| - \ln|\sec x| + C \qquad \begin{bmatrix} \text{by (1) and the boxed} \\ \text{formula above it} \end{bmatrix}$

$\qquad = \ln|(\sec x + \tan x) \cos x| + C = \ln|1 + \sin x| + C$

$\qquad = \ln(1 + \sin x) + C \quad \text{since } 1 + \sin x \geq 0$

Or: $\displaystyle\int \frac{1 - \sin x}{\cos x} \, dx = \int \frac{1 - \sin x}{\cos x} \cdot \frac{1 + \sin x}{1 + \sin x} \, dx = \int \frac{(1 - \sin^2 x) \, dx}{\cos x \, (1 + \sin x)} = \int \frac{\cos x \, dx}{1 + \sin x}$

$\qquad = \int \frac{dw}{w} \qquad [\text{where } w = 1 + \sin x, \, dw = \cos x \, dx]$

$\qquad = \ln|w| + C = \ln|1 + \sin x| + C = \ln(1 + \sin x) + C$

21. Let $u = \tan x$, $du = \sec^2 x \, dx$. Then $\int \sec^2 x \tan x \, dx = \int u \, du = \frac{1}{2} u^2 + C = \frac{1}{2} \tan^2 x + C$.

$\qquad$ Or: Let $v = \sec x$, $dv = \sec x \tan x \, dx$. Then $\int \sec^2 x \tan x \, dx = \int v \, dv = \frac{1}{2} v^2 + C = \frac{1}{2} \sec^2 x + C$.

23. $\int \tan^2 x \, dx = \int (\sec^2 x - 1) \, dx = \tan x - x + C$

25. $\int \sec^6 t \, dt = \int \sec^4 t \cdot \sec^2 t \, dt = \int (\tan^2 t + 1)^2 \sec^2 t \, dt = \int (u^2 + 1)^2 \, du \qquad [u = \tan t, \, du = \sec^2 t \, dt]$

$\qquad = \int (u^4 + 2u^2 + 1) \, du = \frac{1}{5} u^5 + \frac{2}{3} u^3 + u + C = \frac{1}{5} \tan^5 t + \frac{2}{3} \tan^3 t + \tan t + C$

27. $\int_0^{\pi/3} \tan^5 x \sec^4 x \, dx = \int_0^{\pi/3} \tan^5 x \, (\tan^2 x + 1) \sec^2 x \, dx$

$\qquad = \int_0^{\sqrt{3}} u^5 (u^2 + 1) \, du \qquad [u = \tan x, \, du = \sec^2 x \, dx]$

$\qquad = \int_0^{\sqrt{3}} (u^7 + u^5) \, du = \left[\frac{1}{8} u^8 + \frac{1}{6} u^6\right]_0^{\sqrt{3}} = \frac{81}{8} + \frac{27}{6} = \frac{81}{8} + \frac{9}{2} = \frac{81}{8} + \frac{36}{8} = \frac{117}{8}$

Alternate solution:

$\int_0^{\pi/3} \tan^5 x \sec^4 x \, dx = \int_0^{\pi/3} \tan^4 x \sec^3 x \sec x \tan x \, dx = \int_0^{\pi/3} (\sec^2 x - 1)^2 \sec^3 x \sec x \tan x \, dx$

$\qquad = \int_1^2 (u^2 - 1)^2 u^3 \, du \qquad [u = \sec x, \, du = \sec x \tan x \, dx]$

$\qquad = \int_1^2 (u^4 - 2u^2 + 1) u^3 \, du = \int_1^2 (u^7 - 2u^5 + u^3) \, du$

$\qquad = \left[\frac{1}{8} u^8 - \frac{1}{3} u^6 + \frac{1}{4} u^4\right]_1^2 = \left(32 - \frac{64}{3} + 4\right) - \left(\frac{1}{8} - \frac{1}{3} + \frac{1}{4}\right) = \frac{117}{8}$

29. $\int \tan^3 x \sec x \, dx = \int \tan^2 x \sec x \tan x \, dx = \int \left(\sec^2 x - 1\right) \sec x \tan x \, dx$

$\qquad = \int (u^2 - 1) \, du \qquad [u = \sec x, \, du = \sec x \tan x \, dx]$

$\qquad = \tfrac{1}{3}u^3 - u + C = \tfrac{1}{3}\sec^3 x - \sec x + C$

31. $\int \tan^5 x \, dx = \int \left(\sec^2 x - 1\right)^2 \tan x \, dx = \int \sec^4 x \tan x \, dx - 2 \int \sec^2 x \tan x \, dx + \int \tan x \, dx$

$\qquad = \int \sec^3 x \sec x \tan x \, dx - 2 \int \tan x \sec^2 x \, dx + \int \tan x \, dx$

$\qquad = \tfrac{1}{4} \sec^4 x - \tan^2 x + \ln |\sec x| + C \quad [\text{or } \tfrac{1}{4} \sec^4 x - \sec^2 x + \ln |\sec x| + C\,]$

33. $\displaystyle\int \frac{\tan^3 \theta}{\cos^4 \theta} \, d\theta = \int \tan^3 \theta \sec^4 \theta \, d\theta = \int \tan^3 \theta \cdot (\tan^2 \theta + 1) \cdot \sec^2 \theta \, d\theta$

$\qquad = \int u^3(u^2 + 1) \, du \qquad [u = \tan \theta, \, du = \sec^2 \theta \, d\theta]$

$\qquad = \int (u^5 + u^3) \, du = \tfrac{1}{6}u^6 + \tfrac{1}{4}u^4 + C = \tfrac{1}{6} \tan^6 \theta + \tfrac{1}{4} \tan^4 \theta + C$

35. $\int_{\pi/6}^{\pi/2} \cot^2 x \, dx = \int_{\pi/6}^{\pi/2} \left(\csc^2 x - 1\right) dx = \left[- \cot x - x\right]_{\pi/6}^{\pi/2} = \left(0 - \tfrac{\pi}{2}\right) - \left(-\sqrt{3} - \tfrac{\pi}{6}\right) = \sqrt{3} - \tfrac{\pi}{3}$

37. $\int \cot^3 \alpha \csc^3 \alpha \, d\alpha = \int \cot^2 \alpha \csc^2 \alpha \cdot \csc \alpha \cot \alpha \, d\alpha = \int (\csc^2 \alpha - 1) \csc^2 \alpha \cdot \csc \alpha \cot \alpha \, d\alpha$

$\qquad = \int (u^2 - 1)u^2 \cdot (-du) \qquad [u = \csc \alpha, \, du = -\csc \alpha \cot \alpha \, d\alpha]$

$\qquad = \int (u^2 - u^4) \, du = \tfrac{1}{3}u^3 - \tfrac{1}{5}u^5 + C = \tfrac{1}{3} \csc^3 \alpha - \tfrac{1}{5} \csc^5 \alpha + C$

39. $I = \displaystyle\int \csc x \, dx = \int \frac{\csc x \, (\csc x - \cot x)}{\csc x - \cot x} \, dx = \int \frac{-\csc x \cot x + \csc^2 x}{\csc x - \cot x} \, dx.$ Let $u = \csc x - \cot x \;\Rightarrow$

$\quad du = \left(-\csc x \cot x + \csc^2 x\right) dx.$ Then $I = \int du/u = \ln |u| = \ln |\csc x - \cot x| + C.$

41. Use Equation 2(b):

$\int \sin 5x \sin 2x \, dx = \int \tfrac{1}{2}[\cos(5x - 2x) - \cos(5x + 2x)] \, dx = \tfrac{1}{2} \int (\cos 3x - \cos 7x) \, dx$

$\qquad = \tfrac{1}{6} \sin 3x - \tfrac{1}{14} \sin 7x + C$

43. Use Equation 2(c):

$\int \cos 7\theta \cos 5\theta \, d\theta = \int \tfrac{1}{2}[\cos(7\theta - 5\theta) + \cos(7\theta + 5\theta)] \, d\theta = \tfrac{1}{2} \int (\cos 2\theta + \cos 12\theta) \, d\theta$

$\qquad = \tfrac{1}{2}\left(\tfrac{1}{2} \sin 2\theta + \tfrac{1}{12} \sin 12\theta\right) + C = \tfrac{1}{4} \sin 2\theta + \tfrac{1}{24} \sin 12\theta + C$

45. $\displaystyle\int \frac{1 - \tan^2 x}{\sec^2 x} \, dx = \int \left(\cos^2 x - \sin^2 x\right) dx = \int \cos 2x \, dx = \frac{1}{2} \sin 2x + C$

47. Let $u = \tan(t^2) \;\Rightarrow\; du = 2t \sec^2(t^2) \, dt.$ Then

$\int t \sec^2(t^2) \tan^4(t^2) \, dt = \int u^4 \left(\tfrac{1}{2} \, du\right) = \tfrac{1}{10}u^5 + C = \tfrac{1}{10} \tan^5(t^2) + C.$

49. Let $u = \cos x \;\Rightarrow\; du = -\sin x \, dx.$ Then

$\int \sin^5 x \, dx = \int \left(1 - \cos^2 x\right)^2 \sin x \, dx = \int \left(1 - u^2\right)^2 (-du)$

$\qquad = \int (-1 + 2u^2 - u^4) \, du = -\tfrac{1}{5}u^5 + \tfrac{2}{3}u^3 - u + C$

$\qquad = -\tfrac{1}{5} \cos^5 x + \tfrac{2}{3} \cos^3 x - \cos x + C$

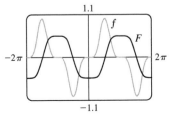

Notice that F is increasing when $f(x) > 0$, so the graphs serve as a
check on our work.

51. $\int \sin 3x \sin 6x \, dx = \int \frac{1}{2}[\cos(3x - 6x) - \cos(3x + 6x)] \, dx$

$$= \frac{1}{2} \int (\cos 3x - \cos 9x) \, dx$$

$$= \frac{1}{6} \sin 3x - \frac{1}{18} \sin 9x + C$$

Notice that $f(x) = 0$ whenever F has a horizontal tangent.

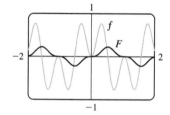

53. $f_{\text{ave}} = \frac{1}{2\pi} \int_{-\pi}^{\pi} \sin^2 x \cos^3 x \, dx = \frac{1}{2\pi} \int_{-\pi}^{\pi} \sin^2 x \left(1 - \sin^2 x\right) \cos x \, dx$

$$= \frac{1}{2\pi} \int_0^0 u^2 \left(1 - u^2\right) du \qquad \text{[where } u = \sin x]$$

$$= 0$$

55. For $0 < x < \frac{\pi}{2}$, we have $0 < \sin x < 1$, so $\sin^3 x < \sin x$. Hence the area is

$\int_0^{\pi/2} \left(\sin x - \sin^3 x\right) dx = \int_0^{\pi/2} \sin x \left(1 - \sin^2 x\right) dx = \int_0^{\pi/2} \cos^2 x \sin x \, dx$. Now let $u = \cos x \Rightarrow$

$du = -\sin x \, dx$. Then area $= \int_1^0 u^2 \left(-du\right) = \int_0^1 u^2 du = \left[\frac{1}{3}u^3\right]_0^1 = \frac{1}{3}$.

57.

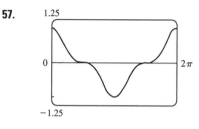

It seems from the graph that $\int_0^{2\pi} \cos^3 x \, dx = 0$, since the area below the

x-axis and above the graph looks about equal to the area above the axis

and below the graph. By Example 1, the integral is

$\left[\sin x - \frac{1}{3} \sin^3 x\right]_0^{2\pi} = 0$. Note that due to symmetry, the integral of

any odd power of $\sin x$ or $\cos x$ between limits which differ by $2n\pi$

(n any integer) is 0.

59. $V = \int_{\pi/2}^{\pi} \pi \sin^2 x \, dx = \pi \int_{\pi/2}^{\pi} \frac{1}{2}(1 - \cos 2x) \, dx = \pi\left[\frac{1}{2}x - \frac{1}{4}\sin 2x\right]_{\pi/2}^{\pi} = \pi\left(\frac{\pi}{2} - 0 - \frac{\pi}{4} + 0\right) = \frac{\pi^2}{4}$

61. Volume $= \pi \int_0^{\pi/2} \left[(1 + \cos x)^2 - 1^2\right] dx = \pi \int_0^{\pi/2} \left(2 \cos x + \cos^2 x\right) dx$

$$= \pi\left[2 \sin x + \frac{1}{2}x + \frac{1}{4}\sin 2x\right]_0^{\pi/2} = \pi\left(2 + \frac{\pi}{4}\right) = 2\pi + \frac{\pi^2}{4}$$

63. $s = f(t) = \int_0^t \sin \omega u \cos^2 \omega u \, du$. Let $y = \cos \omega u \Rightarrow dy = -\omega \sin \omega u \, du$. Then

$s = -\frac{1}{\omega} \int_1^{\cos \omega t} y^2 dy = -\frac{1}{\omega}\left[\frac{1}{3}y^3\right]_1^{\cos \omega t} = \frac{1}{3\omega}\left(1 - \cos^3 \omega t\right)$.

65. Just note that the integrand is odd $[f(-x) = -f(x)]$.

Or: If $m \neq n$, calculate

$$\int_{-\pi}^{\pi} \sin mx \cos nx \, dx = \int_{-\pi}^{\pi} \frac{1}{2}[\sin(m - n)x + \sin(m + n)x] \, dx$$

$$= \frac{1}{2}\left[-\frac{\cos(m - n)x}{m - n} - \frac{\cos(m + n)x}{m + n}\right]_{-\pi}^{\pi} = 0$$

If $m = n$, then the first term in each set of brackets is zero.

67. $\int_{-\pi}^{\pi} \cos mx \cos nx \, dx = \int_{-\pi}^{\pi} \frac{1}{2}[\cos(m - n)x + \cos(m + n)x] \, dx$. If $m \neq n$,

this is equal to $\frac{1}{2}\left[\frac{\sin(m - n)x}{m - n} + \frac{\sin(m + n)x}{m + n}\right]_{-\pi}^{\pi} = 0$. If $m = n$, we get

$\int_{-\pi}^{\pi} \frac{1}{2}[1 + \cos(m + n)x] \, dx = \left[\frac{1}{2}x\right]_{-\pi}^{\pi} + \left[\frac{\sin(m + n)x}{2(m + n)}\right]_{-\pi}^{\pi} = \pi + 0 = \pi$.

7.3 Trigonometric Substitution

1. Let $x = 3 \sec \theta$, where $0 \le \theta < \frac{\pi}{2}$ or $\pi \le \theta < \frac{3\pi}{2}$. Then

$dx = 3 \sec \theta \tan \theta \, d\theta$ and

$$\sqrt{x^2 - 9} = \sqrt{9 \sec^2 \theta - 9} = \sqrt{9(\sec^2 \theta - 1)} = \sqrt{9 \tan^2 \theta}$$
$$= 3 |\tan \theta| = 3 \tan \theta \text{ for the relevant values of } \theta.$$

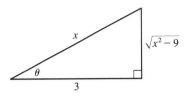

$$\int \frac{1}{x^2 \sqrt{x^2 - 9}} \, dx = \int \frac{1}{9 \sec^2 \theta \cdot 3 \tan \theta} 3 \sec \theta \tan \theta \, d\theta = \frac{1}{9} \int \cos \theta \, d\theta = \frac{1}{9} \sin \theta + C = \frac{1}{9} \frac{\sqrt{x^2 - 9}}{x} + C$$

Note that $-\sec(\theta + \pi) = \sec \theta$, so the figure is sufficient for the case $\pi \le \theta < \frac{3\pi}{2}$.

3. Let $x = 3 \tan \theta$, where $-\frac{\pi}{2} < \theta < \frac{\pi}{2}$. Then $dx = 3 \sec^2 \theta \, d\theta$ and

$$\sqrt{x^2 + 9} = \sqrt{9 \tan^2 \theta + 9} = \sqrt{9(\tan^2 \theta + 1)} = \sqrt{9 \sec^2 \theta}$$
$$= 3 |\sec \theta| = 3 \sec \theta \text{ for the relevant values of } \theta.$$

$$\int \frac{x^3}{\sqrt{x^2 + 9}} \, dx = \int \frac{3^3 \tan^3 \theta}{3 \sec \theta} 3 \sec^2 \theta \, d\theta = 3^3 \int \tan^3 \theta \sec \theta \, d\theta = 3^3 \int \tan^2 \theta \tan \theta \sec \theta \, d\theta$$
$$= 3^3 \int (\sec^2 \theta - 1) \tan \theta \sec \theta \, d\theta = 3^3 \int (u^2 - 1) \, du \qquad [u = \sec \theta, \, du = \sec \theta \tan \theta \, d\theta]$$
$$= 3^3 \left(\frac{1}{3} u^3 - u \right) + C = 3^3 \left(\frac{1}{3} \sec^3 \theta - \sec \theta \right) + C = 3^3 \left[\frac{1}{3} \frac{(x^2 + 9)^{3/2}}{3^3} - \frac{\sqrt{x^2 + 9}}{3} \right] + C$$
$$= \frac{1}{3} (x^2 + 9)^{3/2} - 9\sqrt{x^2 + 9} + C \quad \text{or} \quad \frac{1}{3} (x^2 - 18) \sqrt{x^2 + 9} + C$$

5. Let $t = \sec \theta$, so $dt = \sec \theta \tan \theta \, d\theta$, $t = \sqrt{2} \Rightarrow \theta = \frac{\pi}{4}$, and $t = 2 \Rightarrow \theta = \frac{\pi}{3}$. Then

$$\int_{\sqrt{2}}^{2} \frac{1}{t^3 \sqrt{t^2 - 1}} \, dt = \int_{\pi/4}^{\pi/3} \frac{1}{\sec^3 \theta \tan \theta} \sec \theta \tan \theta \, d\theta = \int_{\pi/4}^{\pi/3} \frac{1}{\sec^2 \theta} \, d\theta = \int_{\pi/4}^{\pi/3} \cos^2 \theta \, d\theta$$
$$= \int_{\pi/4}^{\pi/3} \frac{1}{2} (1 + \cos 2\theta) \, d\theta = \frac{1}{2} \left[\theta + \frac{1}{2} \sin 2\theta \right]_{\pi/4}^{\pi/3}$$
$$= \frac{1}{2} \left[\left(\frac{\pi}{3} + \frac{1}{2} \frac{\sqrt{3}}{2} \right) - \left(\frac{\pi}{4} + \frac{1}{2} \cdot 1 \right) \right] = \frac{1}{2} \left(\frac{\pi}{12} + \frac{\sqrt{3}}{4} - \frac{1}{2} \right) = \frac{\pi}{24} + \frac{\sqrt{3}}{8} - \frac{1}{4}$$

7. Let $x = 5 \sin \theta$, so $dx = 5 \cos \theta \, d\theta$. Then

$$\int \frac{1}{x^2 \sqrt{25 - x^2}} \, dx = \int \frac{1}{5^2 \sin^2 \theta \cdot 5 \cos \theta} 5 \cos \theta \, d\theta$$
$$= \frac{1}{25} \int \csc^2 \theta \, d\theta = -\frac{1}{25} \cot \theta + C$$
$$= -\frac{1}{25} \frac{\sqrt{25 - x^2}}{x} + C$$

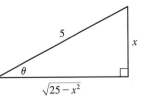

9. Let $x = 4\tan\theta$, where $-\frac{\pi}{2} < \theta < \frac{\pi}{2}$. Then $dx = 4\sec^2\theta\,d\theta$ and

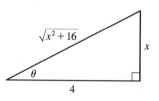

$$\sqrt{x^2 + 16} = \sqrt{16\tan^2\theta + 16} = \sqrt{16(\tan^2\theta + 1)}$$
$$= \sqrt{16\sec^2\theta} = 4\,|\sec\theta|$$
$$= 4\sec\theta \text{ for the relevant values of } \theta.$$

$$\int \frac{dx}{\sqrt{x^2 + 16}} = \int \frac{4\sec^2\theta\,d\theta}{4\sec\theta} = \int \sec\theta\,d\theta = \ln|\sec\theta + \tan\theta| + C_1$$
$$= \ln\left|\frac{\sqrt{x^2 + 16}}{4} + \frac{x}{4}\right| + C_1 = \ln\left|\sqrt{x^2 + 16} + x\right| - \ln|4| + C_1$$
$$= \ln\left(\sqrt{x^2 + 16} + x\right) + C, \text{ where } C = C_1 - \ln 4.$$

(Since $\sqrt{x^2 + 16} + x > 0$, we don't need the absolute value.)

11. Let $2x = \sin\theta$, where $-\frac{\pi}{2} \le \theta \le \frac{\pi}{2}$. Then $x = \frac{1}{2}\sin\theta$,

$dx = \frac{1}{2}\cos\theta\,d\theta$, and $\sqrt{1 - 4x^2} = \sqrt{1 - (2x)^2} = \cos\theta$.

$$\int \sqrt{1 - 4x^2}\,dx = \int \cos\theta\left(\tfrac{1}{2}\cos\theta\right)d\theta = \tfrac{1}{4}\int (1 + \cos 2\theta)\,d\theta$$
$$= \tfrac{1}{4}\left(\theta + \tfrac{1}{2}\sin 2\theta\right) + C = \tfrac{1}{4}(\theta + \sin\theta\cos\theta) + C$$
$$= \tfrac{1}{4}\left[\sin^{-1}(2x) + 2x\sqrt{1 - 4x^2}\right] + C$$

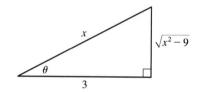

13. Let $x = 3\sec\theta$, where $0 \le \theta < \frac{\pi}{2}$ or $\pi \le \theta < \frac{3\pi}{2}$. Then

$dx = 3\sec\theta\tan\theta\,d\theta$ and $\sqrt{x^2 - 9} = 3\tan\theta$, so

$$\int \frac{\sqrt{x^2 - 9}}{x^3}\,dx = \int \frac{3\tan\theta}{27\sec^3\theta}\,3\sec\theta\tan\theta\,d\theta = \frac{1}{3}\int \frac{\tan^2\theta}{\sec^2\theta}\,d\theta$$

$$= \tfrac{1}{3}\int \sin^2\theta\,d\theta = \tfrac{1}{3}\int \tfrac{1}{2}(1 - \cos 2\theta)\,d\theta = \tfrac{1}{6}\theta - \tfrac{1}{12}\sin 2\theta + C = \tfrac{1}{6}\theta - \tfrac{1}{6}\sin\theta\cos\theta + C$$

$$= \frac{1}{6}\sec^{-1}\left(\frac{x}{3}\right) - \frac{1}{6}\frac{\sqrt{x^2 - 9}}{x}\frac{3}{x} + C = \frac{1}{6}\sec^{-1}\left(\frac{x}{3}\right) - \frac{\sqrt{x^2 - 9}}{2x^2} + C$$

15. Let $x = a\sin\theta$, where $-\frac{\pi}{2} \le \theta \le \frac{\pi}{2}$. Then $dx = a\cos\theta\,d\theta$ and

$$\int \frac{x^2\,dx}{(a^2 - x^2)^{3/2}} = \int \frac{a^2\sin^2\theta\,a\cos\theta\,d\theta}{a^3\cos^3\theta} = \int \tan^2\theta\,d\theta$$

$$= \int (\sec^2\theta - 1)\,d\theta = \tan\theta - \theta + C$$

$$= \frac{x}{\sqrt{a^2 - x^2}} - \sin^{-1}\frac{x}{a} + C$$

17. Let $u = x^2 - 7$, so $du = 2x\,dx$. Then $\displaystyle\int \frac{x}{\sqrt{x^2 - 7}}\,dx = \frac{1}{2}\int \frac{1}{\sqrt{u}}\,du = \frac{1}{2}\cdot 2\sqrt{u} + C = \sqrt{x^2 - 7} + C.$

19. Let $x = \tan\theta$, where $-\frac{\pi}{2} < \theta < \frac{\pi}{2}$. Then $dx = \sec^2\theta\,d\theta$

and $\sqrt{1+x^2} = \sec\theta$, so

$$\int \frac{\sqrt{1+x^2}}{x}\,dx = \int \frac{\sec\theta}{\tan\theta}\sec^2\theta\,d\theta = \int \frac{\sec\theta}{\tan\theta}(1+\tan^2\theta)\,d\theta$$

$$= \int (\csc\theta + \sec\theta\tan\theta)\,d\theta$$

$$= \ln|\csc\theta - \cot\theta| + \sec\theta + C \quad \text{[by Exercise 7.2.39]}$$

$$= \ln\left|\frac{\sqrt{1+x^2}}{x} - \frac{1}{x}\right| + \frac{\sqrt{1+x^2}}{1} + C = \ln\left|\frac{\sqrt{1+x^2}-1}{x}\right| + \sqrt{1+x^2} + C$$

21. Let $u = 4 - 9x^2 \;\Rightarrow\; du = -18x\,dx$. Then $x^2 = \frac{1}{9}(4-u)$ and

$$\int_0^{2/3} x^3\sqrt{4-9x^2}\,dx = \int_4^0 \frac{1}{9}(4-u)u^{1/2}\left(-\frac{1}{18}\right)du = \frac{1}{162}\int_0^4\left(4u^{1/2} - u^{3/2}\right)du$$

$$= \frac{1}{162}\left[\frac{8}{3}u^{3/2} - \frac{2}{5}u^{5/2}\right]_0^4 = \frac{1}{162}\left[\frac{64}{3} - \frac{64}{5}\right] = \frac{64}{1215}$$

Or: Let $3x = 2\sin\theta$, where $-\frac{\pi}{2} \le \theta \le \frac{\pi}{2}$.

23. $5 + 4x - x^2 = -(x^2 - 4x + 4) + 9 = -(x-2)^2 + 9$. Let

$x - 2 = 3\sin\theta$, $-\frac{\pi}{2} \le \theta \le \frac{\pi}{2}$, so $dx = 3\cos\theta\,d\theta$. Then

$$\int \sqrt{5+4x-x^2}\,dx = \int \sqrt{9-(x-2)^2}\,dx = \int \sqrt{9-9\sin^2\theta}\,3\cos\theta\,d\theta$$

$$= \int \sqrt{9\cos^2\theta}\,3\cos\theta\,d\theta = \int 9\cos^2\theta\,d\theta$$

$$= \frac{9}{2}\int (1+\cos 2\theta)\,d\theta = \frac{9}{2}\left(\theta + \frac{1}{2}\sin 2\theta\right) + C$$

$$= \frac{9}{2}\theta + \frac{9}{4}\sin 2\theta + C = \frac{9}{2}\theta + \frac{9}{4}(2\sin\theta\cos\theta) + C$$

$$= \frac{9}{2}\sin^{-1}\left(\frac{x-2}{3}\right) + \frac{9}{2}\cdot\frac{x-2}{3}\cdot\frac{\sqrt{5+4x-x^2}}{3} + C$$

$$= \frac{9}{2}\sin^{-1}\left(\frac{x-2}{3}\right) + \frac{1}{2}(x-2)\sqrt{5+4x-x^2} + C$$

25. $9x^2 + 6x - 8 = (3x+1)^2 - 9$, so let $u = 3x+1$, $du = 3dx$. Then $\displaystyle\int \frac{dx}{\sqrt{9x^2+6x-8}} = \int \frac{\frac{1}{3}\,du}{\sqrt{u^2-9}}$. Now

let $u = 3\sec\theta$, where $0 \le \theta < \frac{\pi}{2}$ or $\pi \le \theta < \frac{3\pi}{2}$. Then $du = 3\sec\theta\tan\theta\,d\theta$ and $\sqrt{u^2-9} = 3\tan\theta$, so

$$\int \frac{\frac{1}{3}\,du}{\sqrt{u^2-9}} = \int \frac{\sec\theta\tan\theta\,d\theta}{3\tan\theta} = \frac{1}{3}\int \sec\theta\,d\theta = \frac{1}{3}\ln|\sec\theta + \tan\theta| + C_1 = \frac{1}{3}\ln\left|\frac{u+\sqrt{u^2-9}}{3}\right| + C_1$$

$$= \frac{1}{3}\ln\left|u + \sqrt{u^2-9}\right| + C = \frac{1}{3}\ln\left|3x+1+\sqrt{9x^2+6x-8}\right| + C$$

27. $x^2 + 2x + 2 = (x+1)^2 + 1$. Let $u = x+1$, $du = dx$. Then

$$\int \frac{dx}{(x^2+2x+2)^2} = \int \frac{du}{(u^2+1)^2} = \int \frac{\sec^2\theta\,d\theta}{\sec^4\theta} \qquad \begin{bmatrix} \text{where } u = \tan\theta,\ du = \sec^2\theta\,d\theta, \\ \text{and } u^2+1 = \sec^2\theta \end{bmatrix}$$

$$= \int \cos^2\theta\,d\theta = \frac{1}{2}\int (1+\cos 2\theta)\,d\theta = \frac{1}{2}(\theta + \sin\theta\cos\theta) + C$$

$$= \frac{1}{2}\left[\tan^{-1}u + \frac{u}{1+u^2}\right] + C = \frac{1}{2}\left[\tan^{-1}(x+1) + \frac{x+1}{x^2+2x+2}\right] + C$$

29. Let $u = x^2$, $du = 2x\,dx$. Then

$$\int x\sqrt{1-x^4}\,dx = \int \sqrt{1-u^2}\left(\tfrac{1}{2}\,du\right) = \tfrac{1}{2}\int \cos\theta \cdot \cos\theta\,d\theta \quad \left[\begin{array}{c}\text{where } u = \sin\theta, \; du = \cos\theta\,d\theta, \\ \text{and } \sqrt{1-u^2} = \cos\theta\end{array}\right]$$

$$= \tfrac{1}{2}\int \tfrac{1}{2}(1+\cos 2\theta)\,d\theta = \tfrac{1}{4}\theta + \tfrac{1}{8}\sin 2\theta + C = \tfrac{1}{4}\theta + \tfrac{1}{4}\sin\theta\cos\theta + C$$

$$= \tfrac{1}{4}\sin^{-1}u + \tfrac{1}{4}u\sqrt{1-u^2} + C = \tfrac{1}{4}\sin^{-1}(x^2) + \tfrac{1}{4}x^2\sqrt{1-x^4} + C$$

31. (a) Let $x = a\tan\theta$, where $-\tfrac{\pi}{2} < \theta < \tfrac{\pi}{2}$. Then $\sqrt{x^2 + a^2} = a\sec\theta$ and

$$\int \frac{dx}{\sqrt{x^2+a^2}} = \int \frac{a\sec^2\theta\,d\theta}{a\sec\theta} = \int \sec\theta\,d\theta = \ln|\sec\theta + \tan\theta| + C_1 = \ln\left|\frac{\sqrt{x^2+a^2}}{a} + \frac{x}{a}\right| + C_1$$

$$= \ln\left(x + \sqrt{x^2+a^2}\right) + C \quad \text{where } C = C_1 - \ln|a|$$

(b) Let $x = a\sinh t$, so that $dx = a\cosh t\,dt$ and $\sqrt{x^2 + a^2} = a\cosh t$. Then

$$\int \frac{dx}{\sqrt{x^2+a^2}} = \int \frac{a\cosh t\,dt}{a\cosh t} = t + C = \sinh^{-1}\frac{x}{a} + C.$$

33. The average value of $f(x) = \sqrt{x^2-1}/x$ on the interval $[1, 7]$ is

$$\frac{1}{7-1}\int_1^7 \frac{\sqrt{x^2-1}}{x}\,dx = \frac{1}{6}\int_0^\alpha \frac{\tan\theta}{\sec\theta}\cdot\sec\theta\tan\theta\,d\theta \quad \left[\begin{array}{c}\text{where } x = \sec\theta, \; dx = \sec\theta\tan\theta\,d\theta, \\ \sqrt{x^2-1} = \tan\theta, \text{ and } \alpha = \sec^{-1}7\end{array}\right]$$

$$= \tfrac{1}{6}\int_0^\alpha \tan^2\theta\,d\theta = \tfrac{1}{6}\int_0^\alpha (\sec^2\theta - 1)\,d\theta$$

$$= \tfrac{1}{6}\Big[\tan\theta - \theta\Big]_0^\alpha = \tfrac{1}{6}(\tan\alpha - \alpha)$$

$$= \tfrac{1}{6}\left(\sqrt{48} - \sec^{-1}7\right)$$

35. Area of $\triangle POQ = \tfrac{1}{2}(r\cos\theta)(r\sin\theta) = \tfrac{1}{2}r^2\sin\theta\cos\theta$. Area of region $PQR = \int_{r\cos\theta}^r \sqrt{r^2 - x^2}\,dx$.
Let $x = r\cos u \;\Rightarrow\; dx = -r\sin u\,du$ for $\theta \le u \le \tfrac{\pi}{2}$. Then we obtain

$$\int \sqrt{r^2 - x^2}\,dx = \int r\sin u\,(-r\sin u)\,du = -r^2\int \sin^2 u\,du = -\tfrac{1}{2}r^2(u - \sin u\cos u) + C$$

$$= -\tfrac{1}{2}r^2\cos^{-1}(x/r) + \tfrac{1}{2}x\sqrt{r^2 - x^2} + C$$

so

$$\text{area of region } PQR = \tfrac{1}{2}\left[-r^2\cos^{-1}(x/r) + x\sqrt{r^2 - x^2}\right]_{r\cos\theta}^r$$

$$= \tfrac{1}{2}\left[0 - (-r^2\theta + r\cos\theta\,r\sin\theta)\right]$$

$$= \tfrac{1}{2}r^2\theta - \tfrac{1}{2}r^2\sin\theta\cos\theta$$

and thus, (area of sector POR) = (area of $\triangle POQ$) + (area of region PQR) = $\tfrac{1}{2}r^2\theta$.

37. From the graph, it appears that the curve $y = x^2\sqrt{4 - x^2}$ and the line

$y = 2 - x$ intersect at about $x = 0.81$ and $x = 2$, with $x^2\sqrt{4 - x^2} > 2 - x$ on

$(0.81, 2)$. So the area bounded by the curve and the line is $A \approx$

$\int_{0.81}^{2} \left[x^2\sqrt{4 - x^2} - (2 - x) \right] dx = \int_{0.81}^{2} x^2\sqrt{4 - x^2}\, dx - \left[2x - \frac{1}{2}x^2 \right]_{0.81}^{2}$.

To evaluate the integral, we put $x = 2\sin\theta$, where $-\frac{\pi}{2} \leq \theta \leq \frac{\pi}{2}$. Then

$dx = 2\cos\theta\, d\theta$, $x = 2 \Rightarrow \theta = \sin^{-1} 1 = \frac{\pi}{2}$, and $x = 0.81 \Rightarrow \theta = \sin^{-1} 0.405 \approx 0.417$. So

$\int_{0.81}^{2} x^2\sqrt{4 - x^2}\, dx \approx \int_{0.417}^{\pi/2} 4\sin^2\theta\,(2\cos\theta)(2\cos\theta\, d\theta) = 4\int_{0.417}^{\pi/2} \sin^2 2\theta\, d\theta = 4\int_{0.417}^{\pi/2} \frac{1}{2}(1 - \cos 4\theta)\, d\theta$

$= 2\left[\theta - \frac{1}{4}\sin 4\theta\right]_{0.417}^{\pi/2} = 2\left[\left(\frac{\pi}{2} - 0 \right) - \left(0.417 - \frac{1}{4}(0.995) \right) \right] \approx 2.81$

Thus, $A \approx 2.81 - \left[\left(2\cdot 2 - \frac{1}{2}\cdot 2^2 \right) - \left(2\cdot 0.81 - \frac{1}{2}\cdot 0.81^2 \right) \right] \approx 2.10$.

39. Let the equation of the large circle be $x^2 + y^2 = R^2$. Then the equation of the small circle is $x^2 + (y - b)^2 = r^2$,

where $b = \sqrt{R^2 - r^2}$ is the distance between the centers of the circles. The desired area is

$A = \int_{-r}^{r} \left[(b + \sqrt{r^2 - x^2}) - \sqrt{R^2 - x^2} \right] dx = 2\int_0^r \left(b + \sqrt{r^2 - x^2} - \sqrt{R^2 - x^2} \right) dx$

$= 2\int_0^r b\, dx + 2\int_0^r \sqrt{r^2 - x^2}\, dx - 2\int_0^r \sqrt{R^2 - x^2}\, dx$

The first integral is just $2br = 2r\sqrt{R^2 - r^2}$. To evaluate the other two integrals, note that

$\int \sqrt{a^2 - x^2}\, dx = \int a^2\cos^2\theta\, d\theta \quad [x = a\sin\theta,\, dx = a\cos\theta\, d\theta] \quad = \frac{1}{2}a^2\int (1 + \cos 2\theta)\, d\theta$

$= \frac{1}{2}a^2\left(\theta + \frac{1}{2}\sin 2\theta\right) + C = \frac{1}{2}a^2(\theta + \sin\theta\cos\theta) + C$

$= \frac{a^2}{2}\arcsin\left(\frac{x}{a}\right) + \frac{a^2}{2}\left(\frac{x}{a}\right)\frac{\sqrt{a^2 - x^2}}{a} + C = \frac{a^2}{2}\arcsin\left(\frac{x}{a}\right) + \frac{x}{2}\sqrt{a^2 - x^2} + C$

so the desired area is

$A = 2r\sqrt{R^2 - r^2} + \left[r^2\arcsin(x/r) + x\sqrt{r^2 - x^2} \right]_0^r - \left[R^2\arcsin(x/R) + x\sqrt{R^2 - x^2} \right]_0^r$

$= 2r\sqrt{R^2 - r^2} + r^2\left(\frac{\pi}{2}\right) - \left[R^2\arcsin(r/R) + r\sqrt{R^2 - r^2} \right] = r\sqrt{R^2 - r^2} + \frac{\pi}{2}r^2 - R^2\arcsin(r/R)$

41. We use cylindrical shells and assume that $R > r$. $x^2 = r^2 - (y - R)^2 \Rightarrow x = \pm\sqrt{r^2 - (y - R)^2}$, so

$g(y) = 2\sqrt{r^2 - (y - R)^2}$ and

$V = \int_{R-r}^{R+r} 2\pi y \cdot 2\sqrt{r^2 - (y - R)^2}\, dy = \int_{-r}^{r} 4\pi(u + R)\sqrt{r^2 - u^2}\, du \quad$ [where $u = y - R$]

$= 4\pi\int_{-r}^{r} u\sqrt{r^2 - u^2}\, du + 4\pi R\int_{-r}^{r} \sqrt{r^2 - u^2}\, du \quad \begin{bmatrix} \text{where } u = r\sin\theta,\, du = r\cos\theta\, d\theta \\ \text{in the second integral} \end{bmatrix}$

$= 4\pi\left[-\frac{1}{3}\left(r^2 - u^2\right)^{3/2} \right]_{-r}^{r} + 4\pi R\int_{-\pi/2}^{\pi/2} r^2\cos^2\theta\, d\theta = -\frac{4\pi}{3}(0 - 0) + 4\pi Rr^2\int_{-\pi/2}^{\pi/2} \cos^2\theta\, d\theta$

$= 2\pi Rr^2\int_{-\pi/2}^{\pi/2}(1 + \cos 2\theta)\, d\theta = 2\pi Rr^2\left[\theta + \frac{1}{2}\sin 2\theta\right]_{-\pi/2}^{\pi/2} = 2\pi^2 Rr^2$

Another method: Use washers instead of shells, so $V = 8\pi R\int_0^r \sqrt{r^2 - y^2}\, dy$ as in Exercise 6.2.61(a), but evaluate

the integral using $y = r\sin\theta$.

7.4 Integration of Rational Functions by Partial Fractions

1. (a) $\dfrac{2x}{(x+3)(3x+1)} = \dfrac{A}{x+3} + \dfrac{B}{3x+1}$

(b) $\dfrac{1}{x^3 + 2x^2 + x} = \dfrac{1}{x(x^2 + 2x + 1)} = \dfrac{1}{x(x+1)^2} = \dfrac{A}{x} + \dfrac{B}{x+1} + \dfrac{C}{(x+1)^2}$

3. (a) $\dfrac{2}{x^2 + 3x - 4} = \dfrac{2}{(x+4)(x-1)} = \dfrac{A}{x+4} + \dfrac{B}{x-1}$

(b) $x^2 + x + 1$ is irreducible, so $\dfrac{x^2}{(x-1)(x^2+x+1)} = \dfrac{A}{x-1} + \dfrac{Bx+C}{x^2+x+1}$.

5. (a) $\dfrac{x^4}{x^4 - 1} = \dfrac{(x^4 - 1) + 1}{x^4 - 1} = 1 + \dfrac{1}{x^4 - 1}$ [or use long division] $= 1 + \dfrac{1}{(x^2 - 1)(x^2 + 1)}$

$= 1 + \dfrac{1}{(x-1)(x+1)(x^2+1)} = 1 + \dfrac{A}{x-1} + \dfrac{B}{x+1} + \dfrac{Cx+D}{x^2+1}$

(b) $\dfrac{t^4 + t^2 + 1}{(t^2 + 1)(t^2 + 4)^2} = \dfrac{At+B}{t^2+1} + \dfrac{Ct+D}{t^2+4} + \dfrac{Et+F}{(t^2+4)^2}$

7. $\displaystyle\int \dfrac{x}{x-6}\,dx = \int \dfrac{(x-6)+6}{x-6}\,dx = \int \left(1 + \dfrac{6}{x-6}\right) dx = x + 6\ln|x-6| + C$

9. $\dfrac{x-9}{(x+5)(x-2)} = \dfrac{A}{x+5} + \dfrac{B}{x-2}$. Multiply both sides by $(x+5)(x-2)$ to get $x - 9 = A(x-2) + B(x+5)$.

Substituting 2 for x gives $-7 = 7B \iff B = -1$. Substituting -5 for x gives $-14 = -7A \iff A = 2$. Thus,

$$\int \dfrac{x-9}{(x+5)(x-2)}\,dx = \int \left(\dfrac{2}{x+5} + \dfrac{-1}{x-2}\right) dx = 2\ln|x+5| - \ln|x-2| + C$$

11. $\dfrac{1}{x^2 - 1} = \dfrac{1}{(x+1)(x-1)} = \dfrac{A}{x+1} + \dfrac{B}{x-1}$. Multiply both sides by $(x+1)(x-1)$ to get

$1 = A(x-1) + B(x+1)$. Substituting 1 for x gives $1 = 2B \iff B = \frac{1}{2}$.

Substituting -1 for x gives $1 = -2A \iff A = -\frac{1}{2}$. Thus,

$$\int_2^3 \dfrac{1}{x^2 - 1}\,dx = \int_2^3 \left(\dfrac{-1/2}{x+1} + \dfrac{1/2}{x-1}\right) dx = \left[-\tfrac{1}{2}\ln|x+1| + \tfrac{1}{2}\ln|x-1|\right]_2^3$$

$$= \left(-\tfrac{1}{2}\ln 4 + \tfrac{1}{2}\ln 2\right) - \left(-\tfrac{1}{2}\ln 3 + \tfrac{1}{2}\ln 1\right) = \tfrac{1}{2}(\ln 2 + \ln 3 - \ln 4) \quad \left[\text{or } \tfrac{1}{2}\ln\tfrac{3}{2}\right]$$

13. $\displaystyle\int \dfrac{ax}{x^2 - bx}\,dx = \int \dfrac{ax}{x(x-b)}\,dx = \int \dfrac{a}{x-b}\,dx = a\ln|x-b| + C$

15. $\dfrac{2x+3}{(x+1)^2} = \dfrac{A}{x+1} + \dfrac{B}{(x+1)^2} \Rightarrow 2x + 3 = A(x+1) + B$. Take $x = -1$ to get $B = 1$, and equate

coefficients of x to get $A = 2$. Now

$$\int_0^1 \dfrac{2x+3}{(x+1)^2}\,dx = \int_0^1 \left[\dfrac{2}{x+1} + \dfrac{1}{(x+1)^2}\right] dx = \left[2\ln(x+1) - \dfrac{1}{x+1}\right]_0^1$$

$$= 2\ln 2 - \tfrac{1}{2} - (2\ln 1 - 1) = 2\ln 2 + \tfrac{1}{2}$$

17. $\dfrac{4y^2 - 7y - 12}{y(y+2)(y-3)} = \dfrac{A}{y} + \dfrac{B}{y+2} + \dfrac{C}{y-3}$ $\Rightarrow$ $4y^2 - 7y - 12 = A(y+2)(y-3) + By(y-3) + Cy(y+2)$.

Setting $y = 0$ gives $-12 = -6A$, so $A = 2$. Setting $y = -2$ gives $18 = 10B$, so $B = \frac{9}{5}$. Setting $y = 3$ gives

$3 = 15C$, so $C = \frac{1}{5}$. Now

$$\int_1^2 \frac{4y^2 - 7y - 12}{y(y+2)(y-3)}\,dy = \int_1^2 \left(\frac{2}{y} + \frac{9/5}{y+2} + \frac{1/5}{y-3} \right) dy = \left[2\ln|y| + \tfrac{9}{5}\ln|y+2| + \tfrac{1}{5}\ln|y-3| \right]_1^2$$

$$= 2\ln 2 + \tfrac{9}{5}\ln 4 + \tfrac{1}{5}\ln 1 - 2\ln 1 - \tfrac{9}{5}\ln 3 - \tfrac{1}{5}\ln 2$$

$$= 2\ln 2 + \tfrac{18}{5}\ln 2 - \tfrac{1}{5}\ln 2 - \tfrac{9}{5}\ln 3 = \tfrac{27}{5}\ln 2 - \tfrac{9}{5}\ln 3 = \tfrac{9}{5}(3\ln 2 - \ln 3) = \tfrac{9}{5}\ln \tfrac{8}{3}$$

19. $\dfrac{1}{(x+5)^2\,(x-1)} = \dfrac{A}{x+5} + \dfrac{B}{(x+5)^2} + \dfrac{C}{x-1}$ $\Rightarrow$ $1 = A(x+5)(x-1) + B(x-1) + C(x+5)^2$. Setting

$x = -5$ gives $1 = -6B$, so $B = -\frac{1}{6}$. Setting $x = 1$ gives $1 = 36C$, so $C = \frac{1}{36}$. Setting $x = -2$ gives

$1 = A(3)(-3) + B(-3) + C(3^2) = -9A - 3B + 9C = -9A + \frac{1}{2} + \frac{1}{4} = -9A + \frac{3}{4}$, so $9A = -\frac{1}{4}$ and

$A = -\frac{1}{36}$. Now

$$\int \frac{1}{(x+5)^2\,(x-1)}\,dx = \int \left[\frac{-1/36}{x+5} - \frac{1/6}{(x+5)^2} + \frac{1/36}{x-1} \right] dx$$

$$= -\frac{1}{36}\ln|x+5| + \frac{1}{6(x+5)} + \frac{1}{36}\ln|x-1| + C$$

21. $\dfrac{5x^2 + 3x - 2}{x^3 + 2x^2} = \dfrac{5x^2 + 3x - 2}{x^2\,(x+2)} = \dfrac{A}{x} + \dfrac{B}{x^2} + \dfrac{C}{x+2}$. Multiply by $x^2(x+2)$ to get

$5x^2 + 3x - 2 = Ax(x+2) + B(x+2) + Cx^2$. Set $x = -2$ to get $C = 3$, and take $x = 0$ to get

$B = -1$. Equating the coefficients of x^2 gives $5 = A + C$ $\Rightarrow$ $A = 2$. So

$$\int \frac{5x^2 + 3x - 2}{x^3 + 2x^2}\,dx = \int \left(\frac{2}{x} - \frac{1}{x^2} + \frac{3}{x+2} \right) dx = 2\ln|x| + \frac{1}{x} + 3\ln|x+2| + C.$$

23. $\dfrac{x^2}{(x+1)^3} = \dfrac{A}{x+1} + \dfrac{B}{(x+1)^2} + \dfrac{C}{(x+1)^3}$. Multiply by $(x+1)^3$ to get $x^2 = A(x+1)^2 + B(x+1) + C$.

Setting $x = -1$ gives $C = 1$. Equating the coefficients of x^2 gives $A = 1$, and setting $x = 0$ gives $B = -2$.

Now $\displaystyle \int \frac{x^2\,dx}{(x+1)^3} = \int \left[\frac{1}{x+1} - \frac{2}{(x+1)^2} + \frac{1}{(x+1)^3} \right] dx = \ln|x+1| + \frac{2}{x+1} - \frac{1}{2(x+1)^2} + C.$

25. $\dfrac{10}{(x-1)(x^2+9)} = \dfrac{A}{x-1} + \dfrac{Bx+C}{x^2+9}$. Multiply both sides by $(x-1)(x^2+9)$ to get

$10 = A(x^2+9) + (Bx+C)(x-1)$ (∗). Substituting 1 for x gives $10 = 10A$ $\Leftrightarrow$ $A = 1$. Substituting 0 for x

gives $10 = 9A - C$ $\Rightarrow$ $C = 9(1) - 10 = -1$. The coefficients of the x^2-terms in (∗) must be equal, so

$0 = A + B$ $\Rightarrow$ $B = -1$. Thus,

$$\int \frac{10}{(x-1)(x^2+9)}\,dx = \int \left(\frac{1}{x-1} + \frac{-x-1}{x^2+9} \right) dx = \int \left(\frac{1}{x-1} - \frac{x}{x^2+9} - \frac{1}{x^2+9} \right) dx$$

$$= \ln|x-1| - \tfrac{1}{2}\ln(x^2+9) \text{ [let } u = x^2+9] - \tfrac{1}{3}\tan^{-1}\left(\tfrac{x}{3}\right) \text{ [Formula 10] } + C$$

27. $\dfrac{x^3 + x^2 + 2x + 1}{(x^2 + 1)(x^2 + 2)} = \dfrac{Ax + B}{x^2 + 1} + \dfrac{Cx + D}{x^2 + 2}$. Multiply both sides by $(x^2 + 1)(x^2 + 2)$ to get

$x^3 + x^2 + 2x + 1 = (Ax + B)(x^2 + 2) + (Cx + D)(x^2 + 1)$ ⇔

$x^3 + x^2 + 2x + 1 = (Ax^3 + Bx^2 + 2Ax + 2B) + (Cx^3 + Dx^2 + Cx + D)$ ⇔

$x^3 + x^2 + 2x + 1 = (A + C)x^3 + (B + D)x^2 + (2A + C)x + (2B + D)$. Comparing coefficients gives us the following system of equations:

$$A + C = 1 \quad \textbf{(1)} \qquad\qquad B + D = 1 \quad \textbf{(2)}$$
$$2A + C = 2 \quad \textbf{(3)} \qquad\qquad 2B + D = 1 \quad \textbf{(4)}$$

Subtracting equation **(1)** from equation **(3)** gives us $A = 1$, so $C = 0$. Subtracting equation **(2)** from equation **(4)**

gives us $B = 0$, so $D = 1$. Thus, $I = \displaystyle\int \dfrac{x^3 + x^2 + 2x + 1}{(x^2 + 1)(x^2 + 2)}\,dx = \int \left(\dfrac{x}{x^2 + 1} + \dfrac{1}{x^2 + 2}\right) dx$. For $\displaystyle\int \dfrac{x}{x^2 + 1}\,dx$,

let $u = x^2 + 1$ so $du = 2x\,dx$ and then $\displaystyle\int \dfrac{x}{x^2 + 1}\,dx = \dfrac{1}{2}\int \dfrac{1}{u}\,du = \dfrac{1}{2}\ln|u| + C = \dfrac{1}{2}\ln(x^2 + 1) + C$. For

$\displaystyle\int \dfrac{1}{x^2 + 2}\,dx$, use Formula 10 with $a = \sqrt{2}$. So $\displaystyle\int \dfrac{1}{x^2 + 2}\,dx = \int \dfrac{1}{x^2 + (\sqrt{2})^2}\,dx = \dfrac{1}{\sqrt{2}}\tan^{-1}\dfrac{x}{\sqrt{2}} + C$.

Thus, $I = \dfrac{1}{2}\ln(x^2 + 1) + \dfrac{1}{\sqrt{2}}\tan^{-1}\dfrac{x}{\sqrt{2}} + C$.

29. $\displaystyle\int \dfrac{x + 4}{x^2 + 2x + 5}\,dx = \int \dfrac{x + 1}{x^2 + 2x + 5}\,dx + \int \dfrac{3}{x^2 + 2x + 5}\,dx = \dfrac{1}{2}\int \dfrac{(2x + 2)\,dx}{x^2 + 2x + 5} + \int \dfrac{3\,dx}{(x + 1)^2 + 4}$

$= \dfrac{1}{2}\ln|x^2 + 2x + 5| + 3\displaystyle\int \dfrac{2\,du}{4(u^2 + 1)} \quad \begin{bmatrix} \text{where } x + 1 = 2u, \\ \text{and } dx = 2\,du \end{bmatrix}$

$= \dfrac{1}{2}\ln(x^2 + 2x + 5) + \dfrac{3}{2}\tan^{-1}u + C = \dfrac{1}{2}\ln(x^2 + 2x + 5) + \dfrac{3}{2}\tan^{-1}\left(\dfrac{x + 1}{2}\right) + C$

31. $\dfrac{1}{x^3 - 1} = \dfrac{1}{(x - 1)(x^2 + x + 1)} = \dfrac{A}{x - 1} + \dfrac{Bx + C}{x^2 + x + 1} \quad\Rightarrow\quad 1 = A(x^2 + x + 1) + (Bx + C)(x - 1)$.

Take $x = 1$ to get $A = \dfrac{1}{3}$. Equating coefficients of x^2 and then comparing the constant terms, we get $0 = \dfrac{1}{3} + B$,

$1 = \dfrac{1}{3} - C$, so $B = -\dfrac{1}{3}$, $C = -\dfrac{2}{3}$ ⇒

$$\int \dfrac{1}{x^3 - 1}\,dx = \int \dfrac{\frac{1}{3}}{x - 1}\,dx + \int \dfrac{-\frac{1}{3}x - \frac{2}{3}}{x^2 + x + 1}\,dx = \tfrac{1}{3}\ln|x - 1| - \dfrac{1}{3}\int \dfrac{x + 2}{x^2 + x + 1}\,dx$$

$$= \tfrac{1}{3}\ln|x - 1| - \dfrac{1}{3}\int \dfrac{x + 1/2}{x^2 + x + 1}\,dx - \dfrac{1}{3}\int \dfrac{(3/2)\,dx}{(x + 1/2)^2 + 3/4}$$

$$= \tfrac{1}{3}\ln|x - 1| - \tfrac{1}{6}\ln(x^2 + x + 1) - \tfrac{1}{2}\left(\dfrac{2}{\sqrt{3}}\right)\tan^{-1}\left(\dfrac{x + \frac{1}{2}}{\sqrt{3}/2}\right) + K$$

$$= \tfrac{1}{3}\ln|x - 1| - \tfrac{1}{6}\ln(x^2 + x + 1) - \dfrac{1}{\sqrt{3}}\tan^{-1}\left(\dfrac{1}{\sqrt{3}}(2x + 1)\right) + K$$

33. Let $u = x^3 + 3x^2 + 4$. Then $du = 3(x^2 + 2x)\,dx$ ⇒

$$\int_2^5 \dfrac{x^2 + 2x}{x^3 + 3x^2 + 4}\,dx = \dfrac{1}{3}\int_{24}^{204} \dfrac{du}{u} = \tfrac{1}{3}[\ln u]_{24}^{204} = \tfrac{1}{3}(\ln 204 - \ln 24) = \tfrac{1}{3}\ln \tfrac{204}{24} = \tfrac{1}{3}\ln \tfrac{17}{2}.$$

35. $\dfrac{1}{x^4 - x^2} = \dfrac{1}{x^2(x-1)(x+1)} = \dfrac{A}{x} + \dfrac{B}{x^2} + \dfrac{C}{x-1} + \dfrac{D}{x+1}$. Multiply by $x^2(x-1)(x+1)$ to get

$1 = Ax(x-1)(x+1) + B(x-1)(x+1) + Cx^2(x+1) + Dx^2(x-1)$. Setting $x = 1$ gives $C = \frac{1}{2}$, taking

$x = -1$ gives $D = -\frac{1}{2}$. Equating the coefficients of x^3 gives $0 = A + C + D = A$. Finally, setting $x = 0$ yields

$B = -1$. Now $\displaystyle\int \dfrac{dx}{x^4 - x^2} = \int \left[\dfrac{-1}{x^2} + \dfrac{1/2}{x-1} - \dfrac{1/2}{x+1} \right] dx = \dfrac{1}{x} + \frac{1}{2} \ln \left| \dfrac{x-1}{x+1} \right| + C.$

37. $\displaystyle\int \dfrac{x-3}{(x^2 + 2x + 4)^2} \, dx = \int \dfrac{x-3}{[(x+1)^2 + 3]^2} \, dx = \int \dfrac{u-4}{(u^2 + 3)^2} \, du$ [with $u = x + 1$]

$= \displaystyle\int \dfrac{u \, du}{(u^2 + 3)^2} - 4 \int \dfrac{du}{(u^2 + 3)^2} = \frac{1}{2} \int \dfrac{dv}{v^2} - 4 \int \dfrac{\sqrt{3} \sec^2 \theta \, d\theta}{9 \sec^4 \theta}$ $\left[\begin{array}{l} v = u^2 + 3 \text{ in the first integral;} \\ u = \sqrt{3} \tan \theta \text{ in the second} \end{array} \right]$

$= \dfrac{-1}{(2v)} - \dfrac{4\sqrt{3}}{9} \displaystyle\int \cos^2 \theta \, d\theta = \dfrac{-1}{2(u^2 + 3)} - \dfrac{2\sqrt{3}}{9} (\theta + \sin \theta \cos \theta) + C$

$= \dfrac{-1}{2(x^2 + 2x + 4)} - \dfrac{2\sqrt{3}}{9} \left[\tan^{-1} \left(\dfrac{x+1}{\sqrt{3}} \right) + \dfrac{\sqrt{3}(x+1)}{x^2 + 2x + 4} \right] + C$

$= \dfrac{-1}{2(x^2 + 2x + 4)} - \dfrac{2\sqrt{3}}{9} \tan^{-1} \left(\dfrac{x+1}{\sqrt{3}} \right) - \dfrac{2(x+1)}{3(x^2 + 2x + 4)} + C$

39. Let $u = \sqrt{x+1}$. Then $x = u^2 - 1$, $dx = 2u \, du$ $\Rightarrow$

$\displaystyle\int \dfrac{dx}{x\sqrt{x+1}} = \int \dfrac{2u \, du}{(u^2 - 1) u} = 2 \int \dfrac{du}{u^2 - 1} = \ln \left| \dfrac{u-1}{u+1} \right| + C = \ln \left| \dfrac{\sqrt{x+1} - 1}{\sqrt{x+1} + 1} \right| + C.$

41. Let $u = \sqrt{x}$, so $u^2 = x$ and $dx = 2u \, du$. Thus,

$\displaystyle\int_9^{16} \dfrac{\sqrt{x}}{x-4} \, dx = \int_3^4 \dfrac{u}{u^2 - 4} 2u \, du = 2 \int_3^4 \dfrac{u^2}{u^2 - 4} \, du = 2 \int_3^4 \left(1 + \dfrac{4}{u^2 - 4} \right) du$ [by long division]

$= 2 + 8 \displaystyle\int_3^4 \dfrac{du}{(u+2)(u-2)}.$ (∗)

Multiply $\dfrac{1}{(u+2)(u-2)} = \dfrac{A}{u+2} + \dfrac{B}{u-2}$ by $(u+2)(u-2)$ to get $1 = A(u-2) + B(u+2)$. Equating

coefficients we get $A + B = 0$ and $-2A + 2B = 1$. Solving gives us $B = \frac{1}{4}$ and $A = -\frac{1}{4}$, so

$\dfrac{1}{(u+2)(u-2)} = \dfrac{-1/4}{u+2} + \dfrac{1/4}{u-2}$ and (∗) is

$2 + 8 \displaystyle\int_3^4 \left(\dfrac{-1/4}{u+2} + \dfrac{1/4}{u-2} \right) du = 2 + 8 \left[-\frac{1}{4} \ln |u+2| + \frac{1}{4} \ln |u-2| \right]_3^4$

$= 2 + \left[2 \ln |u-2| - 2 \ln |u+2| \right]_3^4 = 2 + 2 \left[\ln \left| \dfrac{u-2}{u+2} \right| \right]_3^4$

$= 2 + 2 \left(\ln \frac{2}{6} - \ln \frac{1}{5} \right) = 2 + 2 \ln \frac{2/6}{1/5}$

$= 2 + 2 \ln \frac{5}{3}$ or $2 + \ln \left(\frac{5}{3} \right)^2 = 2 + \ln \frac{25}{9}$

43. Let $u = \sqrt[3]{x^2 + 1}$. Then $x^2 = u^3 - 1$, $2x \, dx = 3u^2 \, du$ $\Rightarrow$

$\displaystyle\int \dfrac{x^3 \, dx}{\sqrt[3]{x^2 + 1}} = \int \dfrac{(u^3 - 1)\frac{3}{2}u^2 \, du}{u} = \frac{3}{2} \int (u^4 - u) \, du = \frac{3}{10} u^5 - \frac{3}{4} u^2 + C$

$= \frac{3}{10} \left(x^2 + 1 \right)^{5/3} - \frac{3}{4} \left(x^2 + 1 \right)^{2/3} + C$

45. If we were to substitute $u = \sqrt{x}$, then the square root would disappear but a cube root would remain. On the other hand, the substitution $u = \sqrt[3]{x}$ would eliminate the cube root but leave a square root. We can eliminate both roots by means of the substitution $u = \sqrt[6]{x}$. (Note that 6 is the least common multiple of 2 and 3.)

Let $u = \sqrt[6]{x}$. Then $x = u^6$, so $dx = 6u^5\,du$ and $\sqrt{x} = u^3$, $\sqrt[3]{x} = u^2$. Thus,

$$\int \frac{dx}{\sqrt{x} - \sqrt[3]{x}} = \int \frac{6u^5\,du}{u^3 - u^2} = 6\int \frac{u^5}{u^2(u-1)}\,du = 6\int \frac{u^3}{u-1}\,du$$

$$= 6\int \left(u^2 + u + 1 + \frac{1}{u-1}\right)du \qquad \text{[by long division]}$$

$$= 6\left(\tfrac{1}{3}u^3 + \tfrac{1}{2}u^2 + u + \ln|u-1|\right) + C = 2\sqrt{x} + 3\sqrt[3]{x} + 6\sqrt[6]{x} + 6\ln\left|\sqrt[6]{x} - 1\right| + C$$

47. Let $u = e^x$. Then $x = \ln u$, $dx = \dfrac{du}{u}$ $\Rightarrow$

$$\int \frac{e^{2x}\,dx}{e^{2x} + 3e^x + 2} = \int \frac{u^2\,(du/u)}{u^2 + 3u + 2} = \int \frac{u\,du}{(u+1)(u+2)} = \int \left[\frac{-1}{u+1} + \frac{2}{u+2}\right]du$$

$$= 2\ln|u+2| - \ln|u+1| + C = \ln\left[(e^x + 2)^2/(e^x + 1)\right] + C$$

49. Let $u = \ln(x^2 - x + 2)$, $dv = dx$. Then $du = \dfrac{2x-1}{x^2 - x + 2}\,dx$, $v = x$, and (by integration by parts)

$$\int \ln(x^2 - x + 2)\,dx = x\ln(x^2 - x + 2) - \int \frac{2x^2 - x}{x^2 - x + 2}\,dx = x\ln(x^2 - x + 2) - \int \left(2 + \frac{x-4}{x^2 - x + 2}\right)dx$$

$$= x\ln(x^2 - x + 2) - 2x - \int \frac{\tfrac{1}{2}(2x - 1)}{x^2 - x + 2}\,dx + \frac{7}{2}\int \frac{dx}{(x - \tfrac{1}{2})^2 + \tfrac{7}{4}}$$

$$= x\ln(x^2 - x + 2) - 2x - \frac{1}{2}\ln(x^2 - x + 2) + \frac{7}{2}\int \frac{\tfrac{\sqrt{7}}{2}\,du}{\tfrac{7}{4}(u^2 + 1)} \qquad \begin{bmatrix} \text{where } x - \tfrac{1}{2} = \tfrac{\sqrt{7}}{2}u, \\ dx = \tfrac{\sqrt{7}}{2}\,du, \\ (x - \tfrac{1}{2})^2 + \tfrac{7}{4} = \tfrac{7}{4}(u^2 + 1) \end{bmatrix}$$

$$= (x - \tfrac{1}{2})\ln(x^2 - x + 2) - 2x + \sqrt{7}\tan^{-1}u + C$$

$$= (x - \tfrac{1}{2})\ln(x^2 - x + 2) - 2x + \sqrt{7}\tan^{-1}\frac{2x - 1}{\sqrt{7}} + C$$

51.

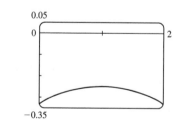

From the graph, we see that the integral will be negative, and we guess that the area is about the same as that of a rectangle with width 2 and height 0.3, so we estimate the integral to be $-(2 \cdot 0.3) = -0.6$. Now

$$\frac{1}{x^2 - 2x - 3} = \frac{1}{(x-3)(x+1)} = \frac{A}{x-3} + \frac{B}{x+1} \quad \Leftrightarrow$$

$1 = (A+B)x + A - 3B$, so $A = -B$ and $A - 3B = 1$ $\Leftrightarrow$ $A = \tfrac{1}{4}$

and $B = -\tfrac{1}{4}$, so the integral becomes

$$\int_0^2 \frac{dx}{x^2 - 2x - 3} = \frac{1}{4}\int_0^2 \frac{dx}{x - 3} - \frac{1}{4}\int_0^2 \frac{dx}{x + 1} = \tfrac{1}{4}\Big[\ln|x - 3| - \ln|x + 1|\Big]_0^2$$

$$= \frac{1}{4}\left[\ln\left|\frac{x-3}{x+1}\right|\right]_0^2 = \tfrac{1}{4}\left(\ln\tfrac{1}{3} - \ln 3\right) = -\tfrac{1}{2}\ln 3 \approx -0.55$$

53. $\displaystyle\int \frac{dx}{x^2 - 2x} = \int \frac{dx}{(x-1)^2 - 1} = \int \frac{du}{u^2 - 1}$ [put $u = x - 1$]

$$= \frac{1}{2} \ln\left|\frac{u-1}{u+1}\right| + C \quad \text{[by Equation 6]} \quad = \frac{1}{2} \ln\left|\frac{x-2}{x}\right| + C$$

55. (a) If $t = \tan\left(\dfrac{x}{2}\right)$, then $\dfrac{x}{2} = \tan^{-1} t$. The figure gives

$$\cos\left(\frac{x}{2}\right) = \frac{1}{\sqrt{1+t^2}} \text{ and } \sin\left(\frac{x}{2}\right) = \frac{t}{\sqrt{1+t^2}}.$$

(b) $\cos x = \cos\left(2 \cdot \dfrac{x}{2}\right) = 2\cos^2\left(\dfrac{x}{2}\right) - 1$

$$= 2\left(\frac{1}{\sqrt{1+t^2}}\right)^2 - 1 = \frac{2}{1+t^2} - 1 = \frac{1-t^2}{1+t^2}$$

$$\sin x = \sin\left(2 \cdot \frac{x}{2}\right) = 2\sin\left(\frac{x}{2}\right)\cos\left(\frac{x}{2}\right) = 2\,\frac{t}{\sqrt{1+t^2}}\,\frac{1}{\sqrt{1+t^2}} = \frac{2t}{1+t^2}$$

(c) $\dfrac{x}{2} = \arctan t \implies x = 2\arctan t \implies dx = \dfrac{2}{1+t^2}\,dt$

57. Let $t = \tan(x/2)$. Then, using the expressions in Exercise 55, we have

$$\int \frac{1}{3\sin x - 4\cos x}\,dx = \int \frac{1}{3\left(\dfrac{2t}{1+t^2}\right) - 4\left(\dfrac{1-t^2}{1+t^2}\right)}\,\frac{2\,dt}{1+t^2} = 2\int \frac{dt}{3(2t) - 4(1-t^2)} = \int \frac{dt}{2t^2 + 3t - 2}$$

$$= \int \frac{dt}{(2t-1)(t+2)} = \int \left[\frac{2}{5}\frac{1}{2t-1} - \frac{1}{5}\frac{1}{t+2}\right]dt \quad \text{[using partial fractions]}$$

$$= \frac{1}{5}\left[\ln|2t-1| - \ln|t+2|\right] + C = \frac{1}{5}\ln\left|\frac{2t-1}{t+2}\right| + C = \frac{1}{5}\ln\left|\frac{2\tan(x/2) - 1}{\tan(x/2) + 2}\right| + C$$

59. Let $t = \tan(x/2)$. Then, by Exercise 55,

$$\int \frac{dx}{2\sin x + \sin 2x} = \frac{1}{2}\int \frac{dx}{\sin x + \sin x\cos x} = \frac{1}{2}\int \frac{2\,dt/(1+t^2)}{2t/(1+t^2) + 2t(1-t^2)/(1+t^2)^2}$$

$$= \frac{1}{2}\int \frac{(1+t^2)\,dt}{t(1+t^2) + t(1-t^2)} = \frac{1}{4}\int \frac{(1+t^2)\,dt}{t} = \frac{1}{4}\int \left(\frac{1}{t} + t\right)dt$$

$$= \frac{1}{4}\ln|t| + \frac{1}{8}t^2 + C = \frac{1}{4}\ln\left|\tan\left(\tfrac{1}{2}x\right)\right| + \frac{1}{8}\tan^2\left(\tfrac{1}{2}x\right) + C$$

61. $\dfrac{x+1}{x-1} = 1 + \dfrac{2}{x-1} > 0$ for $2 \le x \le 3$, so

$$\text{area} = \int_2^3 \left[1 + \frac{2}{x-1}\right]dx = \Big[x + 2\ln|x-1|\Big]_2^3 = (3 + 2\ln 2) - (2 + 2\ln 1) = 1 + 2\ln 2.$$

63. $\dfrac{P+S}{P\left[(r-1)P-S\right]} = \dfrac{A}{P} + \dfrac{B}{(r-1)P-S} \quad \Rightarrow \quad P+S = A\left[(r-1)P-S\right] + BP = \left[(r-1)A+B\right]P - AS$

$\Rightarrow \quad (r-1)A+B = 1,\ -A = 1 \quad \Rightarrow \quad A = -1,\ B = r.$ Now

$$t = \int \frac{P+S}{P\left[(r-1)P-S\right]}\, dP = \int \left[\frac{-1}{P} + \frac{r}{(r-1)P-S}\right] dP = -\int \frac{dP}{P} + \frac{r}{r-1}\int \frac{r-1}{(r-1)P-S}\, dP$$

so $t = -\ln P + \dfrac{r}{r-1}\ln|(r-1)P-S| + C$. Here $r = 0.10$ and $S = 900$, so

$$t = -\ln P + \tfrac{0.1}{-0.9}\ln|{-0.9P} - 900| + C = -\ln P - \tfrac{1}{9}\ln(|-1|\,|0.9P+900|)$$

$$= -\ln P - \tfrac{1}{9}\ln(0.9P+900) + C$$

When $t = 0$, $P = 10{,}000$, so $0 = -\ln 10{,}000 - \tfrac{1}{9}\ln(9900) + C$. Thus, $C = \ln 10{,}000 + \tfrac{1}{9}\ln 9900$ [≈ 10.2326],

so our equation becomes

$$t = \ln 10{,}000 - \ln P + \tfrac{1}{9}\ln 9900 - \tfrac{1}{9}\ln(0.9P+900) = \ln\frac{10{,}000}{P} + \frac{1}{9}\ln\frac{9900}{0.9P+900}$$

$$= \ln\frac{10{,}000}{P} + \frac{1}{9}\ln\frac{1100}{0.1P+100} = \ln\frac{10{,}000}{P} + \frac{1}{9}\ln\frac{11{,}000}{P+1000}$$

65. (a) In Maple, we define $f(x)$, and then use `convert(f,parfrac,x);` to obtain

$$f(x) = \frac{24{,}110/4879}{5x+2} - \frac{668/323}{2x+1} - \frac{9438/80{,}155}{3x-7} + \frac{(22{,}098x + 48{,}935)/260{,}015}{x^2+x+5}.$$

In Mathematica, we use the command `Apart`, and in Derive, we use `Expand`.

(b) $\displaystyle \int f(x)\, dx = \frac{24{,}110}{4879}\cdot\frac{1}{5}\ln|5x+2| - \frac{668}{323}\cdot\frac{1}{2}\ln|2x+1| - \frac{9438}{80{,}155}\cdot\frac{1}{3}\ln|3x-7|$

$$+ \frac{1}{260{,}015}\int \frac{22{,}098\left(x+\frac{1}{2}\right) + 37{,}886}{\left(x+\frac{1}{2}\right)^2 + \frac{19}{4}}\, dx + C$$

$$= \frac{24{,}110}{4879}\cdot\frac{1}{5}\ln|5x+2| - \frac{668}{323}\cdot\frac{1}{2}\ln|2x+1| - \frac{9438}{80{,}155}\cdot\frac{1}{3}\ln|3x-7|$$

$$+ \frac{1}{260{,}015}\left[22{,}098\cdot\frac{1}{2}\ln\left(x^2+x+5\right) + 37{,}886\cdot\sqrt{\frac{4}{19}}\tan^{-1}\left(\frac{1}{\sqrt{19/4}}\left(x+\frac{1}{2}\right)\right)\right] + C$$

$$= \frac{4822}{4879}\ln|5x+2| - \frac{334}{323}\ln|2x+1| - \frac{3146}{80{,}155}\ln|3x-7| + \frac{11{,}049}{260{,}015}\ln\left(x^2+x+5\right)$$

$$+ \frac{75{,}772}{260{,}015\sqrt{19}}\tan^{-1}\left[\frac{1}{\sqrt{19}}(2x+1)\right] + C$$

Using a CAS, we get

$$\frac{4822\ln(5x+2)}{4879} - \frac{334\ln(2x+1)}{323} - \frac{3146\ln(3x-7)}{80{,}155}$$

$$+ \frac{11{,}049\ln\left(x^2+x+5\right)}{260{,}015} + \frac{3988\sqrt{19}}{260{,}015}\tan^{-1}\left[\frac{\sqrt{19}}{19}(2x+1)\right]$$

The main difference in this answer is that the absolute value signs and the constant of integration have been omitted. Also, the fractions have been reduced and the denominators rationalized.

67. There are only finitely many values of x where $Q(x) = 0$ (assuming that Q is not the zero polynomial). At all other values of x, $F(x)/Q(x) = G(x)/Q(x)$, so $F(x) = G(x)$. In other words, the values of F and G agree at all except perhaps finitely many values of x. By continuity of F and G, the polynomials F and G must agree at those values of x too.

More explicitly: if a is a value of x such that $Q(a) = 0$, then $Q(x) \neq 0$ for all x sufficiently close to a. Thus,

$$F(a) = \lim_{x \to a} F(x) \ \ \text{[by continuity of } F] \ = \lim_{x \to a} G(x) \ \ \text{[whenever } Q(x) \neq 0]$$

$$= G(a) \ \ \text{[by continuity of } G]$$

7.5 Strategy for Integration

1. $\displaystyle\int \frac{\sin x + \sec x}{\tan x}\, dx = \int \left(\frac{\sin x}{\tan x} + \frac{\sec x}{\tan x}\right) dx = \int (\cos x + \csc x)\, dx = \sin x + \ln|\csc x - \cot x| + C$

3. $\displaystyle\int_0^2 \frac{2t}{(t-3)^2}\, dt = \int_{-3}^{-1} \frac{2(u+3)}{u^2}\, du \ \ [u = t - 3,\ du = dt] \ = \int_{-3}^{-1} \left(\frac{2}{u} + \frac{6}{u^2}\right) du = \left[2\ln|u| - \frac{6}{u}\right]_{-3}^{-1}$

$$= (2\ln 1 + 6) - (2\ln 3 + 2) = 4 - 2\ln 3 \ \text{or} \ 4 - \ln 9$$

5. Let $u = \arctan y$. Then $du = \dfrac{dy}{1 + y^2}$ $\Rightarrow$ $\displaystyle\int_{-1}^1 \frac{e^{\arctan y}}{1 + y^2}\, dy = \int_{-\pi/4}^{\pi/4} e^u\, du = [e^u]_{-\pi/4}^{\pi/4} = e^{\pi/4} - e^{-\pi/4}.$

7. $\displaystyle\int_1^3 r^4 \ln r\, dr$ $\begin{bmatrix} u = \ln r, & dv = r^4\, dr, \\ du = \dfrac{dr}{r} & v = \dfrac{1}{5}r^5 \end{bmatrix} = \left[\frac{1}{5}r^5 \ln r\right]_1^3 - \int_1^3 \frac{1}{5}r^4\, dr = \frac{243}{5}\ln 3 - 0 - \left[\frac{1}{25}r^5\right]_1^3$

$$= \frac{243}{5}\ln 3 - \left(\frac{243}{25} - \frac{1}{25}\right) = \frac{243}{5}\ln 3 - \frac{242}{25}$$

9. $\displaystyle\int \frac{x-1}{x^2 - 4x + 5}\, dx = \int \frac{(x-2)+1}{(x-2)^2 + 1}\, dx = \int \left(\frac{u}{u^2 + 1} + \frac{1}{u^2 + 1}\right) du \ \ [u = x - 2,\ du = dx]$

$$= \frac{1}{2}\ln(u^2 + 1) + \tan^{-1} u + C = \frac{1}{2}\ln(x^2 - 4x + 5) + \tan^{-1}(x - 2) + C$$

11. $\int \sin^3 \theta \, \cos^5 \theta \, d\theta = \int \cos^5 \theta \sin^2 \theta \sin \theta \, d\theta = -\int \cos^5 \theta \, (1 - \cos^2 \theta)(-\sin \theta)\, d\theta$

$$= -\int u^5 (1 - u^2)\, du \qquad \begin{bmatrix} u = \cos \theta, \\ du = -\sin \theta \, d\theta \end{bmatrix}$$

$$= \int (u^7 - u^5)\, du = \frac{1}{8}u^8 - \frac{1}{6}u^6 + C = \frac{1}{8}\cos^8 \theta - \frac{1}{6}\cos^6 \theta + C$$

Another solution:

$\int \sin^3 \theta \, \cos^5 \theta \, d\theta = \int \sin^3 \theta \, (\cos^2 \theta)^2 \cos \theta \, d\theta = \int \sin^3 \theta \, (1 - \sin^2 \theta)^2 \cos \theta \, d\theta$

$$= \int u^3 (1 - u^2)^2\, du \qquad \begin{bmatrix} u = \sin \theta, \\ du = \cos \theta \, d\theta \end{bmatrix} = \int u^3 (1 - 2u^2 + u^4)\, du$$

$$= \int (u^3 - 2u^5 + u^7)\, du = \frac{1}{4}u^4 - \frac{1}{3}u^6 + \frac{1}{8}u^8 + C = \frac{1}{4}\sin^4 \theta - \frac{1}{3}\sin^6 \theta + \frac{1}{8}\sin^8 \theta + C$$

13. Let $x = \sin \theta$, where $-\frac{\pi}{2} \leq \theta \leq \frac{\pi}{2}$. Then $dx = \cos \theta \, d\theta$ and

$(1 - x^2)^{1/2} = \cos \theta$, so

$$\int \frac{dx}{(1 - x^2)^{3/2}} = \int \frac{\cos \theta \, d\theta}{(\cos \theta)^3} = \int \sec^2 \theta \, d\theta$$

$$= \tan \theta + C = \frac{x}{\sqrt{1 - x^2}} + C$$

15. Let $u = 1 - x^2 \;\Rightarrow\; du = -2x\,dx$. Then

$$\int_0^{1/2} \frac{x}{\sqrt{1-x^2}}\,dx = -\frac{1}{2}\int_1^{3/4}\frac{1}{\sqrt{u}}\,du = \frac{1}{2}\int_{3/4}^1 u^{-1/2}\,du = \frac{1}{2}\Big[2u^{1/2}\Big]_{3/4}^1 = \big[\sqrt{u}\big]_{3/4}^1 = 1 - \frac{\sqrt{3}}{2}$$

17. $\int x\sin^2 x\,dx$ $\quad\begin{bmatrix} u = x, & dv = \sin^2 x\,dx, \\ du = dx & v = \int \sin^2 x\,dx = \int \frac{1}{2}(1 - \cos 2x)\,dx = \frac{1}{2}x - \frac{1}{2}\sin x\cos x \end{bmatrix}$

$\quad = \frac{1}{2}x^2 - \frac{1}{2}x\sin x\cos x - \int\left(\frac{1}{2}x - \frac{1}{2}\sin x\cos x\right)dx$

$\quad = \frac{1}{2}x^2 - \frac{1}{2}x\sin x\cos x - \frac{1}{4}x^2 + \frac{1}{4}\sin^2 x + C = \frac{1}{4}x^2 - \frac{1}{2}x\sin x\cos x + \frac{1}{4}\sin^2 x + C$

Note: $\int \sin x\cos x\,dx = \int s\,ds = \frac{1}{2}s^2 + C$ [where $s = \sin x$, $ds = \cos x\,dx$].

A slightly different method is to write $\int x\sin^2 x\,dx = \int x\cdot\frac{1}{2}(1 - \cos 2x)\,dx = \frac{1}{2}\int x\,dx - \frac{1}{2}\int x\cos 2x\,dx$. If we

evaluate the second integral by parts, we arrive at the equivalent answer $\frac{1}{4}x^2 - \frac{1}{4}x\sin 2x - \frac{1}{8}\cos 2x + C$.

19. Let $u = e^x$. Then $\int e^{x+e^x}\,dx = \int e^{e^x}e^x\,dx = \int e^u\,du = e^u + C = e^{e^x} + C$.

21. Integrate by parts three times, first with $u = t^3$, $dv = e^{-2t}\,dt$:

$$\int t^3 e^{-2t}\,dt = -\frac{1}{2}t^3 e^{-2t} + \frac{1}{2}\int 3t^2 e^{-2t}\,dt = -\frac{1}{2}t^3 e^{-2t} - \frac{3}{4}t^2 e^{-2t} + \frac{1}{2}\int 3t e^{-2t}\,dt$$

$$= -e^{-2t}\left[\tfrac{1}{2}t^3 + \tfrac{3}{4}t^2\right] - \tfrac{3}{4}t e^{-2t} + \tfrac{3}{4}\int e^{-2t}\,dt = -e^{-2t}\left[\tfrac{1}{2}t^3 + \tfrac{3}{4}t^2 + \tfrac{3}{4}t + \tfrac{3}{8}\right] + C$$

$$= -\tfrac{1}{8}e^{-2t}\left(4t^3 + 6t^2 + 6t + 3\right) + C$$

23. Let $u = 1 + \sqrt{x}$. Then $x = (u-1)^2$, $dx = 2(u-1)\,du \;\Rightarrow\;$

$$\int_0^1 (1 + \sqrt{x})^8\,dx = \int_1^2 u^8\cdot 2(u-1)\,du = 2\int_1^2 (u^9 - u^8)\,du = \left[\tfrac{1}{5}u^{10} - 2\cdot\tfrac{1}{9}u^9\right]_1^2$$

$$= \tfrac{1024}{5} - \tfrac{1024}{9} - \tfrac{1}{5} + \tfrac{2}{9} = \tfrac{4097}{45}$$

25. $\dfrac{3x^2 - 2}{x^2 - 2x - 8} = 3 + \dfrac{6x + 22}{(x-4)(x+2)} = 3 + \dfrac{A}{x-4} + \dfrac{B}{x+2} \;\Rightarrow\; 6x + 22 = A(x+2) + B(x-4)$. Setting

$x = 4$ gives $46 = 6A$, so $A = \frac{23}{3}$. Setting $x = -2$ gives $10 = -6B$, so $B = -\frac{5}{3}$. Now

$$\int \frac{3x^2 - 2}{x^2 - 2x - 8}\,dx = \int\left(3 + \frac{23/3}{x-4} - \frac{5/3}{x+2}\right)dx = 3x + \tfrac{23}{3}\ln|x-4| - \tfrac{5}{3}\ln|x+2| + C.$$

27. Let $u = \ln(\sin x)$. Then $du = \cot x\,dx \;\Rightarrow\; \int \cot x\ln(\sin x)\,dx = \int u\,du = \frac{1}{2}u^2 + C = \frac{1}{2}[\ln(\sin x)]^2 + C.$

29. $\displaystyle\int_0^5 \frac{3w - 1}{w + 2}\,dw = \int_0^5\left(3 - \frac{7}{w+2}\right)dw = \Big[3w - 7\ln|w+2|\Big]_0^5$

$\qquad = 15 - 7\ln 7 + 7\ln 2 = 15 + 7(\ln 2 - \ln 7) = 15 + 7\ln\tfrac{2}{7}$

31. As in Example 5,

$$\int \sqrt{\frac{1+x}{1-x}}\,dx = \int \frac{\sqrt{1+x}}{\sqrt{1-x}}\cdot\frac{\sqrt{1+x}}{\sqrt{1+x}}\,dx = \int \frac{1+x}{\sqrt{1-x^2}}\,dx = \int \frac{dx}{\sqrt{1-x^2}} + \int \frac{x\,dx}{\sqrt{1-x^2}}$$

$$= \sin^{-1} x - \sqrt{1-x^2} + C$$

Another method: Substitute $u = \sqrt{(1+x)/(1-x)}$.

33. $3 - 2x - x^2 = -(x^2 + 2x + 1) + 4 = 4 - (x + 1)^2$. Let

$x + 1 = 2 \sin \theta$, where $-\frac{\pi}{2} \le \theta \le \frac{\pi}{2}$. Then $dx = 2 \cos \theta \, d\theta$ and

$\int \sqrt{3 - 2x - x^2} \, dx = \int \sqrt{4 - (x + 1)^2} \, dx = \int \sqrt{4 - 4 \sin^2 \theta} \, 2 \cos \theta \, d\theta$

$\qquad = 4 \int \cos^2 \theta \, d\theta = 2 \int (1 + \cos 2\theta) \, d\theta$

$\qquad = 2\theta + \sin 2\theta + C = 2\theta + 2 \sin \theta \cos \theta + C$

$\qquad = 2 \sin^{-1} \left(\dfrac{x + 1}{2} \right) + 2 \cdot \dfrac{x + 1}{2} \cdot \dfrac{\sqrt{3 - 2x - x^2}}{2} + C$

$\qquad = 2 \sin^{-1} \left(\dfrac{x + 1}{2} \right) + \dfrac{x + 1}{2} \sqrt{3 - 2x - x^2} + C$

(triangle figure: hypotenuse 2, vertical side $x + 1$, angle θ, base $\sqrt{4 - (x + 1)^2} = \sqrt{3 - 2x - x^2}$)

35. Because $f(x) = x^8 \sin x$ is the product of an even function and an odd function, it is odd. Therefore,

$\int_{-1}^{1} x^8 \sin x \, dx = 0$ [by (5.5.7)(b)].

37. $\int_0^{\pi/4} \cos^2 \theta \tan^2 \theta \, d\theta = \int_0^{\pi/4} \sin^2 \theta \, d\theta = \int_0^{\pi/4} \frac{1}{2}(1 - \cos 2\theta) \, d\theta = \left[\frac{1}{2}\theta - \frac{1}{4} \sin 2\theta \right]_0^{\pi/4}$

$\qquad = \left(\frac{\pi}{8} - \frac{1}{4} \right) - (0 - 0) = \frac{\pi}{8} - \frac{1}{4}$

39. Let $u = 1 - x^2$. Then $du = -2x \, dx$ $\Rightarrow$

$\int \dfrac{x \, dx}{1 - x^2 + \sqrt{1 - x^2}} = -\dfrac{1}{2} \int \dfrac{du}{u + \sqrt{u}} = -\int \dfrac{v \, dv}{v^2 + v}$ $[v = \sqrt{u}, u = v^2, du = 2v \, dv]$

$\qquad = -\int \dfrac{dv}{v + 1} = -\ln|v + 1| + C = -\ln \left(\sqrt{1 - x^2} + 1 \right) + C$

41. Let $u = \theta$, $dv = \tan^2 \theta \, d\theta = (\sec^2 \theta - 1) \, d\theta$ $\Rightarrow$ $du = d\theta$ and $v = \tan \theta - \theta$. So

$\int \theta \tan^2 \theta \, d\theta = \theta(\tan \theta - \theta) - \int (\tan \theta - \theta) \, d\theta = \theta \tan \theta - \theta^2 - \ln|\sec \theta| + \frac{1}{2}\theta^2 + C$

$\qquad = \theta \tan \theta - \frac{1}{2}\theta^2 - \ln|\sec \theta| + C$

43. Let $u = 1 + e^x$, so that $du = e^x \, dx$. Then

$\int e^x \sqrt{1 + e^x} \, dx = \int u^{1/2} \, du = \frac{2}{3} u^{3/2} + C = \frac{2}{3}(1 + e^x)^{3/2} + C$.

Or: Let $u = \sqrt{1 + e^x}$, so that $u^2 = 1 + e^x$ and $2u \, du = e^x \, dx$. Then

$\int e^x \sqrt{1 + e^x} \, dx = \int u \cdot 2u \, du = \int 2u^2 \, du = \frac{2}{3} u^3 + C = \frac{2}{3}(1 + e^x)^{3/2} + C$.

45. Let $t = x^3$. Then $dt = 3x^2 \, dx$ $\Rightarrow$ $I = \int x^5 e^{-x^3} \, dx = \frac{1}{3} \int t e^{-t} \, dt$. Now integrate by parts with $u = t$,

$dv = e^{-t} \, dt$: $I = -\frac{1}{3} t e^{-t} + \frac{1}{3} \int e^{-t} \, dt = -\frac{1}{3} t e^{-t} - \frac{1}{3} e^{-t} + C = -\frac{1}{3} e^{-x^3} (x^3 + 1) + C$.

47. $\displaystyle \int \dfrac{x + a}{x^2 + a^2} \, dx = \dfrac{1}{2} \int \dfrac{2x \, dx}{x^2 + a^2} + a \int \dfrac{dx}{x^2 + a^2} = \frac{1}{2} \ln(x^2 + a^2) + a \cdot \dfrac{1}{a} \tan^{-1} \left(\dfrac{x}{a} \right) + C$

$\qquad = \ln \sqrt{x^2 + a^2} + \tan^{-1}(x/a) + C$

49. Let $u = \sqrt{4x + 1}$ $\Rightarrow$ $u^2 = 4x + 1$ $\Rightarrow$ $2u \, du = 4 \, dx$ $\Rightarrow$ $dx = \frac{1}{2} u \, du$. So

$\displaystyle \int \dfrac{1}{x \sqrt{4x + 1}} \, dx = \int \dfrac{\frac{1}{2} u \, du}{\frac{1}{4}(u^2 - 1) u} = 2 \int \dfrac{du}{u^2 - 1} = 2 \left(\frac{1}{2} \right) \ln \left| \dfrac{u - 1}{u + 1} \right| + C$ [by Formula 19]

$\qquad = \ln \left| \dfrac{\sqrt{4x + 1} - 1}{\sqrt{4x + 1} + 1} \right| + C$

51. Let $2x = \tan\theta \Rightarrow x = \frac{1}{2}\tan\theta$, $dx = \frac{1}{2}\sec^2\theta\,d\theta$, $\sqrt{4x^2+1} = \sec\theta$, so

$$\int \frac{dx}{x\sqrt{4x^2+1}} = \int \frac{\frac{1}{2}\sec^2\theta\,d\theta}{\frac{1}{2}\tan\theta\sec\theta} = \int \frac{\sec\theta}{\tan\theta}\,d\theta = \int \csc\theta\,d\theta$$

$$= -\ln|\csc\theta + \cot\theta| + C \qquad \text{[or } \ln|\csc\theta - \cot\theta| + C]$$

$$= -\ln\left|\frac{\sqrt{4x^2+1}}{2x} + \frac{1}{2x}\right| + C \quad \left[\text{or } \ln\left|\frac{\sqrt{4x^2+1}}{2x} - \frac{1}{2x}\right| + C\right]$$

A right triangle with hypotenuse $\sqrt{4x^2+1}$, opposite side $2x$, adjacent side 1, and angle θ.

53. $\displaystyle\int x^2 \sinh(mx)\,dx = \frac{1}{m}x^2\cosh(mx) - \frac{2}{m}\int x\cosh(mx)\,dx \qquad \begin{bmatrix} u = x^2, & dv = \sinh(mx)\,dx, \\ du = 2x\,dx & v = \frac{1}{m}\cosh(mx) \end{bmatrix}$

$$= \frac{1}{m}x^2\cosh(mx) - \frac{2}{m}\left(\frac{1}{m}x\sinh(mx) - \frac{1}{m}\int \sinh(mx)\,dx\right) \qquad \begin{bmatrix} U = x, & dV = \cosh(mx)\,dx, \\ dU = dx & V = \frac{1}{m}\sinh(mx) \end{bmatrix}$$

$$= \frac{1}{m}x^2\cosh(mx) - \frac{2}{m^2}x\sinh(mx) + \frac{2}{m^3}\cosh(mx) + C$$

55. Let $u = \sqrt{x+1}$. Then $x = u^2 - 1 \Rightarrow$

$$\int \frac{dx}{x+4+4\sqrt{x+1}} = \int \frac{2u\,du}{u^2+3+4u} = \int \left[\frac{-1}{u+1} + \frac{3}{u+3}\right]du$$

$$= 3\ln|u+3| - \ln|u+1| + C = 3\ln\left(\sqrt{x+1}+3\right) - \ln\left(\sqrt{x+1}+1\right) + C$$

57. Let $u = \sqrt[3]{x+c}$. Then $x = u^3 - c \Rightarrow$

$$\int x\sqrt[3]{x+c}\,dx = \int (u^3 - c)u \cdot 3u^2\,du = 3\int (u^6 - cu^3)\,du = \frac{3}{7}u^7 - \frac{3}{4}cu^4 + C$$

$$= \frac{3}{7}(x+c)^{7/3} - \frac{3}{4}c(x+c)^{4/3} + C$$

59. Let $u = e^x$. Then $x = \ln u$, $dx = du/u \Rightarrow$

$$\int \frac{dx}{e^{3x} - e^x} = \int \frac{du/u}{u^3 - u} = \int \frac{du}{(u-1)u^2(u+1)} = \int \left[\frac{1/2}{u-1} - \frac{1}{u^2} - \frac{1/2}{u+1}\right]du$$

$$= \frac{1}{u} + \frac{1}{2}\ln\left|\frac{u-1}{u+1}\right| + C = e^{-x} + \frac{1}{2}\ln\left|\frac{e^x-1}{e^x+1}\right| + C$$

61. Let $u = x^5$. Then $du = 5x^4\,dx \Rightarrow$

$$\int \frac{x^4\,dx}{x^{10}+16} = \int \frac{\frac{1}{5}du}{u^2+16} = \frac{1}{5}\cdot\frac{1}{4}\tan^{-1}\left(\frac{1}{4}u\right) + C = \frac{1}{20}\tan^{-1}\left(\frac{1}{4}x^5\right) + C.$$

63. Let $y = \sqrt{x}$ so that $dy = \frac{1}{2\sqrt{x}}dx \Rightarrow dx = 2\sqrt{x}\,dy = 2y\,dy$. Then

$$\int \sqrt{x}\,e^{\sqrt{x}}\,dx = \int ye^y(2y\,dy) = \int 2y^2e^y\,dy \qquad \begin{bmatrix} u = 2y^2, & dv = e^y\,dy, \\ du = 4y\,dy & v = e^y \end{bmatrix}$$

$$= 2y^2e^y - \int 4ye^y\,dy \qquad \begin{bmatrix} U = 4y, & dV = e^y\,dy, \\ dU = 4\,dy & V = e^y \end{bmatrix}$$

$$= 2y^2e^y - \left(4ye^y - \int 4e^y\,dy\right) = 2y^2e^y - 4ye^y + 4e^y + C$$

$$= 2(y^2 - 2y + 2)e^y + C = 2(x - 2\sqrt{x} + 2)e^{\sqrt{x}} + C$$

65. $\int \dfrac{dx}{\sqrt{x+1}+\sqrt{x}} = \int \left(\dfrac{1}{\sqrt{x+1}+\sqrt{x}} \cdot \dfrac{\sqrt{x+1}-\sqrt{x}}{\sqrt{x+1}-\sqrt{x}} \right) dx = \int \left(\sqrt{x+1} - \sqrt{x} \right) dx$

$\qquad\qquad = \frac{2}{3}\left[(x+1)^{3/2} - x^{3/2} \right] + C$

67. Let $u = \sqrt{t}$. Then $du = dt/(2\sqrt{t})$ $\Rightarrow$

$\displaystyle\int_1^3 \dfrac{\arctan\sqrt{t}}{\sqrt{t}}\, dt = \int_1^{\sqrt{3}} \tan^{-1}u\,(2\,du) = 2\left[u\tan^{-1}u - \frac{1}{2}\ln(1+u^2) \right]_1^{\sqrt{3}}$ [Example 5 in Section 7.1]

$\qquad\qquad = 2\left[\left(\sqrt{3}\tan^{-1}\sqrt{3} - \frac{1}{2}\ln 4\right) - \left(\tan^{-1}1 - \frac{1}{2}\ln 2\right) \right]$

$\qquad\qquad = 2\left[\left(\sqrt{3}\cdot\frac{\pi}{3} - \ln 2\right) - \left(\frac{\pi}{4} - \frac{1}{2}\ln 2\right) \right] = \frac{2}{3}\sqrt{3}\,\pi - \frac{1}{2}\pi - \ln 2$

69. Let $u = e^x$. Then $x = \ln u$, $dx = du/u$ $\Rightarrow$

$$\int \dfrac{e^{2x}}{1+e^x}\, dx = \int \dfrac{u^2}{1+u}\dfrac{du}{u} = \int \dfrac{u}{1+u}\, du = \int \left(1 - \dfrac{1}{1+u} \right) du$$

$$= u - \ln|1+u| + C = e^x - \ln(1+e^x) + C$$

71. $\dfrac{x}{x^4+4x^2+3} = \dfrac{x}{(x^2+3)(x^2+1)} = \dfrac{Ax+B}{x^2+3} + \dfrac{Cx+D}{x^2+1}$ $\Rightarrow$

$\quad x = (Ax+B)(x^2+1) + (Cx+D)(x^2+3) = \left(Ax^3 + Bx^2 + Ax + B\right) + \left(Cx^3 + Dx^2 + 3Cx + 3D\right)$

$\quad = (A+C)x^3 + (B+D)x^2 + (A+3C)x + (B+3D)$ $\Rightarrow$

$\quad A+C = 0$, $B+D = 0$, $A+3C = 1$, $B+3D = 0$ $\Rightarrow$ $A = -\frac{1}{2}$, $C = \frac{1}{2}$, $B = 0$, $D = 0$. Thus,

$$\int \dfrac{x}{x^4+4x^2+3}\, dx = \int \left(\dfrac{-\frac{1}{2}x}{x^2+3} + \dfrac{\frac{1}{2}x}{x^2+1} \right) dx$$

$$= -\frac{1}{4}\ln\left(x^2+3\right) + \frac{1}{4}\ln\left(x^2+1\right) + C \quad\text{or}\quad \frac{1}{4}\ln\left(\dfrac{x^2+1}{x^2+3} \right) + C$$

73. $\dfrac{1}{(x-2)(x^2+4)} = \dfrac{A}{x-2} + \dfrac{Bx+C}{x^2+4}$ $\Rightarrow$

$\quad 1 = A(x^2+4) + (Bx+C)(x-2) = (A+B)x^2 + (C-2B)x + (4A-2C)$. So $0 = A+B = C-2B$,

$\quad 1 = 4A - 2C$. Setting $x = 2$ gives $A = \frac{1}{8}$ $\Rightarrow$ $B = -\frac{1}{8}$ and $C = -\frac{1}{4}$. So

$$\int \dfrac{1}{(x-2)(x^2+4)}\, dx = \int \left(\dfrac{\frac{1}{8}}{x-2} + \dfrac{-\frac{1}{8}x - \frac{1}{4}}{x^2+4} \right) dx = \frac{1}{8}\int \dfrac{dx}{x-2} - \frac{1}{16}\int \dfrac{2x\,dx}{x^2+4} - \frac{1}{4}\int \dfrac{dx}{x^2+4}$$

$$= \frac{1}{8}\ln|x-2| - \frac{1}{16}\ln\left(x^2+4\right) - \frac{1}{8}\tan^{-1}(x/2) + C$$

75. $\int \sin x \sin 2x \sin 3x\, dx = \int \sin x \cdot \frac{1}{2}\left[\cos(2x-3x) - \cos(2x+3x) \right] dx = \frac{1}{2}\int \left(\sin x \cos x - \sin x \cos 5x \right) dx$

$\qquad\qquad = \frac{1}{4}\int \sin 2x\, dx - \frac{1}{2}\int \frac{1}{2}\left[\sin(x+5x) + \sin(x-5x) \right] dx$

$\qquad\qquad = -\frac{1}{8}\cos 2x - \frac{1}{4}\int \left(\sin 6x - \sin 4x \right) dx = -\frac{1}{8}\cos 2x + \frac{1}{24}\cos 6x - \frac{1}{16}\cos 4x + C$

77. Let $u = x^{3/2}$ so that $u^2 = x^3$ and $du = \frac{3}{2}x^{1/2}\, dx$ $\Rightarrow$ $\sqrt{x}\, dx = \frac{2}{3}\, du$. Then

$$\int \dfrac{\sqrt{x}}{1+x^3}\, dx = \int \dfrac{\frac{2}{3}}{1+u^2}\, du = \frac{2}{3}\tan^{-1}u + C = \frac{2}{3}\tan^{-1}\left(x^{3/2}\right) + C.$$

79. Let $u = x$, $dv = \sin^2 x \cos x \, dx$ $\Rightarrow$ $du = dx$, $v = \frac{1}{3} \sin^3 x$. Then

$$\int x \sin^2 x \cos x \, dx = \frac{1}{3} x \sin^3 x - \int \frac{1}{3} \sin^3 x \, dx = \frac{1}{3} x \sin^3 x - \frac{1}{3} \int (1 - \cos^2 x) \sin x \, dx$$

$$= \frac{1}{3} x \sin^3 x + \frac{1}{3} \int (1 - y^2) \, dy \quad \begin{bmatrix} y = \cos x, \\ dy = -\sin x \, dx \end{bmatrix}$$

$$= \frac{1}{3} x \sin^3 x + \frac{1}{3} y - \frac{1}{9} y^3 + C = \frac{1}{3} x \sin^3 x + \frac{1}{3} \cos x - \frac{1}{9} \cos^3 x + C$$

81. The function $y = 2xe^{x^2}$ *does* have an elementary antiderivative, so we'll use this fact to help evaluate the integral.

$$\int (2x^2 + 1)e^{x^2} \, dx = \int 2x^2 e^{x^2} \, dx + \int e^{x^2} \, dx = \int x(2xe^{x^2}) \, dx + \int e^{x^2} \, dx$$

$$= xe^{x^2} - \int e^{x^2} \, dx + \int e^{x^2} \, dx \quad \begin{bmatrix} u = x, & dv = 2xe^{x^2} \, dx, \\ du = dx & v = e^{x^2} \end{bmatrix} = xe^{x^2} + C$$

7.6 Integration Using Tables and Computer Algebra Systems

Keep in mind that there are several ways to approach many of these exercises, and different methods can lead to different forms of the answer.

1. We could make the substitution $u = \sqrt{2}\,x$ to obtain the radical $\sqrt{7 - u^2}$ and then use Formula 33 with $a = \sqrt{7}$. Alternatively, we will factor $\sqrt{2}$ out of the radical and use $a = \sqrt{\frac{7}{2}}$.

$$\int \frac{\sqrt{7 - 2x^2}}{x^2} \, dx = \sqrt{2} \int \frac{\sqrt{\frac{7}{2} - x^2}}{x^2} \, dx \overset{33}{=} \sqrt{2} \left[-\frac{1}{x}\sqrt{\frac{7}{2} - x^2} - \sin^{-1} \frac{x}{\sqrt{\frac{7}{2}}} \right] + C$$

$$= -\frac{1}{x}\sqrt{7 - 2x^2} - \sqrt{2} \sin^{-1}\left(\sqrt{\frac{2}{7}}\,x \right) + C$$

3. Let $u = \pi x$ $\Rightarrow$ $du = \pi \, dx$, so

$$\int \sec^3 (\pi x) \, dx = \frac{1}{\pi} \int \sec^3 u \, du \overset{71}{=} \frac{1}{\pi} \left(\frac{1}{2} \sec u \tan u + \frac{1}{2} \ln |\sec u + \tan u| \right) + C$$

$$= \frac{1}{2\pi} \sec \pi x \tan \pi x + \frac{1}{2\pi} \ln |\sec \pi x + \tan \pi x| + C$$

5. $\int_0^1 2x \cos^{-1} x \, dx \overset{91}{=} 2 \left[\frac{2x^2 - 1}{4} \cos^{-1} x - \frac{x\sqrt{1 - x^2}}{4} \right]_0^1 = 2[(\frac{1}{4} \cdot 0 - 0) - (-\frac{1}{4} \cdot \frac{\pi}{2} - 0)] = 2(\frac{\pi}{8}) = \frac{\pi}{4}$

7. By Formula 99 with $a = -3$ and $b = 4$,

$$\int e^{-3x} \cos 4x \, dx = \frac{e^{-3x}}{(-3)^2 + 4^2} (-3 \cos 4x + 4 \sin 4x) + C = \frac{e^{-3x}}{25} (-3 \cos 4x + 4 \sin 4x) + C.$$

9. Let $u = 2x$ and $a = 3$. Then $du = 2 \, dx$ and

$$\int \frac{dx}{x^2\sqrt{4x^2 + 9}} = \int \frac{\frac{1}{2} \, du}{\frac{u^2}{4}\sqrt{u^2 + a^2}} = 2 \int \frac{du}{u^2\sqrt{a^2 + u^2}} \overset{28}{=} -2 \frac{\sqrt{a^2 + u^2}}{a^2 u} + C$$

$$= -2 \frac{\sqrt{4x^2 + 9}}{9 \cdot 2x} + C = -\frac{\sqrt{4x^2 + 9}}{9x} + C$$

11. $\int_{-1}^0 t^2 e^{-t} \, dt \overset{97}{=} \left[\frac{1}{-1} t^2 e^{-t} \right]_{-1}^0 - \frac{2}{-1} \int_{-1}^0 te^{-t} \, dt = e + 2 \int_{-1}^0 te^{-t} \, dt \overset{96}{=} e + 2 \left[\frac{1}{(-1)^2} (-t - 1) e^{-t} \right]_{-1}^0$

$$= e + 2[-e^0 + 0] = e - 2$$

13. $\int \dfrac{\tan^3(1/z)}{z^2}\, dz$ $\begin{bmatrix} u = 1/z, \\ du = -dz/z^2 \end{bmatrix} = -\int \tan^3 u\, du \overset{69}{=} -\dfrac{1}{2}\tan^2 u - \ln|\cos u| + C$

$$= -\dfrac{1}{2}\tan^2\left(\dfrac{1}{z}\right) - \ln\left|\cos\left(\dfrac{1}{z}\right)\right| + C$$

15. Let $u = e^x$. Then $du = e^x\, dx$, so $\int e^x \operatorname{sech}(e^x)\, dx = \int \operatorname{sech} u\, du \overset{107}{=} \tan^{-1}|\sinh u| + C = \tan^{-1}[\sinh(e^x)] + C$

17. Let $z = 6 + 4y - 4y^2 = 6 - (4y^2 - 4y + 1) + 1 = 7 - (2y - 1)^2$, $u = 2y - 1$, and $a = \sqrt{7}$. Then $z = a^2 - u^2$, $du = 2\, dy$, and

$$\int y\sqrt{6 + 4y - 4y^2}\, dy = \int y\sqrt{z}\, dy = \int \tfrac{1}{2}(u + 1)\sqrt{a^2 - u^2}\,\tfrac{1}{2}\, du$$
$$= \tfrac{1}{4}\int u\sqrt{a^2 - u^2}\, du + \tfrac{1}{4}\int \sqrt{a^2 - u^2}\, du$$
$$= \tfrac{1}{4}\int \sqrt{a^2 - u^2}\, du - \tfrac{1}{8}\int(-2u)\sqrt{a^2 - u^2}\, du$$
$$\overset{30}{=} \dfrac{u}{8}\sqrt{a^2 - u^2} + \dfrac{a^2}{8}\sin^{-1}\left(\dfrac{u}{a}\right) - \dfrac{1}{8}\int \sqrt{w}\, dw \quad \begin{bmatrix} w = a^2 - u^2, \\ dw = -2u\, du \end{bmatrix}$$
$$= \dfrac{2y - 1}{8}\sqrt{6 + 4y - 4y^2} + \dfrac{7}{8}\sin^{-1}\dfrac{2y - 1}{\sqrt{7}} - \dfrac{1}{8}\cdot\dfrac{2}{3}w^{3/2} + C$$
$$= \dfrac{2y - 1}{8}\sqrt{6 + 4y - 4y^2} + \dfrac{7}{8}\sin^{-1}\dfrac{2y - 1}{\sqrt{7}} - \dfrac{1}{12}(6 + 4y - 4y^2)^{3/2} + C.$$

This can be rewritten as

$$\sqrt{6 + 4y - 4y^2}\left[\dfrac{1}{8}(2y - 1) - \dfrac{1}{12}(6 + 4y - 4y^2)\right] + \dfrac{7}{8}\sin^{-1}\dfrac{2y - 1}{\sqrt{7}} + C$$
$$= \left(\dfrac{1}{3}y^2 - \dfrac{1}{12}y - \dfrac{5}{8}\right)\sqrt{6 + 4y - 4y^2} + \dfrac{7}{8}\sin^{-1}\left(\dfrac{2y - 1}{\sqrt{7}}\right) + C$$
$$= \dfrac{1}{24}(8y^2 - 2y - 15)\sqrt{6 + 4y - 4y^2} + \dfrac{7}{8}\sin^{-1}\left(\dfrac{2y - 1}{\sqrt{7}}\right) + C$$

19. Let $u = \sin x$. Then $du = \cos x\, dx$, so

$$\int \sin^2 x \cos x \ln(\sin x)\, dx = \int u^2 \ln u\, du \overset{101}{=} \dfrac{u^{2+1}}{(2 + 1)^2}[(2 + 1)\ln u - 1] + C = \tfrac{1}{9}u^3(3\ln u - 1) + C$$
$$= \tfrac{1}{9}\sin^3 x[3\ln(\sin x) - 1] + C$$

21. Let $u = e^x$ and $a = \sqrt{3}$. Then $du = e^x\, dx$ and

$$\int \dfrac{e^x}{3 - e^{2x}}\, dx = \int \dfrac{du}{a^2 - u^2} \overset{19}{=} \dfrac{1}{2a}\ln\left|\dfrac{u + a}{u - a}\right| + C = \dfrac{1}{2\sqrt{3}}\ln\left|\dfrac{e^x + \sqrt{3}}{e^x - \sqrt{3}}\right| + C.$$

23. $\int \sec^5 x\, dx \overset{77}{=} \tfrac{1}{4}\tan x \sec^3 x + \tfrac{3}{4}\int \sec^3 x\, dx \overset{77}{=} \tfrac{1}{4}\tan x \sec^3 x + \tfrac{3}{4}\left(\tfrac{1}{2}\tan x \sec x + \tfrac{1}{2}\int \sec x\, dx\right)$

$$\overset{14}{=} \tfrac{1}{4}\tan x \sec^3 x + \tfrac{3}{8}\tan x \sec x + \tfrac{3}{8}\ln|\sec x + \tan x| + C$$

25. Let $u = \ln x$ and $a = 2$. Then $du = \dfrac{dx}{x}$ and

$$\int \dfrac{\sqrt{4 + (\ln x)^2}}{x}\, dx = \int \sqrt{a^2 + u^2}\, du \overset{21}{=} \dfrac{u}{2}\sqrt{a^2 + u^2} + \dfrac{a^2}{2}\ln\left(u + \sqrt{a^2 + u^2}\right) + C$$
$$= \tfrac{1}{2}(\ln x)\sqrt{4 + (\ln x)^2} + 2\ln\left[\ln x + \sqrt{4 + (\ln x)^2}\right] + C$$

27. Let $u = e^x$. Then $x = \ln u$, $dx = du/u$, so

$$\int \sqrt{e^{2x} - 1}\, dx = \int \dfrac{\sqrt{u^2 - 1}}{u}\, du \overset{41}{=} \sqrt{u^2 - 1} - \cos^{-1}(1/u) + C = \sqrt{e^{2x} - 1} - \cos^{-1}(e^{-x}) + C.$$

29. $\displaystyle\int \frac{x^4\,dx}{\sqrt{x^{10}-2}} = \int \frac{x^4\,dx}{\sqrt{(x^5)^2-2}} = \frac{1}{5}\int \frac{du}{\sqrt{u^2-2}}$ $\qquad [u=x^5,\ du=5x^4\,dx]$

$\qquad\qquad \overset{43}{=} \frac{1}{5}\ln\left|u+\sqrt{u^2-2}\right| + C = \frac{1}{5}\ln\left|x^5+\sqrt{x^{10}-2}\right| + C$

31. Using cylindrical shells, we get

$$V = 2\pi \int_0^2 x\cdot x\sqrt{4-x^2}\,dx = 2\pi \int_0^2 x^2\sqrt{4-x^2}\,dx \overset{31}{=} 2\pi\left[\frac{x}{8}(2x^2-4)\sqrt{4-x^2} + \frac{16}{8}\sin^{-1}\frac{x}{2}\right]_0^2$$

$$= 2\pi\left[(0+2\sin^{-1}1)-(0+2\sin^{-1}0)\right] = 2\pi\left(2\cdot\frac{\pi}{2}\right) = 2\pi^2$$

33. (a) $\displaystyle\frac{d}{du}\left[\frac{1}{b^3}\left(a+bu-\frac{a^2}{a+bu}-2a\ln|a+bu|\right)+C\right] = \frac{1}{b^3}\left[b+\frac{ba^2}{(a+bu)^2}-\frac{2ab}{(a+bu)}\right]$

$\qquad\qquad = \frac{1}{b^3}\left[\frac{b(a+bu)^2+ba^2-(a+bu)2ab}{(a+bu)^2}\right] = \frac{1}{b^3}\left[\frac{b^3u^2}{(a+bu)^2}\right] = \frac{u^2}{(a+bu)^2}$

(b) Let $t=a+bu \;\Rightarrow\; dt=b\,du$. Note that $u=\dfrac{t-a}{b}$ and $du=\dfrac{1}{b}\,dt$.

$$\int \frac{u^2\,du}{(a+bu)^2} = \frac{1}{b^3}\int \frac{(t-a)^2}{t^2}\,dt = \frac{1}{b^3}\int \frac{t^2-2at+a^2}{t^2}\,dt$$

$$= \frac{1}{b^3}\int\left(1-\frac{2a}{t}+\frac{a^2}{t^2}\right)dt = \frac{1}{b^3}\left(t-2a\ln|t|-\frac{a^2}{t}\right)+C$$

$$= \frac{1}{b^3}\left(a+bu-\frac{a^2}{a+bu}-2a\ln|a+bu|\right)+C$$

35. Maple, Mathematica and Derive all give $\displaystyle\int x^2\sqrt{5-x^2}\,dx = -\frac{1}{4}x(5-x^2)^{3/2} + \frac{5}{8}x\sqrt{5-x^2} + \frac{25}{8}\sin^{-1}\left(\frac{1}{\sqrt{5}}x\right)$.

Using Formula 31, we get $\displaystyle\int x^2\sqrt{5-x^2}\,dx = \frac{1}{8}x(2x^2-5)\sqrt{5-x^2} + \frac{1}{8}(5^2)\sin^{-1}\left(\frac{1}{\sqrt{5}}x\right)+C$. But

$-\frac{1}{4}x(5-x^2)^{3/2} + \frac{5}{8}x\sqrt{5-x^2} = \frac{1}{8}x\sqrt{5-x^2}\left[5-2(5-x^2)\right] = \frac{1}{8}x(2x^2-5)\sqrt{5-x^2}$, and the $\sin^{-1}$ terms

are the same in each expression, so the answers are equivalent.

37. Maple and Derive both give $\displaystyle\int \sin^3 x\,\cos^2 x\,dx = -\frac{1}{5}\sin^2 x\,\cos^3 x - \frac{2}{15}\cos^3 x$ (although Derive factors the

expression), and Mathematica gives $\displaystyle\int \sin^3 x\,\cos^2 x\,dx = -\frac{1}{8}\cos x - \frac{1}{48}\cos 3x + \frac{1}{80}\cos 5x$. We can use a CAS to

show that both of these expressions are equal to $-\frac{1}{3}\cos^3 x + \frac{1}{5}\cos^5 x$. Using Formula 86, we write

$$\int \sin^3 x\,\cos^2 x\,dx = -\frac{1}{5}\sin^2 x\,\cos^3 x + \frac{2}{5}\int \sin x\,\cos^2 x\,dx = -\frac{1}{5}\sin^2 x\,\cos^3 x + \frac{2}{5}\left(-\frac{1}{3}\cos^3 x\right)+C$$

$$= -\frac{1}{5}\sin^2 x\,\cos^3 x - \frac{2}{15}\cos^3 x + C$$

39. Maple gives $\displaystyle\int x\sqrt{1+2x}\,dx = \frac{1}{10}(1+2x)^{5/2} - \frac{1}{6}(1+2x)^{3/2}$, Mathematica gives $\sqrt{1+2x}\left(\frac{2}{5}x^2+\frac{1}{15}x-\frac{1}{15}\right)$,

and Derive gives $\frac{1}{15}(1+2x)^{3/2}(3x-1)$. The first two expressions can be simplified to Derive's result. If we use

Formula 54, we get

$$\int x\sqrt{1+2x}\,dx = \frac{2}{15(2)^2}(3\cdot 2x - 2\cdot 1)(1+2x)^{3/2} + C = \frac{1}{30}(6x-2)(1+2x)^{3/2} + C$$

$$= \frac{1}{15}(3x-1)(1+2x)^{3/2}$$

41. Maple gives $\int \tan^5 x\,dx = \frac{1}{4}\tan^4 x - \frac{1}{2}\tan^2 x + \frac{1}{2}\ln(1+\tan^2 x)$, Mathematica

gives $\int \tan^5 x\,dx = \frac{1}{4}[-1-2\cos(2x)]\sec^4 x - \ln(\cos x)$, and Derive gives

$\int \tan^5 x\,dx = \frac{1}{4}\tan^4 x - \frac{1}{2}\tan^2 x - \ln(\cos x)$. These expressions are equivalent, and none includes absolute

value bars or a constant of integration. Note that Mathematica's and Derive's expressions suggest that the integral is

undefined where $\cos x < 0$, which is not the case.

Using Formula 75, $\int \tan^5 x\,dx = \frac{1}{5-1}\tan^{5-1}x - \int \tan^{5-2}x\,dx = \frac{1}{4}\tan^4 x - \int \tan^3 x\,dx$. Using Formula 69,

$\int \tan^3 x\,dx = \frac{1}{2}\tan^2 x + \ln|\cos x| + C$, so $\int \tan^5 x\,dx = \frac{1}{4}\tan^4 x - \frac{1}{2}\tan^2 x - \ln|\cos x| + C$.

43. Derive gives $I = \int 2^x\sqrt{4^x - 1}\,dx = \dfrac{2^{x-1}\sqrt{2^{2x}-1}}{\ln 2} - \dfrac{\ln\left(\sqrt{2^{2x}-1}+2^x\right)}{2\ln 2}$ immediately. Neither Maple nor

Mathematica is able to evaluate I in its given form. However, if we instead write I as $\int 2^x\sqrt{(2^x)^2 - 1}\,dx$, both

systems give the same answer as Derive (after minor simplification). Our trick works because the CAS now

recognizes 2^x as a promising substitution.

45. Maple gives the antiderivative

$$F(x) = \int \frac{x^2 - 1}{x^4 + x^2 + 1}\,dx = -\frac{1}{2}\ln(x^2 + x + 1) + \frac{1}{2}\ln(x^2 - x + 1).$$

We can see that at 0, this antiderivative is 0. From the graphs, it appears

that F has a maximum at $x = -1$ and a minimum at $x = 1$ [since

$F'(x) = f(x)$ changes sign at these x-values], and that F has inflection

points at $x \approx -1.7$, $x = 0$, and $x \approx 1.7$ [since $f(x)$ has extrema at these

x-values].

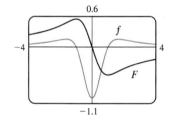

47. Since $f(x) = \sin^4 x \cos^6 x$ is everywhere positive, we know that its antiderivative F is increasing. Maple gives

$\int f(x)\,dx = -\frac{1}{10}\sin^3 x \cos^7 x - \frac{3}{80}\sin x \cos^7 x + \frac{1}{160}\cos^5 x \sin x + \frac{1}{128}\cos^3 x \sin x + \frac{3}{256}\cos x \sin x + \frac{3}{256}x$

and this expression is 0 at $x = 0$.

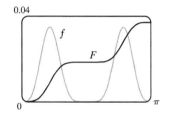

F has a minimum at $x = 0$ and a maximum at $x = \pi$. F has inflection points where f' changes sign, that is, at

$x \approx 0.7$, $x = \pi/2$, and $x \approx 2.5$.

7.7 Approximate Integration

1. (a) $\Delta x = (b - a)/n = (4 - 0)/2 = 2$

$$L_2 = \sum_{i=1}^{2} f(x_{i-1}) \Delta x = f(x_0) \cdot 2 + f(x_1) \cdot 2 = 2[f(0) + f(2)] = 2(0.5 + 2.5) = 6$$

$$R_2 = \sum_{i=1}^{2} f(x_i) \Delta x = f(x_1) \cdot 2 + f(x_2) \cdot 2 = 2[f(2) + f(4)] = 2(2.5 + 3.5) = 12$$

$$M_2 = \sum_{i=1}^{2} f(\overline{x}_i) \Delta x = f(\overline{x}_1) \cdot 2 + f(\overline{x}_2) \cdot 2 = 2[f(1) + f(3)] \approx 2(1.6 + 3.2) = 9.6$$

(b)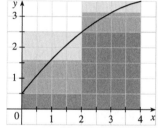

L_2 is an underestimate, since the area under the small rectangles is less than the area under the curve, and R_2 is an overestimate, since the area under the large rectangles is greater than the area under the curve. It appears that M_2 is an overestimate, though it is fairly close to I. See the solution to Exercise 45 for a proof of the fact that if f is concave down on $[a, b]$, then the Midpoint Rule is an overestimate of $\int_a^b f(x)\,dx$.

(c) $T_2 = \left(\frac{1}{2}\,\Delta x\right)[f(x_0) + 2f(x_1) + f(x_2)] = \frac{2}{2}[f(0) + 2f(2) + f(4)] = 0.5 + 2(2.5) + 3.5 = 9$.

This approximation is an underestimate, since the graph is concave down. Thus, $T_2 = 9 < I$. See the solution to Exercise 45 for a general proof of this conclusion.

(d) For any n, we will have $L_n < T_n < I < M_n < R_n$.

3. $f(x) = \cos(x^2)$, $\Delta x = \frac{1-0}{4} = \frac{1}{4}$

(a) $T_4 = \frac{1}{4 \cdot 2}\left[f(0) + 2f\left(\frac{1}{4}\right) + 2f\left(\frac{2}{4}\right) + 2f\left(\frac{3}{4}\right) + f(1)\right] \approx 0.895759$

(b) $M_4 = \frac{1}{4}\left[f\left(\frac{1}{8}\right) + f\left(\frac{3}{8}\right) + f\left(\frac{5}{8}\right) + f\left(\frac{7}{8}\right)\right] \approx 0.908907$

The graph shows that f is concave down on $[0, 1]$. So T_4 is an underestimate and M_4 is an overestimate. We can conclude that $0.895759 < \int_0^1 \cos(x^2)\,dx < 0.908907$.

5. $f(x) = x^2 \sin x$, $\Delta x = \dfrac{b - a}{n} = \dfrac{\pi - 0}{8} = \dfrac{\pi}{8}$

(a) $M_8 = \frac{\pi}{8}\left[f\left(\frac{\pi}{16}\right) + f\left(\frac{3\pi}{16}\right) + f\left(\frac{5\pi}{16}\right) + \cdots + f\left(\frac{15\pi}{16}\right)\right] \approx 5.932957$

(b) $S_8 = \frac{\pi}{8 \cdot 3}\left[f(0) + 4f\left(\frac{\pi}{8}\right) + 2f\left(\frac{2\pi}{8}\right) + 4f\left(\frac{3\pi}{8}\right) + 2f\left(\frac{4\pi}{8}\right) + 4f\left(\frac{5\pi}{8}\right) + 2f\left(\frac{6\pi}{8}\right) + 4f\left(\frac{7\pi}{8}\right) + f(\pi)\right]$

≈ 5.869247

Actual: $\int_0^\pi x^2 \sin x\,dx \overset{84}{=} \left[-x^2 \cos x\right]_0^\pi + 2\int_0^\pi x \cos x\,dx \overset{83}{=} \left[-\pi^2 (-1) - 0\right] + 2[\cos x + x \sin x]_0^\pi$

$\qquad = \pi^2 + 2[(-1 + 0) - (1 + 0)] = \pi^2 - 4 \approx 5.869604$

Errors: $E_M = \text{actual} - M_8 = \int_0^\pi x^2 \sin x\,dx - M_8 \approx -0.063353$

$\qquad E_S = \text{actual} - S_8 = \int_0^\pi x^2 \sin x\,dx - S_8 \approx 0.000357$

7. $f(x) = \sqrt[4]{1 + x^2}$, $\Delta x = \dfrac{2 - 0}{8} = \dfrac{1}{4}$

(a) $T_8 = \frac{1}{4 \cdot 2} \left[f(0) + 2f\left(\frac{1}{4}\right) + 2f\left(\frac{1}{2}\right) + \cdots + 2f\left(\frac{3}{2}\right) + 2f\left(\frac{7}{4}\right) + f(2) \right] \approx 2.413790$

(b) $M_8 = \frac{1}{4} \left[f\left(\frac{1}{8}\right) + f\left(\frac{3}{8}\right) + \cdots + f\left(\frac{13}{8}\right) + f\left(\frac{15}{8}\right) \right] \approx 2.411453$

(c) $S_8 = \frac{1}{4 \cdot 3} \left[f(0) + 4f\left(\frac{1}{4}\right) + 2f\left(\frac{1}{2}\right) + 4f\left(\frac{3}{4}\right) + 2f(1) + 4f\left(\frac{5}{4}\right) + 2f\left(\frac{3}{2}\right) + 4f\left(\frac{7}{4}\right) + f(2) \right] \approx 2.412232$

9. $f(x) = \dfrac{\ln x}{1 + x}$, $\Delta x = \dfrac{2 - 1}{10} = \dfrac{1}{10}$

(a) $T_{10} = \frac{1}{10 \cdot 2}[f(1) + 2f(1.1) + 2f(1.2) + \cdots + 2f(1.8) + 2f(1.9) + f(2)] \approx 0.146879$

(b) $M_{10} = \frac{1}{10}[f(1.05) + f(1.15) + \cdots + f(1.85) + f(1.95)] \approx 0.147391$

(c) $S_{10} = \frac{1}{10 \cdot 3}[f(1) + 4f(1.1) + 2f(1.2) + 4f(1.3) + 2f(1.4) + 4f(1.5) + 2f(1.6) + 4f(1.7)$
$\qquad\qquad + 2f(1.8) + 4f(1.9) + f(2)]$

$\qquad \approx 0.147219$

11. $f(t) = \sin(e^{t/2})$, $\Delta t = \dfrac{\frac{1}{2} - 0}{8} = \dfrac{1}{16}$

(a) $T_8 = \frac{1}{16 \cdot 2} \left[f(0) + 2f\left(\frac{1}{16}\right) + 2f\left(\frac{2}{16}\right) + \cdots + 2f\left(\frac{7}{16}\right) + f\left(\frac{1}{2}\right) \right] \approx 0.451948$

(b) $M_8 = \frac{1}{16} \left[f\left(\frac{1}{32}\right) + f\left(\frac{3}{32}\right) + f\left(\frac{5}{32}\right) + \cdots + f\left(\frac{13}{32}\right) + f\left(\frac{15}{32}\right) \right] \approx 0.451991$

(c) $S_8 = \frac{1}{16 \cdot 3} \left[f(0) + 4f\left(\frac{1}{16}\right) + 2f\left(\frac{2}{16}\right) + \cdots + 4f\left(\frac{7}{16}\right) + f\left(\frac{1}{2}\right) \right] \approx 0.451976$

13. $f(x) = e^{1/x}$, $\Delta x = \dfrac{2 - 1}{4} = \dfrac{1}{4}$

(a) $T_4 = \frac{1}{4 \cdot 2}[f(1) + 2f(1.25) + 2f(1.5) + 2f(1.75) + f(2)] \approx 2.031893$

(b) $M_4 = \frac{1}{4}[f(1.125) + f(1.375) + f(1.625) + f(1.875)] \approx 2.014207$

(c) $S_4 = \frac{1}{4 \cdot 3}[f(1) + 4f(1.25) + 2f(1.5) + 4f(1.75) + f(2)] \approx 2.020651$

15. $f(x) = \dfrac{\cos x}{x}$, $\Delta x = \dfrac{5 - 1}{8} = \dfrac{1}{2}$

(a) $T_8 = \frac{1}{2 \cdot 2} \left[f(1) + 2f\left(\frac{3}{2}\right) + 2f(2) + \cdots + 2f(4) + 2f\left(\frac{9}{2}\right) + f(5) \right] \approx -0.495333$

(b) $M_8 = \frac{1}{2} \left[f\left(\frac{5}{4}\right) + f\left(\frac{7}{4}\right) + f\left(\frac{9}{4}\right) + f\left(\frac{11}{4}\right) + f\left(\frac{13}{4}\right) + f\left(\frac{15}{4}\right) + f\left(\frac{17}{4}\right) + f\left(\frac{19}{4}\right) \right] \approx -0.543321$

(c) $S_8 = \frac{1}{2 \cdot 3} \left[f(1) + 4f\left(\frac{3}{2}\right) + 2f(2) + 4f\left(\frac{5}{2}\right) + 2f(3) + 4f\left(\frac{7}{2}\right) + 2f(4) + 4f\left(\frac{9}{2}\right) + f(5) \right]$
$\qquad \approx -0.526123$

17. $f(y) = \dfrac{1}{1 + y^5}$, $\Delta y = \dfrac{3 - 0}{6} = \dfrac{1}{2}$

(a) $T_6 = \frac{1}{2 \cdot 2} \left[f(0) + 2f\left(\frac{1}{2}\right) + 2f\left(\frac{2}{2}\right) + 2f\left(\frac{3}{2}\right) + 2f\left(\frac{4}{2}\right) + 2f\left(\frac{5}{2}\right) + f(3) \right] \approx 1.064275$

(b) $M_6 = \frac{1}{2} \left[f\left(\frac{1}{4}\right) + f\left(\frac{3}{4}\right) + f\left(\frac{5}{4}\right) + f\left(\frac{7}{4}\right) + f\left(\frac{9}{4}\right) + f\left(\frac{11}{4}\right) \right] \approx 1.067416$

(c) $S_6 = \frac{1}{2 \cdot 3} \left[f(0) + 4f\left(\frac{1}{2}\right) + 2f\left(\frac{2}{2}\right) + 4f\left(\frac{3}{2}\right) + 2f\left(\frac{4}{2}\right) + 4f\left(\frac{5}{2}\right) + f(3) \right] \approx 1.074915$

19. $f(x) = e^{-x^2}$, $\Delta x = \dfrac{2-0}{10} = \dfrac{1}{5}$

(a) $T_{10} = \frac{1}{5 \cdot 2}\{f(0) + 2[f(0.2) + f(0.4) + \cdots + f(1.8)] + f(2)\} \approx 0.881839$

$M_{10} = \frac{1}{5}[f(0.1) + f(0.3) + f(0.5) + \cdots + f(1.7) + f(1.9)] \approx 0.882202$

(b) $f(x) = e^{-x^2}$, $f'(x) = -2xe^{-x^2}$, $f''(x) = (4x^2 - 2)e^{-x^2}$, $f'''(x) = 4x(3 - 2x^2)e^{-x^2}$.

$f'''(x) = 0 \iff x = 0$ or $x = \pm\sqrt{\frac{3}{2}}$. So to find the maximum value of $|f''(x)|$ on $[0, 2]$, we need only

consider its values at $x = 0$, $x = 2$, and $x = \sqrt{\frac{3}{2}}$. $|f''(0)| = 2$, $|f''(2)| \approx 0.2564$ and $\left|f''\left(\sqrt{\frac{3}{2}}\right)\right| \approx 0.8925$.

Thus, taking $K = 2$, $a = 0$, $b = 2$, and $n = 10$ in Theorem 3, we get $|E_T| \le 2 \cdot 2^3/(12 \cdot 10^2) = \frac{1}{75} = 0.01\overline{3}$,

and $|E_M| \le |E_T|/2 \le 0.00\overline{6}$.

(c) Take $K = 2$ [as in part (b)] in Theorem 3. $|E_T| \le \dfrac{K(b-a)^3}{12n^2} \le 10^{-5} \iff \dfrac{2(2-0)^3}{12n^2} \le 10^{-5} \iff$

$\frac{3}{4}n^2 \ge 10^5 \iff n \ge 365.1\ldots \iff n \ge 366$. Take $n = 366$ for T_n. For E_M, again take $K = 2$ in

Theorem 3 to get $|E_M| \le 10^{-5} \iff \frac{3}{2}n^2 \ge 10^5 \iff n \ge 258.2 \Rightarrow n \ge 259$. Take $n = 259$ for M_n.

21. (a) $T_{10} = \frac{1}{10 \cdot 2}\{f(0) + 2[f(0.1) + f(0.2) + \cdots + f(0.9)] + f(1)\} \approx 1.71971349$

$S_{10} = \frac{1}{10 \cdot 3}[f(0) + 4f(0.1) + 2f(0.2) + 4f(0.3) + \cdots + 4f(0.9) + f(1)] \approx 1.71828278$

Since $I = \int_0^1 e^x\,dx = [e^x]_0^1 = e - 1 \approx 1.71828183$, $E_T = I - T_{10} \approx -0.00143166$ and

$E_S = I - S_{10} \approx -0.00000095$.

(b) $f(x) = e^x \Rightarrow f''(x) = e^x \le e$ for $0 \le x \le 1$. Taking $K = e$, $a = 0$, $b = 1$, and $n = 10$ in Theorem 3, we

get $|E_T| \le e(1)^3/(12 \cdot 10^2) \approx 0.002265 > 0.00143166$ [actual $|E_T|$ from (a)]. $f^{(4)}(x) = e^x < e$ for

$0 \le x \le 1$. Using Theorem 4, we have $|E_S| \le e(1)^5/(180 \cdot 10^4) \approx 0.0000015 > 0.00000095$ [actual $|E_S|$

from (a)]. We see that the actual errors are about two-thirds the size of the error estimates.

(c) From part (b), we take $K = e$ to get $|E_T| \le \dfrac{K(b-a)^3}{12n^2} \le 0.00001 \Rightarrow n^2 \ge \dfrac{e(1^3)}{12(0.00001)} \Rightarrow$

$n \ge 150.5$. Take $n = 151$ for T_n. Now $|E_M| \le \dfrac{K(b-a)^3}{24n^2} \le 0.00001 \Rightarrow n \ge 106.4$. Take $n = 107$ for

M_n. Finally, $|E_S| \le \dfrac{K(b-a)^5}{180n^4} \le 0.00001 \Rightarrow n^4 \ge \dfrac{e(1^5)}{180(0.00001)} \Rightarrow n \ge 6.23$. Take $n = 8$ for S_n

(since n has to be even for Simpson's Rule).

23. (a) Using a CAS, we differentiate $f(x) = e^{\cos x}$ twice, and find that

$f''(x) = e^{\cos x}(\sin^2 x - \cos x)$. From the graph, we see that the

maximum value of $|f''(x)|$ occurs at the endpoints of the

interval $[0, 2\pi]$. Since $f''(0) = -e$, we can use $K = e$ or $K = 2.8$.

(b) A CAS gives $M_{10} \approx 7.954926518$. (In Maple, use student[middlesum].)

(c) Using Theorem 3 for the Midpoint Rule, with $K = e$, we get $|E_M| \leq \dfrac{e(2\pi - 0)^3}{24 \cdot 10^2} \approx 0.280945995$. With

$K = 2.8$, we get $|E_M| \leq \dfrac{2.8(2\pi - 0)^3}{24 \cdot 10^2} = 0.289391916$.

(d) A CAS gives $I \approx 7.954926521$.

(e) The actual error is only about 3×10^{-9}, much less than the estimate in part (c).

(f) We use the CAS to differentiate twice more, and then graph

$$f^{(4)}(x) = e^{\cos x}\left(\sin^4 x - 6\sin^2 x \cos x + 3 - 7\sin^2 x + \cos x\right).$$

From the graph, we see that the maximum value of $\left|f^{(4)}(x)\right|$ occurs

at the endpoints of the interval $[0, 2\pi]$. Since $f^{(4)}(0) = 4e$, we can use

$K = 4e$ or $K = 10.9$.

(g) A CAS gives $S_{10} \approx 7.953789422$. (In Maple, use student[simpson].)

(h) Using Theorem 4 with $K = 4e$, we get $|E_S| \leq \dfrac{4e(2\pi - 0)^5}{180 \cdot 10^4} \approx 0.059153618$. With $K = 10.9$, we get

$|E_S| \leq \dfrac{10.9(2\pi - 0)^5}{180 \cdot 10^4} \approx 0.059299814$.

(i) The actual error is about $7.954926521 - 7.953789422 \approx 0.00114$. This is quite a bit smaller than the estimate in part (h), though the difference is not nearly as great as it was in the case of the Midpoint Rule.

(j) To ensure that $|E_S| \leq 0.0001$, we use Theorem 4: $|E_S| \leq \dfrac{4e(2\pi)^5}{180 \cdot n^4} \leq 0.0001 \quad \Rightarrow \quad \dfrac{4e(2\pi)^5}{180 \cdot 0.0001} \leq n^4 \quad \Rightarrow$

$n^4 \geq 5{,}915{,}362 \quad \Leftrightarrow \quad n \geq 49.3$. So we must take $n \geq 50$ to ensure that $|I - S_n| \leq 0.0001$. ($K = 10.9$ leads to the same value of n.)

25. $I = \int_0^1 x^3 \, dx = \left[\frac{1}{4}x^4\right]_0^1 = 0.25.$ $f(x) = x^3$.

$n = 4$: $L_4 = \frac{1}{4}\left[0^3 + \left(\frac{1}{4}\right)^3 + \left(\frac{2}{4}\right)^3 + \left(\frac{3}{4}\right)^3\right] = 0.140625$

$\qquad R_4 = \frac{1}{4}\left[\left(\frac{1}{4}\right)^3 + \left(\frac{2}{4}\right)^3 + \left(\frac{3}{4}\right)^3 + 1^3\right] = 0.390625$

$\qquad T_4 = \frac{1}{4 \cdot 2}\left[0^3 + 2\left(\frac{1}{4}\right)^3 + 2\left(\frac{2}{4}\right)^3 + 2\left(\frac{3}{4}\right)^3 + 1^3\right] = 0.265625,$

$\qquad M_4 = \frac{1}{4}\left[\left(\frac{1}{8}\right)^3 + \left(\frac{3}{8}\right)^3 + \left(\frac{5}{8}\right)^3 + \left(\frac{7}{8}\right)^3\right] = 0.2421875,$

$\qquad E_L = I - L_4 = \frac{1}{4} - 0.140625 = 0.109375,\ E_R = \frac{1}{4} - 0.390625 = -0.140625,$

$\qquad E_T = \frac{1}{4} - 0.265625 = -0.015625,\ E_M = \frac{1}{4} - 0.2421875 = 0.0078125$

$n = 8$: $L_8 = \frac{1}{8}\left[f(0) + f\left(\frac{1}{8}\right) + f\left(\frac{2}{8}\right) + \cdots + f\left(\frac{7}{8}\right)\right] \approx 0.191406$

$\qquad R_8 = \frac{1}{8}\left[f\left(\frac{1}{8}\right) + f\left(\frac{2}{8}\right) + \cdots + f\left(\frac{7}{8}\right) + f(1)\right] \approx 0.316406$

$\qquad T_8 = \frac{1}{8 \cdot 2}\left\{f(0) + 2\left[f\left(\frac{1}{8}\right) + f\left(\frac{2}{8}\right) + \cdots + f\left(\frac{7}{8}\right)\right] + f(1)\right\} \approx 0.253906$

$\qquad M_8 = \frac{1}{8}\left[f\left(\frac{1}{16}\right) + f\left(\frac{3}{16}\right) + \cdots + f\left(\frac{13}{16}\right) + f\left(\frac{15}{16}\right)\right] = 0.248047$

$\qquad E_L \approx \frac{1}{4} - 0.191406 \approx 0.058594,\ E_R \approx \frac{1}{4} - 0.316406 \approx -0.066406,$

$\qquad E_T \approx \frac{1}{4} - 0.253906 \approx -0.003906,\ E_M \approx \frac{1}{4} - 0.248047 \approx 0.001953.$

[continued]

$n = 16$: $L_{16} = \frac{1}{16}\left[f(0) + f\left(\frac{1}{16}\right) + f\left(\frac{2}{16}\right) + \cdots + f\left(\frac{15}{16}\right)\right] \approx 0.219727$

$R_{16} = \frac{1}{16}\left[f\left(\frac{1}{16}\right) + f\left(\frac{2}{16}\right) + \cdots + f\left(\frac{15}{16}\right) + f(1)\right] \approx 0.282227$

$T_{16} = \frac{1}{16 \cdot 2}\left\{f(0) + 2\left[f\left(\frac{1}{16}\right) + f\left(\frac{2}{16}\right) + \cdots + f\left(\frac{15}{16}\right)\right] + f(1)\right\} \approx 0.250977$

$M_{16} = \frac{1}{16}\left[f\left(\frac{1}{32}\right) + f\left(\frac{3}{32}\right) + \cdots + f\left(\frac{31}{32}\right)\right] \approx 0.249512$

$E_L \approx \frac{1}{4} - 0.219727 \approx 0.030273,\ E_R \approx \frac{1}{4} - 0.282227 \approx -0.032227,$

$E_T \approx \frac{1}{4} - 0.250977 \approx -0.000977,\ E_M \approx \frac{1}{4} - 0.249512 \approx 0.000488.$

n	L_n	R_n	T_n	M_n
4	0.140625	0.390625	0.265625	0.242188
8	0.191406	0.316406	0.253906	0.248047
16	0.219727	0.282227	0.250977	0.249512

n	E_L	E_R	E_T	E_M
4	0.109375	−0.140625	−0.015625	0.007813
8	0.058594	−0.066406	−0.003906	0.001953
16	0.030273	−0.032227	−0.000977	0.000488

Observations:

1. E_L and E_R are always opposite in sign, as are E_T and E_M.

2. As n is doubled, E_L and E_R are decreased by about a factor of 2, and E_T and E_M are decreased by a factor of about 4.

3. The Midpoint approximation is about twice as accurate as the Trapezoidal approximation.

4. All the approximations become more accurate as the value of n increases.

5. The Midpoint and Trapezoidal approximations are much more accurate than the endpoint approximations.

27. $\int_1^4 \sqrt{x}\,dx = \left[\frac{2}{3}x^{3/2}\right]_1^4 = \frac{2}{3}(8-1) = \frac{14}{3} \approx 4.666667$

$n = 6$: $\Delta x = (4-1)/6 = \frac{1}{2}$

$T_6 = \frac{1}{2 \cdot 2}\left[\sqrt{1} + 2\sqrt{1.5} + 2\sqrt{2} + 2\sqrt{2.5} + 2\sqrt{3} + 2\sqrt{3.5} + \sqrt{4}\right] \approx 4.661488$

$M_6 = \frac{1}{2}\left[\sqrt{1.25} + \sqrt{1.75} + \sqrt{2.25} + \sqrt{2.75} + \sqrt{3.25} + \sqrt{3.75}\right] \approx 4.669245$

$S_6 = \frac{1}{2 \cdot 3}\left[\sqrt{1} + 4\sqrt{1.5} + 2\sqrt{2} + 4\sqrt{2.5} + 2\sqrt{3} + 4\sqrt{3.5} + \sqrt{4}\right] \approx 4.666563$

$E_T \approx \frac{14}{3} - 4.661488 \approx 0.005178,\quad E_M \approx \frac{14}{3} - 4.669245 \approx -0.002578,$

$E_S \approx \frac{14}{3} - 4.666563 \approx 0.000104.$

$n = 12$: $\Delta x = (4-1)/12 = \frac{1}{4}$

$T_{12} = \frac{1}{4 \cdot 2}\left(f(1) + 2\left[f(1.25) + f(1.5) + \cdots + f(3.5) + f(3.75)\right] + f(4)\right) \approx 4.665367$

$M_{12} = \frac{1}{4}\left[f(1.125) + f(1.375) + f(1.625) + \cdots + f(3.875)\right] \approx 4.667316$

$S_{12} = \frac{1}{4 \cdot 3}\left[f(1) + 4f(1.25) + 2f(1.5) + 4f(1.75) + \cdots + 4f(3.75) + f(4)\right] \approx 4.666659$

$E_T \approx \frac{14}{3} - 4.665367 \approx 0.001300,\quad E_M \approx \frac{14}{3} - 4.667316 \approx -0.000649,$

$E_S \approx \frac{14}{3} - 4.666659 \approx 0.000007.$

Note: These errors were computed more precisely and then rounded to six places. That is, they were not computed by comparing the rounded values of T_n, M_n, and S_n with the rounded value of the actual integral.

n	T_n	M_n	S_n
6	4.661488	4.669245	4.666563
12	4.665367	4.667316	4.666659

n	E_T	E_M	E_S
6	0.005178	−0.002578	0.000104
12	0.001300	−0.000649	0.000007

Observations:

1. E_T and E_M are opposite in sign and decrease by a factor of about 4 as n is doubled.

2. The Simpson's approximation is much more accurate than the Midpoint and Trapezoidal approximations, and seems to decrease by a factor of about 16 as n is doubled.

29. $\Delta x = (4 - 0)/4 = 1$

(a) $T_4 = \frac{1}{2}[f(0) + 2f(1) + 2f(2) + 2f(3) + f(4)] \approx \frac{1}{2}[0 + 2(3) + 2(5) + 2(3) + 1] = 11.5$

(b) $M_4 = 1 \cdot [f(0.5) + f(1.5) + f(2.5) + f(3.5)] \approx 1 + 4.5 + 4.5 + 2 = 12$

(c) $S_4 = \frac{1}{3}[f(0) + 4f(1) + 2f(2) + 4f(3) + f(4)] \approx \frac{1}{3}[0 + 4(3) + 2(5) + 4(3) + 1] = 11.\overline{6}$

31. (a) We are given the function values at the endpoints of 8 intervals of length 0.4, so we'll use the Midpoint Rule with $n = 8/2 = 4$ and $\Delta x = (3.2 - 0)/4 = 0.8$.

$$\int_0^{3.2} f(x)\, dx \approx M_4 = 0.8[f(0.4) + f(1.2) + f(2.0) + f(2.8)]$$
$$= 0.8[6.5 + 6.4 + 7.6 + 8.8]$$
$$= 0.8(29.3) = 23.44$$

(b) $-4 \le f''(x) \le 1 \Rightarrow |f''(x)| \le 4$, so use $K = 4$, $a = 0$, $b = 3.2$, and $n = 4$ in Theorem 3. So

$|E_M| \le \dfrac{4(3.2 - 0)^3}{24(4)^2} = \dfrac{128}{375} = 0.341\overline{3}$.

33. By the Net Change Theorem, the increase in velocity is equal to $\int_0^6 a(t)\, dt$. We use Simpson's Rule with $n = 6$ and $\Delta t = (6 - 0)/6 = 1$ to estimate this integral:

$$\int_0^6 a(t)\, dt \approx S_6 = \frac{1}{3}[a(0) + 4a(1) + 2a(2) + 4a(3) + 2a(4) + 4a(5) + a(6)]$$
$$\approx \frac{1}{3}[0 + 4(0.5) + 2(4.1) + 4(9.8) + 2(12.9) + 4(9.5) + 0] = \frac{1}{3}(113.2) = 37.7\overline{3} \text{ ft/s}$$

35. By the Net Change Theorem, the energy used is equal to $\int_0^6 P(t)\, dt$. We use Simpson's Rule with $n = 12$ and $\Delta t = (6 - 0)/12 = \frac{1}{2}$ to estimate this integral:

$$\int_0^6 P(t)\, dt \approx S_{12} = \frac{1/2}{3}[P(0) + 4P(0.5) + 2P(1) + 4P(1.5) + 2P(2) + 4P(2.5)$$
$$+ 2P(3) + 4P(3.5) + 2P(4) + 4P(4.5) + 2P(5) + 4P(5.5) + P(6)]$$
$$= \frac{1}{6}[1814 + 4(1735) + 2(1686) + 4(1646) + 2(1637) + 4(1609) + 2(1604)$$
$$+ 4(1611) + 2(1621) + 4(1666) + 2(1745) + 4(1886) + 2052]$$
$$= \frac{1}{6}(61{,}064) = 10{,}177.\overline{3} \text{ megawatt-hours.}$$

37. Let $y = f(x)$ denote the curve. Using cylindrical shells, $V = \int_2^{10} 2\pi x f(x)\,dx = 2\pi \int_2^{10} x f(x)\,dx = 2\pi I$.

Now use Simpson's Rule to approximate I:

$$I \approx S_8 = \tfrac{10-2}{3(8)}\left[2f(2) + 4\cdot 3f(3) + 2\cdot 4f(4) + 4\cdot 5f(5) + 2\cdot 6f(6)\right.$$
$$\left. + 4\cdot 7f(7) + 2\cdot 8f(8) + 4\cdot 9f(9) + 10f(10)\right]$$
$$\approx \tfrac{1}{3}[2(0) + 12(1.5) + 8(1.9) + 20(2.2) + 12(3.0) + 28(3.8) + 16(4.0) + 36(3.1) + 10(0)]$$
$$= \tfrac{1}{3}(395.2)$$

Thus, $V \approx 2\pi \cdot \tfrac{1}{3}(395.2) \approx 827.7$ or 828 cubic units.

39. Volume $= \pi \int_0^2 \left(\sqrt[3]{1+x^3}\right)^2 dx = \pi \int_0^2 \left(1+x^3\right)^{2/3} dx$. $V \approx \pi \cdot S_{10}$ where $f(x) = \left(1+x^3\right)^{2/3}$ and $\Delta x = (2-0)/10 = \tfrac{1}{5}$. Therefore,

$$V \approx \pi \cdot S_{10} = \pi \tfrac{1}{5\cdot 3}\left[f(0) + 4f(0.2) + 2f(0.4) + 4f(0.6) + 2f(0.8) + 4f(1)\right.$$
$$\left. + 2f(1.2) + 4f(1.4) + 2f(1.6) + 4f(1.8) + f(2)\right] \approx 12.325078$$

41. $I(\theta) = \dfrac{N^2 \sin^2 k}{k^2}$, where $k = \dfrac{\pi N d \sin\theta}{\lambda}$, $N = 10{,}000$, $d = 10^{-4}$, and $\lambda = 632.8 \times 10^{-9}$. So

$$I(\theta) = \dfrac{\left(10^4\right)^2 \sin^2 k}{k^2}, \text{ where } k = \dfrac{\pi\left(10^4\right)\left(10^{-4}\right)\sin\theta}{632.8 \times 10^{-9}}. \text{ Now } n = 10 \text{ and } \Delta\theta = \dfrac{10^{-6} - (-10^{-6})}{10} = 2 \times 10^{-7},$$

so $M_{10} = 2 \times 10^{-7}[I(-0.0000009) + I(-0.0000007) + \cdots + I(0.0000009)] \approx 59.4$.

43. Consider the function f whose graph is shown. The area $\int_0^2 f(x)\,dx$

is close to 2. The Trapezoidal Rule gives

$$T_2 = \tfrac{2-0}{2\cdot 2}\left[f(0) + 2f(1) + f(2)\right] = \tfrac{1}{2}\left[1 + 2\cdot 1 + 1\right] = 2.$$

The Midpoint Rule gives

$$M_2 = \tfrac{2-0}{2}\left[f(0.5) + f(1.5)\right] = 1[0+0] = 0,$$

so the Trapezoidal Rule is more accurate.

45. Since the Trapezoidal and Midpoint approximations on the interval $[a, b]$ are the sums of the Trapezoidal and Midpoint approximations on the subintervals $[x_{i-1}, x_i]$, $i = 1, 2, \ldots, n$, we can focus our attention on one such interval. The condition $f''(x) < 0$ for $a \le x \le b$ means that the graph of f is concave down as in Figure 5. In that figure, T_n is the area of the trapezoid $AQRD$, $\int_a^b f(x)\,dx$ is the area of the region $AQPRD$, and M_n is the area of the trapezoid $ABCD$, so $T_n < \int_a^b f(x)\,dx < M_n$. In general, the condition $f'' < 0$ implies that the graph of f on $[a, b]$ lies above the chord joining the points $(a, f(a))$ and $(b, f(b))$. Thus, $\int_a^b f(x)\,dx > T_n$. Since M_n is the area under a tangent to the graph, and since $f'' < 0$ implies that the tangent lies above the graph, we also have $M_n > \int_a^b f(x)\,dx$. Thus, $T_n < \int_a^b f(x)\,dx < M_n$.

47. $T_n = \frac{1}{2} \Delta x \left[f(x_0) + 2f(x_1) + \cdots + 2f(x_{n-1}) + f(x_n) \right]$ and

$M_n = \Delta x \left[f(\overline{x}_1) + f(\overline{x}_2) + \cdots + f(\overline{x}_{n-1}) + f(\overline{x}_n) \right]$, where $\overline{x}_i = \frac{1}{2}(x_{i-1} + x_i)$. Now

$$T_{2n} = \frac{1}{2} \left(\frac{1}{2} \Delta x \right) \left[f(x_0) + 2f(\overline{x}_1) + 2f(x_1) + 2f(\overline{x}_2) + 2f(x_2) + \cdots \right.$$
$$\left. + 2f(\overline{x}_{n-1}) + 2f(x_{n-1}) + 2f(\overline{x}_n) + f(x_n) \right]$$

so $\frac{1}{2}(T_n + M_n) = \frac{1}{2} T_n + \frac{1}{2} M_n$

$$= \frac{1}{4} \Delta x \left[f(x_0) + 2f(x_1) + \cdots + 2f(x_{n-1}) + f(x_n) \right]$$
$$+ \frac{1}{4} \Delta x \left[2f(\overline{x}_1) + 2f(\overline{x}_2) + \cdots + 2f(\overline{x}_{n-1}) + 2f(\overline{x}_n) \right]$$

$$= T_{2n}$$

7.8 Improper Integrals

1. (a) Since $\int_1^{\infty} x^4 e^{-x^4} \, dx$ has an infinite interval of integration, it is an improper integral of Type I.

(b) Since $y = \sec x$ has an infinite discontinuity at $x = \frac{\pi}{2}$, $\int_0^{\pi/2} \sec x \, dx$ is a Type II improper integral.

(c) Since $y = \dfrac{x}{(x-2)(x-3)}$ has an infinite discontinuity at $x = 2$, $\displaystyle \int_0^2 \dfrac{x}{x^2 - 5x + 6} \, dx$ is a Type II improper integral.

(d) Since $\displaystyle \int_{-\infty}^0 \dfrac{1}{x^2 + 5} \, dx$ has an infinite interval of integration, it is an improper integral of Type I.

3. The area under the graph of $y = 1/x^3 = x^{-3}$ between $x = 1$ and $x = t$ is

$A(t) = \int_1^t x^{-3} \, dx = \left[-\frac{1}{2} x^{-2} \right]_1^t = -\frac{1}{2} t^{-2} - \left(-\frac{1}{2} \right) = \frac{1}{2} - 1/(2t^2)$. So the area for $1 \le x \le 10$ is

$A(10) = 0.5 - 0.005 = 0.495$, the area for $1 \le x \le 100$ is $A(100) = 0.5 - 0.00005 = 0.49995$, and the area for

$1 \le x \le 1000$ is $A(1000) = 0.5 - 0.0000005 = 0.4999995$. The total area under the curve for $x \ge 1$ is

$\displaystyle \lim_{t \to \infty} A(t) = \lim_{t \to \infty} \left[\frac{1}{2} - 1/(2t^2) \right] = \frac{1}{2}$.

5. $I = \displaystyle \int_1^{\infty} \frac{1}{(3x+1)^2} \, dx = \lim_{t \to \infty} \int_1^t \frac{1}{(3x+1)^2} \, dx$. Now

$$\int \frac{1}{(3x+1)^2} \, dx = \frac{1}{3} \int \frac{1}{u^2} \, du \quad [u = 3x + 1, \, du = 3 \, dx]$$

$$= -\frac{1}{3u} + C = -\frac{1}{3(3x+1)} + C,$$

so $I = \displaystyle \lim_{t \to \infty} \left[-\frac{1}{3(3x+1)} \right]_1^t = \lim_{t \to \infty} \left[-\frac{1}{3(3t+1)} + \frac{1}{12} \right] = 0 + \frac{1}{12} = \frac{1}{12}$.   Convergent

7. $\displaystyle \int_{-\infty}^{-1} \frac{1}{\sqrt{2-w}} \, dw = \lim_{t \to -\infty} \int_t^{-1} \frac{1}{\sqrt{2-w}} \, dw = \lim_{t \to -\infty} \left[-2\sqrt{2-w} \right]_t^{-1} \quad [u = 2 - w, \, du = -dw]$

$$= \lim_{t \to -\infty} \left[-2\sqrt{3} + 2\sqrt{2-t} \right] = \infty. \quad \text{Divergent}$$

9. $\int_4^\infty e^{-y/2}\,dy = \lim_{t\to\infty} \int_4^t e^{-y/2}\,dy = \lim_{t\to\infty}\left[-2e^{-y/2}\right]_4^t = \lim_{t\to\infty}\left(-2e^{-t/2}+2e^{-2}\right) = 0+2e^{-2} = 2e^{-2}$.

Convergent

11. $\displaystyle\int_{-\infty}^\infty \frac{x\,dx}{1+x^2} = \int_{-\infty}^0 \frac{x\,dx}{1+x^2} + \int_0^\infty \frac{x\,dx}{1+x^2}$ and

$\displaystyle\int_{-\infty}^0 \frac{x\,dx}{1+x^2} = \lim_{t\to-\infty}\left[\tfrac12\ln\left(1+x^2\right)\right]_t^0 = \lim_{t\to-\infty}\left[0-\tfrac12\ln\left(1+t^2\right)\right] = -\infty.$ Divergent

13. $\int_{-\infty}^\infty xe^{-x^2}\,dx = \int_{-\infty}^0 xe^{-x^2}\,dx + \int_0^\infty xe^{-x^2}\,dx.$

$\int_{-\infty}^0 xe^{-x^2}\,dx = \lim_{t\to-\infty}\left(-\tfrac12\right)\left[e^{-x^2}\right]_t^0 = \lim_{t\to-\infty}\left(-\tfrac12\right)\left(1-e^{-t^2}\right) = -\tfrac12\cdot 1 = -\tfrac12$, and

$\int_0^\infty xe^{-x^2}\,dx = \lim_{t\to\infty}\left(-\tfrac12\right)\left[e^{-x^2}\right]_0^t = \lim_{t\to\infty}\left(-\tfrac12\right)\left(e^{-t^2}-1\right) = -\tfrac12\cdot(-1) = \tfrac12.$

Therefore, $\int_{-\infty}^\infty xe^{-x^2}\,dx = -\tfrac12 + \tfrac12 = 0$. Convergent

15. $\int_{2\pi}^\infty \sin\theta\,d\theta = \lim_{t\to\infty}\int_{2\pi}^t \sin\theta\,d\theta = \lim_{t\to\infty}\left[-\cos\theta\right]_{2\pi}^t = \lim_{t\to\infty}\left(-\cos t + 1\right).$ This limit does not exist, so the integral

is divergent. Divergent

17. $\displaystyle\int_1^\infty \frac{x+1}{x^2+2x}\,dx = \lim_{t\to\infty}\int_1^t \frac{\tfrac12(2x+2)}{x^2+2x}\,dx = \tfrac12\lim_{t\to\infty}\left[\ln(x^2+2x)\right]_1^t = \tfrac12\lim_{t\to\infty}\left[\ln(t^2+2t)-\ln 3\right]$

$= \infty.$ Divergent

19. $\displaystyle\int_0^\infty se^{-5s}\,ds = \lim_{t\to\infty}\int_0^t se^{-5s}\,ds = \lim_{t\to\infty}\left[-\frac15 se^{-5s} - \frac{1}{25}e^{-5s}\right]_0^t \quad \begin{bmatrix}\text{by integration by}\\ \text{parts with } u = s\end{bmatrix}$

$= \lim_{t\to\infty}\left(-\tfrac15 te^{-5t} - \tfrac{1}{25}e^{-5t} + \tfrac{1}{25}\right) = 0 - 0 + \tfrac{1}{25}$ [by l'Hospital's Rule]

$= \tfrac{1}{25}.$ Convergent

21. $\displaystyle\int_1^\infty \frac{\ln x}{x}\,dx = \lim_{t\to\infty}\left[\frac{(\ln x)^2}{2}\right]_1^t$ (by substitution with $u = \ln x,\ du = dx/x$) $= \lim_{t\to\infty}\frac{(\ln t)^2}{2} = \infty.$ Divergent

23. $\displaystyle\int_{-\infty}^\infty \frac{x^2}{9+x^6}\,dx = \int_{-\infty}^0 \frac{x^2}{9+x^6}\,dx + \int_0^\infty \frac{x^2}{9+x^6}\,dx = 2\int_0^\infty \frac{x^2}{9+x^6}\,dx$ [since the integrand is even].

Now $\displaystyle\int \frac{x^2\,dx}{9+x^6}\ \begin{bmatrix}u = x^3\\ du = 3x^2\,dx\end{bmatrix} = \int \frac{\tfrac13\,du}{9+u^2}\ \begin{bmatrix}u = 3v\\ du = 3\,dv\end{bmatrix} = \int \frac{\tfrac13(3\,dv)}{9+9v^2} = \frac19\int \frac{dv}{1+v^2}$

$= \frac19\tan^{-1} v + C = \frac19\tan^{-1}\left(\frac u3\right) + C = \frac19\tan^{-1}\left(\frac{x^3}{3}\right) + C,$

so $\displaystyle 2\int_0^\infty \frac{x^2}{9+x^6}\,dx = 2\lim_{t\to\infty}\int_0^t \frac{x^2}{9+x^6}\,dx = 2\lim_{t\to\infty}\left[\frac19\tan^{-1}\left(\frac{x^3}{3}\right)\right]_0^t$

$= 2\lim_{t\to\infty}\frac19\tan^{-1}\left(\frac{t^3}{3}\right) = \frac29\cdot\frac{\pi}{2} = \frac{\pi}{9}.$ Convergent

25. Integrate by parts with $u = \ln x,\ dv = dx/x^2 \ \Rightarrow\ du = dx/x,\ v = -1/x.$

$$\int_1^\infty \frac{\ln x}{x^2}\,dx = \lim_{t\to\infty}\int_1^t \frac{\ln x}{x^2}\,dx = \lim_{t\to\infty}\left[-\frac{\ln x}{x} - \frac1x\right]_1^t = \lim_{t\to\infty}\left(-\frac{\ln t}{t} - \frac1t + 0 + 1\right)$$

$$= -0 - 0 + 0 + 1 = 1$$

since $\displaystyle\lim_{t\to\infty}\frac{\ln t}{t} \overset{\text{H}}{=} \lim_{t\to\infty}\frac{1/t}{1} = 0.$ Convergent

27. There is an infinite discontinuity at the left endpoint of $[0, 3]$.

$$\int_0^3 \frac{dx}{\sqrt{x}} = \lim_{t \to 0^+} \int_t^3 \frac{dx}{\sqrt{x}} = \lim_{t \to 0^+} \left[2\sqrt{x}\right]_t^3 = \lim_{t \to 0^+} \left(2\sqrt{3} - 2\sqrt{t}\right) = 2\sqrt{3}. \quad \text{Convergent}$$

29. There is an infinite discontinuity at the right endpoint of $[-1, 0]$.

$$\int_{-1}^0 \frac{dx}{x^2} = \lim_{t \to 0^-} \int_{-1}^t \frac{dx}{x^2} = \lim_{t \to 0^-} \left[\frac{-1}{x}\right]_{-1}^t = \lim_{t \to 0^-} \left[-\frac{1}{t} + \frac{1}{-1}\right] = \infty. \quad \text{Divergent}$$

31. $\int_{-2}^3 \frac{dx}{x^4} = \int_{-2}^0 \frac{dx}{x^4} + \int_0^3 \frac{dx}{x^4}$, but $\int_{-2}^0 \frac{dx}{x^4} = \lim_{t \to 0^-} \left[-\frac{x^{-3}}{3}\right]_{-2}^t = \lim_{t \to 0^-} \left[-\frac{1}{3t^3} - \frac{1}{24}\right] = \infty. \quad \text{Divergent}$

33. There is an infinite discontinuity at $x = 1$. $\int_0^{33}(x-1)^{-1/5}\,dx = \int_0^1 (x-1)^{-1/5}\,dx + \int_1^{33}(x-1)^{-1/5}\,dx.$ Here

$\int_0^1 (x-1)^{-1/5}\,dx = \lim_{t \to 1^-} \int_0^t (x-1)^{-1/5}\,dx = \lim_{t \to 1^-} \left[\frac{5}{4}(x-1)^{4/5}\right]_0^t = \lim_{t \to 1^-} \left[\frac{5}{4}(t-1)^{4/5} - \frac{5}{4}\right] = -\frac{5}{4}$ and

$\int_1^{33}(x-1)^{-1/5}\,dx = \lim_{t \to 1^+} \int_t^{33}(x-1)^{-1/5}\,dx = \lim_{t \to 1^+} \left[\frac{5}{4}(x-1)^{4/5}\right]_t^{33} = \lim_{t \to 1^+} \left[\frac{5}{4}\cdot 16 - \frac{5}{4}(t-1)^{4/5}\right] = 20.$

Thus, $\int_0^{33}(x-1)^{-1/5}\,dx = -\frac{5}{4} + 20 = \frac{75}{4}. \quad \text{Convergent}$

35. $\int_0^\pi \sec x\,dx = \int_0^{\pi/2} \sec x\,dx + \int_{\pi/2}^\pi \sec x\,dx. \quad \int_0^{\pi/2}\sec x\,dx = \lim_{t \to \pi/2^-} \int_0^t \sec x\,dx$

$$= \lim_{t \to \pi/2^-} \left[\ln|\sec x + \tan x|\right]_0^t = \lim_{t \to \pi/2^-} \ln|\sec t + \tan t| = \infty. \text{ Divergent}$$

37. There is an infinite discontinuity at $x = 0$. $\int_{-1}^1 \frac{e^x}{e^x - 1}\,dx = \int_{-1}^0 \frac{e^x}{e^x - 1}\,dx + \int_0^1 \frac{e^x}{e^x - 1}\,dx.$

$$\int_{-1}^0 \frac{e^x}{e^x - 1}\,dx = \lim_{t \to 0^-} \int_{-1}^t \frac{e^x}{e^x - 1}\,dx = \lim_{t \to 0^-} \left[\ln|e^x - 1|\right]_{-1}^t = \lim_{t \to 0^-} \left[\ln|e^t - 1| - \ln|e^{-1} - 1|\right] = -\infty,$$

so $\int_{-1}^1 \frac{e^x}{e^x - 1}\,dx$ is divergent. The integral $\int_0^1 \frac{e^x}{e^x - 1}\,dx$ also diverges since

$$\int_0^1 \frac{e^x}{e^x - 1}\,dx = \lim_{t \to 0^+} \int_t^1 \frac{e^x}{e^x - 1}\,dx = \lim_{t \to 0^+} \left[\ln|e^x - 1|\right]_t^1 = \lim_{t \to 0^+} \left[\ln|e - 1| - \ln|e^t - 1|\right] = \infty.$$

Divergent

39. $I = \int_0^2 z^2 \ln z\,dz = \lim_{t \to 0^+} \int_t^2 z^2 \ln z\,dz \overset{101}{=} \lim_{t \to 0^+} \left[\frac{z^3}{3^2}(3\ln z - 1)\right]_t^2$

$= \lim_{t \to 0^+} \left[\frac{8}{9}(3\ln 2 - 1) - \frac{1}{9}t^3(3\ln t - 1)\right] = \frac{8}{3}\ln 2 - \frac{8}{9} - \frac{1}{9}\lim_{t \to 0^+}\left[t^3(3\ln t - 1)\right] = \frac{8}{3}\ln 2 - \frac{8}{9} - \frac{1}{9}L.$

Now $L = \lim_{t \to 0^+} \left[t^3(3\ln t - 1)\right] = \lim_{t \to 0^+} \frac{3\ln t - 1}{t^{-3}} \overset{\text{H}}{=} \lim_{t \to 0^+} \frac{3/t}{-3/t^4} = \lim_{t \to 0^+} (-t^3) = 0.$ Thus, $L = 0$ and

$I = \frac{8}{3}\ln 2 - \frac{8}{9}. \quad \text{Convergent}$

41.

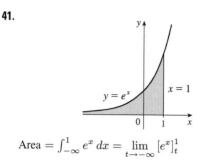

$$\text{Area} = \int_{-\infty}^{1} e^x \, dx = \lim_{t \to -\infty} [e^x]_t^1$$

$$= e - \lim_{t \to -\infty} e^t = e$$

43.

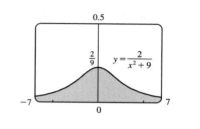

$$\text{Area} = \int_{-\infty}^{\infty} \frac{2}{x^2 + 9} \, dx = 2 \cdot 2 \int_0^{\infty} \frac{1}{x^2 + 9} \, dx$$

$$= 4 \lim_{t \to \infty} \int_0^t \frac{1}{x^2 + 9} \, dx = 4 \lim_{t \to \infty} \left[\frac{1}{3} \tan^{-1} \frac{x}{3} \right]_0^t$$

$$= \frac{4}{3} \lim_{t \to \infty} \left[\tan^{-1} \frac{t}{3} - 0 \right] = \frac{4}{3} \cdot \frac{\pi}{2} = \frac{2\pi}{3}$$

45.

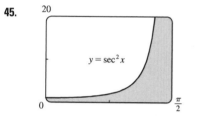

$$\text{Area} = \int_0^{\pi/2} \sec^2 x \, dx = \lim_{t \to (\pi/2)^-} \int_0^t \sec^2 x \, dx$$

$$= \lim_{t \to (\pi/2)^-} [\tan x]_0^t = \lim_{t \to (\pi/2)^-} (\tan t - 0)$$

$$= \infty$$

Infinite area

47. (a)

t	$\int_1^t g(x) \, dx$
2	0.447453
5	0.577101
10	0.621306
100	0.668479
1000	0.672957
10,000	0.673407

$$g(x) = \frac{\sin^2 x}{x^2}.$$

It appears that the integral is convergent.

(b) $-1 \le \sin x \le 1 \implies 0 \le \sin^2 x \le 1 \implies 0 \le \frac{\sin^2 x}{x^2} \le \frac{1}{x^2}$. Since $\int_1^{\infty} \frac{1}{x^2} \, dx$ is convergent

(Equation 2 with $p = 2 > 1$), $\int_1^{\infty} \frac{\sin^2 x}{x^2} \, dx$ is convergent by the Comparison Theorem.

(c)

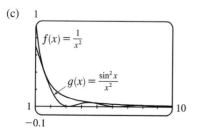

Since $\int_1^{\infty} f(x) \, dx$ is finite and the area under $g(x)$ is less than the area under $f(x)$ on any interval $[1, t]$, $\int_1^{\infty} g(x) \, dx$ must be finite; that is, the integral is convergent.

49. For $x \geq 1$, $\dfrac{\cos^2 x}{1 + x^2} \leq \dfrac{1}{1 + x^2} < \dfrac{1}{x^2}$. $\displaystyle\int_1^\infty \dfrac{1}{x^2}\, dx$ is convergent by Equation 2 with $p = 2 > 1$, so $\displaystyle\int_1^\infty \dfrac{\cos^2 x}{1 + x^2}\, dx$

is convergent by the Comparison Theorem.

51. For $x \geq 1$, $x + e^{2x} > e^{2x} > 0$ $\Rightarrow$ $\dfrac{1}{x + e^{2x}} \leq \dfrac{1}{e^{2x}} = e^{-2x}$ on $[1, \infty)$.

$$\int_1^\infty e^{-2x}\, dx = \lim_{t \to \infty} \left[-\frac{1}{2}e^{-2x}\right]_1^t = \lim_{t \to \infty} \left[-\frac{1}{2}e^{-2t} + \frac{1}{2}e^{-2}\right] = \frac{1}{2}e^{-2}. \text{ Therefore, } \int_1^\infty e^{-2x}\, dx \text{ is convergent,}$$

and by the Comparison Theorem, $\displaystyle\int_1^\infty \dfrac{dx}{x + e^{2x}}$ is also convergent.

53. $\dfrac{1}{x \sin x} \geq \dfrac{1}{x}$ on $\left(0, \frac{\pi}{2}\right]$ since $0 \leq \sin x \leq 1$. $\displaystyle\int_0^{\pi/2} \dfrac{dx}{x} = \lim_{t \to 0^+} \int_t^{\pi/2} \dfrac{dx}{x} = \lim_{t \to 0^+} \left[\ln x\right]_t^{\pi/2}$.

But $\ln t \to -\infty$ as $t \to 0^+$, so $\displaystyle\int_0^{\pi/2} \dfrac{dx}{x}$ is divergent, and by the Comparison Theorem, $\displaystyle\int_0^{\pi/2} \dfrac{dx}{x \sin x}$ is also

divergent.

55. $\displaystyle\int_0^\infty \dfrac{dx}{\sqrt{x}\,(1 + x)} = \int_0^1 \dfrac{dx}{\sqrt{x}\,(1 + x)} + \int_1^\infty \dfrac{dx}{\sqrt{x}\,(1 + x)} = \lim_{t \to 0^+} \int_t^1 \dfrac{dx}{\sqrt{x}\,(1 + x)} + \lim_{t \to \infty} \int_1^t \dfrac{dx}{\sqrt{x}\,(1 + x)}$. Now

$$\int \dfrac{dx}{\sqrt{x}\,(1 + x)} = \int \dfrac{2u\, du}{u(1 + u^2)} \quad [u = \sqrt{x},\, x = u^2,\, dx = 2u\, du]$$

$$= 2\int \dfrac{du}{1 + u^2} = 2\tan^{-1} u + C = 2\tan^{-1}\sqrt{x} + C,$$

so $\displaystyle\int_0^\infty \dfrac{dx}{\sqrt{x}\,(1 + x)} = \lim_{t \to 0^+} \left[2\tan^{-1}\sqrt{x}\right]_t^1 + \lim_{t \to \infty} \left[2\tan^{-1}\sqrt{x}\right]_1^t$

$$= \lim_{t \to 0^+} \left[2\left(\tfrac{\pi}{4}\right) - 2\tan^{-1}\sqrt{t}\right] + \lim_{t \to \infty} \left[2\tan^{-1}\sqrt{t} - 2\left(\tfrac{\pi}{4}\right)\right] = \tfrac{\pi}{2} - 0 + 2\left(\tfrac{\pi}{2}\right) - \tfrac{\pi}{2} = \pi.$$

57. If $p = 1$, then $\displaystyle\int_0^1 \dfrac{dx}{x^p} = \lim_{t \to 0^+} \int_t^1 \dfrac{dx}{x} = \lim_{t \to 0^+} \left[\ln x\right]_t^1 = \infty$. Divergent.

If $p \neq 1$, then $\displaystyle\int_0^1 \dfrac{dx}{x^p} = \lim_{t \to 0^+} \int_t^1 \dfrac{dx}{x^p}$ (note that the integral is not improper if $p < 0$)

$$= \lim_{t \to 0^+} \left[\dfrac{x^{-p+1}}{-p + 1}\right]_t^1 = \lim_{t \to 0^+} \dfrac{1}{1 - p}\left[1 - \dfrac{1}{t^{p-1}}\right]$$

If $p > 1$, then $p - 1 > 0$, so $\dfrac{1}{t^{p-1}} \to \infty$ as $t \to 0^+$, and the integral diverges.

If $p < 1$, then $p - 1 < 0$, so $\dfrac{1}{t^{p-1}} \to 0$ as $t \to 0^+$ and $\displaystyle\int_0^1 \dfrac{dx}{x^p} = \dfrac{1}{1 - p}\left[\lim_{t \to 0^+}\left(1 - t^{1-p}\right)\right] = \dfrac{1}{1 - p}$.

Thus, the integral converges if and only if $p < 1$, and in that case its value is $\dfrac{1}{1 - p}$.

59. First suppose $p = -1$. Then

$$\int_0^1 x^p \ln x \, dx = \int_0^1 \frac{\ln x}{x} \, dx = \lim_{t \to 0^+} \int_t^1 \frac{\ln x}{x} \, dx = \lim_{t \to 0^+} \left[\tfrac{1}{2}(\ln x)^2 \right]_t^1 = -\tfrac{1}{2} \lim_{t \to 0^+} (\ln t)^2 = -\infty,$$

so the integral diverges. Now suppose $p \neq -1$. Then integration by parts gives

$$\int x^p \ln x \, dx = \frac{x^{p+1}}{p+1} \ln x - \int \frac{x^p}{p+1} \, dx = \frac{x^{p+1}}{p+1} \ln x - \frac{x^{p+1}}{(p+1)^2} + C. \text{ If } p < -1, \text{ then } p+1 < 0, \text{ so}$$

$$\int_0^1 x^p \ln x \, dx = \lim_{t \to 0^+} \left[\frac{x^{p+1}}{p+1} \ln x - \frac{x^{p+1}}{(p+1)^2} \right]_t^1 = \frac{-1}{(p+1)^2} - \left(\frac{1}{p+1} \right) \lim_{t \to 0^+} \left[t^{p+1} \left(\ln t - \frac{1}{p+1} \right) \right] = \infty.$$

If $p > -1$, then $p + 1 > 0$ and

$$\int_0^1 x^p \ln x \, dx = \frac{-1}{(p+1)^2} - \left(\frac{1}{p+1} \right) \lim_{t \to 0^+} \frac{\ln t - 1/(p+1)}{t^{-(p+1)}} \overset{\text{H}}{=} \frac{-1}{(p+1)^2} - \left(\frac{1}{p+1} \right) \lim_{t \to 0^+} \frac{1/t}{-(p+1)t^{-(p+2)}}$$

$$= \frac{-1}{(p+1)^2} + \frac{1}{(p+1)^2} \lim_{t \to 0^+} t^{p+1} = \frac{-1}{(p+1)^2}$$

Thus, the integral converges to $-\dfrac{1}{(p+1)^2}$ if $p > -1$ and diverges otherwise.

61. (a) $I = \int_{-\infty}^{\infty} x \, dx = \int_{-\infty}^{0} x \, dx + \int_0^{\infty} x \, dx$, and

$\int_0^{\infty} x \, dx = \lim_{t \to \infty} \int_0^t x \, dx = \lim_{t \to \infty} \left[\tfrac{1}{2} x^2 \right]_0^t = \lim_{t \to \infty} \left[\tfrac{1}{2} t^2 - 0 \right] = \infty$, so I is divergent.

(b) $\int_{-t}^t x \, dx = \left[\tfrac{1}{2} x^2 \right]_{-t}^t = \tfrac{1}{2} t^2 - \tfrac{1}{2} t^2 = 0$, so $\lim_{t \to \infty} \int_{-t}^t x \, dx = 0$. Therefore, $\int_{-\infty}^{\infty} x \, dx \neq \lim_{t \to \infty} \int_{-t}^t x \, dx$.

63. Volume $= \int_1^{\infty} \pi \left(\frac{1}{x} \right)^2 dx = \pi \lim_{t \to \infty} \int_1^t \frac{dx}{x^2} = \pi \lim_{t \to \infty} \left[-\frac{1}{x} \right]_1^t = \pi \lim_{t \to \infty} \left(1 - \frac{1}{t} \right) = \pi < \infty.$

65. Work $= \int_R^{\infty} F \, dr = \lim_{t \to \infty} \int_R^t \frac{GmM}{r^2} \, dr = \lim_{t \to \infty} GmM \left(\frac{1}{R} - \frac{1}{t} \right) = \frac{GmM}{R}$. The initial kinetic energy

provides the work, so $\tfrac{1}{2} m v_0^2 = \dfrac{GmM}{R} \implies v_0 = \sqrt{\dfrac{2GM}{R}}.$

67. (a) We would expect a small percentage of bulbs to burn out in the first few hundred hours, most of the bulbs to burn

out after close to 700 hours, and a few overachievers to burn on and on.

(b) $r(t) = F'(t)$ is the rate at which the fraction $F(t)$ of burnt-out bulbs increases as t increases. This could be interpreted as a fractional burnout rate.

(c) $\int_0^{\infty} r(t) \, dt = \lim_{x \to \infty} F(x) = 1$, since all of the bulbs will eventually burn out.

69. $I = \displaystyle\int_a^\infty \dfrac{1}{x^2+1}\,dx = \lim_{t\to\infty}\int_a^t \dfrac{1}{x^2+1}\,dx = \lim_{t\to\infty}\left[\tan^{-1}x\right]_a^t = \lim_{t\to\infty}\left(\tan^{-1}t - \tan^{-1}a\right) = \frac{\pi}{2} - \tan^{-1}a.$

$I < 0.001 \;\Rightarrow\; \frac{\pi}{2} - \tan^{-1}a < 0.001 \;\Rightarrow\; \tan^{-1}a > \frac{\pi}{2} - 0.001 \;\Rightarrow\; a > \tan\!\left(\frac{\pi}{2} - 0.001\right) \approx 1000.$

71. (a) $F(s) = \displaystyle\int_0^\infty f(t)e^{-st}\,dt = \int_0^\infty e^{-st}\,dt = \lim_{n\to\infty}\left[-\dfrac{e^{-st}}{s}\right]_0^n = \lim_{n\to\infty}\left(\dfrac{e^{-sn}}{-s} + \dfrac{1}{s}\right).$ This converges to $\dfrac{1}{s}$

only if $s > 0$. Therefore $F(s) = \dfrac{1}{s}$ with domain $\{s \mid s > 0\}$.

(b) $F(s) = \displaystyle\int_0^\infty f(t)e^{-st}\,dt = \int_0^\infty e^t e^{-st}\,dt = \lim_{n\to\infty}\int_0^n e^{t(1-s)}\,dt = \lim_{n\to\infty}\left[\dfrac{1}{1-s}e^{t(1-s)}\right]_0^n$

$= \displaystyle\lim_{n\to\infty}\left(\dfrac{e^{(1-s)n}}{1-s} - \dfrac{1}{1-s}\right)$

This converges only if $1 - s < 0 \;\Rightarrow\; s > 1$, in which case $F(s) = \dfrac{1}{s-1}$ with domain $\{s \mid s > 1\}$.

(c) $F(s) = \int_0^\infty f(t)e^{-st}\,dt = \lim_{n\to\infty}\int_0^n te^{-st}\,dt.$ Use integration by parts: let $u = t,\ dv = e^{-st}\,dt \;\Rightarrow$

$du = dt,\ v = -\dfrac{e^{-st}}{s}.$ Then $F(s) = \displaystyle\lim_{n\to\infty}\left[-\dfrac{t}{s}e^{-st} - \dfrac{1}{s^2}e^{-st}\right]_0^n = \lim_{n\to\infty}\left(\dfrac{-n}{se^{sn}} - \dfrac{1}{s^2 e^{sn}} + 0 + \dfrac{1}{s^2}\right) = \dfrac{1}{s^2}$

only if $s > 0$. Therefore, $F(s) = \dfrac{1}{s^2}$ and the domain of F is $\{s \mid s > 0\}$.

73. $G(s) = \int_0^\infty f'(t)e^{-st}\,dt.$ Integrate by parts with $u = e^{-st},\ dv = f'(t)\,dt \;\Rightarrow\; du = -se^{-st},\ v = f(t)$:

$$G(s) = \lim_{n\to\infty}\left[f(t)e^{-st}\right]_0^n + s\int_0^\infty f(t)e^{-st}\,dt = \lim_{n\to\infty}f(n)e^{-sn} - f(0) + sF(s)$$

But $0 \le f(t) \le Me^{at} \;\Rightarrow\; 0 \le f(t)e^{-st} \le Me^{at}e^{-st}$ and $\lim_{t\to\infty}Me^{t(a-s)} = 0$ for $s > a$. So by the Squeeze

Theorem, $\lim_{t\to\infty}f(t)e^{-st} = 0$ for $s > a \;\Rightarrow\; G(s) = 0 - f(0) + sF(s) = sF(s) - f(0)$ for $s > a$.

75. We use integration by parts: let $u = x,\ dv = xe^{-x^2}\,dx \;\Rightarrow\; du = dx,\ v = -\frac{1}{2}e^{-x^2}.$ So

$$\int_0^\infty x^2 e^{-x^2}\,dx = \lim_{t\to\infty}\left[-\tfrac{1}{2}xe^{-x^2}\right]_0^t + \tfrac{1}{2}\int_0^\infty e^{-x^2}\,dx$$

$$= \lim_{t\to\infty}\left[-t\Big/\!\left(2e^{t^2}\right)\right] + \tfrac{1}{2}\int_0^\infty e^{-x^2}\,dx = \tfrac{1}{2}\int_0^\infty e^{-x^2}\,dx$$

(The limit is 0 by l'Hospital's Rule.)

77. For the first part of the integral, let $x = 2\tan\theta \Rightarrow dx = 2\sec^2\theta\,d\theta$.

$$\int \frac{1}{\sqrt{x^2+4}}\,dx = \int \frac{2\sec^2\theta}{2\sec\theta}\,d\theta = \int \sec\theta\,d\theta = \ln|\sec\theta + \tan\theta|. \text{ From the}$$

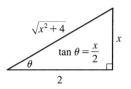

figure, $\tan\theta = \dfrac{x}{2}$, and $\sec\theta = \dfrac{\sqrt{x^2+4}}{2}$. So

$$I = \int_0^\infty \left(\frac{1}{\sqrt{x^2+4}} - \frac{C}{x+2}\right)dx = \lim_{t\to\infty}\left[\ln\left|\frac{\sqrt{x^2+4}}{2} + \frac{x}{2}\right| - C\ln|x+2|\right]_0^t$$

$$= \lim_{t\to\infty}\left[\ln\frac{\sqrt{t^2+4}+t}{2} - C\ln(t+2) - (\ln 1 - C\ln 2)\right]$$

$$= \lim_{t\to\infty}\left[\ln\left(\frac{\sqrt{t^2+4}+t}{2(t+2)^C}\right) + \ln 2^C\right]$$

$$= \ln\left(\lim_{t\to\infty}\frac{t+\sqrt{t^2+4}}{(t+2)^C}\right) + \ln 2^{C-1}$$

Now $L = \lim_{t\to\infty} \dfrac{t+\sqrt{t^2+4}}{(t+2)^C} \overset{\text{H}}{=} \lim_{t\to\infty} \dfrac{1+t/\sqrt{t^2+4}}{C(t+2)^{C-1}} = \dfrac{2}{C\lim_{t\to\infty}(t+2)^{C-1}}$.

If $C < 1$, $L = \infty$ and I diverges. If $C = 1$, $L = 2$ and I converges to $\ln 2 + \ln 2^0 = \ln 2$. If $C > 1$, $L = 0$ and I diverges to $-\infty$.

7 Review

CONCEPT CHECK

1. See Formula 7.1.1 or 7.1.2. We try to choose $u = f(x)$ to be a function that becomes simpler when differentiated (or at least not more complicated) as long as $dv = g'(x)\,dx$ can be readily integrated to give v.

2. See the Strategy for Evaluating $\int \sin^m x \cos^n x\,dx$ on page 484.

3. If $\sqrt{a^2 - x^2}$ occurs, try $x = a\sin\theta$; if $\sqrt{a^2 + x^2}$ occurs, try $x = a\tan\theta$, and if $\sqrt{x^2 - a^2}$ occurs, try $x = a\sec\theta$. See the Table of Trigonometric Substitutions on page 490.

4. See Equation 2 and Expressions 7, 9, and 11 in Section 7.4.

5. See the Midpoint Rule, the Trapezoidal Rule, and Simpson's Rule, as well as their associated error bounds, all in Section 7.7. We would expect the best estimate to be given by Simpson's Rule.

6. See Definitions 1(a), (b), and (c) in Section 7.8.

7. See Definitions 3(b), (a), and (c) in Section 7.8.

8. See the Comparison Theorem after Example 8 in Section 7.8.

──────────────────────── TRUE-FALSE QUIZ ────────────────────────

1. False.   Since the numerator has a higher degree than the denominator,

$$\frac{x\left(x^2+4\right)}{x^2-4} = x + \frac{8x}{x^2-4} = x + \frac{A}{x+2} + \frac{B}{x-2}.$$

3. False.   It can be put in the form $\dfrac{A}{x} + \dfrac{B}{x^2} + \dfrac{C}{x-4}.$

5. False.   This is an improper integral, since the denominator vanishes at $x = 1$.

$$\int_0^4 \frac{x}{x^2-1}\,dx = \int_0^1 \frac{x}{x^2-1}\,dx + \int_1^4 \frac{x}{x^2-1}\,dx \text{ and}$$

$$\int_0^1 \frac{x}{x^2-1}\,dx = \lim_{t\to 1^-}\int_0^t \frac{x}{x^2-1}\,dx = \lim_{t\to 1^-}\left[\tfrac{1}{2}\ln\left|x^2-1\right|\right]_0^t = \lim_{t\to 1^-}\tfrac{1}{2}\ln\left|t^2-1\right| = \infty$$

So the integral diverges.

7. False.   See Exercise 61 in Section 7.8.

9. (a) True.   See the end of Section 7.5.

   (b) False.   Examples include the functions $f(x) = e^{x^2}$, $g(x) = \sin(x^2)$, and $h(x) = \dfrac{\sin x}{x}$.

11. False.   If $f(x) = 1/x$, then f is continuous and decreasing on $[1, \infty)$ with $\lim\limits_{x\to\infty} f(x) = 0$, but $\int_1^\infty f(x)\,dx$
            is divergent.

13. False.   Take $f(x) = 1$ for all x and $g(x) = -1$ for all x. Then $\int_a^\infty f(x)\,dx = \infty$ [divergent]   and
            $\int_a^\infty g(x)\,dx = -\infty$ [divergent],   but $\int_a^\infty [f(x) + g(x)]\,dx = 0$ [convergent].

──────────────────────── EXERCISES ────────────────────────

1. $\displaystyle\int_0^5 \frac{x}{x+10}\,dx = \int_0^5 \left(1 - \frac{10}{x+10}\right)dx = \Big[x - 10\ln(x+10)\Big]_0^5$

$$= 5 - 10\ln 15 + 10\ln 10 = 5 + 10\ln\tfrac{10}{15} = 5 + 10\ln\tfrac{2}{3}$$

3. $\displaystyle\int_0^{\pi/2} \frac{\cos\theta}{1+\sin\theta}\,d\theta = \Big[\ln(1+\sin\theta)\Big]_0^{\pi/2} = \ln 2 - \ln 1 = \ln 2$

5. Let $u = \sec x$. Then $du = \sec x \tan x\,dx$, so

$$\int \tan^7 x \sec^3 x\,dx = \int \tan^6 x \sec^2 x \sec x \tan x\,dx = \int \left(u^2-1\right)^3 u^2\,du = \int \left(u^8 - 3u^6 + 3u^4 - u^2\right)du$$

$$= \tfrac{1}{9}u^9 - \tfrac{3}{7}u^7 + \tfrac{3}{5}u^5 - \tfrac{1}{3}u^3 + C = \tfrac{1}{9}\sec^9 x - \tfrac{3}{7}\sec^7 x + \tfrac{3}{5}\sec^5 x - \tfrac{1}{3}\sec^3 x + C$$

7. Let $u = \ln t$, $du = dt/t$. Then $\displaystyle\int \frac{\sin(\ln t)}{t}\,dt = \int \sin u\,du = -\cos u + C = -\cos(\ln t) + C.$

9. $\displaystyle\int_1^4 x^{3/2}\ln x\,dx \quad \begin{bmatrix} u = \ln x, & dv = x^{3/2}\,dx, \\ du = dx/x & v = \tfrac{2}{5}x^{5/2} \end{bmatrix} = \tfrac{2}{5}\Big[x^{5/2}\ln x\Big]_1^4 - \tfrac{2}{5}\int_1^4 x^{3/2}\,dx$

$$= \tfrac{2}{5}(32\ln 4 - \ln 1) - \tfrac{2}{5}\Big[\tfrac{2}{5}x^{5/2}\Big]_1^4$$

$$= \tfrac{2}{5}(64\ln 2) - \tfrac{4}{25}(32 - 1)$$

$$= \tfrac{128}{5}\ln 2 - \tfrac{124}{25} \quad \left(\text{or } \tfrac{64}{5}\ln 4 - \tfrac{124}{25}\right)$$

11. Let $x = \sec\theta$. Then

$$\int_1^2 \frac{\sqrt{x^2-1}}{x}\,dx = \int_0^{\pi/3} \frac{\tan\theta}{\sec\theta}\sec\theta\tan\theta\,d\theta = \int_0^{\pi/3}\tan^2\theta\,d\theta = \int_0^{\pi/3}(\sec^2\theta - 1)\,d\theta$$

$$= [\tan\theta - \theta]_0^{\pi/3} = \sqrt{3} - \tfrac{\pi}{3}$$

13. $\displaystyle\int \frac{dx}{x^3+x} = \int\left(\frac{1}{x} - \frac{x}{x^2+1}\right)dx = \ln|x| - \tfrac{1}{2}\ln(x^2+1) + C$

15. $\int \sin^2\theta\cos^5\theta\,d\theta = \int\sin^2\theta\left(\cos^2\theta\right)^2\cos\theta\,d\theta = \int\sin^2\theta\left(1 - \sin^2\theta\right)^2\cos\theta\,d\theta$

$$= \int u^2\left(1 - u^2\right)^2 du \quad [u = \sin\theta,\, du = \cos\theta\,d\theta] \;= \int u^2\left(1 - 2u^2 + u^4\right)du$$

$$= \int\left(u^2 - 2u^4 + u^6\right)du = \tfrac{1}{3}u^3 - \tfrac{2}{5}u^5 + \tfrac{1}{7}u^7 + C = \tfrac{1}{3}\sin^3\theta - \tfrac{2}{5}\sin^5\theta + \tfrac{1}{7}\sin^7\theta + C$$

17. Integrate by parts with $u = x$, $dv = \sec x\tan x\,dx \;\Rightarrow\; du = dx$, $v = \sec x$:

$\int x\sec x\tan x\,dx = x\sec x - \int\sec x\,dx \overset{14}{=} x\sec x - \ln|\sec x + \tan x| + C.$

19. $\displaystyle\int \frac{x+1}{9x^2+6x+5}\,dx = \int\frac{x+1}{(9x^2+6x+1)+4}\,dx = \int\frac{x+1}{(3x+1)^2+4}\,dx \quad\begin{bmatrix}u = 3x+1,\\ du = 3\,dx\end{bmatrix}$

$$= \int\frac{\left[\tfrac{1}{3}(u-1)\right]+1}{u^2+4}\left(\tfrac{1}{3}\,du\right) = \tfrac{1}{3}\cdot\tfrac{1}{3}\int\frac{(u-1)+3}{u^2+4}\,du$$

$$= \tfrac{1}{9}\int\frac{u}{u^2+4}\,du + \tfrac{1}{9}\int\frac{2}{u^2+2^2}\,du = \tfrac{1}{9}\cdot\tfrac{1}{2}\ln(u^2+4) + \tfrac{2}{9}\cdot\tfrac{1}{2}\tan^{-1}\left(\tfrac{1}{2}u\right) + C$$

$$= \tfrac{1}{18}\ln(9x^2+6x+5) + \tfrac{1}{9}\tan^{-1}\left[\tfrac{1}{2}(3x+1)\right] + C$$

21.

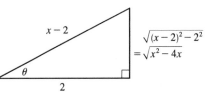

$$\int\frac{dx}{\sqrt{x^2-4x}} = \int\frac{dx}{\sqrt{(x^2-4x+4)-4}} = \int\frac{dx}{\sqrt{(x-2)^2-2^2}}$$

$$= \int\frac{2\sec\theta\tan\theta\,d\theta}{2\tan\theta}\quad\begin{bmatrix}x-2 = 2\sec\theta,\\ dx = 2\sec\theta\tan\theta\,d\theta\end{bmatrix} = \int\sec\theta\,d\theta = \ln|\sec\theta + \tan\theta| + C_1$$

$$= \ln\left|\frac{x-2}{2} + \frac{\sqrt{x^2-4x}}{2}\right| + C_1 = \ln\left|x - 2 + \sqrt{x^2-4x}\right| + C,\text{ where } C = C_1 - \ln 2$$

23. Let $u = \cot 4x$. Then $du = -4\csc^2 4x\,dx \;\Rightarrow$

$$\int\csc^4 4x\,dx = \int\left(\cot^2 4x + 1\right)\csc^2 4x\,dx = \int\left(u^2+1\right)\left(-\tfrac{1}{4}\,du\right)$$

$$= -\tfrac{1}{4}\left(\tfrac{1}{3}u^3 + u\right) + C = -\tfrac{1}{12}\left(\cot^3 4x + 3\cot 4x\right) + C$$

25. $\dfrac{3x^3 - x^2 + 6x - 4}{(x^2 + 1)(x^2 + 2)} = \dfrac{Ax + B}{x^2 + 1} + \dfrac{Cx + D}{x^2 + 2} \quad \Rightarrow$

$3x^3 - x^2 + 6x - 4 = (Ax + B)(x^2 + 2) + (Cx + D)(x^2 + 1)$. Equating the coefficients gives $A + C = 3$,

$B + D = -1$, $2A + C = 6$, and $2B + D = -4 \quad \Rightarrow \quad A = 3, C = 0, B = -3$, and $D = 2$. Now

$$\int \frac{3x^3 - x^2 + 6x - 4}{(x^2 + 1)(x^2 + 2)}\, dx = 3\int \frac{x - 1}{x^2 + 1}\, dx + 2\int \frac{dx}{x^2 + 2}$$

$$= \tfrac{3}{2}\ln(x^2 + 1) - 3\tan^{-1} x + \sqrt{2}\tan^{-1}\left(\tfrac{1}{\sqrt{2}}x\right) + C$$

27. $\int_0^{\pi/2} \cos^3 x \sin 2x\, dx = \int_0^{\pi/2} \cos^3 x\, (2\sin x \cos x)\, dx = \int_0^{\pi/2} 2\cos^4 x \sin x\, dx = \left[-\tfrac{2}{5}\cos^5 x\right]_0^{\pi/2} = \tfrac{2}{5}$

29. The product of an odd function and an even function is an odd function, so $f(x) = x^5 \sec x$ is an odd function.

By Theorem 5.5.7(b), $\int_{-1}^{1} x^5 \sec x\, dx = 0$.

31. Let $u = \sqrt{e^x - 1}$. Then $u^2 = e^x - 1$ and $2u\, du = e^x\, dx$. Also, $e^x + 8 = u^2 + 9$. Thus,

$$\int_0^{\ln 10} \frac{e^x \sqrt{e^x - 1}}{e^x + 8}\, dx = \int_0^3 \frac{u \cdot 2u\, du}{u^2 + 9} = 2\int_0^3 \frac{u^2}{u^2 + 9}\, du = 2\int_0^3 \left(1 - \frac{9}{u^2 + 9}\right) du$$

$$= 2\left[u - \frac{9}{3}\tan^{-1}\left(\frac{u}{3}\right)\right]_0^3 = 2\left[(3 - 3\tan^{-1} 1) - 0\right] = 2\left(3 - 3 \cdot \frac{\pi}{4}\right) = 6 - \frac{3\pi}{2}$$

33. Let $x = 2\sin\theta \quad \Rightarrow \quad (4 - x^2)^{3/2} = (2\cos\theta)^3$, $dx = 2\cos\theta\, d\theta$, so

$$\int \frac{x^2}{(4 - x^2)^{3/2}}\, dx = \int \frac{4\sin^2\theta}{8\cos^3\theta}\, 2\cos\theta\, d\theta$$

$$= \int \tan^2\theta\, d\theta = \int (\sec^2\theta - 1)\, d\theta$$

$$= \tan\theta - \theta + C = \frac{x}{\sqrt{4 - x^2}} - \sin^{-1}\left(\frac{x}{2}\right) + C$$

35. $\displaystyle\int \frac{1}{\sqrt{x + x^{3/2}}}\, dx = \int \frac{dx}{\sqrt{x\,(1 + \sqrt{x}\,)}} = \int \frac{dx}{\sqrt{x}\sqrt{1 + \sqrt{x}}} \quad \begin{bmatrix} u = 1 + \sqrt{x}, \\ du = \dfrac{dx}{2\sqrt{x}} \end{bmatrix} = \int \frac{2\, du}{\sqrt{u}} = \int 2u^{-1/2}\, du$

$$= 4\sqrt{u} + C = 4\sqrt{1 + \sqrt{x}} + C$$

37. $\int (\cos x + \sin x)^2 \cos 2x\, dx = \int (\cos^2 x + 2\sin x \cos x + \sin^2 x) \cos 2x\, dx$

$\quad = \int (1 + \sin 2x) \cos 2x\, dx = \int \cos 2x\, dx + \tfrac{1}{2}\int \sin 4x\, dx = \tfrac{1}{2}\sin 2x - \tfrac{1}{8}\cos 4x + C$

Or: $\int (\cos x + \sin x)^2 \cos 2x\, dx = \int (\cos x + \sin x)^2 (\cos^2 x - \sin^2 x)\, dx$

$\quad = \int (\cos x + \sin x)^3 (\cos x - \sin x)\, dx = \tfrac{1}{4}(\cos x + \sin x)^4 + C_1$

39. We'll integrate $I = \displaystyle\int \frac{xe^{2x}}{(1+2x)^2}\, dx$ by parts with $u = xe^{2x}$ and $dv = \dfrac{dx}{(1+2x)^2}$. Then

$du = (x \cdot 2e^{2x} + e^{2x} \cdot 1)\, dx$ and $v = -\dfrac{1}{2} \cdot \dfrac{1}{1+2x}$, so

$$I = -\frac{1}{2} \cdot \frac{xe^{2x}}{1+2x} - \int \left[-\frac{1}{2} \cdot \frac{e^{2x}(2x+1)}{1+2x} \right] dx = -\frac{xe^{2x}}{4x+2} + \frac{1}{2} \cdot \frac{1}{2} e^{2x} + C$$

$$= e^{2x} \left(\frac{1}{4} - \frac{x}{4x+2} \right) + C.$$

Thus, $\displaystyle\int_0^{1/2} \frac{xe^{2x}}{(1+2x)^2}\, dx = \left[e^{2x} \left(\frac{1}{4} - \frac{x}{4x+2} \right) \right]_0^{1/2} = e\left(\frac{1}{4} - \frac{1}{8} \right) - 1\left(\frac{1}{4} - 0 \right) = \frac{1}{8}e - \frac{1}{4}.$

41. $\displaystyle\int_1^\infty \frac{1}{(2x+1)^3}\, dx = \lim_{t\to\infty} \int_1^t \frac{1}{(2x+1)^3}\, dx = \lim_{t\to\infty} \int_1^t \frac{1}{2}(2x+1)^{-3}\, 2\, dx$

$$= \lim_{t\to\infty} \left[-\frac{1}{4(2x+1)^2} \right]_1^t = -\frac{1}{4} \lim_{t\to\infty} \left[\frac{1}{(2t+1)^2} - \frac{1}{9} \right] = -\frac{1}{4}\left(0 - \frac{1}{9} \right) = \frac{1}{36}$$

43. $\displaystyle\int \frac{dx}{x \ln x}\quad \begin{bmatrix} u = \ln x, \\ du = \dfrac{dx}{x} \end{bmatrix} = \int \frac{du}{u} = \ln|u| + C = \ln|\ln x| + C,$ so

$\displaystyle\int_2^\infty \frac{dx}{x \ln x} = \lim_{t\to\infty} \int_2^t \frac{dx}{x \ln x} = \lim_{t\to\infty} \left[\ln|\ln x| \right]_2^t = \lim_{t\to\infty} [\ln(\ln t) - \ln(\ln 2)] = \infty,$ so the integral is divergent.

45. $\displaystyle\int_0^4 \frac{\ln x}{\sqrt{x}}\, dx = \lim_{t\to 0^+} \int_t^4 \frac{\ln x}{\sqrt{x}}\, dx \overset{*}{=} \lim_{t\to 0^+} \left[2\sqrt{x} \ln x - 4\sqrt{x} \right]_t^4$

$$= \lim_{t\to 0^+} \left[(2 \cdot 2\ln 4 - 4 \cdot 2) - (2\sqrt{t}\ln t - 4\sqrt{t}) \right] \overset{**}{=} (4\ln 4 - 8) - (0 - 0) = 4\ln 4 - 8$$

(∗)   Let $u = \ln x,\ dv = \dfrac{1}{\sqrt{x}}\, dx \Rightarrow du = \dfrac{1}{x}\, dx,\ v = 2\sqrt{x}$. Then

$$\int \frac{\ln x}{\sqrt{x}}\, dx = 2\sqrt{x} \ln x - 2 \int \frac{dx}{\sqrt{x}} = 2\sqrt{x} \ln x - 4\sqrt{x} + C$$

(∗∗) $\displaystyle\lim_{t\to 0^+} (2\sqrt{t} \ln t) = \lim_{t\to 0^+} \frac{2\ln t}{t^{-1/2}} \overset{H}{=} \lim_{t\to 0^+} \frac{2/t}{-\frac{1}{2}t^{-3/2}} = \lim_{t\to 0^+} (-4\sqrt{t}) = 0$

47. $\displaystyle\int_0^3 \frac{dx}{x^2 - x - 2} = \int_0^3 \frac{dx}{(x+1)(x-2)} = \int_0^2 \frac{dx}{(x+1)(x-2)} + \int_2^3 \frac{dx}{(x+1)(x-2)},$ and

$$\int_2^3 \frac{dx}{x^2 - x - 2} = \lim_{t\to 2^+} \int_t^3 \left[\frac{-1/3}{x+1} + \frac{1/3}{x-2} \right] dx = \lim_{t\to 2^+} \left[\frac{1}{3} \ln \left| \frac{x-2}{x+1} \right| \right]_t^3$$

$$= \lim_{t\to 2^+} \left[\frac{1}{3} \ln \frac{1}{4} - \frac{1}{3} \ln \left| \frac{t-2}{t+1} \right| \right] = \infty$$

so $\displaystyle\int_0^3 \frac{dx}{x^2 - x - 2}$ diverges.

49. Let $u = 2x + 1$. Then

$$\int_{-\infty}^{\infty} \frac{dx}{4x^2 + 4x + 5} = \int_{-\infty}^{\infty} \frac{\frac{1}{2}\,du}{u^2 + 4} = \frac{1}{2}\int_{-\infty}^{0} \frac{du}{u^2 + 4} + \frac{1}{2}\int_{0}^{\infty} \frac{du}{u^2 + 4}$$

$$= \tfrac{1}{2}\lim_{t\to-\infty}\left[\tfrac{1}{2}\tan^{-1}\left(\tfrac{1}{2}u\right)\right]_{t}^{0} + \tfrac{1}{2}\lim_{t\to\infty}\left[\tfrac{1}{2}\tan^{-1}\left(\tfrac{1}{2}u\right)\right]_{0}^{t}$$

$$= \tfrac{1}{4}\left[0 - \left(-\tfrac{\pi}{2}\right)\right] + \tfrac{1}{4}\left[\tfrac{\pi}{2} - 0\right] = \tfrac{\pi}{4}$$

51. We first make the substitution $t = x + 1$, so $\ln(x^2 + 2x + 2) = \ln\left[(x+1)^2 + 1\right] = \ln(t^2 + 1)$. Then we use parts with $u = \ln(t^2 + 1)$, $dv = dt$:

$$\int \ln(t^2 + 1)\,dt = t\,\ln(t^2 + 1) - \int \frac{t(2t)\,dt}{t^2 + 1} = t\,\ln(t^2 + 1) - 2\int \frac{t^2\,dt}{t^2 + 1}$$

$$= t\,\ln(t^2 + 1) - 2\int\left(1 - \frac{1}{t^2 + 1}\right)dt = t\,\ln(t^2 + 1) - 2t + 2\arctan t + C$$

$$= (x + 1)\,\ln(x^2 + 2x + 2) - 2x + 2\arctan(x + 1) + K, \quad \text{where } K = C - 2$$

[Alternatively, we could have integrated by parts immediately with $u = \ln(x^2 + 2x + 2)$.] Notice from the graph that $f = 0$ where F has a horizontal tangent. Also, F is always increasing, and $f \geq 0$.

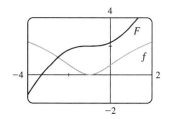

53. From the graph, it seems as though $\int_{0}^{2\pi} \cos^2 x \sin^3 x\,dx$ is equal to 0.
To evaluate the integral, we write the integral as

$I = \int_{0}^{2\pi} \cos^2 x\,(1 - \cos^2 x) \sin x\,dx$ and let $u = \cos x$ $\Rightarrow$

$du = -\sin x\,dx$. Thus, $I = \int_{1}^{1} u^2(1 - u^2)(-du) = 0$.

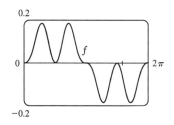

55. $u = e^x$ $\Rightarrow$ $du = e^x\,dx$, so

$$\int e^x\sqrt{1 - e^{2x}}\,dx = \int \sqrt{1 - u^2}\,du \stackrel{30}{=} \tfrac{1}{2}u\sqrt{1 - u^2} + \tfrac{1}{2}\sin^{-1}u + C = \tfrac{1}{2}\left[e^x\sqrt{1 - e^{2x}} + \sin^{-1}(e^x)\right] + C$$

57. $\displaystyle\int \sqrt{x^2 + x + 1}\,dx = \int \sqrt{x^2 + x + \tfrac{1}{4} + \tfrac{3}{4}}\,dx = \int \sqrt{\left(x + \tfrac{1}{2}\right)^2 + \tfrac{3}{4}}\,dx$

$$= \int \sqrt{u^2 + \left(\tfrac{\sqrt{3}}{2}\right)^2}\,du \quad [u = x + \tfrac{1}{2},\, du = dx]$$

$$\stackrel{21}{=} \tfrac{1}{2}u\sqrt{u^2 + \tfrac{3}{4}} + \tfrac{3}{8}\ln\left(u + \sqrt{u^2 + \tfrac{3}{4}}\right) + C$$

$$= \frac{2x + 1}{4}\sqrt{x^2 + x + 1} + \tfrac{3}{8}\ln\left(x + \tfrac{1}{2} + \sqrt{x^2 + x + 1}\right) + C$$

59. (a) $\dfrac{d}{du}\left[-\dfrac{1}{u}\sqrt{a^2-u^2}-\sin^{-1}\left(\dfrac{u}{a}\right)+C\right]=\dfrac{1}{u^2}\sqrt{a^2-u^2}+\dfrac{1}{\sqrt{a^2-u^2}}-\dfrac{1}{\sqrt{1-u^2/a^2}}\cdot\dfrac{1}{a}$

$=\left(a^2-u^2\right)^{-1/2}\left[\dfrac{1}{u^2}\left(a^2-u^2\right)+1-1\right]=\dfrac{\sqrt{a^2-u^2}}{u^2}$

(b) Let $u=a\sin\theta\;\Rightarrow\;du=a\cos\theta\,d\theta,\,a^2-u^2=a^2\left(1-\sin^2\theta\right)=a^2\cos^2\theta.$

$\displaystyle\int\dfrac{\sqrt{a^2-u^2}}{u^2}\,du=\int\dfrac{a^2\cos^2\theta}{a^2\sin^2\theta}\,d\theta=\int\dfrac{1-\sin^2\theta}{\sin^2\theta}\,d\theta=\int\left(\csc^2\theta-1\right)d\theta=-\cot\theta-\theta+C$

$=-\dfrac{\sqrt{a^2-u^2}}{u}-\sin^{-1}\left(\dfrac{u}{a}\right)+C$

61. For $n\geq0$, $\int_0^\infty x^n\,dx=\lim\limits_{t\to\infty}\left[x^{n+1}/(n+1)\right]_0^t=\infty$. For $n<0$, $\int_0^\infty x^n\,dx=\int_0^1 x^n\,dx+\int_1^\infty x^n\,dx$. Both

integrals are improper. By (7.8.2), the second integral diverges if $-1\leq n<0$. By Exercise 7.8.57, the first integral

diverges if $n\leq-1$. Thus, $\int_0^\infty x^n\,dx$ is divergent for all values of n.

63. $f(x)=\sqrt{1+x^4},\;\Delta x=\dfrac{b-a}{n}=\dfrac{1-0}{10}=\dfrac{1}{10}.$

(a) $T_{10}=\dfrac{1}{10\cdot2}\{f(0)+2\left[f(0.1)+f(0.2)+\cdots+f(0.9)\right]+f(1)\}\approx1.090608$

(b) $M_{10}=\dfrac{1}{10}\left[f\left(\dfrac{1}{20}\right)+f\left(\dfrac{3}{20}\right)+f\left(\dfrac{5}{20}\right)+\cdots+f\left(\dfrac{19}{20}\right)\right]\approx1.088840$

(c) $S_{10}=\dfrac{1}{10\cdot3}\left[f(0)+4f(0.1)+2f(0.2)+\cdots+4f(0.9)+f(1)\right]\approx1.089429$

f is concave upward, so the Trapezoidal Rule gives us an overestimate, the Midpoint Rule gives an underestimate,

and we cannot tell whether Simpson's Rule gives us an overestimate or an underestimate.

65. $f(x)=\left(1+x^4\right)^{1/2},\,f'(x)=\dfrac{1}{2}\left(1+x^4\right)^{-1/2}\left(4x^3\right)=2x^3\left(1+x^4\right)^{-1/2},\,f''(x)=\left(2x^6+6x^2\right)\left(1+x^4\right)^{-3/2}.$

A graph of f'' on $[0,1]$ shows that it has its maximum at $x=1$, so $|f''(x)|\leq f''(1)=\sqrt{8}$ on $[0,1]$. By taking

$K=\sqrt{8}$, we find that the error in Exercise 63(a) is bounded by $\dfrac{K(b-a)^3}{12n^2}=\dfrac{\sqrt{8}}{1200}\approx0.0024$, and in (b) by about

$\dfrac{1}{2}(0.0024)=0.0012.$

Note: Another way to estimate K is to let $x=1$ in the factor $2x^6+6x^2$ (maximizing the numerator) and let $x=0$

in the factor $\left(1+x^4\right)^{-3/2}$ (minimizing the denominator). Doing so gives us $K=8$ and errors of $0.00\overline{6}$ and $0.00\overline{3}$.

Using $K=8$ for the Trapezoidal Rule, we have $|E_T|\leq\dfrac{K(b-a)^3}{12n^2}\leq0.00001\;\Leftrightarrow\;\dfrac{8(1-0)^3}{12n^2}\leq\dfrac{1}{100,000}$

$\Leftrightarrow\;n^2\geq\dfrac{800,000}{12}\;\Leftrightarrow\;n\gtrsim258.2$, so we should take $n=259$.

For the Midpoint Rule, $|E_M|\leq\dfrac{K(b-a)^3}{24n^2}\leq0.00001\;\Leftrightarrow\;n^2\geq\dfrac{800,000}{24}\;\Leftrightarrow\;n\gtrsim182.6$, so we should

take $n=183$.

67. $\Delta t = \left(\frac{10}{60} - 0\right)/10 = \frac{1}{60}$.

Distance traveled $= \int_0^{10} v\, dt \approx S_{10}$

$$= \frac{1}{60 \cdot 3}[40 + 4(42) + 2(45) + 4(49) + 2(52) + 4(54) + 2(56) + 4(57) + 2(57) + 4(55) + 56]$$

$$= \frac{1}{180}(1544) = 8.5\overline{7} \text{ mi}$$

69. (a) $f(x) = \sin(\sin x)$. A CAS gives

$$f^{(4)}(x) = \sin(\sin x)\left[\cos^4 x + 7\cos^2 x - 3\right]$$

$$+ \cos(\sin x)\left[6\cos^2 x \sin x + \sin x\right]$$

From the graph, we see that $\left|f^{(4)}(x)\right| < 3.8$ for $x \in [0, \pi]$.

(b) We use Simpson's Rule with $f(x) = \sin(\sin x)$ and $\Delta x = \frac{\pi}{10}$:

$$\int_0^\pi f(x)\, dx \approx \frac{\pi}{10 \cdot 3}\left[f(0) + 4f\left(\frac{\pi}{10}\right) + 2f\left(\frac{2\pi}{10}\right) + \cdots + 4f\left(\frac{9\pi}{10}\right) + f(\pi)\right] \approx 1.786721$$

From part (a), we know that $\left|f^{(4)}(x)\right| < 3.8$ on $[0, \pi]$, so we use Theorem 7.7.4 with $K = 3.8$, and estimate the error as $|E_S| \leq \dfrac{3.8(\pi - 0)^5}{180(10)^4} \approx 0.000646$.

(c) If we want the error to be less than 0.00001, we must have $|E_S| \leq \frac{3.8\pi^5}{180n^4} \leq 0.00001$, so

$$n^4 \geq \frac{3.8\pi^5}{180(0.00001)} \approx 646{,}041.6 \quad \Rightarrow \quad n \geq 28.35. \text{ Since } n \text{ must be even for Simpson's Rule, we must have}$$

$n \geq 30$ to ensure the desired accuracy.

71. $\dfrac{x^3}{x^5 + 2} \leq \dfrac{x^3}{x^5} = \dfrac{1}{x^2}$ for x in $[1, \infty)$. $\displaystyle\int_1^\infty \frac{1}{x^2}\, dx$ is convergent by (7.8.2) with $p = 2 > 1$. Therefore,

$\displaystyle\int_1^\infty \frac{x^3}{x^5 + 2}\, dx$ is convergent by the Comparison Theorem.

73. For x in $\left[0, \frac{\pi}{2}\right]$, $0 \leq \cos^2 x \leq \cos x$. For x in $\left[\frac{\pi}{2}, \pi\right]$, $\cos x \leq 0 \leq \cos^2 x$. Thus,

$$\text{area} = \int_0^{\pi/2}\left(\cos x - \cos^2 x\right) dx + \int_{\pi/2}^\pi \left(\cos^2 x - \cos x\right) dx$$

$$= \left[\sin x - \tfrac{1}{2}x - \tfrac{1}{4}\sin 2x\right]_0^{\pi/2} + \left[\tfrac{1}{2}x + \tfrac{1}{4}\sin 2x - \sin x\right]_{\pi/2}^\pi$$

$$= \left[\left(1 - \tfrac{\pi}{4}\right) - 0\right] + \left[\tfrac{\pi}{2} - \left(\tfrac{\pi}{4} - 1\right)\right] = 2$$

75. Using the formula for disks, the volume is

$$V = \int_0^{\pi/2} \pi\, [f(x)]^2\, dx = \pi \int_0^{\pi/2}\left(\cos^2 x\right)^2 dx = \pi \int_0^{\pi/2}\left[\tfrac{1}{2}(1 + \cos 2x)\right]^2 dx$$

$$= \tfrac{\pi}{4}\int_0^{\pi/2}(1 + \cos^2 2x + 2\cos 2x)\, dx = \tfrac{\pi}{4}\int_0^{\pi/2}\left[1 + \tfrac{1}{2}(1 + \cos 4x) + 2\cos 2x\right] dx$$

$$= \tfrac{\pi}{4}\left[\tfrac{3}{2}x + \tfrac{1}{2}\left(\tfrac{1}{4}\sin 4x\right) + 2\left(\tfrac{1}{2}\sin 2x\right)\right]_0^{\pi/2} = \tfrac{\pi}{4}\left[\left(\tfrac{3\pi}{4} + \tfrac{1}{8}\cdot 0 + 0\right) - 0\right] = \tfrac{3\pi^2}{16}$$

77. By the Fundamental Theorem of Calculus,

$$\int_0^\infty f'(x)\,dx = \lim_{t\to\infty} \int_0^t f'(x)\,dx = \lim_{t\to\infty} [f(t) - f(0)] = \lim_{t\to\infty} f(t) - f(0) = 0 - f(0) = -f(0).$$

79. Let $u = 1/x \;\Rightarrow\; x = 1/u \;\Rightarrow\; dx = -\left(1/u^2\right) du.$

$$\int_0^\infty \frac{\ln x}{1+x^2}\,dx = \int_\infty^0 \frac{\ln (1/u)}{1+1/u^2}\left(-\frac{du}{u^2}\right) = \int_\infty^0 \frac{-\ln u}{u^2+1}\,(-du) = \int_\infty^0 \frac{\ln u}{1+u^2}\,du = -\int_0^\infty \frac{\ln u}{1+u^2}\,du$$

Therefore, $\displaystyle\int_0^\infty \frac{\ln x}{1+x^2}\,dx = -\int_0^\infty \frac{\ln x}{1+x^2}\,dx = 0.$

☐ PROBLEMS PLUS

1.

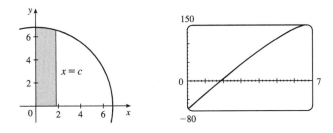

By symmetry, the problem can be reduced to finding the line $x = c$ such that the shaded area is one-third of the area

of the quarter-circle. The equation of the circle is $y = \sqrt{49 - x^2}$, so we require that $\int_0^c \sqrt{49 - x^2}\, dx = \frac{1}{3} \cdot \frac{1}{4}\pi(7)^2$

$\Leftrightarrow \ \left[\frac{1}{2}x\sqrt{49 - x^2} + \frac{49}{2}\sin^{-1}(x/7)\right]_0^c = \frac{49}{12}\pi$ [by Formula 30] $\Leftrightarrow \ \frac{1}{2}c\sqrt{49 - c^2} + \frac{49}{2}\sin^{-1}(c/7) = \frac{49}{12}\pi.$

This equation would be difficult to solve exactly, so we plot the left-hand side as a function of c, and find that the

equation holds for $c \approx 1.85$. So the cuts should be made at distances of about 1.85 inches from the center of the

pizza.

3. The given integral represents the difference of the shaded areas, which appears

to be 0. It can be calculated by integrating with respect to either x or y, so we

find x in terms of y for each curve: $y = \sqrt[3]{1 - x^7} \ \Rightarrow \ x = \sqrt[7]{1 - y^3}$ and

$y = \sqrt[7]{1 - x^3} \ \Rightarrow \ x = \sqrt[3]{1 - y^7}$, so

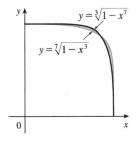

$\int_0^1 \left(\sqrt[3]{1 - y^7} - \sqrt[7]{1 - y^3}\right) dy = \int_0^1 \left(\sqrt[7]{1 - x^3} - \sqrt[3]{1 - x^7}\right) dx.$ But this

equation is of the form $z = -z$. So $\int_0^1 \left(\sqrt[3]{1 - x^7} - \sqrt[7]{1 - x^3}\right) dx = 0.$

5. Recall that $\cos A \cos B = \frac{1}{2}[\cos(A + B) + \cos(A - B)]$. So

$$f(x) = \int_0^\pi \cos t \cos(x - t)\, dt = \frac{1}{2}\int_0^\pi [\cos(t + x - t) + \cos(t - x + t)]\, dt$$

$$= \frac{1}{2}\int_0^\pi [\cos x + \cos(2t - x)]\, dt = \frac{1}{2}\left[t\cos x + \frac{1}{2}\sin(2t - x)\right]_0^\pi$$

$$= \frac{\pi}{2}\cos x + \frac{1}{4}\sin(2\pi - x) - \frac{1}{4}\sin(-x) = \frac{\pi}{2}\cos x + \frac{1}{4}\sin(-x) - \frac{1}{4}\sin(-x)$$

$$= \frac{\pi}{2}\cos x$$

The minimum of $\cos x$ on this domain is -1, so the minimum value of $f(x)$ is $f(\pi) = -\frac{\pi}{2}$.

7. In accordance with the hint, we let $I_k = \int_0^1 (1 - x^2)^k \, dx$, and we find an expression for I_{k+1} in terms of I_k. We integrate I_{k+1} by parts with $u = (1 - x^2)^{k+1}$ $\Rightarrow$ $du = (k+1)(1 - x^2)^k(-2x)$, $dv = dx$ $\Rightarrow$ $v = x$, and then split the remaining integral into identifiable quantities:

$$I_{k+1} = x(1 - x^2)^{k+1}\Big|_0^1 + 2(k+1) \int_0^1 x^2 (1 - x^2)^k \, dx = (2k+2) \int_0^1 (1 - x^2)^k \big[1 - (1 - x^2)\big] \, dx$$

$$= (2k+2)(I_k - I_{k+1})$$

So $I_{k+1}[1 + (2k+2)] = (2k+2)I_k$ $\Rightarrow$ $I_{k+1} = \frac{2k+2}{2k+3} I_k$. Now to complete the proof, we use induction:

$I_0 = 1 = \dfrac{2^0 (0!)^2}{1!}$, so the formula holds for $n = 0$. Now suppose it holds for $n = k$. Then

$$I_{k+1} = \frac{2k+2}{2k+3} I_k = \frac{2k+2}{2k+3} \left[\frac{2^{2k}(k!)^2}{(2k+1)!} \right] = \frac{2(k+1)2^{2k}(k!)^2}{(2k+3)(2k+1)!} = \frac{2(k+1)}{2k+2} \cdot \frac{2(k+1)2^{2k}(k!)^2}{(2k+3)(2k+1)!}$$

$$= \frac{[2(k+1)]^2 \, 2^{2k}(k!)^2}{(2k+3)(2k+2)(2k+1)!} = \frac{2^{2(k+1)} \, [(k+1)!]^2}{[2(k+1)+1]!}$$

So by induction, the formula holds for all integers $n \geq 0$.

9. $0 < a < b$. Now

$$\int_0^1 [bx + a(1 - x)]^t \, dx = \int_a^b \frac{u^t}{(b-a)} \, du \quad [\text{put } u = bx + a(1 - x)] \quad = \left[\frac{u^{t+1}}{(t+1)(b-a)} \right]_a^b = \frac{b^{t+1} - a^{t+1}}{(t+1)(b-a)}.$$

Now let $y = \lim_{t \to 0} \left[\frac{b^{t+1} - a^{t+1}}{(t+1)(b-a)} \right]^{1/t}$. Then $\ln y = \lim_{t \to 0} \left[\frac{1}{t} \ln \frac{b^{t+1} - a^{t+1}}{(t+1)(b-a)} \right]$. This limit is of the form $0/0$,

so we can apply l'Hospital's Rule to get

$$\ln y = \lim_{t \to 0} \left[\frac{b^{t+1} \ln b - a^{t+1} \ln a}{b^{t+1} - a^{t+1}} - \frac{1}{t+1} \right] = \frac{b \ln b - a \ln a}{b - a} - 1 = \frac{b \ln b}{b - a} - \frac{a \ln a}{b - a} - \ln e = \ln \frac{b^{b/(b-a)}}{ea^{a/(b-a)}}.$$

Therefore, $y = e^{-1} \left(\dfrac{b^b}{a^a} \right)^{1/(b-a)}$.

11.

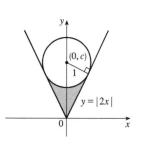

An equation of the circle with center $(0, c)$ and radius 1 is $x^2 + (y - c)^2 = 1^2$, so an equation of the lower semicircle is $y = c - \sqrt{1 - x^2}$. At the points of tangency, the slopes of the line and semicircle must be equal. For $x \geq 0$, we must have

$$y' = 2 \quad \Rightarrow \quad \frac{x}{\sqrt{1 - x^2}} = 2 \quad \Rightarrow \quad x = 2\sqrt{1 - x^2} \quad \Rightarrow$$

$$x^2 = 4(1 - x^2) \quad \Rightarrow \quad 5x^2 = 4 \quad \Rightarrow \quad x^2 = \tfrac{4}{5} \quad \Rightarrow \quad x = \tfrac{2}{5}\sqrt{5}$$

and so $y = 2\left(\tfrac{2}{5}\sqrt{5}\right) = \tfrac{4}{5}\sqrt{5}$. The slope of the perpendicular line segment is $-\tfrac{1}{2}$, so an equation of the line segment is $y - \frac{4}{5}\sqrt{5} = -\frac{1}{2}\left(x - \frac{2}{5}\sqrt{5}\right)$ $\Leftrightarrow$ $y = -\frac{1}{2}x + \frac{1}{5}\sqrt{5} + \frac{4}{5}\sqrt{5}$ $\Leftrightarrow$ $y = -\frac{1}{2}x + \sqrt{5}$, so $c = \sqrt{5}$ and

an equation of the lower semicircle is $y = \sqrt{5} - \sqrt{1 - x^2}$. Thus, the shaded area is

$$2\int_0^{(2/5)\sqrt{5}} \left[\left(\sqrt{5} - \sqrt{1 - x^2}\right) - 2x\right] dx \overset{30}{=} 2\left[\sqrt{5}\,x - \frac{x}{2}\sqrt{1 - x^2} - \frac{1}{2}\sin^{-1}x - x^2\right]_0^{(2/5)\sqrt{5}}$$

$$= 2\left[2 - \frac{\sqrt{5}}{5}\cdot\frac{1}{\sqrt{5}} - \frac{1}{2}\sin^{-1}\left(\frac{2}{\sqrt{5}}\right) - \frac{4}{5}\right] - 2(0)$$

$$= 2\left[1 - \frac{1}{2}\sin^{-1}\left(\frac{2}{\sqrt{5}}\right)\right] = 2 - \sin^{-1}\left(\frac{2}{\sqrt{5}}\right)$$

13. We integrate by parts with $u = \dfrac{1}{\ln(1 + x + t)}$, $dv = \sin t\,dt$, so $du = \dfrac{-1}{(1 + x + t)[\ln(1 + x + t)]^2}$ and

$v = -\cos t$. The integral becomes

$$I = \int_0^\infty \frac{\sin t\,dt}{\ln(1 + x + t)} = \lim_{b\to\infty}\left(\left[\frac{-\cos t}{\ln(1 + x + t)}\right]_0^b - \int_0^b \frac{\cos t\,dt}{(1 + x + t)[\ln(1 + x + t)]^2}\right)$$

$$= \lim_{b\to\infty}\frac{-\cos b}{\ln(1 + x + b)} + \frac{1}{\ln(1 + x)} + \int_0^\infty \frac{-\cos t\,dt}{(1 + x + t)[\ln(1 + x + t)]^2} = \frac{1}{\ln(1 + x)} + J$$

where $J = \displaystyle\int_0^\infty \frac{-\cos t\,dt}{(1 + x + t)[\ln(1 + x + t)]^2}$. Now $-1 \le -\cos t \le 1$ for all t; in fact, the inequality is strict

except at isolated points. So $-\displaystyle\int_0^\infty \frac{dt}{(1 + x + t)[\ln(1 + x + t)]^2} < J < \int_0^\infty \frac{dt}{(1 + x + t)[\ln(1 + x + t)]^2}$ $\Leftrightarrow$

$-\dfrac{1}{\ln(1 + x)} < J < \dfrac{1}{\ln(1 + x)}$ $\Leftrightarrow$ $0 < I < \dfrac{2}{\ln(1 + x)}$.

8 ☐ FURTHER APPLICATIONS OF INTEGRATION

8.1 Arc Length

1. $y = 2 - 3x \implies L = \int_{-2}^{1} \sqrt{1 + (dy/dx)^2}\, dx = \int_{-2}^{1} \sqrt{1 + (-3)^2}\, dx = \sqrt{10}\, [1 - (-2)] = 3\sqrt{10}.$

The arc length can be calculated using the distance formula, since the curve is a line segment, so

$$L = [\text{distance from } (-2, 8) \text{ to } (1, -1)] = \sqrt{[1 - (-2)]^2 + [(-1) - 8]^2} = \sqrt{90} = 3\sqrt{10}$$

3.

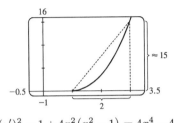

From the figure, the length of the curve is slightly larger than the hypotenuse of the triangle formed by the points $(1, 0)$, $(3, 0)$, and $(3, f(3)) \approx (3, 15)$, where $y = f(x) = \frac{2}{3}(x^2 - 1)^{3/2}$. This length is about $\sqrt{15^2 + 2^2} \approx 15$, so we might estimate the length to be 15.5. $y = \frac{2}{3}(x^2 - 1)^{3/2} \implies y' = (x^2 - 1)^{1/2}(2x) \implies$

$1 + (y')^2 = 1 + 4x^2(x^2 - 1) = 4x^4 - 4x^2 + 1 = (2x^2 - 1)^2$, so, using the fact that $2x^2 - 1 > 0$ for $1 \le x \le 3$,

$$L = \int_1^3 \sqrt{(2x^2 - 1)^2}\, dx = \int_1^3 |2x^2 - 1|\, dx = \int_1^3 (2x^2 - 1)\, dx = \left[\tfrac{2}{3}x^3 - x \right]_1^3$$

$$= (18 - 3) - \left(\tfrac{2}{3} - 1 \right) = \tfrac{46}{3} = 15.\overline{3}$$

5. $y = 1 + 6x^{3/2} \implies dy/dx = 9x^{1/2} \implies 1 + (dy/dx)^2 = 1 + 81x.$ So

$L = \int_0^1 \sqrt{1 + 81x}\, dx = \int_1^{82} u^{1/2} \left(\tfrac{1}{81}\, du \right)$ [where $u = 1 + 81x$ and $du = 81\, dx$]

$$= \tfrac{1}{81} \cdot \tfrac{2}{3} \left[u^{3/2} \right]_1^{82} = \tfrac{2}{243} \left(82\sqrt{82} - 1 \right)$$

7. $y = \dfrac{x^5}{6} + \dfrac{1}{10x^3} \implies \dfrac{dy}{dx} = \dfrac{5}{6}x^4 - \dfrac{3}{10}x^{-4} \implies$

$1 + (dy/dx)^2 = 1 + \tfrac{25}{36}x^8 - \tfrac{1}{2} + \tfrac{9}{100}x^{-8} = \tfrac{25}{36}x^8 + \tfrac{1}{2} + \tfrac{9}{100}x^{-8} = \left(\tfrac{5}{6}x^4 + \tfrac{3}{10}x^{-4} \right)^2.$ So

$$L = \int_1^2 \sqrt{\left(\tfrac{5}{6}x^4 + \tfrac{3}{10}x^{-4} \right)^2}\, dx = \int_1^2 \left(\tfrac{5}{6}x^4 + \tfrac{3}{10}x^{-4} \right) dx = \left[\tfrac{1}{6}x^5 - \tfrac{1}{10}x^{-3} \right]_1^2$$

$$= \left(\tfrac{32}{6} - \tfrac{1}{80} \right) - \left(\tfrac{1}{6} - \tfrac{1}{10} \right) = \tfrac{31}{6} + \tfrac{7}{80} = \tfrac{1261}{240}$$

9. $x = \tfrac{1}{3}\sqrt{y}\,(y - 3) = \tfrac{1}{3}y^{3/2} - y^{1/2} \implies dx/dy = \tfrac{1}{2}y^{1/2} - \tfrac{1}{2}y^{-1/2} \implies$

$1 + (dx/dy)^2 = 1 + \tfrac{1}{4}y - \tfrac{1}{2} + \tfrac{1}{4}y^{-1} = \tfrac{1}{4}y + \tfrac{1}{2} + \tfrac{1}{4}y^{-1} = \left(\tfrac{1}{2}y^{1/2} + \tfrac{1}{2}y^{-1/2} \right)^2.$ So

$L = \int_1^9 \left(\tfrac{1}{2}y^{1/2} + \tfrac{1}{2}y^{-1/2} \right) dy = \tfrac{1}{2} \left[\tfrac{2}{3}y^{3/2} + 2y^{1/2} \right]_1^9 = \tfrac{1}{2} \left[\left(\tfrac{2}{3} \cdot 27 + 2 \cdot 3 \right) - \left(\tfrac{2}{3} \cdot 1 + 2 \cdot 1 \right) \right]$

$$= \tfrac{1}{2} \left(24 - \tfrac{8}{3} \right) = \tfrac{1}{2} \left(\tfrac{64}{3} \right) = \tfrac{32}{3}$$

11. $y = \ln(\sec x) \implies \dfrac{dy}{dx} = \dfrac{\sec x \tan x}{\sec x} = \tan x \implies 1 + \left(\dfrac{dy}{dx} \right)^2 = 1 + \tan^2 x = \sec^2 x$, so

$L = \int_0^{\pi/4} \sqrt{\sec^2 x}\, dx = \int_0^{\pi/4} |\sec x|\, dx = \int_0^{\pi/4} \sec x\, dx = \left[\ln(\sec x + \tan x) \right]_0^{\pi/4}$

$$= \ln(\sqrt{2} + 1) - \ln(1 + 0) = \ln(\sqrt{2} + 1)$$

13. $y = \cosh x \Rightarrow y' = \sinh x \Rightarrow 1 + (y')^2 = 1 + \sinh^2 x = \cosh^2 x.$

So $L = \int_0^1 \cosh x \, dx = [\sinh x]_0^1 = \sinh 1 = \frac{1}{2}(e - 1/e).$

15. $y = e^x \Rightarrow y' = e^x \Rightarrow 1 + (y')^2 = 1 + e^{2x}.$ So

$$L = \int_0^1 \sqrt{1 + e^{2x}} \, dx = \int_1^e \sqrt{1 + u^2} \frac{du}{u} \qquad [u = e^x, \text{ so } x = \ln u, \, dx = du/u]$$

$$= \int_1^e \frac{\sqrt{1 + u^2}}{u^2} u \, du = \int_{\sqrt{2}}^{\sqrt{1+e^2}} \frac{v}{v^2 - 1} v \, dv \qquad \left[v = \sqrt{1 + u^2}, \text{ so } v^2 = 1 + u^2, \, v \, dv = u \, du\right]$$

$$= \int_{\sqrt{2}}^{\sqrt{1+e^2}} \left(1 + \frac{1/2}{v - 1} - \frac{1/2}{v + 1}\right) dv = \left[v + \frac{1}{2} \ln \frac{v - 1}{v + 1}\right]_{\sqrt{2}}^{\sqrt{1+e^2}}$$

$$= \sqrt{1 + e^2} + \frac{1}{2} \ln \frac{\sqrt{1 + e^2} - 1}{\sqrt{1 + e^2} + 1} - \sqrt{2} - \frac{1}{2} \ln \frac{\sqrt{2} - 1}{\sqrt{2} + 1}$$

$$= \sqrt{1 + e^2} - \sqrt{2} + \ln\left(\sqrt{1 + e^2} - 1\right) - 1 - \ln\left(\sqrt{2} - 1\right)$$

Or: Use Formula 23 for $\int \left(\sqrt{1 + u^2}/u\right) du$, or substitute $u = \tan \theta$.

17. $y = \cos x \Rightarrow dy/dx = -\sin x \Rightarrow 1 + (dy/dx)^2 = 1 + \sin^2 x.$ So $L = \int_0^{2\pi} \sqrt{1 + \sin^2 x} \, dx.$

19. $x = y + y^3 \Rightarrow dx/dy = 1 + 3y^2 \Rightarrow 1 + (dx/dy)^2 = 1 + (1 + 3y^2)^2 = 9y^4 + 6y^2 + 2.$

So $L = \int_1^4 \sqrt{9y^4 + 6y^2 + 2} \, dy.$

21. $y = xe^{-x} \Rightarrow dy/dx = e^{-x} - xe^{-x} = e^{-x}(1 - x) \Rightarrow 1 + (dy/dx)^2 = 1 + e^{-2x}(1 - x)^2.$ Let

$f(x) = \sqrt{1 + (dy/dx)^2} = \sqrt{1 + e^{-2x}(1 - x)^2}.$ Then $L = \int_0^5 f(x) \, dx.$ Since $n = 10, \Delta x = \frac{5 - 0}{10} = \frac{1}{2}.$ Now

$$L \approx S_{10} = \frac{1/2}{3}[f(0) + 4f\left(\tfrac{1}{2}\right) + 2f(1) + 4f\left(\tfrac{3}{2}\right) + 2f(2) + 4f\left(\tfrac{5}{2}\right) + 2f(3)$$
$$+ 4f\left(\tfrac{7}{2}\right) + 2f(4) + 4f\left(\tfrac{9}{2}\right) + f(5)] \approx 5.115840$$

The value of the integral produced by a calculator is 5.113568 (to six decimal places).

23. $y = \sec x \Rightarrow dy/dx = \sec x \tan x \Rightarrow L = \int_0^{\pi/3} f(x) \, dx,$ where $f(x) = \sqrt{1 + \sec^2 x \tan^2 x}.$

Since $n = 10, \Delta x = \dfrac{\pi/3 - 0}{10} = \dfrac{\pi}{30}.$ Now

$$L \approx S_{10} = \frac{\pi/30}{3}\left[f(0) + 4f\left(\frac{\pi}{30}\right) + 2f\left(\frac{2\pi}{30}\right) + 4f\left(\frac{3\pi}{30}\right) + 2f\left(\frac{4\pi}{30}\right) + 4f\left(\frac{5\pi}{30}\right)\right.$$
$$\left. + 2f\left(\frac{6\pi}{30}\right) + 4f\left(\frac{7\pi}{30}\right) + 2f\left(\frac{8\pi}{30}\right) + 4f\left(\frac{9\pi}{30}\right) + f\left(\frac{\pi}{3}\right)\right] \approx 1.569619.$$

The value of the integral produced by a calculator is 1.569259 (to six decimal places).

25. (a)

(b)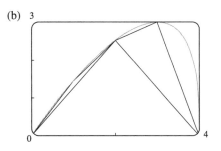

Let $f(x) = y = x\sqrt[3]{4-x}$. The polygon with one side is just the line segment joining the points $(0, f(0)) = (0,0)$ and $(4, f(4)) = (4,0)$, and its length is 4. The polygon with two sides joins the points $(0,0)$, $(2, f(2)) = (2, 2\sqrt[3]{2})$ and $(4,0)$.

Its length is

$$\sqrt{(2-0)^2 + \left(2\sqrt[3]{2} - 0\right)^2} + \sqrt{(4-2)^2 + \left(0 - 2\sqrt[3]{2}\right)^2} = 2\sqrt{4 + 2^{8/3}} \approx 6.43$$

Similarly, the inscribed polygon with four sides joins the points $(0,0)$, $(1, \sqrt[3]{3})$, $(2, 2\sqrt[3]{2})$, $(3,3)$, and $(4,0)$, so its length is

$$\sqrt{1 + \left(\sqrt[3]{3}\right)^2} + \sqrt{1 + \left(2\sqrt[3]{2} - \sqrt[3]{3}\right)^2} + \sqrt{1 + \left(3 - 2\sqrt[3]{2}\right)^2} + \sqrt{1+9} \approx 7.50$$

(c) Using the arc length formula with $\dfrac{dy}{dx} = x\left[\frac{1}{3}(4-x)^{-2/3}(-1)\right] + \sqrt[3]{4-x} = \dfrac{12 - 4x}{3(4-x)^{2/3}}$, the length of the

curve is $L = \displaystyle\int_0^4 \sqrt{1 + \left(\dfrac{dy}{dx}\right)^2}\, dx = \int_0^4 \sqrt{1 + \left[\dfrac{12 - 4x}{3(4-x)^{2/3}}\right]^2}\, dx$.

(d) According to a CAS, the length of the curve is $L \approx 7.7988$. The actual value is larger than any of the approximations in part (b). This is always true, since any approximating straight line between two points on the curve is shorter than the length of the curve between the two points.

27. $x = \ln(1 - y^2) \ \Rightarrow \ \dfrac{dx}{dy} = \dfrac{-2y}{1 - y^2} \ \Rightarrow \ 1 + \left(\dfrac{dx}{dy}\right)^2 = 1 + \dfrac{4y^2}{(1-y^2)^2} = \dfrac{(1+y^2)^2}{(1-y^2)^2}$. So

$$L = \int_0^{1/2} \sqrt{\dfrac{(1+y^2)^2}{(1-y^2)^2}}\, dy = \int_0^{1/2} \dfrac{1 + y^2}{1 - y^2}\, dy = \ln 3 - \tfrac{1}{2} \ \text{[from a CAS]} \ \approx 0.599$$

29. $y^{2/3} = 1 - x^{2/3} \ \Rightarrow \ y = \left(1 - x^{2/3}\right)^{3/2} \ \Rightarrow$

$\dfrac{dy}{dx} = \tfrac{3}{2}\left(1 - x^{2/3}\right)^{1/2}\left(-\tfrac{2}{3}x^{-1/3}\right) = -x^{-1/3}\left(1 - x^{2/3}\right)^{1/2} \ \Rightarrow$

$\left(\dfrac{dy}{dx}\right)^2 = x^{-2/3}\left(1 - x^{2/3}\right) = x^{-2/3} - 1$. Thus

$L = 4\int_0^1 \sqrt{1 + (x^{-2/3} - 1)}\, dx = 4\int_0^1 x^{-1/3}\, dx = 4\displaystyle\lim_{t \to 0^+}\left[\tfrac{3}{2}x^{2/3}\right]_t^1 = 6$.

31. $y = 2x^{3/2} \ \Rightarrow \ y' = 3x^{1/2} \ \Rightarrow \ 1 + (y')^2 = 1 + 9x$. The arc length function with starting point $P_0(1,2)$ is

$$s(x) = \int_1^x \sqrt{1 + 9t}\, dt = \left[\tfrac{2}{27}(1 + 9t)^{3/2}\right]_1^x = \tfrac{2}{27}\left[(1 + 9x)^{3/2} - 10\sqrt{10}\right]$$

33. The prey hits the ground when $y = 0$ $\Leftrightarrow$ $180 - \frac{1}{45}x^2 = 0$ $\Leftrightarrow$ $x^2 = 45 \cdot 180$ $\Rightarrow$ $x = \sqrt{8100} = 90$, since

x must be positive. $y' = -\frac{2}{45}x$ $\Rightarrow$ $1 + (y')^2 = 1 + \frac{4}{45^2}x^2$, so the distance traveled by the prey is

$$L = \int_0^{90} \sqrt{1 + \frac{4}{45^2}x^2}\, dx = \int_0^4 \sqrt{1 + u^2}\left(\frac{45}{2}\, du\right) \qquad [u = \frac{2}{45}x, \ du = \frac{2}{45}\, dx]$$

$$\overset{21}{=} \frac{45}{2}\left[\frac{1}{2}u\sqrt{1 + u^2} + \frac{1}{2}\ln\left(u + \sqrt{1 + u^2}\right)\right]_0^4$$

$$= \frac{45}{2}\left[2\sqrt{17} + \frac{1}{2}\ln\left(4 + \sqrt{17}\right)\right] = 45\sqrt{17} + \frac{45}{4}\ln\left(4 + \sqrt{17}\right) \approx 209.1 \text{ m}$$

35. The sine wave has amplitude 1 and period 14, since it goes through two periods in a distance of 28 in., so its

equation is $y = 1\sin\left(\frac{2\pi}{14}x\right) = \sin\left(\frac{\pi}{7}x\right)$. The width w of the flat metal sheet needed to make the panel is the arc

length of the sine curve from $x = 0$ to $x = 28$. We set up the integral to evaluate w using the arc length formula

with $\frac{dy}{dx} = \frac{\pi}{7}\cos\left(\frac{\pi}{7}x\right)$: $L = \int_0^{28} \sqrt{1 + \left[\frac{\pi}{7}\cos\left(\frac{\pi}{7}x\right)\right]^2}\, dx = 2\int_0^{14} \sqrt{1 + \left[\frac{\pi}{7}\cos\left(\frac{\pi}{7}x\right)\right]^2}\, dx$. This integral would

be very difficult to evaluate exactly, so we use a CAS, and find that $L \approx 29.36$ inches.

37. $y = \int_1^x \sqrt{t^3 - 1}\, dt$ $\Rightarrow$ $\frac{dy}{dx} = \sqrt{x^3 - 1}$ [by FTC1] $\Rightarrow$ $1 + \left(\frac{dy}{dx}\right)^2 = 1 + \left(\sqrt{x^3 - 1}\right)^2 = x^3$ $\Rightarrow$

$$L = \int_1^4 \sqrt{x^3}\, dx = \int_1^4 x^{3/2}\, dx = \frac{2}{5}\left[x^{5/2}\right]_1^4 = \frac{2}{5}(32 - 1) = \frac{62}{5} = 12.4$$

8.2 Area of a Surface of Revolution

1. $y = \ln x$ $\Rightarrow$ $ds = \sqrt{1 + (dy/dx)^2}\, dx = \sqrt{1 + (1/x)^2}\, dx$ $\Rightarrow$ $S = \int_1^3 2\pi(\ln x)\sqrt{1 + (1/x)^2}\, dx$ [by (7)]

3. $y = \sec x$ $\Rightarrow$ $ds = \sqrt{1 + (dy/dx)^2}\, dx = \sqrt{1 + (\sec x \tan x)^2}\, dx$ $\Rightarrow$

$S = \int_0^{\pi/4} 2\pi x \sqrt{1 + (\sec x \tan x)^2}\, dx$ [by (8)]

5. $y = x^3$ $\Rightarrow$ $y' = 3x^2$. So

$$S = \int_0^2 2\pi y \sqrt{1 + (y')^2}\, dx = 2\pi \int_0^2 x^3\sqrt{1 + 9x^4}\, dx \qquad [u = 1 + 9x^4, \ du = 36x^3\, dx]$$

$$= \frac{2\pi}{36}\int_1^{145} \sqrt{u}\, du = \frac{\pi}{18}\left[\frac{2}{3}u^{3/2}\right]_1^{145} = \frac{\pi}{27}\left(145\sqrt{145} - 1\right)$$

7. $y = \sqrt{x}$ $\Rightarrow$ $1 + (dy/dx)^2 = 1 + [1/(2\sqrt{x})]^2 = 1 + 1/(4x)$. So

$$S = \int_4^9 2\pi y \sqrt{1 + \left(\frac{dy}{dx}\right)^2}\, dx = \int_4^9 2\pi \sqrt{x}\sqrt{1 + \frac{1}{4x}}\, dx = 2\pi \int_4^9 \sqrt{x + \frac{1}{4}}\, dx$$

$$= 2\pi\left[\frac{2}{3}\left(x + \frac{1}{4}\right)^{3/2}\right]_4^9 = \frac{4\pi}{3}\left[\frac{1}{8}(4x + 1)^{3/2}\right]_4^9 = \frac{\pi}{6}\left(37\sqrt{37} - 17\sqrt{17}\right)$$

9. $y = \cosh x$ $\Rightarrow$ $1 + (dy/dx)^2 = 1 + \sinh^2 x = \cosh^2 x$. So

$$S = 2\pi \int_0^1 \cosh x \cosh x\, dx = 2\pi \int_0^1 \frac{1}{2}(1 + \cosh 2x)\, dx = \pi\left[x + \frac{1}{2}\sinh 2x\right]_0^1$$

$$= \pi\left(1 + \frac{1}{2}\sinh 2\right) \quad \text{or} \quad \pi\left[1 + \frac{1}{4}\left(e^2 - e^{-2}\right)\right]$$

11. $x = \frac{1}{3}\left(y^2 + 2\right)^{3/2}$ $\Rightarrow$ $dx/dy = \frac{1}{2}\left(y^2 + 2\right)^{1/2}(2y) = y\sqrt{y^2 + 2}$ $\Rightarrow$

$1 + (dx/dy)^2 = 1 + y^2\left(y^2 + 2\right) = \left(y^2 + 1\right)^2$. So

$$S = 2\pi \int_1^2 y\left(y^2 + 1\right) dy = 2\pi\left[\frac{1}{4}y^4 + \frac{1}{2}y^2\right]_1^2 = 2\pi\left(4 + 2 - \frac{1}{4} - \frac{1}{2}\right) = \frac{21\pi}{2}$$

13. $y = \sqrt[3]{x}$ $\Rightarrow$ $x = y^3$ $\Rightarrow$ $1 + (dx/dy)^2 = 1 + 9y^4$. So

$$S = 2\pi \int_1^2 x \sqrt{1 + (dx/dy)^2}\, dy = 2\pi \int_1^2 y^3 \sqrt{1 + 9y^4}\, dy = \frac{2\pi}{36} \int_1^2 \sqrt{1 + 9y^4}\, 36y^3\, dy$$

$$= \frac{\pi}{18} \left[\frac{2}{3}(1 + 9y^4)^{3/2} \right]_1^2 = \frac{\pi}{27}\left(145\sqrt{145} - 10\sqrt{10} \right)$$

15. $x = \sqrt{a^2 - y^2}$ $\Rightarrow$ $dx/dy = \frac{1}{2}(a^2 - y^2)^{-1/2}(-2y) = -y/\sqrt{a^2 - y^2}$ $\Rightarrow$

$$1 + (dx/dy)^2 = 1 + \frac{y^2}{a^2 - y^2} = \frac{a^2 - y^2}{a^2 - y^2} + \frac{y^2}{a^2 - y^2} = \frac{a^2}{a^2 - y^2} \quad \Rightarrow$$

$$S = \int_0^{a/2} 2\pi \sqrt{a^2 - y^2} \frac{a}{\sqrt{a^2 - y^2}}\, dy = 2\pi \int_0^{a/2} a\, dy = 2\pi a\left[y\right]_0^{a/2} = 2\pi a\left(\frac{a}{2} - 0 \right) = \pi a^2. \text{ Note that this is}$$

$\frac{1}{4}$ the surface area of a sphere of radius a, and the length of the interval $y = 0$ to $y = a/2$ is $\frac{1}{4}$ the length of the interval $y = -a$ to $y = a$.

17. $y = \ln x$ $\Rightarrow$ $dy/dx = 1/x$ $\Rightarrow$ $1 + (dy/dx)^2 = 1 + 1/x^2$ $\Rightarrow$ $S = \int_1^3 2\pi \ln x \sqrt{1 + 1/x^2}\, dx$.

Let $f(x) = \ln x \sqrt{1 + 1/x^2}$. Since $n = 10$, $\Delta x = \frac{3 - 1}{10} = \frac{1}{5}$. Then

$$S \approx S_{10} = 2\pi \cdot \frac{1/5}{3} \left[f(1) + 4f(1.2) + 2f(1.4) + \cdots + 2f(2.6) + 4f(2.8) + f(3) \right] \approx 9.023754.$$

The value of the integral produced by a calculator is 9.024262 (to six decimal places).

19. $y = \sec x$ $\Rightarrow$ $dy/dx = \sec x \tan x$ $\Rightarrow$ $1 + (dy/dx)^2 = 1 + \sec^2 x \tan^2 x$ $\Rightarrow$

$S = \int_0^{\pi/3} 2\pi \sec x \sqrt{1 + \sec^2 x \tan^2 x}\, dx.$ Let $f(x) = \sec x \sqrt{1 + \sec^2 x \tan^2 x}$.

Since $n = 10$, $\Delta x = \dfrac{\pi/3 - 0}{10} = \dfrac{\pi}{30}$. Then

$$S \approx S_{10} = 2\pi \cdot \frac{\pi/30}{3}\left[f(0) + 4f\left(\frac{\pi}{30}\right) + 2f\left(\frac{2\pi}{30}\right) + \cdots + 2f\left(\frac{8\pi}{30}\right) + 4f\left(\frac{9\pi}{30}\right) + f\left(\frac{\pi}{3}\right) \right] \approx 13.527296.$$

The value of the integral produced by a calculator is 13.516987 (to six decimal places).

21. $y = 1/x$ $\Rightarrow$ $ds = \sqrt{1 + (dy/dx)^2}\, dx = \sqrt{1 + (-1/x^2)^2}\, dx = \sqrt{1 + 1/x^4}\, dx$ $\Rightarrow$

$$S = \int_1^2 2\pi \cdot \frac{1}{x} \sqrt{1 + \frac{1}{x^4}}\, dx = 2\pi \int_1^2 \frac{\sqrt{x^4 + 1}}{x^3}\, dx = 2\pi \int_1^4 \frac{\sqrt{u^2 + 1}}{u^2} \left(\frac{1}{2}\, du \right) \qquad [u = x^2,\, du = 2x\, dx]$$

$$= \pi \int_1^4 \frac{\sqrt{1 + u^2}}{u^2}\, du \overset{24}{=} \pi \left[-\frac{\sqrt{1 + u^2}}{u} + \ln\left(u + \sqrt{1 + u^2} \right) \right]_1^4$$

$$= \pi\left[-\frac{\sqrt{17}}{4} + \ln\left(4 + \sqrt{17} \right) + \frac{\sqrt{2}}{1} - \ln\left(1 + \sqrt{2} \right) \right] = \pi\left[\sqrt{2} - \frac{\sqrt{17}}{4} + \ln\left(\frac{4 + \sqrt{17}}{1 + \sqrt{2}} \right) \right]$$

23. $y = x^3$ and $0 \le y \le 1$ $\Rightarrow$ $y' = 3x^2$ and $0 \le x \le 1$.

$$S = \int_0^1 2\pi x \sqrt{1 + (3x^2)^2}\, dx = 2\pi \int_0^3 \sqrt{1 + u^2}\, \frac{1}{6}\, du \qquad [u = 3x^2,\, du = 6x\, dx]$$

$$= \frac{\pi}{3} \int_0^3 \sqrt{1 + u^2}\, du \overset{21}{=} \text{ [or use CAS]} \quad \frac{\pi}{3}\left[\frac{1}{2}u\sqrt{1 + u^2} + \frac{1}{2}\ln\left(u + \sqrt{1 + u^2} \right) \right]_0^3$$

$$= \frac{\pi}{3}\left[\frac{3}{2}\sqrt{10} + \frac{1}{2}\ln\left(3 + \sqrt{10} \right) \right] = \frac{\pi}{6}\left[3\sqrt{10} + \ln\left(3 + \sqrt{10} \right) \right]$$

25. $S = 2\pi \displaystyle\int_1^\infty y \sqrt{1 + \left(\dfrac{dy}{dx}\right)^2}\, dx = 2\pi \displaystyle\int_1^\infty \dfrac{1}{x} \sqrt{1 + \dfrac{1}{x^4}}\, dx = 2\pi \displaystyle\int_1^\infty \dfrac{\sqrt{x^4 + 1}}{x^3}\, dx.$ Rather than trying to

evaluate this integral, note that $\sqrt{x^4 + 1} > \sqrt{x^4} = x^2$ for $x > 0$. Thus, if the area is finite,

$$S = 2\pi \int_1^\infty \frac{\sqrt{x^4 + 1}}{x^3}\, dx > 2\pi \int_1^\infty \frac{x^2}{x^3}\, dx = 2\pi \int_1^\infty \frac{1}{x}\, dx$$

But we know that this integral diverges, so the area S is infinite.

27. Since $a > 0$, the curve $3ay^2 = x(a - x)^2$ only has points with

$x \geq 0$. ($3ay^2 \geq 0 \;\Rightarrow\; x(a - x)^2 \geq 0 \;\Rightarrow\; x \geq 0$.) The

curve is symmetric about the x-axis (since the equation is

unchanged when y is replaced by $-y$). $y = 0$ when $x = 0$ or a,

so the curve's loop extends from $x = 0$ to $x = a$.

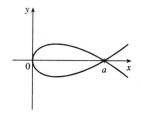

$\dfrac{d}{dx}(3ay^2) = \dfrac{d}{dx}[x(a - x)^2] \;\Rightarrow\; 6ay\dfrac{dy}{dx} = x \cdot 2(a - x)(-1) + (a - x)^2 \;\Rightarrow\;$

$\dfrac{dy}{dx} = \dfrac{(a - x)[-2x + a - x]}{6ay} \;\Rightarrow\; \left(\dfrac{dy}{dx}\right)^2 = \dfrac{(a - x)^2(a - 3x)^2}{36a^2y^2} = \dfrac{(a - x)^2(a - 3x)^2}{36a^2} \cdot \dfrac{3a}{x(a - x)^2}$

$\begin{bmatrix}\text{the last fraction} \\ \text{is } 1/y^2\end{bmatrix} = \dfrac{(a - 3x)^2}{12ax} \;\Rightarrow\;$

$1 + \left(\dfrac{dy}{dx}\right)^2 = 1 + \dfrac{a^2 - 6ax + 9x^2}{12ax} = \dfrac{12ax}{12ax} + \dfrac{a^2 - 6ax + 9x^2}{12ax} = \dfrac{a^2 + 6ax + 9x^2}{12ax} = \dfrac{(a + 3x)^2}{12ax}$ for $x \neq 0$.

(a) $S = \displaystyle\int_{x=0}^a 2\pi y\, ds = 2\pi \int_0^a \dfrac{\sqrt{x}(a - x)}{\sqrt{3a}} \cdot \dfrac{a + 3x}{\sqrt{12ax}}\, dx = 2\pi \int_0^a \dfrac{(a - x)(a + 3x)}{6a}\, dx$

$= \dfrac{\pi}{3a} \displaystyle\int_0^a (a^2 + 2ax - 3x^2)\, dx = \dfrac{\pi}{3a}\left[a^2x + ax^2 - x^3\right]_0^a = \dfrac{\pi}{3a}(a^3 + a^3 - a^3) = \dfrac{\pi}{3a} \cdot a^3 = \dfrac{\pi a^2}{3}.$

Note that we have rotated the top half of the loop about the x-axis. This generates the full surface.

(b) We must rotate the full loop about the y-axis, so we get double the area obtained by rotating the top half of

the loop:

$S = 2 \cdot 2\pi \displaystyle\int_{x=0}^a x\, ds = 4\pi \int_0^a x \dfrac{a + 3x}{\sqrt{12ax}}\, dx = \dfrac{4\pi}{2\sqrt{3a}} \int_0^a x^{1/2}(a + 3x)\, dx$

$= \dfrac{2\pi}{\sqrt{3a}} \displaystyle\int_0^a (ax^{1/2} + 3x^{3/2})\, dx = \dfrac{2\pi}{\sqrt{3a}}\left[\dfrac{2}{3}ax^{3/2} + \dfrac{6}{5}x^{5/2}\right]_0^a = \dfrac{2\pi\sqrt{3}}{3\sqrt{a}}\left(\dfrac{2}{3}a^{5/2} + \dfrac{6}{5}a^{5/2}\right)$

$= \dfrac{2\pi\sqrt{3}}{3}\left(\dfrac{2}{3} + \dfrac{6}{5}\right)a^2 = \dfrac{2\pi\sqrt{3}}{3}\left(\dfrac{28}{15}\right)a^2 = \dfrac{56\pi\sqrt{3}\,a^2}{45}$

29. $\dfrac{x^2}{a^2} + \dfrac{y^2}{b^2} = 1 \quad \Rightarrow \quad \dfrac{y\,(dy/dx)}{b^2} = -\dfrac{x}{a^2} \quad \Rightarrow \quad \dfrac{dy}{dx} = -\dfrac{b^2 x}{a^2 y} \quad \Rightarrow$

$$1 + \left(\frac{dy}{dx}\right)^2 = 1 + \frac{b^4 x^2}{a^4 y^2} = \frac{b^4 x^2 + a^4 y^2}{a^4 y^2} = \frac{b^4 x^2 + a^4 b^2 \left(1 - x^2/a^2\right)}{a^4 b^2 \left(1 - x^2/a^2\right)} = \frac{a^4 b^2 + b^4 x^2 - a^2 b^2 x^2}{a^4 b^2 - a^2 b^2 x^2}$$

$$= \frac{a^4 + b^2 x^2 - a^2 x^2}{a^4 - a^2 x^2} = \frac{a^4 - \left(a^2 - b^2\right) x^2}{a^2 (a^2 - x^2)}$$

The ellipsoid's surface area is twice the area generated by rotating the first quadrant portion of the ellipse about the x-axis. Thus,

$$S = 2 \int_0^a 2\pi y \sqrt{1 + \left(\frac{dy}{dx}\right)^2}\, dx = 4\pi \int_0^a \frac{b}{a} \sqrt{a^2 - x^2}\, \frac{\sqrt{a^4 - (a^2 - b^2) x^2}}{a\sqrt{a^2 - x^2}}\, dx$$

$$= \frac{4\pi b}{a^2} \int_0^a \sqrt{a^4 - (a^2 - b^2) x^2}\, dx = \frac{4\pi b}{a^2} \int_0^{a\sqrt{a^2 - b^2}} \sqrt{a^4 - u^2}\, \frac{du}{\sqrt{a^2 - b^2}} \qquad [u = \sqrt{a^2 - b^2}\, x]$$

$$\overset{30}{=} \frac{4\pi b}{a^2 \sqrt{a^2 - b^2}} \left[\frac{u}{2} \sqrt{a^4 - u^2} + \frac{a^4}{2} \sin^{-1} \frac{u}{a^2} \right]_0^{a\sqrt{a^2 - b^2}}$$

$$= \frac{4\pi b}{a^2 \sqrt{a^2 - b^2}} \left[\frac{a\sqrt{a^2 - b^2}}{2} \sqrt{a^4 - a^2(a^2 - b^2)} + \frac{a^4}{2} \sin^{-1} \frac{\sqrt{a^2 - b^2}}{a} \right] = 2\pi \left[b^2 + \frac{a^2 b \sin^{-1} \dfrac{\sqrt{a^2 - b^2}}{a}}{\sqrt{a^2 - b^2}} \right]$$

31. The analogue of $f(x_i^*)$ in the derivation of (4) is now $c - f(x_i^*)$, so

$$S = \lim_{n \to \infty} \sum_{i=1}^n 2\pi [c - f(x_i^*)] \sqrt{1 + [f'(x_i^*)]^2}\, \Delta x = \int_a^b 2\pi [c - f(x)] \sqrt{1 + [f'(x)]^2}\, dx.$$

33. For the upper semicircle, $f(x) = \sqrt{r^2 - x^2}$, $f'(x) = -x/\sqrt{r^2 - x^2}$. The surface area generated is

$$S_1 = \int_{-r}^r 2\pi \left(r - \sqrt{r^2 - x^2} \right) \sqrt{1 + \frac{x^2}{r^2 - x^2}}\, dx = 4\pi \int_0^r \left(r - \sqrt{r^2 - x^2} \right) \frac{r}{\sqrt{r^2 - x^2}}\, dx$$

$$= 4\pi \int_0^r \left(\frac{r^2}{\sqrt{r^2 - x^2}} - r \right) dx$$

For the lower semicircle, $f(x) = -\sqrt{r^2 - x^2}$ and $f'(x) = \dfrac{x}{\sqrt{r^2 - x^2}}$, so $S_2 = 4\pi \int_0^r \left(\dfrac{r^2}{\sqrt{r^2 - x^2}} + r \right) dx$.

Thus, the total area is $S = S_1 + S_2 = 8\pi \int_0^r \left(\dfrac{r^2}{\sqrt{r^2 - x^2}} \right) dx = 8\pi \left[r^2 \sin^{-1} \left(\dfrac{x}{r} \right) \right]_0^r = 8\pi r^2 \left(\dfrac{\pi}{2} \right) = 4\pi^2 r^2$.

35. In the derivation of (4), we computed a typical contribution to the surface area to be $2\pi \dfrac{y_{i-1} + y_i}{2} |P_{i-1} P_i|$, the

area of a frustum of a cone. When $f(x)$ is not necessarily positive, the approximations $y_i = f(x_i) \approx f(x_i^*)$ and

$y_{i-1} = f(x_{i-1}) \approx f(x_i^*)$ must be replaced by $y_i = |f(x_i)| \approx |f(x_i^*)|$ and $y_{i-1} = |f(x_{i-1})| \approx |f(x_i^*)|$. Thus,

$2\pi \dfrac{y_{i-1} + y_i}{2} |P_{i-1} P_i| \approx 2\pi |f(x_i^*)| \sqrt{1 + [f'(x_i^*)]^2} \Delta x$. Continuing with the rest of the derivation as before, we

obtain $S = \int_a^b 2\pi |f(x)| \sqrt{1 + [f'(x)]^2}\, dx$.

8.3 Applications to Physics and Engineering

1. The weight density of water is $\delta = 62.5 \text{ lb/ft}^3$.

(a) $P = \delta d \approx (62.5 \text{ lb/ft}^3)(3 \text{ ft}) = 187.5 \text{ lb/ft}^2$

(b) $F = PA \approx (187.5 \text{ lb/ft}^2)(5 \text{ ft})(2 \text{ ft}) = 1875 \text{ lb}$. ($A$ is the area of the bottom of the tank.)

(c) As in Example 1, the area of the ith strip is $2(\Delta x)$ and the pressure is $\delta d = \delta x_i$. Thus,

$$F = \int_0^3 \delta x \cdot 2\, dx \approx (62.5)(2) \int_0^3 x\, dx = 125\left[\tfrac{1}{2}x^2\right]_0^3 = 125\left(\tfrac{9}{2}\right) = 562.5 \text{ lb}$$

In Exercises 3–9, n is the number of subintervals of length Δx and x_i^* is a sample point in the ith subinterval $[x_{i-1}, x_i]$.

3. Set up a vertical x-axis as shown, with $x = 0$ at the water's surface and x

increasing in the downward direction. Then the area of the ith rectangular

strip is $6\,\Delta x$ and the pressure on the strip is δx_i^* (where $\delta \approx 62.5 \text{ lb/ft}^3$).

Thus, the hydrostatic force on the strip is $\delta x_i^* \cdot 6\,\Delta x$ and the total

hydrostatic force $\approx \sum\limits_{i=1}^{n} \delta x_i^* \cdot 6\,\Delta x$. The total force

$$F = \lim_{n \to \infty} \sum_{i=1}^{n} \delta x_i^* \cdot 6\,\Delta x = \int_2^6 \delta x \cdot 6\, dx = 6\delta \int_2^6 x\, dx$$

$$= 6\delta\left[\tfrac{1}{2}x^2\right]_2^6 = 6\delta(18 - 2) = 96\delta \approx 6000 \text{ lb}$$

5. Since an equation for the shape is $x^2 + y^2 = 10^2$ ($x \geq 0$), we have

$y = \sqrt{100 - x^2}$. Thus, the area of the ith strip is $2\sqrt{100 - (x_i^*)^2}\,\Delta x$

and the pressure on the strip is $\rho g x_i^*$, so the hydrostatic force on the

strip is $\rho g x_i^* \cdot 2\sqrt{100 - (x_i^*)^2}\,\Delta x$ and the total force on the

plate $\approx \sum\limits_{i=1}^{n} \rho g x_i^* \cdot 2\sqrt{100 - (x_i^*)^2}\,\Delta x$. The total force

$$F = \lim_{n \to \infty} \sum_{i=1}^{n} \rho g x_i^* \cdot 2\sqrt{100 - (x_i^*)^2}\,\Delta x = \int_0^{10} 2\rho g x \sqrt{100 - x^2}\, dx$$

$$= -\rho g \int_0^{10} \left(100 - x^2\right)^{1/2}(-2x)\, dx = -\rho g\left[\tfrac{2}{3}(100 - x^2)^{3/2}\right]_0^{10} = -\tfrac{2}{3}\rho g(0 - 1000)$$

$$= \tfrac{2000}{3}\rho g \approx \tfrac{2000}{3} \cdot 1000 \cdot 9.8 \approx 6.5 \times 10^6 \text{ N} \quad \left[\rho \approx 1000 \text{ kg/m}^3 \text{ and } g \approx 9.8 \text{ m/s}^2.\right]$$

7. Using similar triangles, $\dfrac{4 \text{ ft wide}}{8 \text{ ft high}} = \dfrac{a \text{ ft wide}}{x_i^* \text{ ft high}}$, so $a = \frac{1}{2}x_i^*$ and the

width of the ith rectangular strip is $12 + 2a = 12 + x_i^*$. The area of the

strip is $(12 + x_i^*)\,\Delta x$. The pressure on the strip is δx_i^*.

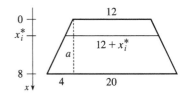

$$F = \lim_{n \to \infty} \sum_{i=1}^{n} \delta x_i^*(12 + x_i^*)\,\Delta x = \int_0^8 \delta x \cdot (12 + x)\,dx$$

$$= \delta \int_0^8 (12x + x^2)\,dx = \delta\left[6x^2 + \frac{x^3}{3}\right]_0^8 = \delta\left(384 + \frac{512}{3}\right)$$

$$= (62.5)\frac{1664}{3} \approx 3.47 \times 10^4 \text{ lb}$$

9. From the figure, the area of the ith rectangular strip is $2\sqrt{r^2 - (x_i^*)^2}\,\Delta x$

and the pressure on it is $\rho g(x_i^* + r)$.

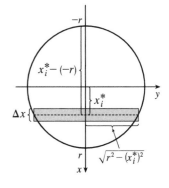

$$F = \lim_{n \to \infty} \sum_{i=1}^{n} \rho g(x_i^* + r)\,2\sqrt{r^2 - (x_i^*)^2}\,\Delta x$$

$$= \int_{-r}^{r} \rho g(x + r) \cdot 2\sqrt{r^2 - x^2}\,dx$$

$$= \rho g \int_{-r}^{r} \sqrt{r^2 - x^2}\,2x\,dx + 2\rho g r \int_{-r}^{r} \sqrt{r^2 - x^2}\,dx$$

The first integral is 0 because the integrand is an odd function. The second

integral can be interpreted as the area of a semicircular disk with radius r,

or we could make the trigonometric substitution $x = r\sin\theta$. Continuing:

$$F = \rho g \cdot 0 + 2\rho g r \cdot \tfrac{1}{2}\pi r^2 = \rho g \pi r^3 = 1000 g \pi r^3 \text{ N (SI units assumed)}.$$

11. By similar triangles, $\dfrac{8}{4\sqrt{3}} = \dfrac{w_i}{x_i^*} \;\Rightarrow\; w_i = \dfrac{2x_i^*}{\sqrt{3}}$. The area of the ith

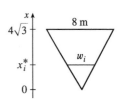

rectangular strip is $\dfrac{2x_i^*}{\sqrt{3}}\,\Delta x$ and the pressure on it is $\rho g\left(4\sqrt{3} - x_i^*\right)$.

$$F = \int_0^{4\sqrt{3}} \rho g\left(4\sqrt{3} - x\right)\frac{2x}{\sqrt{3}}\,dx = 8\rho g \int_0^{4\sqrt{3}} x\,dx - \frac{2\rho g}{\sqrt{3}} \int_0^{4\sqrt{3}} x^2\,dx$$

$$= 4\rho g\left[x^2\right]_0^{4\sqrt{3}} - \frac{2\rho g}{3\sqrt{3}}\left[x^3\right]_0^{4\sqrt{3}} = 192\rho g - \frac{2\rho g}{3\sqrt{3}}\,64 \cdot 3\sqrt{3}$$

$$= 192\rho g - 128\rho g = 64\rho g \approx 64(840)(9.8) \approx 5.27 \times 10^5 \text{ N}$$

13. (a) The top of the cube has depth $d = 1$ m -20 cm $= 80$ cm $= 0.8$ m.

$$F = \rho g d A \approx (1000)(9.8)(0.8)(0.2)^2 = 313.6 \approx 314 \text{ N}$$

(b) The area of a strip is $0.2\,\Delta x$ and the pressure on it is $\rho g x_i^*$.

$$F = \int_{0.8}^1 \rho g x (0.2)\, dx = 0.2\rho g \left[\tfrac{1}{2}x^2\right]_{0.8}^1 = (0.2\rho g)(0.18) = 0.036\rho g = 0.036(1000)(9.8)$$

$$= 352.8 \approx 353 \text{ N}$$

15. (a) The area of a strip is $20\,\Delta x$ and the pressure on it is δx_i.

$$F = \int_0^3 \delta x 20\, dx = 20\delta \left[\tfrac{1}{2}x^2\right]_0^3 = 20\delta \cdot \frac{9}{2} = 90\delta$$

$$= 90(62.5) = 5625 \text{ lb} \approx 5.63 \times 10^3 \text{ lb}$$

(b) $F = \int_0^9 \delta x 20\, dx = 20\delta \left[\tfrac{1}{2}x^2\right]_0^9 = 20\delta \cdot \frac{81}{2} = 810\delta = 810(62.5) = 50{,}625 \text{ lb} \approx 5.06 \times 10^4 \text{ lb}.$

(c) For the first 3 ft, the length of the side is constant at 40 ft. For $3 < x \leq 9$, we can use similar triangles to find the

length a: $\dfrac{a}{40} = \dfrac{9-x}{6} \;\Rightarrow\; a = 40 \cdot \dfrac{9-x}{6}.$

$$F = \int_0^3 \delta x 40\, dx + \int_3^9 \delta x (40)\tfrac{9-x}{6}\, dx = 40\delta \left[\tfrac{1}{2}x^2\right]_0^3 + \tfrac{20}{3}\delta \int_3^9 (9x - x^2)\, dx$$

$$= 180\delta + \tfrac{20}{3}\delta \left[\tfrac{9}{2}x^2 - \tfrac{1}{3}x^3\right]_3^9 = 180\delta + \tfrac{20}{3}\delta \left[\left(\tfrac{729}{2} - 243\right) - \left(\tfrac{81}{2} - 9\right)\right]$$

$$= 180\delta + 600\delta = 780\delta = 780(62.5) = 48{,}750 \text{ lb} \approx 4.88 \times 10^4 \text{ lb}$$

(d) For any right triangle with hypotenuse on the bottom,

$\csc\theta = \dfrac{\Delta x}{\text{hypotenuse}} \;\Rightarrow\;$

$\text{hypotenuse} = \Delta x\,\csc\theta = \Delta x\,\dfrac{\sqrt{40^2+6^2}}{6} = \dfrac{\sqrt{409}}{3}\,\Delta x.$

$F = \int_3^9 \delta x 20\,\tfrac{\sqrt{409}}{3}\, dx = \tfrac{1}{3}\left(20\sqrt{409}\right)\delta\left[\tfrac{1}{2}x^2\right]_3^9$

$= \tfrac{1}{3}\cdot 10\sqrt{409}\,\delta(81 - 9)$

$\approx 303{,}356 \text{ lb} \approx 3.03 \times 10^5 \text{ lb}$

17. $F = \int_2^5 \rho g x \cdot w(x)\, dx$, where $w(x)$ is the width of the plate at depth x. Since $n = 6$, $\Delta x = \frac{5-2}{6} = \frac{1}{2}$, and

$F \approx S_6 = \rho g \cdot \tfrac{1/2}{3}[2 \cdot w(2) + 4 \cdot 2.5 \cdot w(2.5) + 2 \cdot 3 \cdot w(3) + 4 \cdot 3.5 \cdot w(3.5)$

$\qquad\qquad + 2 \cdot 4 \cdot w(4) + 4 \cdot 4.5 \cdot w(4.5) + 5 \cdot w(5)]$

$= \tfrac{1}{6}\rho g(2 \cdot 0 + 10 \cdot 0.8 + 6 \cdot 1.7 + 14 \cdot 2.4 + 8 \cdot 2.9 + 18 \cdot 3.3 + 5 \cdot 3.6)$

$= \tfrac{1}{6}(1000)(9.8)(152.4) \approx 2.5 \times 10^5 \text{ N}$

19. The moment M of the system about the origin is $M = \sum_{i=1}^{2} m_i x_i = m_1 x_1 + m_2 x_2 = 40 \cdot 2 + 30 \cdot 5 = 230$.

The mass m of the system is $m = \sum_{i=1}^{2} m_i = m_1 + m_2 = 40 + 30 = 70$. The center of mass of the system is

$M/m = \frac{230}{70} = \frac{23}{7}$.

21. $m = \sum_{i=1}^{3} m_i = 6 + 5 + 10 = 21$. $M_x = \sum_{i=1}^{3} m_i y_i = 6(5) + 5(-2) + 10(-1) = 10$;

$M_y = \sum_{i=1}^{3} m_i x_i = 6(1) + 5(3) + 10(-2) = 1$. $\bar{x} = \dfrac{M_y}{m} = \dfrac{1}{21}$ and $\bar{y} = \dfrac{M_x}{m} = \dfrac{10}{21}$, so the center of mass of the

system is $\left(\frac{1}{21}, \frac{10}{21}\right)$.

23. Since the region in the figure is symmetric about the y-axis, we know

that $\bar{x} = 0$. The region is "bottom-heavy," so we know that $\bar{y} < 2$,

and we might guess that $\bar{y} = 1.5$.

$A = \int_{-2}^{2} (4 - x^2)\, dx = 2 \int_{0}^{2} (4 - x^2)\, dx = 2 \left[4x - \frac{1}{3} x^3 \right]_{0}^{2}$

$= 2 \left(8 - \frac{8}{3} \right) = \frac{32}{3}$

$\bar{x} = \dfrac{1}{A} \displaystyle\int_{-2}^{2} x(4 - x^2)\, dx = 0$ since $f(x) = x(4 - x^2)$ is an odd

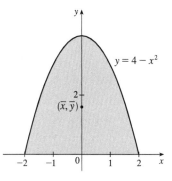

function (or since the region is symmetric about the y-axis).

$$\bar{y} = \frac{1}{A} \int_{-2}^{2} \frac{1}{2}(4 - x^2)^2\, dx = \frac{3}{32} \cdot \frac{1}{2} \cdot 2 \int_{0}^{2} (16 - 8x^2 + x^4)\, dx = \frac{3}{32} \left[16x - \frac{8}{3} x^3 + \frac{1}{5} x^5 \right]_{0}^{2}$$

$$= \frac{3}{32} \left(32 - \frac{64}{3} + \frac{32}{5} \right) = 3 \left(1 - \frac{2}{3} + \frac{1}{5} \right) = 3 \left(\frac{8}{15} \right) = \frac{8}{5}$$

Thus, the centroid is $(\bar{x}, \bar{y}) = \left(0, \frac{8}{5} \right)$.

25. The region in the figure is "right-heavy" and "bottom-heavy," so we know

$\bar{x} > 0.5$ and $\bar{y} < 1$, and we might guess that $\bar{x} = 0.6$ and $\bar{y} = 0.9$.

$A = \int_{0}^{1} e^x\, dx = [e^x]_{0}^{1} = e - 1$,

$\bar{x} = \frac{1}{A} \int_{0}^{1} xe^x\, dx = \frac{1}{e-1} [xe^x - e^x]_{0}^{1}$ [by parts]

$= \frac{1}{e-1} [0 - (-1)] = \frac{1}{e-1}$,

$\bar{y} = \frac{1}{A} \int_{0}^{1} \frac{1}{2} (e^x)^2\, dx = \frac{1}{e-1} \cdot \frac{1}{4} [e^{2x}]_{0}^{1} = \frac{1}{4(e-1)} (e^2 - 1) = \frac{e+1}{4}$.

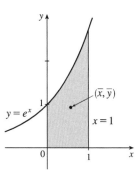

Thus, the centroid is $(\bar{x}, \bar{y}) = \left(\frac{1}{e-1}, \frac{e+1}{4} \right) \approx (0.58, 0.93)$.

27. $A = \int_0^1 (\sqrt{x} - x)\, dx = \left[\frac{2}{3}x^{3/2} - \frac{1}{2}x^2\right]_0^1 = \frac{2}{3} - \frac{1}{2} = \frac{1}{6}$.

$\bar{x} = \frac{1}{A} \int_0^1 x(\sqrt{x} - x)\, dx = 6\int_0^1 (x^{3/2} - x^2)\, dx$

$\quad = 6\left[\frac{2}{5}x^{5/2} - \frac{1}{3}x^3\right]_0^1 = 6\left(\frac{2}{5} - \frac{1}{3}\right) = 6\left(\frac{1}{15}\right) = \frac{2}{5}$;

$\bar{y} = \frac{1}{A} \int_0^1 \frac{1}{2}\left[(\sqrt{x})^2 - x^2\right] dx = 6 \cdot \frac{1}{2}\int_0^1 (x - x^2)\, dx$

$\quad = 3\left[\frac{1}{2}x^2 - \frac{1}{3}x^3\right]_0^1 = 3\left(\frac{1}{2} - \frac{1}{3}\right) = \frac{1}{2}$.

Thus, the centroid is $(\bar{x}, \bar{y}) = \left(\frac{2}{5}, \frac{1}{2}\right)$.

29. $A = \int_0^{\pi/4} (\cos x - \sin x)\, dx = [\sin x + \cos x]_0^{\pi/4} = \sqrt{2} - 1$,

$\bar{x} = A^{-1} \int_0^{\pi/4} x(\cos x - \sin x)\, dx$

$\quad = A^{-1}[x(\sin x + \cos x) + \cos x - \sin x]_0^{\pi/4}$ [integration by parts]

$\quad = A^{-1}\left(\frac{\pi}{4}\sqrt{2} - 1\right) = \dfrac{\frac{1}{4}\pi\sqrt{2} - 1}{\sqrt{2} - 1}$

$\bar{y} = A^{-1}\int_0^{\pi/4} \frac{1}{2}\left(\cos^2 x - \sin^2 x\right) dx = \frac{1}{2A}\int_0^{\pi/4} \cos 2x\, dx = \frac{1}{4A}[\sin 2x]_0^{\pi/4} = \frac{1}{4A} = \dfrac{1}{4(\sqrt{2}-1)}$

Thus, the centroid is $(\bar{x}, \bar{y}) = \left(\dfrac{\pi\sqrt{2} - 4}{4(\sqrt{2} - 1)}, \dfrac{1}{4(\sqrt{2} - 1)}\right) \approx (0.27, 0.60)$.

31. From the figure we see that $\bar{y} = 0$. Now

$$A = \int_0^5 2\sqrt{5 - x}\, dx = 2\left[-\frac{2}{3}(5 - x)^{3/2}\right]_0^5$$

$$= 2\left(0 + \frac{2}{3} \cdot 5^{3/2}\right) = \frac{20}{3}\sqrt{5}$$

so

$\bar{x} = \frac{1}{A}\int_0^5 x\left[\sqrt{5 - x} - (-\sqrt{5 - x})\right] dx = \frac{1}{A}\int_0^5 2x\sqrt{5 - x}\, dx$

$\quad = \frac{1}{A}\int_{\sqrt{5}}^0 2(5 - u^2)u(-2u)\, du$ $[u = \sqrt{5 - x},\ x = 5 - u^2,\ u^2 = 5 - x,\ dx = -2u\, du]$

$\quad = \frac{4}{A}\int_0^{\sqrt{5}} u^2(5 - u^2)\, du = \frac{4}{A}\left[\frac{5}{3}u^3 - \frac{1}{5}u^5\right]_0^{\sqrt{5}} = \frac{3}{5\sqrt{5}}\left(\frac{25}{3}\sqrt{5} - 5\sqrt{5}\right) = 5 - 3 = 2$

Thus, the centroid is $(\bar{x}, \bar{y}) = (2, 0)$.

33. By symmetry, $M_y = 0$ and $\bar{x} = 0$. $A = \frac{1}{2}bh = \frac{1}{2} \cdot 2 \cdot 2 = 2$.

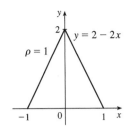

$$M_x = \rho \int_{-1}^{1} \frac{1}{2}(2 - 2x)^2 \, dx = 2\rho \int_0^1 \frac{1}{2}(2 - 2x)^2 \, dx$$

$$= \left(2 \cdot 1 \cdot \frac{1}{2} \cdot 2^2\right) \int_0^1 (1 - x)^2 \, dx$$

$$= 4 \int_1^0 u^2 (-du) \qquad [u = 1 - x, \, du = -dx]$$

$$= -4\left[\frac{1}{3}u^3\right]_1^0 = -4\left(-\frac{1}{3}\right) = \frac{4}{3}$$

$\bar{y} = \frac{1}{m}M_x = \frac{1}{\rho A}M_x = \frac{1}{1 \cdot 2} \cdot \frac{4}{3} = \frac{2}{3}$. Thus, the centroid is $(\bar{x}, \bar{y}) = \left(0, \frac{2}{3}\right)$.

35. $A = \displaystyle\int_0^2 (2^x - x^2) \, dx = \left[\dfrac{2^x}{\ln 2} - \dfrac{x^3}{3}\right]_0^2$

$= \left(\dfrac{4}{\ln 2} - \dfrac{8}{3}\right) - \dfrac{1}{\ln 2} = \dfrac{3}{\ln 2} - \dfrac{8}{3} \approx 1.661418.$

$\bar{x} = \dfrac{1}{A}\displaystyle\int_0^2 x(2^x - x^2) \, dx = \dfrac{1}{A}\displaystyle\int_0^2 (x2^x - x^3) \, dx$

$= \dfrac{1}{A}\left[\dfrac{x2^x}{\ln 2} - \dfrac{2^x}{(\ln 2)^2} - \dfrac{x^4}{4}\right]_0^2 \qquad$ [use parts]

$= \dfrac{1}{A}\left[\dfrac{8}{\ln 2} - \dfrac{4}{(\ln 2)^2} - 4 + \dfrac{1}{(\ln 2)^2}\right]$

$= \dfrac{1}{A}\left[\dfrac{8}{\ln 2} - \dfrac{3}{(\ln 2)^2} - 4\right] \approx \dfrac{1}{A}(1.297453) \approx 0.781$

$\bar{y} = \dfrac{1}{A}\displaystyle\int_0^2 \frac{1}{2}\left[(2^x)^2 - (x^2)^2\right] dx = \dfrac{1}{A}\displaystyle\int_0^2 \frac{1}{2}(2^{2x} - x^4) dx = \dfrac{1}{A} \cdot \dfrac{1}{2}\left[\dfrac{2^{2x}}{2\ln 2} - \dfrac{x^5}{5}\right]_0^2$

$= \dfrac{1}{A} \cdot \dfrac{1}{2}\left(\dfrac{16}{2\ln 2} - \dfrac{32}{5} - \dfrac{1}{2\ln 2}\right) = \dfrac{1}{A}\left(\dfrac{15}{4\ln 2} - \dfrac{16}{5}\right) \approx \dfrac{1}{A}(2.210106) \approx 1.330$

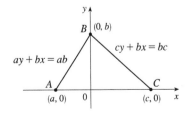

Since the position of a centroid is independent of density when the density is constant, we will assume for convenience that $\rho = 1$ in Exercises 36 and 37.

37. Choose x- and y-axes so that the base (one side of the triangle) lies along the x-axis with the other vertex along the positive y-axis as shown. From geometry, we know the medians intersect at a point $\frac{2}{3}$ of the way from each vertex (along the median) to the opposite side. The median from B goes to the midpoint $\left(\frac{1}{2}(a + c), 0\right)$ of side AC, so the point of intersection of the medians is $\left(\frac{2}{3} \cdot \frac{1}{2}(a + c), \frac{1}{3}b\right) = \left(\frac{1}{3}(a + c), \frac{1}{3}b\right)$.

[continued]

This can also be verified by finding the equations of two medians, and solving them simultaneously to find their point of intersection. Now let us compute the location of the centroid of the triangle. The area is $A = \frac{1}{2}(c - a)b$.

$$\bar{x} = \frac{1}{A}\left[\int_a^0 x \cdot \frac{b}{a}(a - x)\,dx + \int_0^c x \cdot \frac{b}{c}(c - x)\,dx\right] = \frac{1}{A}\left[\frac{b}{a}\int_a^0 (ax - x^2)\,dx + \frac{b}{c}\int_0^c (cx - x^2)\,dx\right]$$

$$= \frac{b}{Aa}\left[\frac{1}{2}ax^2 - \frac{1}{3}x^3\right]_a^0 + \frac{b}{Ac}\left[\frac{1}{2}cx^2 - \frac{1}{3}x^3\right]_0^c = \frac{b}{Aa}\left[-\frac{1}{2}a^3 + \frac{1}{3}a^3\right] + \frac{b}{Ac}\left[\frac{1}{2}c^3 - \frac{1}{3}c^3\right]$$

$$= \frac{2}{a(c-a)} \cdot \frac{-a^3}{6} + \frac{2}{c(c-a)} \cdot \frac{c^3}{6} = \frac{1}{3(c-a)}(c^2 - a^2) = \frac{a+c}{3}$$

and

$$\bar{y} = \frac{1}{A}\left[\int_a^0 \frac{1}{2}\left(\frac{b}{a}(a - x)\right)^2 dx + \int_0^c \frac{1}{2}\left(\frac{b}{c}(c - x)\right)^2 dx\right]$$

$$= \frac{1}{A}\left[\frac{b^2}{2a^2}\int_a^0 (a^2 - 2ax + x^2)\,dx + \frac{b^2}{2c^2}\int_0^c (c^2 - 2cx + x^2)\,dx\right]$$

$$= \frac{1}{A}\left[\frac{b^2}{2a^2}\left[a^2x - ax^2 + \frac{1}{3}x^3\right]_a^0 + \frac{b^2}{2c^2}\left[c^2x - cx^2 + \frac{1}{3}x^3\right]_0^c\right]$$

$$= \frac{1}{A}\left[\frac{b^2}{2a^2}\left(-a^3 + a^3 - \frac{1}{3}a^3\right) + \frac{b^2}{2c^2}\left(c^3 - c^3 + \frac{1}{3}c^3\right)\right] = \frac{1}{A}\left[\frac{b^2}{6}(-a + c)\right] = \frac{2}{(c-a)b} \cdot \frac{(c-a)b^2}{6} = \frac{b}{3}$$

Thus, the centroid is $(\bar{x}, \bar{y}) = \left(\dfrac{a+c}{3}, \dfrac{b}{3}\right)$, as claimed.

Remarks: Actually the computation of $\bar{y}$ is all that is needed. By considering each side of the triangle in turn to be the base, we see that the centroid is $\frac{1}{3}$ of the way from each side to the opposite vertex and must therefore be the intersection of the medians.

The computation of $\bar{y}$ in this problem (and many others) can be simplified by using horizontal rather than vertical approximating rectangles. If the length of a thin rectangle at coordinate y is $\ell(y)$, then its area is $\ell(y)\,\Delta y$, its mass is $\rho\ell(y)\,\Delta y$, and its moment about the x-axis is $\Delta M_x = \rho y\ell(y)\,\Delta y$. Thus,

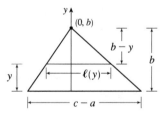

$$M_x = \int \rho y\ell(y)\,dy \quad \text{and} \quad \bar{y} = \frac{\int \rho y\ell(y)\,dy}{\rho A} = \frac{1}{A}\int y\ell(y)\,dy$$

In this problem, $\ell(y) = \dfrac{c - a}{b}(b - y)$ by similar triangles, so

$$\bar{y} = \frac{1}{A}\int_0^b \frac{c - a}{b}\,y(b - y)\,dy = \frac{2}{b^2}\int_0^b (by - y^2)\,dy = \frac{2}{b^2}\left[\frac{1}{2}by^2 - \frac{1}{3}y^3\right]_0^b = \frac{2}{b^2} \cdot \frac{b^3}{6} = \frac{b}{3}$$

Notice that only one integral is needed when this method is used.

39. Divide the lamina into two triangles and one rectangle with respective masses of 2, 2 and 4, so that the total mass

is 8. Using the result of Exercise 37, the triangles have centroids $\left(-1, \frac{2}{3}\right)$ and $\left(1, \frac{2}{3}\right)$. The

centroid of the rectangle (its center) is $\left(0, -\frac{1}{2}\right)$. So, using Formulas 5 and 7, we have

$$\overline{y} = \frac{M_x}{m} = \frac{1}{m}\sum_{i=1}^{3} m_i y_i = \frac{1}{8}\left[2\left(\frac{2}{3}\right) + 2\left(\frac{2}{3}\right) + 4\left(-\frac{1}{2}\right)\right] = \frac{1}{8}\left(\frac{2}{3}\right) = \frac{1}{12}, \text{ and } \overline{x} = 0, \text{ since the lamina is symmetric}$$

about the line $x = 0$. Thus, the centroid is $(\overline{x}, \overline{y}) = \left(0, \frac{1}{12}\right)$.

41. A cone of height h and radius r can be generated by rotating a right

triangle about one of its legs as shown. By Exercise 37, $\overline{x} = \frac{1}{3}r$, so by the

Theorem of Pappus, the volume of the cone is

$$V = Ad = \left(\frac{1}{2} \cdot \text{base} \cdot \text{height}\right) \cdot (2\pi\overline{x}) = \frac{1}{2}rh \cdot 2\pi\left(\frac{1}{3}r\right) = \frac{1}{3}\pi r^2 h.$$

43. Suppose the region lies between two curves $y = f(x)$ and $y = g(x)$ where $f(x) \geq g(x)$, as illustrated in Figure 13.

Choose points x_i with $a = x_0 < x_1 < \cdots < x_n = b$ and choose x_i^* to be the midpoint of the ith subinterval; that

is, $x_i^* = \overline{x}_i = \frac{1}{2}(x_{i-1} + x_i)$. Then the centroid of the ith approximating rectangle R_i is its

center $C_i = \left(\overline{x}_i, \frac{1}{2}[f(\overline{x}_i) + g(\overline{x}_i)]\right)$. Its area is $[f(\overline{x}_i) - g(\overline{x}_i)]\,\Delta x$, so its mass is

$\rho[f(\overline{x}_i) - g(\overline{x}_i)]\,\Delta x$. Thus, $M_y(R_i) = \rho[f(\overline{x}_i) - g(\overline{x}_i)]\,\Delta x \cdot \overline{x}_i = \rho\overline{x}_i\,[f(\overline{x}_i) - g(\overline{x}_i)]\,\Delta x$ and

$M_x(R_i) = \rho[f(\overline{x}_i) - g(\overline{x}_i)]\,\Delta x \cdot \frac{1}{2}[f(\overline{x}_i) + g(\overline{x}_i)] = \rho \cdot \frac{1}{2}\left[f(\overline{x}_i)^2 - g(\overline{x}_i)^2\right]\Delta x$. Summing over i and taking

the limit as $n \to \infty$, we get $M_y = \lim_{n\to\infty} \sum_i \rho\overline{x}_i\,[f(\overline{x}_i) - g(\overline{x}_i)]\,\Delta x = \rho\int_a^b x[f(x) - g(x)]\,dx$ and

$M_x = \lim_{n\to\infty} \sum_i \rho \cdot \frac{1}{2}\left[f(\overline{x}_i)^2 - g(\overline{x}_i)^2\right]\Delta x = \rho\int_a^b \frac{1}{2}\left[f(x)^2 - g(x)^2\right]dx$. Thus,

$$\overline{x} = \frac{M_y}{m} = \frac{M_y}{\rho A} = \frac{1}{A}\int_a^b x[f(x) - g(x)]\,dx \quad \text{and} \quad \overline{y} = \frac{M_x}{m} = \frac{M_x}{\rho A} = \frac{1}{A}\int_a^b \frac{1}{2}\left[f(x)^2 - g(x)^2\right]dx$$

8.4 Applications to Economics and Biology

1. By the Net Change Theorem, $C(2000) - C(0) = \int_0^{2000} C'(x)\,dx \quad \Rightarrow$

$C(2000) = 20{,}000 + \int_0^{2000}(5 - 0.008x + 0.000009x^2)\,dx = 20{,}000 + \left[5x - 0.004x^2 + 0.000003x^3\right]_0^{2000}$

$= 20{,}000 + 10{,}000 - 0.004(4{,}000{,}000) + 0.000003(8{,}000{,}000{,}000) = 30{,}000 - 16{,}000 + 24{,}000$

$= \$38{,}000$

3. If the production level is raised from 1200 units to 1600 units, then the increase in cost is

$$C(1600) - C(1200) = \int_{1200}^{1600} C'(x)\, dx = \int_{1200}^{1600} (74 + 1.1x - 0.002x^2 + 0.00004x^3)\, dx$$

$$= \left[74x + 0.55x^2 - \tfrac{0.002}{3}x^3 + 0.00001x^4\right]_{1200}^{1600}$$

$$= 64{,}331{,}733.33 - 20{,}464{,}800 = \$43{,}866{,}933.33$$

5. $p(x) = 10 \quad \Rightarrow \quad \dfrac{450}{x+8} = 10 \quad \Rightarrow \quad x + 8 = 45 \quad \Rightarrow \quad x = 37.$

$$\text{Consumer surplus} = \int_0^{37} [p(x) - 10]\, dx = \int_0^{37} \left(\frac{450}{x+8} - 10\right) dx$$

$$= [450 \ln(x+8) - 10x]_0^{37}$$

$$= (450 \ln 45 - 370) - 450 \ln 8$$

$$= 450 \ln\left(\tfrac{45}{8}\right) - 370 \approx \$407.25$$

7. $P = p_S(x) \quad \Rightarrow \quad 400 = 200 + 0.2x^{3/2} \quad \Rightarrow \quad 200 = 0.2x^{3/2} \quad \Rightarrow \quad 1000 = x^{3/2} \quad \Rightarrow \quad x = 1000^{2/3} = 100.$

$$\text{Producer surplus} = \int_0^{100} [P - p_S(x)]\, dx = \int_0^{100} \left[400 - \left(200 + 0.2x^{3/2}\right)\right] dx = \int_0^{100} \left(200 - \tfrac{1}{5}x^{3/2}\right) dx$$

$$= \left[200x - \tfrac{2}{25}x^{5/2}\right]_0^{100} = 20{,}000 - 8{,}000 = \$12{,}000$$

9. $p(x) = \dfrac{800{,}000 e^{-x/5000}}{x + 20{,}000} = 16 \quad \Rightarrow \quad x = x_1 \approx 3727.04.$

$$\text{Consumer surplus} = \int_0^{x_1} [p(x) - 16]\, dx \approx \$37{,}753$$

11. $f(8) - f(4) = \int_4^8 f'(t)\, dt = \int_4^8 \sqrt{t}\, dt = \left[\tfrac{2}{3}t^{3/2}\right]_4^8 = \tfrac{2}{3}\left(16\sqrt{2} - 8\right) \approx \$9.75 \text{ million}$

13. $F = \dfrac{\pi P R^4}{8\eta l} = \dfrac{\pi(4000)(0.008)^4}{8(0.027)(2)} \approx 1.19 \times 10^{-4} \text{ cm}^3/\text{s}$

15. $\int_0^{12} c(t)\, dt = \int_0^{12} \tfrac{1}{4}t(12 - t)\, dt = \int_0^{12} \left(3t - \tfrac{1}{4}t^2\right) dt = \left[\tfrac{3}{2}t^2 - \tfrac{1}{12}t^3\right]_0^{12} = (216 - 144) = 72 \text{ mg} \cdot \text{s/L}.$

Thus, the cardiac output is $F = \dfrac{A}{\int_0^{12} c(t)\, dt} = \dfrac{8 \text{ mg}}{72 \text{ mg} \cdot \text{s/L}} = \dfrac{1}{9} \text{ L/s} = \dfrac{60}{9} \text{ L/min.}$

8.5 Probability

1. (a) $\int_{30,000}^{40,000} f(x)\,dx$ is the probability that a randomly chosen tire will have a lifetime between 30,000 and 40,000 miles.

(b) $\int_{25,000}^{\infty} f(x)\,dx$ is the probability that a randomly chosen tire will have a lifetime of at least 25,000 miles.

3. (a) In general, we must satisfy the two conditions that are mentioned before Example 1—namely, (1) $f(x) \geq 0$ for all x, and (2) $\int_{-\infty}^{\infty} f(x)\,dx = 1$. For $0 \leq x \leq 4$, we have $f(x) = \frac{3}{64}x\sqrt{16 - x^2} \geq 0$, so $f(x) \geq 0$ for all x.

Also, $\int_{-\infty}^{\infty} f(x)\,dx = \int_0^4 \frac{3}{64}x\sqrt{16 - x^2}\,dx = -\frac{3}{128}\int_0^4 (16 - x^2)^{1/2}(-2x)\,dx = -\frac{3}{128}\left[\frac{2}{3}(16 - x^2)^{3/2}\right]_0^4$

$= -\frac{1}{64}\left[(16 - x^2)^{3/2}\right]_0^4 = -\frac{1}{64}(0 - 64) = 1.$

Therefore, f is a probability density function.

(b) $P(X < 2) = \int_{-\infty}^2 f(x)\,dx = \int_0^2 \frac{3}{64}x\sqrt{16 - x^2}\,dx = -\frac{3}{128}\int_0^2 (16 - x^2)^{1/2}(-2x)\,dx$

$= -\frac{3}{128}\left[\frac{2}{3}(16 - x^2)^{3/2}\right]_0^2 = -\frac{1}{64}\left[(16 - x^2)^{3/2}\right]_0^2 = -\frac{1}{64}(12^{3/2} - 16^{3/2})$

$= \frac{1}{64}(64 - 12\sqrt{12}) = \frac{1}{64}(64 - 24\sqrt{3}) = 1 - \frac{3}{8}\sqrt{3} \approx 0.350481$

5. (a) In general, we must satisfy the two conditions that are mentioned before Example 1—namely, (1) $f(x) \geq 0$ for all x, and (2) $\int_{-\infty}^{\infty} f(x)\,dx = 1$. Since $f(x) = 0$ or $f(x) = 0.1$, condition (1) is satisfied. For condition (2), we

see that $\int_{-\infty}^{\infty} f(x)\,dx = \int_0^{10} 0.1\,dx = \left[\frac{1}{10}x\right]_0^{10} = 1$. Thus, $f(x)$ is a probability density function for the spinner's values.

(b) Since all the numbers between 0 and 10 are equally likely to be selected, we expect the mean to be halfway between the endpoints of the interval; that is, $x = 5$.

$\mu = \int_{-\infty}^{\infty} xf(x)\,dx = \int_0^{10} x(0.1)\,dx = \left[\frac{1}{20}x^2\right]_0^{10} = \frac{100}{20} = 5$, as expected.

7. We need to find m so that $\int_m^{\infty} f(t)\,dt = \frac{1}{2}$ $\Rightarrow$ $\lim_{x \to \infty} \int_m^x \frac{1}{5}e^{-t/5}\,dt = \frac{1}{2}$ $\Rightarrow$ $\lim_{x \to \infty} \left[\frac{1}{5}(-5)e^{-t/5}\right]_m^x = \frac{1}{2}$ $\Rightarrow$

$(-1)(0 - e^{-m/5}) = \frac{1}{2}$ $\Rightarrow$ $e^{-m/5} = \frac{1}{2}$ $\Rightarrow$ $-m/5 = \ln\frac{1}{2}$ $\Rightarrow$ $m = -5\ln\frac{1}{2} = 5\ln 2 \approx 3.47$ min.

9. We use an exponential density function with $\mu = 2.5$ min.

(a) $P(X > 4) = \int_4^{\infty} f(t)\,dt = \lim_{x \to \infty} \int_4^x \frac{1}{2.5}e^{-t/2.5}\,dt = \lim_{x \to \infty} \left[-e^{-t/2.5}\right]_4^x = 0 + e^{-4/2.5} \approx 0.202$

(b) $P(0 \leq X \leq 2) = \int_0^2 f(t)\,dt = \left[-e^{-t/2.5}\right]_0^2 = -e^{-2/2.5} + 1 \approx 0.551$

(c) We need to find a value a so that $P(X \geq a) = 0.02$, or, equivalently, $P(0 \leq X \leq a) = 0.98$ $\Leftrightarrow$

$\int_0^a f(t)\,dt = 0.98$ $\Leftrightarrow$ $\left[-e^{-t/2.5}\right]_0^a = 0.98$ $\Leftrightarrow$ $-e^{-a/2.5} + 1 = 0.98$ $\Leftrightarrow$ $e^{-a/2.5} = 0.02$ $\Leftrightarrow$

$-a/2.5 = \ln 0.02$ $\Leftrightarrow$ $a = -2.5\ln\frac{1}{50} = 2.5\ln 50 \approx 9.78$ min ≈ 10 min. The ad should say that if you aren't served within 10 minutes, you get a free hamburger.

11. $P(X \geq 10) = \int_{10}^{\infty} \frac{1}{4.2\sqrt{2\pi}} \exp\left(-\frac{(x-9.4)^2}{2 \cdot 4.2^2}\right) dx$. To avoid the improper integral we approximate it by the

integral from 10 to 100. Thus, $P(X \geq 10) \approx \int_{10}^{100} \frac{1}{4.2\sqrt{2\pi}} \exp\left(-\frac{(x-9.4)^2}{2 \cdot 4.2^2}\right) dx \approx 0.443$ (using a calculator

or computer to estimate the integral), so about 44 percent of the households throw out at least 10 lb of paper a week.

Note: We can't evaluate $1 - P(0 \leq X \leq 10)$ for this problem since a significant amount of area lies to the left of

$X = 0$.

13. $P(\mu - 2\sigma \leq X \leq \mu + 2\sigma) = \int_{\mu-2\sigma}^{\mu+2\sigma} \frac{1}{\sigma\sqrt{2\pi}} \exp\left(-\frac{(x-\mu)^2}{2\sigma^2}\right) dx$. Substituting $t = \frac{x-\mu}{\sigma}$ and $dt = \frac{1}{\sigma} dx$

gives us

$$\int_{-2}^{2} \frac{1}{\sigma\sqrt{2\pi}} e^{-t^2/2}(\sigma\, dt) = \frac{1}{\sqrt{2\pi}} \int_{-2}^{2} e^{-t^2/2}\, dt \approx 0.9545$$

15. (a) First $p(r) = \frac{4}{a_0^3} r^2 e^{-2r/a_0} \geq 0$ for $r \geq 0$. Next,

$$\int_{-\infty}^{\infty} p(r)\, dr = \int_{0}^{\infty} \frac{4}{a_0^3} r^2 e^{-2r/a_0}\, dr = \frac{4}{a_0^3} \lim_{t \to \infty} \int_{0}^{t} r^2 e^{-2r/a_0}\, dr$$

By using parts, tables, or a CAS , we find that $\int x^2 e^{bx}\, dx = (e^{bx}/b^3)(b^2 x^2 - 2bx + 2)$. (*)

Next, we use (*) (with $b = -2/a_0$) and l'Hospital's Rule to get $\frac{4}{a_0^3}\left[\frac{a_0^3}{-8}(-2)\right] = 1$. This satisfies the second

condition for a function to be a probability density function.

(b) Using l'Hospital's Rule, $\frac{4}{a_0^3} \lim_{r \to \infty} \frac{r^2}{e^{2r/a_0}} = \frac{4}{a_0^3} \lim_{r \to \infty} \frac{2r}{(2/a_0)e^{2r/a_0}} = \frac{2}{a_0^2} \lim_{r \to \infty} \frac{2}{(2/a_0)e^{2r/a_0}} = 0$.

To find the maximum of p, we differentiate:

$$p'(r) = \frac{4}{a_0^3}\left[r^2 e^{-2r/a_0}\left(-\frac{2}{a_0}\right) + e^{-2r/a_0}(2r)\right] = \frac{4}{a_0^3} e^{-2r/a_0}(2r)\left(-\frac{r}{a_0} + 1\right)$$

$p'(r) = 0 \iff r = 0$ or $1 = \frac{r}{a_0} \iff r = a_0$ [$a_0 \approx 5.59 \times 10^{-11}$ m]. $p'(r)$ changes from positive to

negative at $r = a_0$, so $p(r)$ has its maximum value at $r = a_0$.

(c) It is fairly difficult to find a viewing rectangle, but knowing the

maximum value from part (b) helps.

$$p(a_0) = \frac{4}{a_0^3} a_0^2 e^{-2a_0/a_0} = \frac{4}{a_0} e^{-2} \approx 9,684,098,979$$

With a maximum of nearly 10 billion and a total area under the curve

of 1, we know that the "hump" in the graph must be extremely

narrow.

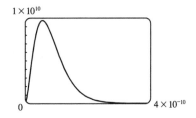

(d) $P(r) = \int_0^r \frac{4}{a_0^3} s^2 e^{-2s/a_0} \, ds \;\; \Rightarrow \;\; P(4a_0) = \int_0^{4a_0} \frac{4}{a_0^3} s^2 e^{-2s/a_0} \, ds.$ Using $(*)$ from part (a)

(with $b = -2/a_0$),

$$P(4a_0) = \frac{4}{a_0^3} \left[\frac{e^{-2s/a_0}}{-8/a_0^3} \left(\frac{4}{a_0^2} s^2 + \frac{4}{a_0} s + 2 \right) \right]_0^{4a_0} = \frac{4}{a_0^3} \left(\frac{a_0^3}{-8} \right) \left[e^{-8}(64 + 16 + 2) - 1(2) \right]$$

$$= -\tfrac{1}{2} \left(82e^{-8} - 2 \right) = 1 - 41e^{-8} \approx 0.986$$

(e) $\mu = \int_{-\infty}^\infty r p(r) \, dr = \frac{4}{a_0^3} \lim_{t \to \infty} \int_0^t r^3 e^{-2r/a_0} \, dr.$ Integrating by parts three times or using a CAS, we find that

$\int x^3 e^{bx} \, dx = \frac{e^{bx}}{b^4} \left(b^3 x^3 - 3b^2 x^2 + 6bx - 6 \right).$ So with $b = -\dfrac{2}{a_0}$, we use l'Hospital's Rule, and get

$$\mu = \frac{4}{a_0^3} \left[-\frac{a_0^4}{16} (-6) \right] = \tfrac{3}{2} a_0.$$

8 Review

CONCEPT CHECK

1. (a) The length of a curve is defined to be the limit of the lengths of the inscribed polygons, as described near Figure 3 in Section 8.1.

   (b) See Equation 8.1.2.

   (c) See Equation 8.1.4.

2. (a) $S = \int_a^b 2\pi f(x) \sqrt{1 + [f'(x)]^2} \, dx$

   (b) If $x = g(y), c \le y \le d$, then $S = \int_c^d 2\pi y \sqrt{1 + [g'(y)]^2} \, dy.$

   (c) $S = \int_a^b 2\pi x \sqrt{1 + [f'(x)]^2} \, dx$ or $S = \int_c^d 2\pi g(y) \sqrt{1 + [g'(y)]^2} \, dy$

3. Let $c(x)$ be the cross-sectional length of the wall (measured parallel to the surface of the fluid) at depth x. Then the hydrostatic force against the wall is given by $F = \int_a^b \delta x c(x) \, dx$, where a and b are the lower and upper limits for x at points of the wall and δ is the weight density of the fluid.

4. (a) The center of mass is the point at which the plate balances horizontally.

   (b) See Equations 8.3.8.

5. If a plane region $\mathcal{R}$ that lies entirely on one side of a line ℓ in its plane is rotated about ℓ, then the volume of the resulting solid is the product of the area of $\mathcal{R}$ and the distance traveled by the centroid of $\mathcal{R}$.

6. See Figure 3 in Section 8.4, and the discussion which precedes it.

7. (a) See the definition in the first paragraph of the subsection *Cardiac Output* in Section 8.4.

   (b) See the discussion in the second paragraph of the subsection *Cardiac Output* in Section 8.4.

8. A probability density function f is a function on the domain of a continuous random variable X such that $\int_a^b f(x) \, dx$ measures the probability that X lies between a and b. Such a function f has nonnegative values and satisfies the relation $\int_D f(x) \, dx = 1$, where D is the domain of the corresponding random variable X. If $D = \mathbb{R}$, or if we define $f(x) = 0$ for real numbers $x \notin D$, then $\int_{-\infty}^\infty f(x) \, dx = 1$. (Of course, to work with f in this way, we must assume that the integrals of f exist.)

9. (a) $\int_0^{100} f(x)\,dx$ represents the probability that the weight of a randomly chosen female college student is less than 100 pounds.

(b) $\mu = \int_{-\infty}^{\infty} x f(x)\,dx = \int_0^{\infty} x f(x)\,dx$

(c) The median of f is the number m such that $\int_m^{\infty} f(x)\,dx = \frac{1}{2}$.

10. See the discussion near Equation 3 in Section 8.5.

──────────────── EXERCISES ────────────────

1. $y = \frac{1}{6}(x^2 + 4)^{3/2} \Rightarrow dy/dx = \frac{1}{4}(x^2 + 4)^{1/2}(2x) \Rightarrow$

$1 + \left(\dfrac{dy}{dx}\right)^2 = 1 + \left[\frac{1}{2}x(x^2 + 4)^{1/2}\right]^2 = 1 + \frac{1}{4}x^2(x^2 + 4) = \frac{1}{4}x^4 + x^2 + 1 = \left(\frac{1}{2}x^2 + 1\right)^2.$

Thus, $L = \int_0^3 \sqrt{\left(\frac{1}{2}x^2 + 1\right)^2}\,dx = \int_0^3 \left(\frac{1}{2}x^2 + 1\right)dx = \left[\frac{1}{6}x^3 + x\right]_0^3 = \frac{15}{2}.$

3. (a) $y = \dfrac{x^4}{16} + \dfrac{1}{2x^2} = \dfrac{1}{16}x^4 + \dfrac{1}{2}x^{-2} \Rightarrow \dfrac{dy}{dx} = \dfrac{1}{4}x^3 - x^{-3} \Rightarrow$

$1 + (dy/dx)^2 = 1 + \left(\frac{1}{4}x^3 - x^{-3}\right)^2 = 1 + \frac{1}{16}x^6 - \frac{1}{2} + x^{-6} = \frac{1}{16}x^6 + \frac{1}{2} + x^{-6} = \left(\frac{1}{4}x^3 + x^{-3}\right)^2.$

Thus, $L = \int_1^2 \left(\frac{1}{4}x^3 + x^{-3}\right)dx = \left[\frac{1}{16}x^4 - \frac{1}{2}x^{-2}\right]_1^2 = \left(1 - \frac{1}{8}\right) - \left(\frac{1}{16} - \frac{1}{2}\right) = \frac{21}{16}.$

(b) $S = \int_1^2 2\pi x\left(\frac{1}{4}x^3 + x^{-3}\right)dx = 2\pi \int_1^2 \left(\frac{1}{4}x^4 + x^{-2}\right)dx = 2\pi\left[\frac{1}{20}x^5 - \frac{1}{x}\right]_1^2$

$= 2\pi\left[\left(\frac{32}{20} - \frac{1}{2}\right) - \left(\frac{1}{20} - 1\right)\right] = 2\pi\left(\frac{8}{5} - \frac{1}{2} - \frac{1}{20} + 1\right) = 2\pi\left(\frac{41}{20}\right) = \frac{41}{10}\pi$

5. $y = e^{-x^2} \Rightarrow dy/dx = -2xe^{-x^2} \Rightarrow 1 + (dy/dx)^2 = 1 + 4x^2 e^{-2x^2}.$

Let $f(x) = \sqrt{1 + 4x^2 e^{-2x^2}}$. Then

$L = \int_0^3 f(x)\,dx \approx S_6 = \dfrac{(3 - 0)/6}{3}[f(0) + 4f(0.5) + 2f(1) + 4f(1.5) + 2f(2) + 4f(2.5) + f(3)]$

≈ 3.292287

7. $y = \int_1^x \sqrt{\sqrt{t} - 1}\,dt \Rightarrow dy/dx = \sqrt{\sqrt{x} - 1} \Rightarrow 1 + (dy/dx)^2 = 1 + \left(\sqrt{x} - 1\right) = \sqrt{x}.$

Thus, $L = \int_1^{16} \sqrt{\sqrt{x}}\,dx = \int_1^{16} x^{1/4}\,dx = \frac{4}{5}\left[x^{5/4}\right]_1^{16} = \frac{4}{5}(32 - 1) = \frac{124}{5}.$

9. As in Example 1 of Section 8.3, $\dfrac{a}{2 - x} = \dfrac{1}{2} \Rightarrow 2a = 2 - x$ and

$w = 2(1.5 + a) = 3 + 2a = 3 + 2 - x = 5 - x$. Thus,

$F = \int_0^2 \rho g x(5 - x)\,dx = \rho g\left[\frac{5}{2}x^2 - \frac{1}{3}x^3\right]_0^2 = \rho g\left(10 - \frac{8}{3}\right) = \frac{22}{3}\delta \ \ [\rho g = \delta] \approx \frac{22}{3} \cdot 62.5 \approx 458 \text{ lb}.$

11. $A = \int_{-2}^{1} \left[(4 - x^2) - (x + 2)\right] dx = \int_{-2}^{1} \left(2 - x - x^2\right) dx = \left[2x - \frac{1}{2}x^2 - \frac{1}{3}x^3\right]_{-2}^{1}$

$\qquad = \left(2 - \frac{1}{2} - \frac{1}{3}\right) - \left(-4 - 2 + \frac{8}{3}\right) = \frac{9}{2} \quad \Rightarrow$

$\qquad\qquad \bar{x} = A^{-1} \int_{-2}^{1} x(2 - x - x^2) dx = \frac{2}{9} \int_{-2}^{1} \left(2x - x^2 - x^3\right) dx = \frac{2}{9}\left[x^2 - \frac{1}{3}x^3 - \frac{1}{4}x^4\right]_{-2}^{1}$

$\qquad\qquad = \frac{2}{9}\left[\left(1 - \frac{1}{3} - \frac{1}{4}\right) - \left(4 + \frac{8}{3} - 4\right)\right] = -\frac{1}{2}$

and $\qquad \bar{y} = A^{-1} \int_{-2}^{1} \frac{1}{2}\left[(4 - x^2)^2 - (x + 2)^2\right] dx = \frac{1}{9} \int_{-2}^{1} \left(x^4 - 9x^2 - 4x + 12\right) dx$

$\qquad\qquad = \frac{1}{9}\left[\frac{1}{5}x^5 - 3x^3 - 2x^2 + 12x\right]_{-2}^{1} = \frac{1}{9}\left[\left(\frac{1}{5} - 3 - 2 + 12\right) - \left(-\frac{32}{5} + 24 - 8 - 24\right)\right] = \frac{12}{5}$

Thus, the centroid is $(\bar{x}, \bar{y}) = \left(-\frac{1}{2}, \frac{12}{5}\right)$.

13. An equation of the line passing through $(0, 0)$ and $(3, 2)$ is $y = \frac{2}{3}x$. $A = \frac{1}{2} \cdot 3 \cdot 2 = 3$. Therefore, using

Equations 8.3.8, $\bar{x} = \frac{1}{3} \int_0^3 x\left(\frac{2}{3}x\right) dx = \frac{2}{27}\left[x^3\right]_0^3 = 2$, and $\bar{y} = \frac{1}{3} \int_0^3 \frac{1}{2}\left(\frac{2}{3}x\right)^2 dx = \frac{2}{81}\left[x^3\right]_0^3 = \frac{2}{3}$.

Thus, the centroid is $(\bar{x}, \bar{y}) = \left(2, \frac{2}{3}\right)$.

15. The centroid of this circle, $(1, 0)$, travels a distance $2\pi(1)$ when the lamina is rotated about the y-axis. The area of

the circle is $\pi(1)^2$. So by the Theorem of Pappus, $V = A(2\pi\bar{x}) = \pi(1)^2 2\pi(1) = 2\pi^2$.

17. $x = 100 \quad \Rightarrow \quad P = 2000 - 0.1(100) - 0.01(100)^2 = 1890$

$\qquad\qquad \text{Consumer surplus} = \int_0^{100} [p(x) - P] dx = \int_0^{100} \left(2000 - 0.1x - 0.01x^2 - 1890\right) dx$

$\qquad\qquad\qquad = \left[110x - 0.05x^2 - \frac{0.01}{3}x^3\right]_0^{100} = 11,000 - 500 - \frac{10,000}{3} \approx \$7166.67$

19. $f(x) = \begin{cases} \frac{\pi}{20} \sin\left(\frac{\pi}{10}x\right) & \text{if } 0 \leq x \leq 10 \\ 0 & \text{if } x < 0 \text{ or } x > 10 \end{cases}$

(a) $f(x) \geq 0$ for all real numbers x and

$\qquad\qquad \int_{-\infty}^{\infty} f(x) dx = \int_0^{10} \frac{\pi}{20} \sin\left(\frac{\pi}{10}x\right) dx = \frac{\pi}{20} \cdot \frac{10}{\pi}\left[-\cos\left(\frac{\pi}{10}x\right)\right]_0^{10}$

$\qquad\qquad\qquad = \frac{1}{2}(-\cos\pi + \cos 0) = \frac{1}{2}(1 + 1) = 1$

Therefore, f is a probability density function.

(b) $P(X < 4) = \int_{-\infty}^{4} f(x) dx = \int_0^4 \frac{\pi}{20} \sin\left(\frac{\pi}{10}x\right) dx = \frac{1}{2}\left[-\cos\left(\frac{\pi}{10}x\right)\right]_0^4 = \frac{1}{2}\left(-\cos\frac{2\pi}{5} + \cos 0\right)$

$\qquad\qquad \approx \frac{1}{2}(-0.309017 + 1) \approx 0.3455$

(c) $\mu = \int_{-\infty}^{\infty} x f(x)\, dx = \int_0^{10} \frac{\pi}{20} x \sin\left(\frac{\pi}{10} x\right) dx$

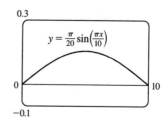

$= \int_0^{\pi} \frac{\pi}{20} \cdot \frac{10}{\pi} u (\sin u)\left(\frac{10}{\pi}\right) du \quad [u = \frac{\pi}{10} x,\, du = \frac{\pi}{10}\, dx]$

$= \frac{5}{\pi} \int_0^{\pi} u \sin u\, du \overset{82}{=} \frac{5}{\pi}\left[\sin u - u \cos u\right]_0^{\pi} = \frac{5}{\pi}[0 - \pi(-1)] = 5$

This answer is expected because the graph of f is symmetric about

the line $x = 5$.

21. (a) The probability density function is $f(t) = \begin{cases} 0 & \text{if } t < 0 \\ \frac{1}{8} e^{-t/8} & \text{if } t \ge 0 \end{cases}$

$P(0 \le X \le 3) = \int_0^3 \frac{1}{8} e^{-t/8}\, dt = \left[-e^{-t/8}\right]_0^3 = -e^{-3/8} + 1 \approx 0.3127$

(b) $P(X > 10) = \int_{10}^{\infty} \frac{1}{8} e^{-t/8}\, dt = \lim_{x \to \infty}\left[-e^{-t/8}\right]_{10}^{x} = \lim_{x \to \infty}\left(-e^{-x/8} + e^{-10/8}\right) = 0 + e^{-5/4} \approx 0.2865$

(c) We need to find m such that $P(X \ge m) = \frac{1}{2} \Rightarrow \int_m^{\infty} \frac{1}{8} e^{-t/8}\, dt = \frac{1}{2} \Rightarrow \lim_{x \to \infty}\left[-e^{-t/8}\right]_m^{x} = \frac{1}{2} \Rightarrow$

$\lim_{x \to \infty}\left(-e^{-x/8} + e^{-m/8}\right) = \frac{1}{2} \Rightarrow e^{-m/8} = \frac{1}{2} \Rightarrow -m/8 = \ln \frac{1}{2} \Rightarrow$

$m = -8 \ln \frac{1}{2} = 8 \ln 2 \approx 5.55$ minutes.

1. $x^2 + y^2 \leq 4y \iff x^2 + (y-2)^2 \leq 4$, so S is part of a circle, as shown in the diagram. The area of S is

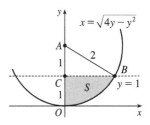

$$\int_0^1 \sqrt{4y - y^2}\, dy \stackrel{113}{=} \left[\frac{y-2}{2}\sqrt{4y-y^2} + 2\cos^{-1}\left(\frac{2-y}{2}\right)\right]_0^1 \quad [a = 2]$$

$$= -\tfrac{1}{2}\sqrt{3} + 2\cos^{-1}\left(\tfrac{1}{2}\right) - 2\cos^{-1} 1$$

$$= -\tfrac{\sqrt{3}}{2} + 2\left(\tfrac{\pi}{3}\right) - 2(0) = \tfrac{2\pi}{3} - \tfrac{\sqrt{3}}{2}$$

Another method (without calculus): Note that $\theta = \angle CAB = \frac{\pi}{3}$, so the area is

$$(\text{area of sector } OAB) - (\text{area of } \triangle ABC) = \tfrac{1}{2}\left(2^2\right)\tfrac{\pi}{3} - \tfrac{1}{2}(1)\sqrt{3} = \tfrac{2\pi}{3} - \tfrac{\sqrt{3}}{2}$$

3. (a) The two spherical zones, whose surface areas we will call S_1 and S_2, are generated by rotation about the y-axis of circular arcs, as indicated in the figure. The arcs are the upper and lower portions of the circle $x^2 + y^2 = r^2$ that are obtained when the circle is cut with the line $y = d$. The portion of the upper arc in the first quadrant is sufficient to generate the upper spherical zone. That portion of the arc can be described by the relation $x = \sqrt{r^2 - y^2}$ for $d \leq y \leq r$. Thus, $dx/dy = -y/\sqrt{r^2 - y^2}$ and

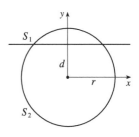

$$ds = \sqrt{1 + \left(\frac{dx}{dy}\right)^2}\, dy = \sqrt{1 + \frac{y^2}{r^2 - y^2}}\, dy = \sqrt{\frac{r^2}{r^2 - y^2}}\, dy = \frac{r\, dy}{\sqrt{r^2 - y^2}}$$

From Formula 8.2.8 we have

$$S_1 = \int_d^r 2\pi x \sqrt{1 + \left(\frac{dx}{dy}\right)^2}\, dy = \int_d^r 2\pi \sqrt{r^2 - y^2}\, \frac{r\, dy}{\sqrt{r^2 - y^2}} = \int_d^r 2\pi r\, dy = 2\pi r(r - d)$$

Similarly, we can compute $S_2 = \int_{-r}^d 2\pi x \sqrt{1 + (dx/dy)^2}\, dy = \int_{-r}^d 2\pi r\, dy = 2\pi r(r + d)$. Note that $S_1 + S_2 = 4\pi r^2$, the surface area of the entire sphere.

(b) $r = 3960$ mi and $d = r(\sin 75°) \approx 3825$ mi, so the surface area of the Arctic Ocean is about

$$2\pi r(r - d) \approx 2\pi(3960)(135) \approx 3.36 \times 10^6 \text{ mi}^2.$$

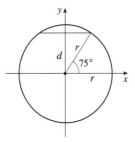

(c) The area on the sphere lies between planes $y = y_1$ and $y = y_2$, where $y_2 - y_1 = h$. Thus, we compute the

surface area on the sphere to be $S = \displaystyle\int_{y_1}^{y_2} 2\pi x \sqrt{1 + \left(\dfrac{dx}{dy}\right)^2}\, dy = \int_{y_1}^{y_2} 2\pi r\, dy = 2\pi r(y_2 - y_1) = 2\pi rh.$

This equals the lateral area of a cylinder of radius r and
height h, since such a cylinder is obtained by rotating the line
$x = r$ about the y-axis, so the surface area of the cylinder
between the planes $y = y_1$ and $y = y_2$ is

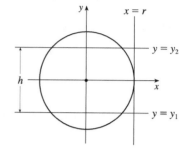

$A = \displaystyle\int_{y_1}^{y_2} 2\pi x \sqrt{1 + \left(\dfrac{dx}{dy}\right)^2}\, dy = \int_{y_1}^{y_2} 2\pi r \sqrt{1 + 0^2}\, dy$

$= 2\pi ry\big|_{y=y_1}^{y_2} = 2\pi r(y_2 - y_1) = 2\pi rh$

(d) $h = 2r\sin 23.45° \approx 3152$ mi, so the surface area of the
Torrid Zone is $2\pi rh \approx 2\pi(3960)(3152) \approx 7.84 \times 10^7$ mi^2.

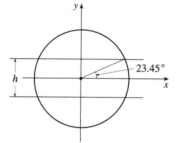

5. (a) Choose a vertical x-axis pointing downward with its origin at the surface. In order to calculate the pressure at
depth z, consider n subintervals of the interval $[0, z]$ by points x_i and choose a point $x_i^* \in [x_{i-1}, x_i]$ for each i.
The thin layer of water lying between depth x_{i-1} and depth x_i has a density of approximately $\rho(x_i^*)$, so the
weight of a piece of that layer with unit cross-sectional area is $\rho(x_i^*)g\,\Delta x$. The total weight of a column of
water extending from the surface to depth z (with unit cross-sectional area) would be approximately

$\displaystyle\sum_{i=1}^{n} \rho(x_i^*)g\,\Delta x.$ The estimate becomes exact if we take the limit as $n \to \infty$; weight (or force) per unit area at

depth z is $W = \displaystyle\lim_{n\to\infty} \sum_{i=1}^{n} \rho(x_i^*)g\,\Delta x.$ In other words, $P(z) = \int_0^z \rho(x)g\,dx.$ More generally, if we make no

assumptions about the location of the origin, then $P(z) = P_0 + \int_0^z \rho(x)g\,dx$, where P_0 is the pressure at $x = 0$.
Differentiating, we get $dP/dz = \rho(z)g.$

(b)

$F = \int_{-r}^{r} P(L+x) \cdot 2\sqrt{r^2 - x^2}\, dx$

$= \int_{-r}^{r} \left(P_0 + \int_0^{L+x} \rho_0 e^{z/H} g\, dz\right) \cdot 2\sqrt{r^2 - x^2}\, dx$

$= P_0 \int_{-r}^{r} 2\sqrt{r^2 - x^2}\, dx + \rho_0 g H \int_{-r}^{r} \left(e^{(L+x)/H} - 1\right) \cdot 2\sqrt{r^2 - x^2}\, dx$

$= (P_0 - \rho_0 gH) \int_{-r}^{r} 2\sqrt{r^2 - x^2}\, dx + \rho_0 gH \int_{-r}^{r} e^{(L+x)/H} \cdot 2\sqrt{r^2 - x^2}\, dx$

$= (P_0 - \rho_0 gH)(\pi r^2) + \rho_0 gH e^{L/H} \int_{-r}^{r} e^{x/H} \cdot 2\sqrt{r^2 - x^2}\, dx$

7. To find the height of the pyramid, we use similar triangles. The first figure shows a cross-section of the pyramid passing through the top and through two opposite corners of the square base. Now $|BD| = b$, since it is a radius of the sphere, which has diameter $2b$ since it is tangent to the opposite sides of the square base. Also, $|AD| = b$ since $\triangle ADB$ is isosceles. So the height is $|AB| = \sqrt{b^2 + b^2} = \sqrt{2}\,b$.

We first observe that the shared volume is equal to half the volume of the sphere, minus the sum of the four equal volumes (caps of the sphere) cut off by the triangular faces of the pyramid. See Exercise 6.2.49 for a derivation of the formula for the volume of a cap of a sphere. To use the formula, we need to find the perpendicular distance h of each triangular face from the surface of the sphere. We first find the distance d from the center of the sphere to one of the triangular faces. The third figure shows a cross-section of the pyramid through the top and through the midpoints of opposite sides of the square base. From similar triangles we find that

$$\frac{d}{b} = \frac{|AB|}{|AC|} = \frac{\sqrt{2}\,b}{\sqrt{b^2 + (\sqrt{2}\,b)^2}} \quad \Rightarrow \quad d = \frac{\sqrt{2}\,b^2}{\sqrt{3b^2}} = \frac{\sqrt{6}}{3}\,b$$

So $h = b - d = b - \frac{\sqrt{6}}{3}\,b = \frac{3 - \sqrt{6}}{3}\,b$. So, using the formula $V = \pi h^2(r - h/3)$ from Exercise 6.2.49 with $r = b$, we find that the volume of each of the caps is

$$\pi\left(\frac{3 - \sqrt{6}}{3}\,b\right)^2\left(b - \frac{3 - \sqrt{6}}{3 \cdot 3}\,b\right) = \frac{15 - 6\sqrt{6}}{9} \cdot \frac{6 + \sqrt{6}}{9}\pi b^3 = \left(\frac{2}{3} - \frac{7}{27}\sqrt{6}\right)\pi b^3.$$ So, using our first observation, the shared volume is $V = \frac{1}{2}\left(\frac{4}{3}\pi b^3\right) - 4\left(\frac{2}{3} - \frac{7}{27}\sqrt{6}\right)\pi b^3 = \left(\frac{28}{27}\sqrt{6} - 2\right)\pi b^3$.

9. We can assume that the cut is made along a vertical line $x = b > 0$, that the disk's boundary is the circle $x^2 + y^2 = 1$, and that the center of mass of the smaller piece (to the right of $x = b$) is $\left(\frac{1}{2}, 0\right)$. We wish to find b to two decimal places. We have

$$\frac{1}{2} = \bar{x} = \frac{\int_b^1 x \cdot 2\sqrt{1 - x^2}\,dx}{\int_b^1 2\sqrt{1 - x^2}\,dx}.$$ Evaluating the numerator gives us

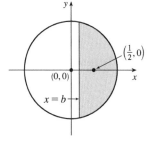

$$-\int_b^1 (1 - x^2)^{1/2}(-2x)\,dx = -\frac{2}{3}\left[(1 - x^2)^{3/2}\right]_b^1 = -\frac{2}{3}\left[0 - (1 - b^2)^{3/2}\right] = \frac{2}{3}(1 - b^2)^{3/2}.$$ Using

Formula 30 in the table of integrals, we find that the denominator is

$$\left[x\sqrt{1-x^2} + \sin^{-1}x\right]_b^1 = \left(0 + \tfrac{\pi}{2}\right) - \left(b\sqrt{1-b^2} + \sin^{-1}b\right). \text{ Thus, we have}$$

$$\frac{1}{2} = \overline{x} = \frac{\tfrac{2}{3}(1-b^2)^{3/2}}{\tfrac{\pi}{2} - b\sqrt{1-b^2} - \sin^{-1}b}, \text{ or, equivalently, } \tfrac{2}{3}(1-b^2)^{3/2} = \tfrac{\pi}{4} - \tfrac{1}{2}b\sqrt{1-b^2} - \tfrac{1}{2}\sin^{-1}b. \text{ Solving this}$$

equation numerically with a calculator or CAS, we obtain $b \approx 0.138173$, or $b = 0.14$ m to two decimal places.

11. If $h = L$, then

$$P = \frac{\text{area under } y = L\sin\theta}{\text{area of rectangle}} = \frac{\int_0^\pi L\sin\theta\, d\theta}{\pi L} = \frac{[-\cos\theta]_0^\pi}{\pi} = \frac{-(-1)+1}{\pi} = \frac{2}{\pi}$$

If $h = L/2$, then

$$P = \frac{\text{area under } y = \tfrac{1}{2}L\sin\theta}{\text{area of rectangle}} = \frac{\int_0^\pi \tfrac{1}{2}L\sin\theta\, d\theta}{\pi L} = \frac{[-\cos\theta]_0^\pi}{2\pi} = \frac{2}{2\pi} = \frac{1}{\pi}$$

9 □ DIFFERENTIAL EQUATIONS

9.1 Modeling with Differential Equations

1. $y = x - x^{-1} \Rightarrow y' = 1 + x^{-2}$. To show that y is a solution of the differential equation, we will substitute the expressions for y and y' in the left-hand side of the equation and show that the left-hand side is equal to the right-hand side.

$$\text{LHS} = xy' + y = x(1 + x^{-2}) + (x - x^{-1}) = x + x^{-1} + x - x^{-1} = 2x = \text{RHS}$$

3. (a) $y = \sin kt \Rightarrow y' = k \cos kt \Rightarrow y'' = -k^2 \sin kt$. $y'' + 9y = 0 \Rightarrow$
$-k^2 \sin kt + 9 \sin kt = 0$ for all $t \Leftrightarrow (9 - k^2) \sin kt = 0$ for all $t \Leftrightarrow 9 - k^2 = 0 \Leftrightarrow k = \pm 3$

   (b) $y = A \sin kt + B \cos kt \Rightarrow y' = Ak \cos kt - Bk \sin kt \Rightarrow y'' = -Ak^2 \sin kt - Bk^2 \cos kt$.
   Thus, $y'' + 9y = 0 \Rightarrow -Ak^2 \sin kt - Bk^2 \cos kt + 9(A \sin kt + B \cos kt) = 0 \Rightarrow$
   $(9 - k^2) A \sin kt + (9 - k^2) B \cos kt = 0$. The last equation is true for all values of A and B if $k = \pm 3$.

5. (a) $y = e^t \Rightarrow y' = e^t \Rightarrow y'' = e^t$. $\text{LHS} = y'' + 2y' + y = e^t + 2e^t + e^t = 4e^t \neq 0$, so $y = e^t$ is not a solution of the differential equation.

   (b) $y = e^{-t} \Rightarrow y' = -e^{-t} \Rightarrow y'' = e^{-t}$. $\text{LHS} = y'' + 2y' + y = e^{-t} - 2e^{-t} + e^{-t} = 0 = \text{RHS}$, so $y = e^{-t}$ is a solution.

   (c) $y = te^{-t} \Rightarrow y' = t(-e^{-t}) + e^{-t}(1) = e^{-t}(1 - t) \Rightarrow y'' = e^{-t}(t - 2)$.

   $$\text{LHS} = y'' + 2y' + y = e^{-t}(t - 2) + 2e^{-t}(1 - t) + te^{-t}$$
   $$= e^{-t}[(t - 2) + 2(1 - t) + t] = e^{-t}(0) = 0 = \text{RHS},$$

   so $y = te^{-t}$ is a solution.

   (d) $y = t^2 e^{-t} \Rightarrow y' = te^{-t}(2 - t) \Rightarrow y'' = e^{-t}(t^2 - 4t + 2)$.

   $$\text{LHS} = y'' + 2y' + y = e^{-t}(t^2 - 4t + 2) + 2te^{-t}(2 - t) + t^2 e^{-t}$$
   $$= e^{-t}[(t^2 - 4t + 2) + 2t(2 - t) + t^2] = e^{-t}(2) \neq 0,$$

   so $y = t^2 e^{-t}$ is not a solution.

7. (a) Since the derivative $y' = -y^2$ is always negative (or 0 if $y = 0$), the function y must be decreasing (or equal to 0) on any interval on which it is defined.

   (b) $y = \dfrac{1}{x + C} \Rightarrow y' = -\dfrac{1}{(x + C)^2}$. $\text{LHS} = y' = -\dfrac{1}{(x + C)^2} = -\left(\dfrac{1}{x + C}\right)^2 = -y^2 = \text{RHS}$

   (c) $y = 0$ is a solution of $y' = -y^2$ that is not a member of the family in part (b).

   (d) If $y(x) = \dfrac{1}{x + C}$, then $y(0) = \dfrac{1}{0 + C} = \dfrac{1}{C}$. Since $y(0) = 0.5$, $\dfrac{1}{C} = \dfrac{1}{2} \Rightarrow C = 2$, so $y = \dfrac{1}{x + 2}$.

9. (a) $\dfrac{dP}{dt} = 1.2P\left(1 - \dfrac{P}{4200}\right)$. Now $\dfrac{dP}{dt} > 0 \;\Rightarrow\; 1 - \dfrac{P}{4200} > 0$ [assuming that $P > 0$] $\;\Rightarrow\; \dfrac{P}{4200} < 1 \;\Rightarrow$

$P < 4200 \;\Rightarrow\;$ the population is increasing for $0 < P < 4200$.

(b) $\dfrac{dP}{dt} < 0 \;\Rightarrow\; P > 4200$

(c) $\dfrac{dP}{dt} = 0 \;\Rightarrow\; P = 4200$ or $P = 0$

11. (a) This function is increasing *and* also decreasing. But $dy/dt = e^t(y-1)^2 \geq 0$ for all t, implying that the graph of the solution of the differential equation cannot be decreasing on any interval.

(b) When $y = 1$, $dy/dt = 0$, but the graph does not have a horizontal tangent line.

13. (a) P increases most rapidly at the beginning, since there are usually many simple, easily-learned sub-skills associated with learning a skill. As t increases, we would expect dP/dt to remain positive, but decrease. This is because as time progresses, the only points left to learn are the more difficult ones.

(c)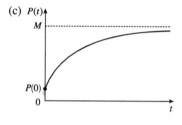

(b) $\dfrac{dP}{dt} = k(M - P)$ is always positive, so the level of performance P is increasing. As P gets close to M, dP/dt gets close to 0; that is, the performance levels off, as explained in part (a).

9.2 Direction Fields and Euler's Method

1. (a)

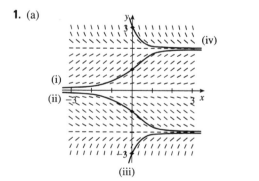

(b) It appears that the constant functions $y = 0$, $y = -2$, and $y = 2$ are equilibrium solutions. Note that these three values of y satisfy the given differential equation $y' = y\left(1 - \frac{1}{4}y^2\right)$.

3. $y' = y - 1$. The slopes at each point are independent of x, so the slopes are the same along each line parallel to the x-axis. Thus, IV is the direction field for this equation. Note that for $y = 1$, $y' = 0$.

5. $y' = y^2 - x^2 = 0 \implies y = \pm x$. There are horizontal tangents on these lines only in graph III, so this equation corresponds to direction field III.

7. (a) $y(0) = 1$ (b) $y(0) = 0$ (c) $y(0) = -1$

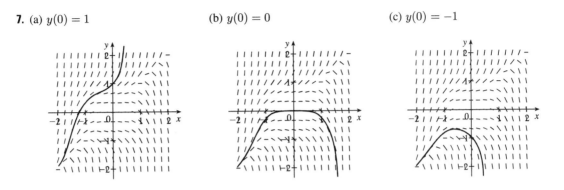

9.

x	y	$y' = 1 + y$
0	0	1
0	1	2
0	2	3
0	-3	-2
0	-2	-1

Note that for $y = -1$, $y' = 0$. The three solution curves sketched go through $(0, 0)$, $(0, -1)$, and $(0, -2)$.

11.

x	y	$y' = y - 2x$
-2	-2	2
-2	2	6
2	2	-2
2	-2	-6

Note that $y' = 0$ for any point on the line $y = 2x$. The slopes are positive to the left of the line and negative to the right of the line. The solution curve in the graph passes through $(1, 0)$.

13.

x	y	$y' = y + xy$
0	± 2	± 2
1	± 2	± 4
-3	± 2	∓ 4

Note that $y' = y(x + 1) = 0$ for any point on $y = 0$ or on $x = -1$. The slopes are positive when the factors y and $x + 1$ have the same sign and negative when they have opposite signs. The solution curve in the graph passes through $(0, 1)$.

15. In Maple, we can use either `directionfield` (in Maple's share library) or `plots[fieldplot]` to plot the direction field. To plot the solution, we can either use the initial-value option in `directionfield`, or actually solve the equation. In Mathematica, we use `PlotVectorField` for the direction field, and the `Plot[Evaluate[...]]` construction to plot the solution, which

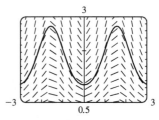

is $y = e^{(1 - \cos 2x)/2}$. In Derive, use `Direction_Field` (in utility file ODE_APPR) to plot the direction field. Then use `DSOLVE1(-y*SIN(2*x),1,x,y,0,1)` (in utility file ODE1) to solve the equation. Simplify each result.

17.

$L = \lim_{t \to \infty} y(t)$ exists for $-2 \le c \le 2$; $L = \pm 2$ for $c = \pm 2$ and $L = 0$ for $-2 < c < 2$. For other values of c, L does not exist.

19. (a) $y' = F(x, y) = y$ and $y(0) = 1$ $\Rightarrow$ $x_0 = 0, y_0 = 1$.

(i) $h = 0.4$ and $y_1 = y_0 + hF(x_0, y_0)$ $\Rightarrow$ $y_1 = 1 + 0.4 \cdot 1 = 1.4$. $x_1 = x_0 + h = 0 + 0.4 = 0.4$, so $y_1 = y(0.4) = 1.4$.

(ii) $h = 0.2$ $\Rightarrow$ $x_1 = 0.2$ and $x_2 = 0.4$, so we need to find y_2.
$y_1 = y_0 + hF(x_0, y_0) = 1 + 0.2y_0 = 1 + 0.2 \cdot 1 = 1.2$,
$y_2 = y_1 + hF(x_1, y_1) = 1.2 + 0.2y_1 = 1.2 + 0.2 \cdot 1.2 = 1.44$.

(iii) $h = 0.1$ $\Rightarrow$ $x_4 = 0.4$, so we need to find y_4. $y_1 = y_0 + hF(x_0, y_0) = 1 + 0.1y_0 = 1 + 0.1 \cdot 1 = 1.1$,
$y_2 = y_1 + hF(x_1, y_1) = 1.1 + 0.1y_1 = 1.1 + 0.1 \cdot 1.1 = 1.21$,
$y_3 = y_2 + hF(x_2, y_2) = 1.21 + 0.1y_2 = 1.21 + 0.1 \cdot 1.21 = 1.331$,
$y_4 = y_3 + hF(x_3, y_3) = 1.331 + 0.1y_3 = 1.331 + 0.1 \cdot 1.331 = 1.4641$.

(b)

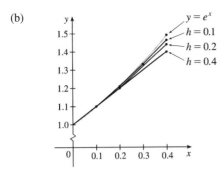

We see that the estimates are underestimates since they are all below the graph of $y = e^x$.

(c) (i) For $h = 0.4$:   (exact value) − (approximate value) $= e^{0.4} − 1.4 \approx 0.0918$

  (ii) For $h = 0.2$:   (exact value) − (approximate value) $= e^{0.4} − 1.44 \approx 0.0518$

  (iii) For $h = 0.1$:   (exact value) − (approximate value) $= e^{0.4} − 1.4641 \approx 0.0277$

Each time the step size is halved, the error estimate also appears to be halved (approximately).

21. $h = 0.5$, $x_0 = 1$, $y_0 = 0$, and $F(x, y) = y − 2x$.

Note that $x_1 = x_0 + h = 1 + 0.5 = 1.5$, $x_2 = 2$, and $x_3 = 2.5$.

$y_1 = y_0 + hF(x_0, y_0) = 0 + 0.5F(1, 0) = 0.5[0 − 2(1)] = −1$.

$y_2 = y_1 + hF(x_1, y_1) = −1 + 0.5F(1.5, −1) = −1 + 0.5[−1 − 2(1.5)] = −3$.

$y_3 = y_2 + hF(x_2, y_2) = −3 + 0.5F(2, −3) = −3 + 0.5[−3 − 2(2)] = −6.5$.

$y_4 = y_3 + hF(x_3, y_3) = −6.5 + 0.5F(2.5, −6.5) = −6.5 + 0.5[−6.5 − 2(2.5)] = −12.25$.

23. $h = 0.1$, $x_0 = 0$, $y_0 = 1$, and $F(x, y) = y + xy$.

Note that $x_1 = x_0 + h = 0 + 0.1 = 0.1$, $x_2 = 0.2$, $x_3 = 0.3$, and $x_4 = 0.4$.

$y_1 = y_0 + hF(x_0, y_0) = 1 + 0.1F(0, 1) = 1 + 0.1[1 + (0)(1)] = 1.1$.

$y_2 = y_1 + hF(x_1, y_1) = 1.1 + 0.1F(0.1, 1.1) = 1.1 + 0.1[1.1 + (0.1)(1.1)] = 1.221$.

$y_3 = y_2 + hF(x_2, y_2) = 1.221 + 0.1F(0.2, 1.221) = 1.221 + 0.1[1.221 + (0.2)(1.221)] = 1.36752$.

$y_4 = y_3 + hF(x_3, y_3) = 1.36752 + 0.1F(0.3, 1.36752) = 1.36752 + 0.1[1.36752 + (0.3)(1.36752)]$

$\qquad = 1.5452976$.

$y_5 = y_4 + hF(x_4, y_4) = 1.5452976 + 0.1F(0.4, 1.5452976)$

$\qquad = 1.5452976 + 0.1[1.5452976 + (0.4)(1.5452976)] = 1.761639264$.

Thus, $y(0.5) \approx 1.7616$.

25. (a) $dy/dx + 3x^2 y = 6x^2 \ \Rightarrow \ y' = 6x^2 − 3x^2 y$. Store this expression in Y_1 and use the following simple program to evaluate $y(1)$ for each part, using $H = h = 1$ and $N = 1$ for part (i), $H = 0.1$ and $N = 10$ for part (ii), and so forth.

$\qquad h \to H$: $0 \to X$: $3 \to Y$:

$\qquad$ For(I, 1, N): $Y + H \times Y_1 \to Y$: $X + H \to X$:

$\qquad$ End(loop):

$\qquad$ Display Y.   [To see all iterations, include this statement in the loop.]

  (i) $H = 1, N = 1 \ \Rightarrow \ y(1) = 3$ (ii) $H = 0.1, N = 10 \ \Rightarrow \ y(1) \approx 2.3928$

  (iii) $H = 0.01, N = 100 \ \Rightarrow \ y(1) \approx 2.3701$ (iv) $H = 0.001, N = 1000 \ \Rightarrow \ y(1) \approx 2.3681$

(b) $y = 2 + e^{-x^3} \quad \Rightarrow \quad y' = -3x^2 e^{-x^3}$

LHS $= y' + 3x^2 y = -3x^2 e^{-x^3} + 3x^2\left(2 + e^{-x^3}\right) = -3x^2 e^{-x^3} + 6x^2 + 3x^2 e^{-x^3} = 6x^2 =$ RHS

$y(0) = 2 + e^{-0} = 2 + 1 = 3$

(c) The exact value of $y(1)$ is $2 + e^{-1^3} = 2 + e^{-1}$.

   (i) For $h = 1$: (exact value) − (approximate value) $= 2 + e^{-1} - 3 \approx -0.6321$

   (ii) For $h = 0.1$: (exact value) − (approximate value) $= 2 + e^{-1} - 2.3928 \approx -0.0249$

   (iii) For $h = 0.01$: (exact value) − (approximate value) $= 2 + e^{-1} - 2.3701 \approx -0.0022$

   (iv) For $h = 0.001$: (exact value) − (approximate value) $= 2 + e^{-1} - 2.3681 \approx -0.0002$

   In (ii)–(iv), it seems that when the step size is divided by 10, the error estimate is also divided by 10 (approximately).

27. (a) $R\dfrac{dQ}{dt} + \dfrac{1}{C}Q = E(t)$ becomes

   $5Q' + \dfrac{1}{0.05}Q = 60$ or $Q' + 4Q = 12$.

(b) From the graph, it appears that the limiting value of the charge Q is about 3.

(c) If $Q' = 0$, then $4Q = 12 \quad \Rightarrow \quad Q = 3$ is an equilibrium solution.

(d)

(e) $Q' + 4Q = 12 \quad \Rightarrow \quad Q' = 12 - 4Q$. Now $Q(0) = 0$, so $t_0 = 0$ and $Q_0 = 0$.

$$Q_1 = Q_0 + hF(t_0, Q_0) = 0 + 0.1(12 - 4 \cdot 0) = 1.2$$
$$Q_2 = Q_1 + hF(t_1, Q_1) = 1.2 + 0.1(12 - 4 \cdot 1.2) = 1.92$$
$$Q_3 = Q_2 + hF(t_2, Q_2) = 1.92 + 0.1(12 - 4 \cdot 1.92) = 2.352$$
$$Q_4 = Q_3 + hF(t_3, Q_3) = 2.352 + 0.1(12 - 4 \cdot 2.352) = 2.6112$$
$$Q_5 = Q_4 + hF(t_4, Q_4) = 2.6112 + 0.1(12 - 4 \cdot 2.6112) = 2.76672$$

Thus, $Q_5 = Q(0.5) \approx 2.77$ C.

9.3 Separable Equations

1. $\dfrac{dy}{dx} = \dfrac{y}{x} \quad \Rightarrow \quad \dfrac{dy}{y} = \dfrac{dx}{x} \quad [y \neq 0] \quad \Rightarrow \quad \displaystyle\int \dfrac{dy}{y} = \int \dfrac{dx}{x} \quad \Rightarrow \quad \ln|y| = \ln|x| + C \quad \Rightarrow$

$|y| = e^{\ln|x| + C} = e^{\ln|x|}e^C = e^C |x| \quad \Rightarrow \quad y = Kx$, where $K = \pm e^C$ is a constant. (In our derivation, K was nonzero, but we can restore the excluded case $y = 0$ by allowing K to be zero.)

3. $(x^2 + 1)y' = xy$ $\Rightarrow$ $\dfrac{dy}{dx} = \dfrac{xy}{x^2+1}$ $\Rightarrow$ $\dfrac{dy}{y} = \dfrac{x\,dx}{x^2+1}$ $\;[y \ne 0]$ $\Rightarrow$ $\displaystyle\int \dfrac{dy}{y} = \int \dfrac{x\,dx}{x^2+1}$ $\Rightarrow$

$\ln|y| = \frac{1}{2}\ln(x^2+1) + C$ $\;[u = x^2+1,\, du = 2x\,dx]$ $= \ln(x^2+1)^{1/2} + \ln e^C = \ln\big(e^C\sqrt{x^2+1}\big)$ $\Rightarrow$

$|y| = e^C\sqrt{x^2+1}$ $\Rightarrow$ $y = K\sqrt{x^2+1}$, where $K = \pm e^C$ is a constant. (In our derivation, K was nonzero, but we can restore the excluded case $y = 0$ by allowing K to be zero.)

5. $(1 + \tan y)\, y' = x^2 + 1$ $\Rightarrow$ $(1 + \tan y)\dfrac{dy}{dx} = x^2 + 1$ $\Rightarrow$ $\left(1 + \dfrac{\sin y}{\cos y}\right)dy = (x^2+1)\,dx$ $\Rightarrow$

$\displaystyle\int \left(1 - \dfrac{-\sin y}{\cos y}\right) dy = \int (x^2+1)\,dx$ $\Rightarrow$ $y - \ln|\cos y| = \frac{1}{3}x^3 + x + C$. Note: The left side is equivalent to

$y + \ln|\sec y|$.

7. $\dfrac{dy}{dt} = \dfrac{te^t}{y\sqrt{1+y^2}}$ $\Rightarrow$ $y\sqrt{1+y^2}\,dy = te^t\,dt$ $\Rightarrow$ $\int y\sqrt{1+y^2}\,dy = \int te^t\,dt$ $\Rightarrow$

$\frac{1}{3}(1+y^2)^{3/2} = te^t - e^t + C$ [where the first integral is evaluated by substitution and the second by parts] $\Rightarrow$

$1 + y^2 = [3(te^t - e^t + C)]^{2/3}$ $\Rightarrow$ $y = \pm\sqrt{[3(te^t - e^t + C)]^{2/3} - 1}$

9. $\dfrac{du}{dt} = 2 + 2u + t + tu$ $\Rightarrow$ $\dfrac{du}{dt} = (1+u)(2+t)$ $\Rightarrow$ $\displaystyle\int \dfrac{du}{1+u} = \int (2+t)dt$ $\;[u \ne -1]$ $\Rightarrow$

$\ln|1 + u| = \frac{1}{2}t^2 + 2t + C$ $\Rightarrow$ $|1+u| = e^{t^2/2 + 2t + C} = Ke^{t^2/2 + 2t}$, where $K = e^C$ $\Rightarrow$

$1 + u = \pm Ke^{t^2/2 + 2t}$ $\Rightarrow$ $u = -1 \pm Ke^{t^2/2 + 2t}$ where $K > 0$. $u = -1$ is also a solution, so

$u = -1 + Ae^{t^2/2 + 2t}$, where A is an arbitrary constant.

11. $\dfrac{dy}{dx} = y^2 + 1$, $y(1) = 0$. $\displaystyle\int \dfrac{dy}{y^2+1} = \int dx$ $\Rightarrow$ $\tan^{-1} y = x + C$. $y = 0$ when $x = 1$, so

$1 + C = \tan^{-1} 0 = 0$ $\Rightarrow$ $C = -1$. Thus, $\tan^{-1} y = x - 1$ and $y = \tan(x-1)$.

13. $x \cos x = (2y + e^{3y})\, y'$ $\Rightarrow$ $x\cos x\,dx = (2y + e^{3y})\,dy$ $\Rightarrow$ $\int(2y + e^{3y})\,dy = \int x\cos x\,dx$ $\Rightarrow$

$y^2 + \frac{1}{3}e^{3y} = x\sin x + \cos x + C$ [where the second integral is evaluated using integration by parts]. Now

$y(0) = 0$ $\Rightarrow$ $0 + \frac{1}{3} = 0 + 1 + C$ $\Rightarrow$ $C = -\frac{2}{3}$. Thus, a solution is $y^2 + \frac{1}{3}e^{3y} = x\sin x + \cos x - \frac{2}{3}$.

We cannot solve explicitly for y.

15. $\dfrac{du}{dt} = \dfrac{2t + \sec^2 t}{2u}$, $u(0) = -5$. $\int 2u\,du = \int(2t + \sec^2 t)\,dt$ $\Rightarrow$ $u^2 = t^2 + \tan t + C$, where

$[u(0)]^2 = 0^2 + \tan 0 + C$ $\Rightarrow$ $C = (-5)^2 = 25$. Therefore, $u^2 = t^2 + \tan t + 25$, so $u = \pm\sqrt{t^2 + \tan t + 25}$.

Since $u(0) = -5$, we must have $u = -\sqrt{t^2 + \tan t + 25}$.

17. $y'\tan x = a + y$, $0 < x < \pi/2$ $\Rightarrow$ $\dfrac{dy}{dx} = \dfrac{a+y}{\tan x}$ $\Rightarrow$ $\dfrac{dy}{a+y} = \cot x\,dx$ $\;[a + y \ne 0]$ $\Rightarrow$

$\displaystyle\int \dfrac{dy}{a+y} = \int \dfrac{\cos x}{\sin x}\,dx$ $\Rightarrow$ $\ln|a + y| = \ln|\sin x| + C$ $\Rightarrow$

$|a+y| = e^{\ln|\sin x| + C} = e^{\ln|\sin x|} \cdot e^C = e^C |\sin x|$ $\Rightarrow$ $a + y = K\sin x$, where $K = \pm e^C$. (In our derivation,

K was nonzero, but we can restore the excluded case $y = -a$ by allowing K to be zero.) $y(\pi/3) = a$ $\Rightarrow$

$a + a = K\sin\left(\dfrac{\pi}{3}\right)$ $\Rightarrow$ $2a = K\dfrac{\sqrt{3}}{2}$ $\Rightarrow$ $K = \dfrac{4a}{\sqrt{3}}$. Thus, $a + y = \dfrac{4a}{\sqrt{3}}\sin x$ and so $y = \dfrac{4a}{\sqrt{3}}\sin x - a$.

19. $\dfrac{dy}{dx} = 4x^3 y$, $y(0) = 7$. $\dfrac{dy}{y} = 4x^3\,dx$ [if $y \ne 0$] $\Rightarrow$ $\displaystyle\int \dfrac{dy}{y} = \int 4x^3\,dx$ $\Rightarrow$ $\ln|y| = x^4 + C$ $\Rightarrow$

$e^{\ln|y|} = e^{x^4 + C}$ $\Rightarrow$ $|y| = e^{x^4} e^C$ $\Rightarrow$ $y = Ae^{x^4}$; $y(0) = 7$ $\Rightarrow$ $A = 7$ $\Rightarrow$ $y = 7e^{x^4}$.

21. (a) $y' = 2x\sqrt{1-y^2}$ $\Rightarrow$ $\dfrac{dy}{dx} = 2x\sqrt{1-y^2}$ $\Rightarrow$ $\dfrac{dy}{\sqrt{1-y^2}} = 2x\,dx$ $\Rightarrow$ $\displaystyle\int \dfrac{dy}{\sqrt{1-y^2}} = \int 2x\,dx$ $\Rightarrow$

$\sin^{-1}y = x^2 + C$ for $-\frac{\pi}{2} \le x^2 + C \le \frac{\pi}{2}$.

(b) $y(0) = 0$ $\Rightarrow$ $\sin^{-1}0 = 0^2 + C$ $\Rightarrow$ $C = 0$, so $\sin^{-1}y = x^2$

and $y = \sin(x^2)$ for $-\sqrt{\pi/2} \le x \le \sqrt{\pi/2}$.

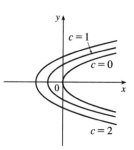

(c) For $\sqrt{1-y^2}$ to be a real number, we must have $-1 \le y \le 1$; that is, $-1 \le y(0) \le 1$. Thus, the initial-value

problem $y' = 2x\sqrt{1-y^2}$, $y(0) = 2$ does *not* have a solution.

23. $\dfrac{dy}{dx} = \dfrac{\sin x}{\sin y}$, $y(0) = \frac{\pi}{2}$. So $\int \sin y\,dy = \int \sin x\,dx$ $\Leftrightarrow$

$-\cos y = -\cos x + C$ $\Leftrightarrow$ $\cos y = \cos x - C$. From the initial

condition, we need $\cos\frac{\pi}{2} = \cos 0 - C$ $\Rightarrow$ $0 = 1 - C$ $\Rightarrow$ $C = 1$,

so the solution is $\cos y = \cos x - 1$. Note that we cannot take $\cos^{-1}$ of

both sides, since that would unnecessarily restrict the solution to the case

where $-1 \le \cos x - 1$ $\Leftrightarrow$ $0 \le \cos x$, as $\cos^{-1}$ is defined only on

$[-1, 1]$. Instead we plot the graph using Maple's

`plots[implicitplot]` or Mathematica's

`Plot[Evaluate[···]]`.

25. (a)

x	y	$y' = 1/y$	x	y	$y' = 1/y$
0	0.5	2	0	-2	-0.5
0	-0.5	-2	0	4	0.25
0	1	1	0	3	$0.\overline{3}$
0	-1	-1	0	0.25	4
0	2	0.5	0	$0.\overline{3}$	3

(b) $y' = 1/y$ $\Rightarrow$ $dy/dx = 1/y$ $\Rightarrow$ (c)

$y\,dy = dx$ $\Rightarrow$ $\int y\,dy = \int dx$ $\Rightarrow$

$\frac{1}{2}y^2 = x + c$ $\Rightarrow$ $y^2 = 2(x + c)$

or $y = \pm\sqrt{2(x + c)}$.

27. The curves $y = kx^2$ form a family of parabolas with axis the y-axis.

Differentiating gives $y' = 2kx$, but $k = y/x^2$, so $y' = 2y/x$. Thus, the

slope of the tangent line at any point (x, y) on one of the parabolas is

$y' = 2y/x$, so the orthogonal trajectories must satisfy $y' = -x/(2y)$

$\Leftrightarrow \ 2y\,dy = -x\,dx \ \Leftrightarrow \ y^2 = -x^2/2 + C_1 \ \Leftrightarrow \ x^2 + 2y^2 = C.$

This is a family of ellipses.

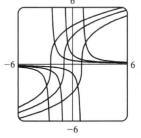

29. Differentiating $y = (x + k)^{-1}$ gives $y' = -\dfrac{1}{(x + k)^2}$, but $k = \dfrac{1}{y} - x$, so

$y' = -\dfrac{1}{(1/y)^2} = -y^2$. Thus, the orthogonal trajectories must satisfy

$y' = -\dfrac{1}{-y^2} = \dfrac{1}{y^2} \ \Leftrightarrow \ y^2\,dy = dx \ \Leftrightarrow \ \dfrac{y^3}{3} = x + C$ or

$y = [3(x + C)]^{1/3}$

31. From Exercise 9.2.27, $\dfrac{dQ}{dt} = 12 - 4Q \ \Leftrightarrow \ \displaystyle\int \dfrac{dQ}{12 - 4Q} = \int dt \ \Leftrightarrow \ -\tfrac{1}{4}\ln|12 - 4Q| = t + C \ \Leftrightarrow$

$\ln|12 - 4Q| = -4t - 4C \ \Leftrightarrow \ |12 - 4Q| = e^{-4t-4C} \ \Leftrightarrow \ 12 - 4Q = Ke^{-4t} \ [K = \pm e^{-4C}] \ \Leftrightarrow$

$4Q = 12 - Ke^{-4t} \ \Leftrightarrow \ Q = 3 - Ae^{-4t} \ [A = K/4].$ $Q(0) = 0 \ \Leftrightarrow \ 0 = 3 - A \ \Leftrightarrow \ A = 3 \ \Leftrightarrow$

$Q(t) = 3 - 3e^{-4t}.$ As $t \to \infty$, $Q(t) \to 3 - 0 = 3$ (the limiting value).

33. $\dfrac{dP}{dt} = k(M - P) \ \Leftrightarrow \ \displaystyle\int \dfrac{dP}{P - M} = \int (-k)\,dt \ \Leftrightarrow \ \ln|P - M| = -kt + C \ \Leftrightarrow \ |P - M| = e^{-kt+C}$

$\Leftrightarrow \ P - M = Ae^{-kt} \ [A = \pm e^C] \ \Leftrightarrow \ P = M + Ae^{-kt}.$ If we assume that performance is at level 0 when

$t = 0$, then $P(0) = 0 \ \Leftrightarrow \ 0 = M + A \ \Leftrightarrow \ A = -M \ \Leftrightarrow \ P(t) = M - Me^{-kt}.$

$\displaystyle\lim_{t \to \infty} P(t) = M - M \cdot 0 = M.$

35. (a) If $a = b$, then $\dfrac{dx}{dt} = k(a - x)(b - x)^{1/2}$ becomes $\dfrac{dx}{dt} = k(a - x)^{3/2} \ \Rightarrow \ (a - x)^{-3/2}dx = k\,dt \ \Rightarrow$

$\int (a - x)^{-3/2}dx = \int k\,dt \ \Rightarrow \ 2(a - x)^{-1/2} = kt + C$ [by substitution] $\Rightarrow \ \dfrac{2}{kt + C} = \sqrt{a - x} \ \Rightarrow$

$\left(\dfrac{2}{kt + C}\right)^2 = a - x \ \Rightarrow \ x(t) = a - \dfrac{4}{(kt + C)^2}.$ The initial concentration of HBr is 0, so $x(0) = 0 \ \Rightarrow$

$0 = a - \dfrac{4}{C^2} \ \Rightarrow \ \dfrac{4}{C^2} = a \ \Rightarrow \ C^2 = \dfrac{4}{a} \ \Rightarrow \ C = 2/\sqrt{a}$ (C is positive since

$kt + C = 2(a - x)^{-1/2} > 0$). Thus, $x(t) = a - \dfrac{4}{(kt + 2/\sqrt{a})^2}.$

(b) $\dfrac{dx}{dt} = k(a - x)(b - x)^{1/2} \ \Rightarrow \ \dfrac{dx}{(a - x)\sqrt{b - x}} = k\,dt \ \Rightarrow \ \displaystyle\int \dfrac{dx}{(a - x)\sqrt{b - x}} = \int k\,dt \ (\star).$ From the

hint, $u = \sqrt{b - x} \ \Rightarrow \ u^2 = b - x \ \Rightarrow \ 2u\,du = -dx,$ so $\displaystyle\int \dfrac{dx}{(a - x)\sqrt{b - x}} = \int \dfrac{-2u\,du}{[a - (b - u^2)]u} =$

$-2\displaystyle\int \dfrac{du}{a - b + u^2} = -2\int \dfrac{du}{(\sqrt{a - b})^2 + u^2} \overset{17}{=} -2\left(\dfrac{1}{\sqrt{a - b}}\tan^{-1}\dfrac{u}{\sqrt{a - b}}\right).$ So $(\star)$ becomes

$\dfrac{-2}{\sqrt{a-b}}\tan^{-1}\dfrac{\sqrt{b-x}}{\sqrt{a-b}} = kt + C$. Now $x(0) = 0 \;\Rightarrow\; C = \dfrac{-2}{\sqrt{a-b}}\tan^{-1}\dfrac{\sqrt{b}}{\sqrt{a-b}}$

and we have $\dfrac{-2}{\sqrt{a-b}}\tan^{-1}\dfrac{\sqrt{b-x}}{\sqrt{a-b}} = kt - \dfrac{2}{\sqrt{a-b}}\tan^{-1}\dfrac{\sqrt{b}}{\sqrt{a-b}} \;\Rightarrow\;$

$\dfrac{2}{\sqrt{a-b}}\left(\tan^{-1}\sqrt{\dfrac{b}{a-b}} - \tan^{-1}\sqrt{\dfrac{b-x}{a-b}}\right) = kt \;\Rightarrow\;$

$t(x) = \dfrac{2}{k\sqrt{a-b}}\left(\tan^{-1}\sqrt{\dfrac{b}{a-b}} - \tan^{-1}\sqrt{\dfrac{b-x}{a-b}}\right).$

37. (a) $\dfrac{dC}{dt} = r - kC \;\Rightarrow\; \dfrac{dC}{dt} = -(kC - r) \;\Rightarrow\; \displaystyle\int\dfrac{dC}{kC-r} = \int -dt \;\Rightarrow\; (1/k)\ln|kC-r| = -t + M_1$

$\Rightarrow\; \ln|kC-r| = -kt + M_2 \;\Rightarrow\; |kC-r| = e^{-kt+M_2} \;\Rightarrow\; kC - r = M_3 e^{-kt} \;\Rightarrow\;$

$kC = M_3 e^{-kt} + r \;\Rightarrow\; C(t) = M_4 e^{-kt} + r/k.\; C(0) = C_0 \;\Rightarrow\; C_0 = M_4 + r/k \;\Rightarrow\;$

$M_4 = C_0 - r/k \;\Rightarrow\; C(t) = (C_0 - r/k)e^{-kt} + r/k.$

(b) If $C_0 < r/k$, then $C_0 - r/k < 0$ and the formula for $C(t)$ shows that $C(t)$ increases and $\lim\limits_{t\to\infty} C(t) = r/k$.

As t increases, the formula for $C(t)$ shows how the role of C_0 steadily diminishes as that of r/k increases.

39. (a) Let $y(t)$ be the amount of salt (in kg) after t minutes. Then $y(0) = 15$. The amount of liquid in the tank is

1000 L at all times, so the concentration at time t (in minutes) is $y(t)/1000$ kg/L and

$\dfrac{dy}{dt} = -\left[\dfrac{y(t)}{1000}\dfrac{\text{kg}}{\text{L}}\right]\left(10\,\dfrac{\text{L}}{\text{min}}\right) = -\dfrac{y(t)}{100}\dfrac{\text{kg}}{\text{min}}.\; \displaystyle\int\dfrac{dy}{y} = -\dfrac{1}{100}\int dt \;\Rightarrow\; \ln y = -\dfrac{t}{100} + C,$ and

$y(0) = 15 \;\Rightarrow\; \ln 15 = C$, so $\ln y = \ln 15 - \dfrac{t}{100}$. It follows that $\ln\left(\dfrac{y}{15}\right) = -\dfrac{t}{100}$ and $\dfrac{y}{15} = e^{-t/100}$, so

$y = 15e^{-t/100}$ kg.

(b) After 20 minutes, $y = 15e^{-20/100} = 15e^{-0.2} \approx 12.3$ kg.

41. Assume that the raindrop begins at rest, so that $v(0) = 0$. $dm/dt = km$ and $(mv)' = gm \;\Rightarrow\;$

$mv' + vm' = gm \;\Rightarrow\; mv' + v(km) = gm \;\Rightarrow\; v' + vk = g \;\Rightarrow\; dv/dt = g - kv \;\Rightarrow\;$

$\displaystyle\int\dfrac{dv}{g-kv} = \int dt \;\Rightarrow\; -(1/k)\ln|g-kv| = t + C \;\Rightarrow\; \ln|g-kv| = -kt - kC \;\Rightarrow\; g - kv = Ae^{-kt}.$

$v(0) = 0 \;\Rightarrow\; A = g.$ So $kv = g - ge^{-kt} \;\Rightarrow\; v = (g/k)\left(1 - e^{-kt}\right)$. Since $k > 0$, as $t \to \infty$, $e^{-kt} \to 0$

and therefore, $\lim\limits_{t\to\infty} v(t) = g/k$.

43. (a) The rate of growth of the area is jointly proportional to $\sqrt{A(t)}$ and $M - A(t)$; that is, the rate is proportional to

the product of those two quantities. So for some constant k, $dA/dt = k\sqrt{A}\,(M - A)$. We are interested in the

maximum of the function dA/dt (when the tissue grows the fastest), so we differentiate, using the Chain Rule

and then substituting for dA/dt from the differential equation:

$$\dfrac{d}{dt}\left(\dfrac{dA}{dt}\right) = k\left[\sqrt{A}(-1)\dfrac{dA}{dt} + (M - A)\cdot\tfrac{1}{2}A^{-1/2}\dfrac{dA}{dt}\right] = \tfrac{1}{2}kA^{-1/2}\dfrac{dA}{dt}[-2A + (M - A)]$$

$$= \tfrac{1}{2}kA^{-1/2}\left[k\sqrt{A}(M - A)\right][M - 3A] = \tfrac{1}{2}k^2(M - A)(M - 3A)$$

This is 0 when $M - A = 0$ [this situation never actually occurs, since the graph of $A(t)$ is asymptotic to the line

$y = M$, as in the logistic model] and when $M - 3A = 0 \;\Leftrightarrow\; A(t) = M/3$. This represents a maximum by

the First Derivative Test, since $\dfrac{d}{dt}\left(\dfrac{dA}{dt}\right)$ goes from positive to negative when $A(t) = M/3$.

(b) From the CAS, we get $A(t) = M\left(\dfrac{Ce^{\sqrt{M}kt} - 1}{Ce^{\sqrt{M}kt} + 1}\right)^2$. To get C in terms of the initial area A_0 and the maximum

area M, we substitute $t = 0$ and $A = A_0 = A(0)$: $A_0 = M\left(\dfrac{C-1}{C+1}\right)^2 \Leftrightarrow (C+1)\sqrt{A_0} = (C-1)\sqrt{M}$

$\Leftrightarrow C\sqrt{A_0} + \sqrt{A_0} = C\sqrt{M} - \sqrt{M} \Leftrightarrow \sqrt{M} + \sqrt{A_0} = C\sqrt{M} - C\sqrt{A_0} \Leftrightarrow$

$\sqrt{M} + \sqrt{A_0} = C\left(\sqrt{M} - \sqrt{A_0}\right) \Leftrightarrow C = \dfrac{\sqrt{M} + \sqrt{A_0}}{\sqrt{M} - \sqrt{A_0}}$. (Notice that if $A_0 = 0$, then $C = 1$.)

9.4 Exponential Growth and Decay

1. The relative growth rate is $\dfrac{1}{P}\dfrac{dP}{dt} = 0.7944$, so $\dfrac{dP}{dt} = 0.7944P$ and, by Theorem 2,

$P(t) = P(0)e^{0.7944t} = 2e^{0.7944t}$. Thus, $P(6) = 2e^{0.7944(6)} \approx 234.99$ or about 235 members.

3. (a) By Theorem 2, $y(t) = y(0)e^{kt} = 500e^{kt}$. Now $y(3) = 500e^{k(3)} = 8000 \Rightarrow e^{3k} = \frac{8000}{500} \Rightarrow$

$3k = \ln 16 \Rightarrow k = (\ln 16)/3$. So $y(t) = 500e^{(\ln 16)t/3} = 500 \cdot 16^{t/3}$

(b) $y(4) = 500 \cdot 16^{4/3} \approx 20{,}159$

(c) $dy/dt = ky \Rightarrow y'(4) = ky(4) = \frac{1}{3}\ln 16\left(500 \cdot 16^{4/3}\right)$ [from part (a)] $\approx 18{,}631$ cells/h

(d) $y(t) = 500 \cdot 16^{t/3} = 30{,}000 \Rightarrow 16^{t/3} = 60 \Rightarrow \frac{1}{3}t \ln 16 = \ln 60 \Rightarrow t = 3(\ln 60)/(\ln 16) \approx 4.4$ h

5. (a) Let the population (in millions) in the year t be $P(t)$. Since the initial time is the year 1750, we substitute

$t - 1750$ for t in Theorem 2, so the exponential model gives $P(t) = P(1750)e^{k(t-1750)}$. Then

$P(1800) = 980 = 790e^{k(1800-1750)} \Rightarrow \frac{980}{790} = e^{k(50)} \Rightarrow \ln\frac{980}{790} = 50k \Rightarrow$

$k = \frac{1}{50}\ln\frac{980}{790} \approx 0.0043104$. So with this model, we have $P(1900) = 790e^{k(1900-1750)} \approx 1508$ million, and

$P(1950) = 790e^{k(1950-1750)} \approx 1871$ million. Both of these estimates are much too low.

(b) In this case, the exponential model gives $P(t) = P(1850)e^{k(t-1850)} \Rightarrow$

$P(1900) = 1650 = 1260e^{k(1900-1850)} \Rightarrow \ln\frac{1650}{1260} = k(50) \Rightarrow k = \frac{1}{50}\ln\frac{1650}{1260} \approx 0.005393$. So with

this model, we estimate $P(1950) = 1260e^{k(1950-1850)} \approx 2161$ million. This is still too low, but closer than the

estimate of $P(1950)$ in part (a).

(c) The exponential model gives $P(t) = P(1900)e^{k(t-1900)} \Rightarrow P(1950) = 2560 = 1650e^{k(1950-1900)} \Rightarrow$

$\ln\frac{2560}{1650} = k(50) \Rightarrow k = \frac{1}{50}\ln\frac{2560}{1650} \approx 0.008785$. With this model, we estimate

$P(2000) = 1650e^{k(2000-1900)} \approx 3972$ million. This is much too low. The discrepancy is explained by the fact

that the world birth rate (average yearly number of births per person) is about the same as always, whereas the

mortality rate (especially the infant mortality rate) is much lower, owing mostly to advances in medical science

and to the wars in the first part of the twentieth century. The exponential model assumes, among other things,

that the birth and mortality rates will remain constant.

7. (a) If $y = [N_2O_5]$ then by Theorem 2, $\dfrac{dy}{dt} = -0.0005y \Rightarrow y(t) = y(0)e^{-0.0005t} = Ce^{-0.0005t}$.

(b) $y(t) = Ce^{-0.0005t} = 0.9C \Rightarrow e^{-0.0005t} = 0.9 \Rightarrow -0.0005t = \ln 0.9 \Rightarrow$

$t = -2000\ln 0.9 \approx 211$ s

9. (a) If $y(t)$ is the mass (in mg) remaining after t years, then $y(t) = y(0)e^{kt} = 100e^{kt}$. $y(30) = 100e^{30k} = \frac{1}{2}(100)$

$\Rightarrow$ $e^{30k} = \frac{1}{2}$ $\Rightarrow$ $k = -(\ln 2)/30$ $\Rightarrow$ $y(t) = 100e^{-(\ln 2)t/30} = 100 \cdot 2^{-t/30}$

(b) $y(100) = 100 \cdot 2^{-100/30} \approx 9.92$ mg

(c) $100e^{-(\ln 2)t/30} = 1$ $\Rightarrow$ $-(\ln 2)t/30 = \ln \frac{1}{100}$ $\Rightarrow$ $t = -30 \frac{\ln 0.01}{\ln 2} \approx 199.3$ years

11. Let $y(t)$ be the level of radioactivity. Thus, $y(t) = y(0)e^{-kt}$ and k is determined by using the half-life:

$y(5730) = \frac{1}{2}y(0)$ $\Rightarrow$ $y(0)e^{-k(5730)} = \frac{1}{2}y(0)$ $\Rightarrow$ $e^{-5730k} = \frac{1}{2}$ $\Rightarrow$

$-5730k = \ln \frac{1}{2}$ $\Rightarrow$ $k = -\dfrac{\ln \frac{1}{2}}{5730} = \dfrac{\ln 2}{5730}$. If 74% of the ^{14}C remains, then we know that $y(t) = 0.74y(0)$

$\Rightarrow$ $0.74 = e^{-t(\ln 2)/5730}$ $\Rightarrow$ $\ln 0.74 = -\dfrac{t \ln 2}{5730}$ $\Rightarrow$ $t = -\dfrac{5730(\ln 0.74)}{\ln 2} \approx 2489 \approx 2500$ years.

13. (a) Using Newton's Law of Cooling, $\dfrac{dT}{dt} = k(T - T_s)$, we have $\dfrac{dT}{dt} = k(T - 75)$.

Now let $y = T - 75$, so $y(0) = T(0) - 75 = 185 - 75 = 110$, so y is a solution of the initial-value

problem $dy/dt = ky$ with $y(0) = 110$ and by Theorem 2 we have $y(t) = y(0)e^{kt} = 110e^{kt}$.

$y(30) = 110e^{30k} = 150 - 75$ $\Rightarrow$ $e^{30k} = \frac{75}{110} = \frac{15}{22}$ $\Rightarrow$ $k = \frac{1}{30} \ln \frac{15}{22}$,

so $y(t) = 110e^{\frac{1}{30}t \ln\left(\frac{15}{22}\right)}$ and $y(45) = 110e^{\frac{45}{30} \ln\left(\frac{15}{22}\right)} \approx 62°$F. Thus, $T(45) \approx 62 + 75 = 137°$F.

(b) $T(t) = 100$ $\Rightarrow$ $y(t) = 25$. $y(t) = 110e^{\frac{1}{30}t \ln\left(\frac{15}{22}\right)} = 25$ $\Rightarrow$ $e^{\frac{1}{30}t \ln\left(\frac{15}{22}\right)} = \frac{25}{110}$ $\Rightarrow$

$\frac{1}{30}t \ln \frac{15}{22} = \ln \frac{25}{110}$ $\Rightarrow$ $t = \dfrac{30 \ln \frac{25}{110}}{\ln \frac{15}{22}} \approx 116$ min.

15. $\dfrac{dT}{dt} = k(T - 20)$. Letting $y = T - 20$, we get $\dfrac{dy}{dt} = ky$, so $y(t) = y(0)e^{kt}$.

$y(0) = T(0) - 20 = 5 - 20 = -15$, so $y(25) = y(0)e^{25k} = -15e^{25k}$, and

$y(25) = T(25) - 20 = 10 - 20 = -10$, so $-15e^{25k} = -10$ $\Rightarrow$ $e^{25k} = \frac{2}{3}$. Thus, $25k = \ln\left(\frac{2}{3}\right)$ and

$k = \frac{1}{25} \ln\left(\frac{2}{3}\right)$, so $y(t) = y(0)e^{kt} = -15e^{(1/25) \ln(2/3)t}$. More simply, $e^{25k} = \frac{2}{3}$ $\Rightarrow$ $e^k = \left(\frac{2}{3}\right)^{1/25}$ $\Rightarrow$

$e^{kt} = \left(\frac{2}{3}\right)^{t/25}$ $\Rightarrow$ $y(t) = -15 \cdot \left(\frac{2}{3}\right)^{t/25}$.

(a) $T(50) = 20 + y(50) = 20 - 15 \cdot \left(\frac{2}{3}\right)^{50/25} = 20 - 15 \cdot \left(\frac{2}{3}\right)^2 = 20 - \frac{20}{3} = 13.\overline{3}°$C

(b) $15 = T(t) = 20 + y(t) = 20 - 15 \cdot \left(\frac{2}{3}\right)^{t/25}$ $\Rightarrow$ $15 \cdot \left(\frac{2}{3}\right)^{t/25} = 5$ $\Rightarrow$ $\left(\frac{2}{3}\right)^{t/25} = \frac{1}{3}$ $\Rightarrow$

$(t/25) \ln\left(\frac{2}{3}\right) = \ln\left(\frac{1}{3}\right)$ $\Rightarrow$ $t = 25 \ln\left(\frac{1}{3}\right) / \ln\left(\frac{2}{3}\right) \approx 67.74$ min.

17. (a) Let $P(h)$ be the pressure at altitude h. Then $dP/dh = kP$ $\Rightarrow$ $P(h) = P(0)e^{kh} = 101.3e^{kh}$.

$P(1000) = 101.3e^{1000k} = 87.14$ $\Rightarrow$ $1000k = \ln\left(\frac{87.14}{101.3}\right)$ $\Rightarrow$

$k = \frac{1}{1000} \ln\left(\frac{87.14}{101.3}\right)$ $\Rightarrow$ $P(h) = 101.3 \, e^{\frac{1}{1000}h \ln\left(\frac{87.14}{101.3}\right)}$, so $P(3000) = 101.3e^{3 \ln\left(\frac{87.14}{101.3}\right)} \approx 64.5$ kPa.

(b) $P(6187) = 101.3 \, e^{\frac{6187}{1000} \ln\left(\frac{87.14}{101.3}\right)} \approx 39.9$ kPa

19. (a) Using $A = A_0\left(1 + \dfrac{r}{n}\right)^{nt}$ with $A_0 = 3000$, $r = 0.05$, and $t = 5$, we have:

(i) Annually: $n = 1$; $A = 3000\left(1 + \frac{0.05}{1}\right)^{1 \cdot 5} = \$3828.84$

(ii) Semiannually: $n = 2$; $A = 3000\left(1 + \frac{0.05}{2}\right)^{2 \cdot 5} = \$3840.25$

(iii) Monthly: $n = 12$; $A = 3000\left(1 + \frac{0.05}{12}\right)^{12 \cdot 5} = \$3850.08$

(iv) Weekly: $n = 52$; $A = 3000\left(1 + \frac{0.05}{52}\right)^{52 \cdot 5} = \$3851.61$

(v) Daily: $n = 365$; $A = 3000\left(1 + \frac{0.05}{365}\right)^{365 \cdot 5} = \$3852.01$

(vi) Continuously: $A = 3000e^{(0.05)5} = \$3852.08$

(b) $dA/dt = 0.05A$ and $A(0) = 3000$.

21. (a) $\dfrac{dP}{dt} = kP - m = k\left(P - \dfrac{m}{k}\right)$. Let $y = P - \dfrac{m}{k}$, so $\dfrac{dy}{dt} = \dfrac{dP}{dt}$ and the differential equation becomes

$\dfrac{dy}{dt} = ky$. The solution is $y = y_0 e^{kt}$ $\Rightarrow$ $P - \dfrac{m}{k} = \left(P_0 - \dfrac{m}{k}\right)e^{kt}$ $\Rightarrow$ $P(t) = \dfrac{m}{k} + \left(P_0 - \dfrac{m}{k}\right)e^{kt}$.

(b) Since $k > 0$, there will be an exponential expansion $\Leftrightarrow$ $P_0 - \dfrac{m}{k} > 0$ $\Leftrightarrow$ $m < kP_0$.

(c) The population will be constant if $P_0 - \dfrac{m}{k} = 0$ $\Leftrightarrow$ $m = kP_0$. It will decline if $P_0 - \dfrac{m}{k} < 0$ $\Leftrightarrow$ $m > kP_0$.

(d) $P_0 = 8{,}000{,}000$, $k = \alpha - \beta = 0.016$, $m = 210{,}000$ $\Rightarrow$ $m > kP_0$ $(= 128{,}000)$, so by part (c), the population was declining.

9.5 The Logistic Equation

1. (a) $dP/dt = 0.05P - 0.0005P^2 = 0.05P(1 - 0.01P) = 0.05P(1 - P/100)$. Comparing to Equation 1,

$dP/dt = kP(1 - P/K)$, we see that the carrying capacity is $K = 100$ and the value of k is 0.05.

(b) The slopes close to 0 occur where P is near 0 or 100. The largest slopes appear to be on the line $P = 50$. The solutions are increasing for $0 < P_0 < 100$ and decreasing for $P_0 > 100$.

(c)

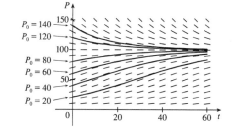

All of the solutions approach $P = 100$ as t increases. As in part (b), the solutions differ since for $0 < P_0 < 100$ they are increasing, and for $P_0 > 100$ they are decreasing. Also, some have an IP and some don't. It appears that the solutions which have $P_0 = 20$ and $P_0 = 40$ have inflection points at $P = 50$.

(d) The equilibrium solutions are $P = 0$ (trivial solution) and $P = 100$. The increasing solutions move away from $P = 0$ and all nonzero solutions approach $P = 100$ as $t \to \infty$.

3. (a) $\dfrac{dy}{dt} = ky\left(1 - \dfrac{y}{K}\right)$ $\Rightarrow$ $y(t) = \dfrac{K}{1 + Ae^{-kt}}$ with $A = \dfrac{K - y(0)}{y(0)}$. With $K = 8 \times 10^7$, $k = 0.71$, and

$y(0) = 2 \times 10^7$, we get the model $y(t) = \dfrac{8 \times 10^7}{1 + 3e^{-0.71t}}$, so $y(1) = \dfrac{8 \times 10^7}{1 + 3e^{-0.71}} \approx 3.23 \times 10^7$ kg.

(b) $y(t) = 4 \times 10^7$ $\Rightarrow$ $\dfrac{8 \times 10^7}{1 + 3e^{-0.71t}} = 4 \times 10^7$ $\Rightarrow$ $2 = 1 + 3e^{-0.71t}$ $\Rightarrow$ $e^{-0.71t} = \frac{1}{3}$ $\Rightarrow$

$-0.71t = \ln \frac{1}{3}$ $\Rightarrow$ $t = \dfrac{\ln 3}{0.71} \approx 1.55$ years

5. (a) We will assume that the difference in the birth and death rates is 20 million/year. Let $t = 0$ correspond to the

year 1990 and use a unit of 1 billion for all calculations. $k \approx \dfrac{1}{P}\dfrac{dP}{dt} = \dfrac{1}{5.3}(0.02) = \dfrac{1}{265}$, so

$$\dfrac{dP}{dt} = kP\left(1 - \dfrac{P}{K}\right) = \dfrac{1}{265}P\left(1 - \dfrac{P}{100}\right),\ P \text{ in billions}$$

(b) $A = \dfrac{K - P_0}{P_0} = \dfrac{100 - 5.3}{5.3} = \dfrac{947}{53} \approx 17.8679$. $P(t) = \dfrac{K}{1 + Ae^{-kt}} = \dfrac{100}{1 + \frac{947}{53}e^{-(1/265)t}}$, so

$P(10) \approx 5.49$ billion.

(c) $P(110) \approx 7.81$, and $P(510) \approx 27.72$. The predictions are 7.81 billion in the year 2100 and 27.72 billion

in 2500.

(d) If $K = 50$, then $P(t) = \dfrac{50}{1 + \frac{447}{53}e^{-(1/265)t}}$. So $P(10) \approx 5.48$, $P(110) \approx 7.61$, and $P(510) \approx 22.41$. The

predictions become 5.48 billion in the year 2000, 7.61 billion in 2100, and 22.41 billion in the year 2500.

7. (a) Our assumption is that $\dfrac{dy}{dt} = ky(1 - y)$, where y is the fraction of the population that has heard the rumor.

(b) Using the logistic equation (1), $\dfrac{dP}{dt} = kP\left(1 - \dfrac{P}{K}\right)$, we substitute $y = \dfrac{P}{K}$, $P = Ky$, and $\dfrac{dP}{dt} = K\dfrac{dy}{dt}$, to

obtain $K\dfrac{dy}{dt} = k(Ky)(1 - y)$ $\Leftrightarrow$ $\dfrac{dy}{dt} = ky(1 - y)$, our equation in part (a). Now the solution to (1) is

$P(t) = \dfrac{K}{1 + Ae^{-kt}}$, where $A = \dfrac{K - P_0}{P_0}$. We use the same substitution to obtain $Ky = \dfrac{K}{1 + \dfrac{K - Ky_0}{Ky_0}e^{-kt}}$

$\Rightarrow$ $y = \dfrac{y_0}{y_0 + (1 - y_0)e^{-kt}}$.

Alternatively, we could use the same steps as outlined in "The Analytic Solution," following Example 2.

(c) Let t be the number of hours since 8 A.M. Then $y_0 = y(0) = \frac{80}{1000} = 0.08$ and $y(4) = \frac{1}{2}$, so

$\dfrac{1}{2} = y(4) = \dfrac{0.08}{0.08 + 0.92e^{-4k}}$. Thus, $0.08 + 0.92e^{-4k} = 0.16$, $e^{-4k} = \dfrac{0.08}{0.92} = \dfrac{2}{23}$, and $e^{-k} = \left(\dfrac{2}{23}\right)^{1/4}$, so

$y = \dfrac{0.08}{0.08 + 0.92(2/23)^{t/4}} = \dfrac{2}{2 + 23(2/23)^{t/4}}$. Solving this equation for t, we get

$2y + 23y\left(\dfrac{2}{23}\right)^{t/4} = 2$ $\Rightarrow$ $\left(\dfrac{2}{23}\right)^{t/4} = \dfrac{2 - 2y}{23y}$ $\Rightarrow$ $\left(\dfrac{2}{23}\right)^{t/4} = \dfrac{2}{23} \cdot \dfrac{1 - y}{y}$ $\Rightarrow$

$\left(\dfrac{2}{23}\right)^{t/4-1} = \dfrac{1 - y}{y}$. It follows that $\dfrac{t}{4} - 1 = \dfrac{\ln[(1 - y)/y]}{\ln \frac{2}{23}}$, so $t = 4\left[1 + \dfrac{\ln((1 - y)/y)}{\ln \frac{2}{23}}\right]$.

When $y = 0.9$, $\dfrac{1 - y}{y} = \frac{1}{9}$, so $t = 4\left(1 - \dfrac{\ln 9}{\ln \frac{2}{23}}\right) \approx 7.6$ h or 7 h 36 min. Thus, 90% of the population will

have heard the rumor by 3:36 P.M.

9. (a) $\dfrac{dP}{dt} = kP\left(1 - \dfrac{P}{K}\right)$ $\Rightarrow$

$$\dfrac{d^2P}{dt^2} = k\left[P\left(-\dfrac{1}{K}\dfrac{dP}{dt}\right) + \left(1 - \dfrac{P}{K}\right)\dfrac{dP}{dt}\right] = k\dfrac{dP}{dt}\left(-\dfrac{P}{K} + 1 - \dfrac{P}{K}\right)$$

$$= k\left[kP\left(1 - \dfrac{P}{K}\right)\right]\left(1 - \dfrac{2P}{K}\right) = k^2P\left(1 - \dfrac{P}{K}\right)\left(1 - \dfrac{2P}{K}\right)$$

(b) P grows fastest when P' has a maximum, that is, when $P'' = 0$. From part (a), $P'' = 0$ $\Leftrightarrow$ $P = 0, P = K$, or $P = K/2$. Since $0 < P < K$, we see that $P'' = 0$ $\Leftrightarrow$ $P = K/2$.

11. (a) The term -15 represents a harvesting of fish at a constant rate—in this case, 15 fish/week. This is the rate at which fish are caught.

(b)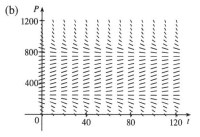

(c) From the graph in part (b), it appears that $P(t) = 250$ and $P(t) = 750$ are the equilibrium solutions. We confirm this analytically by solving the equation $dP/dt = 0$ as follows: $0.08P(1 - P/1000) - 15 = 0$ $\Rightarrow$
$0.08P - 0.00008P^2 - 15 = 0$ $\Rightarrow$ $-0.00008\left(P^2 - 1000P + 187{,}500\right) = 0$ $\Rightarrow$
$(P - 250)(P - 750) = 0$ $\Rightarrow$ $P = 250$ or 750.

(d)

$P_0 = 1000$
$P_0 = 600$
$P_0 = 400$
$P_0 = 300$
$P_0 = 200$

For $0 < P_0 < 250$, $P(t)$ decreases to 0. For $P_0 = 250$, $P(t)$ remains constant. For $250 < P_0 < 750$, $P(t)$ increases and approaches 750. For $P_0 = 750$, $P(t)$ remains constant. For $P_0 > 750$, $P(t)$ decreases and approaches 750.

(e) $\dfrac{dP}{dt} = 0.08P\left(1 - \dfrac{P}{1000}\right) - 15$ $\Leftrightarrow$ $-\dfrac{100{,}000}{8}\cdot\dfrac{dP}{dt} = \left(0.08P - 0.00008P^2 - 15\right)\cdot\left(-\dfrac{100{,}000}{8}\right)$ $\Leftrightarrow$

$-12{,}500\,\dfrac{dP}{dt} = P^2 - 1000P + 187{,}500$ $\Leftrightarrow$ $\dfrac{dP}{(P - 250)(P - 750)} = -\dfrac{1}{12{,}500}\,dt$ $\Leftrightarrow$

$\displaystyle\int\left(\dfrac{-1/500}{P - 250} + \dfrac{1/500}{P - 750}\right)dP = -\dfrac{1}{12{,}500}\,dt$ $\Leftrightarrow$ $\displaystyle\int\left(\dfrac{1}{P - 250} - \dfrac{1}{P - 750}\right)dP = \tfrac{1}{25}\,dt$ $\Leftrightarrow$

$\ln|P - 250| - \ln|P - 750| = \tfrac{1}{25}t + C$ $\Leftrightarrow$ $\ln\left|\dfrac{P - 250}{P - 750}\right| = \tfrac{1}{25}t + C$ $\Leftrightarrow$

$\left|\dfrac{P - 250}{P - 750}\right| = e^{t/25+C} = ke^{t/25}$ $\Leftrightarrow$ $\dfrac{P - 250}{P - 750} = ke^{t/25}$ $\Leftrightarrow$ $P - 250 = Pke^{t/25} - 750ke^{t/25}$ $\Leftrightarrow$

$$P - Pke^{t/25} = 250 - 750ke^{t/25} \quad \Leftrightarrow \quad P(t) = \frac{250 - 750ke^{t/25}}{1 - ke^{t/25}}. \text{ If } t = 0 \text{ and } P = 200, \text{ then}$$

$$200 = \frac{250 - 750k}{1 - k} \quad \Leftrightarrow \quad 200 - 200k = 250 - 750k \quad \Leftrightarrow$$

$$550k = 50 \quad \Leftrightarrow \quad k = \tfrac{1}{11}. \text{ Similarly, if } t = 0 \text{ and } P = 300, \text{ then}$$

$$k = -\tfrac{1}{9}. \text{ Simplifying } P \text{ with these two values of } k \text{ gives us}$$

$$P(t) = \frac{250\left(3e^{t/25} - 11\right)}{e^{t/25} - 11} \text{ and } P(t) = \frac{750\left(e^{t/25} + 3\right)}{e^{t/25} + 9}.$$

13. (a) $\dfrac{dP}{dt} = (kP)\left(1 - \dfrac{P}{K}\right)\left(1 - \dfrac{m}{P}\right)$. If $m < P < K$, then $dP/dt = (+)(+)(+) = + \quad \Rightarrow \quad P$ is increasing.

If $0 < P < m$, then $dP/dt = (+)(+)(-) = - \quad \Rightarrow \quad P$ is decreasing.

(b)

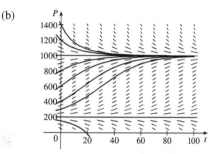

$k = 0.08$, $K = 1000$, and $m = 200 \quad \Rightarrow$

$$\frac{dP}{dt} = 0.08P\left(1 - \frac{P}{1000}\right)\left(1 - \frac{200}{P}\right)$$

For $0 < P_0 < 200$, the population dies out. For $P_0 = 200$, the population is steady. For $200 < P_0 < 1000$, the population increases and approaches 1000. For $P_0 > 1000$, the population decreases and approaches 1000.

The equilibrium solutions are $P(t) = 200$ and $P(t) = 1000$.

(c) $\dfrac{dP}{dt} = kP\left(1 - \dfrac{P}{K}\right)\left(1 - \dfrac{m}{P}\right) = kP\left(\dfrac{K - P}{K}\right)\left(\dfrac{P - m}{P}\right) = \dfrac{k}{K}(K - P)(P - m) \quad \Leftrightarrow$

$$\int \frac{dP}{(K - P)(P - m)} = \int \frac{k}{K}\, dt.$$

By partial fractions, $\dfrac{1}{(K - P)(P - m)} = \dfrac{A}{K - P} + \dfrac{B}{P - m}$, so $A(P - m) + B(K - P) = 1$.

If $P = m$, $B = \dfrac{1}{K - m}$; if $P = K$, $A = \dfrac{1}{K - m}$, so $\dfrac{1}{K - m}\displaystyle\int \left(\dfrac{1}{K - P} + \dfrac{1}{P - m}\right) dP = \int \dfrac{k}{K}\, dt \quad \Rightarrow$

$$\frac{1}{K - m}\left(-\ln|K - P| + \ln|P - m|\right) = \frac{k}{K}t + M \quad \Rightarrow \quad \frac{1}{K - m}\ln\left|\frac{P - m}{K - P}\right| = \frac{k}{K}t + M \quad \Rightarrow$$

$$\ln\left|\frac{P - m}{K - P}\right| = (K - m)\frac{k}{K}t + M_1 \quad \Leftrightarrow \quad \frac{P - m}{K - P} = De^{(K-m)(k/K)t} \quad [D = \pm e^{M_1}].$$

Let $t = 0$: $\dfrac{P_0 - m}{K - P_0} = D$. So $\dfrac{P - m}{K - P} = \dfrac{P_0 - m}{K - P_0}e^{(K-m)(k/K)t}$. Solving for P, we get

$$P(t) = \frac{m(K - P_0) + K(P_0 - m)e^{(K-m)(k/K)t}}{K - P_0 + (P_0 - m)e^{(K-m)(k/K)t}}.$$

(d) If $P_0 < m$, then $P_0 - m < 0$. Let $N(t)$ be the numerator of the expression for $P(t)$ in part (c). Then

$N(0) = P_0(K - m) > 0$, and $P_0 - m < 0 \quad \Leftrightarrow \quad \displaystyle\lim_{t \to \infty} K(P_0 - m)\, e^{(K-m)(k/K)t} = -\infty \quad \Rightarrow$

$\displaystyle\lim_{t \to \infty} N(t) = -\infty$. Since N is continuous, there is a number t such that $N(t) = 0$ and thus $P(t) = 0$. So the species will become extinct.

15. (a) $dP/dt = kP\cos(rt - \phi)$ $\Rightarrow$ $(dP)/P = k\cos(rt - \phi)\,dt$ $\Rightarrow$ $\int (dP)/P = k\int \cos(rt - \phi)\,dt$ $\Rightarrow$

$\ln P = (k/r)\sin(rt - \phi) + C$. (Since this is a growth model, $P > 0$ and we can write $\ln P$ instead of $\ln|P|$.)

Since $P(0) = P_0$, we obtain $\ln P_0 = (k/r)\sin(-\phi) + C = -(k/r)\sin\phi + C$ $\Rightarrow$

$C = \ln P_0 + (k/r)\sin\phi$. Thus, $\ln P = (k/r)\sin(rt - \phi) + \ln P_0 + (k/r)\sin\phi$, which we can rewrite as

$\ln(P/P_0) = (k/r)[\sin(rt - \phi) + \sin\phi]$ or, after exponentiation, $P(t) = P_0 e^{(k/r)[\sin(rt-\phi)+\sin\phi]}$.

(b) As k increases, the amplitude increases, but the minimum value stays the same.

As r increases, the amplitude and the period decrease.

A change in ϕ produces slight adjustments in the phase shift and amplitude.

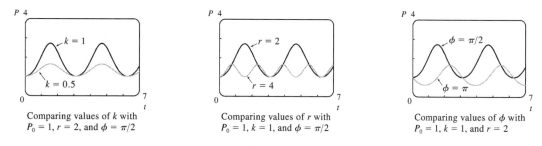

Comparing values of k with $P_0 = 1$, $r = 2$, and $\phi = \pi/2$

Comparing values of r with $P_0 = 1$, $k = 1$, and $\phi = \pi/2$

Comparing values of ϕ with $P_0 = 1$, $k = 1$, and $r = 2$

$P(t)$ oscillates between $P_0 e^{(k/r)(1+\sin\phi)}$ and $P_0 e^{(k/r)(-1+\sin\phi)}$ (the extreme values are attained when $rt - \phi$

is an odd multiple of $\frac{\pi}{2}$), so $\lim\limits_{t\to\infty} P(t)$ does not exist.

17. By Equation (4), $P(t) = \dfrac{K}{1 + Ae^{-kt}}$. By comparison, if $c = (\ln A)/k$ and $u = \frac{1}{2}k(t - c)$, then

$$1 + \tanh u = 1 + \frac{e^u - e^{-u}}{e^u + e^{-u}} = \frac{e^u + e^{-u}}{e^u + e^{-u}} + \frac{e^u - e^{-u}}{e^u + e^{-u}} = \frac{2e^u}{e^u + e^{-u}} \cdot \frac{e^{-u}}{e^{-u}} = \frac{2}{1 + e^{-2u}}$$

and $e^{-2u} = e^{-k(t-c)} = e^{kc}e^{-kt} = e^{\ln A}e^{-kt} = Ae^{-kt}$, so

$$\tfrac{1}{2}K\big[1 + \tanh\big(\tfrac{1}{2}k(t - c)\big)\big] = \frac{K}{2}[1 + \tanh u] = \frac{K}{2}\cdot\frac{2}{1 + e^{-2u}} = \frac{K}{1 + e^{-2u}} = \frac{K}{1 + Ae^{-kt}} = P(t).$$

9.6 Linear Equations

1. $y' + e^x y = x^2 y^2$ is not linear since it cannot be put into the standard linear form (1), $y' + P(x)\,y = Q(x)$.

3. $xy' + \ln x - x^2 y = 0$ $\Rightarrow$ $xy' - x^2 y = -\ln x$ $\Rightarrow$ $y' + (-x)\,y = -\dfrac{\ln x}{x}$, which is in the standard linear

form (1), so this equation is linear.

5. Comparing the given equation, $y' + 2y = 2e^x$, with the general form, $y' + P(x)y = Q(x)$, we see that $P(x) = 2$

and the integrating factor is $I(x) = e^{\int P(x)\,dx} = e^{\int 2\,dx} = e^{2x}$. Multiplying the differential equation by $I(x)$ gives

$e^{2x}y' + 2e^{2x}y = 2e^{3x}$ $\Rightarrow$ $\big(e^{2x}y\big)' = 2e^{3x}$ $\Rightarrow$ $e^{2x}y = \int 2e^{3x}\,dx$ $\Rightarrow$ $e^{2x}y = \tfrac{2}{3}e^{3x} + C$ $\Rightarrow$

$y = \tfrac{2}{3}e^x + Ce^{-2x}$.

7. $xy' - 2y = x^2$ [divide by x] $\Rightarrow$ $y' + \left(-\dfrac{2}{x}\right)y = x$ (∗).

$I(x) = e^{\int P(x)\,dx} = e^{\int(-2/x)\,dx} = e^{-2\,\ln|x|} = e^{\ln|x|^{-2}} = e^{\ln(1/x^2)} = 1/x^2$. Multiplying the differential

equation (∗) by $I(x)$ gives $\dfrac{1}{x^2}\,y' - \dfrac{2}{x^3}\,y = \dfrac{1}{x}$ $\Rightarrow$ $\left(\dfrac{1}{x^2}\,y\right)' = \dfrac{1}{x}$ $\Rightarrow$ $\dfrac{1}{x^2}\,y = \ln|x| + C$ $\Rightarrow$

$y = x^2(\ln|x| + C) = x^2\ln|x| + Cx^2$.

9. Since $P(x)$ is the derivative of the coefficient of y' [$P(x) = 1$ and the coefficient is x], we can write the

differential equation $xy' + y = \sqrt{x}$ in the easily integrable form $(xy)' = \sqrt{x}$ $\Rightarrow$ $xy = \frac{2}{3}x^{3/2} + C$ $\Rightarrow$

$y = \frac{2}{3}\sqrt{x} + C/x$.

11. $I(x) = e^{\int 2x\,dx} = e^{x^2}$. Multiplying the differential equation $y' + 2xy = x^2$ by $I(x)$ gives

$e^{x^2}y' + 2xe^{x^2}y = x^2e^{x^2}$ $\Rightarrow$ $\left(e^{x^2}y\right)' = x^2e^{x^2}$. Thus

$y = e^{-x^2}\left[\int x^2e^{x^2}\,dx + C\right] = e^{-x^2}\left[\frac{1}{2}xe^{x^2} - \int \frac{1}{2}e^{x^2}\,dx + C\right] = \frac{1}{2}x + Ce^{-x^2} - e^{-x^2}\int \frac{1}{2}e^{x^2}\,dx$.

13. $(1+t)\dfrac{du}{dt} + u = 1 + t, t > 0$ [divide by $1 + t$] $\Rightarrow$ $\dfrac{du}{dt} + \dfrac{1}{1+t}\,u = 1$ (∗), which has the form

$u' + P(t)\,u = Q(t)$. The integrating factor is $I(t) = e^{\int P(t)\,dt} = e^{\int[1/(1+t)]\,dt} = e^{\ln(1+t)} = 1 + t$.

Multiplying (∗) by $I(t)$ gives us our original equation back. We rewrite it as $[(1+t)\,u]' = 1 + t$. Thus,

$(1+t)\,u = \int (1+t)\,dt = t + \frac{1}{2}t^2 + C$ $\Rightarrow$ $u = \dfrac{t + \frac{1}{2}t^2 + C}{1+t}$ or $u = \dfrac{t^2 + 2t + 2C}{2(t+1)}$.

15. $y' = x + y$ $\Rightarrow$ $y' + (-1)y = x$. $I(x) = e^{\int(-1)\,dx} = e^{-x}$. Multiplying by e^{-x} gives $e^{-x}y' - e^{-x}y = xe^{-x}$

$\Rightarrow$ $(e^{-x}y)' = xe^{-x}$ $\Rightarrow$ $e^{-x}y = \int xe^{-x}\,dx = -xe^{-x} - e^{-x} + C$ [integration by parts with $u = x$,

$dv = e^{-x}\,dx$] $\Rightarrow$ $y = -x - 1 + Ce^{x}$. $y(0) = 2$ $\Rightarrow$ $-1 + C = 2$ $\Rightarrow$ $C = 3$, so $y = -x - 1 + 3e^{x}$.

17. $\dfrac{dv}{dt} - 2tv = 3t^2e^{t^2}$, $v(0) = 5$. $I(t) = e^{\int(-2t)\,dt} = e^{-t^2}$. Multiply the differential equation by $I(t)$ to get

$e^{-t^2}\dfrac{dv}{dt} - 2te^{-t^2}v = 3t^2$ $\Rightarrow$ $\left(e^{-t^2}v\right)' = 3t^2$ $\Rightarrow$ $e^{-t^2}v = \int 3t^2\,dt = t^3 + C$ $\Rightarrow$ $v = t^3e^{t^2} + Ce^{t^2}$.

$5 = v(0) = 0 \cdot 1 + C \cdot 1 = C$, so $v = t^3e^{t^2} + 5e^{t^2}$.

19. $xy' = y + x^2\sin x$ $\Rightarrow$ $y' - \dfrac{1}{x}\,y = x\sin x$. $I(x) = e^{\int(-1/x)\,dx} = e^{-\ln x} = e^{\ln x^{-1}} = \dfrac{1}{x}$.

Multiplying by $\dfrac{1}{x}$ gives $\dfrac{1}{x}\,y' - \dfrac{1}{x^2}\,y = \sin x$ $\Rightarrow$ $\left(\dfrac{1}{x}\,y\right)' = \sin x$ $\Rightarrow$ $\dfrac{1}{x}\,y = -\cos x + C$ $\Rightarrow$

$y = -x\cos x + Cx$. $y(\pi) = 0$ $\Rightarrow$ $-\pi \cdot (-1) + C\pi = 0$ $\Rightarrow$ $C = -1$, so $y = -x\cos x - x$.

21. $y' + \dfrac{1}{x}y = \cos x$ $(x \neq 0)$, so $I(x) = e^{\int (1/x)\,dx} = e^{\ln|x|} = x$ (for

$x > 0$). Multiplying the differential equation by $I(x)$ gives

$xy' + y = x\cos x \quad \Rightarrow \quad (xy)' = x\cos x$. Thus,

$$y = \frac{1}{x}\left[\int x\cos x\,dx + C\right] = \frac{1}{x}\left[x\sin x + \cos x + C\right]$$

$$= \sin x + \frac{\cos x}{x} + \frac{C}{x}$$

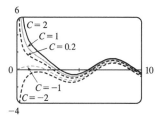

The solutions are asymptotic to the y-axis (except for $C = -1$). In fact, for $C > -1$, $y \to \infty$ as $x \to 0^+$, whereas

for $C < -1$, $y \to -\infty$ as $x \to 0^+$. As x gets larger, the solutions approximate $y = \sin x$ more closely. The

graphs for larger C lie above those for smaller C. The distance between the graphs lessens as x increases.

23. Setting $u = y^{1-n}$, $\dfrac{du}{dx} = (1-n)\,y^{-n}\dfrac{dy}{dx}$ or $\dfrac{dy}{dx} = \dfrac{y^n}{1-n}\dfrac{du}{dx} = \dfrac{u^{n/(1-n)}}{1-n}\dfrac{du}{dx}$. Then the Bernoulli differential

equation becomes $\dfrac{u^{n/(1-n)}}{1-n}\dfrac{du}{dx} + P(x)u^{1/(1-n)} = Q(x)u^{n/(1-n)}$ or $\dfrac{du}{dx} + (1-n)P(x)u = Q(x)(1-n)$.

25. $y' + \dfrac{2}{x}y = \dfrac{y^3}{x^2}$. Here $n = 3$, $P(x) = \dfrac{2}{x}$, $Q(x) = \dfrac{1}{x^2}$ and setting $u = y^{-2}$, u satisfies $u' - \dfrac{4u}{x} = -\dfrac{2}{x^2}$.

Then $I(x) = e^{\int (-4/x)\,dx} = x^{-4}$ and $u = x^4\left(\displaystyle\int -\dfrac{2}{x^6}\,dx + C\right) = x^4\left(\dfrac{2}{5x^5} + C\right) = Cx^4 + \dfrac{2}{5x}$.

Thus, $y = \pm\left(Cx^4 + \dfrac{2}{5x}\right)^{-1/2}$.

27. (a) $2\dfrac{dI}{dt} + 10I = 40$ or $\dfrac{dI}{dt} + 5I = 20$. Then the integrating factor is $e^{\int 5\,dt} = e^{5t}$. Multiplying the differential

   equation by the integrating factor gives $e^{5t}\dfrac{dI}{dt} + 5Ie^{5t} = 20e^{5t} \quad \Rightarrow \quad (e^{5t}I)' = 20e^{5t} \quad \Rightarrow$

   $I(t) = e^{-5t}\left[\int 20e^{5t}\,dt + C\right] = 4 + Ce^{-5t}$. But $0 = I(0) = 4 + C$, so $I(t) = 4 - 4e^{-5t}$.

(b) $I(0.1) = 4 - 4e^{-0.5} \approx 1.57$ A

29. $5\dfrac{dQ}{dt} + 20Q = 60$ with $Q(0) = 0$ C. Then the integrating factor is $e^{\int 4\,dt} = e^{4t}$, and multiplying the differential

equation by the integrating factor gives $e^{4t}\dfrac{dQ}{dt} + 4e^{4t}Q = 12e^{4t} \quad \Rightarrow \quad (e^{4t}Q)' = 12e^{4t} \quad \Rightarrow$

$Q(t) = e^{-4t}\left[\int 12e^{4t}\,dt + C\right] = 3 + Ce^{-4t}$. But $0 = Q(0) = 3 + C$ so $Q(t) = 3\left(1 - e^{-4t}\right)$ is the charge at time

t and $I = dQ/dt = 12e^{-4t}$ is the current at time t.

31. $\dfrac{dP}{dt} + kP = kM$, so $I(t) = e^{\int k\,dt} = e^{kt}$. Multiplying the differential

equation by $I(t)$ gives $e^{kt}\dfrac{dP}{dt} + kPe^{kt} = kMe^{kt}$ $\Rightarrow$

$(e^{kt}P)' = kMe^{kt}$ $\Rightarrow$

$P(t) = e^{-kt}\left(\int kMe^{kt}dt + C\right) = M + Ce^{-kt}$, $k > 0$. Furthermore, it is

reasonable to assume that $0 \le P(0) \le M$, so $-M \le C \le 0$.

33. $y(0) = 0$ kg. Salt is added at a rate of $\left(0.4\,\dfrac{\text{kg}}{\text{L}}\right)\left(5\,\dfrac{\text{L}}{\text{min}}\right) = 2\,\dfrac{\text{kg}}{\text{min}}$. Since solution is drained from the tank at a

rate of 3 L/min, but salt solution is added at a rate of 5 L/min, the tank, which starts out with 100 L of water,

contains $(100 + 2t)$ L of liquid after t min. Thus, the salt concentration at time t is $\dfrac{y(t)}{100 + 2t}\,\dfrac{\text{kg}}{\text{L}}$. Salt therefore

leaves the tank at a rate of $\left(\dfrac{y(t)}{100 + 2t}\,\dfrac{\text{kg}}{\text{L}}\right)\left(3\,\dfrac{\text{L}}{\text{min}}\right) = \dfrac{3y}{100 + 2t}\,\dfrac{\text{kg}}{\text{min}}$. Combining the rates at which salt enters

and leaves the tank, we get $\dfrac{dy}{dt} = 2 - \dfrac{3y}{100 + 2t}$. Rewriting this equation as $\dfrac{dy}{dt} + \left(\dfrac{3}{100 + 2t}\right)y = 2$, we see that

it is linear. $I(t) = \exp\left(\int \dfrac{3\,dt}{100 + 2t}\right) = \exp\left(\tfrac{3}{2}\ln(100 + 2t)\right) = (100 + 2t)^{3/2}$. Multiplying the differential

equation by $I(t)$ gives $(100 + 2t)^{3/2}\dfrac{dy}{dt} + 3(100 + 2t)^{1/2}y = 2(100 + 2t)^{3/2}$ $\Rightarrow$

$\left[(100 + 2t)^{3/2}y\right]' = 2(100 + 2t)^{3/2}$ $\Rightarrow$ $(100 + 2t)^{3/2}y = \tfrac{2}{5}(100 + 2t)^{5/2} + C$ $\Rightarrow$

$y = \tfrac{2}{5}(100 + 2t) + C(100 + 2t)^{-3/2}$. Now $0 = y(0) = \tfrac{2}{5}(100) + C \cdot 100^{-3/2} = 40 + \tfrac{1}{1000}C$ $\Rightarrow$

$C = -40{,}000$, so $y = \left[\tfrac{2}{5}(100 + 2t) - 40{,}000(100 + 2t)^{-3/2}\right]$ kg. From this solution (no pun intended), we

calculate the salt concentration at time t to be $C(t) = \dfrac{y(t)}{100 + 2t} = \left[\dfrac{-40{,}000}{(100 + 2t)^{5/2}} + \dfrac{2}{5}\right]\dfrac{\text{kg}}{\text{L}}$. In particular,

$C(20) = \dfrac{-40{,}000}{140^{5/2}} + \dfrac{2}{5} \approx 0.2275\,\dfrac{\text{kg}}{\text{L}}$ and $y(20) = \tfrac{2}{5}(140) - 40{,}000(140)^{-3/2} \approx 31.85$ kg.

35. (a) $\dfrac{dv}{dt} + \dfrac{c}{m}v = g$ and $I(t) = e^{\int(c/m)dt} = e^{(c/m)t}$, and multiplying the differential equation by $I(t)$ gives

$e^{(c/m)t}\dfrac{dv}{dt} + \dfrac{vce^{(c/m)t}}{m} = ge^{(c/m)t}$ $\Rightarrow$ $\left[e^{(c/m)t}v\right]' = ge^{(c/m)t}$. Hence,

$v(t) = e^{-(c/m)t}\left[\int ge^{(c/m)t}\,dt + K\right] = mg/c + Ke^{-(c/m)t}$. But the object is dropped from rest, so $v(0) = 0$

and $K = -mg/c$. Thus, the velocity at time t is $v(t) = (mg/c)\left[1 - e^{-(c/m)t}\right]$.

(b) $\lim\limits_{t\to\infty} v(t) = mg/c$

(c) $s(t) = \int v(t)\,dt = (mg/c)\left[t + (m/c)e^{-(c/m)t}\right] + c_1$ where $c_1 = s(0) - m^2g/c^2$. $s(0)$ is the initial position,

so $s(0) = 0$ and $s(t) = (mg/c)\left[t + (m/c)e^{-(c/m)t}\right] - m^2g/c^2$.

9.7 Predator-Prey Systems

1. (a) $dx/dt = -0.05x + 0.0001xy$. If $y = 0$, we have $dx/dt = -0.05x$, which indicates that in the absence of y, x

declines at a rate proportional to itself. So x represents the predator population and y represents the prey

population. The growth of the prey population, $0.1y$ (from $dy/dt = 0.1y - 0.005xy$), is restricted only by

encounters with predators (the term $-0.005xy$). The predator population increases only through the term

$0.0001xy$; that is, by encounters with the prey and not through additional food sources.

(b) $dy/dt = -0.015y + 0.00008xy$. If $x = 0$, we have $dy/dt = -0.015y$, which indicates that in the absence of

x, y would decline at a rate proportional to itself. So y represents the predator population and x

represents the prey population. The growth of the prey population, $0.2x$ (from

$dx/dt = 0.2x - 0.0002x^2 - 0.006xy = 0.2x(1 - 0.001x) - 0.006xy$), is restricted by a carrying capacity of

1000 [from the term $1 - 0.001x = 1 - x/1000$] and by encounters with predators (the term $-0.006xy$). The

predator population increases only through the term $0.00008xy$; that is, by encounters with the prey and not

through additional food sources.

3. (a) At $t = 0$, there are about 300 rabbits and 100 foxes. At $t = t_1$, the number of foxes reaches a minimum of about

20 while the number of rabbits is about 1000. At $t = t_2$, the number of rabbits reaches a maximum of about

2400, while the number of foxes rebounds to 100. At $t = t_3$, the number of rabbits decreases to about 1000 and

the number of foxes reaches a maximum of about 315. As t increases, the number of foxes decreases greatly to

100, and the number of rabbits decreases to 300 (the initial populations), and the cycle starts again.

(b)

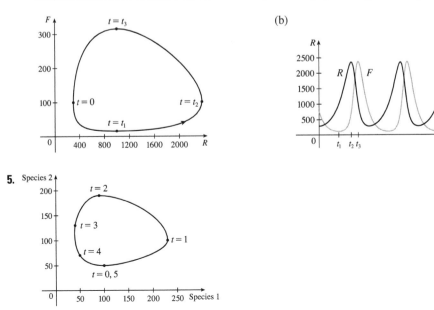

5.

7. $\dfrac{dW}{dR} = \dfrac{-0.02W + 0.00002RW}{0.08R - 0.001RW}$ $\Leftrightarrow$ $(0.08 - 0.001W)R\,dW = (-0.02 + 0.00002R)W\,dR$ $\Leftrightarrow$

$\dfrac{0.08 - 0.001W}{W}\,dW = \dfrac{-0.02 + 0.00002R}{R}\,dR$ $\Leftrightarrow$ $\displaystyle\int\left(\dfrac{0.08}{W} - 0.001\right)dW = \int\left(-\dfrac{0.02}{R} + 0.00002\right)dR$

$\Leftrightarrow$ $0.08\ln|W| - 0.001W = -0.02\ln|R| + 0.00002R + K$ $\Leftrightarrow$

$0.08\ln W + 0.02\ln R = 0.001W + 0.00002R + K$ $\Leftrightarrow$ $\ln\big(W^{0.08}R^{0.02}\big) = 0.00002R + 0.001W + K$ $\Leftrightarrow$

$W^{0.08}R^{0.02} = e^{0.00002R + 0.001W + K}$ $\Leftrightarrow$ $R^{0.02}W^{0.08} = Ce^{0.00002R}e^{0.001W}$ $\Leftrightarrow$ $\dfrac{R^{0.02}W^{0.08}}{e^{0.00002R}e^{0.001W}} = C.$

In general, if $\dfrac{dy}{dx} = \dfrac{-ry + bxy}{kx - axy}$, then $C = \dfrac{x^r y^k}{e^{bx}e^{ay}}$.

9. (a) Letting $W = 0$ gives us $dR/dt = 0.08R(1 - 0.0002R)$. $dR/dt = 0$ $\Leftrightarrow$ $R = 0$ or 5000. Since $dR/dt > 0$

for $0 < R < 5000$, we would expect the rabbit population to *increase* to 5000 for these values of R. Since

$dR/dt < 0$ for $R > 5000$, we would expect the rabbit population to *decrease* to 5000 for these values of R.

Hence, in the absence of wolves, we would expect the rabbit population to stabilize at 5000.

(b) R and W are constant $\Rightarrow$ $R' = 0$ and $W' = 0$ $\Rightarrow$

$$\begin{cases} 0 = 0.08R(1 - 0.0002R) - 0.001RW \\ 0 = -0.02W + 0.00002RW \end{cases} \Rightarrow \begin{cases} 0 = R[0.08(1 - 0.0002R) - 0.001W] \\ 0 = W(-0.02 + 0.00002R) \end{cases}$$

The second equation is true if $W = 0$ or $R = \frac{0.02}{0.00002} = 1000$. If $W = 0$ in the first equation, then either $R = 0$

or $R = \frac{1}{0.0002} = 5000$ [as in part (a)]. If $R = 1000$, then $0 = 1000[0.08(1 - 0.0002 \cdot 1000) - 0.001W]$ $\Leftrightarrow$

$0 = 80(1 - 0.2) - W$ $\Leftrightarrow$ $W = 64$.

Case (i): $W = 0$, $R = 0$: both populations are zero

Case (ii): $W = 0$, $R = 5000$: see part (a)

Case (iii): $R = 1000$, $W = 64$: the predator/prey interaction balances and the populations are stable.

(c) The populations of wolves and rabbits fluctuate

around 64 and 1000, respectively, and eventually

stabilize at those values.

(d)

9 Review

——————————————————— CONCEPT CHECK ———————————————————

1. (a) A differential equation is an equation that contains an unknown function and one or more of its derivatives.

(b) The order of a differential equation is the order of the highest derivative that occurs in the equation.

(c) An initial condition is a condition of the form $y(t_0) = y_0$.

2. $y' = x^2 + y^2 \geq 0$ for all x and y. $y' = 0$ only at the origin, so there is a horizontal tangent at $(0, 0)$, but nowhere else. The graph of the solution is increasing on every interval.

3. See the paragraph preceding Example 1 in Section 9.2.

4. See the paragraph after Figure 14 in Section 9.2.

5. A separable equation is a first-order differential equation in which the expression for dy/dx can be factored as a function of x times a function of y, that is, $dy/dx = g(x)f(y)$. We can solve the equation by integrating both sides of the equation $dy/f(y) = g(x)dx$ and solving for y.

6. A first-order linear differential equation is a differential equation that can be put in the form $\dfrac{dy}{dx} + P(x)\,y = Q(x)$, where P and Q are continuous functions on a given interval. To solve such an equation, multiply it by the integrating factor $I(x) = e^{\int P(x)\,dx}$ to put it in the form $[I(x)\,y]' = I(x)\,Q(x)$ and then integrate both sides to get $I(x)\,y = \int I(x)\,Q(x)\,dx$, that is, $e^{\int P(x)\,dx}y = \int e^{\int P(x)\,dx}Q(x)\,dx$. Solving for y gives us $y = e^{-\int P(x)\,dx}\int e^{\int P(x)\,dx}Q(x)\,dx$.

7. (a) $\dfrac{dy}{dt} = ky$; the relative growth rate, $\dfrac{1}{y}\dfrac{dy}{dt}$, is constant.

(b) The equation in part (a) is an appropriate model for population growth, assuming that there is enough room and nutrition to support the growth.

(c) If $y(0) = y_0$, then the solution is $y(t) = y_0 e^{kt}$.

8. (a) $dP/dt = kP(1 - P/K)$, where K is the carrying capacity.

(b) The equation in part (a) is an appropriate model for population growth, assuming that the population grows at a rate proportional to the size of the population in the beginning, but eventually levels off and approaches its carrying capacity because of limited resources.

9. (a) $dF/dt = kF - aFS$ and $dS/dt = -rS + bFS$.

(b) In the absence of sharks, an ample food supply would support exponential growth of the fish population, that is, $dF/dt = kF$, where k is a positive constant. In the absence of fish, we assume that the shark population would decline at a rate proportional to itself, that is, $dS/dt = -rS$, where r is a positive constant.

——————————————————— TRUE-FALSE QUIZ ———————————————————

1. True. Since $y^4 \geq 0$, $y' = -1 - y^4 < 0$ and the solutions are decreasing functions.

3. False. $x + y$ cannot be written in the form $g(x)f(y)$.

5. True. $e^x y' = y \;\Rightarrow\; y' = e^{-x}y \;\Rightarrow\; y' + (-e^{-x})y = 0$, which is of the form $y' + P(x)\,y = Q(x)$, so the equation is linear.

7. True. By comparing $\dfrac{dy}{dt} = 2y\left(1 - \dfrac{y}{5}\right)$ with the logistic differential equation (9.5.1), we see that the carrying capacity is 5; that is, $\lim\limits_{t \to \infty} y = 5$.

———————————————————— EXERCISES ————————————————————

1. (a)

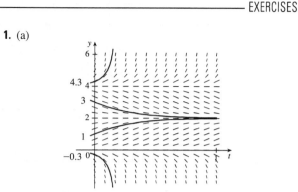

(b) $\lim\limits_{t \to \infty} y(t)$ appears to be finite for $0 \le c \le 4$. In fact $\lim\limits_{t \to \infty} y(t) = 4$ for $c = 4$, $\lim\limits_{t \to \infty} y(t) = 2$ for $0 < c < 4$, and $\lim\limits_{t \to \infty} y(t) = 0$ for $c = 0$. The equilibrium solutions are $y(t) = 0$, $y(t) = 2$, and $y(t) = 4$.

3. (a)

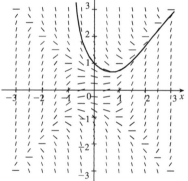

We estimate that when $x = 0.3$, $y = 0.8$, so $y(0.3) \approx 0.8$.

(b) $h = 0.1$, $x_0 = 0$, $y_0 = 1$ and $F(x, y) = x^2 - y^2$. So $y_n = y_{n-1} + 0.1\left(x_{n-1}^2 - y_{n-1}^2\right)$. Thus,
$y_1 = 1 + 0.1\left(0^2 - 1^2\right) = 0.9$,
$y_2 = 0.9 + 0.1\left(0.1^2 - 0.9^2\right) = 0.82$,
$y_3 = 0.82 + 0.1\left(0.2^2 - 0.82^2\right) = 0.75676$. This is close to our graphical estimate of $y(0.3) \approx 0.8$.

(c) The centers of the horizontal line segments of the direction field are located on the lines $y = x$ and $y = -x$. When a solution curve crosses one of these lines, it has a local maximum or minimum.

5. $y' = xe^{-\sin x} - y \cos x \quad \Rightarrow \quad y' + (\cos x)\, y = xe^{-\sin x}$ (*). This is a linear equation and the integrating factor is $I(x) = e^{\int \cos x\, dx} = e^{\sin x}$. Multiplying (*) by $e^{\sin x}$ gives $e^{\sin x}\, y' + e^{\sin x}(\cos x)\, y = x \quad \Rightarrow \quad \left(e^{\sin x}\, y\right)' = x$ $\Rightarrow \quad e^{\sin x}\, y = \frac{1}{2}x^2 + C \quad \Rightarrow \quad y = \left(\frac{1}{2}x^2 + C\right) e^{-\sin x}$.

7. $(3y^2 + 2y)\, y' = x \cos x \quad \Rightarrow \quad (3y^2 + 2y)\, dy = (x \cos x)\, dx \quad \Rightarrow \quad \int (3y^2 + 2y)\, dy = \int (x \cos x)\, dx \quad \Rightarrow$ $y^3 + y^2 = \cos x + x \sin x + C$. For the last step, use integration by parts or Formula 83 in the Table of Integrals.

9. $xyy' = \ln x \quad \Rightarrow \quad y\, dy = \dfrac{\ln x}{x}\, dx \quad \Rightarrow \quad \displaystyle\int y\, dy = \int \dfrac{\ln x}{x}\, dx$ (Make the substitution $u = \ln x$; then $du = dx/x$.) So $\int y\, dy = \int u\, du \quad \Rightarrow \quad \frac{1}{2}y^2 = \frac{1}{2}u^2 + C \quad \Rightarrow \frac{1}{2}y^2 = \frac{1}{2}(\ln x)^2 + C$. $y(1) = 2 \quad \Rightarrow$ $\frac{1}{2}2^2 = \frac{1}{2}(\ln 1)^2 + C = C \quad \Leftrightarrow \quad C = 2$. Therefore, $\frac{1}{2}y^2 = \frac{1}{2}(\ln x)^2 + 2$, or $y = \sqrt{(\ln x)^2 + 4}$. The negative square root is inadmissible, since $y(1) > 0$.

11. Since the equation is linear, let $I(x) = e^{\int dx} = e^x$. Then multiplying by $I(x)$ gives $e^x y' + e^x y = \sqrt{x} \quad \Rightarrow$ $(e^x y)' = \sqrt{x} \quad \Rightarrow \quad y(x) = e^{-x}\left(\int \sqrt{x}\, dx + c\right) = e^{-x}\left(\frac{2}{3}x^{3/2} + c\right)$. But $3 = y(0) = c$, so the solution to the initial-value problem is $y(x) = e^{-x}\left(\frac{2}{3}x^{3/2} + 3\right)$.

13. The curves $kx^2 + y^2 = 1$ form a family of ellipses for $k > 0$, a family of hyperbolas for $k < 0$, and two parallel

lines $y = \pm 1$ for $k = 0$. Solving $kx^2 + y^2 = 1$ for k gives $k = \dfrac{1 - y^2}{x^2}$. Differentiating gives $2kx + 2yy' = 0$ $\Leftrightarrow$

$y' = -\dfrac{kx}{y} = -\left(1 - y^2\right)\dfrac{x}{yx^2} = \dfrac{y^2 - 1}{xy}$. Thus, for $k \neq 0$ the orthogonal trajectories must satisfy $y' = -\dfrac{xy}{y^2 - 1}$

$\Rightarrow \dfrac{y^2 - 1}{y}\,dy = -x\,dx \Rightarrow \dfrac{y^2}{2} - \ln|y| = \dfrac{-x^2}{2} + K \Rightarrow y^2 - 2\ln|y| + x^2 = C$. For $k = 0$, the

orthogonal trajectories are given by $x = C_1$ for C_1 an arbitrary constant.

15. (a) $y(t) = y(0)e^{kt} = 1000e^{kt} \Rightarrow y(2) = 1000e^{2k} = 9000 \Rightarrow e^{2k} = 9 \Rightarrow 2k = \ln 9 \Rightarrow$

$k = \frac{1}{2}\ln 9 = \ln 3 \Rightarrow y(t) = 1000e^{(\ln 3)t} = 1000 \cdot 3^t$

(b) $y(3) = 1000 \cdot 3^3 = 27{,}000$

(c) $y'(t) = 1000 \cdot 3^t \cdot \ln 3$, so $y'(3) = 27{,}000 \ln 3 \approx 29{,}663$ bacteria per hour

(d) $1000 \cdot 3^t = 2 \cdot 1000 \Rightarrow 3^t = 2 \Rightarrow t\ln 3 = \ln 2 \Rightarrow t = (\ln 2)/\ln 3 \approx 0.63$ h

17. (a) $C'(t) = -kC(t) \Rightarrow C(t) = C(0)e^{-kt}$ by Theorem 9.4.2. But $C(0) = C_0$, so $C(t) = C_0e^{-kt}$.

(b) $C(30) = \frac{1}{2}C_0$ since the concentration is reduced by half. Thus, $\frac{1}{2}C_0 = C_0e^{-30k} \Rightarrow \ln\frac{1}{2} = -30k \Rightarrow$

$k = -\frac{1}{30}\ln\frac{1}{2} = \frac{1}{30}\ln 2$. Since 10% of the original concentration remains if 90% is eliminated, we want the

value of t such that $C(t) = \frac{1}{10}C_0$. Therefore, $\frac{1}{10}C_0 = C_0e^{-t(\ln 2)/30} \Rightarrow \ln 0.1 = -t(\ln 2)/30 \Rightarrow$

$t = -\frac{30}{\ln 2}\ln 0.1 \approx 100$ h.

19. (a) $\dfrac{dL}{dt} \propto L_\infty - L \Rightarrow \dfrac{dL}{dt} = k(L_\infty - L) \Rightarrow \displaystyle\int \dfrac{dL}{L_\infty - L} = \int k\,dt \Rightarrow -\ln|L_\infty - L| = kt + C \Rightarrow$

$\ln|L_\infty - L| = -kt - C \Rightarrow |L_\infty - L| = e^{-kt-C} \Rightarrow L_\infty - L = Ae^{-kt} \Rightarrow L = L_\infty - Ae^{-kt}$.

At $t = 0$, $L = L(0) = L_\infty - A \Rightarrow A = L_\infty - L(0) \Rightarrow L(t) = L_\infty - [L_\infty - L(0)]e^{-kt}$.

(b) $L_\infty = 53$ cm, $L(0) = 10$ cm, and $k = 0.2 \Rightarrow L(t) = 53 - (53 - 10)e^{-0.2t} = 53 - 43e^{-0.2t}$.

21. Let P be the population and I be the number of infected people. The rate of spread dI/dt is jointly proportional to

I and to $P - I$, so for some constant k, $dI/dt = kI(P - I) \Rightarrow I = \dfrac{I_0 P}{I_0 + (P - I_0)e^{-kPt}}$ (from the

discussion of logistic growth in Section 9.5).

Now, measuring t in days, we substitute $t = 7$, $P = 5000$, $I_0 = 160$ and $I(7) = 1200$ to find k:

$1200 = \dfrac{160 \cdot 5000}{160 + (5000 - 160)e^{-5000 \cdot 7 \cdot k}} \Leftrightarrow k \approx 0.00006448$. So, putting $I = 5000 \times 80\% = 4000$, we solve

for t: $4000 = \dfrac{160 \cdot 5000}{160 + (5000 - 160)e^{-0.00006448 \cdot 5000 \cdot t}} \Leftrightarrow 160 + 4840e^{-0.3224t} = 200 \Leftrightarrow$

$-0.3224t = \ln\frac{40}{4840} \Leftrightarrow t \approx 14.9$. So it takes about 15 days for 80% of the population to be infected.

23. $\dfrac{dh}{dt} = -\dfrac{R}{V}\left(\dfrac{h}{k + h}\right) \Rightarrow \displaystyle\int \dfrac{k + h}{h}\,dh = \int\left(-\dfrac{R}{V}\right)dt \Rightarrow \int\left(1 + \dfrac{k}{h}\right)dh = -\dfrac{R}{V}\int 1\,dt \Rightarrow$

$h + k\ln h = -\dfrac{R}{V}t + C$. This equation gives a relationship between h and t, but it is not possible to isolate h and

express it in terms of t.

25. (a) $dx/dt = 0.4x(1 - 0.000005x) - 0.002xy$, $dy/dt = -0.2y + 0.000008xy$. If $y = 0$, then

$dx/dt = 0.4x(1 - 0.000005x)$, so $dx/dt = 0 \Leftrightarrow x = 0$ or $x = 200{,}000$, which shows that the insect

population increases logistically with a carrying capacity of 200,000. Since $dx/dt > 0$ for $0 < x < 200{,}000$

and $dx/dt < 0$ for $x > 200{,}000$, we expect the insect population to stabilize at 200,000.

(b) x and y are constant $\Rightarrow x' = 0$ and $y' = 0 \Rightarrow$

$$\begin{cases} 0 = 0.4x(1 - 0.000005x) - 0.002xy \\ 0 = -0.2y + 0.000008xy \end{cases} \Rightarrow \begin{cases} 0 = 0.4x[(1 - 0.000005x) - 0.005y] \\ 0 = y(-0.2 + 0.000008x) \end{cases}$$

The second equation is true if $y = 0$ or $x = \frac{0.2}{0.000008} = 25{,}000$. If $y = 0$ in the first equation, then either $x = 0$

or $x = \frac{1}{0.000005} = 200{,}000$. If $x = 25{,}000$, then $0 = 0.4(25{,}000)[(1 - 0.000005 \cdot 25{,}000) - 0.005y] \Rightarrow$

$0 = 10{,}000[(1 - 0.125) - 0.005y] \Rightarrow 0 = 8750 - 50y \Rightarrow y = 175$.

Case (i): $y = 0$, $x = 0$: Zero populations

Case (ii): $y = 0$, $x = 200{,}000$: In the absence of birds, the insect population is always 200,000.

Case (iii): $x = 25{,}000$, $y = 175$: The predator/prey interaction balances and the populations are stable.

(c) The populations of the birds and insects fluctuate

around 175 and 25,000, respectively, and

eventually stabilize at those values.

(d)

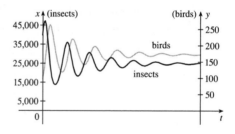

27. (a) $\dfrac{d^2y}{dx^2} = k\sqrt{1 + \left(\dfrac{dy}{dx}\right)^2}$. Setting $z = \dfrac{dy}{dx}$, we get $\dfrac{dz}{dx} = k\sqrt{1 + z^2} \Rightarrow \dfrac{dz}{\sqrt{1 + z^2}} = k\,dx$. Using

Formula 25 gives $\ln\left(z + \sqrt{1 + z^2}\right) = kx + c \Rightarrow z + \sqrt{1 + z^2} = Ce^{kx}$ (where $C = e^c$) $\Rightarrow$

$\sqrt{1 + z^2} = Ce^{kx} - z \Rightarrow 1 + z^2 = C^2e^{2kx} - 2Ce^{kx}z + z^2 \Rightarrow 2Ce^{kx}z = C^2e^{2kx} - 1 \Rightarrow$

$z = \dfrac{C}{2}e^{kx} - \dfrac{1}{2C}e^{-kx}$. Now $\dfrac{dy}{dx} = \dfrac{C}{2}e^{kx} - \dfrac{1}{2C}e^{-kx} \Rightarrow y = \dfrac{C}{2k}e^{kx} + \dfrac{1}{2Ck}e^{-kx} + C'$. From the

diagram in the text, we see that $y(0) = a$ and $y(\pm b) = h$. $a = y(0) = \dfrac{C}{2k} + \dfrac{1}{2Ck} + C' \Rightarrow$

$C' = a - \dfrac{C}{2k} - \dfrac{1}{2Ck} \Rightarrow y = \dfrac{C}{2k}\left(e^{kx} - 1\right) + \dfrac{1}{2Ck}\left(e^{-kx} - 1\right) + a$. From $h = y(\pm b)$, we find

$h = \dfrac{C}{2k}\left(e^{kb} - 1\right) + \dfrac{1}{2Ck}\left(e^{-kb} - 1\right) + a$ and $h = \dfrac{C}{2k}\left(e^{-kb} - 1\right) + \dfrac{1}{2Ck}\left(e^{kb} - 1\right) + a$. Subtracting the

second equation from the first, we get $0 = \dfrac{C}{k}\dfrac{e^{kb} - e^{-kb}}{2} - \dfrac{1}{Ck}\dfrac{e^{kb} - e^{-kb}}{2} = \dfrac{1}{k}\left(C - \dfrac{1}{C}\right)\sinh kb$.

Now $k > 0$ and $b > 0$, so $\sinh kb > 0$ and $C = \pm 1$. If $C = 1$, then

$y = \dfrac{1}{2k}\left(e^{kx} - 1\right) + \dfrac{1}{2k}\left(e^{-kx} - 1\right) + a = \dfrac{1}{k}\dfrac{e^{kx} + e^{-kx}}{2} - \dfrac{1}{k} + a = a + \dfrac{1}{k}(\cosh kx - 1)$. If $C = -1$, then $y = -\dfrac{1}{2k}\left(e^{kx} - 1\right) - \dfrac{1}{2k}\left(e^{-kx} - 1\right) + a = \dfrac{-1}{k}\dfrac{e^{kx} + e^{-kx}}{2} + \dfrac{1}{k} + a = a - \dfrac{1}{k}(\cosh kx - 1).$

Since $k > 0$, $\cosh kx \geq 1$, and $y \geq a$, we conclude that $C = 1$ and $y = a + \dfrac{1}{k}(\cosh kx - 1)$, where

$h = y(b) = a + \dfrac{1}{k}(\cosh kb - 1)$. Since $\cosh(kb) = \cosh(-kb)$, there is no further information to extract from

the condition that $y(b) = y(-b)$. However, we could replace a with the expression $h - \dfrac{1}{k}(\cosh kb - 1)$,

obtaining $y = h + \dfrac{1}{k}(\cosh kx - \cosh kb)$. It would be better still to keep a in the expression for y, and use the

expression for h to solve for k in terms of a, b, and h. That would enable us to express y in terms of x and the

given parameters a, b, and h. Sadly, it is not possible to solve for k in closed form. That would have to be done

by numerical methods when specific parameter values are given.

(b) The length of the cable is

$$L = \int_{-b}^{b} \sqrt{1 + (dy/dx)^2}\, dx = \int_{-b}^{b} \sqrt{1 + \sinh^2 kx}\, dx = \int_{-b}^{b} \cosh kx\, dx = 2\int_{0}^{b} \cosh kx\, dx$$

$$= 2\left[(1/k)\sinh kx\right]_{0}^{b} = (2/k)\sinh kb$$

☐ PROBLEMS PLUS

1. We use the Fundamental Theorem of Calculus to differentiate the given equation:

$$[f(x)]^2 = 100 + \int_0^x \left\{[f(t)]^2 + [f'(t)]^2\right\} dt \quad \Rightarrow \quad 2f(x)f'(x) = [f(x)]^2 + [f'(x)]^2 \quad \Rightarrow$$

$[f(x)]^2 + [f'(x)]^2 - 2f(x)f'(x) = 0 \quad \Rightarrow \quad [f(x) - f'(x)]^2 = 0 \quad \Leftrightarrow \quad f(x) = f'(x)$. We can solve this as a

separable equation, or else use Theorem 9.4.2 with $k = 1$, which says that the solutions are $f(x) = Ce^x$. Now

$[f(0)]^2 = 100$, so $f(0) = C = \pm 10$, and hence $f(x) = \pm 10e^x$ are the only functions satisfying the given

equation.

3. $f'(x) = \lim\limits_{h \to 0} \dfrac{f(x+h) - f(x)}{h} = \lim\limits_{h \to 0} \dfrac{f(x)\,[f(h) - 1]}{h}$ [since $f(x+h) = f(x)f(h)$]

$= f(x) \lim\limits_{h \to 0} \dfrac{f(h) - 1}{h} = f(x) \lim\limits_{h \to 0} \dfrac{f(h) - f(0)}{h - 0} = f(x)f'(0) = f(x)$

Therefore, $f'(x) = f(x)$ for all x and from Theorem 9.4.2 we get $f(x) = Ae^x$. Now $f(0) = 1 \quad \Rightarrow \quad A = 1 \quad \Rightarrow$

$f(x) = e^x$.

5. Let $y(t)$ denote the temperature of the peach pie t minutes after 5:00 P.M. and R the temperature of the room.

Newton's Law of Cooling gives us $dy/dt = k(y - R)$. Solving for y we get $\dfrac{dy}{y - R} = k\,dt \quad \Rightarrow$

$\ln|y - R| = kt + C \quad \Rightarrow \quad |y - R| = e^{kt+C} \quad \Rightarrow \quad y - R = \pm e^{kt} \cdot e^C \quad \Rightarrow \quad y = Me^{kt} + R$, where M is a

nonzero constant. We are given temperatures at three times.

$$\begin{aligned} y(0) &= 100 & \Rightarrow & \quad 100 = M + R & \Rightarrow & \quad R = 100 - M \\ y(10) &= 80 & \Rightarrow & \quad 80 = Me^{10k} + R & \textbf{(1)} \\ y(20) &= 65 & \Rightarrow & \quad 65 = Me^{20k} + R & \textbf{(2)} \end{aligned}$$

Substituting $100 - M$ for R in **(1)** and **(2)** gives us

$$-20 = Me^{10k} - M \quad \textbf{(3)} \quad \text{and} \quad -35 = Me^{20k} - M \quad \textbf{(4)}$$

Dividing **(3)** by **(4)** gives us $\dfrac{-20}{-35} = \dfrac{M(e^{10k} - 1)}{M(e^{20k} - 1)} \quad \Rightarrow \quad \dfrac{4}{7} = \dfrac{e^{10k} - 1}{e^{20k} - 1} \quad \Rightarrow \quad 4e^{20k} - 4 = 7e^{10k} - 7 \quad \Rightarrow$

$4e^{20k} - 7e^{10k} + 3 = 0$. This is a quadratic equation in e^{10k}. $\left(4e^{10k} - 3\right)\left(e^{10k} - 1\right) = 0 \quad \Rightarrow \quad e^{10k} = \frac{3}{4}$ or 1

$\Rightarrow \quad 10k = \ln \frac{3}{4}$ or $\ln 1 \quad \Rightarrow \quad k = \frac{1}{10} \ln \frac{3}{4}$ since k is a nonzero constant of proportionality. Substituting $\frac{3}{4}$ for e^{10k}

in **(3)** gives us $-20 = M \cdot \frac{3}{4} - M \quad \Rightarrow \quad -20 = -\frac{1}{4}M \quad \Rightarrow \quad M = 80$. Now $R = 100 - M$ so $R = 20\,^\circ$C.

7. (a) While running from $(L, 0)$ to (x, y), the dog travels a distance

$$s = \int_x^L \sqrt{1 + (dy/dx)^2}\, dx = -\int_L^x \sqrt{1 + (dy/dx)^2}\, dx, \text{ so } \frac{ds}{dx} = -\sqrt{1 + (dy/dx)^2}.$$ The dog and rabbit

run at the same speed, so the rabbit's position when the dog has traveled a distance s is $(0, s)$. Since the dog runs

straight for the rabbit, $\dfrac{dy}{dx} = \dfrac{s - y}{0 - x}$ (see the figure).

Thus, $s = y - x\dfrac{dy}{dx} \;\Rightarrow\; \dfrac{ds}{dx} = \dfrac{dy}{dx} - \left(x\dfrac{d^2y}{dx^2} + 1 \cdot \dfrac{dy}{dx} \right) = -x\dfrac{d^2y}{dx^2}.$ Equating the two expressions for $\dfrac{ds}{dx}$

gives us $x\dfrac{d^2y}{dx^2} = \sqrt{1 + \left(\dfrac{dy}{dx} \right)^2}$, as claimed.

(b) Letting $z = \dfrac{dy}{dx}$, we obtain the differential equation $x\dfrac{dz}{dx} = \sqrt{1 + z^2}$, or $\dfrac{dz}{\sqrt{1 + z^2}} = \dfrac{dx}{x}$. Integrating:

$\ln x = \displaystyle\int \dfrac{dz}{\sqrt{1 + z^2}} \overset{25}{=} \ln\left(z + \sqrt{1 + z^2} \right) + C.$ When $x = L$, $z = dy/dx = 0$, so $\ln L = \ln 1 + C.$

Therefore, $C = \ln L$, so $\ln x = \ln\left(\sqrt{1 + z^2} + z\right) + \ln L = \ln\left[L\left(\sqrt{1 + z^2} + z\right)\right] \;\Rightarrow$

$x = L\left(\sqrt{1 + z^2} + z\right) \;\Rightarrow\; \sqrt{1 + z^2} = \dfrac{x}{L} - z \;\Rightarrow\; 1 + z^2 = \left(\dfrac{x}{L}\right)^2 - \dfrac{2xz}{L} + z^2 \;\Rightarrow$

$\left(\dfrac{x}{L}\right)^2 - 2z\left(\dfrac{x}{L}\right) - 1 = 0 \;\Rightarrow\; z = \dfrac{(x/L)^2 - 1}{2(x/L)} = \dfrac{x^2 - L^2}{2Lx} = \dfrac{x}{2L} - \dfrac{L}{2}\dfrac{1}{x}$ [for $x > 0$]. Since $z = \dfrac{dy}{dx}$,

$y = \dfrac{x^2}{4L} - \dfrac{L}{2}\ln x + C_1.$ Since $y = 0$ when $x = L$, $0 = \dfrac{L}{4} - \dfrac{L}{2}\ln L + C_1 \;\Rightarrow\; C_1 = \dfrac{L}{2}\ln L - \dfrac{L}{4}.$ Thus,

$y = \dfrac{x^2}{4L} - \dfrac{L}{2}\ln x + \dfrac{L}{2}\ln L - \dfrac{L}{4} = \dfrac{x^2 - L^2}{4L} - \dfrac{L}{2}\ln\left(\dfrac{x}{L}\right).$

(c) As $x \to 0^+$, $y \to \infty$, so the dog never catches the rabbit.

9. (a) We are given that $V = \frac{1}{3}\pi r^2 h$, $dV/dt = 60{,}000\pi$ ft³/h, and $r = 1.5h = \frac{3}{2}h$. So $V = \frac{1}{3}\pi\left(\frac{3}{2}h\right)^2 h = \frac{3}{4}\pi h^3$

$\Rightarrow \dfrac{dV}{dt} = \frac{3}{4}\pi \cdot 3h^2\dfrac{dh}{dt} = \frac{9}{4}\pi h^2\dfrac{dh}{dt}.$ Therefore, $\dfrac{dh}{dt} = \dfrac{4(dV/dt)}{9\pi h^2} = \dfrac{240{,}000\pi}{9\pi h^2} = \dfrac{80{,}000}{3h^2}$ (⋆) $\Rightarrow$

$\int 3h^2\, dh = \int 80{,}000\, dt \;\Rightarrow\; h^3 = 80{,}000t + C.$ When $t = 0$, $h = 60$. Thus, $C = 60^3 = 216{,}000$, so

$h^3 = 80,000t + 216,000$. Let $h = 100$. Then $100^3 = 1,000,000 = 80,000t + 216,000$ ⇒

$80,000t = 784,000$ ⇒ $t = 9.8$, so the time required is 9.8 hours.

(b) The floor area of the silo is $F = \pi \cdot 200^2 = 40,000\pi$ ft^2, and the area of the base of the pile is

$A = \pi r^2 = \pi \left(\frac{3}{2}h\right)^2 = \frac{9\pi}{4}h^2$. So the area of the floor which is not covered when $h = 60$ is

$F - A = 40,000\pi - 8100\pi = 31,900\pi \approx 100,217$ ft^2. Now $A = \frac{9\pi}{4}h^2$ ⇒ $dA/dt = \frac{9\pi}{4} \cdot 2h \, (dh/dt)$,

and from (⋆) in part (a) we know that when $h = 60$, $dh/dt = \frac{80,000}{3(60)^2} = \frac{200}{27}$ ft/h. Therefore,

$dA/dt = \frac{9\pi}{4}(2)(60)\left(\frac{200}{27}\right) = 2000\pi \approx 6283$ ft^2/h.

(c) At $h = 90$ ft, $dV/dt = 60,000\pi - 20,000\pi = 40,000\pi$ ft^3/h. From (⋆) in part (a),

$\dfrac{dh}{dt} = \dfrac{4(dV/dt)}{9\pi h^2} = \dfrac{4(40,000\pi)}{9\pi h^2} = \dfrac{160,000}{9h^2}$ ⇒ $\int 9h^2 \, dh = \int 160,000 \, dt$ ⇒ $3h^3 = 160,000t + C$.

When $t = 0$, $h = 90$; therefore, $C = 3 \cdot 729,000 = 2,187,000$. So $3h^3 = 160,000t + 2,187,000$. At the top,

$h = 100$ ⇒ $3(100)^3 = 160,000t + 2,187,000$ ⇒ $t = \frac{813,000}{160,000} \approx 5.1$. The pile reaches the top after

about 5.1 h.

11. Let $P(a, b)$ be any point on the curve. If m is the slope of the tangent line at P, then $m = y'(a)$, and

an equation of the normal line at P is $y - b = -\dfrac{1}{m}(x - a)$, or equivalently, $y = -\dfrac{1}{m}x + b + \dfrac{a}{m}$.

The y-intercept is always 6, so $b + \dfrac{a}{m} = 6$ ⇒ $\dfrac{a}{m} = 6 - b$ ⇒ $m = \dfrac{a}{6 - b}$.

We will solve the equivalent differential equation $\dfrac{dy}{dx} = \dfrac{x}{6 - y}$ ⇒ $(6 - y) \, dy = x \, dx$ ⇒

$\displaystyle\int (6 - y) \, dy = \int x \, dx$ ⇒ $6y - \frac{1}{2}y^2 = \frac{1}{2}x^2 + C$ ⇒ $12y - y^2 = x^2 + K$. Since $(3, 2)$ is on the curve,

$12(2) - 2^2 = 3^2 + K$ ⇒ $K = 11$. So the curve is given by $12y - y^2 = x^2 + 11$ ⇒

$x^2 + y^2 - 12y + 36 = -11 + 36$ ⇒ $x^2 + (y - 6)^2 = 25$, a circle with center $(0, 6)$ and radius 5.

10 □ PARAMETRIC EQUATIONS AND POLAR COORDINATES

10.1 Curves Defined by Parametric Equations

1. $x = 1 + \sqrt{t}, \quad y = t^2 - 4t, \quad 0 \le t \le 5$

t	0	1	2	3	4	5
x	1	2	$1 + \sqrt{2}$	$1 + \sqrt{3}$	3	$1 + \sqrt{5}$
			2.41	2.73		3.24
y	0	-3	-4	-3	0	5

3. $x = 5\sin t, \quad y = t^2, \quad -\pi \le t \le \pi$

t	$-\pi$	$-\pi/2$	0	$\pi/2$	π
x	0	-5	0	5	0
y	π^2	$\pi^2/4$	0	$\pi^2/4$	π^2
	9.87	2.47		2.47	9.87

5. $x = 3t - 5, \quad y = 2t + 1$

(a)
t	-2	-1	0	1	2	3	4
x	-11	-8	-5	-2	1	4	7
y	-3	-1	1	3	5	7	9

(b) $x = 3t - 5 \quad \Rightarrow \quad 3t = x + 5 \quad \Rightarrow \quad t = \frac{1}{3}(x + 5) \quad \Rightarrow$

$y = 2 \cdot \frac{1}{3}(x + 5) + 1$, so $y = \frac{2}{3}x + \frac{13}{3}$.

7. $x = t^2 - 2, \quad y = 5 - 2t, \quad -3 \le t \le 4$

(a)
t	-3	-2	-1	0	1	2	3	4
x	7	2	-1	-2	-1	2	7	14
y	11	9	7	5	3	1	-1	-3

(b) $y = 5 - 2t \quad \Rightarrow \quad 2t = 5 - y \quad \Rightarrow \quad t = \frac{1}{2}(5 - y) \quad \Rightarrow$

$x = \left[\frac{1}{2}(5 - y)\right]^2 - 2$, so $x = \frac{1}{4}(5 - y)^2 - 2$,

$-3 \le y \le 11$.

9. $x = \sqrt{t}, y = 1 - t$

(a)

t	0	1	2	3	4
x	0	1	1.414	1.732	2
y	1	0	−1	−2	−3

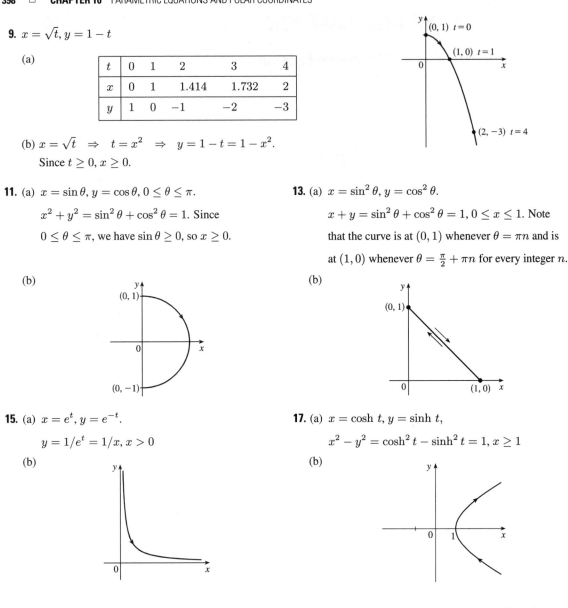

(b) $x = \sqrt{t} \;\Rightarrow\; t = x^2 \;\Rightarrow\; y = 1 - t = 1 - x^2.$

Since $t \geq 0, x \geq 0.$

11. (a) $x = \sin\theta, y = \cos\theta, 0 \leq \theta \leq \pi.$

$x^2 + y^2 = \sin^2\theta + \cos^2\theta = 1.$ Since

$0 \leq \theta \leq \pi,$ we have $\sin\theta \geq 0,$ so $x \geq 0.$

(b)

13. (a) $x = \sin^2\theta, y = \cos^2\theta.$

$x + y = \sin^2\theta + \cos^2\theta = 1, 0 \leq x \leq 1.$ Note

that the curve is at $(0, 1)$ whenever $\theta = \pi n$ and is

at $(1, 0)$ whenever $\theta = \frac{\pi}{2} + \pi n$ for every integer $n.$

(b)

15. (a) $x = e^t, y = e^{-t}.$

$y = 1/e^t = 1/x, x > 0$

(b)

17. (a) $x = \cosh t, y = \sinh t,$

$x^2 - y^2 = \cosh^2 t - \sinh^2 t = 1, x \geq 1$

(b)

19. $x^2 + y^2 = \cos^2 \pi t + \sin^2 \pi t = 1, 1 \leq t \leq 2,$ so the particle moves counterclockwise along the circle $x^2 + y^2 = 1$

from $(-1, 0)$ to $(1, 0),$ along the lower half of the circle.

21. $\left(\frac{1}{2}x\right)^2 + \left(\frac{1}{3}y\right)^2 = \sin^2 t + \cos^2 t = 1,$ so the particle moves once clockwise along the ellipse $\frac{1}{4}x^2 + \frac{1}{9}y^2 = 1,$

starting and ending at $(0, 3).$

23. We must have $1 \leq x \leq 4$ and $2 \leq y \leq 3.$ So the graph of the curve must be contained in the rectangle $[1, 4]$

by $[2, 3].$

25. When $t = -1$, $(x, y) = (0, -1)$. As t increases to 0, x decreases to

-1 and y increases to 0. As t increases from 0 to 1, x increases to 0

and y increases to 1. As t increases beyond 1, both x and y increase.

For $t < -1$, x is positive and decreasing and y is negative and

increasing. We could achieve greater accuracy by estimating x- and

y-values for selected values of t from the given graphs and plotting

the corresponding points.

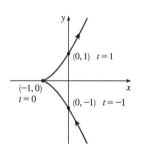

27. When $t = 0$ we see that $x = 0$ and $y = 0$, so the curve starts at the

origin. As t increases from 0 to $\frac{1}{2}$, the graphs show that y increases

from 0 to 1 while x increases from 0 to 1, decreases to 0 and to -1,

then increases back to 0, so we arrive at the point $(0, 1)$. Similarly, as t

increases from $\frac{1}{2}$ to 1, y decreases from 1 to 0 while x repeats its

pattern, and we arrive back at the origin. We could achieve greater accuracy by estimating x- and y-values for

selected values of t from the given graphs and plotting the corresponding points.

29. As in Example 5, we let $y = t$ and $x = t - 3t^3 + t^5$ and use a t-interval of $[-2\pi, 2\pi]$.

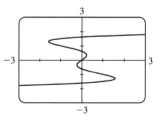

31. (a) $x = x_1 + (x_2 - x_1)t$, $y = y_1 + (y_2 - y_1)t$, $0 \le t \le 1$. Clearly the curve passes through $P_1(x_1, y_1)$ when

$t = 0$ and through $P_2(x_2, y_2)$ when $t = 1$. For $0 < t < 1$, x is strictly between x_1 and x_2 and y is strictly

between y_1 and y_2. For every value of t, x and y satisfy the relation $y - y_1 = \dfrac{y_2 - y_1}{x_2 - x_1}(x - x_1)$, which is the

equation of the line through $P_1(x_1, y_1)$ and $P_2(x_2, y_2)$.

   Finally, any point (x, y) on that line satisfies $\dfrac{y - y_1}{y_2 - y_1} = \dfrac{x - x_1}{x_2 - x_1}$; if we call that common value t, then the

given parametric equations yield the point (x, y); and any (x, y) on the line between $P_1(x_1, y_1)$ and $P_2(x_2, y_2)$

yields a value of t in $[0, 1]$. So the given parametric equations exactly specify the line segment from $P_1(x_1, y_1)$

to $P_2(x_2, y_2)$.

   (b) $x = -2 + [3 - (-2)]t = -2 + 5t$ and $y = 7 + (-1 - 7)t = 7 - 8t$ for $0 \le t \le 1$.

33. The circle $x^2 + y^2 = 4$ can be represented parametrically by $x = 2\cos t$, $y = 2\sin t$; $0 \le t \le 2\pi$. The circle $x^2 + (y-1)^2 = 4$ can be represented by $x = 2\cos t$, $y = 1 + 2\sin t$; $0 \le t \le 2\pi$. This representation gives us the circle with a counterclockwise orientation starting at $(2, 1)$.

(a) To get a clockwise orientation, we could change the equations to $x = 2\cos t$, $y = 1 - 2\sin t$, $0 \le t \le 2\pi$.

(b) To get three times around in the counterclockwise direction, we use the original equations $x = 2\cos t$, $y = 1 + 2\sin t$ with the domain expanded to $0 \le t \le 6\pi$.

(c) To start at $(0, 3)$ using the original equations, we must have $x_1 = 0$; that is, $2\cos t = 0$. Hence, $t = \frac{\pi}{2}$. So we use $x = 2\cos t$, $y = 1 + 2\sin t$; $\frac{\pi}{2} \le t \le \frac{3\pi}{2}$.

Alternatively, if we want t to start at 0, we could change the equations of the curve. For example, we could use $x = -2\sin t$, $y = 1 + 2\cos t$, $0 \le t \le \pi$.

35. (a) Let $x^2/a^2 = \sin^2 t$ and $y^2/b^2 = \cos^2 t$ to obtain $x = a\sin t$ and $y = b\cos t$ with $0 \le t \le 2\pi$ as possible parametric equations for the ellipse $x^2/a^2 + y^2/b^2 = 1$.

(c) As b increases, the ellipse stretches vertically.

(b) The equations are $x = 3\sin t$ and $y = b\cos t$ for $b \in \{1, 2, 4, 8\}$.

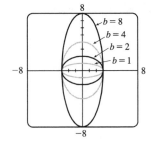

37. The case $\frac{\pi}{2} < \theta < \pi$ is illustrated. C has coordinates $(r\theta, r)$ as in Example 6, and Q has coordinates $(r\theta, r + r\cos(\pi - \theta)) = (r\theta, r(1 - \cos\theta))$ [since $\cos(\pi - \alpha) = \cos\pi\cos\alpha + \sin\pi\sin\alpha = -\cos\alpha$], so P has coordinates $(r\theta - r\sin(\pi - \theta), r(1 - \cos\theta)) = (r(\theta - \sin\theta), r(1 - \cos\theta))$ [since $\sin(\pi - \alpha) = \sin\pi\cos\alpha - \cos\pi\sin\alpha = \sin\alpha$]. Again we have the parametric equations $x = r(\theta - \sin\theta)$, $y = r(1 - \cos\theta)$.

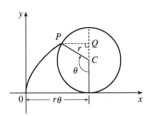

39. It is apparent that $x = |OQ|$ and $y = |QP| = |ST|$. From the diagram, $x = |OQ| = a\cos\theta$ and $y = |ST| = b\sin\theta$. Thus, the parametric equations are $x = a\cos\theta$ and $y = b\sin\theta$. To eliminate θ we rearrange: $\sin\theta = y/b \Rightarrow \sin^2\theta = (y/b)^2$ and $\cos\theta = x/a \Rightarrow \cos^2\theta = (x/a)^2$. Adding the two equations: $\sin^2\theta + \cos^2\theta = 1 = x^2/a^2 + y^2/b^2$. Thus, we have an ellipse.

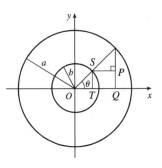

41. $C = (2a \cot\theta, 2a)$, so the x-coordinate of P is $x = 2a \cot\theta$. Let $B = (0, 2a)$. Then $\angle OAB$ is a right angle and $\angle OBA = \theta$, so $|OA| = 2a \sin\theta$ and $A = ((2a \sin\theta)\cos\theta, (2a \sin\theta)\sin\theta)$. Thus, the y-coordinate of P is $y = 2a \sin^2\theta$.

43. (a)

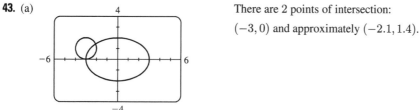

There are 2 points of intersection:
$(-3, 0)$ and approximately $(-2.1, 1.4)$.

(b) A collision point occurs when $x_1 = x_2$ and $y_1 = y_2$ for the same t. So solve the equations:

$$3\sin t = -3 + \cos t \quad \textbf{(1)}$$
$$2\cos t = 1 + \sin t \quad \textbf{(2)}$$

From **(2)**, $\sin t = 2\cos t - 1$. Substituting into **(1)**, we get $3(2\cos t - 1) = -3 + \cos t \implies$ $5\cos t = 0$ (*) $\implies \cos t = 0 \implies t = \frac{\pi}{2}$ or $\frac{3\pi}{2}$. We check that $t = \frac{3\pi}{2}$ satisfies **(1)** and **(2)** but $t = \frac{\pi}{2}$ does not. So the only collision point occurs when $t = \frac{3\pi}{2}$, and this gives the point $(-3, 0)$. [We could check our work by graphing x_1 and x_2 together as functions of t and, on another plot, y_1 and y_2 as functions of t. If we do so, we see that the only value of t for which *both* pairs of graphs intersect is $t = \frac{3\pi}{2}$.]

(c) The circle is centered at $(3, 1)$ instead of $(-3, 1)$. There are still 2 intersection points: $(3, 0)$ and $(2.1, 1.4)$, but there are no collision points, since (*) in part (b) becomes $5\cos t = 6 \implies \cos t = \frac{6}{5} > 1$.

45. $x = t^2$, $y = t^3 - ct$. We use a graphing device to produce the graphs for various values of c with $-\pi \le t \le \pi$. Note that all the members of the family are symmetric about the x-axis. For $c < 0$, the graph does not cross itself, but for $c = 0$ it has a cusp at $(0, 0)$ and for $c > 0$ the graph crosses itself at $x = c$, so the loop grows larger as c increases.

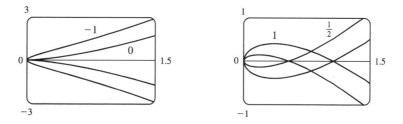

47. Note that all the Lissajous figures are symmetric about the x-axis. The parameters a and b simply stretch the graph in the x- and y-directions respectively. For $a = b = n = 1$ the graph is simply a circle with radius 1. For $n = 2$ the graph crosses itself at the origin and there are loops above and below the x-axis. In general, the figures have $n - 1$ points of intersection, all of which are on the y-axis, and a total of n closed loops.

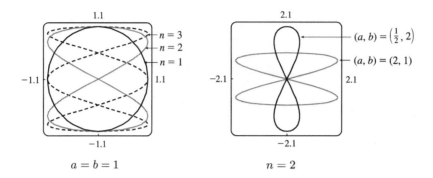

$$a = b = 1 \qquad\qquad n = 2$$

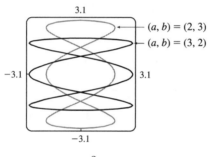

$$n = 3$$

10.2 Calculus with Parametric Curves

1. $x = t - t^3$, $y = 2 - 5t \quad\Rightarrow\quad \dfrac{dy}{dt} = -5$, $\dfrac{dx}{dt} = 1 - 3t^2$, and $\dfrac{dy}{dx} = \dfrac{dy/dt}{dx/dt} = \dfrac{-5}{1 - 3t^2}$ or $\dfrac{5}{3t^2 - 1}$.

3. $x = t^4 + 1$, $y = t^3 + t$; $t = -1$. $\quad \dfrac{dy}{dt} = 3t^2 + 1$, $\dfrac{dx}{dt} = 4t^3$, and $\dfrac{dy}{dx} = \dfrac{dy/dt}{dx/dt} = \dfrac{3t^2 + 1}{4t^3}$.

When $t = -1$, $(x, y) = (2, -2)$ and $dy/dx = \frac{4}{-4} = -1$, so an equation of the tangent to the curve at the point corresponding to $t = -1$ is $y - (-2) = (-1)(x - 2)$, or $y = -x$.

5. $x = e^{\sqrt{t}}$, $y = t - \ln t^2$; $t = 1$. $\quad \dfrac{dy}{dt} = 1 - \dfrac{2t}{t^2} = 1 - \dfrac{2}{t}$, $\dfrac{dx}{dt} = \dfrac{e^{\sqrt{t}}}{2\sqrt{t}}$, and

$\dfrac{dy}{dx} = \dfrac{dy/dt}{dx/dt} = \dfrac{1 - 2/t}{e^{\sqrt{t}}/(2\sqrt{t})} \cdot \dfrac{2t}{2t} = \dfrac{2t - 4}{\sqrt{t}e^{\sqrt{t}}}$. When $t = 1$, $(x, y) = (e, 1)$ and $\dfrac{dy}{dx} = -\dfrac{2}{e}$, so an equation of the

tangent line is $y - 1 = -\frac{2}{e}(x - e)$, or $y = -\frac{2}{e}x + 3$.

7. (a) $x = e^t$, $y = (t-1)^2$; $(1,1)$. $\dfrac{dy}{dt} = 2(t-1)$, $\dfrac{dx}{dt} = e^t$, and $\dfrac{dy}{dx} = \dfrac{dy/dt}{dx/dt} = \dfrac{2(t-1)}{e^t}$.

At $(1,1)$, $t = 0$ and $\dfrac{dy}{dx} = -2$, so an equation of the tangent is $y - 1 = -2(x-1)$, or $y = -2x + 3$.

(b) $x = e^t \Rightarrow t = \ln x$, so $y = (t-1)^2 = (\ln x - 1)^2$ and $\dfrac{dy}{dx} = 2(\ln x - 1)\left(\dfrac{1}{x}\right)$. When $x = 1$,

$\dfrac{dy}{dx} = 2(-1)(1) = -2$, so an equation of the tangent is $y = -2x + 3$, as in part (a).

9. $x = 2\sin 2t$, $y = 2\sin t$; $(\sqrt{3}, 1)$.

$\dfrac{dy}{dx} = \dfrac{dy/dt}{dx/dt} = \dfrac{2\cos t}{2\cdot 2\cos 2t} = \dfrac{\cos t}{2\cos 2t}$. The point $(\sqrt{3}, 1)$ corresponds

to $t = \frac{\pi}{6}$, so the slope of the tangent at that point is

$\dfrac{\cos\frac{\pi}{6}}{2\cos\frac{\pi}{3}} = \dfrac{\frac{\sqrt{3}}{2}}{2\cdot\frac{1}{2}} = \dfrac{\sqrt{3}}{2}$. An equation of the tangent is therefore

$(y - 1) = \frac{\sqrt{3}}{2}\left(x - \sqrt{3}\right)$, or $y = \frac{\sqrt{3}}{2}x - \frac{1}{2}$.

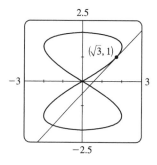

11. $x = 4 + t^2$, $y = t^2 + t^3 \Rightarrow \dfrac{dy}{dx} = \dfrac{dy/dt}{dx/dt} = \dfrac{2t + 3t^2}{2t} = 1 + \dfrac{3}{2}t \Rightarrow$

$\dfrac{d^2y}{dx^2} = \dfrac{d}{dx}\left(\dfrac{dy}{dx}\right) = \dfrac{d(dy/dx)/dt}{dx/dt} = \dfrac{(d/dt)(1 + \frac{3}{2}t)}{2t} = \dfrac{3/2}{2t} = \dfrac{3}{4t}$. The curve is CU when $\dfrac{d^2y}{dx^2} > 0$, that is,

when $t > 0$.

13. $x = t - e^t$, $y = t + e^{-t} \Rightarrow$

$\dfrac{dy}{dx} = \dfrac{dy/dt}{dx/dt} = \dfrac{1 - e^{-t}}{1 - e^t} = \dfrac{1 - \dfrac{1}{e^t}}{1 - e^t} = \dfrac{\dfrac{e^t - 1}{e^t}}{1 - e^t} = -e^{-t} \Rightarrow \dfrac{d^2y}{dx^2} = \dfrac{\dfrac{d}{dt}\left(\dfrac{dy}{dx}\right)}{dx/dt} = \dfrac{\dfrac{d}{dt}(-e^{-t})}{dx/dt} = \dfrac{e^{-t}}{1 - e^t}$.

The curve is CU when $e^t < 1$ [since $e^{-t} > 0$] $\Rightarrow t < 0$.

15. $x = 2\sin t$, $y = 3\cos t$, $0 < t < 2\pi$.

$\dfrac{dy}{dx} = \dfrac{dy/dt}{dx/dt} = \dfrac{-3\sin t}{2\cos t} = -\dfrac{3}{2}\tan t$, so $\dfrac{d^2y}{dx^2} = \dfrac{\dfrac{d}{dt}\left(\dfrac{dy}{dx}\right)}{dx/dt} = \dfrac{-\frac{3}{2}\sec^2 t}{2\cos t} = -\dfrac{3}{4}\sec^3 t$.

The curve is CU when $\sec^3 t < 0 \Rightarrow \sec t < 0 \Rightarrow \cos t < 0 \Rightarrow \frac{\pi}{2} < t < \frac{3\pi}{2}$.

17. $x = 10 - t^2$, $y = t^3 - 12t$.

$dy/dt = 3t^2 - 12 = 3(t+2)(t-2)$, so

$dy/dt = 0 \Leftrightarrow t = \pm 2 \Leftrightarrow$

$(x, y) = (6, \mp 16)$. $dx/dt = -2t$, so $dx/dt = 0$

$\Leftrightarrow t = 0 \Leftrightarrow (x, y) = (10, 0)$. The curve has

horizontal tangents at $(6, \pm 16)$ and a vertical

tangent at $(10, 0)$.

19. $x = 2\cos\theta$, $y = \sin 2\theta$.

$dy/d\theta = 2\cos 2\theta$, so $dy/d\theta = 0$ $\Leftrightarrow$

$2\theta = \frac{\pi}{2} + n\pi$ (n an integer) $\Leftrightarrow$ $\theta = \frac{\pi}{4} + \frac{\pi}{2}n$

$\Leftrightarrow$ $(x, y) = \left(\pm\sqrt{2}, \pm 1\right)$. Also,

$dx/d\theta = -2\sin\theta$, so $dx/d\theta = 0$ $\Leftrightarrow$ $\theta = n\pi$

$\Leftrightarrow$ $(x, y) = (\pm 2, 0)$. The curve has horizontal

tangents at $\left(\pm\sqrt{2}, \pm 1\right)$ (four points), and vertical

tangents at $(\pm 2, 0)$.

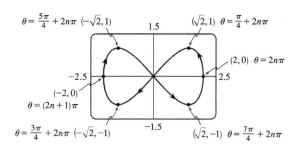

21. From the graph, it appears that the leftmost point on the curve $x = t^4 - t^2$,

$y = t + \ln t$ is about $(-0.25, 0.36)$. To find the exact coordinates, we find

the value of t for which the graph has a vertical tangent, that is,

$0 = dx/dt = 4t^3 - 2t$ $\Leftrightarrow$ $2t(2t^2 - 1) = 0$ $\Leftrightarrow$

$2t\left(\sqrt{2}\,t + 1\right)\left(\sqrt{2}\,t - 1\right) = 0$ $\Leftrightarrow$ $t = 0$ or $\pm\frac{1}{\sqrt{2}}$. The negative and

0 roots are inadmissible since $y(t)$ is only defined for $t > 0$, so the leftmost point must be

$$\left(x\left(\tfrac{1}{\sqrt{2}}\right), y\left(\tfrac{1}{\sqrt{2}}\right)\right) = \left(\left(\tfrac{1}{\sqrt{2}}\right)^4 - \left(\tfrac{1}{\sqrt{2}}\right)^2, \tfrac{1}{\sqrt{2}} + \ln\tfrac{1}{\sqrt{2}}\right) = \left(-\tfrac{1}{4}, \tfrac{1}{\sqrt{2}} - \tfrac{1}{2}\ln 2\right)$$

23. We graph the curve $x = t^4 - 2t^3 - 2t^2$,

$y = t^3 - t$ in the viewing rectangle

$[-2, 1.1]$ by $[-0.5, 0.5]$. This rectangle

corresponds approximately to

$t \in [-1, 0.8]$. We estimate that the curve

has horizontal tangents at about

$(-1, -0.4)$ and $(-0.17, 0.39)$ and vertical tangents at about $(0, 0)$ and $(-0.19, 0.37)$. We calculate

$\dfrac{dy}{dx} = \dfrac{dy/dt}{dx/dt} = \dfrac{3t^2 - 1}{4t^3 - 6t^2 - 4t}$. The horizontal tangents occur when $dy/dt = 3t^2 - 1 = 0$ $\Leftrightarrow$ $t = \pm\frac{1}{\sqrt{3}}$, so

both horizontal tangents are shown in our graph. The vertical tangents occur when $dx/dt = 2t(2t^2 - 3t - 2) = 0$

$\Leftrightarrow$ $2t(2t + 1)(t - 2) = 0$ $\Leftrightarrow$ $t = 0, -\frac{1}{2}$ or 2. It seems that we have missed one vertical tangent, and indeed if

we plot the curve on the t-interval $[-1.2, 2.2]$ we see that there is another vertical tangent at $(-8, 6)$.

25. $x = \cos t$, $y = \sin t \cos t$. $\frac{dx}{dt} = -\sin t$,

$\frac{dy}{dt} = -\sin^2 t + \cos^2 t = \cos 2t$. $(x, y) = (0, 0)$ ⟺ $\cos t = 0$ ⟺

t is an odd multiple of $\frac{\pi}{2}$. When $t = \frac{\pi}{2}$, $\frac{dx}{dt} = -1$ and $\frac{dy}{dt} = -1$, so

$\frac{dy}{dx} = 1$. When $t = \frac{3\pi}{2}$, $\frac{dx}{dt} = 1$ and $\frac{dy}{dt} = -1$. So $\frac{dy}{dx} = -1$. Thus,

$y = x$ and $y = -x$ are both tangent to the curve at $(0, 0)$.

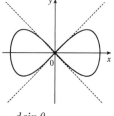

27. (a) $x = r\theta - d\sin\theta$, $y = r - d\cos\theta$; $\dfrac{dx}{d\theta} = r - d\cos\theta$, $\dfrac{dy}{d\theta} = d\sin\theta$. So $\dfrac{dy}{dx} = \dfrac{d\sin\theta}{r - d\cos\theta}$.

(b) If $0 < d < r$, then $|d\cos\theta| \le d < r$, so $r - d\cos\theta \ge r - d > 0$. This shows that $dx/d\theta$ never vanishes, so
the trochoid can have no vertical tangent if $d < r$.

29. The line with parametric equations $x = -7t$, $y = 12t - 5$ is $y = 12\left(-\frac{1}{7}x\right) - 5$, which has slope $-\frac{12}{7}$. The curve

$x = t^3 + 4t$, $y = 6t^2$ has slope $\dfrac{dy}{dx} = \dfrac{dy/dt}{dx/dt} = \dfrac{12t}{3t^2 + 4}$. This equals $-\frac{12}{7}$ ⟺ $3t^2 + 4 = -7t$ ⟺

$(3t + 4)(t + 1) = 0$ ⟺ $t = -1$ or $t = -\frac{4}{3}$ ⟺ $(x, y) = (-5, 6)$ or $\left(-\frac{208}{27}, \frac{32}{3}\right)$.

31. By symmetry of the ellipse about the x- and y-axes,

$$A = 4\int_0^a y\, dx = 4\int_{\pi/2}^0 b\sin\theta\,(-a\sin\theta)\, d\theta = 4ab\int_0^{\pi/2}\sin^2\theta\, d\theta = 4ab\int_0^{\pi/2}\tfrac{1}{2}(1 - \cos 2\theta)\, d\theta$$

$$= 2ab\left[\theta - \tfrac{1}{2}\sin 2\theta\right]_0^{\pi/2} = 2ab\left(\tfrac{\pi}{2}\right) = \pi ab$$

33. $A = \int_0^1 (y - 1)\, dx = \int_{\pi/2}^0 (e^t - 1)(-\sin t)\, dt = \int_0^{\pi/2}(e^t\sin t - \sin t)\, dt \overset{98}{=} \left[\tfrac{1}{2}e^t(\sin t - \cos t) + \cos t\right]_0^{\pi/2}$

$= \tfrac{1}{2}(e^{\pi/2} - 1)$

35. $A = \int_0^{2\pi r} y\, dx = \int_0^{2\pi}(r - d\cos\theta)(r - d\cos\theta)\, d\theta = \int_0^{2\pi}(r^2 - 2dr\cos\theta + d^2\cos^2\theta)\, d\theta$

$= \left[r^2\theta - 2dr\sin\theta + \tfrac{1}{2}d^2\left(\theta + \tfrac{1}{2}\sin 2\theta\right)\right]_0^{2\pi} = 2\pi r^2 + \pi d^2$

37. $x = t - t^2$, $y = \frac{4}{3}t^{3/2}$, $1 \le t \le 2$. $dx/dt = 1 - 2t$ and $dy/dt = 2t^{1/2}$, so

$(dx/dt)^2 + (dy/dt)^2 = (1 - 2t)^2 + (2t^{1/2})^2 = 1 - 4t + 4t^2 + 4t = 1 + 4t^2$. Thus,

$L = \int_a^b \sqrt{(dx/dt)^2 + (dy/dt)^2}\, dt = \int_1^2 \sqrt{1 + 4t^2}\, dt$.

39. $x = t + \cos t$, $y = t - \sin t$, $0 \le t \le 2\pi$. $dx/dt = 1 - \sin t$ and $dy/dt = 1 - \cos t$, so

$(dx/dt)^2 + (dy/dt)^2 = (1 - \sin t)^2 + (1 - \cos t)^2 = (1 - 2\sin t + \sin^2 t) + (1 - 2\cos t + \cos^2 t)$

$$= 3 - 2\sin t - 2\cos t$$

Thus, $L = \int_a^b \sqrt{(dx/dt)^2 + (dy/dt)^2}\, dt = \int_0^{2\pi} \sqrt{3 - 2\sin t - 2\cos t}\, dt$.

41. $x = 1 + 3t^2$, $y = 4 + 2t^3$, $0 \le t \le 1$. $dx/dt = 6t$ and $dy/dt = 6t^2$, so $(dx/dt)^2 + (dy/dt)^2 = 36t^2 + 36t^4$.

Thus, $L = \int_0^1 \sqrt{36t^2 + 36t^4} \, dt = \int_0^1 6t \sqrt{1 + t^2} \, dt = 6 \int_1^2 \sqrt{u} \left(\frac{1}{2} du\right)$ $[u = 1 + t^2, \, du = 2t \, dt]$

$= 3 \left[\frac{2}{3} u^{3/2}\right]_1^2 = 2(2^{3/2} - 1) = 2(2\sqrt{2} - 1)$

43. $x = \dfrac{t}{1+t}$, $y = \ln(1+t)$, $0 \le t \le 2$. $\dfrac{dx}{dt} = \dfrac{(1+t) \cdot 1 - t \cdot 1}{(1+t)^2} = \dfrac{1}{(1+t)^2}$ and $\dfrac{dy}{dt} = \dfrac{1}{1+t}$, so

$\left(\dfrac{dx}{dt}\right)^2 + \left(\dfrac{dy}{dt}\right)^2 = \dfrac{1}{(1+t)^4} + \dfrac{1}{(1+t)^2} = \dfrac{1}{(1+t)^4}[1 + (1+t)^2] = \dfrac{t^2 + 2t + 2}{(1+t)^4}$. Thus,

$L = \int_0^2 \dfrac{\sqrt{t^2 + 2t + 2}}{(1+t)^2} \, dt = \int_1^3 \dfrac{\sqrt{u^2 + 1}}{u^2} \, du$ $[u = t + 1, \, du = dt]$ $\overset{24}{=} \left[-\dfrac{\sqrt{u^2 + 1}}{u} + \ln\left(u + \sqrt{u^2 + 1}\right)\right]_1^3$

$= -\dfrac{\sqrt{10}}{3} + \ln\left(3 + \sqrt{10}\right) + \sqrt{2} - \ln\left(1 + \sqrt{2}\right)$

45. $x = e^t \cos t$, $y = e^t \sin t$, $0 \le t \le \pi$.

$\left(\dfrac{dx}{dt}\right)^2 + \left(\dfrac{dy}{dt}\right)^2 = [e^t(\cos t - \sin t)]^2 + [e^t(\sin t + \cos t)]^2$

$= (e^t)^2 (\cos^2 t - 2\cos t \sin t + \sin^2 t)$

$+ (e^t)^2 (\sin^2 t + 2\sin t \cos t + \cos^2 t)$

$= e^{2t}(2\cos^2 t + 2\sin^2 t) = 2e^{2t}$

Thus, $L = \int_0^\pi \sqrt{2e^{2t}} \, dt = \int_0^\pi \sqrt{2}\, e^t \, dt = \sqrt{2}\left[e^t\right]_0^\pi = \sqrt{2}\,(e^\pi - 1)$.

47. $x = e^t - t$, $y = 4e^{t/2}$, $-8 \le t \le 3$.

$(dx/dt)^2 + (dy/dt)^2 = (e^t - 1)^2 + \left(2e^{t/2}\right)^2 = e^{2t} - 2e^t + 1 + 4e^t$

$= e^{2t} + 2e^t + 1 = (e^t + 1)^2$

Thus,

$L = \int_{-8}^3 \sqrt{(e^t + 1)^2} \, dt = \int_{-8}^3 (e^t + 1) \, dt = \left[e^t + t\right]_{-8}^3$

$= (e^3 + 3) - (e^{-8} - 8) = e^3 - e^{-8} + 11.$

49. $x = t - e^t$, $y = t + e^t$, $-6 \le t \le 6$.

$\left(\frac{dx}{dt}\right)^2 + \left(\frac{dy}{dt}\right)^2 = (1 - e^t)^2 + (1 + e^t)^2 = (1 - 2e^t + e^{2t}) + (1 + 2e^t + e^{2t}) = 2 + 2e^{2t}$, so

$L = \int_{-6}^6 \sqrt{2 + 2e^{2t}} \, dt$. Set $f(t) = \sqrt{2 + 2e^{2t}}$. Then by Simpson's Rule with $n = 6$ and $\Delta t = \frac{6 - (-6)}{6} = 2$,

we get $L \approx \frac{2}{3}[f(-6) + 4f(-4) + 2f(-2) + 4f(0) + 2f(2) + 4f(4) + f(6)] \approx 612.3053.$

51. $x = \sin^2 t$, $y = \cos^2 t$, $0 \le t \le 3\pi$.

$(dx/dt)^2 + (dy/dt)^2 = (2 \sin t \cos t)^2 + (-2 \cos t \sin t)^2 = 8 \sin^2 t \cos^2 t = 2 \sin^2 2t \Rightarrow$

$$\text{Distance} = \int_0^{3\pi} \sqrt{2} \, |\sin 2t| \, dt = 6\sqrt{2} \int_0^{\pi/2} \sin 2t \, dt \text{ [by symmetry]} = -3\sqrt{2} \left[\cos 2t\right]_0^{\pi/2}$$

$$= -3\sqrt{2}(-1-1) = 6\sqrt{2}$$

The full curve is traversed as t goes from 0 to $\frac{\pi}{2}$, because the curve is the segment of $x + y = 1$ that lies in the first

quadrant (since x, $y \ge 0$), and this segment is completely traversed as t goes from 0 to $\frac{\pi}{2}$.

Thus, $L = \int_0^{\pi/2} \sin 2t \, dt = \sqrt{2}$, as above.

53. $x = a \sin \theta$, $y = b \cos \theta$, $0 \le \theta \le 2\pi$.

$$\left(\frac{dx}{d\theta}\right)^2 + \left(\frac{dy}{d\theta}\right)^2 = (a \cos \theta)^2 + (-b \sin \theta)^2 = a^2 \cos^2 \theta + b^2 \sin^2 \theta = a^2(1 - \sin^2 \theta) + b^2 \sin^2 \theta$$

$$= a^2 - (a^2 - b^2) \sin^2 \theta = a^2 - c^2 \sin^2 \theta = a^2\left(1 - \frac{c^2}{a^2} \sin^2 \theta\right) = a^2(1 - e^2 \sin^2 \theta)$$

So $L = 4 \int_0^{\pi/2} \sqrt{a^2 (1 - e^2 \sin^2 \theta)} \, d\theta$ [by symmetry] $= 4a \int_0^{\pi/2} \sqrt{1 - e^2 \sin^2 \theta} \, d\theta$.

55. (a) Notice that $0 \le t \le 2\pi$ does not give the complete curve

because $x(0) \ne x(2\pi)$. In fact, we must take $t \in [0, 4\pi]$ in

order to obtain the complete curve, since the first term in each

of the parametric equations has period 2π and the second has

period $\frac{2\pi}{11/2} = \frac{4\pi}{11}$, and the least common integer multiple of

these two numbers is 4π.

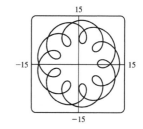

(b) We use the CAS to find the derivatives dx/dt and dy/dt, and then use Formula 1 to find the arc length. Recent

versions of Maple express the integral $\int_0^{4\pi} \sqrt{(dx/dt)^2 + (dy/dt)^2} \, dt$ as $88E(2\sqrt{2}\,i)$, where $E(x)$ is the

elliptic integral $\int_0^1 \frac{\sqrt{1 - x^2 t^2}}{\sqrt{1 - t^2}} \, dt$ and i is the imaginary number $\sqrt{-1}$. Some earlier versions of Maple (as

well as Mathematica) cannot do the integral exactly, so we use the command

`evalf(Int(sqrt(diff(x,t)^2+diff(y,t)^2),t=0..4*Pi));` to estimate the length, and find

that the arc length is approximately 294.03. Derive's `Para_arc_length` function in the utility file

`Int_apps` simplifies the integral to $11 \int_0^{4\pi} \sqrt{-4 \cos t \, \cos\left(\frac{11t}{2}\right) - 4 \sin t \, \sin\left(\frac{11t}{2}\right) + 5} \, dt$.

57. $x = t - t^2$, $y = \frac{4}{3}t^{3/2}$, $1 \le t \le 2$. $\left(\dfrac{dx}{dt}\right)^2 + \left(\dfrac{dy}{dt}\right)^2 = (1-2t)^2 + (2t^{1/2})^2 = 1 - 4t + 4t^2 + 4t = 1 + 4t^2$,

so $S = \int_1^2 2\pi y\, ds = \int_1^2 2\pi \cdot \frac{4}{3}t^{3/2}\sqrt{1+4t^2}\, dt = \int_1^2 \frac{8\pi}{3}t^{3/2}\sqrt{1+4t^2}\, dt.$

59. $x = t^3$, $y = t^2$, $0 \le t \le 1$. $\left(\dfrac{dx}{dt}\right)^2 + \left(\dfrac{dy}{dt}\right)^2 = (3t^2)^2 + (2t)^2 = 9t^4 + 4t^2.$

$$S = \int_0^1 2\pi y \sqrt{\left(\dfrac{dx}{dt}\right)^2 + \left(\dfrac{dy}{dt}\right)^2}\, dt = \int_0^1 2\pi t^2 \sqrt{9t^4 + 4t^2}\, dt = 2\pi \int_0^1 t^2 \sqrt{t^2(9t^2+4)}\, dt$$

$$= 2\pi \int_4^{13} \left(\dfrac{u-4}{9}\right)\sqrt{u}\left(\tfrac{1}{18}du\right) \quad \begin{bmatrix} u = 9t^2+4,\ t^2 = (u-4)/9 \\ du = 18t\, dt,\ \text{so } t\, dt = \tfrac{1}{18}\, du \end{bmatrix} = \dfrac{2\pi}{9\cdot18}\int_4^{13}\left(u^{3/2} - 4u^{1/2}\right)du$$

$$= \dfrac{\pi}{81}\left[\tfrac{2}{5}u^{5/2} - \tfrac{8}{3}u^{3/2}\right]_4^{13} = \dfrac{\pi}{81}\cdot\dfrac{2}{15}\left[3u^{5/2} - 20u^{3/2}\right]_4^{13}$$

$$= \dfrac{2\pi}{1215}\left[\left(3\cdot13^2\sqrt{13} - 20\cdot13\sqrt{13}\right) - (3\cdot32 - 20\cdot8)\right]$$

$$= \dfrac{2\pi}{1215}\left(247\sqrt{13} + 64\right)$$

61. $x = a\cos^3\theta$, $y = a\sin^3\theta$, $0 \le \theta \le \frac{\pi}{2}$.

$\left(\dfrac{dx}{d\theta}\right)^2 + \left(\dfrac{dy}{d\theta}\right)^2 = \left(-3a\cos^2\theta\sin\theta\right)^2 + \left(3a\sin^2\theta\cos\theta\right)^2 = 9a^2\sin^2\theta\cos^2\theta.$

$S = \int_0^{\pi/2} 2\pi \cdot a\sin^3\theta \cdot 3a\sin\theta\cos\theta\, d\theta = 6\pi a^2 \int_0^{\pi/2} \sin^4\theta\cos\theta\, d\theta = \frac{6}{5}\pi a^2\left[\sin^5\theta\right]_0^{\pi/2} = \frac{6}{5}\pi a^2$

63. $x = t + t^3$, $y = t - \dfrac{1}{t^2}$, $1 \le t \le 2$. $\dfrac{dx}{dt} = 1 + 3t^2$ and $\dfrac{dy}{dt} = 1 + \dfrac{2}{t^3}$, so

$\left(\dfrac{dx}{dt}\right)^2 + \left(\dfrac{dy}{dt}\right)^2 = (1+3t^2)^2 + \left(1 + \dfrac{2}{t^3}\right)^2$ and

$S = \displaystyle\int 2\pi y\, ds = \int_1^2 2\pi\left(t - \dfrac{1}{t^2}\right)\sqrt{(1+3t^2)^2 + \left(1 + \dfrac{2}{t^3}\right)^2}\, dt \approx 59.101.$

65. $x = 3t^2$, $y = 2t^3$, $0 \le t \le 5$ $\Rightarrow$ $\left(\dfrac{dx}{dt}\right)^2 + \left(\dfrac{dy}{dt}\right)^2 = (6t)^2 + (6t^2)^2 = 36t^2(1+t^2)$ $\Rightarrow$

$$S = \int_0^5 2\pi x \sqrt{(dx/dt)^2 + (dy/dt)^2}\, dt = \int_0^5 2\pi(3t^2)6t\sqrt{1+t^2}\, dt = 18\pi\int_0^5 t^2\sqrt{1+t^2}\,2t\, dt$$

$$= 18\pi\int_1^{26}(u-1)\sqrt{u}\, du \ \text{[where } u = 1+t^2,\ du = 2t\, dt] \ = 18\pi\int_1^{26}\left(u^{3/2} - u^{1/2}\right)du$$

$$= 18\pi\left[\tfrac{2}{5}u^{5/2} - \tfrac{2}{3}u^{3/2}\right]_1^{26} = 18\pi\left[\left(\tfrac{2}{5}\cdot676\sqrt{26} - \tfrac{2}{3}\cdot26\sqrt{26}\right) - \left(\tfrac{2}{5} - \tfrac{2}{3}\right)\right]$$

$$= \tfrac{24}{5}\pi\left(949\sqrt{26} + 1\right)$$

67. If f' is continuous and $f'(t) \neq 0$ for $a \leq t \leq b$, then either $f'(t) > 0$ for all t in $[a, b]$ or $f'(t) < 0$ for all t in $[a, b]$.

Thus, f is monotonic (in fact, strictly increasing or strictly decreasing) on $[a, b]$. It follows that f has an inverse.

Set $F = g \circ f^{-1}$, that is, define F by $F(x) = g(f^{-1}(x))$. Then $x = f(t) \Rightarrow f^{-1}(x) = t$, so

$y = g(t) = g(f^{-1}(x)) = F(x)$.

69. (a) $\phi = \tan^{-1}\left(\dfrac{dy}{dx}\right) \Rightarrow \dfrac{d\phi}{dt} = \dfrac{d}{dt}\tan^{-1}\left(\dfrac{dy}{dx}\right) = \dfrac{1}{1 + (dy/dx)^2}\left[\dfrac{d}{dt}\left(\dfrac{dy}{dx}\right)\right]$. But $\dfrac{dy}{dx} = \dfrac{dy/dt}{dx/dt} = \dfrac{\dot{y}}{\dot{x}} \Rightarrow$

$\dfrac{d}{dt}\left(\dfrac{dy}{dx}\right) = \dfrac{d}{dt}\left(\dfrac{\dot{y}}{\dot{x}}\right) = \dfrac{\ddot{y}\dot{x} - \ddot{x}\dot{y}}{\dot{x}^2} \Rightarrow \dfrac{d\phi}{dt} = \dfrac{1}{1 + (\dot{y}/\dot{x})^2}\left(\dfrac{\ddot{y}\dot{x} - \ddot{x}\dot{y}}{\dot{x}^2}\right) = \dfrac{\dot{x}\ddot{y} - \ddot{x}\dot{y}}{\dot{x}^2 + \dot{y}^2}$.

Using the Chain Rule, and the fact that $s = \displaystyle\int_0^t \sqrt{\left(\dfrac{dx}{dt}\right)^2 + \left(\dfrac{dy}{dt}\right)^2}\, dt \Rightarrow$

$\dfrac{ds}{dt} = \sqrt{\left(\dfrac{dx}{dt}\right)^2 + \left(\dfrac{dy}{dt}\right)^2} = (\dot{x}^2 + \dot{y}^2)^{1/2}$, we have that

$\dfrac{d\phi}{ds} = \dfrac{d\phi/dt}{ds/dt} = \left(\dfrac{\dot{x}\ddot{y} - \ddot{x}\dot{y}}{\dot{x}^2 + \dot{y}^2}\right)\dfrac{1}{(\dot{x}^2 + \dot{y}^2)^{1/2}} = \dfrac{\dot{x}\ddot{y} - \ddot{x}\dot{y}}{(\dot{x}^2 + \dot{y}^2)^{3/2}}$. So

$\kappa = \left|\dfrac{d\phi}{ds}\right| = \left|\dfrac{\dot{x}\ddot{y} - \ddot{x}\dot{y}}{(\dot{x}^2 + \dot{y}^2)^{3/2}}\right| = \dfrac{|\dot{x}\ddot{y} - \ddot{x}\dot{y}|}{(\dot{x}^2 + \dot{y}^2)^{3/2}}$.

(b) $x = x$ and $y = f(x) \Rightarrow \dot{x} = 1, \ddot{x} = 0$ and $\dot{y} = \dfrac{dy}{dx}, \ddot{y} = \dfrac{d^2y}{dx^2}$.

So $\kappa = \dfrac{\left|1 \cdot (d^2y/dx^2) - 0 \cdot (dy/dx)\right|}{\left[1 + (dy/dx)^2\right]^{3/2}} = \dfrac{|d^2y/dx^2|}{\left[1 + (dy/dx)^2\right]^{3/2}}$.

71. $x = \theta - \sin\theta \Rightarrow \dot{x} = 1 - \cos\theta \Rightarrow \ddot{x} = \sin\theta$, and $y = 1 - \cos\theta \Rightarrow \dot{y} = \sin\theta \Rightarrow \ddot{y} = \cos\theta$.

Therefore, $\kappa = \dfrac{|\cos\theta - \cos^2\theta - \sin^2\theta|}{\left[(1 - \cos\theta)^2 + \sin^2\theta\right]^{3/2}} = \dfrac{|\cos\theta - (\cos^2\theta + \sin^2\theta)|}{(1 - 2\cos\theta + \cos^2\theta + \sin^2\theta)^{3/2}} = \dfrac{|\cos\theta - 1|}{(2 - 2\cos\theta)^{3/2}}$. The top

of the arch is characterized by a horizontal tangent, and from Example 2(b) in Section 10.2, the tangent is horizontal

when $\theta = (2n - 1)\pi$, so take $n = 1$ and substitute $\theta = \pi$ into the expression for κ:

$\kappa = \dfrac{|\cos\pi - 1|}{(2 - 2\cos\pi)^{3/2}} = \dfrac{|-1 - 1|}{[2 - 2(-1)]^{3/2}} = \dfrac{1}{4}$.

73. The coordinates of T are $(r \cos \theta, r \sin \theta)$. Since TP was unwound from

arc TA, TP has length $r\theta$. Also $\angle PTQ = \angle PTR - \angle QTR = \frac{1}{2}\pi - \theta$,

so P has coordinates $x = r \cos \theta + r\theta \cos\left(\frac{1}{2}\pi - \theta\right) = r(\cos \theta + \theta \sin \theta)$,

$y = r \sin \theta - r\theta \sin\left(\frac{1}{2}\pi - \theta\right) = r(\sin \theta - \theta \cos \theta)$.

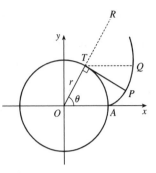

10.3 Polar Coordinates

1. (a) By adding 2π to $\frac{\pi}{2}$, we obtain the

point $\left(1, \frac{5\pi}{2}\right)$. The direction

opposite $\frac{\pi}{2}$ is $\frac{3\pi}{2}$, so $\left(-1, \frac{3\pi}{2}\right)$ is a

point that satisfies the $r < 0$

requirement.

    (b) $\left(-2, \frac{\pi}{4}\right)$

    (c) $(3, 2)$

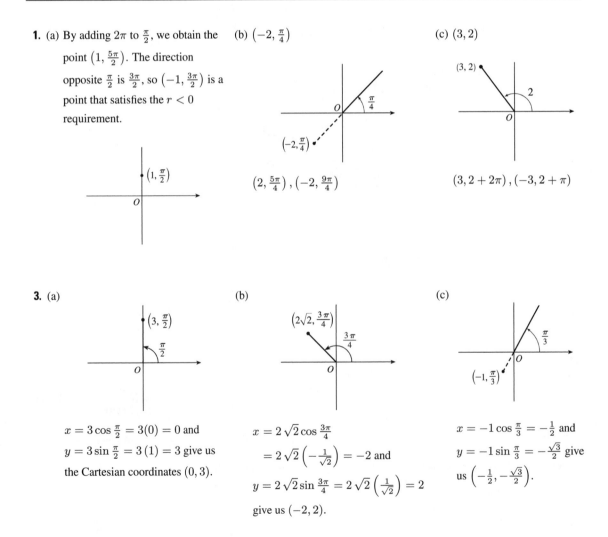

3. (a)

    (b)

    (c)

$x = 3 \cos \frac{\pi}{2} = 3(0) = 0$ and

$y = 3 \sin \frac{\pi}{2} = 3(1) = 3$ give us

the Cartesian coordinates $(0, 3)$.

$x = 2\sqrt{2} \cos \frac{3\pi}{4}$

$= 2\sqrt{2}\left(-\frac{1}{\sqrt{2}}\right) = -2$ and

$y = 2\sqrt{2} \sin \frac{3\pi}{4} = 2\sqrt{2}\left(\frac{1}{\sqrt{2}}\right) = 2$

give us $(-2, 2)$.

$x = -1 \cos \frac{\pi}{3} = -\frac{1}{2}$ and

$y = -1 \sin \frac{\pi}{3} = -\frac{\sqrt{3}}{2}$ give

us $\left(-\frac{1}{2}, -\frac{\sqrt{3}}{2}\right)$.

5. (a) $x = 1$ and $y = 1$ $\Rightarrow$ $r = \sqrt{1^2 + 1^2} = \sqrt{2}$ and $\theta = \tan^{-1}\left(\frac{1}{1}\right) = \frac{\pi}{4}$. Since $(1, 1)$ is in the first quadrant, the

polar coordinates are (i) $\left(\sqrt{2}, \frac{\pi}{4}\right)$ and (ii) $\left(-\sqrt{2}, \frac{5\pi}{4}\right)$.

(b) $x = 2\sqrt{3}$ and $y = -2$ $\Rightarrow$ $r = \sqrt{\left(2\sqrt{3}\right)^2 + (-2)^2} = \sqrt{12 + 4} = \sqrt{16} = 4$ and

$\theta = \tan^{-1}\left(-\frac{2}{2\sqrt{3}}\right) = \tan^{-1}\left(-\frac{1}{\sqrt{3}}\right) = -\frac{\pi}{6}$. Since $\left(2\sqrt{3}, -2\right)$ is in the fourth quadrant and $0 \le \theta \le 2\pi$, the

polar coordinates are (i) $\left(4, \frac{11\pi}{6}\right)$ and (ii) $\left(-4, \frac{5\pi}{6}\right)$.

7. The curves $r = 1$ and $r = 2$ represent circles with center O

and radii 1 and 2. The region in the plane satisfying

$1 \le r \le 2$ consists of both circles and the shaded region

between them in the figure.

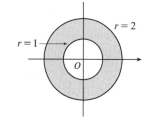

9. The region satisfying $0 \le r < 4$ and

$-\pi/2 \le \theta < \pi/6$ does not include the circle

$r = 4$ nor the line $\theta = \frac{\pi}{6}$.

11. $2 < r < 3$, $\frac{5\pi}{3} \le \theta \le \frac{7\pi}{3}$

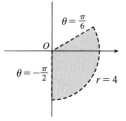

13. $(r, \theta) = \left(1, \frac{\pi}{6}\right)$ $\Rightarrow$ $x = 1 \cos \frac{\pi}{6} = \frac{\sqrt{3}}{2}$ and $y = 1 \sin \frac{\pi}{6} = \frac{1}{2}$.

$(r, \theta) = \left(3, \frac{3\pi}{4}\right)$ $\Rightarrow$ $x = 3 \cos \frac{3\pi}{4} = -\frac{3\sqrt{2}}{2}$ and $y = 3 \sin \frac{3\pi}{4} = \frac{3\sqrt{2}}{2}$. The distance between them is

15. $r = 2$ $\Leftrightarrow$ $\sqrt{x^2 + y^2} = 2$ $\Leftrightarrow$ $x^2 + y^2 = 4$, a circle of radius 2 centered at the origin.

17. $r = 3 \sin \theta$ $\Rightarrow$ $r^2 = 3r \sin \theta$ $\Leftrightarrow$ $x^2 + y^2 = 3y$ $\Leftrightarrow$ $x^2 + \left(y - \frac{3}{2}\right)^2 = \left(\frac{3}{2}\right)^2$, a circle of radius $\frac{3}{2}$ centered

at $\left(0, \frac{3}{2}\right)$. The first two equations are actually equivalent since $r^2 = 3r \sin \theta$ $\Rightarrow$ $r(r - 3 \sin \theta) = 0$ $\Rightarrow$ $r = 0$

or $r = 3 \sin \theta$. But $r = 3 \sin \theta$ gives the point $r = 0$ (the pole) when $\theta = 0$. Thus, the single equation $r = 3 \sin \theta$ is

equivalent to the compound condition ($r = 0$ or $r = 3 \sin \theta$).

19. $r = \csc\theta$ ⇔ $r = \dfrac{1}{\sin\theta}$ ⇔ $r\sin\theta = 1$ ⇔ $y = 1$, a horizontal line 1 unit above the x-axis.

21. $x = 3$ ⇔ $r\cos\theta = 3$ ⇔ $r = 3/\cos\theta$ ⇔ $r = 3\sec\theta$.

23. $x = -y^2$ ⇔ $r\cos\theta = -r^2\sin^2\theta$ ⇔ $\cos\theta = -r\sin^2\theta$ ⇔ $r = -\dfrac{\cos\theta}{\sin^2\theta} = -\cot\theta\csc\theta$.

25. $x^2 + y^2 = 2cx$ ⇔ $r^2 = 2cr\cos\theta$ ⇔ $r^2 - 2cr\cos\theta = 0$ ⇔ $r(r - 2c\cos\theta) = 0$ ⇔ $r = 0$ or $r = 2c\cos\theta$. $r = 0$ is included in $r = 2c\cos\theta$ when $\theta = \frac{\pi}{2} + n\pi$, so the curve is represented by the single equation $r = 2c\cos\theta$.

27. (a) The description leads immediately to the polar equation $\theta = \frac{\pi}{6}$, and the Cartesian equation $\tan\theta = y/x$ ⇒ $y = \left(\tan\frac{\pi}{6}\right)x = \frac{1}{\sqrt{3}}x$ is slightly more difficult to derive.

(b) The easier description here is the Cartesian equation $x = 3$.

29. $\theta = -\pi/6$

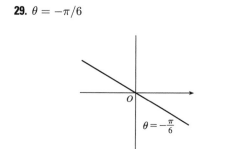

31. $r = \sin\theta$ ⇔ $r^2 = r\sin\theta$ ⇔ $x^2 + y^2 = y$ ⇔ $x^2 + \left(y - \frac{1}{2}\right)^2 = \left(\frac{1}{2}\right)^2$. The reasoning here is the same as in Exercise 17. This is a circle of radius $\frac{1}{2}$ centered at $\left(0, \frac{1}{2}\right)$.

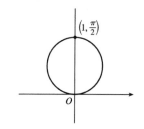

33. $r = 2(1 - \sin\theta)$. This curve is a cardioid.

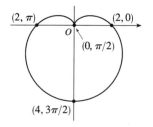

35. $r = \theta$, $\theta \geq 0$

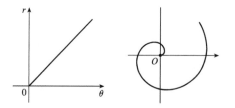

37. $r = \sin 2\theta$

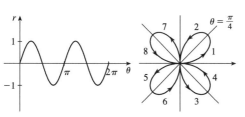

39. $r = 2\cos 4\theta$

41. $r^2 = 4\cos 2\theta$

43. $r = 2\cos\left(\frac{3}{2}\theta\right)$

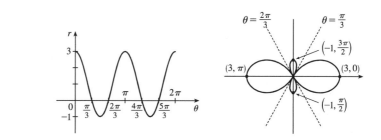

45. $r = 1 + 2\cos 2\theta$

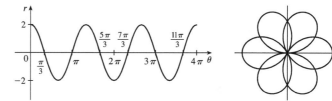

47. For $\theta = 0$, π, and 2π, r has its minimum value of about 0.5. For $\theta = \frac{\pi}{2}$ and $\frac{3\pi}{2}$, r attains its maximum value of 2.

We see that the graph has a similar shape for $0 \le \theta \le \pi$ and $\pi \le \theta \le 2\pi$.

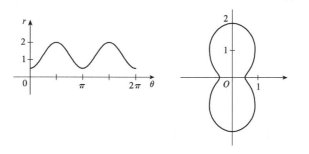

49. $x = r\cos\theta = (4 + 2\sec\theta)\cos\theta = 4\cos\theta + 2$. Now, $r \to \infty$ $\Rightarrow$

$(4 + 2\sec\theta) \to \infty$ $\Rightarrow$ $\theta \to \left(\frac{\pi}{2}\right)^{-}$ or $\theta \to \left(\frac{3\pi}{2}\right)^{+}$ (since we need only

consider $0 \le \theta < 2\pi$), so $\displaystyle\lim_{r\to\infty} x = \lim_{\theta\to\pi/2^{-}} (4\cos\theta + 2) = 2$. Also, $r \to -\infty$

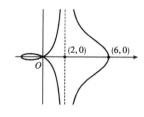

$\Rightarrow$ $(4 + 2\sec\theta) \to -\infty$ $\Rightarrow$ $\theta \to \left(\frac{\pi}{2}\right)^{+}$ or $\theta \to \left(\frac{3\pi}{2}\right)^{-}$, so

$\displaystyle\lim_{r\to-\infty} x = \lim_{\theta\to\pi/2^{+}} (4\cos\theta + 2) = 2$. Therefore, $\displaystyle\lim_{r\to\pm\infty} x = 2$ $\Rightarrow$ $x = 2$ is a

vertical asymptote.

51. To show that $x = 1$ is an asymptote we must prove $\displaystyle\lim_{r\to\pm\infty} x = 1$.

$x = r\cos\theta = (\sin\theta\tan\theta)\cos\theta = \sin^2\theta$. Now, $r \to \infty$ $\Rightarrow$ $\sin\theta\tan\theta \to \infty$

$\Rightarrow$ $\theta \to \left(\frac{\pi}{2}\right)^{-}$, so $\displaystyle\lim_{r\to\infty} x = \lim_{\theta\to\pi/2^{-}} \sin^2\theta = 1$. Also, $r \to -\infty$ $\Rightarrow$

$\sin\theta\tan\theta \to -\infty$ $\Rightarrow$ $\theta \to \left(\frac{\pi}{2}\right)^{+}$, so $\displaystyle\lim_{r\to-\infty} x = \lim_{\theta\to\pi/2^{+}} \sin^2\theta = 1$.

Therefore, $\displaystyle\lim_{r\to\pm\infty} x = 1$ $\Rightarrow$ $x = 1$ is a vertical asymptote. Also notice that $x = \sin^2\theta \ge 0$ for all θ, and

$x = \sin^2\theta \le 1$ for all θ. And $x \ne 1$, since the curve is not defined at odd multiples of $\frac{\pi}{2}$. Therefore, the curve lies

entirely within the vertical strip $0 \le x < 1$.

53. (a) We see that the curve crosses itself at the origin, where $r = 0$ (in fact the inner loop corresponds to negative

r-values), so we solve the equation of the limaçon for $r = 0$ $\Leftrightarrow$ $c\sin\theta = -1$ $\Leftrightarrow$ $\sin\theta = -1/c$. Now if

$|c| < 1$, then this equation has no solution and hence there is no inner loop. But if $c < -1$, then on the interval

$(0, 2\pi)$ the equation has the two solutions $\theta = \sin^{-1}(-1/c)$ and $\theta = \pi - \sin^{-1}(-1/c)$, and if $c > 1$, the

solutions are $\theta = \pi + \sin^{-1}(1/c)$ and $\theta = 2\pi - \sin^{-1}(1/c)$. In each case, $r < 0$ for θ between the two

solutions, indicating a loop.

(b) For $0 < c < 1$, the dimple (if it exists) is characterized by the fact that y has a local maximum at $\theta = \frac{3\pi}{2}$. So we

determine for what c-values $\dfrac{d^2y}{d\theta^2}$ is negative at $\theta = \frac{3\pi}{2}$, since by the Second Derivative Test this indicates a

maximum: $y = r\sin\theta = \sin\theta + c\sin^2\theta$ $\Rightarrow$ $\dfrac{dy}{d\theta} = \cos\theta + 2c\sin\theta\cos\theta = \cos\theta + c\sin 2\theta$ $\Rightarrow$

$\dfrac{d^2y}{d\theta^2} = -\sin\theta + 2c\cos 2\theta$. At $\theta = \frac{3\pi}{2}$, this is equal to $-(-1) + 2c(-1) = 1 - 2c$, which is negative only for

$c > \frac{1}{2}$. A similar argument shows that for $-1 < c < 0$, y only has a local minimum at $\theta = \frac{\pi}{2}$ (indicating a

dimple) for $c < -\frac{1}{2}$.

55. $r = 2\sin\theta$ $\Rightarrow$ $x = r\cos\theta = 2\sin\theta\cos\theta = \sin 2\theta$, $y = r\sin\theta = 2\sin^2\theta$ $\Rightarrow$

$$\frac{dy}{dx} = \frac{dy/d\theta}{dx/d\theta} = \frac{2\cdot 2\sin\theta\cos\theta}{\cos 2\theta \cdot 2} = \frac{\sin 2\theta}{\cos 2\theta} = \tan 2\theta$$

When $\theta = \dfrac{\pi}{6}$, $\dfrac{dy}{dx} = \tan\left(2\cdot\dfrac{\pi}{6}\right) = \tan\dfrac{\pi}{3} = \sqrt{3}$.

57. $r = 1/\theta \quad \Rightarrow \quad x = r\cos\theta = (\cos\theta)/\theta,\ y = r\sin\theta = (\sin\theta)/\theta \quad \Rightarrow$

$$\frac{dy}{dx} = \frac{dy/d\theta}{dx/d\theta} = \frac{\sin\theta(-1/\theta^2) + (1/\theta)\cos\theta}{\cos\theta(-1/\theta^2) - (1/\theta)\sin\theta} \cdot \frac{\theta^2}{\theta^2} = \frac{-\sin\theta + \theta\cos\theta}{-\cos\theta - \theta\sin\theta}$$

When $\theta = \pi$, $\dfrac{dy}{dx} = \dfrac{-0 + \pi(-1)}{-(-1) - \pi(0)} = \dfrac{-\pi}{1} = -\pi$.

59. $r = 1 + \cos\theta \quad \Rightarrow \quad x = r\cos\theta = \cos\theta + \cos^2\theta,\ y = r\sin\theta = \sin\theta + \sin\theta\cos\theta \quad \Rightarrow$

$$\frac{dy}{dx} = \frac{dy/d\theta}{dx/d\theta} = \frac{\cos\theta + \cos^2\theta - \sin^2\theta}{-\sin\theta - 2\cos\theta\sin\theta} = \frac{\cos\theta + \cos 2\theta}{-\sin\theta - \sin 2\theta}$$

When $\theta = \dfrac{\pi}{6}$, $\dfrac{dy}{dx} = \dfrac{\frac{\sqrt{3}}{2} + \frac{1}{2}}{-\frac{1}{2} - \frac{\sqrt{3}}{2}} = \dfrac{\frac{\sqrt{3}}{2} + \frac{1}{2}}{-\left(\frac{1}{2} + \frac{\sqrt{3}}{2}\right)} = -1$.

61. $r = 3\cos\theta \quad \Rightarrow \quad x = r\cos\theta = 3\cos\theta\cos\theta,\ y = r\sin\theta = 3\cos\theta\sin\theta \quad \Rightarrow$

$dy/d\theta = -3\sin^2\theta + 3\cos^2\theta = 3\cos 2\theta = 0 \quad \Rightarrow \quad 2\theta = \frac{\pi}{2}$ or $\frac{3\pi}{2} \quad \Leftrightarrow \quad \theta = \frac{\pi}{4}$ or $\frac{3\pi}{4}$. So the tangent is

horizontal at $\left(\frac{3}{\sqrt{2}}, \frac{\pi}{4}\right)$ and $\left(-\frac{3}{\sqrt{2}}, \frac{3\pi}{4}\right)$ $\left[\text{same as } \left(\frac{3}{\sqrt{2}}, -\frac{\pi}{4}\right)\right]$. $dx/d\theta = -6\sin\theta\cos\theta = -3\sin 2\theta = 0 \quad \Rightarrow$

$2\theta = 0$ or $\pi \quad \Leftrightarrow \quad \theta = 0$ or $\frac{\pi}{2}$. So the tangent is vertical at $(3, 0)$ and $\left(0, \frac{\pi}{2}\right)$.

63. $r = 1 + \cos\theta \quad \Rightarrow \quad x = r\cos\theta = \cos\theta(1 + \cos\theta),\ y = r\sin\theta = \sin\theta(1 + \cos\theta) \quad \Rightarrow$

$dy/d\theta = (1 + \cos\theta)\cos\theta - \sin^2\theta = 2\cos^2\theta + \cos\theta - 1 = (2\cos\theta - 1)(\cos\theta + 1) = 0 \quad \Rightarrow \quad \cos\theta = \frac{1}{2}$ or

$-1 \quad \Rightarrow \quad \theta = \frac{\pi}{3}, \pi$, or $\frac{5\pi}{3} \quad \Rightarrow \quad$ horizontal tangent at $\left(\frac{3}{2}, \frac{\pi}{3}\right)$, $(0, \pi)$ [the pole], and $\left(\frac{3}{2}, \frac{5\pi}{3}\right)$.

$dx/d\theta = -(1 + \cos\theta)\sin\theta - \cos\theta\sin\theta = -\sin\theta(1 + 2\cos\theta) = 0 \quad \Rightarrow \quad \sin\theta = 0$ or $\cos\theta = -\frac{1}{2} \quad \Rightarrow$

$\theta = 0, \pi, \frac{2\pi}{3}$, or $\frac{4\pi}{3} \quad \Rightarrow \quad$ vertical tangent at $(2, 0)$, $\left(\frac{1}{2}, \frac{2\pi}{3}\right)$, and $\left(\frac{1}{2}, \frac{4\pi}{3}\right)$. Note that the tangent is horizontal, not

vertical when $\theta = \pi$, since $\lim\limits_{\theta \to \pi} \dfrac{dy/d\theta}{dx/d\theta} = 0$.

65. $r = \cos 2\theta \quad \Rightarrow \quad x = r\cos\theta = \cos 2\theta\cos\theta,\ y = r\sin\theta = \cos 2\theta\sin\theta \quad \Rightarrow$

$$dy/d\theta = -2\sin 2\theta\sin\theta + \cos 2\theta\cos\theta = -4\sin^2\theta\cos\theta + \left(\cos^3\theta - \sin^2\theta\cos\theta\right)$$

$$= \cos\theta\left(\cos^2\theta - 5\sin^2\theta\right) = \cos\theta\left(1 - 6\sin^2\theta\right) = 0 \quad \Rightarrow$$

$\cos\theta = 0$ or $\sin\theta = \pm\frac{1}{\sqrt{6}} \quad \Rightarrow \quad \theta = \frac{\pi}{2}, \frac{3\pi}{2}, \alpha, \pi - \alpha, \pi + \alpha$, or $2\pi - \alpha$ (where $\alpha = \sin^{-1}\frac{1}{\sqrt{6}}$).

So the tangent is horizontal at $\left(-1, \frac{\pi}{2}\right)$, $\left(-1, \frac{3\pi}{2}\right)$, $\left(\frac{2}{3}, \alpha\right)$, $\left(\frac{2}{3}, \pi - \alpha\right)$, $\left(\frac{2}{3}, \pi + \alpha\right)$, and $\left(\frac{2}{3}, 2\pi - \alpha\right)$.

$$dx/d\theta = -2\sin 2\theta\cos\theta - \cos 2\theta\sin\theta = -4\sin\theta\cos^2\theta - (2\cos^2\theta - 1)\sin\theta$$

$$= \sin\theta(1 - 6\cos^2\theta) = 0 \quad \Rightarrow$$

$\sin\theta = 0$ or $\cos\theta = \pm\frac{1}{\sqrt{6}} \quad \Rightarrow \quad \theta = 0, \pi, \beta, \pi - \beta, \pi + \beta$, or $2\pi - \beta$ (where $\beta = \cos^{-1}\frac{1}{\sqrt{6}}$).

So the tangent is vertical at $(1, 0)$, $(1, \pi)$, $\left(-\frac{2}{3}, \beta\right)$, $\left(-\frac{2}{3}, \pi - \beta\right)$, $\left(-\frac{2}{3}, \pi + \beta\right)$, and $\left(-\frac{2}{3}, 2\pi - \beta\right)$.

67. $r = a \sin\theta + b \cos\theta \;\Rightarrow\; r^2 = ar \sin\theta + br \cos\theta \;\Rightarrow\; x^2 + y^2 = ay + bx \;\Rightarrow$

$x^2 - bx + \left(\frac{1}{2}b\right)^2 + y^2 - ay + \left(\frac{1}{2}a\right)^2 = \left(\frac{1}{2}b\right)^2 + \left(\frac{1}{2}a\right)^2 \;\Rightarrow\; \left(x - \frac{1}{2}b\right)^2 + \left(y - \frac{1}{2}a\right)^2 = \frac{1}{4}\left(a^2 + b^2\right)$, and this

is a circle with center $\left(\frac{1}{2}b, \frac{1}{2}a\right)$ and radius $\frac{1}{2}\sqrt{a^2 + b^2}$.

Note for Exercises 69–74: Maple is able to plot polar curves using the `polarplot` command, or using the `coords=polar` option in a regular `plot` command. In Mathematica, use `PolarPlot`. In Derive, change to `Polar` under `Options State`. If your graphing device cannot plot polar equations, you must convert to parametric equations. For example, in Exercise 69, $x = r \cos\theta = [1 + 2\sin(\theta/2)]\cos\theta$, $y = r \sin\theta = [1 + 2\sin(\theta/2)]\sin\theta$.

69. $r = 1 + 2\sin(\theta/2)$. The parameter interval is $[0, 4\pi]$.

71. $r = e^{\sin\theta} - 2\cos(4\theta)$. The parameter interval is $[0, 2\pi]$.

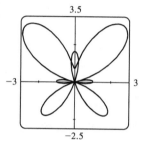

73. $r = 2 - 5\sin(\theta/6)$. The parameter interval is $[-6\pi, 6\pi]$.

75.

$r = 1 + \sin\theta$
$r = 1 + \sin\left(\theta - \frac{\pi}{6}\right)$
$r = 1 + \sin\left(\theta - \frac{\pi}{3}\right)$

It appears that the graph of $r = 1 + \sin\left(\theta - \frac{\pi}{6}\right)$ is the same shape as the graph of $r = 1 + \sin\theta$, but rotated

counterclockwise about the origin by $\frac{\pi}{6}$. Similarly, the graph of $r = 1 + \sin\left(\theta - \frac{\pi}{3}\right)$ is rotated by $\frac{\pi}{3}$. In general, the graph of $r = f(\theta - \alpha)$ is the same shape as that of $r = f(\theta)$, but rotated counterclockwise through α about the origin. That is, for any point (r_0, θ_0) on the curve $r = f(\theta)$, the point $(r_0, \theta_0 + \alpha)$ is on the curve $r = f(\theta - \alpha)$, since $r_0 = f(\theta_0) = f((\theta_0 + \alpha) - \alpha)$.

77. (a) $r = \sin n\theta$. From the graphs, it seems that when n is even, the number of loops in the curve (called a rose) is $2n$, and when n is odd, the number of loops is simply n.

This is because in the case of n odd, every point on the graph is traversed twice, due to the fact that

$$r(\theta + \pi) = \sin\left[n(\theta + \pi)\right] = \sin n\theta \cos n\pi + \cos n\theta \sin n\pi = \begin{cases} \sin n\theta & \text{if } n \text{ is even} \\ -\sin n\theta & \text{if } n \text{ is odd} \end{cases}$$

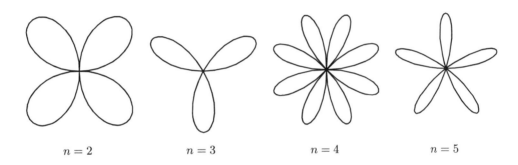

$n = 2$ $n = 3$ $n = 4$ $n = 5$

(b) The graph of $r = |\sin n\theta|$ has $2n$ loops whether n is odd or even, since $r(\theta + \pi) = r(\theta)$.

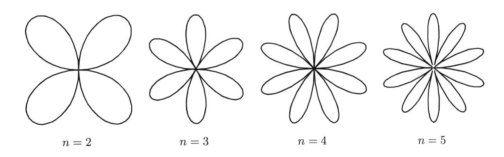

$n = 2$ $n = 3$ $n = 4$ $n = 5$

79. $r = \dfrac{1 - a\cos\theta}{1 + a\cos\theta}$. We start with $a = 0$, since in this case the curve is simply the circle $r = 1$.

As a increases, the graph moves to the left, and its right side becomes flattened. As a increases through about 0.4, the right side seems to grow a dimple, which upon closer investigation (with narrower θ-ranges) seems to appear at $a \approx 0.42$ (the actual value is $\sqrt{2} - 1$). As $a \to 1$, this dimple becomes more pronounced, and the curve begins to stretch out horizontally, until at $a = 1$ the denominator vanishes at $\theta = \pi$, and the dimple becomes an actual cusp. For $a > 1$ we must choose our parameter interval carefully, since $r \to \infty$ as $1 + a\cos\theta \to 0$. ⇔

$\theta \to \pm \cos^{-1}(-1/a)$. As a increases from 1, the curve splits into two parts. The left part has a loop, which grows larger as a increases, and the right part grows broader vertically, and its left tip develops a dimple when $a \approx 2.42$ (actually, $\sqrt{2}+1$). As a increases, the dimple grows more and more pronounced. If $a < 0$, we get the same graph as we do for the corresponding positive a-value, but with a rotation through π about the pole, as happened when c was replaced with $-c$ in Exercise 78.

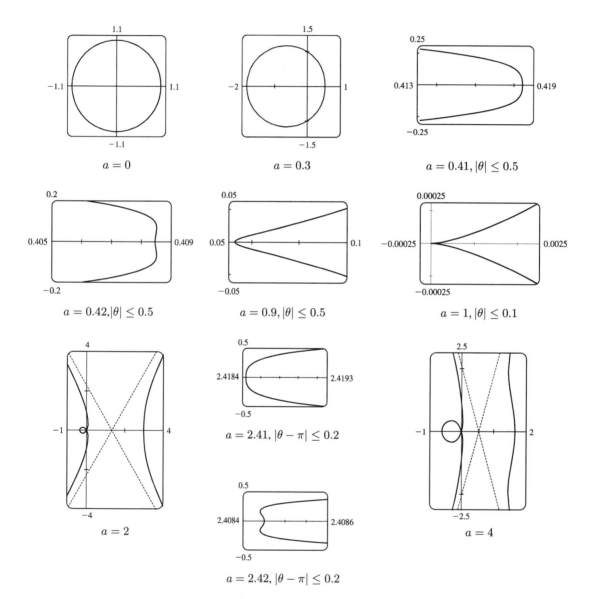

81.
$$\tan \psi = \tan(\phi - \theta) = \frac{\tan \phi - \tan \theta}{1 + \tan \phi \, \tan \theta} = \frac{\dfrac{dy}{dx} - \tan \theta}{1 + \dfrac{dy}{dx} \tan \theta} = \frac{\dfrac{dy/d\theta}{dx/d\theta} - \tan \theta}{1 + \dfrac{dy/d\theta}{dx/d\theta} \tan \theta}$$

$$= \frac{\dfrac{dy}{d\theta} - \dfrac{dx}{d\theta} \tan \theta}{\dfrac{dx}{d\theta} + \dfrac{dy}{d\theta} \tan \theta} = \frac{\left(\dfrac{dr}{d\theta} \sin \theta + r \cos \theta\right) - \tan \theta \left(\dfrac{dr}{d\theta} \cos \theta - r \sin \theta\right)}{\left(\dfrac{dr}{d\theta} \cos \theta - r \sin \theta\right) + \tan \theta \left(\dfrac{dr}{d\theta} \sin \theta + r \cos \theta\right)}$$

$$= \frac{r \cos \theta + r \cdot \dfrac{\sin^2 \theta}{\cos \theta}}{\dfrac{dr}{d\theta} \cos \theta + \dfrac{dr}{d\theta} \cdot \dfrac{\sin^2 \theta}{\cos \theta}} = \frac{r \cos^2 \theta + r \sin^2 \theta}{\dfrac{dr}{d\theta} \cos^2 \theta + \dfrac{dr}{d\theta} \sin^2 \theta} = \frac{r}{dr/d\theta}$$

10.4 Areas and Lengths in Polar Coordinates

1. $r = \sqrt{\theta}, 0 \le \theta \le \frac{\pi}{4}$. $A = \int_0^{\pi/4} \frac{1}{2} r^2 \, d\theta = \int_0^{\pi/4} \frac{1}{2} \left(\sqrt{\theta}\right)^2 d\theta = \int_0^{\pi/4} \frac{1}{2} \theta \, d\theta = \left[\frac{1}{4} \theta^2\right]_0^{\pi/4} = \frac{1}{64} \pi^2$

3. $r = \sin \theta, \frac{\pi}{3} \le \theta \le \frac{2\pi}{3}$.

$$A = \int_{\pi/3}^{2\pi/3} \frac{1}{2} \sin^2 \theta \, d\theta = \frac{1}{4} \int_{\pi/3}^{2\pi/3} (1 - \cos 2\theta) \, d\theta = \frac{1}{4} \left[\theta - \frac{1}{2} \sin 2\theta\right]_{\pi/3}^{2\pi/3}$$

$$= \frac{1}{4} \left[\frac{2\pi}{3} - \frac{1}{2} \sin \frac{4\pi}{3} - \frac{\pi}{3} + \frac{1}{2} \sin \frac{2\pi}{3}\right] = \frac{1}{4}\left[\frac{2\pi}{3} - \frac{1}{2}\left(-\frac{\sqrt{3}}{2}\right) - \frac{\pi}{3} + \frac{1}{2}\left(\frac{\sqrt{3}}{2}\right)\right] = \frac{1}{4}\left(\frac{\pi}{3} + \frac{\sqrt{3}}{2}\right) = \frac{\pi}{12} + \frac{\sqrt{3}}{8}$$

5. $r = \theta, 0 \le \theta \le \pi$. $A = \int_0^\pi \frac{1}{2} \theta^2 d\theta = \left[\frac{1}{6} \theta^3\right]_0^\pi = \frac{1}{6} \pi^3$

7. $r = 4 + 3 \sin \theta, -\frac{\pi}{2} \le \theta \le \frac{\pi}{2}$.

$$A = \int_{-\pi/2}^{\pi/2} \frac{1}{2} (4 + 3 \sin \theta)^2 d\theta = \frac{1}{2} \int_{-\pi/2}^{\pi/2} (16 + 24 \sin \theta + 9 \sin^2 \theta) \, d\theta$$

$$= \frac{1}{2} \int_{-\pi/2}^{\pi/2} (16 + 9 \sin^2 \theta) \, d\theta \quad \text{[by Theorem 5.5.7(b)]}$$

$$= \frac{1}{2} \cdot 2 \int_0^{\pi/2} \left[16 + 9 \cdot \frac{1}{2}(1 - \cos 2\theta)\right] d\theta \quad \text{[by Theorem 5.5.7(a)]}$$

$$= \int_0^{\pi/2} \left(\frac{41}{2} - \frac{9}{2} \cos 2\theta\right) d\theta = \left[\frac{41}{2} \theta - \frac{9}{4} \sin 2\theta\right]_0^{\pi/2} = \left(\frac{41\pi}{4} - 0\right) - (0 - 0) = \frac{41\pi}{4}$$

9. The area above the polar axis is bounded by $r = 3 \cos \theta$ for
$\theta = 0$ to $\theta = \pi/2$ (*not* π). By symmetry,

$$A = 2 \int_0^{\pi/2} \frac{1}{2} r^2 \, d\theta = \int_0^{\pi/2} (3 \cos \theta)^2 d\theta$$

$$= 3^2 \int_0^{\pi/2} \cos^2 \theta \, d\theta = 9 \int_0^{\pi/2} \frac{1}{2}(1 + \cos 2\theta) \, d\theta$$

$$= \frac{9}{2} \left[\theta + \frac{1}{2} \sin 2\theta\right]_0^{\pi/2} = \frac{9}{2}\left[\left(\frac{\pi}{2} + 0\right) - (0 + 0)\right] = \frac{9\pi}{4}.$$

$r = 3 \cos \theta$ with the circle passing through $(3, 0)$ and origin O.

Also, note that this is a circle with radius $\frac{3}{2}$, so its area is $\pi\left(\frac{3}{2}\right)^2 = \frac{9\pi}{4}$.

11. The curve $r^2 = 4\cos 2\theta$ goes through the pole when

$\theta = \pi/4$, so we'll find the area for $0 \le \theta \le \pi/4$ and

multiply it by 4.

$A = 4 \int_0^{\pi/4} \frac{1}{2} r^2 \, d\theta = 2 \int_0^{\pi/4} (4\cos 2\theta) \, d\theta$

$= 8 \int_0^{\pi/4} \cos 2\theta \, d\theta = 4 \left[\sin 2\theta\right]_0^{\pi/4} = 4(1 - 0) = 4$

13. One-sixth of the area lies above the polar axis and is bounded

by the curve $r = 2\cos 3\theta$ for $\theta = 0$ to $\theta = \pi/6$.

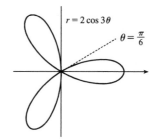

$A = 6 \int_0^{\pi/6} \frac{1}{2} (2\cos 3\theta)^2 \, d\theta = 12 \int_0^{\pi/6} \cos^2 3\theta \, d\theta$

$= \frac{12}{2} \int_0^{\pi/6} (1 + \cos 6\theta) \, d\theta$

$= 6\left[\theta + \frac{1}{6}\sin 6\theta\right]_0^{\pi/6} = 6\left(\frac{\pi}{6}\right) = \pi$

15. $A = \int_0^{2\pi} \frac{1}{2}(1 + 2\sin 6\theta)^2 \, d\theta = \frac{1}{2} \int_0^{2\pi} (1 + 4\sin 6\theta + 4\sin^2 6\theta) \, d\theta$

$= \frac{1}{2} \int_0^{2\pi} \left[1 + 4\sin 6\theta + 4 \cdot \frac{1}{2}(1 - \cos 12\theta)\right] d\theta$

$= \frac{1}{2} \int_0^{2\pi} (3 + 4\sin 6\theta - 2\cos 12\theta) \, d\theta$

$= \frac{1}{2}\left[3\theta - \frac{2}{3}\cos 6\theta - \frac{1}{6}\sin 12\theta\right]_0^{2\pi}$

$= \frac{1}{2}\left[\left(6\pi - \frac{2}{3} - 0\right) - \left(0 - \frac{2}{3} - 0\right)\right] = 3\pi.$

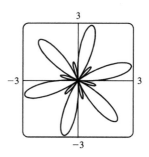

17. The shaded loop is traced out from $\theta = 0$

to $\theta = \pi/2$.

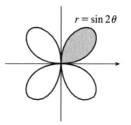

$A = \int_0^{\pi/2} \frac{1}{2} r^2 \, d\theta = \frac{1}{2} \int_0^{\pi/2} \sin^2 2\theta \, d\theta$

$= \frac{1}{2} \int_0^{\pi/2} \frac{1}{2}(1 - \cos 4\theta) \, d\theta$

$= \frac{1}{4}\left[\theta - \frac{1}{4}\sin 4\theta\right]_0^{\pi/2} = \frac{1}{4}\left(\frac{\pi}{2}\right) = \frac{\pi}{8}$

19. $r = 0 \implies 3\cos 5\theta = 0 \implies 5\theta = \frac{\pi}{2} \implies \theta = \frac{\pi}{10}.$

$A = \int_{-\pi/10}^{\pi/10} \frac{1}{2}(3\cos 5\theta)^2 \, d\theta = \int_0^{\pi/10} 9\cos^2 5\theta \, d\theta = \frac{9}{2} \int_0^{\pi/10} (1 + \cos 10\theta) \, d\theta = \frac{9}{2}\left[\theta + \frac{1}{10}\sin 10\theta\right]_0^{\pi/10} = \frac{9\pi}{20}$

21.

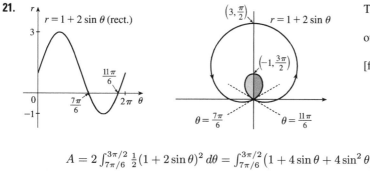

This is a limaçon, with inner loop traced out between $\theta = \frac{7\pi}{6}$ and $\frac{11\pi}{6}$ [found by solving $r = 0$].

$$A = 2 \int_{7\pi/6}^{3\pi/2} \tfrac{1}{2}(1 + 2\sin\theta)^2 \, d\theta = \int_{7\pi/6}^{3\pi/2} (1 + 4\sin\theta + 4\sin^2\theta) \, d\theta$$

$$= \int_{7\pi/6}^{3\pi/2} \left[1 + 4\sin\theta + 4 \cdot \tfrac{1}{2}(1 - \cos 2\theta)\right] d\theta = [\theta - 4\cos\theta + 2\theta - \sin 2\theta]_{7\pi/6}^{3\pi/2}$$

$$= \left(\tfrac{9\pi}{2}\right) - \left(\tfrac{7\pi}{2} + 2\sqrt{3} - \tfrac{\sqrt{3}}{2}\right) = \pi - \tfrac{3\sqrt{3}}{2}$$

23. $4\sin\theta = 2 \Leftrightarrow \sin\theta = \tfrac{1}{2} \Leftrightarrow \theta = \tfrac{\pi}{6}$ or $\tfrac{5\pi}{6}$ (for $0 \le \theta \le 2\pi$). We'll subtract the unshaded area from the shaded area for $\pi/6 \le \theta \le \pi/2$ and double that value.

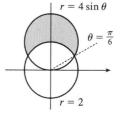

$$A = 2 \int_{\pi/6}^{\pi/2} \tfrac{1}{2}(4\sin\theta)^2 \, d\theta - 2 \int_{\pi/6}^{\pi/2} \tfrac{1}{2}(2)^2 \, d\theta = 2 \int_{\pi/6}^{\pi/2} \tfrac{1}{2}\left[(4\sin\theta)^2 - 2^2\right] d\theta$$

$$= \int_{\pi/6}^{\pi/2} \left(16\sin^2\theta - 4\right) d\theta = \int_{\pi/6}^{\pi/2} \left[8(1 - \cos 2\theta) - 4\right] d\theta$$

$$= \int_{\pi/6}^{\pi/2} (4 - 8\cos 2\theta) \, d\theta = [4\theta - 4\sin 2\theta]_{\pi/6}^{\pi/2}$$

$$= (2\pi - 0) - \left(\tfrac{2\pi}{3} - 4 \cdot \tfrac{\sqrt{3}}{2}\right) = \tfrac{4}{3}\pi + 2\sqrt{3}$$

25. To find the area inside the leminiscate $r^2 = 8\cos 2\theta$ and outside the circle $r = 2$, we first note that the two curves intersect when $r^2 = 8\cos 2\theta$ and $r = 2$; i.e., when $\cos 2\theta = \tfrac{1}{2}$. For $-\pi < \theta \le \pi$,

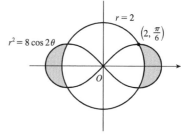

$\cos 2\theta = \tfrac{1}{2} \Leftrightarrow 2\theta = \pm\pi/3$ or $\pm 5\pi/3 \Leftrightarrow \theta = \pm\pi/6$ or $\pm 5\pi/6$. The figure shows that the desired area is 4 times the area between the curves from 0 to $\pi/6$. Thus,

$$A = 4 \int_0^{\pi/6} \left[\tfrac{1}{2}(8\cos 2\theta) - \tfrac{1}{2}(2)^2\right] d\theta = 8 \int_0^{\pi/6} (2\cos 2\theta - 1) \, d\theta$$

$$= 8 \left[\sin 2\theta - \theta\right]_0^{\pi/6} = 8\left(\sqrt{3}/2 - \pi/6\right) = 4\sqrt{3} - 4\pi/3$$

27. $3\cos\theta = 1 + \cos\theta \iff \cos\theta = \frac{1}{2} \implies \theta = \frac{\pi}{3}$ or $-\frac{\pi}{3}$.

$$A = 2\int_0^{\pi/3} \frac{1}{2}\left[(3\cos\theta)^2 - (1+\cos\theta)^2\right] d\theta$$

$$= \int_0^{\pi/3}\left(8\cos^2\theta - 2\cos\theta - 1\right) d\theta$$

$$= \int_0^{\pi/3}\left[4(1+\cos 2\theta) - 2\cos\theta - 1\right] d\theta$$

$$= \int_0^{\pi/3}\left(3 + 4\cos 2\theta - 2\cos\theta\right) d\theta$$

$$= \left[3\theta + 2\sin 2\theta - 2\sin\theta\right]_0^{\pi/3}$$

$$= \pi + \sqrt{3} - \sqrt{3} = \pi$$

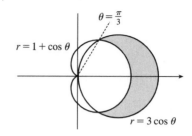

29. $A = 2\int_0^{\pi/4} \frac{1}{2}\sin^2\theta\, d\theta = \int_0^{\pi/4} \frac{1}{2}\left(1 - \cos 2\theta\right) d\theta$

$$= \frac{1}{2}\left[\theta - \frac{1}{2}\sin 2\theta\right]_0^{\pi/4} = \frac{1}{2}\left[\left(\frac{\pi}{4} - \frac{1}{2}\cdot 1\right) - (0-0)\right]$$

$$= \frac{1}{8}\pi - \frac{1}{4}$$

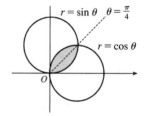

31. $\sin 2\theta = \cos 2\theta \implies \dfrac{\sin 2\theta}{\cos 2\theta} = 1 \implies \tan 2\theta = 1 \implies$

$\quad 2\theta = \frac{\pi}{4} \implies \theta = \frac{\pi}{8} \implies$

$$A = 8\cdot 2\int_0^{\pi/8} \frac{1}{2}\sin^2 2\theta\, d\theta = 8\int_0^{\pi/8} \frac{1}{2}(1 - \cos 4\theta)\, d\theta$$

$$= 4\left[\theta - \frac{1}{4}\sin 4\theta\right]_0^{\pi/8} = 4\left(\frac{\pi}{8} - \frac{1}{4}\cdot 1\right) = \frac{1}{2}\pi - 1$$

33. $A = 2\left[\int_{-\pi/2}^{-\pi/6} \frac{1}{2}(3 + 2\sin\theta)^2\, d\theta + \int_{-\pi/6}^{\pi/2} \frac{1}{2}2^2\, d\theta\right]$

$$= \int_{-\pi/2}^{-\pi/6}\left(9 + 12\sin\theta + 4\sin^2\theta\right) d\theta + \left[4\theta\right]_{-\pi/6}^{\pi/2}$$

$$= \left[9\theta - 12\cos\theta + 2\theta - \sin 2\theta\right]_{-\pi/2}^{-\pi/6} + \frac{8\pi}{3} = \frac{19\pi}{3} - \frac{11\sqrt{3}}{2}$$

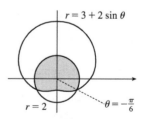

35. The darker shaded region (from $\theta = 0$ to $\theta = 2\pi/3$) represents $\frac{1}{2}$ of the desired area plus $\frac{1}{2}$ of the area of the inner loop. From this area, we'll subtract $\frac{1}{2}$ of the area of the inner loop (the lighter shaded region from $\theta = 2\pi/3$

to $\theta = \pi$), and then double that difference to obtain the desired area.

$A = 2\left[\int_0^{2\pi/3} \frac{1}{2}\left(\frac{1}{2} + \cos\theta\right)^2 d\theta - \int_{2\pi/3}^{\pi} \frac{1}{2}\left(\frac{1}{2} + \cos\theta\right)^2 d\theta\right]$

$= \int_0^{2\pi/3}\left(\frac{1}{4} + \cos\theta + \cos^2\theta\right)d\theta - \int_{2\pi/3}^{\pi}\left(\frac{1}{4} + \cos\theta + \cos^2\theta\right)d\theta$

$= \int_0^{2\pi/3}\left[\frac{1}{4} + \cos\theta + \frac{1}{2}(1 + \cos 2\theta)\right]d\theta$

$\qquad\qquad - \int_{2\pi/3}^{\pi}\left[\frac{1}{4} + \cos\theta + \frac{1}{2}(1 + \cos 2\theta)\right]d\theta$

$= \left[\frac{\theta}{4} + \sin\theta + \frac{\theta}{2} + \frac{\sin 2\theta}{4}\right]_0^{2\pi/3} - \left[\frac{\theta}{4} + \sin\theta + \frac{\theta}{2} + \frac{\sin 2\theta}{4}\right]_{2\pi/3}^{\pi}$

$= \left(\frac{\pi}{6} + \frac{\sqrt{3}}{2} + \frac{\pi}{3} - \frac{\sqrt{3}}{8}\right) - \left(\frac{\pi}{4} + \frac{\pi}{2}\right) + \left(\frac{\pi}{6} + \frac{\sqrt{3}}{2} + \frac{\pi}{3} - \frac{\sqrt{3}}{8}\right)$

$= \frac{\pi}{4} + \frac{3}{4}\sqrt{3} = \frac{1}{4}\left(\pi + 3\sqrt{3}\right)$

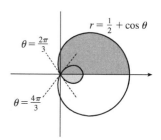

37. The two circles intersect at the pole since $(0, 0)$ satisfies the first equation and $\left(0, \frac{\pi}{2}\right)$ the second. The other intersection point $\left(\frac{1}{\sqrt{2}}, \frac{\pi}{4}\right)$ occurs where $\sin\theta = \cos\theta$.

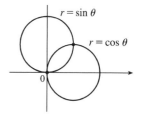

39. The curves intersect at the pole since $\left(0, \frac{\pi}{2}\right)$ satisfies $r = \cos\theta$ and $(0, 0)$ satisfies $r = 1 - \cos\theta$. Now $\cos\theta = 1 - \cos\theta \;\Rightarrow\; 2\cos\theta = 1 \;\Rightarrow\; \cos\theta = \frac{1}{2} \;\Rightarrow\; \theta = \frac{\pi}{3}$ or $\frac{5\pi}{3} \;\Rightarrow$ the other intersection points are $\left(\frac{1}{2}, \frac{\pi}{3}\right)$ and $\left(\frac{1}{2}, \frac{5\pi}{3}\right)$.

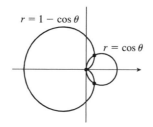

41. The pole is a point of intersection. $\sin\theta = \sin 2\theta = 2\sin\theta\cos\theta \;\Leftrightarrow\; \sin\theta\left(1 - 2\cos\theta\right) = 0 \;\Leftrightarrow\; \sin\theta = 0$ or $\cos\theta = \frac{1}{2} \;\Rightarrow\; \theta = 0, \pi, \frac{\pi}{3}, -\frac{\pi}{3} \;\Rightarrow\; \left(\frac{\sqrt{3}}{2}, \frac{\pi}{3}\right)$ and $\left(\frac{\sqrt{3}}{2}, \frac{2\pi}{3}\right)$ (by symmetry) are the other intersection points.

43.

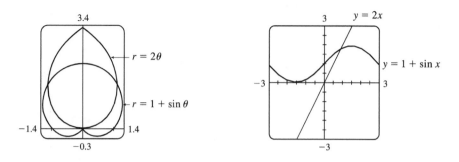

From the first graph, we see that the pole is one point of intersection. By zooming in or using the cursor, we find the θ-values of the intersection points to be $\alpha \approx 0.88786 \approx 0.89$ and $\pi - \alpha \approx 2.25$. (The first of these values may be more easily estimated by plotting $y = 1 + \sin x$ and $y = 2x$ in rectangular coordinates; see the second graph.) By symmetry, the total area contained is twice the area contained in the first quadrant, that is,

$$A = 2 \int_0^\alpha \tfrac{1}{2}(2\theta)^2 \, d\theta + 2 \int_\alpha^{\pi/2} \tfrac{1}{2}(1 + \sin\theta)^2 \, d\theta = \int_0^\alpha 4\theta^2 \, d\theta + \int_\alpha^{\pi/2} \left[1 + 2\sin\theta + \tfrac{1}{2}(1 - \cos 2\theta)\right] d\theta$$

$$= \left[\tfrac{4}{3}\theta^3\right]_0^\alpha + \left[\theta - 2\cos\theta + \left(\tfrac{1}{2}\theta - \tfrac{1}{4}\sin 2\theta\right)\right]_\alpha^{\pi/2}$$

$$= \tfrac{4}{3}\alpha^3 + \left[\left(\tfrac{\pi}{2} + \tfrac{\pi}{4}\right) - \left(\alpha - 2\cos\alpha + \tfrac{1}{2}\alpha - \tfrac{1}{4}\sin 2\alpha\right)\right] \approx 3.4645$$

45. $L = \int_a^b \sqrt{r^2 + (dr/d\theta)^2} \, d\theta = \int_0^{\pi/3} \sqrt{(3\sin\theta)^2 + (3\cos\theta)^2} \, d\theta = \int_0^{\pi/3} \sqrt{9(\sin^2\theta + \cos^2\theta)} \, d\theta$

$= 3 \int_0^{\pi/3} d\theta = 3[\theta]_0^{\pi/3} = 3\left(\tfrac{\pi}{3}\right) = \pi.$

As a check, note that the circumference of a circle with radius $\tfrac{3}{2}$ is $2\pi\left(\tfrac{3}{2}\right) = 3\pi$, and since $\theta = 0$ to $\pi = \tfrac{\pi}{3}$ traces out $\tfrac{1}{3}$ of the circle (from $\theta = 0$ to $\theta = \pi$), $\tfrac{1}{3}(3\pi) = \pi$.

47. $L = \int_a^b \sqrt{r^2 + (dr/d\theta)^2} \, d\theta = \int_0^{2\pi} \sqrt{\left(\theta^2\right)^2 + (2\theta)^2} \, d\theta = \int_0^{2\pi} \sqrt{\theta^4 + 4\theta^2} \, d\theta$

$= \int_0^{2\pi} \sqrt{\theta^2\left(\theta^2 + 4\right)} \, d\theta = \int_0^{2\pi} \theta\sqrt{\theta^2 + 4} \, d\theta$

Now let $u = \theta^2 + 4$, so that $du = 2\theta \, d\theta \; \left[\theta \, d\theta = \tfrac{1}{2} \, du\right]$ and

$$\int_0^{2\pi} \theta\sqrt{\theta^2 + 4} \, d\theta = \int_4^{4\pi^2 + 4} \tfrac{1}{2}\sqrt{u} \, du = \tfrac{1}{2} \cdot \tfrac{2}{3} \left[u^{3/2}\right]_4^{4(\pi^2 + 1)} = \tfrac{1}{3}\left[4^{3/2}\left(\pi^2 + 1\right)^{3/2} - 4^{3/2}\right]$$

$$= \tfrac{8}{3}\left[\left(\pi^2 + 1\right)^{3/2} - 1\right]$$

49. The curve $r = 3\sin 2\theta$ is completely traced with $0 \le \theta \le 2\pi$. $r^2 + \left(\tfrac{dr}{d\theta}\right)^2 = (3\sin 2\theta)^2 + (6\cos 2\theta)^2 \quad \Rightarrow$

$L = \int_0^{2\pi} \sqrt{9\sin^2 2\theta + 36\cos^2 2\theta} \, d\theta \approx 29.0653$

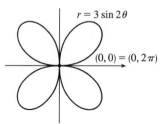

$r = 3\sin 2\theta$

$(0, 0) = (0, 2\pi)$

51. The curve $r = \sin\left(\frac{\theta}{2}\right)$ is completely traced with $0 \le \theta \le 4\pi$. $r^2 + \left(\frac{dr}{d\theta}\right)^2 = \sin^2\left(\frac{\theta}{2}\right) + \left[\frac{1}{2}\cos\left(\frac{\theta}{2}\right)\right]^2 \Rightarrow$

$$L = \int_0^{4\pi} \sqrt{\sin^2\left(\frac{\theta}{2}\right) + \frac{1}{4}\cos^2\left(\frac{\theta}{2}\right)}\, d\theta$$

$$\approx 9.6884$$

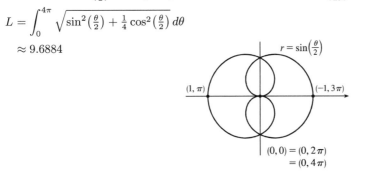

53. The curve $r = \cos^4(\theta/4)$ is completely traced with $0 \le \theta \le 4\pi$.

$$r^2 + (dr/d\theta)^2 = [\cos^4(\theta/4)]^2 + \left[4\cos^3(\theta/4)\cdot(-\sin(\theta/4))\cdot\frac{1}{4}\right]^2$$

$$= \cos^8(\theta/4) + \cos^6(\theta/4)\sin^2(\theta/4)$$

$$= \cos^6(\theta/4)[\cos^2(\theta/4) + \sin^2(\theta/4)]$$

$$= \cos^6(\theta/4)$$

$$L = \int_0^{4\pi}\sqrt{\cos^6(\theta/4)}\, d\theta = \int_0^{4\pi}\left|\cos^3(\theta/4)\right|\, d\theta$$

$$= 2\int_0^{2\pi}\cos^3(\theta/4)\, d\theta \quad [\text{since } \cos^3(\theta/4) \ge 0 \text{ for } 0 \le \theta \le 2\pi] \quad = 8\int_0^{\pi/2}\cos^3 u\, du \; \left[u = \frac{1}{4}\theta\right]$$

$$\overset{68}{=} 8\left[\frac{1}{3}(2 + \cos^2 u)\sin u\right]_0^{\pi/2} = \frac{8}{3}[(2\cdot 1) - (3\cdot 0)] = \frac{16}{3}$$

55. (a) From (10.2.7),

$$S = \int_a^b 2\pi y\sqrt{(dx/d\theta)^2 + (dy/d\theta)^2}\, d\theta$$

$$= \int_a^b 2\pi y\sqrt{r^2 + (dr/d\theta)^2}\, d\theta \qquad [\text{from the derivation of Equation 10.4.5}]$$

$$= \int_a^b 2\pi r\sin\theta\sqrt{r^2 + (dr/d\theta)^2}\, d\theta$$

(b) The curve $r^2 = \cos 2\theta$ goes through the pole when

$\cos 2\theta = 0 \quad\Rightarrow\quad 2\theta = \dfrac{\pi}{2} \quad\Rightarrow\quad \theta = \dfrac{\pi}{4}$. We'll rotate the curve

from $\theta = 0$ to $\theta = \dfrac{\pi}{4}$ and double this value to obtain the total surface

area generated. $r^2 = \cos 2\theta \quad\Rightarrow\quad 2r\dfrac{dr}{d\theta} = -2\sin 2\theta \quad\Rightarrow$

$$\left(\frac{dr}{d\theta}\right)^2 = \frac{\sin^2 2\theta}{r^2} = \frac{\sin^2 2\theta}{\cos 2\theta}.$$

$$S = 2\int_0^{\pi/4} 2\pi\sqrt{\cos 2\theta}\sin\theta\sqrt{\cos 2\theta + (\sin^2 2\theta)/\cos 2\theta}\, d\theta$$

$$= 4\pi\int_0^{\pi/4}\sqrt{\cos 2\theta}\sin\theta\sqrt{\frac{\cos^2 2\theta + \sin^2 2\theta}{\cos 2\theta}}\, d\theta = 4\pi\int_0^{\pi/4}\sqrt{\cos 2\theta}\sin\theta\frac{1}{\sqrt{\cos 2\theta}}\, d\theta$$

$$= 4\pi\int_0^{\pi/4}\sin\theta\, d\theta = 4\pi\left[-\cos\theta\right]_0^{\pi/4} = -4\pi\left(\frac{\sqrt{2}}{2} - 1\right) = 2\pi\left(2 - \sqrt{2}\right)$$

10.5 Conic Sections

1. $x = 2y^2 \;\Rightarrow\; y^2 = \frac{1}{2}x$. $4p = \frac{1}{2}$, so $p = \frac{1}{8}$. The
vertex is $(0,0)$, the focus is $\left(\frac{1}{8}, 0\right)$, and the
directrix is $x = -\frac{1}{8}$.

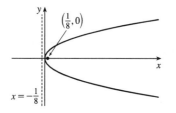

3. $4x^2 = -y \;\Rightarrow\; x^2 = -\frac{1}{4}y$. $4p = -\frac{1}{4}$, so
$p = -\frac{1}{16}$. The vertex is $(0,0)$, the focus is
$\left(0, -\frac{1}{16}\right)$, and the directrix is $y = \frac{1}{16}$.

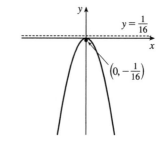

5. $(x + 2)^2 = 8 (y - 3)$. $4p = 8$, so $p = 2$. The vertex is $(-2, 3)$, the focus is $(-2, 5)$, and the directrix is $y = 1$.

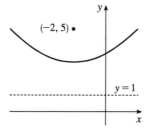

7. $y^2 + 2y + 12x + 25 = 0 \;\Rightarrow$

$y^2 + 2y + 1 = -12x - 24 \;\Rightarrow\; (y + 1)^2 = -12(x + 2)$. $4p = -12$, so $p = -3$. The vertex is $(-2, -1)$, the
focus is $(-5, -1)$, and the directrix is $x = 1$.

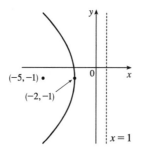

9. The equation has the form $y^2 = 4px$, where $p < 0$. Since the parabola passes through $(-1, 1)$, we have

$1^2 = 4p(-1)$, so $4p = -1$ and an equation is $y^2 = -x$ or $x = -y^2$. $4p = -1$, so $p = -\frac{1}{4}$ and the focus is

$\left(-\frac{1}{4}, 0\right)$ while the directrix is $x = \frac{1}{4}$.

11. $\dfrac{x^2}{9} + \dfrac{y^2}{5} = 1 \Rightarrow a = \sqrt{9} = 3, b = \sqrt{5}, c = \sqrt{a^2 - b^2} = \sqrt{9-5} = 2.$ The ellipse is centered at $(0,0)$, with

vertices at $(\pm 3, 0)$. The foci are $(\pm 2, 0)$.

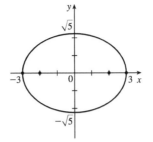

13. $4x^2 + y^2 = 16 \Rightarrow \dfrac{x^2}{4} + \dfrac{y^2}{16} = 1 \Rightarrow a = \sqrt{16} = 4, b = \sqrt{4} = 2, c = \sqrt{a^2 - b^2} = \sqrt{16-4} = 2\sqrt{3}.$ The

ellipse is centered at $(0,0)$, with vertices at $(0, \pm 4)$. The foci are $\left(0, \pm 2\sqrt{3}\right)$.

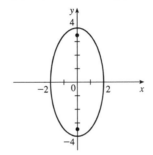

15. $9x^2 - 18x + 4y^2 = 27 \Leftrightarrow 9(x^2 - 2x + 1) + 4y^2 = 27 + 9 \Leftrightarrow$

$9(x-1)^2 + 4y^2 = 36 \Leftrightarrow \dfrac{(x-1)^2}{4} + \dfrac{y^2}{9} = 1 \Rightarrow a = 3, b = 2,$

$c = \sqrt{5} \Rightarrow$ center $(1,0)$, vertices $(1, \pm 3)$, foci $\left(1, \pm\sqrt{5}\right)$

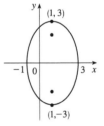

17. The center is $(0,0)$, $a = 3$, and $b = 2$, so an equation is $\dfrac{x^2}{4} + \dfrac{y^2}{9} = 1.$ $c = \sqrt{a^2 - b^2} = \sqrt{5}$, so the foci are

$\left(0, \pm\sqrt{5}\right)$.

19. $\dfrac{x^2}{144} - \dfrac{y^2}{25} = 1 \Rightarrow a = 12, b = 5, c = \sqrt{144 + 25} = 13 \Rightarrow$

center $(0,0)$, vertices $(\pm 12, 0)$, foci $(\pm 13, 0)$,

asymptotes $y = \pm\frac{5}{12}x$.

Note: It is helpful to draw a $2a$-by-$2b$ rectangle whose center is the

center of the hyperbola. The asymptotes are the extended diagonals

of the rectangle.

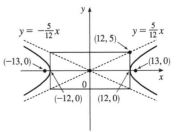

21. $y^2 - x^2 = 4$ ⟺ $\dfrac{y^2}{4} - \dfrac{x^2}{4} = 1$ ⟹ $a = \sqrt{4} = 2 = b$,

$c = \sqrt{4+4} = 2\sqrt{2}$ ⟹ center $(0,0)$, vertices $(0, \pm 2)$,

foci $\left(0, \pm 2\sqrt{2}\right)$, asymptotes $y = \pm x$

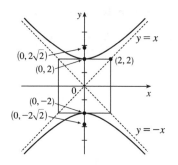

23. $2y^2 - 4y - 3x^2 + 12x = -8$ ⟺

$2(y^2 - 2y + 1) - 3(x^2 - 4x + 4) = -8 + 2 - 12$ ⟺

$2(y-1)^2 - 3(x-2)^2 = -18$ ⟺ $\dfrac{(x-2)^2}{6} - \dfrac{(y-1)^2}{9} = 1$

⟹ $a = \sqrt{6}, b = 3, c = \sqrt{15}$ ⟹ center $(2,1)$, vertices

$\left(2 \pm \sqrt{6}, 1\right)$, foci $\left(2 \pm \sqrt{15}, 1\right)$, asymptotes $y - 1 = \pm \dfrac{3}{\sqrt{6}} (x - 2)$

or $y - 1 = \pm \dfrac{\sqrt{6}}{2}(x - 2)$

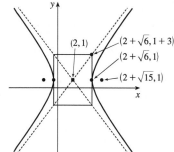

25. $x^2 = y + 1$ ⟺ $x^2 = 1(y+1)$. This is an equation of a *parabola* with $4p = 1$, so $p = \frac{1}{4}$. The vertex is $(0, -1)$

and the focus is $\left(0, -\frac{3}{4}\right)$.

27. $x^2 = 4y - 2y^2$ ⟺ $x^2 + 2y^2 - 4y = 0$ ⟺ $x^2 + 2(y^2 - 2y + 1) = 2$ ⟺ $x^2 + 2(y-1)^2 = 2$ ⟺

$\dfrac{x^2}{2} + \dfrac{(y-1)^2}{1} = 1$. This is an equation of an *ellipse* with vertices at $\left(\pm\sqrt{2}, 1\right)$. The foci are at

$\left(\pm\sqrt{2-1}, 1\right) = (\pm 1, 1)$.

29. $y^2 + 2y = 4x^2 + 3$ ⟺ $y^2 + 2y + 1 = 4x^2 + 4$ ⟺ $(y+1)^2 - 4x^2 = 4$ ⟺ $\dfrac{(y+1)^2}{4} - x^2 = 1$. This is

an equation of a *hyperbola* with vertices $(0, -1 \pm 2) = (0, 1)$ and $(0, -3)$. The foci are at

$\left(0, -1 \pm \sqrt{4+1}\right) = \left(0, -1 \pm \sqrt{5}\right)$.

31. The parabola with vertex $(0,0)$ and focus $(0, -2)$ opens downward and has $p = -2$, so its equation is

$x^2 = 4py = -8y$.

33. The distance from the focus $(-4, 0)$ to the directrix $x = 2$ is $2 - (-4) = 6$, so the distance from the focus to the

vertex is $\frac{1}{2}(6) = 3$ and the vertex is $(-1, 0)$. Since the focus is to the left of the vertex, $p = -3$. An equation is

$y^2 = 4p(x+1)$ ⟹ $y^2 = -12(x+1)$.

35. The parabola must have equation $y^2 = 4px$, so $(-4)^2 = 4p(1)$ ⟹ $p = 4$ ⟹ $y^2 = 16x$.

37. The ellipse with foci $(\pm 2, 0)$ and vertices $(\pm 5, 0)$ has center $(0, 0)$ and a horizontal major axis, with $a = 5$ and

$c = 2$, so $b = \sqrt{a^2 - c^2} = \sqrt{21}$. An equation is $\dfrac{x^2}{25} + \dfrac{y^2}{21} = 1$.

39. Since the vertices are $(0, 0)$ and $(0, 8)$, the ellipse has center $(0, 4)$ with a vertical axis and $a = 4$. The foci at $(0, 2)$

and $(0, 6)$ are 2 units from the center, so $c = 2$ and $b = \sqrt{a^2 - c^2} = \sqrt{4^2 - 2^2} = \sqrt{12}$. An equation is

$\dfrac{(x-0)^2}{b^2} + \dfrac{(y-4)^2}{a^2} = 1$ ⟹ $\dfrac{x^2}{12} + \dfrac{(y-4)^2}{16} = 1$.

41. Center $(2, 2)$, $c = 2$, $a = 3$ $\Rightarrow$ $b = \sqrt{5}$ $\Rightarrow$ $\frac{1}{9}(x - 2)^2 + \frac{1}{5}(y - 2)^2 = 1$

43. Center $(0, 0)$, vertical axis, $c = 3$, $a = 1$ $\Rightarrow$ $b = \sqrt{8} = 2\sqrt{2}$ $\Rightarrow$ $y^2 - \frac{1}{8}x^2 = 1$

45. Center $(4, 3)$, horizontal axis, $c = 3$, $a = 2$ $\Rightarrow$ $b = \sqrt{5}$ $\Rightarrow$ $\frac{1}{4}(x - 4)^2 - \frac{1}{5}(y - 3)^2 = 1$

47. Center $(0, 0)$, horizontal axis, $a = 3$, $\frac{b}{a} = 2$ $\Rightarrow$ $b = 6$ $\Rightarrow$ $\frac{1}{9}x^2 - \frac{1}{36}y^2 = 1$

49. In Figure 8, we see that the point on the ellipse closest to a focus is the closer vertex (which is a distance $a - c$ from it) while the farthest point is the other vertex (at a distance of $a + c$). So for this lunar orbit,

$(a - c) + (a + c) = 2a = (1728 + 110) + (1728 + 314)$, or $a = 1940$; and

$(a + c) - (a - c) = 2c = 314 - 110$, or $c = 102$. Thus, $b^2 = a^2 - c^2 = 3{,}753{,}196$, and the equation is

$$\frac{x^2}{3{,}763{,}600} + \frac{y^2}{3{,}753{,}196} = 1.$$

51. (a) Set up the coordinate system so that A is $(-200, 0)$ and B is $(200, 0)$.

$|PA| - |PB| = (1200)(980) = 1{,}176{,}000 \text{ ft} = \frac{2450}{11} \text{ mi} = 2a$ $\Rightarrow$ $a = \frac{1225}{11}$, and $c = 200$ so

$b^2 = c^2 - a^2 = \frac{3{,}339{,}375}{121}$ $\Rightarrow$ $\frac{121x^2}{1{,}500{,}625} - \frac{121y^2}{3{,}339{,}375} = 1$.

(b) Due north of B $\Rightarrow$ $x = 200$ $\Rightarrow$ $\frac{(121)(200)^2}{1{,}500{,}625} - \frac{121y^2}{3{,}339{,}375} = 1$ $\Rightarrow$ $y = \frac{133{,}575}{539} \approx 248 \text{ mi}$

53. The function whose graph is the upper branch of this hyperbola is concave upward. The

function is $y = f(x) = a\sqrt{1 + \dfrac{x^2}{b^2}} = \dfrac{a}{b}\sqrt{b^2 + x^2}$, so $y' = \dfrac{a}{b}x(b^2 + x^2)^{-1/2}$ and

$y'' = \dfrac{a}{b}\left[(b^2 + x^2)^{-1/2} - x^2(b^2 + x^2)^{-3/2}\right] = ab(b^2 + x^2)^{-3/2} > 0$ for all x, and so f is concave upward.

55. (a) If $k > 16$, then $k - 16 > 0$, and $\dfrac{x^2}{k} + \dfrac{y^2}{k - 16} = 1$ is an *ellipse* since it is the sum of two squares on the left side.

(b) If $0 < k < 16$, then $k - 16 < 0$, and $\dfrac{x^2}{k} + \dfrac{y^2}{k - 16} = 1$ is a *hyperbola* since it is the difference of two squares on the left side.

(c) If $k < 0$, then $k - 16 < 0$, and there is *no curve* since the left side is the sum of two negative terms, which cannot equal 1.

(d) In case (a), $a^2 = k$, $b^2 = k - 16$, and $c^2 = a^2 - b^2 = 16$, so the foci are at $(\pm 4, 0)$. In case (b), $k - 16 < 0$, so $a^2 = k$, $b^2 = 16 - k$, and $c^2 = a^2 + b^2 = 16$, and so again the foci are at $(\pm 4, 0)$.

57. Use the parametrization $x = 2\cos t$, $y = \sin t$, $0 \le t \le 2\pi$ to get

$$L = 4\int_0^{\pi/2} \sqrt{(dx/dt)^2 + (dy/dt)^2}\, dt = 4\int_0^{\pi/2}\sqrt{4\sin^2 t + \cos^2 t}\, dt = 4\int_0^{\pi/2}\sqrt{3\sin^2 t + 1}\, dt$$

Using Simpson's Rule with $n = 10$, $\Delta t = \frac{\pi/2 - 0}{10} = \frac{\pi}{20}$, and $f(t) = \sqrt{3\sin^2 t + 1}$, we get

$$L \approx \frac{4}{3}\left(\frac{\pi}{20}\right)\left[f(0) + 4f\left(\frac{\pi}{20}\right) + 2f\left(\frac{2\pi}{20}\right) + \cdots + 2f\left(\frac{8\pi}{20}\right) + 4f\left(\frac{9\pi}{20}\right) + f\left(\frac{\pi}{2}\right)\right] \approx 9.69$$

59. $\dfrac{x^2}{a^2} + \dfrac{y^2}{b^2} = 1 \;\Rightarrow\; \dfrac{2x}{a^2} + \dfrac{2yy'}{b^2} = 0 \;\Rightarrow\; y' = -\dfrac{b^2x}{a^2y}$ $(y \ne 0)$. Thus, the slope of the tangent line at P is

$-\dfrac{b^2x_1}{a^2y_1}$. The slope of F_1P is $\dfrac{y_1}{x_1+c}$ and of F_2P is $\dfrac{y_1}{x_1-c}$. By the formula from Problems Plus, we have

$$\tan\alpha = \frac{\dfrac{y_1}{x_1+c} + \dfrac{b^2x_1}{a^2y_1}}{1 - \dfrac{b^2x_1y_1}{a^2y_1(x_1+c)}} = \frac{a^2y_1^2 + b^2x_1(x_1+c)}{a^2y_1(x_1+c) - b^2x_1y_1} = \frac{a^2b^2 + b^2cx_1}{c^2x_1y_1 + a^2cy_1} \quad \left[\begin{array}{l}\text{using } b^2x_1^2 + a^2y_1^2 = a^2b^2 \\ \text{and } a^2 - b^2 = c^2\end{array}\right]$$

$$= \frac{b^2(cx_1+a^2)}{cy_1(cx_1+a^2)} = \frac{b^2}{cy_1}$$

and

$$\tan\beta = \frac{-\dfrac{y_1}{x_1-c} - \dfrac{b^2x_1}{a^2y_1}}{1 - \dfrac{b^2x_1y_1}{a^2y_1(x_1-c)}} = \frac{-a^2y_1^2 - b^2x_1(x_1-c)}{a^2y_1(x_1-c) - b^2x_1y_1} = \frac{-a^2b^2 + b^2cx_1}{c^2x_1y_1 - a^2cy_1} = \frac{b^2(cx_1-a^2)}{cy_1(cx_1-a^2)} = \frac{b^2}{cy_1}$$

So $\alpha = \beta$.

10.6 Conic Sections in Polar Coordinates

1. The directrix $y = 6$ is above the focus at the origin, so we use the form with "$+ e\sin\theta$" in the denominator. (See Theorem 6 and Figure 2.) $r = \dfrac{ed}{1 + e\sin\theta} = \dfrac{\frac{7}{4}\cdot 6}{1 + \frac{7}{4}\sin\theta} = \dfrac{42}{4 + 7\sin\theta}$

3. The directrix $x = -5$ is to the left of the focus at the origin, so we use the form with "$- e\cos\theta$" in the denominator. $r = \dfrac{ed}{1 - e\cos\theta} = \dfrac{\frac{3}{4}\cdot 5}{1 - \frac{3}{4}\cos\theta} = \dfrac{15}{4 - 3\cos\theta}$

5. The vertex $(4, 3\pi/2)$ is 4 units below the focus at the origin, so the directrix is 8 units below the focus $(d = 8)$, and we use the form with "$-e\sin\theta$" in the denominator. $e = 1$ for a parabola, so an equation is
$$r = \frac{ed}{1 - e\sin\theta} = \frac{1(8)}{1 - 1\sin\theta} = \frac{8}{1 - \sin\theta}.$$

7. The directrix $r = 4\sec\theta$ (equivalent to $r\cos\theta = 4$ or $x = 4$) is to the right of the focus at the origin, so we will use the form with "$+e\cos\theta$" in the denominator. The distance from the focus to the directrix is $d = 4$, so an equation is
$$r = \frac{ed}{1 + e\cos\theta} = \frac{0.5(4)}{1 + 0.5\cos\theta} \cdot \frac{2}{2} = \frac{4}{2 + \cos\theta}.$$

9. $r = \dfrac{1}{1 + \sin\theta} = \dfrac{ed}{1 + e\sin\theta}$, where $d = e = 1$.

(a) Eccentricity $= e = 1$

(b) Since $e = 1$, the conic is a parabola.

(c) Since "$+e\sin\theta$" appears in the denominator, the directrix is above the focus at the origin. $d = |Fl| = 1$, so an equation of the directrix is $y = 1$.

(d) The vertex is at $\left(\frac{1}{2}, \frac{\pi}{2}\right)$, midway between the focus and the directrix.

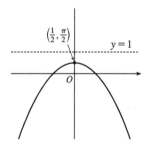

11. $r = \dfrac{12}{4 - \sin \theta} \cdot \dfrac{1/4}{1/4} = \dfrac{3}{1 - \frac{1}{4} \sin \theta}$, where $e = \frac{1}{4}$ and $ed = 3 \Rightarrow d = 12$.

(a) Eccentricity $= e = \frac{1}{4}$

(b) Since $e = \frac{1}{4} < 1$, the conic is an ellipse.

(c) Since "$-e \sin \theta$" appears in the denominator, the directrix is below
the focus at the origin. $d = |Fl| = 12$, so an equation of the directrix
is $y = -12$.

(d) The vertices are $\left(4, \frac{\pi}{2}\right)$ and $\left(\frac{12}{5}, \frac{3\pi}{2}\right)$, so the center is midway
between them, that is, $\left(\frac{4}{5}, \frac{\pi}{2}\right)$.

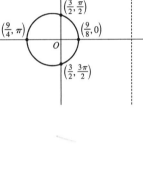

13. $r = \dfrac{9}{6 + 2 \cos \theta} \cdot \dfrac{1/6}{1/6} = \dfrac{3/2}{1 + \frac{1}{3} \cos \theta}$, where $e = \frac{1}{3}$ and $ed = \frac{3}{2} \Rightarrow d = \frac{9}{2}$.

(a) Eccentricity $= e = \frac{1}{3}$

(b) Since $e = \frac{1}{3} < 1$, the conic is an ellipse.

(c) Since "$+e \cos \theta$" appears in the denominator, the directrix is to the
right of the focus at the origin. $d = |Fl| = \frac{9}{2}$, so an equation of the
directrix is $x = \frac{9}{2}$.

(d) The vertices are $\left(\frac{9}{8}, 0\right)$ and $\left(\frac{9}{4}, \pi\right)$, so the center is midway between
them, that is, $\left(\frac{9}{16}, \pi\right)$.

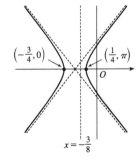

15. $r = \dfrac{3}{4 - 8 \cos \theta} \cdot \dfrac{1/4}{1/4} = \dfrac{3/4}{1 - 2 \cos \theta}$, where $e = 2$ and $ed = \frac{3}{4} \Rightarrow d = \frac{3}{8}$.

(a) Eccentricity $= e = 2$

(b) Since $e = 2 > 1$, the conic is a hyperbola.

(c) Since "$-e \cos \theta$" appears in the denominator, the directrix is to the
left of the focus at the origin. $d = |Fl| = \frac{3}{8}$, so an equation of the
directrix is $x = -\frac{3}{8}$.

(d) The vertices are $\left(-\frac{3}{4}, 0\right)$ and $\left(\frac{1}{4}, \pi\right)$, so the center is midway
between them, that is, $\left(\frac{1}{2}, \pi\right)$.

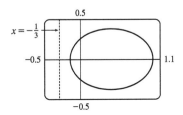

17. (a) The equation is $r = \dfrac{1}{4 - 3 \cos \theta} = \dfrac{1/4}{1 - \frac{3}{4} \cos \theta}$, so $e = \frac{3}{4}$ and

$ed = \frac{1}{4} \Rightarrow d = \frac{1}{3}$. The conic is an ellipse, and the equation of its

directrix is $x = r \cos \theta = -\frac{1}{3} \Rightarrow r = -\dfrac{1}{3 \cos \theta}$. We must be

careful in our choice of parameter values in this equation
$(-1 \le \theta \le 1$ works well$)$.

(b) The equation is obtained by replacing θ with $\theta - \frac{\pi}{3}$ in the equation of

the original conic (see Example 4), so $r = \dfrac{1}{4 - 3\cos\left(\theta - \frac{\pi}{3}\right)}$.

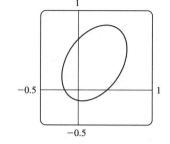

19. For $e < 1$ the curve is an ellipse. It is nearly circular when e is close to 0.
As e increases, the graph is stretched out to the right, and grows larger
(that is, its right-hand focus moves to the right while its left-hand focus
remains at the origin.) At $e = 1$, the curve becomes a parabola with focus
at the origin.

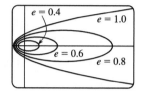

21. $|PF| = e\,|Pl| \quad\Rightarrow\quad r = e[d - r\cos(\pi - \theta)] = e(d + r\cos\theta) \quad\Rightarrow$

$r(1 - e\cos\theta) = ed \quad\Rightarrow\quad r = \dfrac{ed}{1 - e\cos\theta}$

23. $|PF| = e\,|Pl| \quad\Rightarrow\quad r = e[d - r\sin(\theta - \pi)] = e(d + r\sin\theta) \quad\Rightarrow$

$r(1 - e\sin\theta) = ed \quad\Rightarrow\quad r = \dfrac{ed}{1 - e\sin\theta}$

25. (a) If the directrix is $x = -d$, then $r = \dfrac{ed}{1 - e\cos\theta}$ [see Figure 2(b)], and, from (4), $a^2 = \dfrac{e^2d^2}{(1 - e^2)^2} \quad\Rightarrow$

$ed = a(1 - e^2)$. Therefore, $r = \dfrac{a(1 - e^2)}{1 - e\cos\theta}$.

(b) $e = 0.017$ and the major axis $= 2a = 2.99 \times 10^8 \quad\Rightarrow\quad a = 1.495 \times 10^8$.

Therefore $r = \dfrac{1.495 \times 10^8\left[1 - (0.017)^2\right]}{1 - 0.017\cos\theta} \approx \dfrac{1.49 \times 10^8}{1 - 0.017\cos\theta}$.

27. Here $2a = $ length of major axis $= 36.18$ AU $\quad\Rightarrow\quad a = 18.09$ AU and $e = 0.97$. By Exercise 25(a), the equation

of the orbit is $r = \dfrac{18.09\left[1 - (0.97)^2\right]}{1 - 0.97\cos\theta} \approx \dfrac{1.07}{1 - 0.97\cos\theta}$. By Exercise 26(a), the maximum distance from the

comet to the sun is $18.09(1 + 0.97) \approx 35.64$ AU or about 3.314 billion miles.

29. The minimum distance is at perihelion, where

$$4.6 \times 10^7 = r = a(1 - e) = a(1 - 0.206) = a(0.794) \quad \Rightarrow \quad a = 4.6 \times 10^7/0.794.$$ So the maximum distance, which is at aphelion, is $r = a(1 + e) = (4.6 \times 10^7/0.794)(1.206) \approx 7.0 \times 10^7$ km.

31. From Exercise 29, we have $e = 0.206$ and $a(1 - e) = 4.6 \times 10^7$ km. Thus, $a = 4.6 \times 10^7/0.794$. From

Exercise 25, we can write the equation of Mercury's orbit as $r = a \dfrac{1 - e^2}{1 - e \cos \theta}$. So since

$$\frac{dr}{d\theta} = \frac{-a(1 - e^2)e \sin \theta}{(1 - e \cos \theta)^2} \quad \Rightarrow$$

$$r^2 + \left(\frac{dr}{d\theta}\right)^2 = \frac{a^2(1 - e^2)^2}{(1 - e \cos \theta)^2} + \frac{a^2(1 - e^2)^2 e^2 \sin^2 \theta}{(1 - e \cos \theta)^4} = \frac{a^2(1 - e^2)^2}{(1 - e \cos \theta)^4}(1 - 2e \cos \theta + e^2)$$

the length of the orbit is

$$L = \int_0^{2\pi} \sqrt{r^2 + (dr/d\theta)^2}\, d\theta = a(1 - e^2) \int_0^{2\pi} \frac{\sqrt{1 + e^2 - 2e \cos \theta}}{(1 - e \cos \theta)^2}\, d\theta \approx 3.6 \times 10^8 \text{ km}$$

This seems reasonable, since Mercury's orbit is nearly circular, and the circumference of a circle of radius a is $2\pi a \approx 3.6 \times 10^8$ km.

10 Review

CONCEPT CHECK

1. (a) A parametric curve is a set of points of the form $(x, y) = (f(t), g(t))$, where f and g are continuous functions of a variable t.

(b) Sketching a parametric curve, like sketching the graph of a function, is difficult to do in general. We can plot points on the curve by finding $f(t)$ and $g(t)$ for various values of t, either by hand or with a calculator or computer. Sometimes, when f and g are given by formulas, we can eliminate t from the equations $x = f(t)$ and $y = g(t)$ to get a Cartesian equation relating x and y. It may be easier to graph that equation than to work with the original formulas for x and y in terms of t.

2. (a) You can find $\dfrac{dy}{dx}$ as a function of t by calculating $\dfrac{dy}{dx} = \dfrac{dy/dt}{dx/dt}$ (if $dx/dt \neq 0$).

(b) Calculate the area as $\int_a^b y\, dx = \int_\alpha^\beta g(t) f'(t)\, dt$ [or $\int_\beta^\alpha g(t) f'(t)\, dt$ if the leftmost point is $(f(\beta), g(\beta))$ rather than $(f(\alpha), g(\alpha))$].

3. (a) $L = \int_\alpha^\beta \sqrt{(dx/dt)^2 + (dy/dt)^2}\, dt = \int_\alpha^\beta \sqrt{[f'(t)]^2 + [g'(t)]^2}\, dt$

(b) $S = \int_\alpha^\beta 2\pi y \sqrt{(dx/dt)^2 + (dy/dt)^2}\, dt = \int_\alpha^\beta 2\pi g(t) \sqrt{[f'(t)]^2 + [g'(t)]^2}\, dt$

4. (a) See Figure 5 in Section 10.3.

(b) $x = r \cos \theta$, $y = r \sin \theta$

(c) To find a polar representation (r, θ) with $r \geq 0$ and $0 \leq \theta < 2\pi$, first calculate $r = \sqrt{x^2 + y^2}$. Then θ is specified by $\cos \theta = x/r$ and $\sin \theta = y/r$.

5. (a) Calculate $\dfrac{dy}{dx} = \dfrac{\frac{dy}{d\theta}}{\frac{dx}{d\theta}} = \dfrac{\frac{d}{d\theta}(y)}{\frac{d}{d\theta}(x)} = \dfrac{\frac{d}{d\theta}(r\sin\theta)}{\frac{d}{d\theta}(r\cos\theta)} = \dfrac{\left(\frac{dr}{d\theta}\right)\sin\theta + r\cos\theta}{\left(\frac{dr}{d\theta}\right)\cos\theta - r\sin\theta}$, where $r = f(\theta)$.

(b) Calculate $A = \int_a^b \frac{1}{2}r^2\,d\theta = \int_a^b \frac{1}{2}[f(\theta)]^2\,d\theta$

(c) $L = \int_a^b \sqrt{(dx/d\theta)^2 + (dy/d\theta)^2}\,d\theta = \int_a^b \sqrt{r^2 + (dr/d\theta)^2}\,d\theta = \int_a^b \sqrt{[f(\theta)]^2 + [f'(\theta)]^2}\,d\theta$

6. (a) A parabola is a set of points in a plane whose distances from a fixed point F (the focus) and a fixed line l (the directrix) are equal.

(b) $x^2 = 4py$; $y^2 = 4px$

7. (a) An ellipse is a set of points in a plane the sum of whose distances from two fixed points (the foci) is a constant.

(b) $\dfrac{x^2}{a^2} + \dfrac{y^2}{a^2 - c^2} = 1$.

8. (a) A hyperbola is a set of points in a plane the difference of whose distances from two fixed points (the foci) is a constant. This difference should be interpreted as the larger distance minus the smaller distance.

(b) $\dfrac{x^2}{a^2} - \dfrac{y^2}{c^2 - a^2} = 1$

(c) $y = \pm\dfrac{\sqrt{c^2 - a^2}}{a}\,x$

9. (a) If a conic section has focus F and corresponding directrix l, then the eccentricity e is the fixed ratio $|PF|\,/\,|Pl|$ for points P of the conic section.

(b) $e < 1$ for an ellipse; $e > 1$ for a hyperbola; $e = 1$ for a parabola.

(c) $x = d$: $r = \dfrac{ed}{1 + e\cos\theta}$. $x = -d$: $r = \dfrac{ed}{1 - e\cos\theta}$. $y = d$: $r = \dfrac{ed}{1 + e\sin\theta}$. $y = -d$: $r = \dfrac{ed}{1 - e\sin\theta}$.

───────────────── TRUE-FALSE QUIZ ─────────────────

1. False. Consider the curve defined by $x = f(t) = (t - 1)^3$ and $y = g(t) = (t - 1)^2$. Then $g'(t) = 2(t - 1)$, so $g'(1) = 0$, but its graph has a *vertical* tangent when $t = 1$. *Note:* The statement is true if $f'(1) \neq 0$ when $g'(1) = 0$.

3. False. For example, if $f(t) = \cos t$ and $g(t) = \sin t$ for $0 \leq t \leq 4\pi$, then the curve is a circle of radius 1, hence its length is 2π, but

$$\int_0^{4\pi} \sqrt{[f'(t)]^2 + [g'(t)]^2}\,dt = \int_0^{4\pi} \sqrt{(-\sin t)^2 + (\cos t)^2}\,dt$$

$$= \int_0^{4\pi} 1\,dt = 4\pi,$$

since as t increases from 0 to 4π, the circle is traversed twice.

5. True. The curve $r = 1 - \sin 2\theta$ is unchanged if we rotate it through $180°$ about O because $1 - \sin 2(\theta + \pi) = 1 - \sin(2\theta + 2\pi) = 1 - \sin 2\theta$. So it's unchanged if we replace r by $-r$. (See the discussion after Example 8 in Section 10.3.) In other words, it's the same curve as $r = -(1 - \sin 2\theta) = \sin 2\theta - 1$.

7. False. The first pair of equations yields the portion of the parabola $y = x^2$ with $x \geq 0$, whereas the second pair of equations traces out the whole parabola $y = x^2$.

9. True. By rotating and translating the parabola, we can assume it has an equation of the form $y = cx^2$, where $c > 0$. The tangent at the point (a, ca^2) is the line $y - ca^2 = 2ca(x - a)$; i.e., $y = 2cax - ca^2$. This tangent meets the parabola at the points (x, cx^2) where $cx^2 = 2cax - ca^2$. This equation is equivalent to $x^2 = 2ax - a^2$ (since $c > 0$). But $x^2 = 2ax - a^2 \Leftrightarrow x^2 - 2ax + a^2 = 0 \Leftrightarrow (x - a)^2 = 0 \Leftrightarrow x = a \Leftrightarrow (x, cx^2) = (a, ca^2)$. This shows that each tangent meets the parabola at exactly one point.

─────────────────────── EXERCISES ───────────────────────

1. $x = t^2 + 4t$, $y = 2 - t$, $-4 \leq t \leq 1$. $t = 2 - y$, so
$x = (2 - y)^2 + 4(2 - y) = 4 - 4y + y^2 + 8 - 4y = y^2 - 8y + 12 \Leftrightarrow$
$x + 4 = y^2 - 8y + 16 = (y - 4)^2$. This is part of a parabola with vertex
$(-4, 4)$, opening to the right.

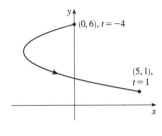

3. $x = \tan\theta$, $y = \cot\theta$. $y = 1/\tan\theta = 1/x$. The whole curve is traced out as θ ranges over the open interval $\left(-\frac{\pi}{2}, \frac{\pi}{2}\right)$ [or any open interval of the form $\left(-\frac{\pi}{2} + n\pi, \frac{\pi}{2} + n\pi\right)$, where n is an integer].

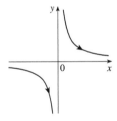

5. Three different sets of parametric equations for the curve $y = \sqrt{x}$ are

(i) $x = t$, $y = \sqrt{t}$, $t \geq 0$

(ii) $x = t^4$, $y = t^2$

(iii) $x = \tan^2 t$, $y = \tan t$, $0 \leq t < \pi/2$
   There are many other sets of equations that also give this curve.

7. $r = 1 - \cos\theta$. This cardiod is symmetric about the polar axis.

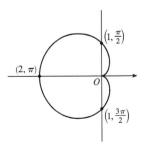

9. $r = 1 + \cos 2\theta$. The curve is symmetric about the pole and both the horizontal and vertical axes.

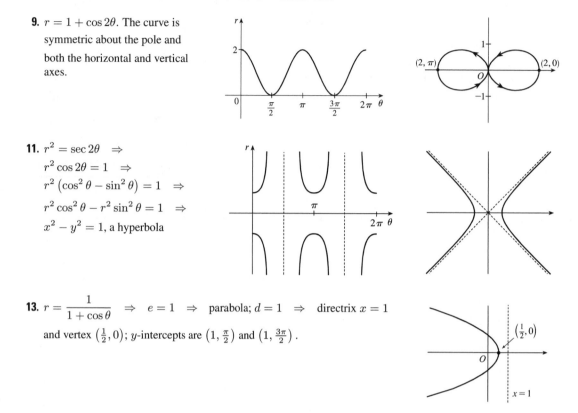

11. $r^2 = \sec 2\theta \quad \Rightarrow$
$r^2 \cos 2\theta = 1 \quad \Rightarrow$
$r^2 (\cos^2 \theta - \sin^2 \theta) = 1 \quad \Rightarrow$
$r^2 \cos^2 \theta - r^2 \sin^2 \theta = 1 \quad \Rightarrow$
$x^2 - y^2 = 1$, a hyperbola

13. $r = \dfrac{1}{1 + \cos \theta} \quad \Rightarrow \quad e = 1 \quad \Rightarrow \quad$ parabola; $d = 1 \quad \Rightarrow \quad$ directrix $x = 1$

and vertex $\left(\frac{1}{2}, 0\right)$; y-intercepts are $\left(1, \frac{\pi}{2}\right)$ and $\left(1, \frac{3\pi}{2}\right)$.

15. $x + y = 2 \quad \Leftrightarrow \quad r \cos \theta + r \sin \theta = 2 \quad \Leftrightarrow \quad r(\cos \theta + \sin \theta) = 2 \quad \Leftrightarrow \quad r = \dfrac{2}{\cos \theta + \sin \theta}$

17. $r = (\sin \theta)/\theta$. As $\theta \to \pm\infty$, $r \to 0$. As $\theta \to 0$, $r \to 1$. In the first figure, there are an infinite number of x-intercepts at $x = \pi n$, n a nonzero integer. These correspond to pole points in the second figure.

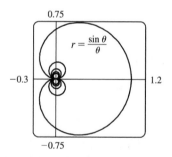

19. $x = \ln t$, $y = 1 + t^2$; $t = 1$. $\dfrac{dy}{dt} = 2t$ and $\dfrac{dx}{dt} = \dfrac{1}{t}$, so $\dfrac{dy}{dx} = \dfrac{dy/dt}{dx/dt} = \dfrac{2t}{1/t} = 2t^2$. When $t = 1$, $(x, y) = (0, 2)$

and $dy/dx = 2$.

21. $r = e^{-\theta} \quad \Rightarrow \quad y = r \sin \theta = e^{-\theta} \sin \theta$ and $x = r \cos \theta = e^{-\theta} \cos \theta \quad \Rightarrow$

$\dfrac{dy}{dx} = \dfrac{dy/d\theta}{dx/d\theta} = \dfrac{\frac{dr}{d\theta} \sin \theta + r \cos \theta}{\frac{dr}{d\theta} \cos \theta - r \sin \theta} = \dfrac{-e^{-\theta} \sin \theta + e^{-\theta} \cos \theta}{-e^{-\theta} \cos \theta - e^{-\theta} \sin \theta} \cdot \dfrac{-e^{\theta}}{-e^{\theta}} = \dfrac{\sin \theta - \cos \theta}{\cos \theta + \sin \theta}$. When $\theta = \pi$,

$\dfrac{dy}{dx} = \dfrac{0 - (-1)}{-1 + 0} = \dfrac{1}{-1} = -1$.

23. $x = t \cos t$, $y = t \sin t$. $\dfrac{dy}{dx} = \dfrac{dy/dt}{dx/dt} = \dfrac{t \cos t + \sin t}{-t \sin t + \cos t}$. $\dfrac{d^2y}{dx^2} = \dfrac{\frac{d}{dt}\left(\frac{dy}{dx}\right)}{dx/dt}$, where

$$\dfrac{d}{dt}\left(\dfrac{dy}{dx}\right) = \dfrac{(-t \sin t + \cos t)(-t \sin t + 2 \cos t) - (t \cos t + \sin t)(-t \cos t - 2 \sin t)}{(-t \sin t + \cos t)^2}$$

$$= \dfrac{t^2 + 2}{(-t \sin t + \cos t)^2} \quad \Rightarrow \quad \dfrac{d^2y}{dx^2} = \dfrac{t^2 + 2}{(-t \sin t + \cos t)^3}.$$

25. We graph the curve $x = t^3 - 3t$, $y = t^2 + t + 1$ for $-2.2 \le t \le 1.2$. By

zooming in or using a cursor, we find that the lowest point is about

$(1.4, 0.75)$. To find the exact values, we find the t-value at which

$dy/dt = 2t + 1 = 0 \ \Leftrightarrow \ t = -\frac{1}{2} \ \Leftrightarrow \ (x, y) = \left(\frac{11}{8}, \frac{3}{4}\right)$.

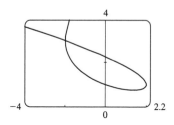

27. $x = 2a \cos t - a \cos 2t \ \Rightarrow \ \dfrac{dx}{dt} = -2a \sin t + 2a \sin 2t = 2a \sin t(2 \cos t - 1) = 0 \ \Leftrightarrow$

$\sin t = 0$ or $\cos t = \frac{1}{2} \ \Rightarrow \ t = 0, \frac{\pi}{3}, \pi$, or $\frac{5\pi}{3}$.

$y = 2a \sin t - a \sin 2t \ \Rightarrow$

$\dfrac{dy}{dt} = 2a \cos t - 2a \cos 2t = 2a\left(1 + \cos t - 2 \cos^2 t\right) = 2a(1 - \cos t)(1 + 2 \cos t) = 0 \ \Rightarrow \ t = 0, \frac{2\pi}{3}$, or $\frac{4\pi}{3}$.

Thus the graph has vertical tangents where

$t = \frac{\pi}{3}, \pi$ and $\frac{5\pi}{3}$, and horizontal tangents where

$t = \frac{2\pi}{3}$ and $\frac{4\pi}{3}$. To determine what the slope is

where $t = 0$, we use l'Hospital's Rule to evaluate

$\lim\limits_{t \to 0} \dfrac{dy/dt}{dx/dt} = 0$, so there is a horizontal tangent

there.

t	x	y
0	a	0
$\frac{\pi}{3}$	$\frac{3}{2}a$	$\frac{\sqrt{3}}{2}a$
$\frac{2\pi}{3}$	$-\frac{1}{2}a$	$\frac{3\sqrt{3}}{2}a$
π	$-3a$	0
$\frac{4\pi}{3}$	$-\frac{1}{2}a$	$-\frac{3\sqrt{3}}{2}a$
$\frac{5\pi}{3}$	$\frac{3}{2}a$	$-\frac{\sqrt{3}}{2}a$

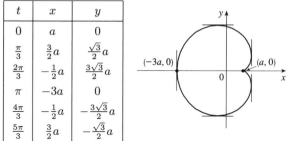

29. The curve $r^2 = 9 \cos 5\theta$ has 10 "petals." For instance, for $-\frac{\pi}{10} \le \theta \le \frac{\pi}{10}$, there are two petals, one with $r > 0$ and one with $r < 0$.

$$A = 10 \int_{-\pi/10}^{\pi/10} \frac{1}{2}r^2 \, d\theta = 5 \int_{-\pi/10}^{\pi/10} 9 \cos 5\theta \, d\theta = 5 \cdot 9 \cdot 2 \int_0^{\pi/10} \cos 5\theta \, d\theta = 18 \left[\sin 5\theta\right]_0^{\pi/10} = 18$$

31. The curves intersect when $4 \cos \theta = 2 \ \Rightarrow \ \cos \theta = \frac{1}{2} \ \Rightarrow \ \theta = \pm\frac{\pi}{3}$

for $-\pi \le \theta \le \pi$. The points of intersection are $\left(2, \frac{\pi}{3}\right)$ and $\left(2, -\frac{\pi}{3}\right)$.

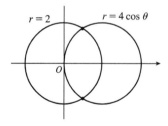

33. The curves intersect where $2 \sin \theta = \sin \theta + \cos \theta \Rightarrow$

$\sin \theta = \cos \theta \Rightarrow \theta = \frac{\pi}{4}$, and also at the origin (at which $\theta = \frac{3\pi}{4}$ on the

second curve).

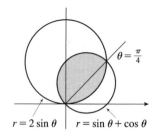

$r = 2 \sin \theta \quad | \quad r = \sin \theta + \cos \theta$

$$A = \int_0^{\pi/4} \tfrac{1}{2}(2\sin\theta)^2 \, d\theta + \int_{\pi/4}^{3\pi/4} \tfrac{1}{2}(\sin\theta + \cos\theta)^2 \, d\theta$$

$$= \int_0^{\pi/4} (1 - \cos 2\theta) \, d\theta + \tfrac{1}{2}\int_{\pi/4}^{3\pi/4}(1 + \sin 2\theta) \, d\theta$$

$$= \left[\theta - \tfrac{1}{2}\sin 2\theta\right]_0^{\pi/4} + \left[\tfrac{1}{2}\theta - \tfrac{1}{4}\cos 2\theta\right]_{\pi/4}^{3\pi/4} = \tfrac{1}{2}(\pi - 1)$$

35. $x = 3t^2$, $y = 2t^3$.

$$L = \int_0^2 \sqrt{(dx/dt)^2 + (dy/dt)^2} \, dt = \int_0^2 \sqrt{(6t)^2 + (6t^2)^2} \, dt = \int_0^2 \sqrt{36t^2 + 36t^4} \, dt$$

$$= \int_0^2 \sqrt{36t^2}\sqrt{1+t^2} \, dt = \int_0^2 6\,|t|\,\sqrt{1+t^2} \, dt = 6\int_0^2 t\sqrt{1+t^2} \, dt$$

$$= 6\int_1^5 u^{1/2}\left(\tfrac{1}{2}du\right) \quad \left[u = 1 + t^2, \, du = 2t\,dt\right]$$

$$= 6 \cdot \tfrac{1}{2} \cdot \tfrac{2}{3}\left[u^{3/2}\right]_1^5 = 2\left(5^{3/2} - 1\right) = 2\left(5\sqrt{5} - 1\right)$$

37. $L = \int_\pi^{2\pi} \sqrt{r^2 + (dr/d\theta)^2} \, d\theta = \int_\pi^{2\pi}\sqrt{(1/\theta)^2 + (-1/\theta^2)^2} \, d\theta = \int_\pi^{2\pi} \dfrac{\sqrt{\theta^2 + 1}}{\theta^2} \, d\theta$

$$\overset{24}{=} \left[-\frac{\sqrt{\theta^2+1}}{\theta} + \ln\left(\theta + \sqrt{\theta^2+1}\right)\right]_\pi^{2\pi} = \frac{\sqrt{\pi^2+1}}{\pi} - \frac{\sqrt{4\pi^2+1}}{2\pi} + \ln\left(\frac{2\pi + \sqrt{4\pi^2+1}}{\pi + \sqrt{\pi^2+1}}\right)$$

$$= \frac{2\sqrt{\pi^2+1} - \sqrt{4\pi^2+1}}{2\pi} + \ln\left(\frac{2\pi + \sqrt{4\pi^2+1}}{\pi + \sqrt{\pi^2+1}}\right)$$

39. $x = 4\sqrt{t}$, $y = \dfrac{t^3}{3} + \dfrac{1}{2t^2}$, $1 \le t \le 4 \Rightarrow$

$S = \int_1^4 2\pi y\sqrt{(dx/dt)^2 + (dy/dt)^2} \, dt = \int_1^4 2\pi\left(\tfrac{1}{3}t^3 + \tfrac{1}{2}t^{-2}\right)\sqrt{\left(2/\sqrt{t}\right)^2 + (t^2 - t^{-3})^2} \, dt$

$= 2\pi\int_1^4\left(\tfrac{1}{3}t^3 + \tfrac{1}{2}t^{-2}\right)\sqrt{(t^2 + t^{-3})^2} \, dt = 2\pi\int_1^4\left(\tfrac{1}{3}t^5 + \tfrac{5}{6} + \tfrac{1}{2}t^{-5}\right) dt = 2\pi\left[\tfrac{1}{18}t^6 + \tfrac{5}{6}t - \tfrac{1}{8}t^{-4}\right]_1^4 = \frac{471{,}295}{1024}\pi$

41. For all c except -1, the curve is asymptotic to the line $x = 1$. For

$c < -1$, the curve bulges to the right near $y = 0$. As c increases, the

bulge becomes smaller, until at $c = -1$ the curve is the straight line $x = 1$.

As c continues to increase, the curve bulges to the left, until at $c = 0$ there

is a cusp at the origin. For $c > 0$, there is a loop to the left of the origin,

whose size and roundness increase as c increases. Note that the x-intercept

of the curve is always $-c$.

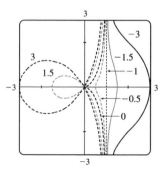

43. $\dfrac{x^2}{9} + \dfrac{y^2}{8} = 1$ is an ellipse with center $(0,0)$.

$a = 3, b = 2\sqrt{2}, c = 1 \;\Rightarrow$

foci $(\pm 1, 0)$, vertices $(\pm 3, 0)$.

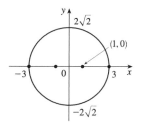

45. $6y^2 + x - 36y + 55 = 0 \;\Leftrightarrow$

$6(y^2 - 6y + 9) = -(x+1) \;\Leftrightarrow$

$(y-3)^2 = -\frac{1}{6}(x+1)$, a parabola with vertex

$(-1, 3)$, opening to the left, $p = -\frac{1}{24} \;\Rightarrow$ focus

$\left(-\frac{25}{24}, 3\right)$ and directrix $x = -\frac{23}{24}$.

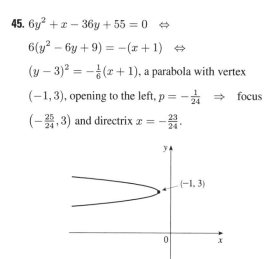

47. The parabola opens upward with vertex $(0, 4)$ [midway between the focus $(0, 6)$ and the directrix $y = 2$] and $p = 2$,

so its equation is $(x - 0)^2 = 4 \cdot 2(y - 4) \;\Leftrightarrow\; x^2 = 8(y - 4)$.

49. The hyperbola has center $(0, 0)$ and foci on the x-axis. $c = 3$ and $b/a = \frac{1}{2}$ (from the asymptotes)

$\Rightarrow \quad 9 = c^2 = a^2 + b^2 = (2b)^2 + b^2 = 5b^2 \;\Rightarrow\; b = \frac{3}{\sqrt{5}} \;\Rightarrow\; a = \frac{6}{\sqrt{5}} \;\Rightarrow$ an equation of the hyperbola is

$\dfrac{x^2}{36/5} - \dfrac{y^2}{9/5} = 1 \;\Leftrightarrow\; 5x^2 - 20y^2 = 36.$

51. $x^2 = -(y - 100)$ has its vertex at $(0, 100)$, so one of the vertices of the ellipse is $(0, 100)$. Another form of the

equation of a parabola is $x^2 = 4p(y - 100)$ so $4p(y - 100) = -(y - 100) \;\Rightarrow\; 4p = -1 \;\Rightarrow\; p = -\frac{1}{4}$.

Therefore the shared focus is found at $\left(0, \frac{399}{4}\right)$ so $2c = \frac{399}{4} - 0 \;\Rightarrow\; c = \frac{399}{8}$ and the center of the ellipse is

$\left(0, \frac{399}{8}\right)$. So $a = 100 - \frac{399}{8} = \frac{401}{8}$ and $b^2 = a^2 - c^2 = \frac{401^2 - 399^2}{8^2} = 25$. So the equation of the ellipse is

$\dfrac{x^2}{b^2} + \dfrac{\left(y - \frac{399}{8}\right)^2}{a^2} = 1 \;\Rightarrow\; \dfrac{x^2}{25} + \dfrac{\left(y - \frac{399}{8}\right)^2}{\left(\frac{401}{8}\right)^2} = 1$ or $\dfrac{x^2}{25} + \dfrac{(8y - 399)^2}{160{,}801} = 1.$

53. Directrix $x = 4 \;\Rightarrow\; d = 4$, so $e = \frac{1}{3} \;\Rightarrow\; r = \dfrac{ed}{1 + e\cos\theta} = \dfrac{4}{3 + \cos\theta}.$

55. In polar coordinates, an equation for the circle is $r = 2a\sin\theta$. Thus, the coordinates of Q are

$x = r\cos\theta = 2a\sin\theta\cos\theta$ and $y = r\sin\theta = 2a\sin^2\theta$. The coordinates of R are $x = 2a\cot\theta$ and $y = 2a$.

Since P is the midpoint of QR, we use the midpoint formula to get $x = a(\sin\theta\cos\theta + \cot\theta)$ and

$y = a\left(1 + \sin^2\theta\right).$

☐ PROBLEMS PLUS

1. $x = \displaystyle\int_1^t \frac{\cos u}{u}\, du$, $y = \displaystyle\int_1^t \frac{\sin u}{u}\, du$, so by FTC1, we have $\dfrac{dx}{dt} = \dfrac{\cos t}{t}$ and $\dfrac{dy}{dt} = \dfrac{\sin t}{t}$. Vertical tangent lines

occur when $\dfrac{dx}{dt} = 0 \iff \cos t = 0$. The parameter value corresponding to $(x, y) = (0, 0)$ is $t = 1$, so the

nearest vertical tangent occurs when $t = \frac{\pi}{2}$. Therefore, the arc length between these points is

$$L = \int_1^{\pi/2} \sqrt{\left(\frac{dx}{dt}\right)^2 + \left(\frac{dy}{dt}\right)^2}\, dt = \int_1^{\pi/2} \sqrt{\frac{\cos^2 t}{t^2} + \frac{\sin^2 t}{t^2}}\, dt = \int_1^{\pi/2} \frac{dt}{t} = [\ln t]_1^{\pi/2} = \ln \frac{\pi}{2}$$

3. In terms of x and y, we have $x = r \cos\theta = (1 + c\sin\theta)\cos\theta = \cos\theta + c\sin\theta\cos\theta = \cos\theta + \frac{1}{2}c\sin 2\theta$ and

$y = r\sin\theta = (1 + c\sin\theta)\sin\theta = \sin\theta + c\sin^2\theta$. Now $-1 \le \sin\theta \le 1 \implies$

$-1 \le \sin\theta + c\sin^2\theta \le 1 + c \le 2$, so $-1 \le y \le 2$. Furthermore, $y = 2$ when $c = 1$ and $\theta = \frac{\pi}{2}$, while $y = -1$

for $c = 0$ and $\theta = \frac{3\pi}{2}$. Therefore, we need a viewing rectangle with $-1 \le y \le 2$.

To find the x-values, look at the equation $x = \cos\theta + \frac{1}{2}c\sin 2\theta$ and use the fact that $\sin 2\theta \ge 0$ for $0 \le \theta \le \frac{\pi}{2}$

and $\sin 2\theta \le 0$ for $-\frac{\pi}{2} \le \theta \le 0$. [Because $r = 1 + c\sin\theta$ is symmetric about the y-axis, we only need to consider

$-\frac{\pi}{2} \le \theta \le \frac{\pi}{2}$.] So for $-\frac{\pi}{2} \le \theta \le 0$, x has a maximum value when $c = 0$ and then $x = \cos\theta$ has a maximum value

of 1 at $\theta = 0$. Thus, the maximum value of x must occur on $\left[0, \frac{\pi}{2}\right]$ with $c = 1$. Then $x = \cos\theta + \frac{1}{2}\sin 2\theta \implies$

$\frac{dx}{d\theta} = -\sin\theta + \cos 2\theta = -\sin\theta + 1 - 2\sin^2\theta \implies \frac{dx}{d\theta} = -(2\sin\theta - 1)(\sin\theta + 1) = 0$ when $\sin\theta = -1$ or $\frac{1}{2}$

(but $\sin\theta \ne -1$ for $0 \le \theta \le \frac{\pi}{2}$). If $\sin\theta = \frac{1}{2}$, then $\theta = \frac{\pi}{6}$ and

$x = \cos\frac{\pi}{6} + \frac{1}{2}\sin\frac{\pi}{3} = \frac{3}{4}\sqrt{3}$. Thus, the maximum value of x is

$\frac{3}{4}\sqrt{3}$, and, by symmetry, the minimum value is $-\frac{3}{4}\sqrt{3}$. Therefore,

the smallest viewing rectangle that contains every member of the

family of polar curves $r = 1 + c\sin\theta$, where $0 \le c \le 1$, is

$\left[-\frac{3}{4}\sqrt{3}, \frac{3}{4}\sqrt{3}\right] \times [-1, 2]$.

5. (a) If (a, b) lies on the curve, then there is some parameter value t_1 such that $\dfrac{3t_1}{1 + t_1^3} = a$ and $\dfrac{3t_1^2}{1 + t_1^3} = b$. If

$t_1 = 0$, the point is $(0, 0)$, which lies on the line $y = x$. If $t_1 \ne 0$, then the point corresponding to $t = \dfrac{1}{t_1}$ is

given by $x = \dfrac{3(1/t_1)}{1 + (1/t_1)^3} = \dfrac{3t_1^2}{t_1^3 + 1} = b$, $y = \dfrac{3(1/t_1)^2}{1 + (1/t_1)^3} = \dfrac{3t_1}{t_1^3 + 1} = a$. So (b, a) also lies on the curve.

[Another way to see this is to do part (e) first; the result is immediate.] The curve intersects the line $y = x$ when

$\dfrac{3t}{1 + t^3} = \dfrac{3t^2}{1 + t^3} \implies t = t^2 \implies t = 0$ or 1, so the points are $(0, 0)$ and $\left(\frac{3}{2}, \frac{3}{2}\right)$.

(b) $\dfrac{dy}{dt} = \dfrac{(1+t^3)(6t) - 3t^2(3t^2)}{(1+t^3)^2} = \dfrac{6t - 3t^4}{(1+t^3)^2} = 0$ when $6t - 3t^4 = 3t(2 - t^3) = 0 \;\Rightarrow\; t = 0$ or $t = \sqrt[3]{2}$,

so there are horizontal tangents at $(0,0)$ and $(\sqrt[3]{2}, \sqrt[3]{4})$. Using the symmetry from part (a), we see that there are vertical tangents at $(0,0)$ and $(\sqrt[3]{4}, \sqrt[3]{2})$.

(c) Notice that as $t \to -1^+$, we have $x \to -\infty$ and $y \to \infty$. As $t \to -1^-$, we have $x \to \infty$ and $y \to -\infty$. Also

$$y - (-x - 1) = y + x + 1 = \frac{3t + 3t^2 + (1 + t^3)}{1 + t^3} = \frac{(t+1)^3}{1+t^3} = \frac{(t+1)^2}{t^2 - t + 1} \to 0 \text{ as } t \to -1. \text{ So}$$

$y = -x - 1$ is a slant asymptote.

(d) $\dfrac{dx}{dt} = \dfrac{(1+t^3)(3) - 3t(3t^2)}{(1+t^3)^2} = \dfrac{3 - 6t^3}{(1+t^3)^2}$ and from part (b) we have $\dfrac{dy}{dt} = \dfrac{6t - 3t^4}{(1+t^3)^2}$.

So $\dfrac{dy}{dx} = \dfrac{dy/dt}{dx/dt} = \dfrac{t(2 - t^3)}{1 - 2t^3}$. Also

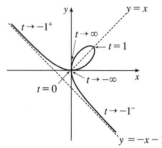

$$\frac{d^2y}{dx^2} = \frac{\dfrac{d}{dt}\left(\dfrac{dy}{dx}\right)}{dx/dt} = \frac{2(1+t^3)^4}{3(1 - 2t^3)^3} > 0 \;\Leftrightarrow\; t < \tfrac{1}{\sqrt[3]{2}}. \text{ So}$$

the curve is concave upward there and has a minimum point at $(0,0)$ and a maximum point at $(\sqrt[3]{2}, \sqrt[3]{4})$. Using this together with the information from parts (a), (b), and (c), we sketch the curve.

(e) $x^3 + y^3 = \left(\dfrac{3t}{1+t^3}\right)^3 + \left(\dfrac{3t^2}{1+t^3}\right)^3 = \dfrac{27t^3 + 27t^6}{(1+t^3)^3} = \dfrac{27t^3(1+t^3)}{(1+t^3)^3} = \dfrac{27t^3}{(1+t^3)^2}$ and

$3xy = 3\left(\dfrac{3t}{1+t^3}\right)\left(\dfrac{3t^2}{1+t^3}\right) = \dfrac{27t^3}{(1+t^3)^2}$, so $x^3 + y^3 = 3xy$.

(f) We start with the equation from part (e) and substitute $x = r\cos\theta$, $y = r\sin\theta$. Then $x^3 + y^3 = 3xy \;\Rightarrow$

$r^3 \cos^3\theta + r^3 \sin^3\theta = 3r^2 \cos\theta \sin\theta$. For $r \neq 0$, this gives $r = \dfrac{3\cos\theta \sin\theta}{\cos^3\theta + \sin^3\theta}$. Dividing numerator and

denominator by $\cos^3\theta$, we obtain $r = \dfrac{3\left(\dfrac{1}{\cos\theta}\right)\dfrac{\sin\theta}{\cos\theta}}{1 + \dfrac{\sin^3\theta}{\cos^3\theta}} = \dfrac{3\sec\theta\tan\theta}{1 + \tan^3\theta}$.

(g) The loop corresponds to $\theta \in \left(0, \tfrac{\pi}{2}\right)$, so its area is

$$A = \int_0^{\pi/2} \frac{r^2}{2}\,d\theta = \frac{1}{2}\int_0^{\pi/2}\left(\frac{3\sec\theta\tan\theta}{1 + \tan^3\theta}\right)^2 d\theta = \frac{9}{2}\int_0^{\pi/2}\frac{\sec^2\theta\tan^2\theta}{(1 + \tan^3\theta)^2}\,d\theta$$

$$= \frac{9}{2}\int_0^{\infty}\frac{u^2\,du}{(1 + u^3)^2} \; \text{ [let } u = \tan\theta] \; = \lim_{b\to\infty}\frac{9}{2}\left[-\frac{1}{3}(1 + u^3)^{-1}\right]_0^b = \frac{3}{2}$$

(h) By symmetry, the area between the folium and the line $y = -x - 1$ is equal to the enclosed area in the third quadrant, plus twice the enclosed area in the fourth quadrant. The area in the third quadrant is $\tfrac{1}{2}$, and since

$y = -x - 1 \;\Rightarrow\; r\sin\theta = -r\cos\theta - 1 \;\Rightarrow\; r = -\dfrac{1}{\sin\theta + \cos\theta}$, the area in the fourth quadrant is

$$\frac{1}{2}\int_{-\pi/2}^{-\pi/4}\left[\left(-\frac{1}{\sin\theta + \cos\theta}\right)^2 - \left(\frac{3\sec\theta\tan\theta}{1 + \tan^3\theta}\right)^2\right] d\theta \overset{\text{CAS}}{=} \frac{1}{2}. \text{ Therefore, the total area is } \tfrac{1}{2} + 2\left(\tfrac{1}{2}\right) = \tfrac{3}{2}.$$

11 □ INFINITE SEQUENCES AND SERIES

11.1 Sequences

1. (a) A sequence is an ordered list of numbers. It can also be defined as a function whose domain is the set of positive integers.

   (b) The terms a_n approach 8 as n becomes large. In fact, we can make a_n as close to 8 as we like by taking n sufficiently large.

   (c) The terms a_n become large as n becomes large. In fact, we can make a_n as large as we like by taking n sufficiently large.

3. $a_n = 1 - (0.2)^n$, so the sequence is $\{0.8, 0.96, 0.992, 0.9984, 0.99968, \ldots\}$.

5. $a_n = \dfrac{3(-1)^n}{n!}$, so the sequence is $\left\{ \dfrac{-3}{1}, \dfrac{3}{2}, \dfrac{-3}{6}, \dfrac{3}{24}, \dfrac{-3}{120}, \ldots \right\} = \left\{ -3, \dfrac{3}{2}, -\dfrac{1}{2}, \dfrac{1}{8}, -\dfrac{1}{40}, \ldots \right\}$.

7. $a_1 = 3$, $a_{n+1} = 2a_n - 1$. Each term is defined in terms of the preceding term.
   $a_2 = 2a_1 - 1 = 2(3) - 1 = 5$. $a_3 = 2a_2 - 1 = 2(5) - 1 = 9$. $a_4 = 2a_3 - 1 = 2(9) - 1 = 17$.
   $a_5 = 2a_4 - 1 = 2(17) - 1 = 33$. The sequence is $\{3, 5, 9, 17, 33, \ldots\}$.

9. The numerators are all 1 and the denominators are powers of 2, so $a_n = \dfrac{1}{2^n}$.

11. $\{2, 7, 12, 17, \ldots\}$. Each term is larger than the preceding one by 5, so
    $a_n = a_1 + d(n - 1) = 2 + 5(n - 1) = 5n - 3$.

13. $\left\{ 1, -\dfrac{2}{3}, \dfrac{4}{9}, -\dfrac{8}{27}, \ldots \right\}$. Each term is $-\dfrac{2}{3}$ times the preceding one, so $a_n = \left(-\dfrac{2}{3} \right)^{n-1}$.

15. $a_n = n(n - 1)$. $a_n \to \infty$ as $n \to \infty$, so the sequence diverges.

17. $a_n = \dfrac{3 + 5n^2}{n + n^2} = \dfrac{(3 + 5n^2)/n^2}{(n + n^2)/n^2} = \dfrac{5 + 3/n^2}{1 + 1/n}$, so $a_n \to \dfrac{5 + 0}{1 + 0} = 5$ as $n \to \infty$. Converges

19. $a_n = \dfrac{2^n}{3^{n+1}} = \dfrac{1}{3} \left(\dfrac{2}{3} \right)^n$, so $\lim\limits_{n \to \infty} a_n = \dfrac{1}{3} \lim\limits_{n \to \infty} \left(\dfrac{2}{3} \right)^n = \dfrac{1}{3} \cdot 0 = 0$ by (8) with $r = \dfrac{2}{3}$. Converges

21. $a_n = \dfrac{(-1)^{n-1} n}{n^2 + 1} = \dfrac{(-1)^{n-1}}{n + 1/n}$, so $0 \le |a_n| = \dfrac{1}{n + 1/n} \le \dfrac{1}{n} \to 0$ as $n \to \infty$, so $a_n \to 0$ by the Squeeze
    Theorem and Theorem 6. Converges

23. $a_n = \cos(n/2)$. This sequence diverges since the terms don't approach any particular real number as $n \to \infty$. The terms take on values between -1 and 1.

25. $a_n = \dfrac{(2n - 1)!}{(2n + 1)!} = \dfrac{(2n - 1)!}{(2n + 1)(2n)(2n - 1)!} = \dfrac{1}{(2n + 1)(2n)} \to 0$ as $n \to \infty$. Converges

27. $a_n = \dfrac{e^n + e^{-n}}{e^{2n} - 1} \cdot \dfrac{e^{-n}}{e^{-n}} = \dfrac{1 + e^{-2n}}{e^n - e^{-n}} \to \dfrac{1 + 0}{e^n - 0} \to 0$ as $n \to \infty$. Converges

29. $a_n = n^2 e^{-n} = \dfrac{n^2}{e^n}$. Since $\lim\limits_{x \to \infty} \dfrac{x^2}{e^x} \overset{\text{H}}{=} \lim\limits_{x \to \infty} \dfrac{2x}{e^x} \overset{\text{H}}{=} \lim\limits_{x \to \infty} \dfrac{2}{e^x} = 0$, it follows from Theorem 3 that $\lim\limits_{n \to \infty} a_n = 0$.
    Converges

31. $0 \le \dfrac{\cos^2 n}{2^n} \le \dfrac{1}{2^n}$ [since $0 \le \cos^2 n \le 1$], so since $\lim\limits_{n \to \infty} \dfrac{1}{2^n} = 0$, $\left\{ \dfrac{\cos^2 n}{2^n} \right\}$ converges to 0 by the Squeeze
    Theorem.

33. $a_n = n\sin(1/n) = \dfrac{\sin(1/n)}{1/n}$. Since $\lim\limits_{x\to\infty}\dfrac{\sin(1/x)}{1/x} = \lim\limits_{t\to 0+}\dfrac{\sin t}{t}$ [where $t = 1/x$] $= 1$, it follows from

Theorem 3 that $\{a_n\}$ converges to 1.

35. $a_n = \left(1 + \dfrac{2}{n}\right)^{1/n} \Rightarrow \ln a_n = \dfrac{1}{n}\ln\left(1 + \dfrac{2}{n}\right)$. As $n \to \infty$, $\dfrac{1}{n} \to 0$ and $\ln\left(1 + \dfrac{2}{n}\right) \to 0$, so $\ln a_n \to 0$.

Thus, $a_n \to e^0 = 1$ as $n \to \infty$. Converges

37. $\{0, 1, 0, 0, 1, 0, 0, 0, 1, \dots\}$ diverges since the sequence takes on only two values, 0 and 1, and never stays arbitrarily close to either one (or any other value) for n sufficiently large.

39. $a_n = \dfrac{n!}{2^n} = \dfrac{1}{2}\cdot\dfrac{2}{2}\cdot\dfrac{3}{2}\cdot\ \cdots\ \cdot\dfrac{(n-1)}{2}\cdot\dfrac{n}{2} \geq \dfrac{1}{2}\cdot\dfrac{n}{2}$ [for $n > 1$] $= \dfrac{n}{4} \to \infty$ as $n \to \infty$, so $\{a_n\}$ diverges.

41.

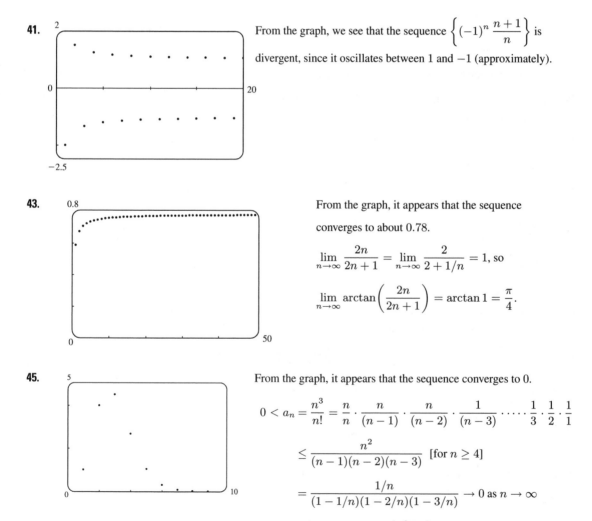

From the graph, we see that the sequence $\left\{(-1)^n\,\dfrac{n+1}{n}\right\}$ is

divergent, since it oscillates between 1 and -1 (approximately).

43.

From the graph, it appears that the sequence converges to about 0.78.

$$\lim_{n\to\infty}\frac{2n}{2n+1} = \lim_{n\to\infty}\frac{2}{2+1/n} = 1,\text{ so}$$

$$\lim_{n\to\infty}\arctan\left(\frac{2n}{2n+1}\right) = \arctan 1 = \frac{\pi}{4}.$$

45.

From the graph, it appears that the sequence converges to 0.

$$0 < a_n = \frac{n^3}{n!} = \frac{n}{n}\cdot\frac{n}{(n-1)}\cdot\frac{n}{(n-2)}\cdot\frac{1}{(n-3)}\ \cdots\ \frac{1}{3}\cdot\frac{1}{2}\cdot\frac{1}{1}$$

$$\leq \frac{n^2}{(n-1)(n-2)(n-3)}\quad\text{[for } n \geq 4\text{]}$$

$$= \frac{1/n}{(1-1/n)(1-2/n)(1-3/n)} \to 0\text{ as } n \to \infty$$

So by the Squeeze Theorem, $\{n^3/n!\}$ converges to 0.

47.

From the graph, it appears that the sequence approaches 0.

$$0 < a_n = \frac{1 \cdot 3 \cdot 5 \cdot \cdots \cdot (2n-1)}{(2n)^n} = \frac{1}{2n} \cdot \frac{3}{2n} \cdot \frac{5}{2n} \cdot \cdots \cdot \frac{2n-1}{2n}$$

$$\leq \frac{1}{2n} \cdot (1) \cdot (1) \cdot \cdots \cdot (1) = \frac{1}{2n} \to 0 \text{ as } n \to \infty$$

So by the Squeeze Theorem, $\left\{ \dfrac{1 \cdot 3 \cdot 5 \cdot \cdots \cdot (2n-1)}{(2n)^n} \right\}$ converges

to 0.

49. (a) $a_n = 1000(1.06)^n$ ⇒ $a_1 = 1060$, $a_2 = 1123.60$, $a_3 = 1191.02$, $a_4 = 1262.48$, and $a_5 = 1338.23$.

(b) $\lim\limits_{n \to \infty} a_n = 1000 \lim\limits_{n \to \infty} (1.06)^n$, so the sequence diverges by (8) with $r = 1.06 > 1$.

51. If $|r| \geq 1$, then $\{r^n\}$ diverges by (8), so $\{nr^n\}$ diverges also, since $|nr^n| = n |r^n| \geq |r^n|$. If $|r| < 1$ then

$$\lim_{x \to \infty} xr^x = \lim_{x \to \infty} \frac{x}{r^{-x}} \overset{\text{H}}{=} \lim_{x \to \infty} \frac{1}{(-\ln r)\, r^{-x}} = \lim_{x \to \infty} \frac{r^x}{-\ln r} = 0, \text{ so } \lim_{n \to \infty} nr^n = 0, \text{ and hence } \{nr^n\} \text{ converges}$$

whenever $|r| < 1$.

53. Since $\{a_n\}$ is a decreasing sequence, $a_n > a_{n+1}$ for all $n \geq 1$. Because all of its terms lie between 5 and 8, $\{a_n\}$ is a bounded sequence. By the Monotonic Sequence Theorem, $\{a_n\}$ is convergent; that is, $\{a_n\}$ has a limit L. L must be less than 8 since $\{a_n\}$ is decreasing, so $5 \leq L < 8$.

55. $a_n = \dfrac{1}{2n+3}$ is decreasing since $a_{n+1} = \dfrac{1}{2(n+1)+3} = \dfrac{1}{2n+5} < \dfrac{1}{2n+3} = a_n$ for each $n \geq 1$. The

sequence is bounded since $0 < a_n \leq \frac{1}{5}$ for all $n \geq 1$. Note that $a_1 = \frac{1}{5}$.

57. $a_n = \cos(n\pi/2)$ is not monotonic. The first few terms are $0, -1, 0, 1, 0, -1, 0, 1, \ldots$. In fact, the sequence consists of the terms $0, -1, 0, 1$ repeated over and over again in that order. The sequence is bounded since $|a_n| \leq 1$ for all $n \geq 1$.

59. $a_n = \dfrac{n}{n^2+1}$ defines a decreasing sequence since for $f(x) = \dfrac{x}{x^2+1}$,

$$f'(x) = \frac{(x^2+1)(1) - x(2x)}{(x^2+1)^2} = \frac{1-x^2}{(x^2+1)^2} \leq 0 \text{ for } x \geq 1. \text{ The sequence is bounded since } 0 < a_n \leq \frac{1}{2} \text{ for all}$$

$n \geq 1$.

61. $a_1 = 2^{1/2}$, $a_2 = 2^{3/4}$, $a_3 = 2^{7/8}$, $\ldots$, so $a_n = 2^{(2^n - 1)/2^n} = 2^{1 - (1/2^n)}$. $\lim\limits_{n \to \infty} a_n = \lim\limits_{n \to \infty} 2^{1 - (1/2^n)} = 2^1 = 2$.

Alternate solution: Let $L = \lim\limits_{n \to \infty} a_n$. (We could show the limit exists by showing that $\{a_n\}$ is bounded and

increasing.) Then L must satisfy $L = \sqrt{2 \cdot L}$ ⇒ $L^2 = 2L$ ⇒ $L(L-2) = 0$. $L \neq 0$ since the sequence increases, so $L = 2$.

63. We show by induction that $\{a_n\}$ is increasing and bounded above by 3.

Let P_n be the proposition that $a_{n+1} > a_n$ and $0 < a_n < 3$. Clearly P_1 is true. Assume that P_n is true.

Then $a_{n+1} > a_n$ ⇒ $\dfrac{1}{a_{n+1}} < \dfrac{1}{a_n}$ ⇒ $-\dfrac{1}{a_{n+1}} > -\dfrac{1}{a_n}$.

Now $a_{n+2} = 3 - \dfrac{1}{a_{n+1}} > 3 - \dfrac{1}{a_n} = a_{n+1}$ ⇔ P_{n+1}. This proves that $\{a_n\}$ is increasing and bounded above

by 3, so $1 = a_1 < a_n < 3$, that is, $\{a_n\}$ is bounded, and hence convergent by the Monotonic Sequence Theorem.

If $L = \lim\limits_{n \to \infty} a_n$, then $\lim\limits_{n \to \infty} a_{n+1} = L$ also, so L must satisfy $L = 3 - 1/L$ ⇒ $L^2 - 3L + 1 = 0$ ⇒

$L = \frac{3 \pm \sqrt{5}}{2}$. But $L > 1$, so $L = \frac{3 + \sqrt{5}}{2}$.

65. (a) Let a_n be the number of rabbit pairs in the nth month. Clearly $a_1 = 1 = a_2$. In the nth month, each pair that is 2 or more months old (that is, a_{n-2} pairs) will produce a new pair to add to the a_{n-1} pairs already present. Thus, $a_n = a_{n-1} + a_{n-2}$, so that $\{a_n\} = \{f_n\}$, the Fibonacci sequence.

(b) $a_n = \dfrac{f_{n+1}}{f_n} \Rightarrow a_{n-1} = \dfrac{f_n}{f_{n-1}} = \dfrac{f_{n-1} + f_{n-2}}{f_{n-1}} = 1 + \dfrac{f_{n-2}}{f_{n-1}} = 1 + \dfrac{1}{f_{n-1}/f_{n-2}} = 1 + \dfrac{1}{a_{n-2}}$. If

$L = \lim\limits_{n\to\infty} a_n$, then $L = \lim\limits_{n\to\infty} a_{n-1}$ and $L = \lim\limits_{n\to\infty} a_{n-2}$, so L must satisfy $L = 1 + \dfrac{1}{L} \Rightarrow$

$L^2 - L - 1 = 0 \Rightarrow L = \dfrac{1+\sqrt{5}}{2}$ (since L must be positive).

67. (a)

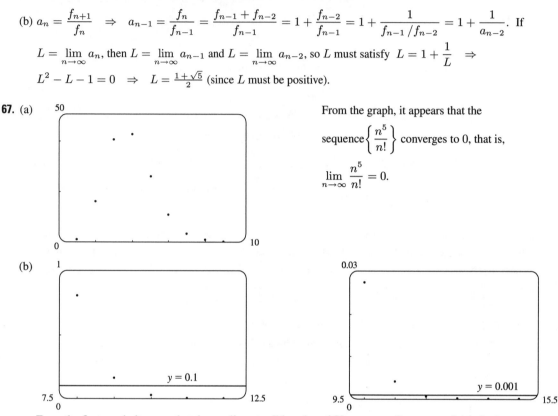

From the graph, it appears that the sequence $\left\{ \dfrac{n^5}{n!} \right\}$ converges to 0, that is,

$$\lim_{n\to\infty} \dfrac{n^5}{n!} = 0.$$

From the first graph, it seems that the smallest possible value of N corresponding to $\varepsilon = 0.1$ is 9, since $n^5/n! < 0.1$ whenever $n \geq 10$, but $9^5/9! > 0.1$. From the second graph, it seems that for $\varepsilon = 0.001$, the smallest possible value for N is 11.

69. If $\lim\limits_{n\to\infty} |a_n| = 0$ then $\lim\limits_{n\to\infty} -|a_n| = 0$, and since $-|a_n| \leq a_n \leq |a_n|$, we have that $\lim\limits_{n\to\infty} a_n = 0$ by the Squeeze Theorem.

71. (a) First we show that $a > a_1 > b_1 > b$.

$a_1 - b_1 = \dfrac{a+b}{2} - \sqrt{ab} = \dfrac{1}{2}\left(a - 2\sqrt{ab} + b\right) = \dfrac{1}{2}\left(\sqrt{a} - \sqrt{b}\right)^2 > 0$ (since $a > b$) $\Rightarrow a_1 > b_1$. Also

$a - a_1 = a - \dfrac{1}{2}(a + b) = \dfrac{1}{2}(a - b) > 0$ and $b - b_1 = b - \sqrt{ab} = \sqrt{b}\left(\sqrt{b} - \sqrt{a}\right) < 0$, so $a > a_1 > b_1 > b$.

In the same way we can show that $a_1 > a_2 > b_2 > b_1$ and so the given assertion is true for $n = 1$. Suppose it is true for $n = k$, that is, $a_k > a_{k+1} > b_{k+1} > b_k$. Then

$$a_{k+2} - b_{k+2} = \dfrac{1}{2}(a_{k+1} + b_{k+1}) - \sqrt{a_{k+1}b_{k+1}} = \dfrac{1}{2}\left(a_{k+1} - 2\sqrt{a_{k+1}b_{k+1}} + b_{k+1}\right)$$

$$= \dfrac{1}{2}\left(\sqrt{a_{k+1}} - \sqrt{b_{k+1}}\right)^2 > 0$$

$$a_{k+1} - a_{k+2} = a_{k+1} - \dfrac{1}{2}(a_{k+1} + b_{k+1}) = \dfrac{1}{2}(a_{k+1} - b_{k+1}) > 0$$

and $b_{k+1} - b_{k+2} = b_{k+1} - \sqrt{a_{k+1}b_{k+1}} = \sqrt{b_{k+1}}\left(\sqrt{b_{k+1}} - \sqrt{a_{k+1}}\right) < 0 \;\Rightarrow$

$a_{k+1} > a_{k+2} > b_{k+2} > b_{k+1}$, so the assertion is true for $n = k + 1$. Thus, it is true for all n by mathematical induction.

(b) From part (a) we have $a > a_n > a_{n+1} > b_{n+1} > b_n > b$, which shows that both sequences, $\{a_n\}$ and $\{b_n\}$, are monotonic and bounded. So they are both convergent by the Monotonic Sequence Theorem.

(c) Let $\lim_{n\to\infty} a_n = \alpha$ and $\lim_{n\to\infty} b_n = \beta$. Then $\lim_{n\to\infty} a_{n+1} = \lim_{n\to\infty} \dfrac{a_n + b_n}{2} \;\Rightarrow\; \alpha = \dfrac{\alpha + \beta}{2} \;\Rightarrow$

$2\alpha = \alpha + \beta \;\Rightarrow\; \alpha = \beta$.

73. (a) Suppose $\{p_n\}$ converges to p. Then $p_{n+1} = \dfrac{bp_n}{a + p_n} \;\Rightarrow\; \lim_{n\to\infty} p_{n+1} = \dfrac{b\,\lim\limits_{n\to\infty} p_n}{a + \lim\limits_{n\to\infty} p_n} \;\Rightarrow$

$p = \dfrac{bp}{a + p} \;\Rightarrow\; p^2 + ap = bp \;\Rightarrow\; p(p + a - b) = 0 \;\Rightarrow\; p = 0 \text{ or } p = b - a.$

(b) $p_{n+1} = \dfrac{bp_n}{a + p_n} = \dfrac{\dfrac{b}{a}p_n}{1 + \dfrac{p_n}{a}} < \dfrac{b}{a}p_n$ since $1 + \dfrac{p_n}{a} > 1$.

(c) By part (b), $p_1 < \left(\dfrac{b}{a}\right)p_0$, $p_2 < \left(\dfrac{b}{a}\right)p_1 < \left(\dfrac{b}{a}\right)^2 p_0$, $p_3 < \left(\dfrac{b}{a}\right)p_2 < \left(\dfrac{b}{a}\right)^3 p_0$, etc. In general,

$p_n < \left(\dfrac{b}{a}\right)^n p_0$, so $\lim_{n\to\infty} p_n \le \lim_{n\to\infty} \left(\dfrac{b}{a}\right)^n \cdot p_0 = 0$ since $b < a$. [By result 8, $\lim_{n\to\infty} r^n = 0$ if $-1 < r < 1$.

Here $r = \dfrac{b}{a} \in (0, 1)$.]

(d) Let $a < b$. We first show, by induction, that if $p_0 < b - a$, then $p_n < b - a$ and $p_{n+1} > p_n$.

For $n = 0$, we have $p_1 - p_0 = \dfrac{bp_0}{a + p_0} - p_0 = \dfrac{p_0(b - a - p_0)}{a + p_0} > 0$ since $p_0 < b - a$. So $p_1 > p_0$.

Now we suppose the assertion is true for $n = k$, that is, $p_k < b - a$ and $p_{k+1} > p_k$. Then

$b - a - p_{k+1} = b - a - \dfrac{bp_k}{a + p_k} = \dfrac{a(b - a) + bp_k - ap_k - bp_k}{a + p_k} = \dfrac{a(b - a - p_k)}{a + p_k} > 0$ because $p_k < b - a$.

So $p_{k+1} < b - a$. And $p_{k+2} - p_{k+1} = \dfrac{bp_{k+1}}{a + p_{k+1}} - p_{k+1} = \dfrac{p_{k+1}(b - a - p_{k+1})}{a + p_{k+1}} > 0$ since $p_{k+1} < b - a$.

Therefore, $p_{k+2} > p_{k+1}$. Thus, the assertion is true for $n = k + 1$. It is therefore true for all n by mathematical induction. A similar proof by induction shows that if $p_0 > b - a$, then $p_n > b - a$ and $\{p_n\}$ is decreasing.

In either case the sequence $\{p_n\}$ is bounded and monotonic, so it is convergent by the Monotonic Sequence Theorem. It then follows from part (a) that $\lim_{n\to\infty} p_n = b - a$.

11.2 Series

1. (a) A sequence is an ordered list of numbers whereas a series is the *sum* of a list of numbers.

(b) A series is convergent if the sequence of partial sums is a convergent sequence. A series is divergent if it is not convergent.

3.

n	s_n
1	-2.40000
2	-1.92000
3	-2.01600
4	-1.99680
5	-2.00064
6	-1.99987
7	-2.00003
8	-1.99999
9	-2.00000
10	-2.00000

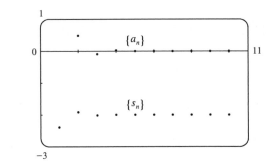

From the graph and the table, it seems that the series converges to -2. In fact, it is a geometric series with $a = -2.4$ and $r = -\frac{1}{5}$, so its sum is

$$\sum_{n=1}^{\infty} \frac{12}{(-5)^n} = \frac{-2.4}{1 - \left(-\frac{1}{5}\right)} = \frac{-2.4}{1.2} = -2.$$ Note that the dot corresponding to

$n = 1$ is part of both $\{a_n\}$ and $\{s_n\}$.

TI-86 Note: To graph $\{a_n\}$ and $\{s_n\}$, set your calculator to Param mode and DrawDot mode. (DrawDot is under GRAPH, MORE, FORMT (F3).) Now under E(t) = make the assignments: xt1=t, yt1=12/(-5)^t, xt2=t, yt2=sum seq(yt1,t,1,t,1). (sum and seq are under LIST, OPS (F5), MORE.) Under WIND use 1,10,1,0,10,1,-3,1,1 to obtain a graph similar to the one above. Then use TRACE (F4) to see the values.

5.

n	s_n
1	1.55741
2	-0.62763
3	-0.77018
4	0.38764
5	-2.99287
6	-3.28388
7	-2.41243
8	-9.21214
9	-9.66446
10	-9.01610

The series $\displaystyle\sum_{n=1}^{\infty} \tan n$ diverges, since its terms do not approach 0.

7.

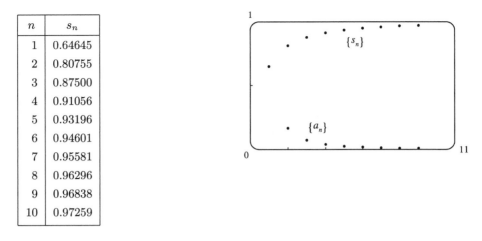

n	s_n
1	0.64645
2	0.80755
3	0.87500
4	0.91056
5	0.93196
6	0.94601
7	0.95581
8	0.96296
9	0.96838
10	0.97259

From the graph, it seems that the series converges to 1. To find the sum, we write

$$s_n = \sum_{i=1}^{n} \left(\frac{1}{i^{1.5}} - \frac{1}{(i+1)^{1.5}} \right)$$

$$= \left(1 - \frac{1}{2^{1.5}} \right) + \left(\frac{1}{2^{1.5}} - \frac{1}{3^{1.5}} \right) + \left(\frac{1}{3^{1.5}} - \frac{1}{4^{1.5}} \right) + \cdots + \left(\frac{1}{n^{1.5}} - \frac{1}{(n+1)^{1.5}} \right) = 1 - \frac{1}{(n+1)^{1.5}}$$

So the sum is $\lim\limits_{n \to \infty} s_n = 1 - 0 = 1$.

9. (a) $\lim\limits_{n \to \infty} a_n = \lim\limits_{n \to \infty} \dfrac{2n}{3n+1} = \dfrac{2}{3}$, so the *sequence* $\{a_n\}$ is convergent by (11.1.1).

(b) Since $\lim\limits_{n \to \infty} a_n = \frac{2}{3} \neq 0$, the *series* $\sum\limits_{n=1}^{\infty} a_n$ is divergent by the Test for Divergence (7).

11. $3 + 2 + \frac{4}{3} + \frac{8}{9} + \cdots$ is a geometric series with first term $a = 3$ and common ratio $r = \frac{2}{3}$. Since $|r| = \frac{2}{3} < 1$, the

series converges to $\frac{a}{1-r} = \frac{3}{1-2/3} = \frac{3}{1/3} = 9$.

13. $-2 + \frac{5}{2} - \frac{25}{8} + \frac{125}{32} - \cdots$ is a geometric series with $a = -2$ and $r = \frac{5/2}{-2} = -\frac{5}{4}$. Since $|r| = \frac{5}{4} > 1$, the series

diverges by (4).

15. $\sum\limits_{n=1}^{\infty} 5 \left(\frac{2}{3} \right)^{n-1}$ is a geometric series with $a = 5$ and $r = \frac{2}{3}$. Since $|r| = \frac{2}{3} < 1$, the series converges to

$\frac{a}{1-r} = \frac{5}{1-2/3} = \frac{5}{1/3} = 15$.

17. $\sum\limits_{n=1}^{\infty} \dfrac{(-3)^{n-1}}{4^n} = \dfrac{1}{4} \sum\limits_{n=1}^{\infty} \left(-\dfrac{3}{4} \right)^{n-1}$. The latter series is geometric with $a = 1$ and $r = -\frac{3}{4}$. Since $|r| = \frac{3}{4} < 1$, it

converges to $\frac{1}{1-(-3/4)} = \frac{4}{7}$. Thus, the given series converges to $\left(\frac{1}{4} \right) \left(\frac{4}{7} \right) = \frac{1}{7}$.

19. $\sum\limits_{n=0}^{\infty} \dfrac{\pi^n}{3^{n+1}} = \dfrac{1}{3} \sum\limits_{n=0}^{\infty} \left(\dfrac{\pi}{3} \right)^n$ is a geometric series with ratio $r = \dfrac{\pi}{3}$. Since $|r| > 1$, the series diverges.

21. $\sum_{n=1}^{\infty} \dfrac{n}{n+5}$ diverges since $\lim_{n\to\infty} a_n = \lim_{n\to\infty} \dfrac{n}{n+5} = 1 \neq 0$. [Use (7), the Test for Divergence.]

23. Using partial fractions, the partial sums are

$$s_n = \sum_{i=2}^{n} \frac{2}{(i-1)(i+1)} = \sum_{i=2}^{n} \left(\frac{1}{i-1} - \frac{1}{i+1} \right)$$

$$= \left(1 - \frac{1}{3}\right) + \left(\frac{1}{2} - \frac{1}{4}\right) + \left(\frac{1}{3} - \frac{1}{5}\right) + \cdots + \left(\frac{1}{n-3} - \frac{1}{n-1}\right) + \left(\frac{1}{n-2} - \frac{1}{n}\right)$$

This sum is a telescoping series and $s_n = 1 + \dfrac{1}{2} - \dfrac{1}{n-1} - \dfrac{1}{n}$.

Thus, $\sum_{n=2}^{\infty} \dfrac{2}{n^2 - 1} = \lim_{n\to\infty} \left(1 + \dfrac{1}{2} - \dfrac{1}{n-1} - \dfrac{1}{n}\right) = \dfrac{3}{2}$.

25. $\sum_{k=2}^{\infty} \dfrac{k^2}{k^2 - 1}$ diverges by the Test for Divergence since $\lim_{k\to\infty} a_k = \lim_{k\to\infty} \dfrac{k^2}{k^2 - 1} = 1 \neq 0$.

27. Converges. $\sum_{n=1}^{\infty} \dfrac{3^n + 2^n}{6^n} = \sum_{n=1}^{\infty} \left(\dfrac{3^n}{6^n} + \dfrac{2^n}{6^n}\right) = \sum_{n=1}^{\infty} \left[\left(\dfrac{1}{2}\right)^n + \left(\dfrac{1}{3}\right)^n\right] = \dfrac{1/2}{1-1/2} + \dfrac{1/3}{1-1/3} = 1 + \dfrac{1}{2} = \dfrac{3}{2}$

29. $\sum_{n=1}^{\infty} \sqrt[n]{2} = 2 + \sqrt{2} + \sqrt[3]{2} + \sqrt[4]{2} + \cdots$ diverges by the Test for Divergence since

$$\lim_{n\to\infty} a_n = \lim_{n\to\infty} \sqrt[n]{2} = \lim_{n\to\infty} 2^{1/n} = 2^0 = 1 \neq 0.$$

31. $\lim_{n\to\infty} a_n = \lim_{n\to\infty} \arctan n = \frac{\pi}{2} \neq 0$, so the series diverges by the Test for Divergence.

33. The first series is a telescoping sum:

$$\sum_{n=1}^{\infty} \frac{3}{n(n+3)} = \sum_{n=1}^{\infty} \left(\frac{1}{n} - \frac{1}{n+3}\right) = \sum_{n=1}^{\infty} \left(\frac{1}{n} - \frac{1}{n+1} + \frac{1}{n+1} - \frac{1}{n+2} + \frac{1}{n+2} - \frac{1}{n+3}\right)$$

$$= \sum_{n=1}^{\infty} \left(\frac{1}{n} - \frac{1}{n+1}\right) + \sum_{n=1}^{\infty} \left(\frac{1}{n+1} - \frac{1}{n+2}\right) + \sum_{n=1}^{\infty} \left(\frac{1}{n+2} - \frac{1}{n+3}\right)$$

$$= 1 + \frac{1}{2} + \frac{1}{3} = \frac{11}{6}$$

The second series is geometric with first term $\dfrac{5}{4}$ and ratio $\dfrac{1}{4}$: $\sum_{n=1}^{\infty} \dfrac{5}{4^n} = \dfrac{5/4}{1 - 1/4} = \dfrac{5}{3}$. Thus,

$$\sum_{n=1}^{\infty} \left(\frac{3}{n(n+3)} + \frac{5}{4^n}\right) = \sum_{n=1}^{\infty} \frac{3}{n(n+3)} + \sum_{n=1}^{\infty} \frac{5}{4^n} \text{ [sum of two convergent series]} = \frac{11}{6} + \frac{5}{3} = \frac{7}{2}.$$

35. $0.\overline{2} = \dfrac{2}{10} + \dfrac{2}{10^2} + \cdots$ is a geometric series with $a = \dfrac{2}{10}$ and $r = \dfrac{1}{10}$. It converges to $\dfrac{a}{1-r} = \dfrac{2/10}{1 - 1/10} = \dfrac{2}{9}$.

37. $3.\overline{417} = 3 + \dfrac{417}{10^3} + \dfrac{417}{10^6} + \cdots = 3 + \dfrac{417/10^3}{1 - 1/10^3} = 3 + \dfrac{417}{999} = \dfrac{3414}{999} = \dfrac{1138}{333}$

39. $0.123\overline{456} = \dfrac{123}{1000} + \dfrac{0.000456}{1 - 0.001} = \dfrac{123}{1000} + \dfrac{456}{999{,}000} = \dfrac{123{,}333}{999{,}000} = \dfrac{41{,}111}{333{,}000}$

41. $\displaystyle\sum_{n=1}^{\infty} \dfrac{x^n}{3^n} = \sum_{n=1}^{\infty} \left(\dfrac{x}{3}\right)^n$ is a geometric series with $r = \dfrac{x}{3}$, so the series converges $\Leftrightarrow |r| < 1 \Leftrightarrow \dfrac{|x|}{3} < 1 \Leftrightarrow$

$|x| < 3$; that is, $-3 < x < 3$. In that case, the sum of the series is $\dfrac{a}{1 - r} = \dfrac{x/3}{1 - x/3} = \dfrac{x/3}{1 - x/3} \cdot \dfrac{3}{3} = \dfrac{x}{3 - x}$.

43. $\displaystyle\sum_{n=0}^{\infty} 4^n x^n = \sum_{n=0}^{\infty} (4x)^n$ is a geometric series with $r = 4x$, so the series converges $\Leftrightarrow |r| < 1 \Leftrightarrow$

$4|x| < 1 \Leftrightarrow |x| < \tfrac{1}{4}$. In that case, the sum of the series is $\dfrac{1}{1 - 4x}$.

45. $\displaystyle\sum_{n=0}^{\infty} \dfrac{\cos^n x}{2^n}$ is a geometric series with first term 1 and ratio $r = \dfrac{\cos x}{2}$, so it converges $\Leftrightarrow |r| < 1$. But

$|r| = \dfrac{|\cos x|}{2} \le \dfrac{1}{2}$ for all x. Thus, the series converges for all real values of x and the sum of the series is

$\dfrac{1}{1 - (\cos x)/2} = \dfrac{2}{2 - \cos x}$.

47. After defining f, We use `convert(f,parfrac);` in Maple, `Apart` in Mathematica, or `Expand Rational`

and `Simplify` in Derive to find that the general term is $\dfrac{1}{(4n + 1)(4n - 3)} = -\dfrac{1/4}{4n + 1} + \dfrac{1/4}{4n - 3}$. So the

nth partial sum is

$$s_n = \sum_{k=1}^{n} \left(-\dfrac{1/4}{4k + 1} + \dfrac{1/4}{4k - 3}\right) = \dfrac{1}{4}\sum_{k=1}^{n}\left(\dfrac{1}{4k - 3} - \dfrac{1}{4k + 1}\right)$$

$$= \dfrac{1}{4}\left[\left(1 - \dfrac{1}{5}\right) + \left(\dfrac{1}{5} - \dfrac{1}{9}\right) + \left(\dfrac{1}{9} - \dfrac{1}{13}\right) + \cdots + \left(\dfrac{1}{4n - 3} - \dfrac{1}{4n + 1}\right)\right] = \dfrac{1}{4}\left(1 - \dfrac{1}{4n + 1}\right)$$

The series converges to $\displaystyle\lim_{n\to\infty} s_n = \tfrac{1}{4}$. This can be confirmed by directly computing the sum using

`sum(f,1..infinity);` (in Maple), `Sum[f,{n,1,Infinity}]` (in Mathematica), or `Calculus Sum`

(from 1 to ∞) and `Simplify` (in Derive).

49. For $n = 1$, $a_1 = 0$ since $s_1 = 0$. For $n > 1$,

$$a_n = s_n - s_{n-1} = \dfrac{n - 1}{n + 1} - \dfrac{(n - 1) - 1}{(n - 1) + 1} = \dfrac{(n - 1)n - (n + 1)(n - 2)}{(n + 1)n} = \dfrac{2}{n(n + 1)}$$

Also, $\displaystyle\sum_{n=1}^{\infty} a_n = \lim_{n\to\infty} s_n = \lim_{n\to\infty} \dfrac{1 - 1/n}{1 + 1/n} = 1$.

51. (a) The first step in the chain occurs when the local government spends D dollars. The people who receive it spend a fraction c of those D dollars, that is, Dc dollars. Those who receive the Dc dollars spend a fraction c of it, that is, Dc^2 dollars. Continuing in this way, we see that the total spending after n transactions is

$$S_n = D + Dc + Dc^2 + \cdots + Dc^{n-1} = \frac{D(1-c^n)}{1-c} \text{ by (3).}$$

(b) $\displaystyle\lim_{n\to\infty} S_n = \lim_{n\to\infty} \frac{D(1-c^n)}{1-c} = \frac{D}{1-c} \lim_{n\to\infty}(1-c^n) = \frac{D}{1-c}$ (since $0 < c < 1 \ \Rightarrow \ \displaystyle\lim_{n\to\infty} c^n = 0$)

$\qquad = \dfrac{D}{s}$ (since $c + s = 1$) $= kD$ (since $k = 1/s$)

If $c = 0.8$, then $s = 1 - c = 0.2$ and the multiplier is $k = 1/s = 5$.

53. $\sum_{n=2}^{\infty}(1+c)^{-n}$ is a geometric series with $a = (1+c)^{-2}$ and $r = (1+c)^{-1}$, so the series converges when

$\left|(1+c)^{-1}\right| < 1 \ \Leftrightarrow \ |1+c| > 1 \ \Leftrightarrow \ 1+c > 1$ or $1+c < -1 \ \Leftrightarrow \ c > 0$ or $c < -2$. We calculate

the sum of the series and set it equal to 2: $\dfrac{(1+c)^{-2}}{1-(1+c)^{-1}} = 2 \ \Leftrightarrow \ \left(\dfrac{1}{1+c}\right)^2 = 2 - 2\left(\dfrac{1}{1+c}\right) \ \Leftrightarrow$

$1 = 2(1+c)^2 - 2(1+c) \ \Leftrightarrow \ 2c^2 + 2c - 1 = 0 \ \Leftrightarrow \ c = \dfrac{-2 \pm \sqrt{12}}{4} = \dfrac{\pm\sqrt{3}-1}{2}$. However, the negative root is

inadmissible because $-2 < \dfrac{-\sqrt{3}-1}{2} < 0$. So $c = \dfrac{\sqrt{3}-1}{2}$.

55. Let d_n be the diameter of C_n. We draw lines from the centers of the C_i to the center of D (or C), and using the Pythagorean Theorem, we can write $1^2 + \left(1 - \frac{1}{2}d_1\right)^2 = \left(1 + \frac{1}{2}d_1\right)^2 \ \Leftrightarrow$

$1 = \left(1 + \frac{1}{2}d_1\right)^2 - \left(1 - \frac{1}{2}d_1\right)^2 = 2d_1$ (difference of squares)

$\Rightarrow \ d_1 = \frac{1}{2}$. Similarly,

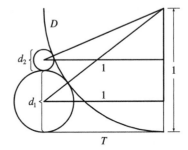

$1 = \left(1 + \frac{1}{2}d_2\right)^2 - \left(1 - d_1 - \frac{1}{2}d_2\right)^2 = 2d_2 + 2d_1 - d_1^2 - d_1 d_2$

$\quad = (2 - d_1)(d_1 + d_2) \ \Leftrightarrow$

$d_2 = \dfrac{1}{2-d_1} - d_1 = \dfrac{(1-d_1)^2}{2-d_1}, \ 1 = \left(1 + \frac{1}{2}d_3\right)^2 - \left(1 - d_1 - d_2 - \frac{1}{2}d_3\right)^2 \ \Leftrightarrow \ d_3 = \dfrac{[1-(d_1+d_2)]^2}{2-(d_1+d_2)}$, and

in general, $d_{n+1} = \dfrac{\left(1 - \sum_{i=1}^{n} d_i\right)^2}{2 - \sum_{i=1}^{n} d_i}$. If we actually calculate d_2 and d_3 from the formulas above, we find that they

are $\dfrac{1}{6} = \dfrac{1}{2\cdot 3}$ and $\dfrac{1}{12} = \dfrac{1}{3\cdot 4}$ respectively, so we suspect that in general, $d_n = \dfrac{1}{n(n+1)}$. To prove this, we use

induction: Assume that for all $k \leq n$, $d_k = \dfrac{1}{k(k+1)} = \dfrac{1}{k} - \dfrac{1}{k+1}$. Then

$$\sum_{i=1}^{n} d_i = 1 - \frac{1}{n+1} = \frac{n}{n+1} \quad \text{(telescoping sum). Substituting this into our formula for } d_{n+1}, \text{ we get}$$

$$d_{n+1} = \frac{\left[1 - \dfrac{n}{n+1}\right]^2}{2 - \left(\dfrac{n}{n+1}\right)} = \frac{\dfrac{1}{(n+1)^2}}{\dfrac{n+2}{n+1}} = \frac{1}{(n+1)(n+2)}, \text{ and the induction is complete.}$$

Now, we observe that the partial sums $\sum_{i=1}^{n} d_i$ of the diameters of the circles approach 1 as $n \to \infty$; that is,

$$\sum_{n=1}^{\infty} a_n = \sum_{n=1}^{\infty} \frac{1}{n(n+1)} = 1, \text{ which is what we wanted to prove.}$$

57. The series $1 - 1 + 1 - 1 + 1 - 1 + \cdots$ diverges (geometric series with $r = -1$) so we cannot say that

$0 = 1 - 1 + 1 - 1 + 1 - 1 + \cdots$.

59. $\sum_{n=1}^{\infty} ca_n = \lim_{n \to \infty} \sum_{i=1}^{n} ca_i = \lim_{n \to \infty} c \sum_{i=1}^{n} a_i = c \lim_{n \to \infty} \sum_{i=1}^{n} a_i = c \sum_{n=1}^{\infty} a_n$, which exists by hypothesis.

61. Suppose on the contrary that $\sum (a_n + b_n)$ converges. Then $\sum (a_n + b_n)$ and $\sum a_n$ are convergent series. So by

Theorem 8, $\sum [(a_n + b_n) - a_n]$ would also be convergent. But $\sum [(a_n + b_n) - a_n] = \sum b_n$, a contradiction,

since $\sum b_n$ is given to be divergent.

63. The partial sums $\{s_n\}$ form an increasing sequence, since $s_n - s_{n-1} = a_n > 0$ for all n. Also, the sequence $\{s_n\}$

is bounded since $s_n \leq 1000$ for all n. So by Theorem 11.1.11, the sequence of partial sums converges, that is, the

series $\sum a_n$ is convergent.

65. (a) At the first step, only the interval $\left(\frac{1}{3}, \frac{2}{3}\right)$ (length $\frac{1}{3}$) is removed. At the second step, we remove the intervals

$\left(\frac{1}{9}, \frac{2}{9}\right)$ and $\left(\frac{7}{9}, \frac{8}{9}\right)$, which have a total length of $2 \cdot \left(\frac{1}{3}\right)^2$. At the third step, we remove 2^2 intervals, each of

length $\left(\frac{1}{3}\right)^3$. In general, at the nth step we remove 2^{n-1} intervals, each of length $\left(\frac{1}{3}\right)^n$, for a length of

$2^{n-1} \cdot \left(\frac{1}{3}\right)^n = \frac{1}{3}\left(\frac{2}{3}\right)^{n-1}$. Thus, the total length of all removed intervals is $\sum_{n=1}^{\infty} \frac{1}{3}\left(\frac{2}{3}\right)^{n-1} = \frac{1/3}{1 - 2/3} = 1$

(geometric series with $a = \frac{1}{3}$ and $r = \frac{2}{3}$). Notice that at the nth step, the leftmost interval that is removed is

$\left(\left(\frac{1}{3}\right)^n, \left(\frac{2}{3}\right)^n\right)$, so we never remove 0, and 0 is in the Cantor set. Also, the rightmost interval removed is

$\left(1 - \left(\frac{2}{3}\right)^n, 1 - \left(\frac{1}{3}\right)^n\right)$, so 1 is never removed. Some other numbers in the Cantor set are $\frac{1}{3}, \frac{2}{3}, \frac{1}{9}, \frac{2}{9}, \frac{7}{9}$, and $\frac{8}{9}$.

(b) The area removed at the first step is $\frac{1}{9}$; at the second step, $8 \cdot \left(\frac{1}{9}\right)^2$; at the third step, $(8)^2 \cdot \left(\frac{1}{9}\right)^3$. In general, the

area removed at the nth step is $(8)^{n-1}\left(\frac{1}{9}\right)^n = \frac{1}{9}\left(\frac{8}{9}\right)^{n-1}$, so the total area of all removed squares is

$$\sum_{n=1}^{\infty} \frac{1}{9}\left(\frac{8}{9}\right)^{n-1} = \frac{1/9}{1 - 8/9} = 1.$$

67. (a) For $\sum\limits_{n=1}^{\infty} \dfrac{n}{(n+1)!}$, $s_1 = \dfrac{1}{1\cdot 2} = \dfrac{1}{2}$, $s_2 = \dfrac{1}{2} + \dfrac{2}{1\cdot 2\cdot 3} = \dfrac{5}{6}$, $s_3 = \dfrac{5}{6} + \dfrac{3}{1\cdot 2\cdot 3\cdot 4} = \dfrac{23}{24}$,

$s_4 = \dfrac{23}{24} + \dfrac{4}{1\cdot 2\cdot 3\cdot 4\cdot 5} = \dfrac{119}{120}$. The denominators are $(n+1)!$, so a guess would be $s_n = \dfrac{(n+1)!-1}{(n+1)!}$.

(b) For $n = 1$, $s_1 = \dfrac{1}{2} = \dfrac{2!-1}{2!}$, so the formula holds for $n = 1$. Assume $s_k = \dfrac{(k+1)!-1}{(k+1)!}$. Then

$$s_{k+1} = \dfrac{(k+1)!-1}{(k+1)!} + \dfrac{k+1}{(k+2)!} = \dfrac{(k+1)!-1}{(k+1)!} + \dfrac{k+1}{(k+1)!(k+2)}$$

$$= \dfrac{(k+2)! - (k+2) + k + 1}{(k+2)!} = \dfrac{(k+2)!-1}{(k+2)!}$$

Thus, the formula is true for $n = k+1$. So by induction, the guess is correct.

(c) $\lim\limits_{n\to\infty} s_n = \lim\limits_{n\to\infty} \dfrac{(n+1)!-1}{(n+1)!} = \lim\limits_{n\to\infty}\left[1 - \dfrac{1}{(n+1)!}\right] = 1$ and so $\sum\limits_{n=1}^{\infty} \dfrac{n}{(n+1)!} = 1$.

11.3 The Integral Test and Estimates of Sums

1. The picture shows that $a_2 = \dfrac{1}{2^{1.3}} < \displaystyle\int_1^2 \dfrac{1}{x^{1.3}}\,dx$,

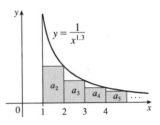

$a_3 = \dfrac{1}{3^{1.3}} < \displaystyle\int_2^3 \dfrac{1}{x^{1.3}}\,dx$, and so on, so $\sum\limits_{n=2}^{\infty} \dfrac{1}{n^{1.3}} < \displaystyle\int_1^\infty \dfrac{1}{x^{1.3}}\,dx$. The

integral converges by (7.8.2) with $p = 1.3 > 1$, so the series converges.

3. The function $f(x) = 1/x^4$ is continuous, positive, and decreasing on $[1, \infty)$, so the Integral Test applies.

$$\int_1^\infty \dfrac{1}{x^4}\,dx = \lim_{t\to\infty} \int_1^t x^{-4}\,dx = \lim_{t\to\infty}\left[\dfrac{x^{-3}}{-3}\right]_1^t = \lim_{t\to\infty}\left(-\dfrac{1}{3t^3} + \dfrac{1}{3}\right) = \dfrac{1}{3}.$$ Since this improper integral is

convergent, the series $\sum\limits_{n=1}^{\infty} \dfrac{1}{n^4}$ is also convergent by the Integral Test.

5. The function $f(x) = 1/(3x+1)$ is continuous, positive, and decreasing on $[1, \infty)$, so the Integral Test applies.

$$\int_1^\infty \dfrac{dx}{3x+1} = \lim_{b\to\infty} \int_1^b \dfrac{dx}{3x+1} = \lim_{b\to\infty}\left[\tfrac{1}{3}\ln(3x+1)\right]_1^b = \lim_{b\to\infty}\left[\tfrac{1}{3}\ln(3b+1) - \tfrac{1}{3}\ln 4\right] = \infty$$

so the improper integral diverges, and so does the series $\sum_{n=1}^{\infty} 1/(3n+1)$.

7. $f(x) = xe^{-x}$ is continuous and positive on $[1, \infty)$. $f'(x) = -xe^{-x} + e^{-x} = e^{-x}(1-x) < 0$ for $x > 1$, so f is decreasing on $[1, \infty)$. Thus, the Integral Test applies.

$$\int_1^\infty xe^{-x}\,dx = \lim_{b\to\infty} \int_1^b xe^{-x}\,dx = \lim_{b\to\infty}\left[-xe^{-x} - e^{-x}\right]_1^b \quad \text{(by parts)}$$

$$= \lim_{b\to\infty}\left[-be^{-b} - e^{-b} + e^{-1} + e^{-1}\right] = 2/e$$

since $\lim\limits_{b\to\infty} be^{-b} = \lim\limits_{b\to\infty} (b/e^b) \overset{\text{H}}{=} \lim\limits_{b\to\infty} (1/e^b) = 0$ and $\lim\limits_{b\to\infty} e^{-b} = 0$. Thus, $\sum\limits_{n=1}^{\infty} ne^{-n}$ converges.

9. The series $\sum_{n=1}^{\infty} \dfrac{1}{n^{0.85}}$ is a p-series with $p = 0.85 \le 1$, so it diverges by (1). Therefore, the series $\sum_{n=1}^{\infty} \dfrac{2}{n^{0.85}}$ must

also diverge, for if it converged, then $\sum_{n=1}^{\infty} \dfrac{1}{n^{0.85}}$ would have to converge (by Theorem 8(i) in Section 11.2).

11. $1 + \dfrac{1}{8} + \dfrac{1}{27} + \dfrac{1}{64} + \dfrac{1}{125} + \cdots = \sum_{n=1}^{\infty} \dfrac{1}{n^3}$. This is a p-series with $p = 3 > 1$, so it converges by (1).

13. $\sum_{n=1}^{\infty} \dfrac{5 - 2\sqrt{n}}{n^3} = 5 \sum_{n=1}^{\infty} \dfrac{1}{n^3} - 2 \sum_{n=1}^{\infty} \dfrac{1}{n^{5/2}}$ by Theorem 11.2.8, since $\sum_{n=1}^{\infty} \dfrac{1}{n^3}$ and $\sum_{n=1}^{\infty} \dfrac{1}{n^{5/2}}$ both converge by (1)

(with $p = 3 > 1$ and $p = \frac{5}{2} > 1$). Thus, $\sum_{n=1}^{\infty} \dfrac{5 - 2\sqrt{n}}{n^3}$ converges.

15. The function $f(x) = \dfrac{1}{x^2 + 4}$ is continuous, positive, and decreasing on $[1, \infty)$, so we can apply the Integral Test.

$$\int_1^{\infty} \frac{1}{x^2 + 4}\, dx = \lim_{t \to \infty} \int_1^t \frac{1}{x^2 + 4}\, dx = \lim_{t \to \infty} \left[\frac{1}{2} \tan^{-1} \frac{x}{2} \right]_1^t = \frac{1}{2} \lim_{t \to \infty} \left[\tan^{-1}\left(\frac{t}{2}\right) - \tan^{-1}\left(\frac{1}{2}\right) \right]$$

$$= \frac{1}{2}\left[\frac{\pi}{2} - \tan^{-1}\left(\frac{1}{2}\right) \right]$$

Therefore, the series $\sum_{n=1}^{\infty} \dfrac{1}{n^2 + 4}$ converges.

17. $f(x) = \dfrac{x}{x^2 + 1}$ is continuous and positive on $[1, \infty)$, and since

$f'(x) = \dfrac{1 - x^2}{(x^2 + 1)^2} < 0$ for $x > 1$, f is also decreasing. Using the Integral Test,

$$\int_1^{\infty} \frac{x}{x^2 + 1}\, dx = \lim_{t \to \infty} \int_1^t \frac{x}{x^2 + 1}\, dx = \lim_{t \to \infty} \left[\frac{\ln(x^2 + 1)}{2} \right]_1^t = \frac{1}{2} \lim_{t \to \infty} [\ln(t^2 + 1) - \ln 2] = \infty, \text{ so the series}$$

diverges.

19. $f(x) = xe^{-x^2}$ is continuous and positive on $[1, \infty)$, and since $f'(x) = e^{-x^2}\left(1 - 2x^2\right) < 0$ for

$x > 1$, f is decreasing as well. Thus, we can use the Integral Test.

$\int_1^{\infty} xe^{-x^2}\, dx = \lim_{t \to \infty} \left[-\frac{1}{2} e^{-x^2} \right]_1^t = 0 - \left(-\frac{1}{2}e^{-1} \right) = 1/(2e)$. Since the integral converges, the series converges.

21. $f(x) = \dfrac{1}{x \ln x}$ is continuous and positive on $[2, \infty)$, and also decreasing since $f'(x) = -\dfrac{1 + \ln x}{x^2 (\ln x)^2} < 0$ for $x > 2$,

so we can use the Integral Test. $\displaystyle\int_2^{\infty} \frac{1}{x \ln x}\, dx = \lim_{t \to \infty} \left[\ln(\ln x)\right]_2^t = \lim_{t \to \infty} [\ln(\ln t) - \ln(\ln 2)] = \infty$, so the series

diverges.

23. The function $f(x) = \dfrac{1}{x^3 + x}$ is continuous, positive, and decreasing on $[1, \infty)$, so the Integral Test applies. We use

partial fractions to evaluate the integral:

$$\int_1^\infty \frac{1}{x^3 + x}\, dx = \lim_{t \to \infty} \int_1^t \left[\frac{1}{x} - \frac{x}{1 + x^2} \right] dx = \lim_{t \to \infty} \left[\ln x - \frac{1}{2}\ln(1 + x^2) \right]_1^t$$

$$= \lim_{t \to \infty} \left[\ln \frac{x}{\sqrt{1 + x^2}} \right]_1^t = \lim_{t \to \infty} \left(\ln \frac{t}{\sqrt{1 + t^2}} - \ln \frac{1}{\sqrt{2}} \right)$$

$$= \lim_{t \to \infty} \left(\ln \frac{1}{\sqrt{1 + 1/t^2}} + \frac{1}{2}\ln 2 \right) = \frac{1}{2}\ln 2$$

so the series $\displaystyle\sum_{n=1}^\infty \frac{1}{n^3 + n}$ converges.

25. We have already shown (in Exercise 21) that when $p = 1$ the series $\displaystyle\sum_{n=2}^\infty \frac{1}{n(\ln n)^p}$ diverges, so assume that $p \neq 1$.

$f(x) = \dfrac{1}{x(\ln x)^p}$ is continuous and positive on $[2, \infty)$, and $f'(x) = -\dfrac{p + \ln x}{x^2(\ln x)^{p+1}} < 0$ if $x > e^{-p}$, so that f is

eventually decreasing and we can use the Integral Test.

$$\int_2^\infty \frac{1}{x(\ln x)^p}\, dx = \lim_{t \to \infty} \left[\frac{(\ln x)^{1-p}}{1 - p} \right]_2^t \quad \text{(for } p \neq 1) \quad = \lim_{t \to \infty} \left[\frac{(\ln t)^{1-p}}{1 - p} \right] - \frac{(\ln 2)^{1-p}}{1 - p}$$

This limit exists whenever $1 - p < 0 \iff p > 1$, so the series converges for $p > 1$.

27. Clearly the series cannot converge if $p \geq -\frac{1}{2}$, because then $\displaystyle\lim_{n \to \infty} n(1 + n^2)^p \neq 0$. Also, if $p = -1$ the series

diverges (see Exercise 17). So assume $p < -\frac{1}{2}$, $p \neq -1$. Then $f(x) = x(1 + x^2)^p$ is continuous,

positive, and eventually decreasing on $[1, \infty)$, and we can use the Integral Test.

$$\int_1^\infty x(1 + x^2)^p\, dx = \lim_{t \to \infty} \left[\frac{1}{2} \cdot \frac{(1 + x^2)^{p+1}}{p + 1} \right]_1^t = \lim_{t \to \infty} \frac{1}{2} \cdot \frac{(1 + t^2)^{p+1}}{p + 1} - \frac{2^p}{p + 1}. \text{ This limit exists and is finite}$$

$\iff p + 1 < 0 \iff p < -1$, so the series converges whenever $p < -1$.

29. Since this is a p-series with $p = x$, $\zeta(x)$ is defined when $x > 1$. Unless specified otherwise, the domain of a

function f is the set of numbers x such that the expression for $f(x)$ makes sense and defines a real number. So, in

the case of a series, it's the set of numbers x such that the series is convergent.

31. (a) $f(x) = \dfrac{1}{x^2}$ is positive and continuous and $f'(x) = -\dfrac{2}{x^3}$ is negative for $x > 0$, and so the Integral

Test applies. $\displaystyle\sum_{n=1}^\infty \frac{1}{n^2} \approx s_{10} = \frac{1}{1^2} + \frac{1}{2^2} + \frac{1}{3^2} + \cdots + \frac{1}{10^2} \approx 1.549768.$

$R_{10} \leq \displaystyle\int_{10}^\infty \frac{1}{x^2}\, dx = \lim_{t \to \infty} \left[\frac{-1}{x} \right]_{10}^t = \lim_{t \to \infty} \left(-\frac{1}{t} + \frac{1}{10} \right) = \frac{1}{10}$, so the error is at most 0.1.

(b) $s_{10} + \int_{11}^{\infty} \frac{1}{x^2} \, dx \le s \le s_{10} + \int_{10}^{\infty} \frac{1}{x^2} \, dx \quad \Rightarrow \quad s_{10} + \frac{1}{11} \le s \le s_{10} + \frac{1}{10} \quad \Rightarrow$

$1.549768 + 0.090909 = 1.640677 \le s \le 1.549768 + 0.1 = 1.649768$, so we get $s \approx 1.64522$ (the average of 1.640677 and 1.649768) with error ≤ 0.005 (the maximum of $1.649768 - 1.64522$ and $1.64522 - 1.640677$, rounded up).

(c) $R_n \le \int_n^{\infty} \frac{1}{x^2} \, dx = \frac{1}{n}$. So $R_n < 0.001$ if $\frac{1}{n} < \frac{1}{1000} \quad \Leftrightarrow \quad n > 1000$.

33. $f(x) = x^{-3/2}$ is positive and continuous and $f'(x) = -\frac{3}{2} x^{-5/2}$ is negative for $x > 0$, so the Integral Test applies. From the end of Example 6, we see that the error is at most half the length of the interval. From (3), the interval is $\left(s_n + \int_{n+1}^{\infty} f(x) \, dx, \, s_n + \int_n^{\infty} f(x) \, dx \right)$, so its length is $\int_n^{\infty} f(x) \, dx - \int_{n+1}^{\infty} f(x) \, dx = \int_n^{n+1} f(x) \, dx$. Thus, we need n such that

$$0.01 > \frac{1}{2} \int_n^{n+1} x^{-3/2} \, dx = \frac{1}{2} \left[\frac{-2}{\sqrt{x}} \right]_n^{n+1} = \frac{1}{\sqrt{n}} - \frac{1}{\sqrt{n+1}}$$

$\Leftrightarrow \quad n > 13.08$ (use a graphing calculator to solve $1/\sqrt{x} - 1/\sqrt{x+1} < 0.01$). Again from the end of Example 6, we approximate s by the midpoint of this interval. In general, the midpoint is

$\frac{1}{2} \left[\left(s_n + \int_{n+1}^{\infty} f(x) \, dx \right) + \left(s_n + \int_n^{\infty} f(x) \, dx \right) \right] = s_n + \frac{1}{2} \left(\int_{n+1}^{\infty} f(x) \, dx + \int_n^{\infty} f(x) \, dx \right)$. So using $n = 14$,

we have $s \approx s_{14} + \frac{1}{2} \left(\int_{14}^{\infty} x^{-3/2} \, dx + \int_{15}^{\infty} x^{-3/2} \, dx \right) \approx 2.0872 + \frac{1}{\sqrt{14}} + \frac{1}{\sqrt{15}} \approx 2.6127 \approx 2.61$. Any larger

value of n will also work. For instance, $s \approx s_{30} + \frac{1}{\sqrt{30}} + \frac{1}{\sqrt{31}} \approx 2.6124$.

35. $\sum_{n=1}^{\infty} n^{-1.001} = \sum_{n=1}^{\infty} \frac{1}{n^{1.001}}$ is a convergent p-series with $p = 1.001 > 1$. Using (2), we get

$R_n \le \int_n^{\infty} x^{-1.001} \, dx = \lim_{t \to \infty} \left[\frac{x^{-0.001}}{-0.001} \right]_n^t = -1000 \lim_{t \to \infty} \left[\frac{1}{x^{0.001}} \right]_n^t = -1000 \left(-\frac{1}{n^{0.001}} \right) = \frac{1000}{n^{0.001}}$. We want

$R_n < 0.000\,000\,005 \quad \Leftrightarrow \quad \frac{1000}{n^{0.001}} < 5 \times 10^{-9} \quad \Leftrightarrow \quad n^{0.001} > \frac{1000}{5 \times 10^{-9}} \quad \Leftrightarrow$

$n > \left(2 \times 10^{11} \right)^{1000} = 2^{1000} \times 10^{11,000} \approx 1.07 \times 10^{301} \times 10^{11,000} = 1.07 \times 10^{11,301}$.

37. (a) From the figure, $a_2 + a_3 + \cdots + a_n \le \int_1^n f(x) \, dx$, so with

$f(x) = \frac{1}{x}, \frac{1}{2} + \frac{1}{3} + \frac{1}{4} + \cdots + \frac{1}{n} \le \int_1^n \frac{1}{x} \, dx = \ln n$. Thus,

$s_n = 1 + \frac{1}{2} + \frac{1}{3} + \frac{1}{4} + \cdots + \frac{1}{n} \le 1 + \ln n$.

(b) By part (a), $s_{10^6} \le 1 + \ln 10^6 \approx 14.82 < 15$ and $s_{10^9} \le 1 + \ln 10^9 \approx 21.72 < 22$.

39. $b^{\ln n} = \left(e^{\ln b} \right)^{\ln n} = \left(e^{\ln n} \right)^{\ln b} = n^{\ln b} = \frac{1}{n^{-\ln b}}$. This is a p-series, which converges for all b such that $-\ln b > 1$

$\Leftrightarrow \quad \ln b < -1 \quad \Leftrightarrow \quad b < e^{-1} \quad \Leftrightarrow \quad b < 1/e$ [with $b > 0$].

11.4 The Comparison Tests

1. (a) We cannot say anything about $\sum a_n$. If $a_n > b_n$ for all n and $\sum b_n$ is convergent, then $\sum a_n$ could be convergent or divergent. (See the note after Example 2.)

(b) If $a_n < b_n$ for all n, then $\sum a_n$ is convergent. [This is part (i) of the Comparison Test.]

3. $\dfrac{1}{n^2 + n + 1} < \dfrac{1}{n^2}$ for all $n \geq 1$, so $\displaystyle\sum_{n=1}^{\infty} \dfrac{1}{n^2 + n + 1}$ converges by comparison with $\displaystyle\sum_{n=1}^{\infty} \dfrac{1}{n^2}$, which converges because it is a p-series with $p = 2 > 1$.

5. $\dfrac{5}{2 + 3^n} < \dfrac{5}{3^n}$ for all $n \geq 1$, so $\displaystyle\sum_{n=1}^{\infty} \dfrac{5}{2 + 3^n}$ converges by comparison with $\displaystyle\sum_{n=1}^{\infty} \dfrac{5}{3^n} = 5 \displaystyle\sum_{n=1}^{\infty} \dfrac{1}{3^n}$, which converges because $\displaystyle\sum_{n=1}^{\infty} \dfrac{1}{3^n}$ is a convergent geometric series with $r = \frac{1}{3}$ ($|r| < 1$).

7. $\dfrac{n+1}{n^2} > \dfrac{n}{n^2} = \dfrac{1}{n}$ for all $n \geq 1$, so $\displaystyle\sum_{n=1}^{\infty} \dfrac{n+1}{n^2}$ diverges by comparison with the harmonic series $\displaystyle\sum_{n=1}^{\infty} \dfrac{1}{n}$.

9. $\dfrac{\cos^2 n}{n^2 + 1} \leq \dfrac{1}{n^2 + 1} < \dfrac{1}{n^2}$, so the series $\displaystyle\sum_{n=1}^{\infty} \dfrac{\cos^2 n}{n^2 + 1}$ converges by comparison with the p-series $\displaystyle\sum_{n=1}^{\infty} \dfrac{1}{n^2}$ ($p = 2 > 1$).

11. If $a_n = \dfrac{n^2 + 1}{n^3 - 1}$ and $b_n = \dfrac{1}{n}$, then $\displaystyle\lim_{n \to \infty} \dfrac{a_n}{b_n} = \lim_{n \to \infty} \dfrac{n^3 + n}{n^3 - 1} = \lim_{n \to \infty} \dfrac{1 + 1/n^2}{1 - 1/n^3} = 1$, so $\displaystyle\sum_{n=2}^{\infty} \dfrac{n^2 + 1}{n^3 - 1}$ diverges by the Limit Comparison Test with the divergent (partial) harmonic series $\displaystyle\sum_{n=2}^{\infty} \dfrac{1}{n}$.

Or: Since $a_n = \dfrac{n^2 + 1}{n^3 - 1} > \dfrac{n^2 + 1}{n^3} > \dfrac{n^2}{n^3} = \dfrac{1}{n} = b_n$, we could use the Comparison Test.

13. $\dfrac{n-1}{n4^n}$ is positive for $n > 1$ and $\dfrac{n-1}{n4^n} < \dfrac{n}{n4^n} = \dfrac{1}{4^n}$, so $\displaystyle\sum_{n=1}^{\infty} \dfrac{n-1}{n4^n}$ converges by comparison with the convergent geometric series $\displaystyle\sum_{n=1}^{\infty} \left(\dfrac{1}{4}\right)^n$.

15. $\dfrac{2 + (-1)^n}{n\sqrt{n}} \leq \dfrac{3}{n\sqrt{n}}$, and $\displaystyle\sum_{n=1}^{\infty} \dfrac{3}{n\sqrt{n}}$ converges because it is a constant multiple of the convergent p-series $\displaystyle\sum_{n=1}^{\infty} \dfrac{1}{n\sqrt{n}}$ ($p = \frac{3}{2} > 1$), so the given series converges by the Comparison Test.

17. Use the Limit Comparison Test with $a_n = \dfrac{1}{\sqrt{n^2 + 1}}$ and $b_n = \dfrac{1}{n}$:

$\displaystyle\lim_{n \to \infty} \dfrac{a_n}{b_n} = \lim_{n \to \infty} \dfrac{n}{\sqrt{n^2 + 1}} = \lim_{n \to \infty} \dfrac{1}{\sqrt{1 + (1/n^2)}} = 1 > 0$. Since the harmonic series $\displaystyle\sum_{n=1}^{\infty} \dfrac{1}{n}$ diverges, so does $\displaystyle\sum_{n=1}^{\infty} \dfrac{1}{\sqrt{n^2 + 1}}$.

19. $\dfrac{2^n}{1 + 3^n} < \dfrac{2^n}{3^n} = \left(\dfrac{2}{3}\right)^n$. $\displaystyle\sum_{n=1}^{\infty} \left(\dfrac{2}{3}\right)^n$ is a convergent geometric series ($|r| = \frac{2}{3} < 1$), so $\displaystyle\sum_{n=1}^{\infty} \dfrac{2^n}{1 + 3^n}$ converges by the Comparison Test.

21. Use the Limit Comparison Test with $a_n = \dfrac{1}{1 + \sqrt{n}}$ and $b_n = \dfrac{1}{\sqrt{n}}$: $\lim\limits_{n \to \infty} \dfrac{a_n}{b_n} = \lim\limits_{n \to \infty} \dfrac{\sqrt{n}}{1 + \sqrt{n}} = 1 > 0$. Since

$\displaystyle\sum_{n=1}^{\infty} \dfrac{1}{\sqrt{n}}$ is a divergent p-series $(p = \frac{1}{2} \le 1)$, $\displaystyle\sum_{n=1}^{\infty} \dfrac{1}{1 + \sqrt{n}}$ also diverges.

23. Use the Limit Comparison Test with $a_n = \dfrac{5 + 2n}{(1 + n^2)^2}$ and $b_n = \dfrac{1}{n^3}$:

$\lim\limits_{n \to \infty} \dfrac{a_n}{b_n} = \lim\limits_{n \to \infty} \dfrac{n^3(5 + 2n)}{(1 + n^2)^2} = \lim\limits_{n \to \infty} \dfrac{5n^3 + 2n^4}{(1 + n^2)^2} \cdot \dfrac{1/n^4}{1/(n^2)^2} = \lim\limits_{n \to \infty} \dfrac{\frac{5}{n} + 2}{\left(\frac{1}{n^2} + 1\right)^2} = 2 > 0$. Since $\displaystyle\sum_{n=1}^{\infty} \dfrac{1}{n^3}$ is a

convergent p-series $(p = 3 > 1)$, the series $\displaystyle\sum_{n=1}^{\infty} \dfrac{5 + 2n}{(1 + n^2)^2}$ also converges.

25. If $a_n = \dfrac{1 + n + n^2}{\sqrt{1 + n^2 + n^6}}$ and $b_n = \dfrac{1}{n}$, then

$\lim\limits_{n \to \infty} \dfrac{a_n}{b_n} = \lim\limits_{n \to \infty} \dfrac{n + n^2 + n^3}{\sqrt{1 + n^2 + n^6}} = \lim\limits_{n \to \infty} \dfrac{1/n^2 + 1/n + 1}{\sqrt{1/n^6 + 1/n^4 + 1}} = 1 > 0$, so $\displaystyle\sum_{n=1}^{\infty} \dfrac{1 + n + n^2}{\sqrt{1 + n^2 + n^6}}$ diverges by the

Limit Comparison Test with the divergent harmonic series $\displaystyle\sum_{n=1}^{\infty} \dfrac{1}{n}$.

27. Use the Limit Comparison Test with $a_n = \left(1 + \dfrac{1}{n}\right)^2 e^{-n}$ and $b_n = e^{-n}$: $\lim\limits_{n \to \infty} \dfrac{a_n}{b_n} = \lim\limits_{n \to \infty} \left(1 + \dfrac{1}{n}\right)^2 = 1 > 0$.

Since $\displaystyle\sum_{n=1}^{\infty} e^{-n} = \sum_{n=1}^{\infty} \dfrac{1}{e^n}$ is a convergent geometric series $\left(|r| = \frac{1}{e} < 1\right)$, the series $\displaystyle\sum_{n=1}^{\infty} \left(1 + \dfrac{1}{n}\right)^2 e^{-n}$ also

converges.

29. Clearly $n! = n(n - 1)(n - 2) \cdots (3)(2) \ge 2 \cdot 2 \cdot 2 \cdots \cdot 2 \cdot 2 = 2^{n-1}$, so $\dfrac{1}{n!} \le \dfrac{1}{2^{n-1}}$. $\displaystyle\sum_{n=1}^{\infty} \dfrac{1}{2^{n-1}}$ is a convergent

geometric series $\left(|r| = \frac{1}{2} < 1\right)$, so $\displaystyle\sum_{n=1}^{\infty} \dfrac{1}{n!}$ converges by the Comparison Test.

31. Use the Limit Comparison Test with $a_n = \sin\left(\dfrac{1}{n}\right)$ and $b_n = \dfrac{1}{n}$. Then $\sum a_n$ and $\sum b_n$ are series with positive

terms and $\lim\limits_{n \to \infty} \dfrac{a_n}{b_n} = \lim\limits_{n \to \infty} \dfrac{\sin(1/n)}{1/n} = \lim\limits_{\theta \to 0} \dfrac{\sin \theta}{\theta} = 1 > 0$. Since $\sum_{n=1}^{\infty} b_n$ is the divergent harmonic series,

$\sum_{n=1}^{\infty} \sin(1/n)$ also diverges. (Note that we could also use l'Hospital's Rule to evaluate the limit:

$\lim\limits_{x \to \infty} \dfrac{\sin(1/x)}{1/x} \overset{\text{H}}{=} \lim\limits_{x \to \infty} \dfrac{\cos(1/x) \cdot (-1/x^2)}{-1/x^2} = \lim\limits_{x \to \infty} \cos \dfrac{1}{x} = \cos 0 = 1$.)

33. $\displaystyle\sum_{n=1}^{10} \dfrac{1}{n^4 + n^2} = \dfrac{1}{2} + \dfrac{1}{20} + \dfrac{1}{90} + \cdots + \dfrac{1}{10,100} \approx 0.567975$. Now $\dfrac{1}{n^4 + n^2} < \dfrac{1}{n^4}$, so using the reasoning and

notation of Example 5, the error is $R_{10} \le T_{10} = \displaystyle\sum_{n=11}^{\infty} \dfrac{1}{n^4} \le \int_{10}^{\infty} \dfrac{dx}{x^4} = \lim\limits_{t \to \infty} \left[-\dfrac{x^{-3}}{3} \right]_{10}^{t} = \dfrac{1}{3000} = 0.000\overline{3}$.

35. $\displaystyle\sum_{n=1}^{10} \dfrac{1}{1 + 2^n} = \dfrac{1}{3} + \dfrac{1}{5} + \dfrac{1}{9} + \cdots + \dfrac{1}{1025} \approx 0.76352$. Now $\dfrac{1}{1 + 2^n} < \dfrac{1}{2^n}$, so the error is

$R_{10} \le T_{10} = \displaystyle\sum_{n=11}^{\infty} \dfrac{1}{2^n} = \dfrac{1/2^{11}}{1 - 1/2}$ (geometric series) ≈ 0.00098.

37. Since $\dfrac{d_n}{10^n} \le \dfrac{9}{10^n}$ for each n, and since $\displaystyle\sum_{n=1}^{\infty} \dfrac{9}{10^n}$ is a convergent geometric series ($|r| = \frac{1}{10} < 1$),

$0.d_1 d_2 d_3 \ldots = \displaystyle\sum_{n=1}^{\infty} \dfrac{d_n}{10^n}$ will always converge by the Comparison Test.

39. Since $\sum a_n$ converges, $\displaystyle\lim_{n \to \infty} a_n = 0$, so there exists N such that $|a_n - 0| < 1$ for all $n > N$ $\Rightarrow$ $0 \le a_n < 1$

for all $n > N$ $\Rightarrow$ $0 \le a_n^2 \le a_n$. Since $\sum a_n$ converges, so does $\sum a_n^2$ by the Comparison Test.

41. (a) Since $\displaystyle\lim_{n \to \infty} \dfrac{a_n}{b_n} = \infty$, there is an integer N such that $\dfrac{a_n}{b_n} > 1$ whenever $n > N$. (Take $M = 1$ in

Definition 11.1.5.) Then $a_n > b_n$ whenever $n > N$ and since $\sum b_n$ is divergent, $\sum a_n$ is also divergent by the Comparison Test.

(b) (i) If $a_n = \dfrac{1}{\ln n}$ and $b_n = \dfrac{1}{n}$ for $n \ge 2$, then

$$\lim_{n \to \infty} \frac{a_n}{b_n} = \lim_{n \to \infty} \frac{n}{\ln n} = \lim_{x \to \infty} \frac{x}{\ln x} \overset{\text{H}}{=} \lim_{x \to \infty} \frac{1}{1/x} = \lim_{x \to \infty} x = \infty, \text{ so by part (a), } \sum_{n=2}^{\infty} \frac{1}{\ln n} \text{ is divergent.}$$

(ii) If $a_n = \dfrac{\ln n}{n}$ and $b_n = \dfrac{1}{n}$, then $\displaystyle\sum_{n=1}^{\infty} b_n$ is the divergent harmonic series and

$$\lim_{n \to \infty} \frac{a_n}{b_n} = \lim_{n \to \infty} \ln n = \lim_{x \to \infty} \ln x = \infty, \text{ so } \sum_{n=1}^{\infty} a_n \text{ diverges by part (a).}$$

43. $\displaystyle\lim_{n \to \infty} n a_n = \lim_{n \to \infty} \dfrac{a_n}{1/n}$, so we apply the Limit Comparison Test with $b_n = \dfrac{1}{n}$. Since $\displaystyle\lim_{n \to \infty} n a_n > 0$ we know that

either both series converge or both series diverge, and we also know that $\displaystyle\sum_{n=0}^{\infty} \dfrac{1}{n}$ diverges (p-series with $p = 1$).

Therefore, $\sum a_n$ must be divergent.

45. Yes. Since $\sum a_n$ is a convergent series with positive terms, $\displaystyle\lim_{n \to \infty} a_n = 0$ by Theorem 11.2.6, and

$\sum b_n = \sum \sin(a_n)$ is a series with positive terms (for large enough n). We have

$\displaystyle\lim_{n \to \infty} \dfrac{b_n}{a_n} = \lim_{n \to \infty} \dfrac{\sin(a_n)}{a_n} = 1 > 0$ by Theorem 3.4.2. Thus, $\sum b_n$ is also convergent by the Limit Comparison

Test.

11.5 Alternating Series

1. (a) An alternating series is a series whose terms are alternately positive and negative.

(b) An alternating series $\sum_{n=1}^{\infty} (-1)^{n-1} b_n$ converges if $0 < b_{n+1} \le b_n$ for all n and $\displaystyle\lim_{n \to \infty} b_n = 0$. (This is the Alternating Series Test.)

(c) The error involved in using the partial sum s_n as an approximation to the total sum s is the remainder $R_n = s - s_n$ and the size of the error is smaller than b_{n+1}; that is, $|R_n| \le b_{n+1}$. (This is the Alternating Series Estimation Theorem.)

3. $\dfrac{4}{7} - \dfrac{4}{8} + \dfrac{4}{9} - \dfrac{4}{10} + \dfrac{4}{11} - \cdots = \displaystyle\sum_{n=1}^{\infty} (-1)^{n-1} \dfrac{4}{n+6}$. Now $b_n = \dfrac{4}{n+6} > 0$, $\{b_n\}$ is decreasing, and

$\displaystyle\lim_{n \to \infty} b_n = 0$, so the series converges by the Alternating Series Test.

5. $b_n = \dfrac{1}{\sqrt{n}} > 0$, $\{b_n\}$ is decreasing, and $\lim\limits_{n\to\infty} b_n = 0$, so the series $\sum\limits_{n=1}^{\infty} \dfrac{(-1)^{n-1}}{\sqrt{n}}$ converges by the Alternating Series Test.

7. $\sum\limits_{n=1}^{\infty} a_n = \sum\limits_{n=1}^{\infty}(-1)^n \dfrac{3n-1}{2n+1} = \sum\limits_{n=1}^{\infty}(-1)^n b_n$. Now $\lim\limits_{n\to\infty} b_n = \lim\limits_{n\to\infty}\dfrac{3-1/n}{2+1/n} = \dfrac{3}{2} \neq 0$. Since $\lim\limits_{n\to\infty} a_n \neq 0$ (in fact the limit does not exist), the series diverges by the Test for Divergence.

9. $b_n = \dfrac{1}{4n^2+1} > 0$, $\{b_n\}$ is decreasing, and $\lim\limits_{n\to\infty} b_n = 0$, so the series $\sum\limits_{n=1}^{\infty} \dfrac{(-1)^{n+1}}{4n^2+1}$ converges by the Alternating Series Test.

11. $b_n = \dfrac{n^2}{n^3+4} > 0$ for $n \geq 1$. $\{b_n\}$ is decreasing for $n \geq 2$ since

$$\left(\frac{x^2}{x^3+4}\right)' = \frac{(x^3+4)(2x)-x^2(3x^2)}{(x^3+4)^2} = \frac{x(2x^3+8-3x^3)}{(x^3+4)^2} = \frac{x(8-x^3)}{(x^3+4)^2} < 0 \text{ for } x > 2. \text{ Also,}$$

$\lim\limits_{n\to\infty} b_n = \lim\limits_{n\to\infty}\dfrac{1/n}{1+4/n^3} = 0$. Thus, the series $\sum\limits_{n=1}^{\infty}(-1)^{n+1}\dfrac{n^2}{n^3+4}$ converges by the Alternating Series Test.

13. $\sum\limits_{n=2}^{\infty}(-1)^n \dfrac{n}{\ln n}$. $\lim\limits_{n\to\infty}\dfrac{n}{\ln n} = \lim\limits_{x\to\infty}\dfrac{x}{\ln x} \overset{\text{H}}{=} \lim\limits_{x\to\infty}\dfrac{1}{1/x} = \infty$, so the series diverges by the Test for Divergence.

15. $\sum\limits_{n=1}^{\infty}\dfrac{\cos n\pi}{n^{3/4}} = \sum\limits_{n=1}^{\infty}\dfrac{(-1)^n}{n^{3/4}}$. $b_n = \dfrac{1}{n^{3/4}}$ is decreasing and positive and $\lim\limits_{n\to\infty}\dfrac{1}{n^{3/4}} = 0$, so the series converges by the Alternating Series Test.

17. $\sum\limits_{n=1}^{\infty}(-1)^n \sin\dfrac{\pi}{n}$. $b_n = \sin\dfrac{\pi}{n} > 0$ for $n \geq 2$ and $\sin\dfrac{\pi}{n} \geq \sin\dfrac{\pi}{n+1}$, and $\lim\limits_{n\to\infty}\sin\dfrac{\pi}{n} = \sin 0 = 0$, so the series converges by the Alternating Series Test.

19. $\dfrac{n^n}{n!} = \dfrac{n \cdot n \cdot \cdots \cdot n}{1 \cdot 2 \cdot \cdots \cdot n} \geq n \Rightarrow \lim\limits_{n\to\infty}\dfrac{n^n}{n!} = \infty \Rightarrow \lim\limits_{n\to\infty}\dfrac{(-1)^n n^n}{n!}$ does not exist. So the series diverges by the Test for Divergence.

21.

n	a_n	s_n
1	1	1
2	-0.35355	0.64645
3	0.19245	0.83890
4	-0.125	0.71390
5	0.08944	0.80334
6	-0.06804	0.73530
7	0.05399	0.78929
8	-0.04419	0.74510
9	0.03704	0.78214
10	-0.03162	0.75051

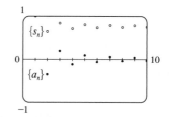

By the Alternating Series Estimation Theorem, the error in the

approximation $\sum\limits_{n=1}^{\infty}\dfrac{(-1)^{n-1}}{n^{3/2}} \approx 0.75051$ is

$|s - s_{10}| \leq b_{11} = 1/(11)^{3/2} \approx 0.0275$ (to four decimal places, rounded up).

23. The series $\sum_{n=1}^{\infty} (-1)^{n-1} \dfrac{1}{n^2}$ satisfies (i) of the Alternating Series Test because $\dfrac{1}{(n+1)^2} < \dfrac{1}{n^2}$ and

(ii) $\lim\limits_{n\to\infty} \dfrac{1}{n^2} = 0$, so the series is convergent. Now $b_{10} = \dfrac{1}{10^2} = 0.01$ and $b_{11} = \dfrac{1}{11^2} = \dfrac{1}{121} \approx 0.008 < 0.01$, so

by the Alternating Series Estimation Theorem, $n = 10$. (That is, since the 11th term is less than the desired error, we need to add the first 10 terms to get the sum to the desired accuracy.)

25. The series $\sum_{n=1}^{\infty} \dfrac{(-2)^n}{n!} = \sum_{n=1}^{\infty} (-1)^n \dfrac{2^n}{n!}$ satisfies (i) of the Alternating Series Test

because $b_{n+1} = \dfrac{2^{n+1}}{(n+1)!} = \dfrac{2 \cdot 2^n}{(n+1)n!} = \dfrac{2}{n+1} \cdot \dfrac{2^n}{n!} = \dfrac{2}{n+1} \cdot b_n \le b_n$ and (ii)

$\lim\limits_{n\to\infty} \dfrac{2^n}{n!} = \dfrac{2}{n} \cdot \dfrac{2}{n-1} \cdot \cdots \cdot \dfrac{2}{2} \cdot \dfrac{2}{1} = 0$, so the series is convergent. Now $b_7 = 2^7/7! \approx 0.025 > 0.01$ and

$b_8 = 2^8/8! \approx 0.006 < 0.01$, so by the Alternating Series Estimation Theorem, $n = 7$. (That is, since the 8th term is less than the desired error, we need to add the first 7 terms to get the sum to the desired accuracy.)

27. $b_7 = \dfrac{1}{7^5} = \dfrac{1}{16,807} \approx 0.000\,059\,5$, so

$\sum_{n=1}^{\infty} \dfrac{(-1)^{n+1}}{n^5} \approx s_6 = \sum_{n=1}^{6} \dfrac{(-1)^{n+1}}{n^5} = 1 - \frac{1}{32} + \frac{1}{243} - \frac{1}{1024} + \frac{1}{3125} - \frac{1}{7776} \approx 0.972\,080$. Adding b_7 to s_6 does

not change the fourth decimal place of s_6, so the sum of the series, correct to four decimal places, is 0.9721.

29. $b_7 = \dfrac{7^2}{10^7} = 0.000\,004\,9$, so

$\sum_{n=1}^{\infty} \dfrac{(-1)^{n-1}n^2}{10^n} \approx s_6 = \sum_{n=1}^{6} \dfrac{(-1)^{n-1}n^2}{10^n} = \frac{1}{10} - \frac{4}{100} + \frac{9}{1000} - \frac{16}{10,000} + \frac{25}{100,000} - \frac{36}{1,000,000} = 0.067\,614$.

Adding b_7 to s_6 does not change the fourth decimal place of s_6, so the sum of the series, correct to four decimal places, is 0.0676.

31. $\sum_{n=1}^{\infty} \dfrac{(-1)^{n-1}}{n} = 1 - \frac{1}{2} + \frac{1}{3} - \frac{1}{4} + \cdots + \frac{1}{49} - \frac{1}{50} + \frac{1}{51} - \frac{1}{52} + \cdots$. The 50th partial sum of this series is an

underestimate, since $\sum_{n=1}^{\infty} \dfrac{(-1)^{n-1}}{n} = s_{50} + \left(\frac{1}{51} - \frac{1}{52} \right) + \left(\frac{1}{53} - \frac{1}{54} \right) + \cdots$, and the terms in parentheses are

all positive. The result can be seen geometrically in Figure 1.

33. Clearly $b_n = \dfrac{1}{n+p}$ is decreasing and eventually positive and $\lim\limits_{n\to\infty} b_n = 0$ for any p. So the series converges (by

the Alternating Series Test) for any p for which every b_n is defined, that is, $n + p \ne 0$ for $n \ge 1$, or p is not a negative integer.

35. $\sum b_{2n} = \sum 1/(2n)^2$ clearly converges (by comparison with the p-series for $p = 2$). So

suppose that $\sum (-1)^{n-1} b_n$ converges. Then by Theorem 11.2.8(ii), so does

$\sum \left[(-1)^{n-1} b_n + b_n \right] = 2 \left(1 + \frac{1}{3} + \frac{1}{5} + \cdots \right) = 2 \sum \dfrac{1}{2n-1}$. But this diverges by comparison with the

harmonic series, a contradiction. Therefore, $\sum (-1)^{n-1} b_n$ must diverge. The Alternating Series Test does not

apply since $\{b_n\}$ is not decreasing.

11.6 Absolute Convergence and the Ratio and Root Tests

1. (a) Since $\lim\limits_{n\to\infty}\left|\dfrac{a_{n+1}}{a_n}\right| = 8 > 1$, part (b) of the Ratio Test tells us that the series $\sum a_n$ is divergent.

(b) Since $\lim\limits_{n\to\infty}\left|\dfrac{a_{n+1}}{a_n}\right| = 0.8 < 1$, part (a) of the Ratio Test tells us that the series $\sum a_n$ is absolutely convergent
(and therefore convergent).

(c) Since $\lim\limits_{n\to\infty}\left|\dfrac{a_{n+1}}{a_n}\right| = 1$, the Ratio Test fails and the series $\sum a_n$ might converge or it might diverge.

3. $\displaystyle\sum_{n=0}^{\infty}\dfrac{(-10)^n}{n!}$. Using the Ratio Test, $\lim\limits_{n\to\infty}\left|\dfrac{a_{n+1}}{a_n}\right| = \lim\limits_{n\to\infty}\left|\dfrac{(-10)^{n+1}}{(n+1)!}\cdot\dfrac{n!}{(-10)^n}\right| = \lim\limits_{n\to\infty}\left|\dfrac{-10}{n+1}\right| = 0 < 1$, so the
series is absolutely convergent.

5. $\displaystyle\sum_{n=1}^{\infty}\dfrac{(-1)^{n+1}}{\sqrt[4]{n}}$ converges by the Alternating Series Test, but $\displaystyle\sum_{n=1}^{\infty}\dfrac{1}{\sqrt[4]{n}}$ is a divergent p-series $\left(p = \frac{1}{4} \leq 1\right)$, so the
given series is conditionally convergent.

7. $\lim\limits_{n\to\infty}|a_n| = \lim\limits_{n\to\infty}\dfrac{n}{5+n} = \lim\limits_{n\to\infty}\dfrac{1}{5/n+1} = 1$, so $\lim\limits_{n\to\infty} a_n \neq 0$. Thus, the given series is divergent by the Test
for Divergence.

9. $\lim\limits_{n\to\infty}\left|\dfrac{a_{n+1}}{a_n}\right| = \lim\limits_{n\to\infty}\dfrac{1/(2n+2)!}{1/(2n)!} = \lim\limits_{n\to\infty}\dfrac{(2n)!}{(2n+2)!} = \lim\limits_{n\to\infty}\dfrac{(2n)!}{(2n+2)(2n+1)(2n)!}$

$= \lim\limits_{n\to\infty}\dfrac{1}{(2n+2)(2n+1)} = 0 < 1$, so the series $\displaystyle\sum_{n=1}^{\infty}\dfrac{1}{(2n)!}$ is absolutely convergent by the Ratio Test. Of course,
absolute convergence is the same as convergence for this series, since all of its terms are positive.

11. Since $0 \leq \dfrac{e^{1/n}}{n^3} \leq \dfrac{e}{n^3} = e\left(\dfrac{1}{n^3}\right)$ and $\displaystyle\sum_{n=1}^{\infty}\dfrac{1}{n^3}$ is a convergent p-series $(p = 3 > 1)$, $\displaystyle\sum_{n=1}^{\infty}\dfrac{e^{1/n}}{n^3}$ converges, and so

$\displaystyle\sum_{n=1}^{\infty}\dfrac{(-1)^n e^{1/n}}{n^3}$ is absolutely convergent.

13. $\lim\limits_{n\to\infty}\left|\dfrac{a_{n+1}}{a_n}\right| = \lim\limits_{n\to\infty}\left[\dfrac{(n+1)\,3^{n+1}}{4^n}\cdot\dfrac{4^{n-1}}{n\cdot 3^n}\right] = \lim\limits_{n\to\infty}\left(\dfrac{3}{4}\cdot\dfrac{n+1}{n}\right) = \dfrac{3}{4} < 1$, so the series $\displaystyle\sum_{n=1}^{\infty}\dfrac{n(-3)^n}{4^{n-1}}$ is
absolutely convergent by the Ratio Test.

15. $\lim\limits_{n\to\infty}\left|\dfrac{a_{n+1}}{a_n}\right| = \lim\limits_{n\to\infty}\left[\dfrac{10^{n+1}}{(n+2)\,4^{2n+3}}\cdot\dfrac{(n+1)\,4^{2n+1}}{10^n}\right] = \lim\limits_{n\to\infty}\left(\dfrac{10}{4^2}\cdot\dfrac{n+1}{n+2}\right) = \dfrac{5}{8} < 1$, so the series

$\displaystyle\sum_{n=1}^{\infty}\dfrac{10^n}{(n+1)4^{2n+1}}$ is absolutely convergent by the Ratio Test. Since the terms of this series are positive, absolute
convergence is the same as convergence.

17. $\displaystyle\sum_{n=2}^{\infty} \frac{(-1)^n}{\ln n}$ converges by the Alternating Series Test since $\displaystyle\lim_{n\to\infty} \frac{1}{\ln n} = 0$ and $\left\{\dfrac{1}{\ln n}\right\}$ is decreasing. Now

$\ln n < n$, so $\dfrac{1}{\ln n} > \dfrac{1}{n}$, and since $\displaystyle\sum_{n=2}^{\infty} \frac{1}{n}$ is the divergent (partial) harmonic series, $\displaystyle\sum_{n=2}^{\infty} \frac{1}{\ln n}$ diverges by the

Comparison Test. Thus, $\displaystyle\sum_{n=2}^{\infty} \frac{(-1)^n}{\ln n}$ is conditionally convergent.

19. $\dfrac{|\cos(n\pi/3)|}{n!} \leq \dfrac{1}{n!}$ and $\displaystyle\sum_{n=1}^{\infty} \frac{1}{n!}$ converges (use the Ratio Test or the result of Exercise 11.4.29), so the series

$\displaystyle\sum_{n=1}^{\infty} \frac{\cos(n\pi/3)}{n!}$ converges absolutely by the Comparison Test.

21. $\displaystyle\lim_{n\to\infty} \sqrt[n]{|a_n|} = \lim_{n\to\infty} \left(\frac{n^n}{3^{1+3n}}\right)^{1/n} = \lim_{n\to\infty} \frac{n}{\sqrt[n]{3} \cdot 3^3} = \infty$, so the series $\displaystyle\sum_{n=1}^{\infty} \frac{n^n}{3^{1+3n}}$ is divergent by the Root Test.

Or: $\displaystyle\lim_{n\to\infty} \left|\frac{a_{n+1}}{a_n}\right| = \lim_{n\to\infty}\left[\frac{(n+1)^{n+1}}{3^{4+3n}} \cdot \frac{3^{1+3n}}{n^n}\right] = \lim_{n\to\infty}\left[\frac{1}{3^3} \cdot \left(\frac{n+1}{n}\right)^n (n+1)\right]$

$= \dfrac{1}{27} \displaystyle\lim_{n\to\infty}\left(1+\frac{1}{n}\right)^n \lim_{n\to\infty}(n+1) = \tfrac{1}{27}e \lim_{n\to\infty}(n+1) = \infty$,

so the series is divergent by the Ratio Test.

23. $\displaystyle\lim_{n\to\infty} \sqrt[n]{|a_n|} = \lim_{n\to\infty} \frac{n^2+1}{2n^2+1} = \lim_{n\to\infty} \frac{1+1/n^2}{2+1/n^2} = \frac{1}{2} < 1$, so the series $\displaystyle\sum_{n=1}^{\infty}\left(\frac{n^2+1}{2n^2+1}\right)^n$ is absolutely

convergent by the Root Test.

25. Use the Ratio Test with the series $1 - \dfrac{1\cdot3}{3!} + \dfrac{1\cdot3\cdot5}{5!} - \dfrac{1\cdot3\cdot5\cdot7}{7!} + \cdots + (-1)^{n-1}\dfrac{1\cdot3\cdot5\cdot\cdots\cdot(2n-1)}{(2n-1)!} + \cdots$

$= \displaystyle\sum_{n=1}^{\infty} (-1)^{n-1}\frac{1\cdot3\cdot5\cdot\cdots\cdot(2n-1)}{(2n-1)!}.$

$\displaystyle\lim_{n\to\infty}\left|\frac{a_{n+1}}{a_n}\right| = \lim_{n\to\infty}\left|\frac{(-1)^n\cdot1\cdot3\cdot5\cdot\cdots\cdot(2n-1)[2(n+1)-1]}{[2(n+1)-1]!} \cdot \frac{(2n-1)!}{(-1)^{n-1}\cdot1\cdot3\cdot5\cdot\cdots\cdot(2n-1)}\right|$

$= \displaystyle\lim_{n\to\infty}\left|\frac{(-1)(2n+1)(2n-1)!}{(2n+1)(2n)(2n-1)!}\right|$

$= \displaystyle\lim_{n\to\infty}\frac{1}{2n} = 0 < 1,$

so the given series is absolutely convergent and therefore convergent.

27. $\displaystyle\sum_{n=1}^{\infty}\frac{2\cdot4\cdot6\cdot\cdots\cdot(2n)}{n!} = \sum_{n=1}^{\infty}\frac{(2\cdot1)\cdot(2\cdot2)\cdot(2\cdot3)\cdot\cdots\cdot(2\cdot n)}{n!} = \sum_{n=1}^{\infty}\frac{2^n n!}{n!} = \sum_{n=1}^{\infty}2^n$, which diverges by the

Test for Divergence since $\displaystyle\lim_{n\to\infty} 2^n = \infty$.

29. By the recursive definition, $\displaystyle\lim_{n\to\infty}\left|\frac{a_{n+1}}{a_n}\right| = \lim_{n\to\infty}\left|\frac{5n+1}{4n+3}\right| = \frac{5}{4} > 1$, so the series diverges by the Ratio Test.

31. (a) $\lim\limits_{n\to\infty}\left|\dfrac{1/(n+1)^3}{1/n^3}\right| = \lim\limits_{n\to\infty}\dfrac{n^3}{(n+1)^3} = \lim\limits_{n\to\infty}\dfrac{1}{(1+1/n)^3} = 1$. Inconclusive.

(b) $\lim\limits_{n\to\infty}\left|\dfrac{(n+1)}{2^{n+1}}\cdot\dfrac{2^n}{n}\right| = \lim\limits_{n\to\infty}\dfrac{n+1}{2n} = \lim\limits_{n\to\infty}\left(\dfrac{1}{2}+\dfrac{1}{2n}\right) = \dfrac{1}{2}$. Conclusive (convergent).

(c) $\lim\limits_{n\to\infty}\left|\dfrac{(-3)^n}{\sqrt{n+1}}\cdot\dfrac{\sqrt{n}}{(-3)^{n-1}}\right| = 3\lim\limits_{n\to\infty}\sqrt{\dfrac{n}{n+1}} = 3\lim\limits_{n\to\infty}\sqrt{\dfrac{1}{1+1/n}} = 3$. Conclusive (divergent).

(d) $\lim\limits_{n\to\infty}\left|\dfrac{\sqrt{n+1}}{1+(n+1)^2}\cdot\dfrac{1+n^2}{\sqrt{n}}\right| = \lim\limits_{n\to\infty}\left[\sqrt{1+\dfrac{1}{n}}\cdot\dfrac{1/n^2+1}{1/n^2+(1+1/n)^2}\right] = 1$. Inconclusive.

33. (a) $\lim\limits_{n\to\infty}\left|\dfrac{a_{n+1}}{a_n}\right| = \lim\limits_{n\to\infty}\left|\dfrac{x^{n+1}}{(n+1)!}\cdot\dfrac{n!}{x^n}\right| = \lim\limits_{n\to\infty}\left|\dfrac{x}{n+1}\right| = |x|\lim\limits_{n\to\infty}\dfrac{1}{n+1} = |x|\cdot 0 = 0 < 1$, so by the Ratio

Test the series $\sum\limits_{n=0}^{\infty}\dfrac{x^n}{n!}$ converges for all x.

(b) Since the series of part (a) always converges, we must have $\lim\limits_{n\to\infty}\dfrac{x^n}{n!} = 0$ by Theorem 11.2.6.

35. (a) $s_5 = \sum\limits_{n=1}^{5}\dfrac{1}{n2^n} = \dfrac{1}{2}+\dfrac{1}{8}+\dfrac{1}{24}+\dfrac{1}{64}+\dfrac{1}{160} = \dfrac{661}{960} \approx 0.68854$. Now the ratios

$r_n = \dfrac{a_{n+1}}{a_n} = \dfrac{n2^n}{(n+1)2^{n+1}} = \dfrac{n}{2(n+1)}$ form an increasing sequence, since

$r_{n+1} - r_n = \dfrac{n+1}{2(n+2)} - \dfrac{n}{2(n+1)} = \dfrac{(n+1)^2 - n(n+2)}{2(n+1)(n+2)} = \dfrac{1}{2(n+1)(n+2)} > 0$. So by Exercise 34(b),

the error in using s_5 is $R_5 \le \dfrac{a_6}{1 - \lim\limits_{n\to\infty} r_n} = \dfrac{1/(6\cdot 2^6)}{1 - 1/2} = \dfrac{1}{192} \approx 0.00521$.

(b) The error in using s_n as an approximation to the sum is $R_n = \dfrac{a_{n+1}}{1 - \frac{1}{2}} = \dfrac{2}{(n+1)2^{n+1}}$. We want

$R_n < 0.00005 \quad\Leftrightarrow\quad \dfrac{1}{(n+1)2^n} < 0.00005 \quad\Leftrightarrow\quad (n+1)2^n > 20{,}000$. To find such an n we can use trial

and error or a graph. We calculate $(11+1)2^{11} = 24{,}576$, so $s_{11} = \sum\limits_{n=1}^{11}\dfrac{1}{n2^n} \approx 0.693109$ is within 0.00005 of

the actual sum.

37. Summing the inequalities $-|a_i| \le a_i \le |a_i|$ for $i = 1, 2, \ldots, n$, we get $-\sum_{i=1}^{n}|a_i| \le \sum_{i=1}^{n} a_i \le \sum_{i=1}^{n}|a_i|$

$\Rightarrow -\lim\limits_{n\to\infty}\sum_{i=1}^{n}|a_i| \le \lim\limits_{n\to\infty}\sum_{i=1}^{n} a_i \le \lim\limits_{n\to\infty}\sum_{i=1}^{n}|a_i| \Rightarrow -\sum_{n=1}^{\infty}|a_n| \le \sum_{n=1}^{\infty} a_n \le \sum_{n=1}^{\infty}|a_n| \Rightarrow$

$\left|\sum_{n=1}^{\infty} a_n\right| \le \sum_{n=1}^{\infty}|a_n|$.

39. (a) Since $\sum a_n$ is absolutely convergent, and since $\left|a_n^+\right| \le |a_n|$ and $\left|a_n^-\right| \le |a_n|$ (because a_n^+ and a_n^- each equal

either a_n or 0), we conclude by the Comparison Test that both $\sum a_n^+$ and $\sum a_n^-$ must be absolutely convergent.

(Or use Theorem 11.2.8.)

(b) We will show by contradiction that both $\sum a_n^+$ and $\sum a_n^-$ must diverge. For suppose that

$\sum a_n^+$ converged. Then so would $\sum\left(a_n^+ - \frac{1}{2}a_n\right)$ by Theorem 11.2.8. But

$\sum\left(a_n^+ - \frac{1}{2}a_n\right) = \sum\left[\frac{1}{2}\left(a_n + |a_n|\right) - \frac{1}{2}a_n\right] = \frac{1}{2}\sum|a_n|$, which diverges because $\sum a_n$ is only conditionally

convergent. Hence, $\sum a_n^+$ can't converge. Similarly, neither can $\sum a_n^-$.

11.7 Strategy for Testing Series

1. $\lim\limits_{n\to\infty} a_n = \lim\limits_{n\to\infty} \dfrac{n^2-1}{n^2+1} = \lim\limits_{n\to\infty} \dfrac{1-1/n^2}{1+1/n} = 1 \neq 0$, so the series $\displaystyle\sum_{n=1}^{\infty} \dfrac{n^2-1}{n^2+1}$ diverges by the Test for Divergence.

3. $\dfrac{1}{n^2+n} < \dfrac{1}{n^2}$ for all $n \geq 1$, so $\displaystyle\sum_{n=1}^{\infty} \dfrac{1}{n^2+n}$ converges by the Comparison Test with $\displaystyle\sum_{n=1}^{\infty} \dfrac{1}{n^2}$, a p-series that converges because $p = 2 > 1$.

5. $\lim\limits_{n\to\infty} \left| \dfrac{a_{n+1}}{a_n} \right| = \lim\limits_{n\to\infty} \left| \dfrac{(-3)^{n+2}}{2^{3(n+1)}} \cdot \dfrac{2^{3n}}{(-3)^{n+1}} \right| = \lim\limits_{n\to\infty} \left| \dfrac{-3 \cdot 2^{3n}}{2^{3n} \cdot 2^3} \right| = \lim\limits_{n\to\infty} \dfrac{3}{2^3} = \dfrac{3}{8} < 1$, so the series

$\displaystyle\sum_{n=1}^{\infty} \dfrac{(-3)^{n+1}}{2^{3n}}$ is absolutely convergent by the Ratio Test.

7. Let $f(x) = \dfrac{1}{x\sqrt{\ln x}}$. Then f is positive, continuous, and decreasing on $[2, \infty)$, so we can apply the Integral Test.

Since $\displaystyle\int \dfrac{1}{x\sqrt{\ln x}}\, dx \begin{bmatrix} u = \ln x, \\ du = dx/x \end{bmatrix} = \int u^{-1/2}\, du = 2u^{1/2} + C = 2\sqrt{\ln x} + C$, we find

$\displaystyle\int_2^{\infty} \dfrac{dx}{x\sqrt{\ln x}} = \lim_{t\to\infty} \int_2^t \dfrac{dx}{x\sqrt{\ln x}} = \lim_{t\to\infty} \left[2\sqrt{\ln x} \right]_2^t = \lim_{t\to\infty} \left(2\sqrt{\ln t} - 2\sqrt{\ln 2} \right) = \infty$. Since the integral

diverges, the given series $\displaystyle\sum_{n=2}^{\infty} \dfrac{1}{n\sqrt{\ln n}}$ diverges.

9. $\displaystyle\sum_{k=1}^{\infty} k^2 e^{-k} = \sum_{k=1}^{\infty} \dfrac{k^2}{e^k}$. Using the Ratio Test, we get

$\lim\limits_{k\to\infty} \left| \dfrac{a_{k+1}}{a_k} \right| = \lim\limits_{k\to\infty} \left| \dfrac{(k+1)^2}{e^{k+1}} \cdot \dfrac{e^k}{k^2} \right| = \lim\limits_{k\to\infty} \left[\left(\dfrac{k+1}{k} \right)^2 \cdot \dfrac{1}{e} \right] = 1^2 \cdot \dfrac{1}{e} = \dfrac{1}{e} < 1$, so the series converges.

11. $b_n = \dfrac{1}{n\ln n} > 0$ for $n \geq 2$, $\{b_n\}$ is decreasing, and $\lim\limits_{n\to\infty} b_n = 0$, so the given series $\displaystyle\sum_{n=2}^{\infty} \dfrac{(-1)^{n+1}}{n\ln n}$ converges by the Alternating Series Test.

13. $\lim\limits_{n\to\infty} \left| \dfrac{a_{n+1}}{a_n} \right| = \lim\limits_{n\to\infty} \left| \dfrac{3^{n+1}(n+1)^2}{(n+1)!} \cdot \dfrac{n!}{3^n n^2} \right| = \lim\limits_{n\to\infty} \left[\dfrac{3(n+1)^2}{(n+1)n^2} \right] = 3\lim\limits_{n\to\infty} \dfrac{n+1}{n^2} = 0 < 1$, so the series

$\displaystyle\sum_{n=1}^{\infty} \dfrac{3^n n^2}{n!}$ converges by the Ratio Test.

15. $\lim\limits_{n\to\infty} \left| \dfrac{a_{n+1}}{a_n} \right| = \lim\limits_{n\to\infty} \left| \dfrac{(n+1)!}{2 \cdot 5 \cdot 8 \cdots (3n+2)[3(n+1)+2]} \cdot \dfrac{2 \cdot 5 \cdot 8 \cdots (3n+2)}{n!} \right|$

$= \lim\limits_{n\to\infty} \dfrac{n+1}{3n+5} = \dfrac{1}{3} < 1$

so the series $\displaystyle\sum_{n=0}^{\infty} \dfrac{n!}{2 \cdot 5 \cdot 8 \cdots (3n+2)}$ converges by the Ratio Test.

17. $\lim\limits_{n\to\infty} 2^{1/n} = 2^0 = 1$, so $\lim\limits_{n\to\infty} (-1)^n 2^{1/n}$ does not exist and the series $\displaystyle\sum_{n=1}^{\infty} (-1)^n 2^{1/n}$ diverges by the Test for Divergence.

19. Let $f(x) = \dfrac{\ln x}{\sqrt{x}}$. Then $f'(x) = \dfrac{2 - \ln x}{2x^{3/2}} < 0$ when $\ln x > 2$ or $x > e^2$, so $\dfrac{\ln n}{\sqrt{n}}$ is decreasing for $n > e^2$.

By l'Hospital's Rule, $\lim\limits_{n\to\infty} \dfrac{\ln n}{\sqrt{n}} = \lim\limits_{n\to\infty} \dfrac{1/n}{1/(2\sqrt{n})} = \lim\limits_{n\to\infty} \dfrac{2}{\sqrt{n}} = 0$, so the series $\sum\limits_{n=1}^{\infty} (-1)^n \dfrac{\ln n}{\sqrt{n}}$ converges by

the Alternating Series Test.

21. $\sum\limits_{n=1}^{\infty} \dfrac{(-2)^{2n}}{n^n} = \sum\limits_{n=1}^{\infty} \left(\dfrac{4}{n}\right)^n$. $\lim\limits_{n\to\infty} \sqrt[n]{|a_n|} = \lim\limits_{n\to\infty} \dfrac{4}{n} = 0 < 1$, so the given series is absolutely convergent by the
Root Test.

23. Using the Limit Comparison Test with $a_n = \tan\left(\dfrac{1}{n}\right)$ and $b_n = \dfrac{1}{n}$, we have

$$\lim_{n\to\infty} \frac{a_n}{b_n} = \lim_{n\to\infty} \frac{\tan(1/n)}{1/n} = \lim_{x\to\infty} \frac{\tan(1/x)}{1/x} \overset{\text{H}}{=} \lim_{x\to\infty} \frac{\sec^2(1/x) \cdot (-1/x^2)}{-1/x^2} = \lim_{x\to\infty} \sec^2(1/x) = 1^2 = 1 > 0.$$

Since $\sum_{n=1}^{\infty} b_n$ is the divergent harmonic series, $\sum_{n=1}^{\infty} a_n$ is also divergent.

25. Use the Ratio Test. $\lim\limits_{n\to\infty} \left|\dfrac{a_{n+1}}{a_n}\right| = \lim\limits_{n\to\infty} \left|\dfrac{(n+1)!}{e^{(n+1)^2}} \cdot \dfrac{e^{n^2}}{n!}\right| = \lim\limits_{n\to\infty} \dfrac{(n+1)n! \cdot e^{n^2}}{e^{n^2+2n+1}n!} = \lim\limits_{n\to\infty} \dfrac{n+1}{e^{2n+1}} = 0 < 1$, so

$\sum\limits_{n=1}^{\infty} \dfrac{n!}{e^{n^2}}$ converges.

27. $\displaystyle\int_2^{\infty} \dfrac{\ln x}{x^2}\,dx = \lim_{t\to\infty} \left[-\dfrac{\ln x}{x} - \dfrac{1}{x}\right]_1^t$ (using integration by parts) $\overset{\text{H}}{=} 1$. So $\sum\limits_{n=1}^{\infty} \dfrac{\ln n}{n^2}$ converges by the Integral Test,

and since $\dfrac{k \ln k}{(k+1)^3} < \dfrac{k \ln k}{k^3} = \dfrac{\ln k}{k^2}$, the given series $\sum\limits_{k=1}^{\infty} \dfrac{k \ln k}{(k+1)^3}$ converges by the Comparison Test.

29. $0 < \dfrac{\tan^{-1} n}{n^{3/2}} < \dfrac{\pi/2}{n^{3/2}}$. $\sum\limits_{n=1}^{\infty} \dfrac{\pi/2}{n^{3/2}} = \dfrac{\pi}{2} \sum\limits_{n=1}^{\infty} \dfrac{1}{n^{3/2}}$ which is a convergent p-series $(p = \frac{3}{2} > 1)$, so

$\sum\limits_{n=1}^{\infty} \dfrac{\tan^{-1} n}{n^{3/2}}$ converges by the Comparison Test.

31. $\lim\limits_{k\to\infty} a_k = \lim\limits_{k\to\infty} \dfrac{5^k}{3^k + 4^k} = [\text{divide by } 4^k] \lim\limits_{k\to\infty} \dfrac{(5/4)^k}{(3/4)^k + 1} = \infty$ since $\lim\limits_{k\to\infty} \left(\dfrac{3}{4}\right)^k = 0$ and $\lim\limits_{k\to\infty} \left(\dfrac{5}{4}\right)^k = \infty$.

Thus, $\sum\limits_{k=1}^{\infty} \dfrac{5^k}{3^k + 4^k}$ diverges by the Test for Divergence.

33. Let $a_n = \dfrac{\sin(1/n)}{\sqrt{n}}$ and $b_n = \dfrac{1}{n\sqrt{n}}$. Then $\lim\limits_{n\to\infty} \dfrac{a_n}{b_n} = \lim\limits_{n\to\infty} \dfrac{\sin(1/n)}{1/n} = 1 > 0$, so $\sum\limits_{n=1}^{\infty} \dfrac{\sin(1/n)}{\sqrt{n}}$ converges by

limit comparison with the convergent p-series $\sum\limits_{n=1}^{\infty} \dfrac{1}{n^{3/2}}$ $(p = 3/2 > 1)$.

35. $\lim\limits_{n\to\infty} \sqrt[n]{|a_n|} = \lim\limits_{n\to\infty} \left(\dfrac{n}{n+1}\right)^{n^2/n} = \lim\limits_{n\to\infty} \dfrac{1}{[(n+1)/n]^n} = \dfrac{1}{\lim\limits_{n\to\infty} (1+1/n)^n} = \dfrac{1}{e} < 1$, so the series

$\sum\limits_{n=1}^{\infty} \left(\dfrac{n}{n+1}\right)^{n^2}$ converges by the Root Test.

37. $\lim\limits_{n\to\infty} \sqrt[n]{|a_n|} = \lim\limits_{n\to\infty} \left(2^{1/n} - 1\right) = 1 - 1 = 0 < 1$, so the series $\sum\limits_{n=1}^{\infty} \left(\sqrt[n]{2} - 1\right)^n$ converges by the Root Test.

11.8 Power Series

1. A power series is a series of the form $\sum_{n=0}^{\infty} c_n x^n = c_0 + c_1 x + c_2 x^2 + c_3 x^3 + \cdots$, where x is a variable and the c_n's are constants called the coefficients of the series.

More generally, a series of the form $\sum_{n=0}^{\infty} c_n (x - a)^n = c_0 + c_1(x - a) + c_2(x - a)^2 + \cdots$ is called a power series in $(x - a)$ or a power series centered at a or a power series about a, where a is a constant.

3. If $a_n = \dfrac{x^n}{\sqrt{n}}$, then $\lim\limits_{n \to \infty} \left| \dfrac{a_{n+1}}{a_n} \right| = \lim\limits_{n \to \infty} \left| \dfrac{x^{n+1}}{\sqrt{n+1}} \cdot \dfrac{\sqrt{n}}{x^n} \right| = \lim\limits_{n \to \infty} \left| \dfrac{x}{\sqrt{n+1}/\sqrt{n}} \right| = \lim\limits_{n \to \infty} \dfrac{|x|}{\sqrt{1+1/n}} = |x|.$

By the Ratio Test, the series $\sum_{n=1}^{\infty} \dfrac{x^n}{\sqrt{n}}$ converges when $|x| < 1$, so the radius of convergence $R = 1$. Now we'll

check the endpoints, that is, $x = \pm 1$. When $x = 1$, the series $\sum_{n=1}^{\infty} \dfrac{1}{\sqrt{n}}$ diverges because it is a p-series with

$p = \frac{1}{2} \leq 1$. When $x = -1$, the series $\sum_{n=1}^{\infty} \dfrac{(-1)^n}{\sqrt{n}}$ converges by the Alternating Series Test. Thus, the interval of

convergence is $I = [-1, 1)$.

5. If $a_n = \dfrac{(-1)^{n-1} x^n}{n^3}$, then $\lim\limits_{n \to \infty} \left| \dfrac{a_{n+1}}{a_n} \right| = \lim\limits_{n \to \infty} \left| \dfrac{(-1)^n x^{n+1}}{(n+1)^3} \cdot \dfrac{n^3}{(-1)^{n-1} x^n} \right| = \lim\limits_{n \to \infty} \left| \dfrac{(-1)x n^3}{(n+1)^3} \right|$

$= \lim\limits_{n \to \infty} \left[\left(\dfrac{n}{n+1} \right)^3 |x| \right] = 1^3 \cdot |x| = |x|.$ By the Ratio Test, the series $\sum_{n=1}^{\infty} \dfrac{(-1)^{n-1} x^n}{n^3}$ converges when $|x| < 1$,

so the radius of convergence $R = 1$. Now we'll check the endpoints, that is, $x = \pm 1$. When $x = 1$, the series

$\sum_{n=1}^{\infty} \dfrac{(-1)^{n-1}}{n^3}$ converges by the Alternating Series Test. When $x = -1$, the series

$\sum_{n=1}^{\infty} \dfrac{(-1)^{n-1}(-1)^n}{n^3} = -\sum_{n=1}^{\infty} \dfrac{1}{n^3}$ converges because it is a constant multiple of a convergent p-series ($p = 3 > 1$).

Thus, the interval of convergence is $I = [-1, 1]$.

7. If $a_n = \dfrac{x^n}{n!}$, then $\lim\limits_{n \to \infty} \left| \dfrac{a_{n+1}}{a_n} \right| = \lim\limits_{n \to \infty} \left| \dfrac{x^{n+1}}{(n+1)!} \cdot \dfrac{n!}{x^n} \right| = \lim\limits_{n \to \infty} \left| \dfrac{x}{n+1} \right| = |x| \lim\limits_{n \to \infty} \dfrac{1}{n+1} = |x| \cdot 0 = 0 < 1$ for

all real x. So, by the Ratio Test, $R = \infty$, and $I = (-\infty, \infty)$.

9. $\lim\limits_{n \to \infty} \left| \dfrac{a_{n+1}}{a_n} \right| = \lim\limits_{n \to \infty} \dfrac{(n+1) 4^{n+1} |x|^{n+1}}{n 4^n |x|^n} = \lim\limits_{n \to \infty} \left(1 + \dfrac{1}{n} \right) 4 |x| = 4 |x|.$ Now $4 |x| < 1 \iff |x| < \frac{1}{4}$, so by

the Ratio Test, $R = \frac{1}{4}$. When $x = \frac{1}{4}$, we get the divergent series $\sum_{n=1}^{\infty} (-1)^n n$, and when $x = -\frac{1}{4}$, we get the

divergent series $\sum_{n=1}^{\infty} n$. Thus, $I = \left(-\frac{1}{4}, \frac{1}{4} \right)$.

11. $a_n = \dfrac{(-2)^n x^n}{\sqrt[4]{n}}$, so $\lim\limits_{n \to \infty} \left| \dfrac{a_{n+1}}{a_n} \right| = \lim\limits_{n \to \infty} \dfrac{2^{n+1} |x|^{n+1}}{\sqrt[4]{n+1}} \cdot \dfrac{\sqrt[4]{n}}{2^n |x|^n} = \lim\limits_{n \to \infty} 2 |x| \sqrt[4]{\dfrac{n}{n+1}} = 2 |x|,$ so by the

Ratio Test, the series converges when $2 |x| < 1 \iff |x| < \frac{1}{2}$, so $R = \frac{1}{2}$. When $x = -\frac{1}{2}$, we get the divergent

p-series $\sum_{n=1}^{\infty} \dfrac{1}{\sqrt[4]{n}}$ ($p = \frac{1}{4} \leq 1$). When $x = \frac{1}{2}$, we get the series $\sum_{n=1}^{\infty} \dfrac{(-1)^n}{\sqrt[4]{n}}$, which converges by the Alternating

Series Test. Thus, $I = \left(-\frac{1}{2}, \frac{1}{2} \right]$.

13. If $a_n = (-1)^n \dfrac{x^n}{4^n \ln n}$, then

$$\lim_{n\to\infty} \left| \frac{a_{n+1}}{a_n} \right| = \lim_{n\to\infty} \left| \frac{x^{n+1}}{4^{n+1} \ln(n+1)} \cdot \frac{4^n \ln n}{x^n} \right| = \frac{|x|}{4} \lim_{n\to\infty} \frac{\ln n}{\ln(n+1)} = \frac{|x|}{4} \cdot 1 \text{ (by l'Hospital's Rule)} = \frac{|x|}{4}.$$

By the Ratio Test, the series converges when $\dfrac{|x|}{4} < 1 \iff |x| < 4$, so $R = 4$. When $x = -4$,

$$\sum_{n=2}^{\infty} (-1)^n \frac{x^n}{4^n \ln n} = \sum_{n=2}^{\infty} \frac{[(-1)(-4)]^n}{4^n \ln n} = \sum_{n=2}^{\infty} \frac{1}{\ln n}. \text{ Since } \ln n < n \text{ for } n \geq 2, \frac{1}{\ln n} > \frac{1}{n} \text{ and } \sum_{n=2}^{\infty} \frac{1}{n} \text{ is the}$$

divergent harmonic series (without the $n = 1$ term), $\displaystyle\sum_{n=2}^{\infty} \frac{1}{\ln n}$ is divergent by the Comparison Test. When $x = 4$,

$$\sum_{n=2}^{\infty} (-1)^n \frac{x^n}{4^n \ln n} = \sum_{n=2}^{\infty} (-1)^n \frac{1}{\ln n}, \text{ which converges by the Alternating Series Test. Thus, } I = (-4, 4].$$

15. If $a_n = \sqrt{n}\,(x-1)^n$, then $\displaystyle\lim_{n\to\infty} \left| \frac{a_{n+1}}{a_n} \right| = \lim_{n\to\infty} \left| \frac{\sqrt{n+1}\,|x-1|^{n+1}}{\sqrt{n}\,|x-1|^n} \right| = \lim_{n\to\infty} \sqrt{1 + \frac{1}{n}}\,|x-1| = |x-1|$. By

the Ratio Test, the series converges when $|x-1| < 1$ [so $R = 1$] $\iff -1 < x-1 < 1 \iff 0 < x < 2$.

When $x = 0$, the series becomes $\sum_{n=0}^{\infty} (-1)^n \sqrt{n}$, which diverges by the Test for Divergence. When $x = 2$, the

series becomes $\sum_{n=0}^{\infty} \sqrt{n}$, which also diverges by the Test for Divergence. Thus, $I = (0, 2)$.

17. If $a_n = (-1)^n \dfrac{(x+2)^n}{n2^n}$, then

$$\lim_{n\to\infty} \left| \frac{a_{n+1}}{a_n} \right| = \lim_{n\to\infty} \left[\frac{|x+2|^{n+1}}{(n+1)2^{n+1}} \cdot \frac{n2^n}{|x+2|^n} \right] = \lim_{n\to\infty} \frac{n}{n+1} \cdot \frac{|x+2|}{2} = \frac{|x+2|}{2}. \text{ By the Ratio Test, the}$$

series converges when $\dfrac{|x+2|}{2} < 1 \iff |x+2| < 2$ [so $R = 2$] $\iff -2 < x+2 < 2 \iff -4 < x < 0$.

When $x = -4$, the series becomes $\displaystyle\sum_{n=1}^{\infty} (-1)^n \frac{(-2)^n}{n2^n} = \sum_{n=1}^{\infty} \frac{2^n}{n2^n} = \sum_{n=1}^{\infty} \frac{1}{n}$, which is the divergent harmonic series.

When $x = 0$, the series is $\displaystyle\sum_{n=1}^{\infty} \frac{(-1)^n}{n}$, the alternating harmonic series, which converges by the Alternating Series

Test. Thus, $I = (-4, 0]$.

19. If $a_n = \dfrac{(x-2)^n}{n^n}$, then $\displaystyle\lim_{n\to\infty} \sqrt[n]{|a_n|} = \lim_{n\to\infty} \frac{|x-2|}{n} = 0$, so the series converges for all x (by the Root Test).

$R = \infty$ and $I = (-\infty, \infty)$.

21. $a_n = \dfrac{n}{b^n}(x-a)^n$, where $b > 0$.

$$\lim_{n\to\infty} \left| \frac{a_{n+1}}{a_n} \right| = \lim_{n\to\infty} \frac{(n+1)\,|x-a|^{n+1}}{b^{n+1}} \cdot \frac{b^n}{n\,|x-a|^n} = \lim_{n\to\infty} \left(1 + \frac{1}{n} \right) \frac{|x-a|}{b} = \frac{|x-a|}{b}.$$

By the Ratio Test, the series converges when $\dfrac{|x-a|}{b} < 1 \iff |x-a| < b$ [so $R = b$] $\iff$

$-b < x-a < b \iff a-b < x < a+b$. When $|x-a| = b$, $\displaystyle\lim_{n\to\infty} |a_n| = \lim_{n\to\infty} n = \infty$, so the series diverges.

Thus, $I = (a-b, a+b)$.

23. If $a_n = n!(2x-1)^n$, then $\displaystyle\lim_{n\to\infty} \left| \frac{a_{n+1}}{a_n} \right| = \lim_{n\to\infty} \left| \frac{(n+1)!(2x-1)^{n+1}}{n!(2x-1)^n} \right| = \lim_{n\to\infty} (n+1)\,|2x-1| \to \infty$

as $n \to \infty$ for all $x \neq \frac{1}{2}$. Since the series diverges for all $x \neq \frac{1}{2}$, $R = 0$ and $I = \{\frac{1}{2}\}$.

25. $\lim\limits_{n\to\infty}\left|\dfrac{a_{n+1}}{a_n}\right| = \lim\limits_{n\to\infty}\left[\dfrac{|4x+1|^{n+1}}{(n+1)^2}\cdot\dfrac{n^2}{|4x+1|^n}\right] = \lim\limits_{n\to\infty}\dfrac{|4x+1|}{(1+1/n)^2} = |4x+1|$, so by the Ratio Test, the

series converges when $|4x+1| < 1 \Leftrightarrow -1 < 4x+1 < 1 \Leftrightarrow -2 < 4x < 0 \Leftrightarrow -\frac{1}{2} < x < 0$, so

$R = \frac{1}{4}$. When $x = -\frac{1}{2}$, the series becomes $\sum\limits_{n=1}^{\infty}\dfrac{(-1)^n}{n^2}$, which converges by the Alternating Series Test. When

$x = 0$, the series becomes $\sum\limits_{n=1}^{\infty}\dfrac{1}{n^2}$, a convergent p-series ($p = 2 > 1$). $I = \left[-\frac{1}{2}, 0\right]$.

27. If $a_n = \dfrac{x^n}{(\ln n)^n}$, then $\lim\limits_{n\to\infty}\sqrt[n]{|a_n|} = \lim\limits_{n\to\infty}\dfrac{|x|}{\ln n} = 0 < 1$ for all x, so $R = \infty$ and $I = (-\infty, \infty)$ by the

Root Test.

29. (a) We are given that the power series $\sum_{n=0}^{\infty} c_n x^n$ is convergent for $x = 4$. So by Theorem 3, it must converge for
at least $-4 < x \leq 4$. In particular, it converges when $x = -2$; that is, $\sum_{n=0}^{\infty} c_n(-2)^n$ is convergent.

(b) It does not follow that $\sum_{n=0}^{\infty} c_n(-4)^n$ is necessarily convergent. [See the comments after Theorem 3 about
convergence at the endpoint of an interval. An example is $c_n = (-1)^n/(n4^n)$.]

31. If $a_n = \dfrac{(n!)^k}{(kn)!}x^n$, then

$$\lim\limits_{n\to\infty}\left|\dfrac{a_{n+1}}{a_n}\right| = \lim\limits_{n\to\infty}\dfrac{[(n+1)!]^k(kn)!}{(n!)^k[k(n+1)]!}|x| = \lim\limits_{n\to\infty}\dfrac{(n+1)^k}{(kn+k)(kn+k-1)\cdots(kn+2)(kn+1)}|x|$$

$$= \lim\limits_{n\to\infty}\left[\dfrac{(n+1)}{(kn+1)}\dfrac{(n+1)}{(kn+2)}\cdots\dfrac{(n+1)}{(kn+k)}\right]|x|$$

$$= \lim\limits_{n\to\infty}\left[\dfrac{n+1}{kn+1}\right]\lim\limits_{n\to\infty}\left[\dfrac{n+1}{kn+2}\right]\cdots\lim\limits_{n\to\infty}\left[\dfrac{n+1}{kn+k}\right]|x| = \left(\dfrac{1}{k}\right)^k|x| < 1 \Leftrightarrow$$

$|x| < k^k$ for convergence, and the radius of convergence is $R = k^k$.

33. (a) If $a_n = \dfrac{(-1)^n x^{2n+1}}{n!(n+1)!\,2^{2n+1}}$, then

$$\lim\limits_{n\to\infty}\left|\dfrac{a_{n+1}}{a_n}\right| = \lim\limits_{n\to\infty}\left|\dfrac{x^{2n+3}}{(n+1)!(n+2)!\,2^{2n+3}}\cdot\dfrac{n!(n+1)!\,2^{2n+1}}{x^{2n+1}}\right| = \left(\dfrac{x}{2}\right)^2\lim\limits_{n\to\infty}\dfrac{1}{(n+1)(n+2)} = 0 \text{ for}$$

all x. So $J_1(x)$ converges for all x and its domain is $(-\infty, \infty)$.

(b), (c) The initial terms of $J_1(x)$ up to $n = 5$ are

$a_0 = \dfrac{x}{2},\ a_1 = -\dfrac{x^3}{16},\ a_2 = \dfrac{x^5}{384},\ a_3 = -\dfrac{x^7}{18{,}432},$

$a_4 = \dfrac{x^9}{1{,}474{,}560},$ and $a_5 = -\dfrac{x^{11}}{176{,}947{,}200}.$ The

partial sums seem to approximate $J_1(x)$ well near

the origin, but as $|x|$ increases, we need to take a

large number of terms to get a good approximation.

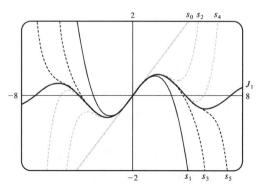

35. $s_{2n-1} = 1 + 2x + x^2 + 2x^3 + x^4 + 2x^5 + \cdots + x^{2n-2} + 2x^{2n-1}$

$\qquad = 1(1 + 2x) + x^2(1 + 2x) + x^4(1 + 2x) + \cdots + x^{2n-2}(1 + 2x)$

$\qquad = (1 + 2x)(1 + x^2 + x^4 + \cdots + x^{2n-2})$

$\qquad = (1 + 2x)\dfrac{1 - x^{2n}}{1 - x^2}$ [by (11.2.3)] with $r = x^2$] $\to \dfrac{1 + 2x}{1 - x^2}$ as $n \to \infty$ [by (11.2.4)],

when $|x| < 1$. Also $s_{2n} = s_{2n-1} + x^{2n} \to \dfrac{1 + 2x}{1 - x^2}$ since $x^{2n} \to 0$ for $|x| < 1$. Therefore,

$s_n \to \dfrac{1 + 2x}{1 - x^2}$ since s_{2n} and s_{2n-1} both approach $\dfrac{1 + 2x}{1 - x^2}$ as $n \to \infty$. Thus, the interval of convergence is $(-1, 1)$

and $f(x) = \dfrac{1 + 2x}{1 - x^2}$.

37. We use the Root Test on the series $\sum c_n x^n$. We need $\displaystyle\lim_{n \to \infty} \sqrt[n]{|c_n x^n|} = |x| \lim_{n \to \infty} \sqrt[n]{|c_n|} = c|x| < 1$ for

convergence, or $|x| < 1/c$, so $R = 1/c$.

39. For $2 < x < 3$, $\sum c_n x^n$ diverges and $\sum d_n x^n$ converges. By Exercise 11.2.61, $\sum (c_n + d_n)x^n$ diverges. Since both series converge for $|x| < 2$, the radius of convergence of $\sum (c_n + d_n)x^n$ is 2.

11.9 Representations of Functions as Power Series

1. If $f(x) = \displaystyle\sum_{n=0}^{\infty} c_n x^n$ has radius of convergence 10, then $f'(x) = \displaystyle\sum_{n=1}^{\infty} n c_n x^{n-1}$ also has radius of convergence 10

by Theorem 2.

3. Our goal is to write the function in the form $\dfrac{1}{1 - r}$, and then use Equation (1) to represent the function as a sum of a

power series. $f(x) = \dfrac{1}{1 + x} = \dfrac{1}{1 - (-x)} = \displaystyle\sum_{n=0}^{\infty} (-x)^n = \sum_{n=0}^{\infty} (-1)^n x^n$ with $|-x| < 1 \Leftrightarrow |x| < 1$, so $R = 1$

and $I = (-1, 1)$.

5. Replacing x with x^3 in (1) gives $f(x) = \dfrac{1}{1 - x^3} = \displaystyle\sum_{n=0}^{\infty} (x^3)^n = \sum_{n=0}^{\infty} x^{3n}$. The series converges when $|x^3| < 1$

$\Leftrightarrow |x|^3 < 1 \Leftrightarrow |x| < \sqrt[3]{1} \Leftrightarrow |x| < 1$. Thus, $R = 1$ and $I = (-1, 1)$.

7. $f(x) = \dfrac{1}{x - 5} = -\dfrac{1}{5}\left(\dfrac{1}{1 - x/5}\right) = -\dfrac{1}{5}\displaystyle\sum_{n=0}^{\infty}\left(\dfrac{x}{5}\right)^n$ or equivalently, $-\sum_{n=0}^{\infty}\dfrac{1}{5^{n+1}}x^n$. The series converges when

$\left|\dfrac{x}{5}\right| < 1$; that is, when $|x| < 5$, so $I = (-5, 5)$.

9. $f(x) = \dfrac{x}{9 + x^2} = \dfrac{x}{9}\left[\dfrac{1}{1 + (x/3)^2}\right] = \dfrac{x}{9}\left[\dfrac{1}{1 - \{-(x/3)^2\}}\right] = \dfrac{x}{9}\displaystyle\sum_{n=0}^{\infty}\left[-\left(\dfrac{x}{3}\right)^2\right]^n$

$\qquad = \dfrac{x}{9}\displaystyle\sum_{n=0}^{\infty}(-1)^n \dfrac{x^{2n}}{9^n} = \sum_{n=0}^{\infty}(-1)^n \dfrac{x^{2n+1}}{9^{n+1}}$. The geometric series $\displaystyle\sum_{n=0}^{\infty}\left[-\left(\dfrac{x}{3}\right)^2\right]^n$ converges when

$\left|-\left(\dfrac{x}{3}\right)^2\right| < 1 \Leftrightarrow \dfrac{|x^2|}{9} < 1 \Leftrightarrow |x|^2 < 9 \Leftrightarrow |x| < 3$, so $R = 3$ and $I = (-3, 3)$.

11. $f(x) = \dfrac{3}{x^2 + x - 2} = \dfrac{3}{(x+2)(x-1)} = \dfrac{A}{x+2} + \dfrac{B}{x-1}$ $\Rightarrow$ $3 = A(x-1) + B(x+2)$. Taking $x = -2$, we

get $A = -1$. Taking $x = 1$, we get $B = 1$. Thus,

$$\frac{3}{x^2 + x - 2} = \frac{1}{x-1} - \frac{1}{x+2} = -\frac{1}{1-x} - \frac{1}{2}\frac{1}{1+x/2} = -\sum_{n=0}^{\infty} x^n - \frac{1}{2}\sum_{n=0}^{\infty}\left(-\frac{x}{2}\right)^n$$

$$= \sum_{n=0}^{\infty}\left[-1 - \tfrac{1}{2}\left(-\tfrac{1}{2}\right)^n\right]x^n = \sum_{n=0}^{\infty}\left[-1 + \left(-\tfrac{1}{2}\right)^{n+1}\right]x^n = \sum_{n=0}^{\infty}\left[\frac{(-1)^{n+1}}{2^{n+1}} - 1\right]x^n$$

We represented the given function as the sum of two geometric series; the first converges for $x \in (-1, 1)$ and the

second converges for $x \in (-2, 2)$. Thus, the sum converges for $x \in (-1, 1) = I$.

13. **(a)** $f(x) = \dfrac{1}{(1+x)^2} = \dfrac{d}{dx}\left(\dfrac{-1}{1+x}\right) = -\dfrac{d}{dx}\left[\displaystyle\sum_{n=0}^{\infty}(-1)^n x^n\right]$ [from Exercise 3]

$$= \sum_{n=1}^{\infty}(-1)^{n+1}nx^{n-1} \text{ [from Theorem 2(i)]} = \sum_{n=0}^{\infty}(-1)^n(n+1)x^n \text{ with } R = 1.$$

In the last step, note that we *decreased* the initial value of the summation variable n by 1, and then *increased*

each occurrence of n in the term by 1 [also note that $(-1)^{n+2} = (-1)^n$].

(b) $f(x) = \dfrac{1}{(1+x)^3} = -\dfrac{1}{2}\dfrac{d}{dx}\left[\dfrac{1}{(1+x)^2}\right] = -\dfrac{1}{2}\dfrac{d}{dx}\left[\displaystyle\sum_{n=0}^{\infty}(-1)^n(n+1)x^n\right]$ [from part (a)]

$$= -\frac{1}{2}\sum_{n=1}^{\infty}(-1)^n(n+1)nx^{n-1} = \frac{1}{2}\sum_{n=0}^{\infty}(-1)^n(n+2)(n+1)x^n \text{ with } R = 1.$$

(c) $f(x) = \dfrac{x^2}{(1+x)^3} = x^2 \cdot \dfrac{1}{(1+x)^3} = x^2 \cdot \dfrac{1}{2}\displaystyle\sum_{n=0}^{\infty}(-1)^n(n+2)(n+1)x^n$ [from part (b)]

$$= \frac{1}{2}\sum_{n=0}^{\infty}(-1)^n(n+2)(n+1)x^{n+2}. \text{ To write the power series with } x^n \text{ rather than } x^{n+2},$$

we will *decrease* each occurrence of n in the term by 2 and *increase* the initial value of the summation variable

by 2. This gives us $\dfrac{1}{2}\displaystyle\sum_{n=2}^{\infty}(-1)^n(n)(n-1)x^n$.

15. $f(x) = \ln(5 - x) = -\displaystyle\int \dfrac{dx}{5-x} = -\dfrac{1}{5}\int \dfrac{dx}{1 - x/5}$

$$= -\frac{1}{5}\int\left[\sum_{n=0}^{\infty}\left(\frac{x}{5}\right)^n\right]dx = C - \frac{1}{5}\sum_{n=0}^{\infty}\frac{x^{n+1}}{5^n(n+1)} = C - \sum_{n=1}^{\infty}\frac{x^n}{n5^n}$$

Putting $x = 0$, we get $C = \ln 5$. The series converges for $|x/5| < 1$ $\Leftrightarrow$ $|x| < 5$, so $R = 5$.

17. $\dfrac{1}{2-x} = \dfrac{1}{2(1-x/2)} = \dfrac{1}{2}\displaystyle\sum_{n=0}^{\infty}\left(\dfrac{x}{2}\right)^n = \sum_{n=0}^{\infty}\dfrac{1}{2^{n+1}}x^n$ for $\left|\dfrac{x}{2}\right| < 1$ $\Leftrightarrow$ $|x| < 2$. Now

$$\frac{1}{(x-2)^2} = \frac{d}{dx}\left(\frac{1}{2-x}\right) = \frac{d}{dx}\left(\sum_{n=0}^{\infty}\frac{1}{2^{n+1}}x^n\right) = \sum_{n=1}^{\infty}\frac{n}{2^{n+1}}x^{n-1} = \sum_{n=0}^{\infty}\frac{n+1}{2^{n+2}}x^n. \text{ So}$$

$$f(x) = \frac{x^3}{(x-2)^2} = x^3\sum_{n=0}^{\infty}\frac{n+1}{2^{n+2}}x^n = \sum_{n=0}^{\infty}\frac{n+1}{2^{n+2}}x^{n+3} \text{ or } \sum_{n=3}^{\infty}\frac{n-2}{2^{n-1}}x^n \text{ for } |x| < 2. \text{ Thus, } R = 2 \text{ and}$$

$I = (-2, 2)$.

19. $f(x) = \ln(3 + x) = \int \dfrac{dx}{3 + x} = \dfrac{1}{3} \int \dfrac{dx}{1 + x/3} = \dfrac{1}{3} \int \dfrac{dx}{1 - (-x/3)} = \dfrac{1}{3} \int \sum\limits_{n=0}^{\infty} \left(-\dfrac{x}{3}\right)^n dx$

$= C + \dfrac{1}{3} \sum\limits_{n=0}^{\infty} \dfrac{(-1)^n}{(n+1)3^n} x^{n+1} = \ln 3 + \dfrac{1}{3} \sum\limits_{n=1}^{\infty} \dfrac{(-1)^{n-1}}{n3^{n-1}} x^n \quad [C = f(0) = \ln 3]$

$= \ln 3 + \sum\limits_{n=1}^{\infty} \dfrac{(-1)^{n-1}}{n3^n} x^n.$ The series converges when $|-x/3| < 1 \iff |x| < 3$, so $R = 3$.

The terms of the series are $a_0 = \ln 3,\ a_1 = \dfrac{x}{3},\ a_2 = -\dfrac{x^2}{18},\ a_3 = \dfrac{x^3}{81},\ a_4 = -\dfrac{x^4}{324},\ a_5 = \dfrac{x^5}{1215}, \ldots.$

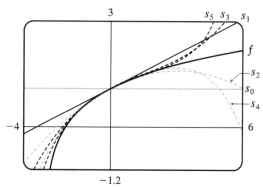

As n increases, $s_n(x)$ approximates f better on the interval of convergence, which is $(-3, 3)$.

21. $f(x) = \ln\left(\dfrac{1 + x}{1 - x}\right) = \ln(1 + x) - \ln(1 - x) = \int \dfrac{dx}{1 + x} + \int \dfrac{dx}{1 - x}$

$= \int \dfrac{dx}{1 - (-x)} + \int \dfrac{dx}{1 - x} = \int \left[\sum\limits_{n=0}^{\infty} (-1)^n x^n + \sum\limits_{n=0}^{\infty} x^n\right] dx$

$= \int \left[(1 - x + x^2 - x^3 + x^4 - \cdots) + (1 + x + x^2 + x^3 + x^4 + \cdots)\right] dx$

$= \int \left(2 + 2x^2 + 2x^4 + \cdots\right) dx = \int \sum\limits_{n=0}^{\infty} 2x^{2n}\, dx = C + \sum\limits_{n=0}^{\infty} \dfrac{2x^{2n+1}}{2n + 1}$

But $f(0) = \ln \dfrac{1}{1} = 0$, so $C = 0$ and we have $f(x) = \sum\limits_{n=0}^{\infty} \dfrac{2x^{2n+1}}{2n + 1}$ with $R = 1$. If $x = \pm 1$, then

$f(x) = \pm 2 \sum\limits_{n=0}^{\infty} \dfrac{1}{2n + 1}$, which both diverge by the Limit Comparison Test with $b_n = \dfrac{1}{n}$.

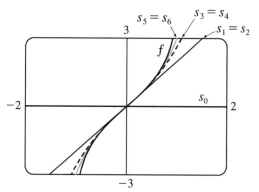

As n increases, $s_n(x)$ approximates f better on the interval of convergence, which is $(-1, 1)$.

23. $\dfrac{t}{1-t^8} = t \cdot \dfrac{1}{1-t^8} = t \displaystyle\sum_{n=0}^{\infty}(t^8)^n = \displaystyle\sum_{n=0}^{\infty}t^{8n+1}$ $\Rightarrow$ $\displaystyle\int \dfrac{t}{1-t^8}\, dt = C + \displaystyle\sum_{n=0}^{\infty}\dfrac{t^{8n+2}}{8n+2}$. The series for $\dfrac{1}{1-t^8}$

converges when $|t^8| < 1$ $\Leftrightarrow$ $|t| < 1$, so $R = 1$ for that series and also the series for $t/(1-t^8)$. By Theorem

2, the series for $\displaystyle\int \dfrac{t}{1-t^8}\, dt$ also has $R = 1$.

25. By Example 7, $\tan^{-1} x = \displaystyle\sum_{n=0}^{\infty}(-1)^n \dfrac{x^{2n+1}}{2n+1}$ with $R = 1$, so

$x - \tan^{-1} x = x - \left(x - \dfrac{x^3}{3} + \dfrac{x^5}{5} - \dfrac{x^7}{7} + \cdots \right) = \dfrac{x^3}{3} - \dfrac{x^5}{5} + \dfrac{x^7}{7} - \cdots = \displaystyle\sum_{n=1}^{\infty}(-1)^{n+1}\dfrac{x^{2n+1}}{2n+1}$

and $\dfrac{x - \tan^{-1} x}{x^3} = \displaystyle\sum_{n=1}^{\infty}(-1)^{n+1}\dfrac{x^{2n-2}}{2n+1}$, so

$\displaystyle\int \dfrac{x - \tan^{-1} x}{x^3}\, dx = C + \displaystyle\sum_{n=1}^{\infty}(-1)^{n+1}\dfrac{x^{2n-1}}{(2n+1)(2n-1)} = C + \displaystyle\sum_{n=1}^{\infty}(-1)^{n+1}\dfrac{x^{2n-1}}{4n^2-1}$. By Theorem 2, $R = 1$.

27. $\dfrac{1}{1+x^5} = \dfrac{1}{1-(-x^5)} = \displaystyle\sum_{n=0}^{\infty}(-x^5)^n = \displaystyle\sum_{n=0}^{\infty}(-1)^n x^{5n}$ $\Rightarrow$

$\displaystyle\int \dfrac{1}{1+x^5}\, dx = \displaystyle\int \displaystyle\sum_{n=0}^{\infty}(-1)^n x^{5n}\, dx = C + \displaystyle\sum_{n=0}^{\infty}(-1)^n \dfrac{x^{5n+1}}{5n+1}$. Thus,

$I = \displaystyle\int_0^{0.2} \dfrac{1}{1+x^5}\, dx = \left[x - \dfrac{x^6}{6} + \dfrac{x^{11}}{11} - \cdots \right]_0^{0.2} = 0.2 - \dfrac{(0.2)^6}{6} + \dfrac{(0.2)^{11}}{11} - \cdots$. The series is alternating, so

if we use the first two terms, the error is at most $(0.2)^{11}/11 \approx 1.9 \times 10^{-9}$. So $I \approx 0.2 - (0.2)^6/6 \approx 0.199989$ to

six decimal places.

29. We substitute x^4 for x in Example 7, and find that

$$\displaystyle\int x^2 \tan^{-1}(x^4)\, dx = \displaystyle\int x^2 \displaystyle\sum_{n=0}^{\infty}(-1)^n \dfrac{(x^4)^{2n+1}}{2n+1}\, dx$$

$$= \displaystyle\int \displaystyle\sum_{n=0}^{\infty}(-1)^n \dfrac{x^{8n+6}}{2n+1}\, dx = C + \displaystyle\sum_{n=0}^{\infty}(-1)^n \dfrac{x^{8n+7}}{(2n+1)(8n+7)}$$

So $\displaystyle\int_0^{1/3} x^2 \tan^{-1}(x^4)\, dx = \left[\dfrac{x^7}{7} - \dfrac{x^{15}}{45} + \cdots \right]_0^{1/3} = \dfrac{1}{7 \cdot 3^7} - \dfrac{1}{45 \cdot 3^{15}} + \cdots$. The series is alternating,

so if we use only one term, the error is at most $1/(45 \cdot 3^{15}) \approx 1.5 \times 10^{-9}$. So

$\displaystyle\int_0^{1/3} x^2 \tan^{-1}(x^4)\, dx \approx 1/(7 \cdot 3^7) \approx 0.000065$ to six decimal places.

31. Using the result of Example 6, $\ln(1-x) = -\sum_{n=1}^{\infty} \dfrac{x^n}{n}$, with $x = -0.1$, we have

$$\ln 1.1 = \ln[1-(-0.1)] = 0.1 - \frac{0.01}{2} + \frac{0.001}{3} - \frac{0.0001}{4} + \frac{0.00001}{5} - \cdots. \text{ The series is alternating, so if}$$

we use only the first four terms, the error is at most $\dfrac{0.00001}{5} = 0.000002$. So

$$\ln 1.1 \approx 0.1 - \frac{0.01}{2} + \frac{0.001}{3} - \frac{0.0001}{4} \approx 0.09531.$$

33. (a) $J_0(x) = \displaystyle\sum_{n=0}^{\infty} \frac{(-1)^n x^{2n}}{2^{2n}(n!)^2}$, $J_0'(x) = \displaystyle\sum_{n=1}^{\infty} \frac{(-1)^n 2n x^{2n-1}}{2^{2n}(n!)^2}$, and $J_0''(x) = \displaystyle\sum_{n=1}^{\infty} \frac{(-1)^n 2n(2n-1)x^{2n-2}}{2^{2n}(n!)^2}$, so

$$x^2 J_0''(x) + x J_0'(x) + x^2 J_0(x) = \sum_{n=1}^{\infty} \frac{(-1)^n 2n(2n-1)x^{2n}}{2^{2n}(n!)^2} + \sum_{n=1}^{\infty} \frac{(-1)^n 2n x^{2n}}{2^{2n}(n!)^2} + \sum_{n=0}^{\infty} \frac{(-1)^n x^{2n+2}}{2^{2n}(n!)^2}$$

$$= \sum_{n=1}^{\infty} \frac{(-1)^n 2n(2n-1)x^{2n}}{2^{2n}(n!)^2} + \sum_{n=1}^{\infty} \frac{(-1)^n 2n x^{2n}}{2^{2n}(n!)^2} + \sum_{n=1}^{\infty} \frac{(-1)^{n-1} x^{2n}}{2^{2n-2}[(n-1)!]^2}$$

$$= \sum_{n=1}^{\infty} \frac{(-1)^n 2n(2n-1)x^{2n}}{2^{2n}(n!)^2} + \sum_{n=1}^{\infty} \frac{(-1)^n 2n x^{2n}}{2^{2n}(n!)^2} + \sum_{n=1}^{\infty} \frac{(-1)^n (-1)^{-1} 2^2 n^2 x^{2n}}{2^{2n}(n!)^2}$$

$$= \sum_{n=1}^{\infty} (-1)^n \left[\frac{2n(2n-1) + 2n - 2^2 n^2}{2^{2n}(n!)^2}\right] x^{2n} = \sum_{n=1}^{\infty} (-1)^n \left[\frac{4n^2 - 2n + 2n - 4n^2}{2^{2n}(n!)^2}\right] x^{2n} = 0$$

(b) $\displaystyle\int_0^1 J_0(x)\,dx = \int_0^1 \left[\sum_{n=0}^{\infty} \frac{(-1)^n x^{2n}}{2^{2n}(n!)^2}\right] dx = \int_0^1 \left(1 - \frac{x^2}{4} + \frac{x^4}{64} - \frac{x^6}{2304} + \cdots\right) dx$

$$= \left[x - \frac{x^3}{3\cdot 4} + \frac{x^5}{5\cdot 64} - \frac{x^7}{7\cdot 2304} + \cdots\right]_0^1 = 1 - \frac{1}{12} + \frac{1}{320} - \frac{1}{16{,}128} + \cdots.$$

Since $\frac{1}{16{,}128} \approx 0.000062$, it follows from The Alternating Series Estimation Theorem that, correct to three

decimal places, $\int_0^1 J_0(x)\,dx \approx 1 - \frac{1}{12} + \frac{1}{320} \approx 0.920$.

35. (a) $f(x) = \displaystyle\sum_{n=0}^{\infty} \frac{x^n}{n!} \;\Rightarrow\; f'(x) = \sum_{n=1}^{\infty} \frac{n x^{n-1}}{n!} = \sum_{n=1}^{\infty} \frac{x^{n-1}}{(n-1)!} = \sum_{n=0}^{\infty} \frac{x^n}{n!} = f(x)$

(b) By Theorem 9.4.2, the only solution to the differential equation $df(x)/dx = f(x)$ is $f(x) = Ke^x$, but

$f(0) = 1$, so $K = 1$ and $f(x) = e^x$.

Or: We could solve the equation $df(x)/dx = f(x)$ as a separable differential equation.

37. If $a_n = \dfrac{x^n}{n^2}$, then by the Ratio Test, $\displaystyle\lim_{n\to\infty}\left|\dfrac{a_{n+1}}{a_n}\right| = \lim_{n\to\infty}\left|\dfrac{x^{n+1}}{(n+1)^2} \cdot \dfrac{n^2}{x^n}\right| = |x| \lim_{n\to\infty}\left(\dfrac{n}{n+1}\right)^2 = |x| < 1$ for

convergence, so $R = 1$. When $x = \pm 1$, $\displaystyle\sum_{n=1}^{\infty}\left|\dfrac{x^n}{n^2}\right| = \sum_{n=1}^{\infty}\dfrac{1}{n^2}$ which is a convergent p-series ($p = 2 > 1$), so the

interval of convergence for f is $[-1, 1]$. By Theorem 2, the radii of convergence of f' and f'' are both 1, so we need

only check the endpoints. $f(x) = \displaystyle\sum_{n=1}^{\infty}\dfrac{x^n}{n^2} \Rightarrow f'(x) = \sum_{n=1}^{\infty}\dfrac{nx^{n-1}}{n^2} = \sum_{n=0}^{\infty}\dfrac{x^n}{n+1}$, and this series diverges for

$x = 1$ (harmonic series) and converges for $x = -1$ (Alternating Series Test), so the interval of convergence

is $[-1, 1)$. $f''(x) = \displaystyle\sum_{n=1}^{\infty}\dfrac{nx^{n-1}}{n+1}$ diverges at both 1 and -1 (Test for Divergence) since $\displaystyle\lim_{n\to\infty}\dfrac{n}{n+1} = 1 \neq 0$, so its

interval of convergence is $(-1, 1)$.

39. By Example 7, $\tan^{-1} x = \displaystyle\sum_{n=0}^{\infty}(-1)^n\dfrac{x^{2n+1}}{2n+1}$ for $|x| < 1$. In particular, for $x = \dfrac{1}{\sqrt{3}}$, we have

$$\dfrac{\pi}{6} = \tan^{-1}\left(\dfrac{1}{\sqrt{3}}\right) = \sum_{n=0}^{\infty}(-1)^n\dfrac{(1/\sqrt{3})^{2n+1}}{2n+1} = \sum_{n=0}^{\infty}(-1)^n\left(\dfrac{1}{3}\right)^n\dfrac{1}{\sqrt{3}}\dfrac{1}{2n+1}, \text{ so}$$

$$\pi = \dfrac{6}{\sqrt{3}}\sum_{n=0}^{\infty}\dfrac{(-1)^n}{(2n+1)3^n} = 2\sqrt{3}\sum_{n=0}^{\infty}\dfrac{(-1)^n}{(2n+1)3^n}.$$

11.10 Taylor and Maclaurin Series

1. Using Theorem 5 with $\displaystyle\sum_{n=0}^{\infty}b_n(x-5)^n$, $b_n = \dfrac{f^{(n)}(a)}{n!}$, so $b_8 = \dfrac{f^{(8)}(5)}{8!}$.

3.

n	$f^{(n)}(x)$	$f^{(n)}(0)$
0	$\cos x$	1
1	$-\sin x$	0
2	$-\cos x$	-1
3	$\sin x$	0
4	$\cos x$	1
$\vdots$	$\vdots$	$\vdots$

We use Equation 7 with $f(x) = \cos x$.

$$\cos x = f(0) + f'(0)x + \dfrac{f''(0)}{2!}x^2 + \dfrac{f^{(3)}(0)}{3!}x^3 + \dfrac{f^{(4)}(0)}{4!}x^4 + \cdots$$

$$= 1 - \dfrac{x^2}{2!} + \dfrac{x^4}{4!} - \cdots = \sum_{n=0}^{\infty}\dfrac{(-1)^n x^{2n}}{(2n)!}$$

If $a_n = \dfrac{(-1)^n x^{2n}}{(2n)!}$, then

$$\lim_{n\to\infty}\left|\dfrac{a_{n+1}}{a_n}\right| = \lim_{n\to\infty}\left|\dfrac{x^{2n+2}}{(2n+2)!} \cdot \dfrac{(2n)!}{x^{2n}}\right| = x^2 \lim_{n\to\infty}\dfrac{1}{(2n+2)(2n+1)} = 0 < 1 \text{ for all } x.$$

So $R = \infty$ (Ratio Test).

5.

n	$f^{(n)}(x)$	$f^{(n)}(0)$
0	$(1+x)^{-3}$	1
1	$-3(1+x)^{-4}$	-3
2	$12(1+x)^{-5}$	12
3	$-60(1+x)^{-6}$	-60
4	$360(1+x)^{-7}$	360
$\vdots$	$\vdots$	$\vdots$

$$(1+x)^{-3} = f(0) + f'(0)x + \frac{f''(0)}{2!}x^2 + \frac{f'''(0)}{3!}x^3 + \frac{f^{(4)}(0)}{4!}x^4 + \cdots$$

$$= 1 - 3x + \frac{4 \cdot 3}{2!}x^2 - \frac{5 \cdot 4 \cdot 3}{3!}x^3 + \frac{6 \cdot 5 \cdot 4 \cdot 3}{4!}x^4 - \cdots$$

$$= 1 - 3x + \frac{4 \cdot 3 \cdot 2}{2 \cdot 2!}x^2 - \frac{5 \cdot 4 \cdot 3 \cdot 2}{2 \cdot 3!}x^3 + \frac{6 \cdot 5 \cdot 4 \cdot 3 \cdot 2}{2 \cdot 4!}x^4 - \cdots$$

$$= \sum_{n=0}^{\infty} \frac{(-1)^n (n+2)! \, x^n}{2(n!)} = \sum_{n=0}^{\infty} \frac{(-1)^n (n+2)(n+1)x^n}{2}$$

$$\lim_{n \to \infty} \left| \frac{a_{n+1}}{a_n} \right| = \lim_{n \to \infty} \left| \frac{(n+3)(n+2)x^{n+1}}{2} \cdot \frac{2}{(n+2)(n+1)x^n} \right| = |x| \lim_{n \to \infty} \frac{n+3}{n+1} = |x| < 1 \text{ for convergence,}$$

so $R = 1$ (Ratio Test).

7.

n	$f^{(n)}(x)$	$f^{(n)}(0)$
0	e^{5x}	1
1	$5e^{5x}$	5
2	$5^2 e^{5x}$	25
3	$5^3 e^{5x}$	125
4	$5^4 e^{5x}$	625
$\vdots$	$\vdots$	$\vdots$

$$e^{5x} = \sum_{n=0}^{\infty} \frac{f^{(n)}(0)}{n!}x^n = \sum_{n=0}^{\infty} \frac{5^n}{n!}x^n.$$

$$\lim_{n \to \infty} \left| \frac{a_{n+1}}{a_n} \right| = \lim_{n \to \infty} \left[\frac{5^{n+1}|x|^{n+1}}{(n+1)!} \cdot \frac{n!}{5^n |x|^n} \right]$$

$$= \lim_{n \to \infty} \frac{5|x|}{n+1} = 0 < 1 \text{ for all } x, \text{ so } R = \infty.$$

9.

n	$f^{(n)}(x)$	$f^{(n)}(0)$
0	$\sinh x$	0
1	$\cosh x$	1
2	$\sinh x$	0
3	$\cosh x$	1
4	$\sinh x$	0
$\vdots$	$\vdots$	$\vdots$

$$f^{(n)}(0) = \begin{cases} 0 & \text{if } n \text{ is even} \\ 1 & \text{if } n \text{ is odd} \end{cases} \text{ so } \sinh x = \sum_{n=0}^{\infty} \frac{x^{2n+1}}{(2n+1)!}.$$

Use the Ratio Test to find R. If $a_n = \dfrac{x^{2n+1}}{(2n+1)!}$, then

$$\lim_{n \to \infty} \left| \frac{a_{n+1}}{a_n} \right| = \lim_{n \to \infty} \left| \frac{x^{2n+3}}{(2n+3)!} \cdot \frac{(2n+1)!}{x^{2n+1}} \right|$$

$$= x^2 \cdot \lim_{n \to \infty} \frac{1}{(2n+3)(2n+2)} = 0 < 1$$

for all x, so $R = \infty$.

11.

n	$f^{(n)}(x)$	$f^{(n)}(2)$
0	$1 + x + x^2$	7
1	$1 + 2x$	5
2	2	2
3	0	0
4	0	0
⋮	⋮	⋮

$$f(x) = 7 + 5(x - 2) + \frac{2}{2!}(x - 2)^2 + \sum_{n=3}^{\infty} \frac{0}{n!}(x - 2)^n$$

$$= 7 + 5(x - 2) + (x - 2)^2$$

Since $a_n = 0$ for large n, $R = \infty$.

13. Clearly, $f^{(n)}(x) = e^x$, so $f^{(n)}(3) = e^3$ and $e^x = \sum_{n=0}^{\infty} \frac{e^3}{n!}(x - 3)^n$. If $a_n = \frac{e^3}{n!}(x - 3)^n$, then

$$\lim_{n \to \infty} \left| \frac{a_{n+1}}{a_n} \right| = \lim_{n \to \infty} \left| \frac{e^3(x - 3)^{n+1}}{(n + 1)!} \cdot \frac{n!}{e^3(x - 3)^n} \right| = \lim_{n \to \infty} \frac{|x - 3|}{n + 1} = 0 < 1 \text{ for all } x, \text{ so } R = \infty.$$

15.

n	$f^{(n)}(x)$	$f^{(n)}(\pi)$
0	$\cos x$	-1
1	$-\sin x$	0
2	$-\cos x$	1
3	$\sin x$	0
4	$\cos x$	-1
⋮	⋮	⋮

$$\cos x = \sum_{k=0}^{\infty} \frac{f^{(k)}(\pi)}{k!}(x - \pi)^k = -1 + \frac{(x - \pi)^2}{2!} - \frac{(x - \pi)^4}{4!} + \frac{(x - \pi)^6}{6!} - \cdots = \sum_{n=0}^{\infty} (-1)^{n+1} \frac{(x - \pi)^{2n}}{(2n)!}.$$

$$\lim_{n \to \infty} \left| \frac{a_{n+1}}{a_n} \right| = \lim_{n \to \infty} \left[\frac{|x - \pi|^{2n+2}}{(2n + 2)!} \cdot \frac{(2n)!}{|x - \pi|^{2n}} \right] = \lim_{n \to \infty} \frac{|x - \pi|^2}{(2n + 2)(2n + 1)} = 0 < 1 \text{ for all } x, \text{ so } R = \infty.$$

17.

n	$f^{(n)}(x)$	$f^{(n)}(9)$
0	$x^{-1/2}$	$\frac{1}{3}$
1	$-\frac{1}{2}x^{-3/2}$	$-\frac{1}{2} \cdot \frac{1}{3^3}$
2	$\frac{3}{4}x^{-5/2}$	$-\frac{1}{2} \cdot \left(-\frac{3}{2}\right) \cdot \frac{1}{3^5}$
3	$-\frac{15}{8}x^{-7/2}$	$-\frac{1}{2} \cdot \left(-\frac{3}{2}\right) \cdot \left(-\frac{5}{2}\right) \cdot \frac{1}{3^7}$
⋮	⋮	⋮

$$\frac{1}{\sqrt{x}} = \frac{1}{3} - \frac{1}{2 \cdot 3^3}(x - 9) + \frac{3}{2^2 \cdot 3^5} \frac{(x - 9)^2}{2!} - \frac{3 \cdot 5}{2^3 \cdot 3^7} \frac{(x - 9)^3}{3!} + \cdots$$

$$= \sum_{n=0}^{\infty} (-1)^n \frac{1 \cdot 3 \cdot 5 \cdot \cdots \cdot (2n - 1)}{2^n \cdot 3^{2n+1} \cdot n!}(x - 9)^n.$$

$$\lim_{n \to \infty} \left| \frac{a_{n+1}}{a_n} \right| = \lim_{n \to \infty} \left[\frac{1 \cdot 3 \cdot 5 \cdot \cdots \cdot (2n-1)[2(n+1)-1] \, |x-9|^{n+1}}{2^{n+1} \cdot 3^{[2(n+1)+1]} \cdot (n+1)!} \cdot \frac{2^n \cdot 3^{2n+1} \cdot n!}{1 \cdot 3 \cdot 5 \cdot \cdots \cdot (2n-1) \, |x-9|^n} \right]$$

$$= \lim_{n \to \infty} \left[\frac{(2n+1) \, |x-9|}{2 \cdot 3^2 \, (n+1)} \right] = \frac{1}{9} |x-9| < 1$$

for convergence, so $|x-9| < 9$ and $R = 9$.

19. If $f(x) = \cos x$, then $f^{(n+1)}(x) = \pm \sin x$ or $\pm \cos x$. In each case, $\left| f^{(n+1)}(x) \right| \le 1$, so by Formula 9 with $a = 0$

and $M = 1$, $|R_n(x)| \le \dfrac{1}{(n+1)!} |x|^{n+1}$. Thus, $|R_n(x)| \to 0$ as $n \to \infty$ by Equation 10. So $\lim\limits_{n \to \infty} R_n(x) = 0$ and,

by Theorem 8, the series in Exercise 3 represents $\cos x$ for all x.

21. If $f(x) = \sinh x$, then for all n, $f^{(n+1)}(x) = \cosh x$ or $\sinh x$. Since $|\sinh x| < |\cosh x| = \cosh x$ for all x, we

have $\left| f^{(n+1)}(x) \right| \le \cosh x$ for all n. If d is any positive number and $|x| \le d$, then $\left| f^{(n+1)}(x) \right| \le \cosh x \le \cosh d$,

so by Formula 9 with $a = 0$ and $M = \cosh d$, we have $|R_n(x)| \le \dfrac{\cosh d}{(n+1)!} |x|^{n+1}$. It follows that $|R_n(x)| \to 0$

as $n \to \infty$ for $|x| \le d$ (by Equation 10). But d was an arbitrary positive number. So by Theorem 8, the series

represents $\sinh x$ for all x.

23. $\cos x = \sum\limits_{n=0}^{\infty} (-1)^n \dfrac{x^{2n}}{(2n)!} \quad \Rightarrow \quad f(x) = \cos(\pi x) = \sum\limits_{n=0}^{\infty} \dfrac{(-1)^n (\pi x)^{2n}}{(2n)!} = \sum\limits_{n=0}^{\infty} \dfrac{(-1)^n \pi^{2n} x^{2n}}{(2n)!}, R = \infty$

25. $\tan^{-1} x = \sum\limits_{n=0}^{\infty} (-1)^n \dfrac{x^{2n+1}}{2n+1} \quad \Rightarrow \quad f(x) = x \tan^{-1} x = x \sum\limits_{n=0}^{\infty} (-1)^n \dfrac{x^{2n+1}}{2n+1} = \sum\limits_{n=0}^{\infty} (-1)^n \dfrac{x^{2n+2}}{2n+1}, R = 1$

27. $e^x = \sum\limits_{n=0}^{\infty} \dfrac{x^n}{n!} \quad \Rightarrow \quad f(x) = x^2 e^{-x} = x^2 \sum\limits_{n=0}^{\infty} \dfrac{(-x)^n}{n!} = \sum\limits_{n=0}^{\infty} \dfrac{(-1)^n \, x^{n+2}}{n!}, R = \infty$

29. $\sin^2 x = \dfrac{1}{2}(1 - \cos 2x) = \dfrac{1}{2}\left[1 - \sum\limits_{n=0}^{\infty} \dfrac{(-1)^n (2x)^{2n}}{(2n)!} \right] = \dfrac{1}{2}\left[1 - 1 - \sum\limits_{n=1}^{\infty} \dfrac{(-1)^n (2x)^{2n}}{(2n)!} \right]$

$$= \sum\limits_{n=1}^{\infty} \dfrac{(-1)^{n+1} 2^{2n-1} x^{2n}}{(2n)!}, R = \infty$$

31. $\dfrac{\sin x}{x} = \dfrac{1}{x} \sum\limits_{n=0}^{\infty} \dfrac{(-1)^n x^{2n+1}}{(2n+1)!} = \sum\limits_{n=0}^{\infty} \dfrac{(-1)^n x^{2n}}{(2n+1)!}$ and this series also gives the required value at $x = 0$ (namely 1);

$R = \infty$.

33.

n	$f^{(n)}(x)$	$f^{(n)}(0)$
0	$(1+x)^{1/2}$	1
1	$\frac{1}{2}(1+x)^{-1/2}$	$\frac{1}{2}$
2	$-\frac{1}{4}(1+x)^{-3/2}$	$-\frac{1}{4}$
3	$\frac{3}{8}(1+x)^{-5/2}$	$\frac{3}{8}$
4	$-\frac{15}{16}(1+x)^{-7/2}$	$-\frac{15}{16}$
$\vdots$	$\vdots$	$\vdots$

So $f^{(n)}(0) = \dfrac{(-1)^{n-1}\,1\cdot3\cdot5\cdot\,\cdots\,\cdot(2n-3)}{2^n}$ for $n \geq 2$, and

$$\sqrt{1+x} = 1 + \frac{x}{2} + \sum_{n=2}^{\infty} \frac{(-1)^{n-1}1\cdot3\cdot5\cdot\,\cdots\,\cdot(2n-3)}{2^n n!}x^n. \text{ If } a_n = \frac{(-1)^{n-1}1\cdot3\cdot5\cdot\,\cdots\,\cdot(2n-3)}{2^n n!}x^n,$$

then $\displaystyle\lim_{n\to\infty}\left|\frac{a_{n+1}}{a_n}\right| = \lim_{n\to\infty}\left|\frac{1\cdot3\cdot5\cdot\,\cdots\,\cdot(2n-3)(2n-1)x^{n+1}}{2^{n+1}(n+1)!} \cdot \frac{2^n n!}{1\cdot3\cdot5\cdot\,\cdots\,\cdot(2n-3)x^n}\right|$

$$= \frac{|x|}{2}\lim_{n\to\infty}\frac{2n-1}{n+1} = \frac{|x|}{2}\cdot2 = |x| < 1 \text{ for convergence, so } R = 1.$$

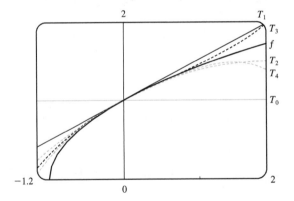

Notice that, as n increases, $T_n(x)$ becomes a better approximation to $f(x)$ for $-1 < x < 1$.

35. $\cos x = \displaystyle\sum_{n=0}^{\infty}(-1)^n\frac{x^{2n}}{(2n)!} \quad \Rightarrow \quad f(x) = \cos(x^2) = \sum_{n=0}^{\infty}\frac{(-1)^n\left(x^2\right)^{2n}}{(2n)!} = \sum_{n=0}^{\infty}\frac{(-1)^n\,x^{4n}}{(2n)!},\ R = \infty$

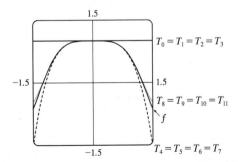

Notice that, as n increases, $T_n(x)$ becomes a better approximation to $f(x)$.

37. $e^x = \sum_{n=0}^{\infty} \dfrac{x^n}{n!}$, so

$$e^{-0.2} = \sum_{n=0}^{\infty} \frac{(-0.2)^n}{n!} = 1 - 0.2 + \frac{1}{2!}(0.2)^2 - \frac{1}{3!}(0.2)^3 + \frac{1}{4!}(0.2)^4 - \frac{1}{5!}(0.2)^5 + \frac{1}{6!}(0.2)^6 - \cdots . \text{ But}$$

$\dfrac{1}{6!}(0.2)^6 = 8.\overline{8} \times 10^{-8}$, so by the Alternating Series Estimation Theorem, $e^{-0.2} \approx \sum_{n=0}^{5} \dfrac{(-0.2)^n}{n!} \approx 0.81873$,

correct to five decimal places.

39. $\cos x \overset{(16)}{=} \sum_{n=0}^{\infty} (-1)^n \dfrac{x^{2n}}{(2n)!} \quad \Rightarrow \quad \cos(x^3) = \sum_{n=0}^{\infty} (-1)^n \dfrac{(x^3)^{2n}}{(2n)!} = \sum_{n=0}^{\infty} (-1)^n \dfrac{x^{6n}}{(2n)!} \quad \Rightarrow$

$x \cos(x^3) = \sum_{n=0}^{\infty} (-1)^n \dfrac{x^{6n+1}}{(2n)!} \quad \Rightarrow \quad \displaystyle\int x \cos(x^3)\, dx = C + \sum_{n=0}^{\infty} (-1)^n \dfrac{x^{6n+2}}{(6n+2)(2n)!}$, with $R = \infty$.

41. Using the series from Exercise 33 and substituting x^3 for x, we get

$$\int \sqrt{x^3 + 1}\, dx = \int \left[1 + \frac{x^3}{2} + \sum_{n=2}^{\infty} \frac{(-1)^{n-1}\, 1 \cdot 3 \cdot 5 \cdots (2n-3)}{2^n n!} x^{3n} \right] dx$$

$$= C + x + \frac{x^4}{8} + \sum_{n=2}^{\infty} \frac{(-1)^{n-1}\, 1 \cdot 3 \cdot 5 \cdots (2n-3)}{2^n n!(3n+1)} x^{3n+1}$$

43. By Exercise 39, $\displaystyle\int x \cos(x^3)\, dx = C + \sum_{n=0}^{\infty} (-1)^n \dfrac{x^{6n+2}}{(6n+2)(2n)!}$, so $\displaystyle\int_0^1 x \cos(x^3)\, dx$

$$= \left[\sum_{n=0}^{\infty} (-1)^n \frac{x^{6n+2}}{(6n+2)(2n)!} \right]_0^1 = \sum_{n=0}^{\infty} \frac{(-1)^n}{(6n+2)(2n)!} = \frac{1}{2} - \frac{1}{8 \cdot 2!} + \frac{1}{14 \cdot 4!} - \frac{1}{20 \cdot 6!} + \cdots, \text{ but}$$

$\dfrac{1}{20 \cdot 6!} = \dfrac{1}{14{,}400} \approx 0.000\,069$, so $\displaystyle\int_0^1 x \cos(x^3)\, dx \approx \dfrac{1}{2} - \dfrac{1}{16} + \dfrac{1}{336} \approx 0.440$ (correct to three decimal places)

by the Alternating Series Estimation Theorem.

45. We first find a series representation for $f(x) = (1 + x)^{-1/2}$, and then substitute.

n	$f^{(n)}(x)$	$f^{(n)}(0)$
0	$(1+x)^{-1/2}$	1
1	$-\frac{1}{2}(1+x)^{-3/2}$	$-\frac{1}{2}$
2	$\frac{3}{4}(1+x)^{-5/2}$	$\frac{3}{4}$
3	$-\frac{15}{8}(1+x)^{-7/2}$	$-\frac{15}{8}$
$\vdots$	$\vdots$	$\vdots$

$$\frac{1}{\sqrt{1+x}} = 1 - \frac{x}{2} + \frac{3}{4}\left(\frac{x^2}{2!}\right) - \frac{15}{8}\left(\frac{x^3}{3!}\right) + \cdots \quad \Rightarrow \quad \frac{1}{\sqrt{1+x^3}} = 1 - \frac{1}{2}x^3 + \frac{3}{8}x^6 - \frac{5}{16}x^9 + \cdots \quad \Rightarrow$$

$$\int_0^{0.1} \frac{dx}{\sqrt{1+x^3}} = \left[x - \frac{1}{8}x^4 + \frac{3}{56}x^7 - \frac{1}{32}x^{10} + \cdots \right]_0^{0.1} \approx (0.1) - \frac{1}{8}(0.1)^4, \text{ by the Alternating Series}$$

Estimation Theorem, since $\frac{3}{56}(0.1)^7 \approx 0.000\,000\,005\,4 < 10^{-8}$, which is the maximum desired error. Therefore,

$$\int_0^{0.1} \frac{dx}{\sqrt{1+x^3}} \approx 0.099\,987\,50.$$

47. $\displaystyle\lim_{x\to0} \frac{x - \tan^{-1}x}{x^3} = \lim_{x\to0} \frac{x - \left(x - \frac{1}{3}x^3 + \frac{1}{5}x^5 - \frac{1}{7}x^7 + \cdots \right)}{x^3} = \lim_{x\to0} \frac{\frac{1}{3}x^3 - \frac{1}{5}x^5 + \frac{1}{7}x^7 - \cdots}{x^3}$

$\displaystyle= \lim_{x\to0} \left(\frac{1}{3} - \frac{1}{5}x^2 + \frac{1}{7}x^4 - \cdots \right) = \frac{1}{3}$

since power series are continuous functions.

49. $\displaystyle\lim_{x\to0} \frac{\sin x - x + \frac{1}{6}x^3}{x^5} = \lim_{x\to0} \frac{\left(x - \frac{1}{3!}x^3 + \frac{1}{5!}x^5 - \frac{1}{7!}x^7 + \cdots \right) - x + \frac{1}{6}x^3}{x^5}$

$\displaystyle= \lim_{x\to0} \frac{\frac{1}{5!}x^5 - \frac{1}{7!}x^7 + \cdots}{x^5} = \lim_{x\to0} \left(\frac{1}{5!} - \frac{x^2}{7!} + \frac{x^4}{9!} - \cdots \right) = \frac{1}{5!} = \frac{1}{120}$

since power series are continuous functions.

51. As in Example 8(a), we have $e^{-x^2} = 1 - \frac{x^2}{1!} + \frac{x^4}{2!} - \frac{x^6}{3!} + \cdots$ and we know that $\cos x = 1 - \frac{x^2}{2!} + \frac{x^4}{4!} - \cdots$

from Equation 16. Therefore, $e^{-x^2}\cos x = \left(1 - x^2 + \frac{1}{2}x^4 - \cdots \right)\left(1 - \frac{1}{2}x^2 + \frac{1}{24}x^4 - \cdots \right)$. Writing only the

terms with degree ≤ 4, we get $e^{-x^2}\cos x = 1 - \frac{1}{2}x^2 + \frac{1}{24}x^4 - x^2 + \frac{1}{2}x^4 + \frac{1}{2}x^4 + \cdots = 1 - \frac{3}{2}x^2 + \frac{25}{24}x^4 + \cdots$.

53.

$$\begin{array}{r} 1 + \frac{1}{6}x^2 + \frac{7}{360}x^4 + \cdots \\ x - \frac{1}{6}x^3 + \frac{1}{120}x^5 - \cdots \, \overline{\big)\, x} \\ \underline{x - \frac{1}{6}x^3 + \frac{1}{120}x^5 - \cdots} \\ \frac{1}{6}x^3 - \frac{1}{120}x^5 + \cdots \\ \underline{\frac{1}{6}x^3 - \frac{1}{36}x^5 + \cdots} \\ \frac{7}{360}x^5 + \cdots \\ \underline{\frac{7}{360}x^5 + \cdots} \\ \cdots \end{array}$$

$\dfrac{x}{\sin x} \overset{(15)}{=} \dfrac{x}{x - \frac{1}{6}x^3 + \frac{1}{120}x^5 - \cdots}$. From the long division above, $\dfrac{x}{\sin x} = 1 + \frac{1}{6}x^2 + \frac{7}{360}x^4 + \cdots$.

55. $\displaystyle\sum_{n=0}^{\infty} (-1)^n \frac{x^{4n}}{n!} = \sum_{n=0}^{\infty} \frac{\left(-x^4 \right)^n}{n!} = e^{-x^4}$, by (11).

57. $\displaystyle\sum_{n=0}^{\infty} \frac{(-1)^n \pi^{2n+1}}{4^{2n+1}(2n+1)!} = \sum_{n=0}^{\infty} \frac{(-1)^n \left(\frac{\pi}{4} \right)^{2n+1}}{(2n+1)!} = \sin\frac{\pi}{4} = \frac{1}{\sqrt{2}}$, by (15).

59. $3 + \dfrac{9}{2!} + \dfrac{27}{3!} + \dfrac{81}{4!} + \cdots = \dfrac{3^1}{1!} + \dfrac{3^2}{2!} + \dfrac{3^3}{3!} + \dfrac{3^4}{4!} + \cdots = \displaystyle\sum_{n=1}^{\infty} \dfrac{3^n}{n!} = \displaystyle\sum_{n=0}^{\infty} \dfrac{3^n}{n!} - 1 = e^3 - 1$, by (11).

61. Assume that $|f'''(x)| \le M$, so $f'''(x) \le M$ for $a \le x \le a + d$. Now $\int_a^x f'''(t)\,dt \le \int_a^x M\,dt$ ⇒

$f''(x) - f''(a) \le M(x - a)$ ⇒ $f''(x) \le f''(a) + M(x - a)$. Thus, $\int_a^x f''(t)\,dt \le \int_a^x [f''(a) + M(t - a)]\,dt$

⇒ $f'(x) - f'(a) \le f''(a)(x - a) + \frac{1}{2}M(x - a)^2$ ⇒ $f'(x) \le f'(a) + f''(a)(x - a) + \frac{1}{2}M(x - a)^2$ ⇒

$\int_a^x f'(t)\,dt \le \int_a^x \left[f'(a) + f''(a)(t - a) + \frac{1}{2}M(t - a)^2\right]dt$ ⇒

$f(x) - f(a) \le f'(a)(x - a) + \frac{1}{2}f''(a)(x - a)^2 + \frac{1}{6}M(x - a)^3$. So

$f(x) - f(a) - f'(a)(x - a) - \frac{1}{2}f''(a)(x - a)^2 \le \frac{1}{6}M(x - a)^3$. But

$R_2(x) = f(x) - T_2(x) = f(x) - f(a) - f'(a)(x - a) - \frac{1}{2}f''(a)(x - a)^2$, so $R_2(x) \le \frac{1}{6}M(x - a)^3$.

A similar argument using $f'''(x) \ge -M$ shows that $R_2(x) \ge -\frac{1}{6}M(x - a)^3$. So $|R_2(x_2)| \le \frac{1}{6}M\,|x - a|^3$.

Although we have assumed that $x > a$, a similar calculation shows that this inequality is also true if $x < a$.

11.11 The Binomial Series

1. The general binomial series in (2) is

$$(1 + x)^k = \sum_{n=0}^{\infty} \binom{k}{n} x^n = 1 + kx + \frac{k(k - 1)}{2!}x^2 + \frac{k(k - 1)(k - 2)}{3!}x^3 + \cdots.$$

$$(1 + x)^{1/2} = \sum_{n=0}^{\infty} \binom{\frac{1}{2}}{n} x^n = 1 + \left(\tfrac{1}{2}\right)x + \frac{\left(\frac{1}{2}\right)\left(-\frac{1}{2}\right)}{2!}x^2 + \frac{\left(\frac{1}{2}\right)\left(-\frac{1}{2}\right)\left(-\frac{3}{2}\right)}{3!}x^3 + \cdots$$

$$= 1 + \frac{x}{2} - \frac{x^2}{2^2 \cdot 2!} + \frac{1 \cdot 3 \cdot x^3}{2^3 \cdot 3!} - \frac{1 \cdot 3 \cdot 5 \cdot x^4}{2^4 \cdot 4!} + \cdots$$

$$= 1 + \frac{x}{2} + \sum_{n=2}^{\infty} \frac{(-1)^{n-1} 1 \cdot 3 \cdot 5 \cdot \cdots \cdot (2n - 3)x^n}{2^n \cdot n!} \quad \text{for } |x| < 1, \text{ so } R = 1$$

3. $\dfrac{1}{(2 + x)^3} = \dfrac{1}{[2(1 + x/2)]^3} = \dfrac{1}{8}\left(1 + \dfrac{x}{2}\right)^{-3} = \dfrac{1}{8}\displaystyle\sum_{n=0}^{\infty}\binom{-3}{n}\left(\dfrac{x}{2}\right)^n$. The binomial coefficient is

$$\binom{-3}{n} = \frac{(-3)(-4)(-5)\cdot \cdots \cdot(-3 - n + 1)}{n!} = \frac{(-3)(-4)(-5)\cdot \cdots \cdot[-(n + 2)]}{n!}$$

$$= \frac{(-1)^n \cdot 2 \cdot 3 \cdot 4 \cdot 5 \cdot \cdots \cdot (n + 1)(n + 2)}{2 \cdot n!} = \frac{(-1)^n(n + 1)(n + 2)}{2}$$

Thus, $\dfrac{1}{(2 + x)^3} = \dfrac{1}{8}\displaystyle\sum_{n=0}^{\infty}\dfrac{(-1)^n(n + 1)(n + 2)}{2}\dfrac{x^n}{2^n} = \displaystyle\sum_{n=0}^{\infty}\dfrac{(-1)^n(n + 1)(n + 2)x^n}{2^{n+4}}$ for $\left|\dfrac{x}{2}\right| < 1$ ⇔

$|x| < 2$, so $R = 2$.

5. $\sqrt[4]{1-8x} = (1-8x)^{1/4} = \sum\limits_{n=0}^{\infty} \binom{\frac{1}{4}}{n} (-8x)^n$

$$= 1 + \tfrac{1}{4}(-8x) + \frac{\frac{1}{4}\left(-\frac{3}{4}\right)}{2!}(-8x)^2 + \frac{\left(\frac{1}{4}\right)\left(-\frac{3}{4}\right)\left(-\frac{7}{4}\right)}{3!}(-8x)^3 + \cdots$$

$$= 1 - 2x + \sum\limits_{n=2}^{\infty} \frac{(-1)^n (-1)^{n-1} \cdot 3 \cdot 7 \cdot \ldots \cdot (4n-5)\,8^n}{4^n \cdot n!} x^n$$

$$= 1 - 2x - \sum\limits_{n=2}^{\infty} \frac{3 \cdot 7 \cdot \ldots \cdot (4n-5)\,2^n}{n!} x^n$$

and $|-8x| < 1 \iff |x| < \tfrac{1}{8}$, so $R = \tfrac{1}{8}$.

7. We must write the binomial in the form $(1+$ expression$)$, so we'll factor out a 4.

$$\frac{x}{\sqrt{4+x^2}} = \frac{x}{\sqrt{4(1+x^2/4)}} = \frac{x}{2\sqrt{1+x^2/4}} = \frac{x}{2}\left(1+\frac{x^2}{4}\right)^{-1/2} = \frac{x}{2}\sum\limits_{n=0}^{\infty}\binom{-\frac{1}{2}}{n}\left(\frac{x^2}{4}\right)^n$$

$$= \frac{x}{2}\left[1 + \left(-\tfrac{1}{2}\right)\frac{x^2}{4} + \frac{\left(-\frac{1}{2}\right)\left(-\frac{3}{2}\right)}{2!}\left(\frac{x^2}{4}\right)^2 + \frac{\left(-\frac{1}{2}\right)\left(-\frac{3}{2}\right)\left(-\frac{5}{2}\right)}{3!}\left(\frac{x^2}{4}\right)^3 + \cdots\right]$$

$$= \frac{x}{2} + \frac{x}{2}\sum\limits_{n=1}^{\infty}(-1)^n\frac{1 \cdot 3 \cdot 5 \cdot \ldots \cdot (2n-1)}{2^n \cdot 4^n \cdot n!}x^{2n}$$

$$= \frac{x}{2} + \sum\limits_{n=1}^{\infty}(-1)^n\frac{1 \cdot 3 \cdot 5 \cdot \ldots \cdot (2n-1)}{n!\,2^{3n+1}}x^{2n+1} \text{ and } \frac{x^2}{4} < 1 \iff \frac{|x|}{2} < 1 \iff$$

$|x| < 2$, so $R = 2$.

9. $(1+2x)^{3/4} = 1 + \frac{3}{4}(2x) + \frac{\left(\frac{3}{4}\right)\left(-\frac{1}{4}\right)}{2!}(2x)^2 + \frac{\left(\frac{3}{4}\right)\left(-\frac{1}{4}\right)\left(-\frac{5}{4}\right)}{3!}(2x)^3 + \cdots$

$$= 1 + \frac{3}{2}x + 3\sum\limits_{n=2}^{\infty}(-1)^{n+1}\frac{1 \cdot 5 \cdot 9 \cdot \ldots \cdot (4n-7)}{4^n \cdot n!} \cdot 2^n x^n$$

$$= 1 + \frac{3}{2}x + 3\sum\limits_{n=2}^{\infty}(-1)^{n+1}\frac{1 \cdot 5 \cdot 9 \cdot \ldots \cdot (4n-7)}{2^n \cdot n!} x^n \text{ and } |2x| < 1 \iff |x| < \frac{1}{2}, \text{ so } R = \frac{1}{2}.$$

The three Taylor polynomials are $T_1(x) = 1 + \frac{3}{2}x$, $T_2(x) = 1 + \frac{3}{2}x - \frac{3}{8}x^2$, and

$T_3(x) = 1 + \frac{3}{2}x - \frac{3}{8}x^2 + \frac{5}{16}x^3$.

11. (a) $1/\sqrt{1-x^2} = \left[1+\left(-x^2\right)\right]^{-1/2}$

$$= 1 + \left(-\tfrac{1}{2}\right)\left(-x^2\right) + \frac{\left(-\tfrac{1}{2}\right)\left(-\tfrac{3}{2}\right)}{2!}\left(-x^2\right)^2 + \frac{\left(-\tfrac{1}{2}\right)\left(-\tfrac{3}{2}\right)\left(-\tfrac{5}{2}\right)}{3!}\left(-x^2\right)^3 + \cdots$$

$$= 1 + \sum_{n=1}^{\infty} \frac{1 \cdot 3 \cdot 5 \cdots \cdots (2n-1)}{2^n \cdot n!} x^{2n}$$

(b) $\sin^{-1} x = \displaystyle\int \frac{1}{\sqrt{1-x^2}}\, dx = C + x + \sum_{n=1}^{\infty} \frac{1 \cdot 3 \cdot 5 \cdots \cdots (2n-1)}{(2n+1)2^n \cdot n!} x^{2n+1}$

$$= x + \sum_{n=1}^{\infty} \frac{1 \cdot 3 \cdot 5 \cdots \cdots (2n-1)}{(2n+1)2^n \cdot n!} x^{2n+1} \quad \text{since } 0 = \sin^{-1} 0 = C.$$

13. (a) $\sqrt[3]{1+x} = (1+x)^{1/3} = \displaystyle\sum_{n=0}^{\infty} \binom{\tfrac{1}{3}}{n} x^n$

$$= 1 + \frac{1}{3}x + \frac{\left(\tfrac{1}{3}\right)\left(-\tfrac{2}{3}\right)}{2!}x^2 + \frac{\left(\tfrac{1}{3}\right)\left(-\tfrac{2}{3}\right)\left(-\tfrac{5}{3}\right)}{3!}x^3 + \cdots$$

$$= 1 + \frac{x}{3} + \sum_{n=2}^{\infty} (-1)^{n+1} \frac{2 \cdot 5 \cdot 8 \cdots \cdots (3n-4)}{3^n \cdot n!} x^n$$

(b) $\sqrt[3]{1+x} = 1 + \frac{1}{3}x - \frac{1}{9}x^2 + \frac{5}{81}x^3 - \cdots$. $\sqrt[3]{1.01} = \sqrt[3]{1+0.01}$, so let $x = 0.01$. The sum of the first two

terms is then $1 + \frac{1}{3}(0.01) \approx 1.0033$. The third term is $\frac{1}{9}(0.01)^2 \approx 0.000\,01$, which does not affect the fourth

decimal place of the sum, so we have $\sqrt[3]{1.01} \approx 1.0033$.

15. (a) $[1+(-x)]^{-2} = 1 + (-2)(-x) + \frac{(-2)(-3)}{2!}(-x)^2 + \frac{(-2)(-3)(-4)}{3!}(-x)^3 + \cdots$

$$= 1 + 2x + 3x^2 + 4x^3 + \cdots = \sum_{n=0}^{\infty} (n+1)x^n,$$

so $\dfrac{x}{(1-x)^2} = x \displaystyle\sum_{n=0}^{\infty} (n+1)x^n = \sum_{n=0}^{\infty} (n+1)x^{n+1} = \sum_{n=1}^{\infty} nx^n.$

(b) With $x = \frac{1}{2}$ in part (a), we have $\displaystyle\sum_{n=1}^{\infty} n\left(\tfrac{1}{2}\right)^n = \sum_{n=1}^{\infty} \frac{n}{2^n} = \frac{\tfrac{1}{2}}{\left(1-\tfrac{1}{2}\right)^2} = \frac{\tfrac{1}{2}}{\tfrac{1}{4}} = 2.$

17. (a) $\left(1+x^2\right)^{1/2} = 1 + \left(\tfrac{1}{2}\right)x^2 + \frac{\left(\tfrac{1}{2}\right)\left(-\tfrac{1}{2}\right)}{2!}\left(x^2\right)^2 + \frac{\left(\tfrac{1}{2}\right)\left(-\tfrac{1}{2}\right)\left(-\tfrac{3}{2}\right)}{3!}\left(x^2\right)^3 + \cdots$

$$= 1 + \frac{x^2}{2} + \sum_{n=2}^{\infty} \frac{(-1)^{n-1} 1 \cdot 3 \cdot 5 \cdots \cdots (2n-3)}{2^n \cdot n!} x^{2n}$$

(b) The coefficient of x^{10} (corresponding to $n = 5$) in the above Maclaurin series is $\dfrac{f^{(10)}(0)}{10!}$, so

$$\frac{f^{(10)}(0)}{10!} = \frac{(-1)^4 \cdot 1 \cdot 3 \cdot 5 \cdot 7}{2^5 \cdot 5!} \quad \Rightarrow \quad f^{(10)}(0) = 10!\left(\frac{1 \cdot 3 \cdot 5 \cdot 7}{2^5 \cdot 5!}\right) = 99{,}225.$$

19. (a) $g(x) = \sum_{n=0}^{\infty} \binom{k}{n} x^n \quad \Rightarrow \quad g'(x) = \sum_{n=1}^{\infty} \binom{k}{n} n x^{n-1}$, so

$$(1+x)g'(x) = (1+x) \sum_{n=1}^{\infty} \binom{k}{n} n x^{n-1} = \sum_{n=1}^{\infty} \binom{k}{n} n x^{n-1} + \sum_{n=1}^{\infty} \binom{k}{n} n x^n$$

$$= \sum_{n=0}^{\infty} \binom{k}{n+1} (n+1)x^n + \sum_{n=0}^{\infty} \binom{k}{n} n x^n \qquad \begin{bmatrix} \text{Replace } n \text{ with } n+1 \\ \text{in the first series} \end{bmatrix}$$

$$= \sum_{n=0}^{\infty} (n+1) \frac{k(k-1)(k-2)\cdots(k-n+1)(k-n)}{(n+1)!} x^n$$

$$+ \sum_{n=0}^{\infty} \left[(n) \frac{k(k-1)(k-2)\cdots(k-n+1)}{n!} \right] x^n$$

$$= \sum_{n=0}^{\infty} \frac{(n+1)k(k-1)(k-2)\cdots(k-n+1)}{(n+1)!} [(k-n)+n] x^n$$

$$= k \sum_{n=0}^{\infty} \frac{k(k-1)(k-2)\cdots(k-n+1)}{n!} x^n = k \sum_{n=0}^{\infty} \binom{k}{n} x^n = kg(x)$$

Thus, $g'(x) = \dfrac{kg(x)}{1+x}$.

(b) $h(x) = (1+x)^{-k} g(x) \quad \Rightarrow$

$h'(x) = -k(1+x)^{-k-1} g(x) + (1+x)^{-k} g'(x)$ [Product Rule]

$$= -k(1+x)^{-k-1} g(x) + (1+x)^{-k} \frac{kg(x)}{1+x} \quad \text{[from part (a)]}$$

$$= -k(1+x)^{-k-1} g(x) + k(1+x)^{-k-1} g(x) = 0$$

(c) From part (b) we see that $h(x)$ must be constant for $x \in (-1, 1)$, so $h(x) = h(0) = 1$ for $x \in (-1, 1)$.

Thus, $h(x) = 1 = (1+x)^{-k} g(x) \quad \Leftrightarrow \quad g(x) = (1+x)^k$ for $x \in (-1, 1)$.

11.12 Applications of Taylor Polynomials

1. (a)

n	$f^{(n)}(x)$	$f^{(n)}(0)$	$T_n(x)$
0	$\cos x$	1	1
1	$-\sin x$	0	1
2	$-\cos x$	-1	$1 - \frac{1}{2}x^2$
3	$\sin x$	0	$1 - \frac{1}{2}x^2$
4	$\cos x$	1	$1 - \frac{1}{2}x^2 + \frac{1}{24}x^4$
5	$-\sin x$	0	$1 - \frac{1}{2}x^2 + \frac{1}{24}x^4$
6	$-\cos x$	-1	$1 - \frac{1}{2}x^2 + \frac{1}{24}x^4 - \frac{1}{720}x^6$

(b)

x	f	$T_0 = T_1$	$T_2 = T_3$	$T_4 = T_5$	T_6
$\frac{\pi}{4}$	0.7071	1	0.6916	0.7074	0.7071
$\frac{\pi}{2}$	0	1	-0.2337	0.0200	-0.0009
π	-1	1	-3.9348	0.1239	-1.2114

(c) As n increases, $T_n(x)$ is a good approximation to $f(x)$ on a larger and larger interval.

3.

n	$f^{(n)}(x)$	$f^{(n)}(1)$
0	$\ln x$	0
1	$1/x$	1
2	$-1/x^2$	-1
3	$2/x^3$	2
4	$-6/x^4$	-6

$$T_4(x) = \sum_{n=0}^{4} \frac{f^{(n)}(1)}{n!}(x-1)^n = 0 + (x-1) - \tfrac{1}{2}(x-1)^2 + \tfrac{1}{3}(x-1)^3 - \tfrac{1}{4}(x-1)^4$$

5.

n	$f^{(n)}(x)$	$f^{(n)}\left(\frac{\pi}{6}\right)$
0	$\sin x$	$\frac{1}{2}$
1	$\cos x$	$\frac{\sqrt{3}}{2}$
2	$-\sin x$	$-\frac{1}{2}$
3	$-\cos x$	$-\frac{\sqrt{3}}{2}$

$$T_3(x) = \sum_{n=0}^{3} \frac{f^{(n)}\left(\frac{\pi}{6}\right)}{n!}\left(x-\tfrac{\pi}{6}\right)^n = \tfrac{1}{2} + \tfrac{\sqrt{3}}{2}\left(x-\tfrac{\pi}{6}\right) - \tfrac{1}{4}\left(x-\tfrac{\pi}{6}\right)^2 - \tfrac{\sqrt{3}}{12}\left(x-\tfrac{\pi}{6}\right)^3$$

7.

n	$f^{(n)}(x)$	$f^{(n)}(0)$
0	$\arcsin x$	0
1	$1/\sqrt{1-x^2}$	1
2	$x/(1-x^2)^{3/2}$	0
3	$(2x^2+1)/(1-x^2)^{5/2}$	1

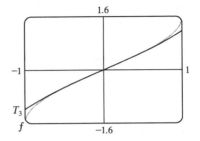

$$T_3(x) = \sum_{n=0}^{3} \frac{f^{(n)}(0)}{n!} x^n = x + \frac{x^3}{6}$$

9.

n	$f^{(n)}(x)$	$f^{(n)}(0)$
0	xe^{-2x}	0
1	$(1-2x)e^{-2x}$	1
2	$4(x-1)e^{-2x}$	-4
3	$4(3-2x)e^{-2x}$	12

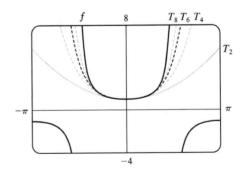

$$T_3(x) = \sum_{n=0}^{3} \frac{f^{(n)}(0)}{n!} x^n = \frac{0}{1}\cdot 1 + \frac{1}{1}x^1 + \frac{-4}{2}x^2 + \frac{12}{6}x^3 = x - 2x^2 + 2x^3$$

11. In Maple, we can find the Taylor polynomials by the following method: first define `f:=sec(x);` and then set

`T2:=convert(taylor(f,x=0,3),polynom);`, `T4:=convert(taylor(f,x=0,5),polynom);`,

etc. (The third argument in the `taylor` function is one more than the degree of the desired polynomial). We must

convert to the type `polynom` because the output of the

`taylor` function contains an error term which we do not

want. In Mathematica, we use

`Tn:=Normal[Series[f,{x,0,n}]]`, with n=2, 4,

etc. Note that in Mathematica, the "degree" argument is the

same as the degree of the desired polynomial. In Derive,

author sec x, then enter `Calculus,Taylor,8,0;` and

then simplify the expression. The eighth Taylor polynomial is

$$T_8(x) = 1 + \tfrac{1}{2}x^2 + \tfrac{5}{24}x^4 + \tfrac{61}{720}x^6 + \tfrac{277}{8064}x^8.$$

13.

n	$f^{(n)}(x)$	$f^{(n)}(4)$
0	$\sqrt{x}$	2
1	$\frac{1}{2}x^{-1/2}$	$\frac{1}{4}$
2	$-\frac{1}{4}x^{-3/2}$	$-\frac{1}{32}$
3	$\frac{3}{8}x^{-5/2}$	

(a) $f(x) = \sqrt{x} \approx T_2(x) = 2 + \frac{1}{4}(x-4) - \frac{1/32}{2!}(x-4)^2 = 2 + \frac{1}{4}(x-4) - \frac{1}{64}(x-4)^2$

(b) $|R_2(x)| \le \frac{M}{3!}|x-4|^3$, where $|f'''(x)| \le M$. Now $4 \le x \le 4.2 \Rightarrow |x-4| \le 0.2 \Rightarrow$

$|x-4|^3 \le 0.008$. Since $f'''(x)$ is decreasing on $[4, 4.2]$, we can take $M = |f'''(4)| = \frac{3}{8}4^{-5/2} = \frac{3}{256}$, so

$|R_2(x)| \le \frac{3/256}{6}(0.008) = \frac{0.008}{512} = 0.000015625$.

(c) From the graph of $|R_2(x)| = |\sqrt{x} - T_2(x)|$, it seems

that the error is less than 1.52×10^{-5} on $[4, 4.2]$.

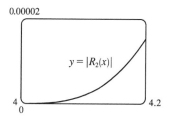

15.

n	$f^{(n)}(x)$	$f^{(n)}(1)$
0	$x^{2/3}$	1
1	$\frac{2}{3}x^{-1/3}$	$\frac{2}{3}$
2	$-\frac{2}{9}x^{-4/3}$	$-\frac{2}{9}$
3	$\frac{8}{27}x^{-7/3}$	$\frac{8}{27}$
4	$-\frac{56}{81}x^{-10/3}$	

(a) $f(x) = x^{2/3} \approx T_3(x) = 1 + \frac{2}{3}(x-1) - \frac{2/9}{2!}(x-1)^2 + \frac{8/27}{3!}(x-1)^3$

$= 1 + \frac{2}{3}(x-1) - \frac{1}{9}(x-1)^2 + \frac{4}{81}(x-1)^3$

(b) $|R_3(x)| \le \frac{M}{4!}|x-1|^4$, where $\left|f^{(4)}(x)\right| \le M$. Now $0.8 \le x \le 1.2 \Rightarrow |x-1| \le 0.2 \Rightarrow$

$|x-1|^4 \le 0.0016$. Since $\left|f^{(4)}(x)\right|$ is decreasing on $[0.8, 1.2]$, we can take $M = \left|f^{(4)}(0.8)\right| = \frac{56}{81}(0.8)^{-10/3}$,

so $|R_3(x)| \le \frac{\frac{56}{81}(0.8)^{-10/3}}{24}(0.0016) \approx 0.00009697$.

(c) From the graph of $|R_3(x)| = \left|x^{2/3} - T_3(x)\right|$, it seems

that the error is less than 0.0000533 on $[0.8, 1.2]$.

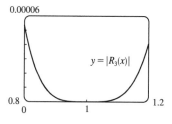

17.

n	$f^{(n)}(x)$	$f^{(n)}(0)$
0	$\tan x$	0
1	$\sec^2 x$	1
2	$2\sec^2 x \tan x$	0
3	$4\sec^2 x \tan^2 x + 2\sec^4 x$	2
4	$8\sec^2 x \tan^3 x + 16\sec^4 x \tan x$	

(a) $f(x) = \tan x \approx T_3(x) = x + \frac{1}{3}x^3$

(b) $|R_3(x)| \le \dfrac{M}{4!}\,|x|^4$, where $\left|f^{(4)}(x)\right| \le M$. Now

$0 \le x \le \frac{\pi}{6} \quad\Rightarrow\quad x^4 \le \left(\frac{\pi}{6}\right)^4$, and letting $x = \frac{\pi}{6}$

$\left[\text{since } f^{(4)} \text{ is increasing on } \left(0, \frac{\pi}{6}\right)\right]$ gives

$$|R_3(x)| \le \dfrac{8\left(\frac{2}{\sqrt{3}}\right)^2\left(\frac{1}{\sqrt{3}}\right)^3 + 16\left(\frac{2}{\sqrt{3}}\right)^4\left(\frac{1}{\sqrt{3}}\right)}{4!}\left(\frac{\pi}{6}\right)^4$$

$$= \tfrac{4\sqrt{3}}{9}\left(\tfrac{\pi}{6}\right)^4 \approx 0.057859$$

(c)

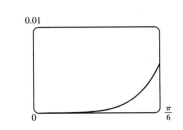

0.01

0

$\frac{\pi}{6}$

From the graph of

$|R_3(x)| = |\tan x - T_3(3)|$, it seems that

the error is less than 0.006 on $[0, \pi/6]$.

19.

n	$f^{(n)}(x)$	$f^{(n)}(0)$
0	e^{x^2}	1
1	$e^{x^2}(2x)$	0
2	$e^{x^2}(2 + 4x^2)$	2
3	$e^{x^2}(12x + 8x^3)$	0
4	$e^{x^2}(12 + 48x^2 + 16x^4)$	

(a) $f(x) = e^{x^2} \approx T_3(x) = 1 + \frac{2}{2!}x^2 = 1 + x^2$

(b) $|R_3(x)| \le \dfrac{M}{4!}\,|x|^4$, where $\left|f^{(4)}(x)\right| \le M$.

Now $0 \le x \le 0.1 \quad\Rightarrow\quad x^4 \le (0.1)^4$, and

letting $x = 0.1$ gives

$$|R_3(x)| \le \dfrac{e^{0.01}(12 + 0.48 + 0.0016)}{24}(0.1)^4 \approx$$

0.00006.

(c)

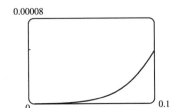

0.00008

0

0.1

From the graph of

$|R_3(x)| = \left|e^{x^2} - (1 + x^2)\right|$, it appears that

the error is less than 0.000051 on $[0, 0.1]$.

21.

n	$f^{(n)}(x)$	$f^{(n)}(0)$
0	$x \sin x$	0
1	$\sin x + x \cos x$	0
2	$2 \cos x - x \sin x$	2
3	$-3 \sin x - x \cos x$	0
4	$-4 \cos x + x \sin x$	-4
5	$5 \sin x + x \cos x$	

(a) $f(x) = x \sin x \approx T_4(x) = \frac{2}{2!}(x-0)^2 + \frac{-4}{4!}(x-0)^4 = x^2 - \frac{1}{6}x^4$

(b) $|R_4(x)| \leq \frac{M}{5!}|x|^5$, where $\left|f^{(5)}(x)\right| \leq M$. Now $-1 \leq x \leq 1 \Rightarrow |x| \leq 1$, and a graph of $f^{(5)}(x)$ shows that

$\left|f^{(5)}(x)\right| \leq 5$ for $-1 \leq x \leq 1$. Thus, we can take $M = 5$ and get $|R_4(x)| \leq \frac{5}{5!} \cdot 1^5 = \dfrac{1}{24} = 0.041\overline{6}$.

(c) From the graph of $|R_4(x)| = |x \sin x - T_4(x)|$, it

seems that the error is less than 0.0082 on $[-1, 1]$.

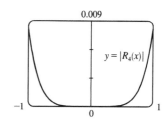

23. From Exercise 5, $\sin x = \frac{1}{2} + \frac{\sqrt{3}}{2}\left(x - \frac{\pi}{6}\right) - \frac{1}{4}\left(x - \frac{\pi}{6}\right)^2 - \frac{\sqrt{3}}{12}\left(x - \frac{\pi}{6}\right)^3 + R_3(x)$, where $|R_3(x)| \leq \dfrac{M}{4!}\left|x - \frac{\pi}{6}\right|^4$

with $\left|f^{(4)}(x)\right| = |\sin x| \leq M = 1$. Now $x = 35° = (30° + 5°) = \left(\frac{\pi}{6} + \frac{\pi}{36}\right)$ radians, so the error is

$\left|R_3\left(\frac{\pi}{36}\right)\right| \leq \dfrac{\left(\frac{\pi}{36}\right)^4}{4!} < 0.000003$. Therefore, to five decimal places,

$\sin 35° \approx \frac{1}{2} + \frac{\sqrt{3}}{2}\left(\frac{\pi}{36}\right) - \frac{1}{4}\left(\frac{\pi}{36}\right)^2 - \frac{\sqrt{3}}{12}\left(\frac{\pi}{36}\right)^3 \approx 0.57358$.

25. All derivatives of e^x are e^x, so $|R_n(x)| \leq \dfrac{e^x}{(n+1)!}|x|^{n+1}$, where $0 < x < 0.1$. Letting $x = 0.1$,

$R_n(0.1) \leq \dfrac{e^{0.1}}{(n+1)!}(0.1)^{n+1} < 0.00001$, and by trial and error we find that $n = 3$ satisfies this inequality since

$R_3(0.1) < 0.0000046$. Thus, by adding the four terms of the Maclaurin series for e^x corresponding to $n = 0, 1, 2,$

and 3, we can estimate $e^{0.1}$ to within 0.00001. (In fact, this sum is $1.1051\overline{6}$ and $e^{0.1} \approx 1.10517$.)

27. $\sin x = x - \frac{1}{3!}x^3 + \frac{1}{5!}x^5 - \cdots$. By the Alternating

Series Estimation Theorem, the error in the

approximation $\sin x = x - \frac{1}{3!}x^3$ is less than

$\left|\frac{1}{5!}x^5\right| < 0.01 \iff |x^5| < 120(0.01) \iff$

$|x| < (1.2)^{1/5} \approx 1.037$. The curves $y = x - \frac{1}{6}x^3$ and

$y = \sin x - 0.01$ intersect at $x \approx 1.043$, so the graph

confirms our estimate. Since both the sine function and

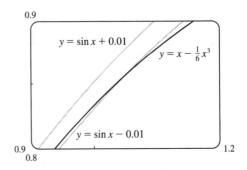

the given approximation are odd functions, we need to check the estimate only for $x > 0$. Thus, the desired range of

values for x is $-1.037 < x < 1.037$.

29. Let $s(t)$ be the position function of the car, and for convenience set $s(0) = 0$. The velocity of the car is

$v(t) = s'(t)$ and the acceleration is $a(t) = s''(t)$, so the second degree Taylor polynomial is

$T_2(t) = s(0) + v(0)t + \frac{a(0)}{2}t^2 = 20t + t^2$. We estimate the distance travelled during the next second to be

$s(1) \approx T_2(1) = 20 + 1 = 21$ m. The function $T_2(t)$ would not be accurate over a full minute, since the car could

not possibly maintain an acceleration of 2 m/s^2 for that long (if it did, its final speed would be

140 m/s ≈ 313 mi/h!)

31. $E = \frac{q}{D^2} - \frac{q}{(D+d)^2} = \frac{q}{D^2} - \frac{q}{D^2(1 + d/D)^2} = \frac{q}{D^2}\left[1 - \left(1 + \frac{d}{D}\right)^{-2}\right]$.

We use the Binomial Series to expand $(1 + d/D)^{-2}$:

$$E = \frac{q}{D^2}\left[1 - \left(1 - 2\left(\frac{d}{D}\right) + \frac{2 \cdot 3}{2!}\left(\frac{d}{D}\right)^2 - \frac{2 \cdot 3 \cdot 4}{3!}\left(\frac{d}{D}\right)^3 + \cdots\right)\right]$$

$$= \frac{q}{D^2}\left[2\left(\frac{d}{D}\right) - 3\left(\frac{d}{D}\right)^2 + 4\left(\frac{d}{D}\right)^3 - \cdots\right] \approx \frac{q}{D^2} \cdot 2\left(\frac{d}{D}\right) = 2qd \cdot \frac{1}{D^3}$$

when D is much larger than d; that is, when P is far away from the dipole.

33. (a) If the water is deep, then $2\pi d/L$ is large, and we know that $\tanh x \to 1$ as $x \to \infty$. So we can approximate

$\tanh(2\pi d/L) \approx 1$, and so $v^2 \approx gL/(2\pi) \iff v \approx \sqrt{gL/(2\pi)}$.

(b) From the table, the first term in the Maclaurin

series of $\tanh x$ is x, so if the water is shallow,

we can approximate $\tanh \frac{2\pi d}{L} \approx \frac{2\pi d}{L}$, and so

$v^2 \approx \frac{gL}{2\pi} \cdot \frac{2\pi d}{L} \iff v \approx \sqrt{gd}$.

n	$f^{(n)}(x)$	$f^{(n)}(0)$
0	$\tanh x$	0
1	$\text{sech}^2 x$	1
2	$-2\,\text{sech}^2 x \tanh x$	0
3	$2\,\text{sech}^2 x\,(3\tanh^2 x - 1)$	-2

(c) Since $\tanh x$ is an odd function, its Maclaurin series is alternating, so the error in the approximation

$$\tanh \frac{2\pi d}{L} \approx \frac{2\pi d}{L} \text{ is less than the first neglected term, which is } \frac{|f'''(0)|}{3!}\left(\frac{2\pi d}{L}\right)^3 = \frac{1}{3}\left(\frac{2\pi d}{L}\right)^3.$$

If $L > 10d$, then $\dfrac{1}{3}\left(\dfrac{2\pi d}{L}\right)^3 < \dfrac{1}{3}\left(2\pi \cdot \dfrac{1}{10}\right)^3 = \dfrac{\pi^3}{375}$, so the error in the approximation $v^2 = gd$ is less

than $\dfrac{gL}{2\pi} \cdot \dfrac{\pi^3}{375} \approx 0.0132gL$.

35. (a) L is the length of the arc subtended by the angle θ, so $L = R\theta$ ⇒

$\theta = L/R$. Now $\sec \theta = (R+C)/R$ ⇒ $R \sec \theta = R + C$ ⇒

$C = R \sec \theta - R = R \sec(L/R) - R$.

(b) From Exercise 11, $\sec x \approx T_4(x) = 1 + \frac{1}{2}x^2 + \frac{5}{24}x^4$. By part (a),

$$C \approx R\left[1 + \frac{1}{2}\left(\frac{L}{R}\right)^2 + \frac{5}{24}\left(\frac{L}{R}\right)^4\right] - R = R + \frac{1}{2}R \cdot \frac{L^2}{R^2} + \frac{5}{24}R \cdot \frac{L^4}{R^4} - R = \frac{L^2}{2R} + \frac{5L^4}{24R^3}.$$

(c) Taking $L = 100$ km and $R = 6370$ km, the formula in part (a) says that

$$C = R\sec(L/R) - R = 6370\sec(100/6370) - 6370 \approx 0.785\,009\,965\,44 \text{ km}$$

The formula in part (b) says that

$$C \approx \frac{L^2}{2R} + \frac{5L^4}{24R^3} = \frac{100^2}{2 \cdot 6370} + \frac{5 \cdot 100^4}{24 \cdot 6370^3} \approx 0.785\,009\,957\,36 \text{ km}.$$

The difference between these two results is only $0.000\,000\,008\,08$ km, or $0.000\,008\,08$ m!

37. Using $f(x) = T_n(x) + R_n(x)$ with $n = 1$ and $x = r$, we have $f(r) = T_1(r) + R_1(r)$, where T_1 is the first-degree

Taylor polynomial of f at a. Because $a = x_n$, $f(r) = f(x_n) + f'(x_n)(r - x_n) + R_1(r)$. But r is a root of f, so

$f(r) = 0$ and we have $0 = f(x_n) + f'(x_n)(r - x_n) + R_1(r)$. Taking the first two terms to the left side gives us

$f'(x_n)(x_n - r) - f(x_n) = R_1(r)$. Dividing by $f'(x_n)$, we get $x_n - r - \dfrac{f(x_n)}{f'(x_n)} = \dfrac{R_1(r)}{f'(x_n)}$. By the formula for

Newton's method, the left side of the preceding equation is $x_{n+1} - r$, so $|x_{n+1} - r| = \left|\dfrac{R_1(r)}{f'(x_n)}\right|$. Taylor's

Inequality gives us $|R_1(r)| \le \dfrac{|f''(r)|}{2!}|r - x_n|^2$. Combining this inequality with the facts $|f''(x)| \le M$ and

$|f'(x)| \ge K$ gives us $|x_{n+1} - r| \le \dfrac{M}{2K}|x_n - r|^2$.

11 Review

1. (a) See Definition 11.1.1.

(b) See Definition 11.2.2.

(c) The terms of the sequence $\{a_n\}$ approach 3 as n becomes large.

(d) By adding sufficiently many terms of the series, we can make the partial sums as close to 3 as we like.

2. (a) See Definition 11.1.10.

(b) A sequence is monotonic if it is either increasing or decreasing.

(c) By Theorem 11.1.11, every bounded, monotonic sequence is convergent.

3. (a) See (4) in Section 11.2.

(b) The p-series $\displaystyle\sum_{n=1}^{\infty} \frac{1}{n^p}$ is convergent if $p > 1$.

4. If $\sum a_n = 3$, then $\displaystyle\lim_{n\to\infty} a_n = 0$ and $\displaystyle\lim_{n\to\infty} s_n = 3$.

5. (a) See the Test for Divergence on page 718.

(b) See the Integral Test on page 724.

(c) See the Comparison Test on page 731.

(d) See the Limit Comparison Test on page 732.

(e) See the Alternating Series Test on page 736.

(f) See the Ratio Test on page 742.

(g) See the Root Test on page 744.

6. (a) A series $\sum a_n$ is called *absolutely convergent* if the series of absolute values $\sum |a_n|$ is convergent.

(b) If a series $\sum a_n$ is absolutely convergent, then it is convergent.

(c) A series $\sum a_n$ is called *conditionally convergent* if it is convergent but not absolutely convergent.

7. (a) Use (3) in Section 11.3.

(b) See Example 5 in Section 11.4.

(c) By adding terms until you reach the desired accuracy given by the Alternating Series Estimation Theorem on page 738.

8. (a) $\sum_{n=0}^{\infty} c_n(x-a)^n$

(b) Given the power series $\sum_{n=0}^{\infty} c_n(x-a)^n$, the radius of convergence is:

(i) 0 if the series converges only when $x = a$

(ii) ∞ if the series converges for all x, or

(iii) a positive number R such that the series converges if $|x - a| < R$ and diverges if $|x - a| > R$.

(c) The interval of convergence of a power series is the interval that consists of all values of x for which the series converges. Corresponding to the cases in part (b), the interval of convergence is: (i) the single point $\{a\}$, (ii) all real numbers, that is, the real number line $(-\infty, \infty)$, or (iii) an interval with endpoints $a - R$ and $a + R$ which can contain neither, either, or both of the endpoints. In this case, we must test the series for convergence at each endpoint to determine the interval of convergence.

9. (a), (b) See Theorem 11.9.2.

10. (a) $T_n(x) = \sum\limits_{i=0}^{n} \dfrac{f^{(i)}(a)}{i!}(x-a)^i$

(b) $\sum\limits_{n=0}^{\infty} \dfrac{f^{(n)}(a)}{n!}(x-a)^n$

(c) $\sum\limits_{n=0}^{\infty} \dfrac{f^{(n)}(0)}{n!}x^n$ [$a = 0$ in part (b)]

(d) See Theorem 11.10.8.

(e) See Taylor's Inequality (11.10.9).

11. (a) – (e) See the table on page 767.

12. See the Binomial Series (11.11.2) for the expansion. The radius of convergence for the binomial series is 1.

───────────────────── TRUE-FALSE QUIZ ─────────────────────

1. False.   See Note 2 after Theorem 11.2.6.

3. True.   If $\lim\limits_{n\to\infty} a_n = L$, then given any $\varepsilon > 0$, we can find a positive integer N such that $|a_n - L| < \varepsilon$ whenever $n > N$. If $n > N$, then $2n + 1 > N$ and $|a_{2n+1} - L| < \varepsilon$. Thus, $\lim\limits_{n\to\infty} a_{2n+1} = L$.

5. False.   For example, take $c_n = (-1)^n / (n6^n)$.

7. False, since $\lim\limits_{n\to\infty} \left| \dfrac{a_{n+1}}{a_n} \right| = \lim\limits_{n\to\infty} \left| \dfrac{1}{(n+1)^3} \cdot \dfrac{n^3}{1} \right| = \lim\limits_{n\to\infty} \left| \dfrac{n^3}{(n+1)^3} \cdot \dfrac{1/n^3}{1/n^3} \right| = \lim\limits_{n\to\infty} \dfrac{1}{(1+1/n)^3} = 1$.

9. False.   See the note after Example 2 in Section 11.4.

11. True.   See (8) in Section 11.1.

13. True.   By Theorem 11.10.5 the coefficient of x^3 is $\dfrac{f'''(0)}{3!} = \dfrac{1}{3} \;\Rightarrow\; f'''(0) = 2$.

Or: Use Theorem 11.9.2 to differentiate f three times.

15. False.   For example, let $a_n = b_n = (-1)^n$. Then $\{a_n\}$ and $\{b_n\}$ are divergent, but $a_n b_n = 1$, so $\{a_n b_n\}$ is convergent.

17. True by Theorem 11.6.3. $\left[\sum (-1)^n a_n \text{ is absolutely convergent and hence convergent.} \right]$

───────────────────── EXERCISES ─────────────────────

1. $\left\{ \dfrac{2+n^3}{1+2n^3} \right\}$ converges since $\lim\limits_{n\to\infty} \dfrac{2+n^3}{1+2n^3} = \lim\limits_{n\to\infty} \dfrac{2/n^3 + 1}{1/n^3 + 2} = \dfrac{1}{2}$.

3. $\lim\limits_{n\to\infty} a_n = \lim\limits_{n\to\infty} \dfrac{n^3}{1+n^2} = \lim\limits_{n\to\infty} \dfrac{n}{1/n^2 + 1} = \infty$, so the sequence diverges.

5. $|a_n| = \left| \dfrac{n\sin n}{n^2 + 1} \right| \le \dfrac{n}{n^2 + 1} < \dfrac{1}{n}$, so $|a_n| \to 0$ as $n \to \infty$. Thus, $\lim\limits_{n\to\infty} a_n = 0$. The sequence $\{a_n\}$ is convergent.

7. $\left\{\left(1+\dfrac{3}{n}\right)^{4n}\right\}$ is convergent. Let $y = \left(1+\dfrac{3}{x}\right)^{4x}$. Then

$$\lim_{x\to\infty} \ln y = \lim_{x\to\infty} 4x \ln(1+3/x) = \lim_{x\to\infty} \frac{\ln(1+3/x)}{1/(4x)} \overset{\text{H}}{=} \lim_{x\to\infty} \frac{\dfrac{1}{1+3/x}\left(-\dfrac{3}{x^2}\right)}{-1/(4x^2)} = \lim_{x\to\infty} \frac{12}{1+3/x} = 12$$

so $\displaystyle\lim_{x\to\infty} y = \lim_{n\to\infty}\left(1+\frac{3}{n}\right)^{4n} = e^{12}$.

Or: Use Exercise 4.4.54 .

9. We use induction, hypothesizing that $a_{n-1} < a_n < 2$. Note first that $1 < a_2 = \frac{1}{3}(1+4) = \frac{5}{3} < 2$, so the hypothesis holds for $n = 2$. Now assume that $a_{k-1} < a_k < 2$. Then

$a_k = \frac{1}{3}(a_{k-1}+4) < \frac{1}{3}(a_k+4) < \frac{1}{3}(2+4) = 2$. So $a_k < a_{k+1} < 2$, and the induction is complete. To find the limit of the sequence, we note that $L = \lim_{n\to\infty} a_n = \lim_{n\to\infty} a_{n+1} \;\Rightarrow\; L = \frac{1}{3}(L+4) \;\Rightarrow\; L = 2$.

11. $\dfrac{n}{n^3+1} < \dfrac{n}{n^3} = \dfrac{1}{n^2}$, so $\displaystyle\sum_{n=1}^{\infty} \frac{n}{n^3+1}$ converges by the Comparison Test with the convergent p-series

$\displaystyle\sum_{n=1}^{\infty} \frac{1}{n^2}$ $(p = 2 > 1)$.

13. $\displaystyle\lim_{n\to\infty}\left|\frac{a_{n+1}}{a_n}\right| = \lim_{n\to\infty}\left[\frac{(n+1)^3}{5^{n+1}} \cdot \frac{5^n}{n^3}\right] = \lim_{n\to\infty}\left(1+\frac{1}{n}\right)^3 \cdot \frac{1}{5} = \frac{1}{5} < 1$, so $\displaystyle\sum_{n=1}^{\infty} \frac{n^3}{5^n}$ converges by the Ratio Test.

15. Let $f(x) = \dfrac{1}{x\sqrt{\ln x}}$. Then f is continuous, positive, and decreasing on $[2,\infty)$, so the Integral Test applies.

$$\int_2^{\infty} f(x)\,dx = \lim_{t\to\infty}\int_2^t \frac{1}{x\sqrt{\ln x}}\,dx \quad \begin{bmatrix} u = \ln x, \\ du = \dfrac{1}{x}\,dx \end{bmatrix} = \lim_{t\to\infty}\int_{\ln 2}^{\ln t} u^{-1/2}\,du = \lim_{t\to\infty}\left[2\sqrt{u}\,\right]_{\ln 2}^{\ln t}$$

$= \lim_{t\to\infty}(2\sqrt{\ln t} - 2\sqrt{\ln 2}) = \infty$, so the series $\displaystyle\sum_{n=2}^{\infty}\frac{1}{n\sqrt{\ln n}}$ diverges.

17. $|a_n| = \left|\dfrac{\cos 3n}{1+(1.2)^n}\right| \le \dfrac{1}{1+(1.2)^n} < \dfrac{1}{(1.2)^n} = \left(\dfrac{5}{6}\right)^n$, so $\displaystyle\sum_{n=1}^{\infty}|a_n|$ converges by comparison with the convergent

geometric series $\displaystyle\sum_{n=1}^{\infty}\left(\frac{5}{6}\right)^n$ $\left(r = \frac{5}{6} < 1\right)$. It follows that $\displaystyle\sum_{n=1}^{\infty} a_n$ converges (by Theorem 3 in Section 11.6).

19. $\displaystyle\lim_{n\to\infty}\left|\frac{a_{n+1}}{a_n}\right| = \lim_{n\to\infty}\frac{1\cdot 3\cdot 5\cdots(2n-1)(2n+1)}{5^{n+1}\,(n+1)!} \cdot \frac{5^n n!}{1\cdot 3\cdot 5\cdots(2n-1)} = \lim_{n\to\infty}\frac{2n+1}{5(n+1)} = \frac{2}{5} < 1$, so

the series converges by the Ratio Test.

21. $b_n = \dfrac{\sqrt{n}}{n+1} > 0$, $\{b_n\}$ is decreasing, and $\lim_{n\to\infty} b_n = 0$, so the series $\displaystyle\sum_{n=1}^{\infty}(-1)^{n-1}\frac{\sqrt{n}}{n+1}$ converges by the

Alternating Series Test.

23. Consider the series of absolute values: $\sum_{n=1}^{\infty} n^{-1/3}$ is a p-series with $p = \frac{1}{3} \le 1$ and is therefore divergent. But if

we apply the Alternating Series Test, we see that $b_n = \dfrac{1}{\sqrt[3]{n}} > 0$, $\{b_n\}$ is decreasing, and $\lim_{n\to\infty} b_n = 0$, so the series

$\displaystyle\sum_{n=1}^{\infty}(-1)^{n-1}n^{-1/3}$ converges. Thus, $\displaystyle\sum_{n=1}^{\infty}(-1)^{n-1}n^{-1/3}$ is conditionally convergent.

25. $\left|\dfrac{a_{n+1}}{a_n}\right| = \left|\dfrac{(-1)^{n+1}\,(n+2)\,3^{n+1}}{2^{2n+3}} \cdot \dfrac{2^{2n+1}}{(-1)^n\,(n+1)\,3^n}\right| = \dfrac{n+2}{n+1} \cdot \dfrac{3}{4} = \dfrac{1+(2/n)}{1+(1/n)} \cdot \dfrac{3}{4} \to \dfrac{3}{4} < 1$ as $n \to \infty$, so

by the Ratio Test, $\displaystyle\sum_{n=1}^{\infty} \dfrac{(-1)^n\,(n+1)\,3^n}{2^{2n+1}}$ is absolutely convergent.

27. $\dfrac{2^{2n+1}}{5^n} = \dfrac{2^{2n} \cdot 2^1}{5^n} = \dfrac{(2^2)^n \cdot 2}{5^n} = 2\left(\dfrac{4}{5}\right)^n$, so $\displaystyle\sum_{n=1}^{\infty} \dfrac{2^{2n+1}}{5^n} = 2\sum_{n=1}^{\infty}\left(\dfrac{4}{5}\right)^n$ is a geometric series with $a = \dfrac{8}{5}$ and

$r = \dfrac{4}{5}$. Since $|r| = \dfrac{4}{5} < 1$, the series converges to $\dfrac{a}{1-r} = \dfrac{8/5}{1-4/5} = \dfrac{8/5}{1/5} = 8$.

29. $\displaystyle\sum_{n=1}^{\infty}\left[\tan^{-1}(n+1) - \tan^{-1} n\right] = \lim_{n\to\infty} s_n = \lim_{n\to\infty}\left[(\tan^{-1} 2 - \tan^{-1} 1) + (\tan^{-1} 3 - \tan^{-1} 2) + \cdots\right.$

$\left. + (\tan^{-1}(n+1) - \tan^{-1} n)\right]$

$= \lim_{n\to\infty}\left[\tan^{-1}(n+1) - \tan^{-1} 1\right] = \dfrac{\pi}{2} - \dfrac{\pi}{4} = \dfrac{\pi}{4}$

31. $1 - e + \dfrac{e^2}{2!} - \dfrac{e^3}{3!} + \dfrac{e^4}{4!} - \cdots = \displaystyle\sum_{n=0}^{\infty} (-1)^n \dfrac{e^n}{n!} = \sum_{n=0}^{\infty} \dfrac{(-e)^n}{n!} = e^{-e}$ since $e^x = \displaystyle\sum_{n=0}^{\infty} \dfrac{x^n}{n!}$ for all x.

33. $\cosh x = \dfrac{1}{2}(e^x + e^{-x}) = \dfrac{1}{2}\left(\displaystyle\sum_{n=0}^{\infty}\dfrac{x^n}{n!} + \sum_{n=0}^{\infty}\dfrac{(-x)^n}{n!}\right)$

$= \dfrac{1}{2}\left[\left(1 + x + \dfrac{x^2}{2!} + \dfrac{x^3}{3!} + \dfrac{x^4}{4!} + \cdots\right) + \left(1 - x + \dfrac{x^2}{2!} - \dfrac{x^3}{3!} + \dfrac{x^4}{4!} - \cdots\right)\right]$

$= \dfrac{1}{2}\left(2 + 2 \cdot \dfrac{x^2}{2!} + 2 \cdot \dfrac{x^4}{4!} + \cdots\right)$

$= 1 + \dfrac{1}{2}x^2 + \displaystyle\sum_{n=2}^{\infty} \dfrac{x^{2n}}{(2n)!}$

$\geq 1 + \dfrac{1}{2}x^2 \quad$ for all x

35. $\displaystyle\sum_{n=1}^{\infty} \dfrac{(-1)^{n+1}}{n^5} = 1 - \dfrac{1}{32} + \dfrac{1}{243} - \dfrac{1}{1024} + \dfrac{1}{3125} - \dfrac{1}{7776} + \dfrac{1}{16,807} - \dfrac{1}{32,768} + \cdots$.

Since $b_8 = \dfrac{1}{8^5} = \dfrac{1}{32,768} < 0.000031$, $\displaystyle\sum_{n=1}^{\infty} \dfrac{(-1)^{n+1}}{n^5} \approx \sum_{n=1}^{7} \dfrac{(-1)^{n+1}}{n^5} \approx 0.9721$.

37. $\displaystyle\sum_{n=1}^{\infty} \dfrac{1}{2+5^n} \approx \sum_{n=1}^{8} \dfrac{1}{2+5^n} \approx 0.18976224$. To estimate the error, note that $\dfrac{1}{2+5^n} < \dfrac{1}{5^n}$, so the remainder term is

$R_8 = \displaystyle\sum_{n=9}^{\infty} \dfrac{1}{2+5^n} < \sum_{n=9}^{\infty} \dfrac{1}{5^n} = \dfrac{1/5^9}{1-1/5} = 6.4 \times 10^{-7}$ (geometric series with $a = \frac{1}{5^9}$ and $r = \frac{1}{5}$).

39. Use the Limit Comparison Test. $\displaystyle\lim_{n\to\infty}\left|\dfrac{\left(\frac{n+1}{n}\right)a_n}{a_n}\right| = \lim_{n\to\infty}\dfrac{n+1}{n} = \lim_{n\to\infty}\left(1 + \dfrac{1}{n}\right) = 1 > 0$.

Since $\sum |a_n|$ is convergent, so is $\displaystyle\sum\left|\left(\dfrac{n+1}{n}\right)a_n\right|$, by the Limit Comparison Test.

41. $\lim\limits_{n\to\infty}\left|\dfrac{a_{n+1}}{a_n}\right| = \lim\limits_{n\to\infty}\left[\dfrac{|x+2|^{n+1}}{(n+1)\,4^{n+1}}\cdot\dfrac{n4^n}{|x+2|^n}\right] = \lim\limits_{n\to\infty}\left[\dfrac{n}{n+1}\,\dfrac{|x+2|}{4}\right] = \dfrac{|x+2|}{4} < 1 \ \Leftrightarrow\ |x+2| < 4$,

so $R = 4$. $|x+2| < 4 \ \Leftrightarrow\ -4 < x+2 < 4 \ \Leftrightarrow\ -6 < x < 2$. If $x = -6$, then the series $\sum\limits_{n=1}^{\infty}\dfrac{(x+2)^n}{n4^n}$

becomes $\sum\limits_{n=1}^{\infty}\dfrac{(-4)^n}{n4^n} = \sum\limits_{n=1}^{\infty}\dfrac{(-1)^n}{n}$, the alternating harmonic series, which converges by the Alternating Series

Test. When $x = 2$, the series becomes the harmonic series $\sum\limits_{n=1}^{\infty}\dfrac{1}{n}$, which diverges. Thus, $I = [-6, 2)$.

43. $\lim\limits_{n\to\infty}\left|\dfrac{a_{n+1}}{a_n}\right| = \lim\limits_{n\to\infty}\left|\dfrac{2^{n+1}\,(x-3)^{n+1}}{\sqrt{n+4}}\cdot\dfrac{\sqrt{n+3}}{2^n\,(x-3)^n}\right| = 2\,|x-3|\,\lim\limits_{n\to\infty}\sqrt{\dfrac{n+3}{n+4}} = 2\,|x-3| < 1 \ \Leftrightarrow$

$|x-3| < \tfrac{1}{2}$, so $R = \tfrac{1}{2}$. $|x-3| < \tfrac{1}{2} \ \Leftrightarrow\ -\tfrac{1}{2} < x-3 < \tfrac{1}{2} \ \Leftrightarrow\ \tfrac{5}{2} < x < \tfrac{7}{2}$. For $x = \tfrac{7}{2}$, the series

$\sum\limits_{n=1}^{\infty}\dfrac{2^n\,(x-3)^n}{\sqrt{n+3}}$ becomes $\sum\limits_{n=0}^{\infty}\dfrac{1}{\sqrt{n+3}} = \sum\limits_{n=3}^{\infty}\dfrac{1}{n^{1/2}}$, which diverges ($p = \tfrac{1}{2} \le 1$), but for $x = \tfrac{5}{2}$, we get

$\sum\limits_{n=0}^{\infty}\dfrac{(-1)^n}{\sqrt{n+3}}$, which is a convergent alternating series, so $I = \left[\tfrac{5}{2}, \tfrac{7}{2}\right)$.

45.

n	$f^{(n)}(x)$	$f^{(n)}\left(\tfrac{\pi}{6}\right)$
0	$\sin x$	$\tfrac{1}{2}$
1	$\cos x$	$\tfrac{\sqrt{3}}{2}$
2	$-\sin x$	$-\tfrac{1}{2}$
3	$-\cos x$	$-\tfrac{\sqrt{3}}{2}$
4	$\sin x$	$\tfrac{1}{2}$
⋮	⋮	⋮

$\sin x = f\left(\tfrac{\pi}{6}\right) + f'\left(\tfrac{\pi}{6}\right)\left(x - \tfrac{\pi}{6}\right) + \dfrac{f''\left(\tfrac{\pi}{6}\right)}{2!}\left(x - \tfrac{\pi}{6}\right)^2 + \dfrac{f^{(3)}\left(\tfrac{\pi}{6}\right)}{3!}\left(x - \tfrac{\pi}{6}\right)^3 + \dfrac{f^{(4)}\left(\tfrac{\pi}{6}\right)}{4!}\left(x - \tfrac{\pi}{6}\right)^4 + \cdots$

$= \dfrac{1}{2}\left[1 - \dfrac{1}{2!}\left(x - \tfrac{\pi}{6}\right)^2 + \dfrac{1}{4!}\left(x - \tfrac{\pi}{6}\right)^4 - \cdots\right] + \dfrac{\sqrt{3}}{2}\left[\left(x - \tfrac{\pi}{6}\right) - \dfrac{1}{3!}\left(x - \tfrac{\pi}{6}\right)^3 + \cdots\right]$

$= \dfrac{1}{2}\sum\limits_{n=0}^{\infty}(-1)^n\,\dfrac{1}{(2n)!}\left(x - \tfrac{\pi}{6}\right)^{2n} + \dfrac{\sqrt{3}}{2}\sum\limits_{n=0}^{\infty}(-1)^n\,\dfrac{1}{(2n+1)!}\left(x - \tfrac{\pi}{6}\right)^{2n+1}$

47. $\dfrac{1}{1+x} = \dfrac{1}{1-(-x)} = \sum\limits_{n=0}^{\infty}(-x)^n = \sum\limits_{n=0}^{\infty}(-1)^n\,x^n$ for $|x| < 1 \ \Rightarrow\ \dfrac{x^2}{1+x} = \sum\limits_{n=0}^{\infty}(-1)^n\,x^{n+2}$ with $R = 1$.

49. $\dfrac{1}{1-x} = \sum\limits_{n=0}^{\infty}x^n$ for $|x| < 1 \ \Rightarrow\ \ln(1-x) = -\displaystyle\int\dfrac{dx}{1-x} = -\int\sum\limits_{n=0}^{\infty}x^n\,dx = C - \sum\limits_{n=0}^{\infty}\dfrac{x^{n+1}}{n+1}$.

$\ln(1-0) = C - 0 \ \Rightarrow\ C = 0 \ \Rightarrow\ \ln(1-x) = -\sum\limits_{n=0}^{\infty}\dfrac{x^{n+1}}{n+1} = \sum\limits_{n=1}^{\infty}\dfrac{-x^n}{n}$ with $R = 1$.

51. $\sin x = \sum\limits_{n=0}^{\infty} \dfrac{(-1)^n x^{2n+1}}{(2n+1)!} \;\Rightarrow\; \sin(x^4) = \sum\limits_{n=0}^{\infty} \dfrac{(-1)^n (x^4)^{2n+1}}{(2n+1)!} = \sum\limits_{n=0}^{\infty} \dfrac{(-1)^n x^{8n+4}}{(2n+1)!}$ for all x, so the radius of

convergence is ∞.

53. $f(x) = 1/\sqrt[4]{16-x} = 1/\left(\sqrt[4]{16}\,\sqrt[4]{1-\tfrac{1}{16}x}\right) = \tfrac{1}{2}\left(1-\tfrac{1}{16}x\right)^{-1/4}$

$$= \frac{1}{2}\left[1 + \left(-\frac{1}{4}\right)\left(-\frac{x}{16}\right) + \frac{\left(-\frac{1}{4}\right)\left(-\frac{5}{4}\right)}{2!}\left(-\frac{x}{16}\right)^2 + \frac{\left(-\frac{1}{4}\right)\left(-\frac{5}{4}\right)\left(-\frac{9}{4}\right)}{3!}\left(-\frac{x}{16}\right)^3 + \cdots\right]$$

$$= \frac{1}{2} + \sum_{n=1}^{\infty} \frac{1\cdot5\cdot9\cdot\cdots\cdot(4n-3)}{2\cdot4^n\cdot n!\cdot16^n}x^n = \frac{1}{2} + \sum_{n=1}^{\infty} \frac{1\cdot5\cdot9\cdot\cdots\cdot(4n-3)}{2^{6n+1}\,n!}x^n$$

for $\left|-\dfrac{x}{16}\right| < 1 \;\Leftrightarrow\; |x| < 16$, so $R = 16$.

55. $e^x = \sum\limits_{n=0}^{\infty} \dfrac{x^n}{n!}$, so $\dfrac{e^x}{x} = \dfrac{1}{x} + \sum\limits_{n=1}^{\infty} \dfrac{x^{n-1}}{n!}$ and $\displaystyle\int \dfrac{e^x}{x}\,dx = C + \ln|x| + \sum\limits_{n=1}^{\infty} \dfrac{x^n}{n\cdot n!}$.

57. (a)

n	$f^{(n)}(x)$	$f^{(n)}(1)$
0	$x^{1/2}$	1
1	$\tfrac{1}{2}x^{-1/2}$	$\tfrac{1}{2}$
2	$-\tfrac{1}{4}x^{-3/2}$	$-\tfrac{1}{4}$
3	$\tfrac{3}{8}x^{-5/2}$	$\tfrac{3}{8}$
4	$-\tfrac{15}{16}x^{-7/2}$	$-\tfrac{15}{16}$
$\vdots$	$\vdots$	$\vdots$

$\sqrt{x} \approx T_3(x) = 1 + \dfrac{1/2}{1!}(x-1) - \dfrac{1/4}{2!}(x-1)^2 + \dfrac{3/8}{3!}(x-1)^3$

$= 1 + \tfrac{1}{2}(x-1) - \tfrac{1}{8}(x-1)^2 + \tfrac{1}{16}(x-1)^3$

(b)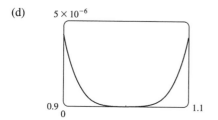

(c) $|R_3(x)| \le \dfrac{M}{4!}|x-1|^4$, where $\left|f^{(4)}(x)\right| \le M$ with

$f^{(4)}(x) = -\tfrac{15}{16}x^{-7/2}$. Now $0.9 \le x \le 1.1 \;\Rightarrow$

$-0.1 \le x-1 \le 0.1 \;\Rightarrow\; (x-1)^4 \le (0.1)^4$,

and letting $x = 0.9$ gives $M = \dfrac{15}{16(0.9)^{7/2}}$, so

$|R_3(x)| \le \dfrac{15}{16(0.9)^{7/2}4!}(0.1)^4 \approx 0.000\,005\,648$

$\approx 0.000\,006$

$= 6 \times 10^{-6}$

(d)

From the graph of $|R_3(x)| = |\sqrt{x} - T_3(x)|$, it appears

that the error is less than 5×10^{-6} on $[0.9, 1.1]$.

59. $\sin x = \sum_{n=0}^{\infty} (-1)^n \dfrac{x^{2n+1}}{(2n+1)!} = x - \dfrac{x^3}{3!} + \dfrac{x^5}{5!} - \dfrac{x^7}{7!} + \cdots$, so $\sin x - x = -\dfrac{x^3}{3!} + \dfrac{x^5}{5!} - \dfrac{x^7}{7!} + \cdots$ and

$\dfrac{\sin x - x}{x^3} = -\dfrac{1}{3!} + \dfrac{x^2}{5!} - \dfrac{x^4}{7!} + \cdots$. Thus, $\lim\limits_{x \to 0} \dfrac{\sin x - x}{x^3} = \lim\limits_{x \to 0} \left(-\dfrac{1}{6} + \dfrac{x^2}{120} - \dfrac{x^4}{5040} + \cdots \right) = -\dfrac{1}{6}$.

61. $f(x) = \sum_{n=0}^{\infty} c_n x^n \quad \Rightarrow \quad f(-x) = \sum_{n=0}^{\infty} c_n(-x)^n = \sum_{n=0}^{\infty} (-1)^n c_n x^n$

(a) If f is an odd function, then $f(-x) = -f(x) \quad \Rightarrow \quad \sum_{n=0}^{\infty}(-1)^n c_n x^n = \sum_{n=0}^{\infty} -c_n x^n$. The coefficients of

any power series are uniquely determined (by Theorem 11.10.5), so $(-1)^n c_n = -c_n$. If n is even, then

$(-1)^n = 1$, so $c_n = -c_n \quad \Rightarrow \quad 2c_n = 0 \quad \Rightarrow \quad c_n = 0$. Thus, all even coefficients are 0, that is,

$c_0 = c_2 = c_4 = \cdots = 0$.

(b) If f is even, then $f(-x) = f(x) \quad \Rightarrow \quad \sum_{n=0}^{\infty}(-1)^n c_n x^n = \sum_{n=0}^{\infty} c_n x^n \quad \Rightarrow \quad (-1)^n c_n = c_n$.

If n is odd, then $(-1)^n = -1$, so $-c_n = c_n \quad \Rightarrow \quad 2c_n = 0 \quad \Rightarrow \quad c_n = 0$. Thus, all odd coefficients are 0,

that is, $c_1 = c_3 = c_5 = \cdots = 0$.

☐ PROBLEMS PLUS

1. It would be far too much work to compute 15 derivatives of f. The key idea is to remember that $f^{(n)}(0)$ occurs in the coefficient of x^n in the Maclaurin series of f. We start with the Maclaurin series for sin:

$$\sin x = x - \frac{x^3}{3!} + \frac{x^5}{5!} - \cdots.$$ Then $\sin(x^3) = x^3 - \frac{x^9}{3!} + \frac{x^{15}}{5!} - \cdots$, and so the coefficient of x^{15} is

$$\frac{f^{(15)}(0)}{15!} = \frac{1}{5!}.$$ Therefore, $f^{(15)}(0) = \frac{15!}{5!} = 6 \cdot 7 \cdot 8 \cdot 9 \cdot 10 \cdot 11 \cdot 12 \cdot 13 \cdot 14 \cdot 15 = 10{,}897{,}286{,}400.$

3. (a) From Formula 14a in Appendix D, with $x = y = \theta$, we get $\tan 2\theta = \frac{2\tan\theta}{1 - \tan^2\theta}$, so $\cot 2\theta = \frac{1 - \tan^2\theta}{2\tan\theta}$ $\Rightarrow$

$$2\cot 2\theta = \frac{1 - \tan^2\theta}{\tan\theta} = \cot\theta - \tan\theta.$$ Replacing θ by $\frac{1}{2}x$, we get $2\cot x = \cot\frac{1}{2}x - \tan\frac{1}{2}x$, or

$$\tan\tfrac{1}{2}x = \cot\tfrac{1}{2}x - 2\cot x$$

(b) From part (a) with $\frac{x}{2^{n-1}}$ in place of x, $\tan\frac{x}{2^n} = \cot\frac{x}{2^n} - 2\cot\frac{x}{2^{n-1}}$, so the nth partial sum of

$\sum_{n=1}^{\infty} \frac{1}{2^n}\tan\frac{x}{2^n}$ is

$$s_n = \frac{\tan(x/2)}{2} + \frac{\tan(x/4)}{4} + \frac{\tan(x/8)}{8} + \cdots + \frac{\tan(x/2^n)}{2^n}$$

$$= \left[\frac{\cot(x/2)}{2} - \cot x\right] + \left[\frac{\cot(x/4)}{4} - \frac{\cot(x/2)}{2}\right] + \left[\frac{\cot(x/8)}{8} - \frac{\cot(x/4)}{4}\right] + \cdots$$

$$+ \left[\frac{\cot(x/2^n)}{2^n} - \frac{\cot(x/2^{n-1})}{2^{n-1}}\right] = -\cot x + \frac{\cot(x/2^n)}{2^n} \quad \text{(telescoping sum)}$$

Now $\dfrac{\cot(x/2^n)}{2^n} = \dfrac{\cos(x/2^n)}{2^n\sin(x/2^n)} = \dfrac{\cos(x/2^n)}{x} \cdot \dfrac{x/2^n}{\sin(x/2^n)} \to \dfrac{1}{x} \cdot 1 = \dfrac{1}{x}$ as $n \to \infty$ since $x/2^n \to 0$

for $x \neq 0$. Therefore, if $x \neq 0$ and $x \neq k\pi$ where k is any integer, then

$$\sum_{n=1}^{\infty}\frac{1}{2^n}\tan\frac{x}{2^n} = \lim_{n\to\infty} s_n = \lim_{n\to\infty}\left(-\cot x + \frac{1}{2^n}\cot\frac{x}{2^n}\right) = -\cot x + \frac{1}{x}$$

If $x = 0$, then all terms in the series are 0, so the sum is 0.

5. (a) At each stage, each side is replaced by four shorter sides, each of length $\frac{1}{3}$ of the side length at the preceding stage. Writing s_0 and ℓ_0 for the number of sides and the length of the side of the initial triangle, we generate the table at right. In general, we have $s_n = 3 \cdot 4^n$ and $\ell_n = \left(\frac{1}{3}\right)^n$, so the length of the perimeter at the nth

$s_0 = 3$	$\ell_0 = 1$
$s_1 = 3 \cdot 4$	$\ell_1 = 1/3$
$s_2 = 3 \cdot 4^2$	$\ell_2 = 1/3^2$
$s_3 = 3 \cdot 4^3$	$\ell_3 = 1/3^3$
$\cdots$	$\cdots$

stage of construction is $p_n = s_n\ell_n = 3 \cdot 4^n \cdot \left(\frac{1}{3}\right)^n = 3 \cdot \left(\frac{4}{3}\right)^n$.

(b) $p_n = \dfrac{4^n}{3^{n-1}} = 4\left(\dfrac{4}{3}\right)^{n-1}$. Since $\frac{4}{3} > 1$, $p_n \to \infty$ as $n \to \infty$.

(c) The area of each of the small triangles added at a given stage is one-ninth of the area of the triangle added at the preceding stage. Let a be the area of the original triangle. Then the area a_n of each of the small triangles added at stage n is $a_n = a \cdot \dfrac{1}{9^n} = \dfrac{a}{9^n}$. Since a small triangle is added to each side at every stage, it follows that the total area A_n added to the figure at the nth stage is

$$A_n = s_{n-1} \cdot a_n = 3 \cdot 4^{n-1} \cdot \frac{a}{9^n} = a \cdot \frac{4^{n-1}}{3^{2n-1}}.$$ Then the total area enclosed by the snowflake curve is

$$A = a + A_1 + A_2 + A_3 + \cdots = a + a \cdot \frac{1}{3} + a \cdot \frac{4}{3^3} + a \cdot \frac{4^2}{3^5} + a \cdot \frac{4^3}{3^7} + \cdots. $$ After the first term, this is a

geometric series with common ratio $\frac{4}{9}$, so $A = a + \dfrac{a/3}{1 - \frac{4}{9}} = a + \dfrac{a}{3} \cdot \dfrac{9}{5} = \dfrac{8a}{5}$. But the area of the original

equilateral triangle with side 1 is $a = \frac{1}{2} \cdot 1 \cdot \sin\frac{\pi}{3} = \frac{\sqrt{3}}{4}$. So the area enclosed by the snowflake curve is

$\frac{8}{5} \cdot \frac{\sqrt{3}}{4} = \frac{2\sqrt{3}}{5}$.

7. (a) Let $a = \arctan x$ and $b = \arctan y$. Then, from Formula 14b in Appendix D,

$$\tan(a - b) = \frac{\tan a - \tan b}{1 + \tan a \tan b} = \frac{\tan(\arctan x) - \tan(\arctan y)}{1 + \tan(\arctan x)\tan(\arctan y)} = \frac{x - y}{1 + xy}.$$

Now $\arctan x - \arctan y = a - b = \arctan(\tan(a-b)) = \arctan\dfrac{x-y}{1+xy}$ since $-\frac{\pi}{2} < a - b < \frac{\pi}{2}$.

(b) From part (a) we have

$$\arctan\tfrac{120}{119} - \arctan\tfrac{1}{239} = \arctan\frac{\frac{120}{119} - \frac{1}{239}}{1 + \frac{120}{119} \cdot \frac{1}{239}} = \arctan\frac{\frac{28,561}{28,441}}{\frac{28,561}{28,441}} = \arctan 1 = \tfrac{\pi}{4}$$

(c) Replacing y by $-y$ in the formula of part (a), we get $\arctan x + \arctan y = \arctan\dfrac{x+y}{1-xy}$. So

$$4\arctan\tfrac{1}{5} = 2\big(\arctan\tfrac{1}{5} + \arctan\tfrac{1}{5}\big) = 2\arctan\frac{\frac{1}{5} + \frac{1}{5}}{1 - \frac{1}{5} \cdot \frac{1}{5}} = 2\arctan\tfrac{5}{12} = \arctan\tfrac{5}{12} + \arctan\tfrac{5}{12}$$

$$= \arctan\frac{\frac{5}{12} + \frac{5}{12}}{1 - \frac{5}{12} \cdot \frac{5}{12}} = \arctan\tfrac{120}{119}$$

Thus, from part (b), we have $4\arctan\tfrac{1}{5} - \arctan\tfrac{1}{239} = \arctan\tfrac{120}{119} - \arctan\tfrac{1}{239} = \tfrac{\pi}{4}$.

(d) From Example 7 in Section 11.9 we have $\arctan x = x - \dfrac{x^3}{3} + \dfrac{x^5}{5} - \dfrac{x^7}{7} + \dfrac{x^9}{9} - \dfrac{x^{11}}{11} + \cdots$, so

$$\arctan\frac{1}{5} = \frac{1}{5} - \frac{1}{3 \cdot 5^3} + \frac{1}{5 \cdot 5^5} - \frac{1}{7 \cdot 5^7} + \frac{1}{9 \cdot 5^9} - \frac{1}{11 \cdot 5^{11}} + \cdots$$

This is an alternating series and the size of the terms decreases to 0, so by the Alternating Series Estimation Theorem, the sum lies between s_5 and s_6, that is, $0.197395560 < \arctan\frac{1}{5} < 0.197395562$.

(e) From the series in part (d) we get $\arctan \frac{1}{239} = \frac{1}{239} - \frac{1}{3 \cdot 239^3} + \frac{1}{5 \cdot 239^5} - \cdots$. The third term is less than

2.6×10^{-13}, so by the Alternating Series Estimation Theorem, we have, to nine decimal places,

$\arctan \frac{1}{239} \approx s_2 \approx 0.004184076$. Thus, $0.004184075 < \arctan \frac{1}{239} < 0.004184077$.

(f) From part (c) we have $\pi = 16 \arctan \frac{1}{5} - 4 \arctan \frac{1}{239}$, so from parts (d) and (e) we have

$16(0.197395560) - 4(0.004184077) < \pi < 16(0.197395562) - 4(0.004184075) \quad \Rightarrow$

$3.141592652 < \pi < 3.141592692$. So, to 7 decimal places, $\pi \approx 3.1415927$.

9. We start with the geometric series $\sum\limits_{n=0}^{\infty} x^n = \frac{1}{1-x}$, $|x| < 1$, and differentiate:

$$\sum_{n=1}^{\infty} n x^{n-1} = \frac{d}{dx}\left(\sum_{n=0}^{\infty} x^n\right) = \frac{d}{dx}\left(\frac{1}{1-x}\right) = \frac{1}{(1-x)^2} \text{ for } |x| < 1 \quad \Rightarrow$$

$$\sum_{n=1}^{\infty} n x^n = x \sum_{n=1}^{\infty} n x^{n-1} = \frac{x}{(1-x)^2} \text{ for } |x| < 1. \text{ Differentiate again:}$$

$$\sum_{n=1}^{\infty} n^2 x^{n-1} = \frac{d}{dx}\frac{x}{(1-x)^2} = \frac{(1-x)^2 - x \cdot 2(1-x)(-1)}{(1-x)^4} = \frac{x+1}{(1-x)^3} \quad \Rightarrow \quad \sum_{n=1}^{\infty} n^2 x^n = \frac{x^2 + x}{(1-x)^3} \quad \Rightarrow$$

$$\sum_{n=1}^{\infty} n^3 x^{n-1} = \frac{d}{dx}\frac{x^2+x}{(1-x)^3} = \frac{(1-x)^3(2x+1) - (x^2+x)3(1-x)^2(-1)}{(1-x)^6} = \frac{x^2 + 4x + 1}{(1-x)^4} \quad \Rightarrow$$

$$\sum_{n=1}^{\infty} n^3 x^n = \frac{x^3 + 4x^2 + x}{(1-x)^4}, \, |x| < 1. \text{ The radius of convergence is 1 because that is the radius of convergence for}$$

the geometric series we started with. If $x = \pm 1$, the series is $\sum n^3 (\pm 1)^n$, which diverges by the Test For

Divergence, so the interval of convergence is $(-1, 1)$.

11. $\ln\left(1 - \frac{1}{n^2}\right) = \ln\left(\frac{n^2-1}{n^2}\right) = \ln\frac{(n+1)(n-1)}{n^2} = \ln[(n+1)(n-1)] - \ln n^2$

$$= \ln(n+1) + \ln(n-1) - 2\ln n$$

$$= \ln(n-1) - \ln n - \ln n + \ln(n+1)$$

$$= \ln\frac{n-1}{n} - [\ln n - \ln(n+1)] = \ln\frac{n-1}{n} - \ln\frac{n}{n+1}.$$

Let $s_k = \sum\limits_{n=2}^{k} \ln\left(1 - \frac{1}{n^2}\right) = \sum\limits_{n=2}^{k}\left(\ln\frac{n-1}{n} - \ln\frac{n}{n+1}\right)$ for $k \geq 2$. Then

$$s_k = \left(\ln\frac{1}{2} - \ln\frac{2}{3}\right) + \left(\ln\frac{2}{3} - \ln\frac{3}{4}\right) + \cdots + \left(\ln\frac{k-1}{k} - \ln\frac{k}{k+1}\right) = \ln\frac{1}{2} - \ln\frac{k}{k+1}, \text{ so}$$

$$\sum_{n=2}^{\infty} \ln\left(1 - \frac{1}{n^2}\right) = \lim_{k\to\infty} s_k = \lim_{k\to\infty}\left(\ln\frac{1}{2} - \ln\frac{k}{k+1}\right) = \ln\frac{1}{2} - \ln 1 = \ln 1 - \ln 2 - \ln 1 = -\ln 2.$$

13. $u = 1 + \dfrac{x^3}{3!} + \dfrac{x^6}{6!} + \dfrac{x^9}{9!} + \cdots, \; v = x + \dfrac{x^4}{4!} + \dfrac{x^7}{7!} + \dfrac{x^{10}}{10!} + \cdots, \; w = \dfrac{x^2}{2!} + \dfrac{x^5}{5!} + \dfrac{x^8}{8!} + \cdots.$

Use the Ratio Test to show that the series for u, v, and w have positive radii of convergence (∞ in each case), so Theorem 11.9.2 applies, and hence, we may differentiate each of these series:

$$\frac{du}{dx} = \frac{3x^2}{3!} + \frac{6x^5}{6!} + \frac{9x^8}{9!} + \cdots = \frac{x^2}{2!} + \frac{x^5}{5!} + \frac{x^8}{8!} + \cdots = w$$

Similarly, $\dfrac{dv}{dx} = 1 + \dfrac{x^3}{3!} + \dfrac{x^6}{6!} + \dfrac{x^9}{9!} + \cdots = u$, and $\dfrac{dw}{dx} = x + \dfrac{x^4}{4!} + \dfrac{x^7}{7!} + \dfrac{x^{10}}{10!} + \cdots = v$.

So $u' = w$, $v' = u$, and $w' = v$. Now differentiate the left hand side of the desired equation:

$$\frac{d}{dx}\left(u^3 + v^3 + w^3 - 3uvw\right) = 3u^2 u' + 3v^2 v' + 3w^2 w' - 3(u'vw + uv'w + uvw')$$

$$= 3u^2 w + 3v^2 u + 3w^2 v - 3(vw^2 + u^2 w + uv^2) = 0 \quad \Rightarrow$$

$u^3 + v^3 + w^3 - 3uvw = C$. To find the value of the constant C, we put $x = 0$ in the last equation and get $1^3 + 0^3 + 0^3 - 3(1 \cdot 0 \cdot 0) = C \quad \Rightarrow \quad C = 1$, so $u^3 + v^3 + w^3 - 3uvw = 1$.

15. If L is the length of a side of the equilateral triangle, then the area is $A = \frac{1}{2}L \cdot \frac{\sqrt{3}}{2}L = \frac{\sqrt{3}}{4}L^2$ and so $L^2 = \frac{4}{\sqrt{3}}A$.

Let r be the radius of one of the circles. When there are n rows of circles, the figure shows that

$$L = \sqrt{3}r + r + (n-2)(2r) + r + \sqrt{3}r = r(2n - 2 + 2\sqrt{3}), \text{ so } r = \frac{L}{2(n + \sqrt{3} - 1)}.$$

The number of circles is $1 + 2 + \cdots + n = \dfrac{n(n+1)}{2}$, and so the total area of the circles is

$$A_n = \frac{n(n+1)}{2}\pi r^2 = \frac{n(n+1)}{2}\pi \frac{L^2}{4(n + \sqrt{3} - 1)^2} = \frac{n(n+1)}{2}\pi \frac{4A/\sqrt{3}}{4(n + \sqrt{3} - 1)^2}$$

$$= \frac{n(n+1)}{(n + \sqrt{3} - 1)^2}\frac{\pi A}{2\sqrt{3}} \quad \Rightarrow$$

$$\frac{A_n}{A} = \frac{n(n+1)}{(n + \sqrt{3} - 1)^2}\frac{\pi}{2\sqrt{3}}$$

$$= \frac{1 + 1/n}{[1 + (\sqrt{3} - 1)/n]^2}\frac{\pi}{2\sqrt{3}} \longrightarrow \frac{\pi}{2\sqrt{3}} \text{ as } n \to \infty$$

$\sqrt{3}r \quad r \quad 2r \quad 2r \quad r \quad \sqrt{3}r$

$\longmapsto \quad\quad L \quad\quad \longrightarrow$

17. As in Section 11.9 we have to integrate the function x^x by integrating series. Writing $x^x = (e^{\ln x})^x = e^{x \ln x}$ and

using the Maclaurin series for e^x, we have $x^x = (e^{\ln x})^x = e^{x \ln x} = \displaystyle\sum_{n=0}^{\infty} \frac{(x \ln x)^n}{n!} = \displaystyle\sum_{n=0}^{\infty} \frac{x^n (\ln x)^n}{n!}.$

As with power series, we can integrate this series term-by-term:

$$\int_0^1 x^x \, dx = \sum_{n=0}^{\infty} \int_0^1 \frac{x^n (\ln x)^n}{n!} \, dx = \sum_{n=0}^{\infty} \frac{1}{n!} \int_0^1 x^n (\ln x)^n \, dx$$

We integrate by parts with $u = (\ln x)^n$, $dv = x^n \, dx$, so $du = \dfrac{n(\ln x)^{n-1}}{x} \, dx$ and $v = \dfrac{x^{n+1}}{n+1}$:

$$\int_0^1 x^n (\ln x)^n \, dx = \lim_{t \to 0^+} \int_t^1 x^n (\ln x)^n \, dx = \lim_{t \to 0^+} \left[\frac{x^{n+1}}{n+1} (\ln x)^n \right]_t^1 - \lim_{t \to 0^+} \int_t^1 \frac{n}{n+1} x^n (\ln x)^{n-1} \, dx$$

$$= 0 - \frac{n}{n+1} \int_0^1 x^n (\ln x)^{n-1} \, dx$$

(where l'Hospital's Rule was used to help evaluate the first limit).

Further integration by parts gives $\displaystyle\int_0^1 x^n (\ln x)^k \, dx = -\frac{k}{n+1} \int_0^1 x^n (\ln x)^{k-1} \, dx$ and, combining

these steps, we get $\displaystyle\int_0^1 x^n (\ln x)^n \, dx = \frac{(-1)^n n!}{(n+1)^n} \int_0^1 x^n \, dx = \frac{(-1)^n n!}{(n+1)^{n+1}} \quad \Rightarrow$

$$\int_0^1 x^x \, dx = \sum_{n=0}^{\infty} \frac{1}{n!} \int_0^1 x^n (\ln x)^n \, dx = \sum_{n=0}^{\infty} \frac{1}{n!} \frac{(-1)^n n!}{(n+1)^{n+1}} = \sum_{n=0}^{\infty} \frac{(-1)^n}{(n+1)^{n+1}} = \sum_{n=1}^{\infty} \frac{(-1)^{n-1}}{n^n}.$$

19. Let $f(x) = \sum_{m=0}^{\infty} c_m x^m$ and $g(x) = e^{f(x)} = \sum_{n=0}^{\infty} d_n x^n$. Then $g'(x) = \sum_{n=0}^{\infty} n d_n x^{n-1}$, so $n d_n$ occurs as the coefficient of x^{n-1}. But also

$$g'(x) = e^{f(x)} f'(x) = \left(\sum_{n=0}^{\infty} d_n x^n \right) \left(\sum_{m=1}^{\infty} m c_m x^{m-1} \right)$$

$$= \left(d_0 + d_1 x + d_2 x^2 + \cdots + d_{n-1} x^{n-1} + \cdots \right) \left(c_1 + 2c_2 x + 3c_3 x^2 + \cdots + n c_n x^{n-1} + \cdots \right)$$

so the coefficient of x^{n-1} is $c_1 d_{n-1} + 2c_2 d_{n-2} + 3c_3 d_{n-3} + \cdots + n c_n d_0 = \sum_{i=1}^{n} i c_i d_{n-i}$. Therefore, $n d_n = \sum_{i=1}^{n} i c_i d_{n-i}$.

21. Call the series S. We group the terms according to the number of digits in their denominators:

$$S = \underbrace{\left(\frac{1}{1} + \frac{1}{2} + \cdots + \frac{1}{8} + \frac{1}{9} \right)}_{g_1} + \underbrace{\left(\frac{1}{11} + \cdots + \frac{1}{99} \right)}_{g_2} + \underbrace{\left(\frac{1}{111} + \cdots + \frac{1}{999} \right)}_{g_3} + \cdots$$

Now in the group g_n, since we have 9 choices for each of the n digits in the denominator, there are 9^n terms.

Furthermore, each term in g_n is less than $\frac{1}{10^{n-1}}$ [except for the first term in g_1]. So $g_n < 9^n \cdot \frac{1}{10^{n-1}} = 9 \left(\frac{9}{10} \right)^{n-1}$.

Now $\sum_{n=1}^{\infty} 9 \left(\frac{9}{10} \right)^{n-1}$ is a geometric series with $a = 9$ and $r = \frac{9}{10} < 1$. Therefore, by the Comparison Test,

$$S = \sum_{n=1}^{\infty} g_n < \sum_{n=1}^{\infty} 9 \left(\frac{9}{10} \right)^{n-1} = \frac{9}{1 - 9/10} = 90.$$

☐ APPENDIXES

A Numbers, Inequalities, and Absolute Values

1. $|5 - 23| = |-18| = 18$

3. $|-\pi| = \pi$ because $\pi > 0$.

5. $\left|\sqrt{5} - 5\right| = -\left(\sqrt{5} - 5\right) = 5 - \sqrt{5}$ because $\sqrt{5} - 5 < 0$.

7. For $x < 2$, $x - 2 < 0$, so $|x - 2| = -(x - 2) = 2 - x$.

9. $|x + 1| = \begin{cases} x + 1 & \text{for } x + 1 \geq 0 \iff x \geq -1 \\ -(x + 1) & \text{for } x + 1 < 0 \iff x < -1 \end{cases}$

11. $\left|x^2 + 1\right| = x^2 + 1$ (since $x^2 + 1 \geq 0$ for all x).

13. $2x + 7 > 3 \iff 2x > -4 \iff x > -2$, so $x \in (-2, \infty)$.

15. $1 - x \leq 2 \iff -x \leq 1 \iff x \geq -1$, so $x \in [-1, \infty)$.

17. $2x + 1 < 5x - 8 \iff 9 < 3x \iff 3 < x$, so $x \in (3, \infty)$.

19. $-1 < 2x - 5 < 7 \iff 4 < 2x < 12 \iff 2 < x < 6$, so $x \in (2, 6)$.

21. $0 \leq 1 - x < 1 \iff -1 \leq -x < 0 \iff 1 \geq x > 0$, so $x \in (0, 1]$.

23. $4x < 2x + 1 \leq 3x + 2$. So $4x < 2x + 1 \iff 2x < 1 \iff x < \frac{1}{2}$, and $2x + 1 \leq 3x + 2 \iff -1 \leq x$. Thus, $x \in \left[-1, \frac{1}{2}\right)$.

25. $(x - 1)(x - 2) > 0$. *Case 1:* (both factors are positive, so their product is positive)

$$x - 1 > 0 \iff x > 1, \text{ and } x - 2 > 0 \iff x > 2, \text{ so } x \in (2, \infty).$$

Case 2: (both factors are negative, so their product is positive)

$$x - 1 < 0 \iff x < 1, \text{ and } x - 2 < 0 \iff x < 2, \text{ so } x \in (-\infty, 1).$$

Thus, the solution set is $(-\infty, 1) \cup (2, \infty)$.

27. $2x^2 + x \leq 1 \iff 2x^2 + x - 1 \leq 0 \iff (2x - 1)(x + 1) \leq 0$. *Case 1:* $2x - 1 \geq 0 \iff x \geq \frac{1}{2}$, and $x + 1 \leq 0 \iff x \leq -1$, which is an impossible combination. *Case 2:* $2x - 1 \leq 0 \iff x \leq \frac{1}{2}$, and $x + 1 \geq 0 \iff x \geq -1$, so $x \in \left[-1, \frac{1}{2}\right]$. Thus, the solution set is $\left[-1, \frac{1}{2}\right]$.

29. $x^2 + x + 1 > 0 \iff x^2 + x + \frac{1}{4} + \frac{3}{4} > 0 \iff \left(x + \frac{1}{2}\right)^2 + \frac{3}{4} > 0$. But since $\left(x + \frac{1}{2}\right)^2 \geq 0$ for every real x, the original inequality will be true for all real x as well. Thus, the solution set is $(-\infty, \infty)$.

31. $x^2 < 3$ $\Leftrightarrow$ $x^2 - 3 < 0$ $\Leftrightarrow$ $\left(x - \sqrt{3}\right)\left(x + \sqrt{3}\right) < 0$. *Case 1:* $x > \sqrt{3}$ and $x < -\sqrt{3}$, which is impossible.

Case 2: $x < \sqrt{3}$ and $x > -\sqrt{3}$. Thus, the solution set is $\left(-\sqrt{3}, \sqrt{3}\right)$.

Another method: $x^2 < 3$ $\Leftrightarrow$ $|x| < \sqrt{3}$ $\Leftrightarrow$ $-\sqrt{3} < x < \sqrt{3}$.

33. $x^3 - x^2 \le 0$ $\Leftrightarrow$ $x^2(x - 1) \le 0$. Since $x^2 \ge 0$ for all x, the inequality is satisfied when $x - 1 \le 0$ $\Leftrightarrow$ $x \le 1$. Thus, the solution set is $(-\infty, 1]$.

35. $x^3 > x$ $\Leftrightarrow$ $x^3 - x > 0$ $\Leftrightarrow$ $x(x^2 - 1) > 0$ $\Leftrightarrow$ $x(x - 1)(x + 1) > 0$. Constructing a table:

Interval	x	$x - 1$	$x + 1$	$x(x - 1)(x + 1)$
$x < -1$	$-$	$-$	$-$	$-$
$-1 < x < 0$	$-$	$-$	$+$	$+$
$0 < x < 1$	$+$	$-$	$+$	$-$
$x > 1$	$+$	$+$	$+$	$+$

Since $x^3 > x$ when the last column is positive, the solution set is $(-1, 0) \cup (1, \infty)$.

37. $1/x < 4$. This is clearly true for $x < 0$. So suppose $x > 0$. then $1/x < 4$ $\Leftrightarrow$ $1 < 4x$ $\Leftrightarrow$ $\frac{1}{4} < x$. Thus, the solution set is $(-\infty, 0) \cup \left(\frac{1}{4}, \infty\right)$.

39. $C = \frac{5}{9}(F - 32)$ $\Rightarrow$ $F = \frac{9}{5}C + 32$. So $50 \le F \le 95$ $\Rightarrow$ $50 \le \frac{9}{5}C + 32 \le 95$ $\Rightarrow$ $18 \le \frac{9}{5}C \le 63$ $\Rightarrow$ $10 \le C \le 35$. So the interval is $[10, 35]$.

41. (a) Let T represent the temperature in degrees Celsius and h the height in km. $T = 20$ when $h = 0$ and T decreases by $10\,°C$ for every km ($1\,°C$ for each 100-m rise). Thus, $T = 20 - 10h$ when $0 \le h \le 12$.

(b) From part (a), $T = 20 - 10h$ $\Rightarrow$ $10h = 20 - T$ $\Rightarrow$ $h = 2 - T/10$. So $0 \le h \le 5$ $\Rightarrow$ $0 \le 2 - T/10 \le 5$ $\Rightarrow$ $-2 \le -T/10 \le 3$ $\Rightarrow$ $-20 \le -T \le 30$ $\Rightarrow$ $20 \ge T \ge -30$ $\Rightarrow$ $-30 \le T \le 20$. Thus, the range of temperatures (in $°C$) to be expected is $[-30, 20]$.

43. $|2x| = 3$ $\Leftrightarrow$ either $2x = 3$ or $2x = -3$ $\Leftrightarrow$ $x = \frac{3}{2}$ or $x = -\frac{3}{2}$.

45. $|x + 3| = |2x + 1|$ $\Leftrightarrow$ either $x + 3 = 2x + 1$ or $x + 3 = -(2x + 1)$. In the first case, $x = 2$, and in the second case, $x + 3 = -2x - 1$ $\Leftrightarrow$ $3x = -4$ $\Leftrightarrow$ $x = -\frac{4}{3}$. So the solutions are $-\frac{4}{3}$ and 2.

47. By Property 5 of absolute values, $|x| < 3$ $\Leftrightarrow$ $-3 < x < 3$, so $x \in (-3, 3)$.

49. $|x - 4| < 1$ $\Leftrightarrow$ $-1 < x - 4 < 1$ $\Leftrightarrow$ $3 < x < 5$, so $x \in (3, 5)$.

51. $|x + 5| \ge 2$ $\Leftrightarrow$ $x + 5 \ge 2$ or $x + 5 \le -2$ $\Leftrightarrow$ $x \ge -3$ or $x \le -7$, so $x \in (-\infty, -7] \cup [-3, \infty)$.

53. $|2x - 3| \le 0.4$ $\Leftrightarrow$ $-0.4 \le 2x - 3 \le 0.4$ $\Leftrightarrow$ $2.6 \le 2x \le 3.4$ $\Leftrightarrow$ $1.3 \le x \le 1.7$, so $x \in [1.3, 1.7]$.

55. $1 \le |x| \le 4$. So either $1 \le x \le 4$ or $1 \le -x \le 4$ $\Leftrightarrow$ $-1 \ge x \ge -4$. Thus, $x \in [-4, -1] \cup [1, 4]$.

57. $a(bx - c) \ge bc$ $\Leftrightarrow$ $bx - c \ge \dfrac{bc}{a}$ $\Leftrightarrow$ $bx \ge \dfrac{bc}{a} + c = \dfrac{bc + ac}{a}$ $\Leftrightarrow$ $x \ge \dfrac{bc + ac}{ab}$

59. $ax + b < c \iff ax < c - b \iff x > \dfrac{c-b}{a}$ (since $a < 0$)

61. $|(x + y) - 5| = |(x - 2) + (y - 3)| \le |x - 2| + |y - 3| < 0.01 + 0.04 = 0.05$

63. If $a < b$ then $a + a < a + b$ and $a + b < b + b$. So $2a < a + b < 2b$. Dividing by 2, we get $a < \frac{1}{2}(a + b) < b$.

65. $|ab| = \sqrt{(ab)^2} = \sqrt{a^2 b^2} = \sqrt{a^2}\,\sqrt{b^2} = |a|\,|b|$

67. If $0 < a < b$, then $a \cdot a < a \cdot b$ and $a \cdot b < b \cdot b$ [using Rule 3 of Inequalities]. So $a^2 < ab < b^2$ and hence $a^2 < b^2$.

69. Observe that the sum, difference and product of two integers is always an integer. Let the rational numbers be represented by $r = m/n$ and $s = p/q$ (where m, n, p and q are integers with $n \ne 0$, $q \ne 0$). Now

$r + s = \dfrac{m}{n} + \dfrac{p}{q} = \dfrac{mq + pn}{nq}$, but $mq + pn$ and nq are both integers, so $\dfrac{mq + pn}{nq} = r + s$ is a rational number by

definition. Similarly, $r - s = \dfrac{m}{n} - \dfrac{p}{q} = \dfrac{mq - pn}{nq}$ is a rational number. Finally, $r \cdot s = \dfrac{m}{n} \cdot \dfrac{p}{q} = \dfrac{mp}{nq}$ but mp and

nq are both integers, so $\dfrac{mp}{nq} = r \cdot s$ is a rational number by definition.

B Coordinate Geometry and Lines

1. From the Distance Formula with $x_1 = 1$, $x_2 = 4$, $y_1 = 1$, $y_2 = 5$, we find the distance from $(1, 1)$ to $(4, 5)$ to be
$\sqrt{(4 - 1)^2 + (5 - 1)^2} = \sqrt{3^2 + 4^2} = \sqrt{25} = 5$.

3. $\sqrt{(-1 - 6)^2 + [3 - (-2)]^2} = \sqrt{(-7)^2 + 5^2} = \sqrt{74}$

5. $\sqrt{(4 - 2)^2 + (-7 - 5)^2} = \sqrt{2^2 + (-12)^2} = \sqrt{148} = 2\sqrt{37}$

7. From (2), the slope is $\dfrac{11 - 5}{4 - 1} = \dfrac{6}{3} = 2$.

9. With $P(-3, 3)$ and $Q(-1, -6)$, the slope m of the line through P and Q is $m = \dfrac{-6 - 3}{-1 - (-3)} = -\dfrac{9}{2}$.

11. Since $|AC| = \sqrt{(-4 - 0)^2 + (3 - 2)^2} = \sqrt{(-4)^2 + 1^2} = \sqrt{17}$ and

$|BC| = \sqrt{[-4 - (-3)]^2 + [3 - (-1)]^2} = \sqrt{(-1)^2 + 4^2} = \sqrt{17}$, the triangle has two sides of equal length, and

so is isosceles.

13. Using $A(-2, 9)$, $B(4, 6)$, $C(1, 0)$, and $D(-5, 3)$, we have

$|AB| = \sqrt{[4 - (-2)]^2 + (6 - 9)^2} = \sqrt{6^2 + (-3)^2} = \sqrt{45} = \sqrt{9}\,\sqrt{5} = 3\sqrt{5}$,

$|BC| = \sqrt{(1 - 4)^2 + (0 - 6)^2} = \sqrt{(-3)^2 + (-6)^2} = \sqrt{45} = \sqrt{9}\,\sqrt{5} = 3\sqrt{5}$,

$|CD| = \sqrt{(-5 - 1)^2 + (3 - 0)^2} = \sqrt{(-6)^2 + 3^2} = \sqrt{45} = \sqrt{9}\,\sqrt{5} = 3\sqrt{5}$, and

$|DA| = \sqrt{[-2 - (-5)]^2 + (9 - 3)^2} = \sqrt{3^2 + 6^2} = \sqrt{45} = \sqrt{9}\,\sqrt{5} = 3\sqrt{5}$. So all sides are of equal length and

we have a rhombus. Moreover, $m_{AB} = \dfrac{6 - 9}{4 - (-2)} = -\dfrac{1}{2}$, $m_{BC} = \dfrac{0 - 6}{1 - 4} = 2$, $m_{CD} = \dfrac{3 - 0}{-5 - 1} = -\dfrac{1}{2}$, and

$m_{DA} = \dfrac{9 - 3}{-2 - (-5)} = 2$, so the sides are perpendicular. Thus, A, B, C, and D are vertices of a square.

15. The slope of the line segment AB is $\dfrac{4-1}{7-1} = \dfrac{1}{2}$, the slope of CD is $\dfrac{7-10}{-1-5} = \dfrac{1}{2}$, the slope of BC is

$\dfrac{10-4}{5-7} = -3$, and the slope of DA is $\dfrac{1-7}{1-(-1)} = -3$. So AB is parallel to CD and BC is parallel to DA.

Hence $ABCD$ is a parallelogram.

17. $x = 3$

19. $xy = 0 \iff x = 0$ or $y = 0$. The graph consists of the coordinate axes.

21. By the point-slope form of the equation of a line, an equation of the line through $(2, -3)$ with slope 6 is

$y - (-3) = 6(x - 2)$ or $y = 6x - 15$.

23. $y - 7 = \frac{2}{3}(x - 1)$ or $y = \frac{2}{3}x + \frac{19}{3}$

25. The slope of the line through $(2, 1)$ and $(1, 6)$ is $m = \dfrac{6-1}{1-2} = -5$, so an equation of the line is

$y - 1 = -5(x - 2)$ or $y = -5x + 11$.

27. By the slope-intercept form of the equation of a line, an equation of the line is $y = 3x - 2$.

29. Since the line passes through $(1, 0)$ and $(0, -3)$, its slope is $m = \dfrac{-3-0}{0-1} = 3$, so an equation is $y = 3x - 3$.

Another method: From Exercise 61, $\dfrac{x}{1} + \dfrac{y}{-3} = 1 \implies -3x + y = -3 \implies y = 3x - 3$.

31. Since $m = 0$, $y - 5 = 0(x - 4)$ or $y = 5$.

33. Putting the line $x + 2y = 6$ into its slope-intercept form gives us $y = -\frac{1}{2}x + 3$, so we see that this line has

slope $-\frac{1}{2}$. Thus, we want the line of slope $-\frac{1}{2}$ that passes through the point $(1, -6)$: $y - (-6) = -\frac{1}{2}(x - 1) \iff$

$y = -\frac{1}{2}x - \frac{11}{2}$.

35. $2x + 5y + 8 = 0 \iff y = -\frac{2}{5}x - \frac{8}{5}$. Since this line has slope $-\frac{2}{5}$, a line perpendicular to it would have slope $\frac{5}{2}$,

so the required line is $y - (-2) = \frac{5}{2}[x - (-1)] \iff y = \frac{5}{2}x + \frac{1}{2}$.

37. $x + 3y = 0 \iff y = -\frac{1}{3}x$, so the slope is $-\frac{1}{3}$ and the y-intercept is 0.

39. $y = -2$ is a horizontal line with slope 0 and y-intercept -2.

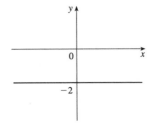

41. $3x - 4y = 12$ $\Leftrightarrow$ $y = \frac{3}{4}x - 3$, so the slope is $\frac{3}{4}$ and the y-intercept is -3.

43. $\{(x, y) \mid x < 0\}$

45. $\{(x, y) \mid xy < 0\} = \{(x, y) \mid x < 0 \text{ and } y > 0\}$
$\quad\quad\quad\quad\quad\quad\quad \cup \{(x, y) \mid x > 0 \text{ and } y < 0\}$

47. $\left\{(x, y) \,\middle|\, |x| \leq 2\right\} = \{(x, y) \mid -2 \leq x \leq 2\}$

49. $\{(x, y) \mid 0 \leq y \leq 4, x \leq 2\}$

51. $\{(x, y) \mid 1 + x \leq y \leq 1 - 2x\}$

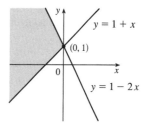

53. Let $P(0, y)$ be a point on the y-axis. The distance from P to $(5, -5)$ is

$\sqrt{(5 - 0)^2 + (-5 - y)^2} = \sqrt{5^2 + (y + 5)^2}$. The distance from P to $(1, 1)$ is

$\sqrt{(1 - 0)^2 + (1 - y)^2} = \sqrt{1^2 + (y - 1)^2}$. We want these distances to be equal:

$\sqrt{5^2 + (y + 5)^2} = \sqrt{1^2 + (y - 1)^2}$ $\Leftrightarrow$ $5^2 + (y + 5)^2 = 1^2 + (y - 1)^2$ $\Leftrightarrow$

$25 + (y^2 + 10y + 25) = 1 + (y^2 - 2y + 1)$ $\Leftrightarrow$ $12y = -48$ $\Leftrightarrow$ $y = -4$. So the desired point is $(0, -4)$.

55. (a) Using the midpoint formula from Exercise 54 with $(1, 3)$ and $(7, 15)$, we get $\left(\frac{1+7}{2}, \frac{3+15}{2}\right) = (4, 9)$.

(b) Using the same formula, we get $\left(\frac{-1+8}{2}, \frac{6-12}{2}\right) = \left(\frac{7}{2}, -3\right)$.

57. $2x - y = 4 \iff y = 2x - 4 \implies m_1 = 2$ and $6x - 2y = 10 \iff 2y = 6x - 10 \iff y = 3x - 5 \implies$
$m_2 = 3$. Since $m_1 \neq m_2$, the two lines are not parallel. To find the point of intersection: $2x - 4 = 3x - 5 \iff$
$x = 1 \implies y = -2$. Thus, the point of intersection is $(1, -2)$.

59. With $A(1, 4)$ and $B(7, -2)$, the slope of segment AB is $\frac{-2-4}{7-1} = -1$, so its perpendicular bisector has slope 1.
The midpoint of AB is $\left(\frac{1+7}{2}, \frac{4+(-2)}{2}\right) = (4, 1)$, so an equation of the perpendicular bisector is $y - 1 = 1(x - 4)$
or $y = x - 3$.

61. (a) Since the x-intercept is a, the point $(a, 0)$ is on the line, and similarly since the y-intercept is b, $(0, b)$ is on the
line. Hence, the slope of the line is $m = \dfrac{b - 0}{0 - a} = -\dfrac{b}{a}$. Substituting into $y = mx + b$ gives $y = -\dfrac{b}{a}x + b \iff$
$\dfrac{b}{a}x + y = b \iff \dfrac{x}{a} + \dfrac{y}{b} = 1$.

(b) Letting $a = 6$ and $b = -8$ gives $\dfrac{x}{6} + \dfrac{y}{-8} = 1 \iff -8x + 6y = -48$ [multiply by -48] $\iff$
$6y = 8x - 48 \iff 3y = 4x - 24 \iff y = \frac{4}{3}x - 8$.

C Graphs of Second-Degree Equations

1. An equation of the circle with center $(3, -1)$ and radius 5 is $(x - 3)^2 + (y + 1)^2 = 5^2 = 25$.

3. The equation has the form $x^2 + y^2 = r^2$. Since $(4, 7)$ lies on the circle, we have $4^2 + 7^2 = r^2 \implies r^2 = 65$. So
the required equation is $x^2 + y^2 = 65$.

5. $x^2 + y^2 - 4x + 10y + 13 = 0 \iff x^2 - 4x + y^2 + 10y = -13 \iff$
$(x^2 - 4x + 4) + (y^2 + 10y + 25) = -13 + 4 + 25 = 16 \iff (x - 2)^2 + (y + 5)^2 = 4^2$. Thus, we have a
circle with center $(2, -5)$ and radius 4.

7. $x^2 + y^2 + x = 0 \iff (x^2 + x + \frac{1}{4}) + y^2 = \frac{1}{4} \iff (x + \frac{1}{2})^2 + y^2 = (\frac{1}{2})^2$. Thus, we have a circle with
center $(-\frac{1}{2}, 0)$ and radius $\frac{1}{2}$.

9. $2x^2 + 2y^2 - x + y = 1 \iff 2(x^2 - \frac{1}{2}x + \frac{1}{16}) + 2(y^2 + \frac{1}{2}y + \frac{1}{16}) = 1 + \frac{1}{8} + \frac{1}{8} \iff$
$2(x - \frac{1}{4})^2 + 2(y + \frac{1}{4})^2 = \frac{5}{4} \iff (x - \frac{1}{4})^2 + (y + \frac{1}{4})^2 = \frac{5}{8}$. Thus, we have a circle with center $(\frac{1}{4}, -\frac{1}{4})$ and
radius $\frac{\sqrt{5}}{2\sqrt{2}} = \frac{\sqrt{10}}{4}$.

11. $y = -x^2$. Parabola

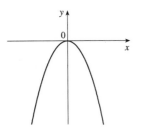

13. $x^2 + 4y^2 = 16 \iff \dfrac{x^2}{16} + \dfrac{y^2}{4} = 1$. Ellipse

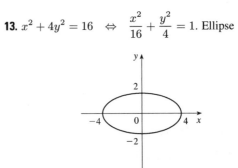

15. $16x^2 - 25y^2 = 400$ $\Leftrightarrow$ $\dfrac{x^2}{25} - \dfrac{y^2}{16} = 1$.

Hyperbola

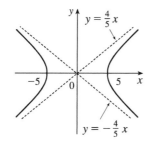

17. $4x^2 + y^2 = 1$ $\Leftrightarrow$ $\dfrac{x^2}{1/4} + y^2 = 1$. Ellipse

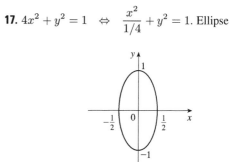

19. $x = y^2 - 1$. Parabola with vertex at $(-1, 0)$

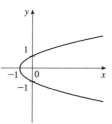

21. $9y^2 - x^2 = 9$ $\Leftrightarrow$ $y^2 - \dfrac{x^2}{9} = 1$. Hyperbola

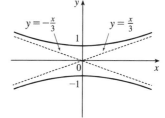

23. $xy = 4$. Hyperbola

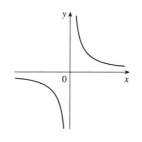

25. $9(x-1)^2 + 4(y-2)^2 = 36$ $\Leftrightarrow$

$\dfrac{(x-1)^2}{4} + \dfrac{(y-2)^2}{9} = 1$. Ellipse centered at $(1, 2)$

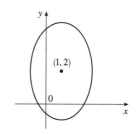

27. $y = x^2 - 6x + 13 = (x^2 - 6x + 9) + 4 =$

$(x-3)^2 + 4$. Parabola with vertex at $(3, 4)$

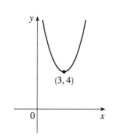

29. $x = -y^2 + 4$. Parabola with vertex at $(4, 0)$

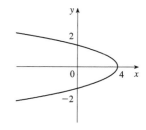

31. $x^2 + 4y^2 - 6x + 5 = 0 \iff$

$(x^2 - 6x + 9) + 4y^2 = -5 + 9 = 4 \iff$

$\dfrac{(x-3)^2}{4} + y^2 = 1.$ Ellipse centered at $(3, 0)$

33. $y = 3x$ and $y = x^2$ intersect where $3x = x^2 \iff$

$0 = x^2 - 3x = x(x - 3)$, that is, at $(0, 0)$

and $(3, 9)$.

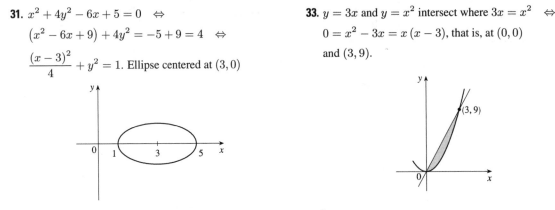

35. The parabola must have an equation of the form $y = a(x-1)^2 - 1$. Substituting $x = 3$ and $y = 3$ into the equation gives $3 = a(3-1)^2 - 1$, so $a = 1$, and the equation is $y = (x-1)^2 - 1 = x^2 - 2x$.

Note that using the other point $(-1, 3)$ would have given the same value for a, and hence the same equation.

37. $\{(x, y) \mid x^2 + y^2 \le 1\}$

39. $\{(x, y) \mid y \ge x^2 - 1\}$

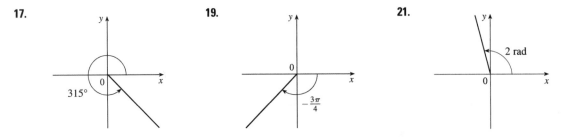

D Trigonometry

1. $210° = 210\left(\frac{\pi}{180}\right) = \frac{7\pi}{6}$ rad

3. $9° = 9\left(\frac{\pi}{180}\right) = \frac{\pi}{20}$ rad

5. $900° = 900\left(\frac{\pi}{180}\right) = 5\pi$ rad

7. 4π rad $= 4\pi\left(\frac{180}{\pi}\right) = 720°$

9. $\frac{5\pi}{12}$ rad $= \frac{5\pi}{12}\left(\frac{180}{\pi}\right) = 75°$

11. $-\frac{3\pi}{8}$ rad $= -\frac{3\pi}{8}\left(\frac{180}{\pi}\right) = -67.5°$

13. Using Formula 3, $a = r\theta = 36 \cdot \frac{\pi}{12} = 3\pi$ cm.

15. Using Formula 3, $\theta = a/r = \frac{1}{1.5} = \frac{2}{3}$ rad $= \frac{2}{3}\left(\frac{180}{\pi}\right) = \left(\frac{120}{\pi}\right)° \approx 38.2°$.

17.

19.

21.

23.

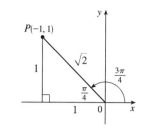

From the diagram we see that a point on the terminal side is $P(-1, 1)$. Therefore, taking $x = -1$, $y = 1$, $r = \sqrt{2}$ in the definitions of the trigonometric ratios, we have $\sin \frac{3\pi}{4} = \frac{1}{\sqrt{2}}$, $\cos \frac{3\pi}{4} = -\frac{1}{\sqrt{2}}$, $\tan \frac{3\pi}{4} = -1$, $\csc \frac{3\pi}{4} = \sqrt{2}$, $\sec \frac{3\pi}{4} = -\sqrt{2}$, and $\cot \frac{3\pi}{4} = -1$.

25.

From the diagram we see that a point on the terminal line is $P(0, 1)$. Therefore taking $x = 0$, $y = 1$, $r = 1$ in the definitions of the trigonometric ratios, we have $\sin \frac{9\pi}{2} = 1$, $\cos \frac{9\pi}{2} = 0$, $\tan \frac{9\pi}{2} = y/x$ is undefined since $x = 0$, $\csc \frac{9\pi}{2} = 1$, $\sec \frac{9\pi}{2} = r/x$ is undefined since $x = 0$, and $\cot \frac{9\pi}{2} = 0$.

27.

Using Figure 8 we see that a point on the terminal line is $P\left(-\sqrt{3}, 1\right)$. Therefore taking $x = -\sqrt{3}$, $y = 1$, $r = 2$ in the definitions of the trigonometric ratios, we have $\sin \frac{5\pi}{6} = \frac{1}{2}$, $\cos \frac{5\pi}{6} = -\frac{\sqrt{3}}{2}$, $\tan \frac{5\pi}{6} = -\frac{1}{\sqrt{3}}$, $\csc \frac{5\pi}{6} = 2$, $\sec \frac{5\pi}{6} = -\frac{2}{\sqrt{3}}$, and $\cot \frac{5\pi}{6} = -\sqrt{3}$.

29. $\sin \theta = y/r = \frac{3}{5}$ $\Rightarrow$ $y = 3$, $r = 5$, and $x = \sqrt{r^2 - y^2} = 4$ (since $0 < \theta < \frac{\pi}{2}$). Therefore taking $x = 4$, $y = 3$, $r = 5$ in the definitions of the trigonometric ratios, we have $\cos \theta = \frac{4}{5}$, $\tan \theta = \frac{3}{4}$, $\csc \theta = \frac{5}{3}$, $\sec \theta = \frac{5}{4}$, and $\cot \theta = \frac{4}{3}$.

31. $\frac{\pi}{2} < \phi < \pi$ $\Rightarrow$ ϕ is in the second quadrant, where x is negative and y is positive. Therefore $\sec \phi = r/x = -1.5 = -\frac{3}{2}$ $\Rightarrow$ $r = 3$, $x = -2$, and $y = \sqrt{r^2 - x^2} = \sqrt{5}$. Taking $x = -2$, $y = \sqrt{5}$, and $r = 3$ in the definitions of the trigonometric ratios, we have $\sin \phi = \frac{\sqrt{5}}{3}$, $\cos \phi = -\frac{2}{3}$, $\tan \phi = -\frac{\sqrt{5}}{2}$, $\csc \phi = \frac{3}{\sqrt{5}}$, and $\cot \theta = -\frac{2}{\sqrt{5}}$.

33. $\pi < \beta < 2\pi$ means that β is in the third or fourth quadrant where y is negative. Also since $\cot \beta = x/y = 3$ which is positive, x must also be negative. Therefore $\cot \beta = x/y = \frac{3}{1}$ $\Rightarrow$ $x = -3$, $y = -1$, and $r = \sqrt{x^2 + y^2} = \sqrt{10}$. Taking $x = -3$, $y = -1$ and $r = \sqrt{10}$ in the definitions of the trigonometric ratios, we have $\sin \beta = -\frac{1}{\sqrt{10}}$, $\cos \beta = -\frac{3}{\sqrt{10}}$, $\tan \beta = \frac{1}{3}$, $\csc \beta = -\sqrt{10}$, and $\sec \beta = -\frac{\sqrt{10}}{3}$.

35. $\sin 35° = \dfrac{x}{10}$ $\Rightarrow$ $x = 10 \sin 35° \approx 5.73576$ cm

37. $\tan \frac{2\pi}{5} = \dfrac{x}{8}$ $\Rightarrow$ $x = 8 \tan \frac{2\pi}{5} \approx 24.62147$ cm

39.

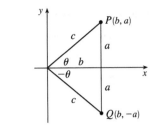

(a) From the diagram we see that $\sin \theta = \dfrac{y}{r} = \dfrac{a}{c}$, and

$$\sin(-\theta) = \frac{-a}{c} = -\frac{a}{c} = -\sin\theta.$$

(b) Again from the diagram we see that $\cos\theta = \dfrac{x}{r} = \dfrac{b}{c} = \cos(-\theta)$.

41. (a) Using (12a) and (13a), we have

$$\tfrac{1}{2}\left[\sin(x+y) + \sin(x-y)\right] = \tfrac{1}{2}\left[\sin x \cos y + \cos x \sin y + \sin x \cos y - \cos x \sin y\right]$$

$$= \tfrac{1}{2}(2\sin x \cos y) = \sin x \cos y$$

(b) This time, using (12b) and (13b), we have

$$\tfrac{1}{2}\left[\cos(x+y) + \cos(x-y)\right] = \tfrac{1}{2}\left[\cos x \cos y - \sin x \sin y + \cos x \cos y + \sin x \sin y\right]$$

$$= \tfrac{1}{2}(2\cos x \cos y) = \cos x \cos y$$

(c) Again using (12b) and (13b), we have

$$\tfrac{1}{2}\left[\cos(x-y) - \cos(x+y)\right] = \tfrac{1}{2}\left[\cos x \cos y + \sin x \sin y - \cos x \cos y + \sin x \sin y\right]$$

$$= \tfrac{1}{2}(2\sin x \sin y) = \sin x \sin y$$

43. Using (12a), we have $\sin\left(\frac{\pi}{2} + x\right) = \sin\frac{\pi}{2}\cos x + \cos\frac{\pi}{2}\sin x = 1\cdot\cos x + 0\cdot\sin x = \cos x$.

45. Using (6), we have $\sin\theta\cot\theta = \sin\theta\cdot\dfrac{\cos\theta}{\sin\theta} = \cos\theta$.

47. $\sec y - \cos y = \dfrac{1}{\cos y} - \cos y$ [by (6)] $= \dfrac{1 - \cos^2 y}{\cos y} = \dfrac{\sin^2 y}{\cos y}$ [by (7)] $= \dfrac{\sin y}{\cos y}\sin y = \tan y \sin y$ [by (6)]

49. $\cot^2\theta + \sec^2\theta = \dfrac{\cos^2\theta}{\sin^2\theta} + \dfrac{1}{\cos^2\theta}$ [by (6)] $= \dfrac{\cos^2\theta\cos^2\theta + \sin^2\theta}{\sin^2\theta\cos^2\theta}$

$$= \dfrac{(1 - \sin^2\theta)(1 - \sin^2\theta) + \sin^2\theta}{\sin^2\theta\cos^2\theta} \text{ [by (7)]} = \dfrac{1 - \sin^2\theta + \sin^4\theta}{\sin^2\theta\cos^2\theta}$$

$$= \dfrac{\cos^2\theta + \sin^4\theta}{\sin^2\theta\cos^2\theta} \text{ [by (7)]} = \dfrac{1}{\sin^2\theta} + \dfrac{\sin^2\theta}{\cos^2\theta} = \csc^2\theta + \tan^2\theta \text{ [by (6)]}$$

51. Using (14a), we have $\tan 2\theta = \tan(\theta + \theta) = \dfrac{\tan\theta + \tan\theta}{1 - \tan\theta\tan\theta} = \dfrac{2\tan\theta}{1 - \tan^2\theta}$.

53. Using (15a) and (16a),

$$\sin x \sin 2x + \cos x \cos 2x = \sin x\,(2\sin x \cos x) + \cos x\,(2\cos^2 x - 1) = 2\sin^2 x \cos x + 2\cos^3 x - \cos x$$

$$= 2(1 - \cos^2 x)\cos x + 2\cos^3 x - \cos x \text{ [by (7)]}$$

$$= 2\cos x - 2\cos^3 x + 2\cos^3 x - \cos x = \cos x$$

$Or:$ $\sin x \sin 2x + \cos x \cos 2x = \cos(2x - x)$ [by 13(b)] $= \cos x$

55. $\dfrac{\sin\phi}{1 - \cos\phi} = \dfrac{\sin\phi}{1 - \cos\phi}\cdot\dfrac{1 + \cos\phi}{1 + \cos\phi} = \dfrac{\sin\phi\,(1 + \cos\phi)}{1 - \cos^2\phi} = \dfrac{\sin\phi\,(1 + \cos\phi)}{\sin^2\phi}$ [by (7)]

$$= \dfrac{1 + \cos\phi}{\sin\phi} = \dfrac{1}{\sin\phi} + \dfrac{\cos\phi}{\sin\phi} = \csc\phi + \cot\phi \text{ [by (6)]}$$

57. Using (12a),

$$\sin 3\theta + \sin \theta = \sin (2\theta + \theta) + \sin \theta = \sin 2\theta \cos \theta + \cos 2\theta \sin \theta + \sin \theta$$

$$= \sin 2\theta \cos \theta + (2\cos^2 \theta - 1) \sin \theta + \sin \theta \quad \text{[by (16a)]}$$

$$= \sin 2\theta \cos \theta + 2\cos^2 \theta \sin \theta - \sin \theta + \sin \theta = \sin 2\theta \cos \theta + \sin 2\theta \cos \theta \quad \text{[by (15a)]}$$

$$= 2\sin 2\theta \cos \theta$$

59. Since $\sin x = \frac{1}{3}$ we can label the opposite side as having length 1, the hypotenuse as having length 3, and use the Pythagorean Theorem to get that the adjacent side has length $\sqrt{8}$. Then, from the diagram, $\cos x = \frac{\sqrt{8}}{3}$. Similarly we have that $\sin y = \frac{3}{5}$. Now use (12a):

$\sin(x + y) = \sin x \cos y + \cos x \sin y = \frac{1}{3} \cdot \frac{4}{5} + \frac{\sqrt{8}}{3} \cdot \frac{3}{5} = \frac{4}{15} + \frac{3\sqrt{8}}{15} = \frac{4 + 6\sqrt{2}}{15}$.

61. Using (13b) and the values for $\cos x$ and $\sin y$ obtained in Exercise 59, we have

$$\cos (x - y) = \cos x \cos y + \sin x \sin y = \frac{\sqrt{8}}{3} \cdot \frac{4}{5} + \frac{1}{3} \cdot \frac{3}{5} = \frac{8\sqrt{2} + 3}{15}$$

63. Using (15a) and the values for $\sin y$ and $\cos y$ obtained in Exercise 59, we have

$$\sin 2y = 2\sin y \cos y = 2 \cdot \frac{3}{5} \cdot \frac{4}{5} = \frac{24}{25}$$

65. $2\cos x - 1 = 0 \iff \cos x = \frac{1}{2} \implies x = \frac{\pi}{3}, \frac{5\pi}{3}$ for $x \in [0, 2\pi]$.

67. $2\sin^2 x = 1 \iff \sin^2 x = \frac{1}{2} \iff \sin x = \pm \frac{1}{\sqrt{2}} \implies x = \frac{\pi}{4}, \frac{3\pi}{4}, \frac{5\pi}{4}, \frac{7\pi}{4}$.

69. Using (15a), we have $\sin 2x = \cos x \iff 2\sin x \cos x - \cos x = 0 \iff \cos x(2\sin x - 1) = 0 \iff$
$\cos x = 0$ or $2\sin x - 1 = 0 \implies x = \frac{\pi}{2}, \frac{3\pi}{2}$ or $\sin x = \frac{1}{2} \implies x = \frac{\pi}{6}$ or $\frac{5\pi}{6}$. Therefore, the solutions are
$x = \frac{\pi}{6}, \frac{\pi}{2}, \frac{5\pi}{6}, \frac{3\pi}{2}$.

71. $\sin x = \tan x \iff \sin x - \tan x = 0 \iff \sin x - \frac{\sin x}{\cos x} = 0 \iff \sin x \left(1 - \frac{1}{\cos x}\right) = 0 \iff \sin x = 0$

or $1 - \frac{1}{\cos x} = 0 \implies x = 0, \pi, 2\pi$ or $1 = \frac{1}{\cos x} \implies \cos x = 1 \implies x = 0, 2\pi$. Therefore the solutions are $x = 0, \pi, 2\pi$.

73. We know that $\sin x = \frac{1}{2}$ when $x = \frac{\pi}{6}$ or $\frac{5\pi}{6}$, and from Figure 13(a), we see that $\sin x \leq \frac{1}{2} \implies 0 \leq x \leq \frac{\pi}{6}$ or $\frac{5\pi}{6} \leq x \leq 2\pi$ for $x \in [0, 2\pi]$.

75. $\tan x = -1$ when $x = \frac{3\pi}{4}, \frac{7\pi}{4}$, and $\tan x = 1$ when $x = \frac{\pi}{4}$ or $\frac{5\pi}{4}$. From Figure 14 we see that $-1 < \tan x < 1$
$\implies 0 \leq x < \frac{\pi}{4}, \frac{3\pi}{4} < x < \frac{5\pi}{4}$, and $\frac{7\pi}{4} < x \leq 2\pi$.

77. $y = \cos\left(x - \frac{\pi}{3}\right)$. We start with the graph of $y = \cos x$ and shift it $\frac{\pi}{3}$ units to the right.

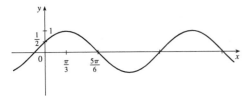

79. $y = \frac{1}{3}\tan\left(x - \frac{\pi}{2}\right)$. We start with the graph of $y = \tan x$, shift it $\frac{\pi}{2}$ units to the right and compress it to $\frac{1}{3}$ of its original vertical size.

81. $y = |\sin x|$. We start with the graph of $y = \sin x$ and reflect the parts below the x-axis about the x-axis.

83. From the figure in the text, we see that $x = b\cos\theta$, $y = b\sin\theta$, and from the distance formula we have that the distance c from (x, y) to $(a, 0)$ is $c = \sqrt{(x-a)^2 + (y-0)^2}$ $\Rightarrow$

$$c^2 = (b\cos\theta - a)^2 + (b\sin\theta)^2 = b^2\cos^2\theta - 2ab\cos\theta + a^2 + b^2\sin^2\theta$$

$$= a^2 + b^2\left(\cos^2\theta + \sin^2\theta\right) - 2ab\cos\theta = a^2 + b^2 - 2ab\cos\theta \quad \text{[by (7)]}$$

85. Using the Law of Cosines, we have $c^2 = 1^2 + 1^2 - 2(1)(1)\cos(\alpha - \beta) = 2\left[1 - \cos(\alpha - \beta)\right]$. Now, using the distance formula, $c^2 = |AB|^2 = (\cos\alpha - \cos\beta)^2 + (\sin\alpha - \sin\beta)^2$. Equating these two expressions for c^2, we get $2\left[1 - \cos(\alpha - \beta)\right] = \cos^2\alpha + \sin^2\alpha + \cos^2\beta + \sin^2\beta - 2\cos\alpha\cos\beta - 2\sin\alpha\sin\beta$ $\Rightarrow$ $1 - \cos(\alpha - \beta) = 1 - \cos\alpha\cos\beta - \sin\alpha\sin\beta$ $\Rightarrow$ $\cos(\alpha - \beta) = \cos\alpha\cos\beta + \sin\alpha\sin\beta$.

87. In Exercise 86 we used the subtraction formula for cosine to prove the addition formula for cosine. Using that formula with $x = \frac{\pi}{2} - \alpha$, $y = \beta$, we get $\cos\left[\left(\frac{\pi}{2} - \alpha\right) + \beta\right] = \cos\left(\frac{\pi}{2} - \alpha\right)\cos\beta - \sin\left(\frac{\pi}{2} - \alpha\right)\sin\beta$ $\Rightarrow$ $\cos\left[\frac{\pi}{2} - (\alpha - \beta)\right] = \cos\left(\frac{\pi}{2} - \alpha\right)\cos\beta - \sin\left(\frac{\pi}{2} - \alpha\right)\sin\beta$. Now we use the identities given in the problem, $\cos\left(\frac{\pi}{2} - \theta\right) = \sin\theta$ and $\sin\left(\frac{\pi}{2} - \theta\right) = \cos\theta$, to get $\sin(\alpha - \beta) = \sin\alpha\cos\beta - \cos\alpha\sin\beta$.

89. Using the formula from part (a), the area of the triangle is $\frac{1}{2}(10)(3)\sin 107° \approx 14.34457$ cm^2.

E Sigma Notation

1. $\displaystyle\sum_{i=1}^{5} \sqrt{i} = \sqrt{1} + \sqrt{2} + \sqrt{3} + \sqrt{4} + \sqrt{5}$

3. $\displaystyle\sum_{i=4}^{6} 3^i = 3^4 + 3^5 + 3^6$

5. $\displaystyle\sum_{k=0}^{4} \frac{2k-1}{2k+1} = -1 + \frac{1}{3} + \frac{3}{5} + \frac{5}{7} + \frac{7}{9}$

7. $\displaystyle\sum_{i=1}^{n} i^{10} = 1^{10} + 2^{10} + 3^{10} + \cdots + n^{10}$

9. $\displaystyle\sum_{j=0}^{n-1} (-1)^j = 1 - 1 + 1 - 1 + \cdots + (-1)^{n-1}$

11. $1 + 2 + 3 + 4 + \cdots + 10 = \displaystyle\sum_{i=1}^{10} i$

13. $\dfrac{1}{2} + \dfrac{2}{3} + \dfrac{3}{4} + \dfrac{4}{5} + \cdots + \dfrac{19}{20} = \displaystyle\sum_{i=1}^{19} \frac{i}{i+1}$

15. $2 + 4 + 6 + 8 + \cdots + 2n = \displaystyle\sum_{i=1}^{n} 2i$

17. $1 + 2 + 4 + 8 + 16 + 32 = \displaystyle\sum_{i=0}^{5} 2^i$

19. $x + x^2 + x^3 + \cdots + x^n = \displaystyle\sum_{i=1}^{n} x^i$

21. $\displaystyle\sum_{i=4}^{8} (3i-2) = [3(4)-2] + [3(5)-2] + [3(6)-2] + [3(7)-2] + [3(8)-2] = 10 + 13 + 16 + 19 + 22 = 80$

23. $\displaystyle\sum_{j=1}^{6} 3^{j+1} = 3^2 + 3^3 + 3^4 + 3^5 + 3^6 + 3^7 = 9 + 27 + 81 + 243 + 729 + 2187 = 3276$

(For a more general method, see Exercise 47.)

25. $\displaystyle\sum_{n=1}^{20} (-1)^n = -1 + 1 - 1 + 1 - 1 + 1 - 1 + 1 - 1 + 1 - 1 + 1 - 1 + 1 - 1 + 1 - 1 + 1 - 1 + 1 = 0$

27. $\displaystyle\sum_{i=0}^{4} \left(2^i + i^2\right) = (1+0) + (2+1) + (4+4) + (8+9) + (16+16) = 61$

29. $\displaystyle\sum_{i=1}^{n} 2i = 2\sum_{i=1}^{n} i = 2 \cdot \frac{n(n+1)}{2}$ [by Theorem 3(c)] $= n(n+1)$

31. $\displaystyle\sum_{i=1}^{n} \left(i^2 + 3i + 4\right) = \sum_{i=1}^{n} i^2 + 3\sum_{i=1}^{n} i + \sum_{i=1}^{n} 4 = \frac{n(n+1)(2n+1)}{6} + \frac{3n(n+1)}{2} + 4n$

$\qquad = \frac{1}{6}\left[(2n^3 + 3n^2 + n) + (9n^2 + 9n) + 24n\right] = \frac{1}{6}\left(2n^3 + 12n^2 + 34n\right)$

$\qquad = \frac{1}{3}n(n^2 + 6n + 17)$

33. $\displaystyle\sum_{i=1}^{n} (i+1)(i+2) = \sum_{i=1}^{n} \left(i^2 + 3i + 2\right) = \sum_{i=1}^{n} i^2 + 3\sum_{i=1}^{n} i + \sum_{i=1}^{n} 2$

$\qquad = \frac{n(n+1)(2n+1)}{6} + \frac{3n(n+1)}{2} + 2n = \frac{n(n+1)}{6}\left[(2n+1) + 9\right] + 2n$

$\qquad = \frac{n(n+1)}{3}(n+5) + 2n = \frac{n}{3}\left[(n+1)(n+5) + 6\right] = \frac{n}{3}\left(n^2 + 6n + 11\right)$

35. $\displaystyle\sum_{i=1}^{n}\left(i^3 - i - 2\right) = \sum_{i=1}^{n} i^3 - \sum_{i=1}^{n} i - \sum_{i=1}^{n} 2 = \left[\frac{n(n+1)}{2}\right]^2 - \frac{n(n+1)}{2} - 2n$

$\qquad\qquad = \frac{1}{4}n(n+1)\left[n(n+1)-2\right] - 2n = \frac{1}{4}n(n+1)(n+2)(n-1) - 2n$

$\qquad\qquad = \frac{1}{4}n\left[(n+1)(n-1)(n+2) - 8\right] = \frac{1}{4}n\left[(n^2-1)(n+2) - 8\right] = \frac{1}{4}n\left(n^3 + 2n^2 - n - 10\right)$

37. By Theorem 2(a) and Example 3, $\displaystyle\sum_{i=1}^{n} c = c\sum_{i=1}^{n} 1 = cn.$

39. $\displaystyle\sum_{i=1}^{n}\left[(i+1)^4 - i^4\right] = \left(2^4 - 1^4\right) + \left(3^4 - 2^4\right) + \left(4^4 - 3^4\right) + \cdots + \left[(n+1)^4 - n^4\right]$

$\qquad\qquad\qquad = (n+1)^4 - 1^4 = n^4 + 4n^3 + 6n^2 + 4n$

On the other hand,

$\qquad\displaystyle\sum_{i=1}^{n}\left[(i+1)^4 - i^4\right] = \sum_{i=1}^{n}\left(4i^3 + 6i^2 + 4i + 1\right) = 4\sum_{i=1}^{n} i^3 + 6\sum_{i=1}^{n} i^2 + 4\sum_{i=1}^{n} i + \sum_{i=1}^{n} 1$

$\qquad\qquad\qquad = 4S + n(n+1)(2n+1) + 2n(n+1) + n \quad\left[\text{where } S = \sum_{i=1}^{n} i^3\right]$

$\qquad\qquad\qquad = 4S + 2n^3 + 3n^2 + n + 2n^2 + 2n + n = 4S + 2n^3 + 5n^2 + 4n$

Thus, $n^4 + 4n^3 + 6n^2 + 4n = 4S + 2n^3 + 5n^2 + 4n$, from which it follows that

$4S = n^4 + 2n^3 + n^2 = n^2\left(n^2 + 2n + 1\right) = n^2(n+1)^2$ and $S = \left[\dfrac{n(n+1)}{2}\right]^2.$

41. (a) $\displaystyle\sum_{i=1}^{n}\left[i^4 - (i-1)^4\right] = \left(1^4 - 0^4\right) + \left(2^4 - 1^4\right) + \left(3^4 - 2^4\right) + \cdots + \left[n^4 - (n-1)^4\right] = n^4 - 0 = n^4$

(b) $\displaystyle\sum_{i=1}^{100}\left(5^i - 5^{i-1}\right) = \left(5^1 - 5^0\right) + \left(5^2 - 5^1\right) + \left(5^3 - 5^2\right) + \cdots + \left(5^{100} - 5^{99}\right) = 5^{100} - 5^0 = 5^{100} - 1$

(c) $\displaystyle\sum_{i=3}^{99}\left(\frac{1}{i} - \frac{1}{i+1}\right) = \left(\frac{1}{3} - \frac{1}{4}\right) + \left(\frac{1}{4} - \frac{1}{5}\right) + \left(\frac{1}{5} - \frac{1}{6}\right) + \cdots + \left(\frac{1}{99} - \frac{1}{100}\right) = \frac{1}{3} - \frac{1}{100} = \frac{97}{300}$

(d) $\displaystyle\sum_{i=1}^{n}(a_i - a_{i-1}) = (a_1 - a_0) + (a_2 - a_1) + (a_3 - a_2) + \cdots + (a_n - a_{n-1}) = a_n - a_0$

43. $\displaystyle\lim_{n\to\infty}\sum_{i=1}^{n}\frac{1}{n}\left(\frac{i}{n}\right)^2 = \lim_{n\to\infty}\frac{1}{n^3}\sum_{i=1}^{n} i^2 = \lim_{n\to\infty}\frac{1}{n^3}\frac{n(n+1)(2n+1)}{6} = \lim_{n\to\infty}\frac{1}{6}\left(1+\frac{1}{n}\right)\left(2+\frac{1}{n}\right)$

$\qquad\qquad = \frac{1}{6}(1)(2) = \frac{1}{3}$

45. $\displaystyle\lim_{n\to\infty}\sum_{i=1}^{n}\frac{2}{n}\left[\left(\frac{2i}{n}\right)^3 + 5\left(\frac{2i}{n}\right)\right] = \lim_{n\to\infty}\sum_{i=1}^{n}\left[\frac{16}{n^4}i^3 + \frac{20}{n^2}i\right] = \lim_{n\to\infty}\left[\frac{16}{n^4}\sum_{i=1}^{n} i^3 + \frac{20}{n^2}\sum_{i=1}^{n} i\right]$

$\qquad\qquad = \lim_{n\to\infty}\left[\frac{16}{n^4}\frac{n^2(n+1)^2}{4} + \frac{20}{n^2}\frac{n(n+1)}{2}\right] = \lim_{n\to\infty}\left[\frac{4(n+1)^2}{n^2} + \frac{10n(n+1)}{n^2}\right]$

$\qquad\qquad = \lim_{n\to\infty}\left[4\left(1+\frac{1}{n}\right)^2 + 10\left(1+\frac{1}{n}\right)\right] = 4\cdot 1 + 10\cdot 1 = 14$

47. Let $S = \sum_{i=1}^{n} ar^{i-1} = a + ar + ar^2 + \cdots + ar^{n-1}$. Multiplying both sides by r gives us

$rS = ar + ar^2 + \cdots + ar^{n-1} + ar^n$. Subtracting the first equation from the second, we find

$(r-1)S = ar^n - a = a(r^n - 1)$, so $S = \dfrac{a(r^n - 1)}{r - 1}$ (since $r \neq 1$).

49. $\sum_{i=1}^{n}\left(2i + 2^i\right) = 2\sum_{i=1}^{n} i + \sum_{i=1}^{n} 2 \cdot 2^{i-1} = 2\dfrac{n(n+1)}{2} + \dfrac{2(2^n - 1)}{2 - 1} = 2^{n+1} + n^2 + n - 2$.

For the first sum we have used Theorem 3(c), and for the second, Exercise 47 with $a = r = 2$.

G Complex Numbers

1. $(5 - 6i) + (3 + 2i) = (5 + 3) + (-6 + 2)i = 8 + (-4)i = 8 - 4i$

3. $(2 + 5i)(4 - i) = 2(4) + 2(-i) + (5i)(4) + (5i)(-i) = 8 - 2i + 20i - 5i^2$
$$= 8 + 18i - 5(-1) = 8 + 18i + 5 = 13 + 18i$$

5. $\overline{12 + 7i} = 12 - 7i$

7. $\dfrac{1 + 4i}{3 + 2i} = \dfrac{1 + 4i}{3 + 2i} \cdot \dfrac{3 - 2i}{3 - 2i} = \dfrac{3 - 2i + 12i - 8(-1)}{3^2 + 2^2} = \dfrac{11 + 10i}{13} = \dfrac{11}{13} + \dfrac{10}{13}i$

9. $\dfrac{1}{1 + i} = \dfrac{1}{1 + i} \cdot \dfrac{1 - i}{1 - i} = \dfrac{1 - i}{1 - (-1)} = \dfrac{1 - i}{2} = \dfrac{1}{2} - \dfrac{1}{2}i$

11. $i^3 = i^2 \cdot i = (-1)i = -i$

13. $\sqrt{-25} = \sqrt{25}\, i = 5i$

15. $\overline{12 - 5i} = 12 + 15i$ and $|12 - 15i| = \sqrt{12^2 + (-5)^2} = \sqrt{144 + 25} = \sqrt{169} = 13$

17. $\overline{-4i} = \overline{0 - 4i} = 0 + 4i = 4i$ and $|-4i| = \sqrt{0^2 + (-4)^2} = \sqrt{16} = 4$

19. $4x^2 + 9 = 0 \Leftrightarrow 4x^2 = -9 \Leftrightarrow x^2 = -\frac{9}{4} \Leftrightarrow x = \pm\sqrt{-\frac{9}{4}} = \pm\sqrt{\frac{9}{4}}\,i = \pm\frac{3}{2}i$.

21. $x^2 + 2x + 5 = 0 \Leftrightarrow x = \dfrac{-2 \pm \sqrt{2^2 - 4 \cdot 1 \cdot 5}}{2 \cdot 1} = \dfrac{-2 \pm \sqrt{-16}}{2} = \dfrac{-2 \pm 4i}{2} = -1 \pm 2i$

23. By the quadratic formula, $z^2 + z + 2 = 0 \Leftrightarrow z = \dfrac{-1 \pm \sqrt{1^2 - 4(1)(2)}}{2(1)} = \dfrac{-1 \pm \sqrt{-7}}{2} = -\dfrac{1}{2} \pm \dfrac{\sqrt{7}}{2}i$.

25. For $z = -3 + 3i$, $r = \sqrt{(-3)^2 + 3^2} = 3\sqrt{2}$ and $\tan\theta = \frac{3}{-3} = -1 \Rightarrow \theta = \frac{3\pi}{4}$ (since z lies in the second quadrant). Therefore, $-3 + 3i = 3\sqrt{2}\left(\cos\frac{3\pi}{4} + i\sin\frac{3\pi}{4}\right)$.

27. For $z = 3 + 4i$, $r = \sqrt{3^2 + 4^2} = 5$ and $\tan\theta = \frac{4}{3} \Rightarrow \theta = \tan^{-1}\left(\frac{4}{3}\right)$ (since z lies in the first quadrant). Therefore, $3 + 4i = 5\left[\cos\left(\tan^{-1}\frac{4}{3}\right) + i\sin\left(\tan^{-1}\frac{4}{3}\right)\right]$.

29. For $z = \sqrt{3} + i$, $r = \sqrt{\left(\sqrt{3}\right)^2 + 1^2} = 2$ and $\tan\theta = \frac{1}{\sqrt{3}}$ $\Rightarrow$ $\theta = \frac{\pi}{6}$ $\Rightarrow$ $z = 2\left(\cos\frac{\pi}{6} + i\sin\frac{\pi}{6}\right)$.

For $w = 1 + \sqrt{3}\,i$, $r = 2$ and $\tan\theta = \sqrt{3}$ $\Rightarrow$ $\theta = \frac{\pi}{3}$ $\Rightarrow$ $w = 2\left(\cos\frac{\pi}{3} + i\sin\frac{\pi}{3}\right)$.

Therefore, $zw = 2 \cdot 2\left[\cos\left(\frac{\pi}{6} + \frac{\pi}{3}\right) + i\sin\left(\frac{\pi}{6} + \frac{\pi}{3}\right)\right] = 4\left(\cos\frac{\pi}{2} + i\sin\frac{\pi}{2}\right)$,

$z/w = \frac{2}{2}\left[\cos\left(\frac{\pi}{6} - \frac{\pi}{3}\right) + i\sin\left(\frac{\pi}{6} - \frac{\pi}{3}\right)\right] = \cos\left(-\frac{\pi}{6}\right) + i\sin\left(-\frac{\pi}{6}\right)$, and $1 = 1 + 0i = 1(\cos 0 + i\sin 0)$ $\Rightarrow$

$1/z = \frac{1}{2}\left[\cos\left(0 - \frac{\pi}{6}\right) + i\sin\left(0 - \frac{\pi}{6}\right)\right] = \frac{1}{2}\left[\cos\left(-\frac{\pi}{6}\right) + i\sin\left(-\frac{\pi}{6}\right)\right]$. For $1/z$, we could also use the formula that

precedes Example 5 to obtain $1/z = \frac{1}{2}\left(\cos\frac{\pi}{6} - i\sin\frac{\pi}{6}\right)$.

31. For $z = 2\sqrt{3} - 2i$, $r = \sqrt{\left(2\sqrt{3}\right)^2 + (-2)^2} = 4$ and $\tan\theta = \frac{-2}{2\sqrt{3}} = -\frac{1}{\sqrt{3}}$

$\Rightarrow$ $\theta = -\frac{\pi}{6}$ $\Rightarrow$ $z = 4\left[\cos\left(-\frac{\pi}{6}\right) + i\sin\left(-\frac{\pi}{6}\right)\right]$. For $w = -1 + i$, $r = \sqrt{2}$,

$\tan\theta = \frac{1}{-1} = -1$ $\Rightarrow$ $\theta = \frac{3\pi}{4}$ $\Rightarrow$ $w = \sqrt{2}\left(\cos\frac{3\pi}{4} + i\sin\frac{3\pi}{4}\right)$. Therefore,

$zw = 4\sqrt{2}\left[\cos\left(-\frac{\pi}{6} + \frac{3\pi}{4}\right) + i\sin\left(-\frac{\pi}{6} + \frac{3\pi}{4}\right)\right] = 4\sqrt{2}\left(\cos\frac{7\pi}{12} + i\sin\frac{7\pi}{12}\right)$,

$z/w = \frac{4}{\sqrt{2}}\left[\cos\left(-\frac{\pi}{6} - \frac{3\pi}{4}\right) + i\sin\left(-\frac{\pi}{6} - \frac{3\pi}{4}\right)\right] = \frac{4}{\sqrt{2}}\left[\cos\left(-\frac{11\pi}{12}\right) + i\sin\left(-\frac{11\pi}{12}\right)\right]$

$= 2\sqrt{2}\left(\cos\frac{13\pi}{12} + i\sin\frac{13\pi}{12}\right)$, and

$1/z = \frac{1}{4}\left[\cos\left(-\frac{\pi}{6}\right) - i\sin\left(-\frac{\pi}{6}\right)\right] = \frac{1}{4}\left(\cos\frac{\pi}{6} + i\sin\frac{\pi}{6}\right)$.

33. For $z = 1 + i$, $r = \sqrt{2}$ and $\tan\theta = \frac{1}{1} = 1$ $\Rightarrow$ $\theta = \frac{\pi}{4}$ $\Rightarrow$ $z = \sqrt{2}\left(\cos\frac{\pi}{4} + i\sin\frac{\pi}{4}\right)$. So by

De Moivre's Theorem,

$$(1 + i)^{20} = \left[\sqrt{2}\left(\cos\frac{\pi}{4} + i\sin\frac{\pi}{4}\right)\right]^{20} = \left(2^{1/2}\right)^{20}\left(\cos\frac{20 \cdot \pi}{4} + i\sin\frac{20 \cdot \pi}{4}\right)$$

$$= 2^{10}(\cos 5\pi + i\sin 5\pi) = 2^{10}[-1 + i(0)] = -2^{10} = -1024$$

35. For $z = 2\sqrt{3} + 2i$, $r = \sqrt{\left(2\sqrt{3}\right)^2 + 2^2} = \sqrt{16} = 4$ and $\tan\theta = \frac{2}{2\sqrt{3}} = \frac{1}{\sqrt{3}}$ $\Rightarrow$ $\theta = \frac{\pi}{6}$ $\Rightarrow$

$z = 4\left(\cos\frac{\pi}{6} + i\sin\frac{\pi}{6}\right)$. So by De Moivre's Theorem,

$$\left(2\sqrt{3} + 2i\right)^5 = \left[4\left(\cos\frac{\pi}{6} + i\sin\frac{\pi}{6}\right)\right]^5 = 4^5\left(\cos\frac{5\pi}{6} + i\sin\frac{5\pi}{6}\right) = 1024\left[-\frac{\sqrt{3}}{2} + \frac{1}{2}i\right] = -512\sqrt{3} + 512i$$

37. $1 = 1 + 0i = 1(\cos 0 + i\sin 0)$. Using Equation 3 with $r = 1$, $n = 8$, and $\theta = 0$, we have

$$w_k = 1^{1/8}\left[\cos\left(\frac{0 + 2k\pi}{8}\right) + i\sin\left(\frac{0 + 2k\pi}{8}\right)\right] = \cos\frac{k\pi}{4} + i\sin\frac{k\pi}{4}, \text{ where } k = 0, 1, 2, \ldots, 7.$$

$w_0 = 1(\cos 0 + i\sin 0) = 1$, $w_1 = 1\left(\cos\frac{\pi}{4} + i\sin\frac{\pi}{4}\right) = \frac{1}{\sqrt{2}} + \frac{1}{\sqrt{2}}i$,

$w_2 = 1\left(\cos\frac{\pi}{2} + i\sin\frac{\pi}{2}\right) = i$, $w_3 = 1\left(\cos\frac{3\pi}{4} + i\sin\frac{3\pi}{4}\right) = -\frac{1}{\sqrt{2}} + \frac{1}{\sqrt{2}}i$,

$w_4 = 1(\cos\pi + i\sin\pi) = -1$, $w_5 = 1\left(\cos\frac{5\pi}{4} + i\sin\frac{5\pi}{4}\right) = -\frac{1}{\sqrt{2}} - \frac{1}{\sqrt{2}}i$,

$w_6 = 1\left(\cos\frac{3\pi}{2} + i\sin\frac{3\pi}{2}\right) = -i$, $w_7 = 1\left(\cos\frac{7\pi}{4} + i\sin\frac{7\pi}{4}\right) = \frac{1}{\sqrt{2}} - \frac{1}{\sqrt{2}}i$

39. $i = 0 + i = 1\left(\cos\frac{\pi}{2} + i\sin\frac{\pi}{2}\right)$. Using Equation 3 with $r = 1$, $n = 3$, and $\theta = \frac{\pi}{2}$, we have

$$w_k = 1^{1/3}\left[\cos\left(\frac{\frac{\pi}{2} + 2k\pi}{3}\right) + i\sin\left(\frac{\frac{\pi}{2} + 2k\pi}{3}\right)\right], \text{ where } k = 0, 1, 2.$$

$$w_0 = \left(\cos\frac{\pi}{6} + i\sin\frac{\pi}{6}\right) = \frac{\sqrt{3}}{2} + \frac{1}{2}i$$

$$w_1 = \left(\cos\frac{5\pi}{6} + i\sin\frac{5\pi}{6}\right) = -\frac{\sqrt{3}}{2} + \frac{1}{2}i$$

$$w_2 = \left(\cos\frac{9\pi}{6} + i\sin\frac{9\pi}{6}\right) = -i$$

41. Using Euler's formula (6) with $y = \frac{\pi}{2}$, we have $e^{i\pi/2} = \cos\frac{\pi}{2} + i\sin\frac{\pi}{2} = 0 + 1i = i$.

43. Using Euler's formula (6) with $y = \frac{\pi}{3}$, we have $e^{i\pi/3} = \cos\frac{\pi}{3} + i\sin\frac{\pi}{3} = \frac{1}{2} + \frac{\sqrt{3}}{2}i$.

45. Using Equation 7 with $x = 2$ and $y = \pi$, we have $e^{2+i\pi} = e^2 e^{i\pi} = e^2(\cos\pi + i\sin\pi) = e^2(-1 + 0) = -e^2$.

47. Take $r = 1$ and $n = 3$ in De Moivre's Theorem to get

$$[1(\cos\theta + i\sin\theta)]^3 = 1^3(\cos 3\theta + i\sin 3\theta)$$

$$(\cos\theta + i\sin\theta)^3 = \cos 3\theta + i\sin 3\theta$$

$$\cos^3\theta + 3(\cos^2\theta)(i\sin\theta) + 3(\cos\theta)(i\sin\theta)^2 + (i\sin\theta)^3 = \cos 3\theta + i\sin 3\theta$$

$$\cos^3\theta + (3\cos^2\theta\sin\theta)i - 3\cos\theta\sin^2\theta - (\sin^3\theta)i = \cos 3\theta + i\sin 3\theta$$

$$(\cos^3\theta - 3\sin^2\theta\cos\theta) + (3\sin\theta\cos^2\theta - \sin^3\theta)i = \cos 3\theta + i\sin 3\theta$$

Equating real and imaginary parts gives

$$\cos 3\theta = \cos^3\theta - 3\sin^2\theta\cos\theta \quad \text{and} \quad \sin 3\theta = 3\sin\theta\cos^2\theta - \sin^3\theta$$

49. $F(x) = e^{rx} = e^{(a+bi)x} = e^{ax+bxi} = e^{ax}(\cos bx + i\sin bx) = e^{ax}\cos bx + i(e^{ax}\sin bx) \quad\Rightarrow$

$F'(x) = (e^{ax}\cos bx)' + i(e^{ax}\sin bx)' = (ae^{ax}\cos bx - be^{ax}\sin bx) + i(ae^{ax}\sin bx + be^{ax}\cos bx)$

$\qquad = a\left[e^{ax}(\cos bx + i\sin bx)\right] + b\left[e^{ax}(-\sin bx + i\cos bx)\right] = ae^{rx} + b\left[e^{ax}(i^2\sin bx + i\cos bx)\right]$

$\qquad = ae^{rx} + bi\left[e^{ax}(\cos bx + i\sin bx)\right] = ae^{rx} + bie^{rx} = (a + bi)e^{rx} = re^{rx}$